D1269301

Amazing Light

Springer

New York
Berlin
Heidelberg
Barcelona
Budapest
Hong Kong
London
Milan
Paris
Santa Clara
Singapore
Tokyo

Charles Hard Townes

Raymond Y. Chiao, Editor

Amazing Light

A Volume Dedicated To Charles Hard Townes On His 80th Birthday

With 189 illustrations

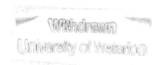

Springer

Raymond Y. Chiao
Department of Physics
University of California at Berkeley
Berkeley, CA 94720
USA

Library of Congress Cataloging-in-Publication Data
Chiao, Raymond Y.
    Amazing light : a volume dedicated to Charles Hard Townes on his
80th birthday / Raymond Y. Chiao.
        p.   cm.
    Includes bibliographical references and index.
    ISBN 0-387-94658-6 (hardcover : alk. paper)
    1. Quantum electronics—Congresses. 2. Lasers—Congresses.
3. Masers—Congresses. 4. Astrophysics—Congresses. 5. Townes,
Charles H.—Congresses. 6. Physicists—United States—Biography–
–Congresses.   I. Title.
QC685.C45   1996
537.5—dc20   95-49220

Printed on acid-free paper.

Production managed by Robert Wexler; manufacturing supervised by Jacqui Ashri.
Photocomposed copy prepared from the author's LaTeX files using Springer's svmult macro.
Printed and bound by Edwards Brothers, Inc., Ann Arbor, MI.
Printed in the United States of America.

9 8 7 6 5 4 3 2 1

IEEE ISBN 0-7803-1181-7   IEEE Order Number PC5658
ISBN 0-387-94658-6 Springer-Verlag New York Berlin Heidelberg SPIN 10524527

To Charles Hard Townes

Preface

This Festschrift is a collection of essays contributed by students, colleagues, and admirers to honor an eminent scholar on a special anniversary: Charles Hard Townes on the occasion of his 80th birthday, July 28, 1995. In 1964, Townes shared the Nobel Prize in physics with Alexander Mikhailovich Prokhorov and Nikolai Gennadyevich Basov "for fundamental work in the field of quantum electronics, which has led to the construction of oscillators and amplifiers based on the maser-laser principle." His contributions have covered a much wider area, however. His fruitful interests spanning several decades have included many scientific subjects, including, microwave spectroscopy and astrophysics (other articles in this volume will expand further on this point). He has also contributed to public service, having served as the chairman of the Science and Technology Advisory Committee for NASA's Apollo program, and as a member and vice chairman of the President's Science Advisory Committee. As the enormous breadth of contributions from his students shows, he has educated scholars who are now in a wide range of fields. The contributions from his many admirers, among whom are nine fellow Nobel laureates, attest to his impact on many disciplines ranging from electrical engineering to medicine. His influence extends even to theology, as is indicated by one essay. The broadly international character of this Festschrift reflects his deep belief in the international, universal nature of science. Contributors have come from Australia, Canada, England, France, Germany, Italy, Japan, Russia, Switzerland, and the United States. His students have also come from countries all over the globe, which, in addition to the above, include China, Greece, India, Iran, and Israel. In light of the breadth of these contributions, it was difficult to decide on their order. In the end, to be as impartial as possible, I have decided to present them in alphabetical order by first author, apart from the Introduction by Arthur Schawlow.

This book would have been impossible without the hard work and assistance of Marnie McElhiney, who helped type, organize, compile, and proofread the manuscripts, and the help of Grant McKinney and David E. Johnson, who assisted in compiling the manuscripts into LaTeX. They made many valuable suggestions along the way while I edited the manuscripts. Many thanks also go to my neighbors in Birge Hall, Dr. Norbert Geis and Paul McEuen, for their expert advice on the editing of some manuscripts. I would especially like to thank Arthur Schawlow for his wise advice and friendly encouragement. I gratefully acknowledge the support of the Office of Naval Research under ONR Grant No. N00014-90-J-1259 during my editorship. Also, the support and encouragement from the Berkeley physics department through its chairman, Roger Falcone, and through its staff, was essential for the production of this volume. Moreover, this Festschrift would not have been possible without the generous donations of the supporters and contributors listed below. Finally, I would like to thank my dear wife Florence for her patience and wise advice during this labor.

R. Y. C.
Berkeley, July 28, 1995

◊   ◊   ◊

Many thanks go to the following societies, corporations, and persons, whose generous donations have made the production of this Festschrift possible:

Supporters:	Contributors:
Optical Society of America	Burleigh Instruments, Inc.
AT&T	Cleveland Crystals, Inc.
Mr. and Mrs. S. D. Bechtel	Mr. and Mrs. David R. Dunlap
Carnegie Institution of Washington	Prof. Thomas E. Everhart
Coherent Laser Group	Mr. Richard C. Gerstenberg
NEC Research Institute, Inc.	Lambda Physik
Optical Coating Laboratory, Inc.	Mr. and Mrs. F. James McDonald
Mr. David Packard	Mr. and Mrs. Thomas A. Murphy
The Perkin-Elmer Corp.	Mr. and Mrs. James M. Roche
Rudolph Research Corp.	Prof. and Mrs. Arthur Rosenfeld
Prof. Arthur L. Schawlow	Mr. and Mrs. Marian O. Scully
	Mr. Casper W. Weinberger
	Mr. and Mrs. Warren L. Wolfson
	Mr. Timothy F. Wullger

Contents

List of Contributors

Vladimir A. Alekseev Quantum Radiophysics Division, P.N. Lebedev Physical Institute, Russian Academy of Sciences, 117924 Moscow, Russia

Nicolai G. Basov Quantum Radiophysics Division, P.N. Lebedev Physical Institute, Russian Academy of Sciences, 117924 Moscow, Russia

Sara C. Beck Department of Physics and Astronomy, University of Tel Aviv, Ramat Aviv, 69978 Israel

Thomas Beckert Institut für Theoretische Astrophysik, D-69120 Heidelberg, Germany

Sergei P. Belov Physikalisches Institut, Universität zu Köln, D-50937 Köln, Germany

William R. Bennett, Jr. Department of Applied Physics, Yale University, New Haven, CT 06520-8284, USA

Frank Bertoldi Max-Planck-Institut für Extraterrestrische Physik, D-85740 Garching, Germany

Manfred Bester Space Sciences Laboratory, University of California at Berkeley, Berkeley, CA 94720-7450, USA

Albert L. Betz Center for Astrophysics and Space Astronomy, University of Colorado, Boulder, CO 80309-0446, USA

Brebis Bleaney Clarendon Laboratory, Garford House, Oxford, United Kingdom

Nicolaas Bloembergen Harvard University, Cambridge, MA 02138, USA

Robert W. Boyd Institute of Optics, University of Rochester, Rochester, NY 14627, USA

Richard G. Brewer IBM Research Division, Almaden Research Center, San Jose, CA 95120-6099, USA

Roland Brodbeck Institute of Quantum Mechanics, Infrared Physics Laboratory, Swiss Federal Institute of Technology (ETH), CH-8093 Zürich, Switzerland

Vieniamin P. Chebotayev Department of Applied Physics, Yale University, New Haven, CT 06520-8284, USA

Raymond Y. Chiao Department of Physics, University of California at Berkeley, Berkeley, CA 94720-7300, USA

Meiro Chiba Institute of Atomic Energy, Kyoto University, Kyoto, Japan

Claude N. Cohen-Tannoudji Physique Atomique et Moleculaire, Collège de France et Laboratoire Kastler Brossel de l'Ecole Normale Supérieure, 75231 Paris, France

Eugene D. Commins Department of Physics, University of California at Berkeley, Berkeley, CA 94720-7300, USA

David A. Coppeta Electro-Optics Technology Center, Tufts University, Medford, MA 02155, USA

David R. Corbin DuPont Experimental Station, Wilmington, DE 19880-0356, USA

Michael K. Crawford DuPont Experimental Station, Wilmington, DE 19880-0356, USA

Herman Z. Cummins Department of Physics, City College of the City University of New York, New York, NY 10031, USA

Izabel Dabrowski Steacie Institute for Molecular Science, National Research Council of Canada, Ottawa, Ontario K1A 0R6, Canada

William C. Danchi Space Sciences Laboratory, University of California at Berkeley, Berkeley, CA 94720-7450, USA

Michael Danos National Institute of Standards, Gaithersburg, MD 20899, USA

Francesco De Martini Dipartimento di Fisica, Università deglis Studi di Roma, Roma 1-00185, Italy

Sidney D. Drell Stanford Linear Accelerator, Stanford University, Stanford, CA 94305-4060, USA

Wolfgang J. Duschl Max-Planck-Institut für Radioastronomie, D-53121 Bonn, Germany

Carolyn L. Ebrahimi Marine Physical Laboratory, Scripps Institution of Oceanography, University of California at San Diego, La Jolla, CA 92152, USA

Gerald Ehrenstein Biophysics Section, National Institutes of Health, Bethesda, MD 20892, USA

Neal J. Evans II Department of Astronomy, University of Texas, Austin, TX 78712-1083, USA

Heidi Fearn Department of Physics, California State University at Fullerton, Fullerton, CA 92634, USA

Uri Feldman Laboratory for Astrophysics, National Air and Space Museum, Smithsonian Institution, Washington, DC 20560, USA

Roger S. Foster Remote Sensing Division, Naval Research Laboratory, Washington, DC 20375-5351, USA

Tserensodnom Gantsog Department of Theoretical Physics, National University of Mongolia, 210646 Ulaanbaatar, Mongolia

Elsa Garmire Dartmouth College, Hanover, NH 03755, USA

Thomas R. Geballe Joint Astronomy Center, University Park, Hilo, HI 96720, USA

Reinhard Genzel Max-Planck Institut für Extraterrestrische Physik, D-85740 Garching, Germany

Marco Giangrasso Dipartimento di Fisica, Università deglis Studi di Roma, Roma 1-00185, Italy

Joseph A. Giordmaine NEC Research Institute, Princeton, NJ 08540, USA

Alfred E. Glassgold Department of Physics, New York University, New York, NY 10003, USA

Paul F. Goldsmith Department of Astronomy, Cornell University, Ithaca, NY 14853, USA

James P. Gordon AT&T Bell Telephone Laboratories, Holmdel, NJ 07733, USA

Matthew A. Greenhouse Laboratory for Astrophysics, National Air and Space Museum, Smithsonian Institution, Washington, DC 20560, USA

Mikhail A. Gubin Quantum Radiophysics Division, P.N. Lebedev Physical Institute, Russian Academy of Sciences, 117924 Moscow, Russia

T. Kenneth Gustafson Department of Electrical Engineering, Computer Sciences, and Electronic Research, University of California at Berkeley, Berkeley, CA 94720, USA

Erwin L. Hahn Department of Physics, University of California at Berkeley, Berkeley, CA 94720-7300, USA

William Happer Department of Physics, Princeton University, Princeton, NJ 08544, USA

Patrick J. Harshman Department of Electrical Engineering, Computer Sciences, and Electronic Research, University of California at Berkeley, Berkeley, CA 94720, USA. Present Address: Harmonic Lightwave, Inc., 3005 Bunker Hill Lane, Santa Clara, CA 95054, USA

Eric Herbst Departments of Physics and Astronomy, Ohio State University, Columbus, OH 43210-1106, USA

Gerhard Herzberg Steacie Institute for Molecular Science, National Research Council of Canada, Ottawa, Ontario K1A 0R6, Canada

John A. Hoffnagle IBM Research Division, Almaden Research Center, San Jose, CA 95120-6099, USA

David J. Hollenbach NASA Ames Research Center, Moffett Field, CA 94035-1000, USA

Daniel Huguenin Institute of Quantum Mechanics, Infrared Physics Laboratory, Swiss Federal Institute of Technology (ETH), CH-8093 Zürich, Switzerland

Ali Javan Department of Physics, Massachusetts Institute of Technology, Cambridge, MA 02130, USA

Paul L. Kelley Electro-Optics Technology Center, Tufts University, Medford, MA 02155, USA

Giyuu Kido Institute of Atomic Energy, Kyoto University, Kyoto, Japan

Thomas Klaus Physikalisches Institut, Universität zu Köln, D-50937 Köln, Germany

Egbert Klisch Physikalisches Institut, Universität zu Köln, D-50937 Köln, Germany

Fritz K. Kneubühl Institute of Quantum Mechanics, Infrared Physics Laboratory, Swiss Federal Institute of Technology (ETH), CH-8093 Zürich, Switzerland

Walter D. Knight Department of Physics, University of California at Berkeley, Berkeley, CA 94720-7300, USA

John H. Lacy Department of Astronomy, University of Texas, Austin, TX 78712-1083, USA

Willis E. Lamb, Jr. Optical Science Center, University of Arizona, Tucson, AZ 85721, USA

Harold Lecar Department of Molecular and Cell Biology, University of California at Berkeley, Berkeley, CA 94720, USA

Leon M. Lederman Enrico Fermi Laboratory and University of Chicago, Batavia, IL 60510, USA

Vladilen S. Letokhov Institute of Spectroscopy, Russian Academy of Sciences, 14292 Moscow Region, Russia

Philip R. Maloney NASA Ames Research Center, Moffett Field, CA 94035-1000, USA

Demetrios N. Matsakis United States Naval Observatory, Washington, DC 20392-5420, USA

Michael May 728 East Angela Street, Pleasanton, CA 94566, USA

Christopher F. McKee Space Sciences Laboratory, University of California at Berkeley, Berkeley, CA 94720-7450, USA

Peter G. Mezger Max-Planck-Institut für Radioastronomie, D-53121 Bonn, Germany

Georg M. Meyer Department of Physics, Texas A&M University, College Station, TX 77843, USA

Takeshi Morimoto Institute of Atomic Energy, Kyoto University, Kyoto, Japan

Elna M. Nagasako Institute of Optics, University of Rochester, Rochester, NY 14627, USA

A. Penzias AT&T Bell Telephone Laboratories, Murray Hill, NJ 07974-0630, USA

Francesco A. Pepe Institute of Quantum Mechanics, Infrared Physics Laboratory, Swiss Federal Institute of Technology (ETH), CH-8093 Zürich, Switzerland

Victor O. Ponomarev P.N. Lebedev Physical Institute, Russian Academy of Sciences, 117924 Moscow, Russia

Alexander M. Prokhorov General Physics Institute, Russian Academy of Sciences, 117942 Moscow, Russia

Norman F. Ramsey Lyman Physics Laboratory, Harvard University, Cambridge, MA 02138, USA

Robert J. Russell Center for Theology in the Natural Sciences, Berkeley, CA 94709, USA

Bernard Sadoulet Center for Particle Astrophysics, Department of Physics, University of California at Berkeley, Berkeley, CA 94720, USA

Goran Sandell Joint Astronomy Center, University Park, Hilo, HI 96720, USA

Arthur L. Schawlow Department of Physics, Stanford University, Stanford, CA 94305-4060, USA

Wolfgang Schleich Max-Planck-Institut für Quantenoptik, D-85748 Garching, Germany

Marlan O. Scully Department of Physics, Texas A&M University, College Station, TX 77843, USA

Eugene Serabyn Division of Physics, Mathematics, and Astronomy, California Institute of Technology, Pasadena, CA 91125, USA

Koichi Shimoda 1-19-15 Kichijoji-Minamicho, 180 Tokyo, Japan

Robert J. Smalley DuPont Experimental Station, Wilmington, DE 19880-0356, USA

Howard A. Smith Laboratory for Astrophysics, National Air and Space Museum, Smithsonian Institution, Washington, DC 20560, USA

Boris P. Stoicheff Department of Physics, University of Toronto, Toronto M5S 1A7, Canada

John W.V. Storey School of Physics, University of New South Wales, Sydney 2052, Australia

Vladimir S. Strelnitski New Mexico Institute of Mining and Technology, Socorro, NM 87801, USA

Edmund C. Sutton Department of Astronomy, University of Illinois, Urbana, IL 61801, USA

Jean-Pierre E. Taran Office National d'Etudes et de Recherches Aérospatiales, 92322 Chatillon, France

Herbert Walther Max-Planck-Institut für Quantenoptik, D-85748 Garching, Germany

Herbert Walther Sektion Physik, Ludwig-Maximilians-Universität, D-00000 München, Germany

Kenneth M. Watson Marine Physical Laboratory, Scripps Institution of Oceanography, University of California at San Diego, La Jolla, CA 92152, USA

David H. Whiffen 16 St. Andrews Road, Sogursey near Bridgewater, Somerset TA5 1TE, United Kingdom

John R. Whinnery Department of Electrical Engineering and Computer Sciences, University of California at Berkeley, Berkeley, CA 94720, USA

Gispert Winnewisser Physikalisches Institut, Universität zu Köln, D-50937 Köln, Germany

Eric R. Wollman Department of Physics and Astronomy, Bates College, Lewiston, ME 04240, USA

Herbert J. Zeiger Department of Electrical Engineering and Computer Science, Massachusetts Institute of Technology, Cambridge, MA 02139, USA

Robert Zylka Max-Planck-Institut für Radioastronomie, D-53121 Bonn, Germany

1

Introduction: Charles Townes as I Have Known Him

Arthur L. Schawlow

Charles Townes's many important discoveries rank him high among the greatest scientists of his generation. Certainly they attest to his extraordinary scientific talents. He has a powerful command of the tools of both theoretical and experimental physics. He also has remarkable personal qualities which have contributed greatly to his success as a scientist, an academic leader, and an advisor to government and industry.

I have been fortunate to have known Charles Townes for more than forty-five years and to have worked with him for several of those years. It began in 1949 when, as the work for my Ph.D. at the University of Toronto was nearing completion, I was awarded a Carbide and Carbon Chemicals Corporation postdoctoral fellowship. This fellowship had been established a year earlier for research on the applications of microwave spectroscopy to chemistry to help support Charles Townes's program at Columbia University. Dr. Helmut Schulz, who instigated the fellowship, hoped that Townes' research along with the millimeter wavelength magnetron tubes being developed at the Columbia Radiation Laboratory might lead to some method of controlling chemical reactions with radiation of wavelengths shorter than radio waves but longer than visible light.

We first met at a meeting of the American Physical Society in Washington, D. C., near the end of April 1949. He greeted me pleasantly, and we had some opportunity to get acquainted. I had not known much about his work until then, but I was greatly impressed by the fact that he was giving two invited papers at that meeting. It is considered an honor to be asked to give such a paper, for which half an hour is allotted rather than the ten minutes allowed for a contributed paper; to give two invited papers was nearly unprecedented. One of these was on molecular microwave spectra of some molecules. The other was on the use of these narrow microwave absorption lines as frequency standards, that is, for controlling an atomic clock.

Because he had arranged to spend the summer at Brookhaven National Laboratory, it was not possible for me to start at Columbia until September. Meanwhile, he suggested that I should read Linus Pauling's book, *The Nature of the Chemical Bond*. This was particularly appropriate because at that time a considerable part of his effort, along with his colleague Benjamin Dailey in Columbia's Chemistry Department, was directed toward understanding the relationship between chemical binding and the hyperfine structures which they could observe in microwave spectra.

At that time, his research group consisted of about ten graduate students and Jan Loubser from South Africa, my predecessor as Carbide and Carbon Chemicals postdoctoral fellow. Soon after, we were joined by Eilif Amble, another postdoctoral fellow from Norway. The group's laboratory occupied about half of the tenth floor of the Pupin Physics Laboratory and a large room on the eleventh floor. The other half of the tenth floor was given over to the molecular and atomic beam experiments supervised by Professors I. I. Rabi and Polycarp Kusch.

I was given a laboratory room adjacent to the atomic beams group. Charles Townes suggested that I try to observe the microwave spectrum of the free radical molecule OH. This molecule, like most of those with odd numbers of electrons, is not stable by itself, but easily forms compounds with many other atoms and molecules. However, OH molecules can be produced in electrical discharges through such gases as water vapor. Even then he realized that OH would be one of the most abundant molecules in astronomical objects such as galaxies and planetary atmospheres. Indeed, the OH spectrum was one of the first observed by microwave telescopes and has led to the rich field of astronomical spectroscopy at centimeter and millimeter wavelengths.

In our early search, the OH spectrum proved to be elusive. Although, with the collaboration of T. Michael Sanders and Wilton A. Hardy, we constructed a spectrometer that used Zeeman modulation so as to be sensitive only to free radicals, for a long time we found nothing. The problem was that we had no direct way of being sure that our gas discharge was producing free radicals. Eventually, a different set of conditions for the discharge was found to be effective in producing OH so that its spectrum was observed and measured. With this information, radio astronomers later could know at what frequencies to look and could use shifts of the line frequencies to infer the motion of the astronomical object relative to the earth. In this, as in so many other instances, Charles Townes's vision proved to be penetrating and far-sighted.

Indeed, he was full of ideas for good experiments and inspired a large group of graduate students working on a wide range of projects. They would meet weekly at a seminar where a student would present research or some recent publication of interest. Good ideas would emerge from these discussions. Both at the seminars and in private conversations Townes had a remarkable way of helping a student to develop a half-formed idea without taking it away from him. This was an excellent way to help students to learn how to formulate their own ideas and have confidence in them. Many of his students have gone on to brilliant research careers. He was a master of all the relevant theory and of experimental techniques. Many a student was surprised by how quickly he could find out what was wrong with the apparatus and get it working. His students were greatly impressed by what he could do with them, but perhaps even more by the fact that in Smythe's book on electromagnetic theory, the author thanked Charles Townes for working out solutions for all of the (notoriously difficult) problems in the book.

The breadth of his knowledge and interests in physics was also striking. He used microwave spectroscopy, not only to study chemical bonding but also to measure nuclear quadrupole moments and masses with great precision.

One of his most remarkable attributes was his ability to concentrate, despite interruptions. If anyone called him or came to see him he would give that person his complete attention. As soon as that was finished, he would return to the task that had been interrupted and immediately continue, apparently without having to spend time picking up the thread of his thoughts. He planned his research thoroughly and in detail. When he, with Jim Gordon and Herb Zeiger were working to build the first ammonia beam maser, two of his distinguished colleagues with extensive experience in atomic and molecular beams tried to persuade him to abandon the project as unworkable. However, he had analyzed the problems carefully and felt confident that the difficulties could be overcome, and they soon were.

He was also concerned for the personal well-being of his students, and entertained all of them at his home from time to time. Usually this was very much appreciated, but there was at least one surprise. One graduate student had come from a distant country, one from which communication had been cut off, and had left his wife and family behind. Frances and Charles invited this man to their home every year for Thanksgiving dinner. But after a few years he told Charles that he could not come any more because his wife did not like it! Everyone was quite surprised that he had acquired a wife, and they never did find out how it happened. Frances Townes was a very gracious hostess, also interested in the people in his research group. I came to know their family of four lovely daughters. It was said around Columbia that physicists had only daughters, not sons, and that was true for the faculty members I knew. Although Charles Townes worked long and hard, he always made time for his family. He was in his office or laboratory six days of the week, but never on Sundays. The family attended the Riverside Church, and both Charles and Frances were active in church affairs. He has always been, and remains, an eloquent spokesman for the compatibility of science and religion. Although he worked hard for six days and nights, Sundays were always kept clear for church and family. Often they would take walks or excursions to renew his childhood interest in nature.

It was Frances Townes who made sure that I became acquainted with Charles's younger sister, Aurelia, when she came to New York in 1950 to continue her musical studies. It didn't take me long to realize what a wonderful person Aurelia was, and we were married in May 1951. Although everything I have done in physics since then has been enormously aided and influenced by what I learned from Charles Townes, I have to say that meeting Aurelia was the best thing that happened to me in New York. After our marriage I became acquainted with his parents, two brothers, and three sisters. Theirs was a truly remarkable family of extremely intelligent and talented people.

I. I. Rabi was the Columbia Physics Department chairman and had attracted an outstanding group of professors, researchers, and students to Columbia. There were no less than eleven future Nobel laureates around during the time I was at Columbia—H. Yukawa, W. Lamb, P. Kusch, C. Townes, J. Rainwater, A. Bohr, V. Fitch, L. Lederman, M. Schwartz, J. Steinberger, and myself. Even in this galaxy Charles Townes was a brilliant star, beginning to receive many honors from the United States and abroad.

Although he has been universally honored and respected, I know of a couple of times that he "didn't get no respect." Once when he was a young man he bought an accordion and began learning to play it. Then he took it with him on a train trip through Mexico. His fellow passsengers urged him to play, and he responded with some lively Mexican songs. Soon most of the people in the car were singing along and having a great time. Then one of the Mexicans said, "Play us an American song." Not having practiced very many songs he responded with "Old Black Joe." This was not the peppy music they were expecting and was met with momentary silence. Then one of the Mexicans said, "You play another one like that we kill you." You cannot please everyone, but Charles does have considerable musical talent, and he and Aurelia would sometimes sing duets.

Another occasion, when I was at lunch in the Columbia Faculty Club with Charles and a few of his colleagues, the discussion turned to two articles which had just appeared in *Fortune* magazine. The science editor, Francis Bello, had picked ten outstanding scientists under forty in universities, and ten in industry, and had tried to draw conclusions about what they had in common. One thing he noted was that they were all oldest sons or only sons. Charles remarked that it did not seem right to him, for he had an older brother and two older sisters. Thereupon Rabi squelched him by saying, "You didn't make the list, did you?" There cannot have been many lists since then of the outstanding scientists of the twentieth century that failed to include Charles Townes.

I left Columbia in September 1951 to begin physics research at Bell Telephone Laboratories in Murray Hill, New Jersey. A few months before that Charles Townes had conceived the idea of the maser, and I had witnessed the disclosure in his notebook (see figures on pages 5 and 6). My research at Bell Labs was on super-conductivity, and I did not participate at all in the exciting developments on masers during the next few years. However, I continued to go to Columbia on most Saturdays to work with Charles on our book, *Microwave Spectroscopy*, which was published in 1955. Then in 1957 he was consulting at Bell Labs, and I was beginning to think seriously about the possibility of extending the maser principle from microwaves to shorter wavelengths, such as the infrared region of the spectrum. It turned out that he was also thinking about this problem, and so we decided to look at the problem together. Without interrupting our other duties, over the next few months we worked on this in odd moments. Again he was enormously helpful in clarifying my ideas as well as contributing his own. By the spring of 1958 we were convinced that it would be possible to build an optical maser, which is now known as a laser. We submitted our work to the *Physical Review* and it was published December 1958.

Charles and I have had no further opportunities for collaboration, but we have kept in close touch even though our research fields have diverged. Masers provided tools for him to begin some innovative microwave astronomy. Subsequently, he has used maser techniques to aid in infrared astronomy.

Despite heavy commitments that would overwhelm anyone else, he is generous with his time when needed. In 1985 it became apparent that the group home where our nonverbal autistic son lives was on the verge of bankruptcy and would close unless someone else took charge of it. Aurelia and I helped to organize a nonprofit corporation, California Vocations, for that purpose and asked Charles to be a director. He has been serving on the board of directors ever since then, and his counsel has been invaluable. He is a wise man, who listens to what everyone says before coming to a conclusion. Quite often he will recognize some aspect of the discussion that others have overlooked.

In all, I feel very fortunate to have known Charles Townes. He is a marvelous scientist and an inspiration to everyone.

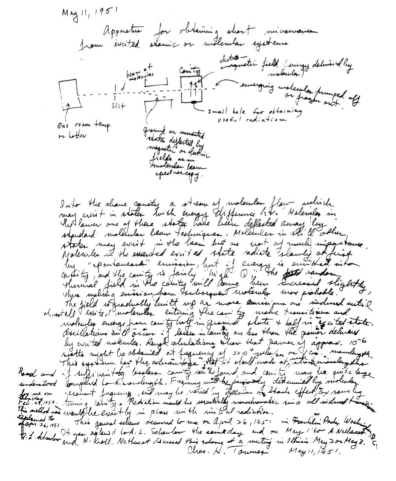

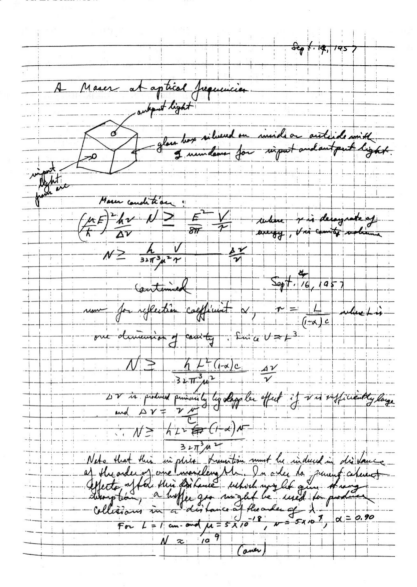

Sept. 14, 1957

A Maser at optical frequencies.

output light

glass box silvered on inside or outside with
windows for input and output light.

input
light
from arc

Maser condition:

$$\left(\frac{\mu E}{h}\right)^2 \frac{h\nu}{\Delta\nu} \quad N \geq \frac{E^2}{8\pi}\frac{V}{r} \qquad \text{where } r \text{ is decay rate of energy, } V \text{ is cavity volume}$$

$$N \geq \frac{h}{32\pi^3\mu^2 r} \frac{V}{\Delta\nu}{\nu}$$

continued   Sept. 16, 1957

now for reflection coefficient α, $\quad r = \frac{L}{(1-\alpha)c}$ where L is
one dimension of cavity. Since $V = L^3$

$$N \geq \frac{h L^2(1-\alpha)c}{32\pi^3\mu^2}\frac{\Delta\nu}{\nu}$$

$\Delta\nu$ is produced primarily by Doppler effect if ν is sufficiently large
and $\Delta\nu = \nu\frac{\nu}{c}$

$$\therefore N \geq \frac{h L^2 (1-\alpha)\nu}{32\pi^3\mu^2}$$

Note that this implies transition must be induced in distance
of the order of one wavelength. In order to prevent coherent
effects after this distance which might give strong
absorption, a buffer gas might be used for producing
collisions in a distance of the order of λ.
For $L = 1$ cm. and $\mu = 5\times10^{-18}$, $\nu = 5\times10^{9}$, $\alpha = 0.90$

$$N \approx 10^9$$

(over)

2

Methane Optical Frequency Standard

Nicolai G. Basov
Vladimir A. Alekseev
and Mikhail A. Gubin

2.1 Introduction

The creation of masers and lasers is one of the most important achievements of the 20th century. The development of quantum electronics began in the middle of the 1950s [1, 2] and led to completely new approaches to the solutions of numerous physical and technical problems. This article is dedicated to Prof. Charles Townes on his 80th birthday and presents a prominent example of the progress connected with quantum electronics.

From its early stage, the investigations at Columbia University and at the Lebedev Institute made it clear that masers and lasers opened a new era in frequency, time, and length measurements, which allow, in particular, the creation of a unified time and length standard. Nowadays the units of time and length are the most precise ones. The frequency reproducibility and accuracy of the established Cs^{133} time (frequency) standard is now as high as $\Delta\omega/\omega_0 = 10^{-13}$, where $\Delta\omega$ is a frequency uncertainty and ω_0 is the central frequency. Together with another microwave source, the hydrogen maser [3], which possesses very narrow spectral width of radiation and very low frequency drift ($\sim 10^{-15}$ per day), the cesium standard sets the international atomic time/frequency scale. The state of the art and further progress of this scale are important for numerous applications in metrology, physics and navigation.

The high accuracy of the cesium standard is based on the small value of transition line width γ and very small ratio $\gamma/\omega_0 \sim 10^{-8}$–$10^{-9}$, which was achieved due to dramatic progress in the methods of radio spectroscopy [4]. Nowadays the methods of laser spectroscopy permit the elimination of the Doppler broadening [5], resulting in resonances with the relative width $\gamma/\omega_0 \sim 10^{-11} \ldots 10^{-12}$ in the optical domain of spectra. On the other hand, the radio-optical frequency chains allow the measurement of the frequency of optical transition in units of the cesium microwave transition frequency [6, 7]. Thus the foundation is prepared for transition to a unified optical standard of length and frequency.

Various methods are used to detect optical nonlinear resonances [8, 9]. This article is devoted to the optical frequency standard which is based on the methane infrared transition ($\lambda = 3.39\,\mu m$) using the so-called "two-mode" method of the resonance detection.

In Sect. 2 we give the description of the experimental setup and the method of resonance detection which yields transitions with minimal relative line widths of $\gamma/\omega_0 \cong 4 \cdot 10^{-12}$. In Sect. 3 the results of the methane resonance frequency measurements carried out by direct comparison of the methane transition frequency with the frequency of a primary cesium standard are given. These measurements were carried out in Germany in collaboration with G. Kramer and B. Lipphardt using a transportable methane standard constructed at the Lebedev Physical Institute and the radio-optical chain and the cesium frequency standard continuously operating at Physikalische-Technische-Bundesanstalt (PTB, Braunschweig, Germany).

In Sect. 4 the procedure and results of the determination of the unperturbed transition frequency of the reference methane line ($\lambda = 3.39\,\mu m$) are described.

In Sect. 5 we give theoretical estimates of the accuracy of the optical standard.

2.2   Two-mode He-Ne Laser with a Methane Absorption Cell

The saturation of the inhomogeneously broadened spectral line by counter-propagating laser waves results in a resonant dependence of all characteristics of the laser field: its intensity, state of polarization, phase and frequency. Thus, in principle, narrow optical resonances can be detected by controlling the variation of any of these values.

An ideal method of frequency registration must exclude the low frequency technical noise of laser radiation, and provide a quantum sensitivity limit, determined by the shot noise of a photodetector and by the natural fluctuations of the laser's frequency (the Schawlow-Townes limit [10]).

In our experiment we used a two-mode technique of saturated dispersion registration which permits us to achieve the quantum limit of sensitivity [11].

The principle of the technique is shown in Fig. 2.1. A laser, containing a non-linear absorption cell, operates in a two-mode regime. If the spacing of mode frequencies ω_1 and ω_2 satisfies the condition

$$\Delta\omega_D \gg \omega_{12} \equiv \omega_1 - \omega_2 \gg \gamma, \qquad (2.1)$$

where $\Delta\omega_D$ is the Doppler line width, a dispersion-shape resonance, caused by the effect of a non-linear frequency pulling towards the line center ω_0, is observed in the mode beat frequency against the slow-varying background. This happens when the frequency of one of the modes coincides with the central frequency ω_0 of the Doppler-broadened absorption line. The second mode ω_2, which does not coincide with the line center, plays the role of an internal heterodyne signal. The frequency fluctuations caused by technical noise, for example, by mechanical vibrations of cavity elements, are practically the same for both modes, so that they are subtracted

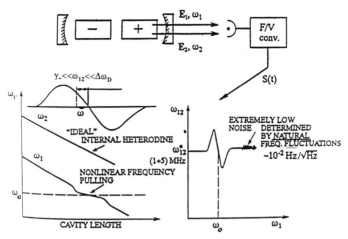

FIGURE 2.1. Principle of frequency registration of the saturated dispersion resonances ("frequency resonances") in a two-mode laser with an absorption cell. ω_{12}^0 is the mean value of the mode beat frequency in the He-Ne/CH$_4$ laser used, which oscillated at two orthogonal-polarized modes, ω_{12}^0 was in the range 1-5 MHz. F/V converter denotes a frequency-to-voltage converter (frequency detector).

and do not influence the beat frequency. The realization of the two-mode technique for He-Ne lasers with a CH$_4$ absorption cell led to an increase of the sensitivity of the frequency registration compared to the resonances of saturated absorption recorded via the intensity of the laser field. As a result a number of relatively compact and reliable He-Ne/CH$_4$ optical frequency standards were developed with the level of reproducibility and accuracy of $\sim 10^{-13}$.

Optical Stark broadening is the main factor limiting the resolution of nonlinear laser spectroscopy with internal absorption cells. It is proportional to laser intensity and can be reduced at low laser intensities but in this case the laser operates near the threshold which results in an increase of both technical and natural noise. For this reason, for obtaining resonances with the width $\gamma \leq 1$ kHz, it is necessary to use the telescopic expansion of the laser beam inside the absorption cell.

Figure 2.2 shows the main components of the experimental setup for high-precision spectrometry of the $F_2^{(2)}$-line of methane based on two-mode He-Ne lasers with intracavity telescopes. It involves the following lasers: a narrow-band reference laser L1, a heterodyne laser L2, a transportable optical frequency standard L3 (TE-30T), and a stationary laser L4 (TE-200) with a beam waist $2w_0 \cong 200$ mm. Lasers L1 and L2 provide high-power reference radiation at a frequency shifted from the line center of methane by some fixed interval (600 kHz in our case). In the frequency standard L3 only a laser with a telescope (beam waist $2w_0 \cong 30$ mm) was used. The full scheme of this standard is described in [12]. For the detection of optical resonances, study of their shapes, and the comparison of the laser frequencies, automatic phase-frequency control systems, based on an IBM PC/AT, were developed. These systems allowed specifically the realization of synchronous fre-

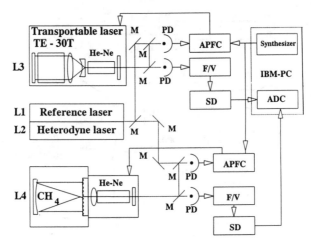

FIGURE 2.2. Scheme of a high precision laser spectrometer using two-mode He-Ne/CH$_4$ lasers: L1-reference two-mode narrowband laser; L2-single-mode heterodyne laser; L3-transportable laser with a telescope; L4-stationary laser with a telescope; M denotes mirror; PD denotes photodectector; APFC denotes system for automatic phase frequency control; F/V denotes frequency-to-voltage converter; SD denotes synchronous detector; ADC denotes analog-to-digital converter.

quency tuning of lasers with telescopes within the absorption line. The board of the analog-to-digital converter includes two channels for the simultaneous recording of spectra of saturated dispersion and mathematical processing.

Figure 2.3 shows saturated dispersion resonances recorded under typical experimental conditions at both lasers with telescopic expanders. At the transportable laser (TE-30T) the magnetic hyperfine structure of the $F_2^{(2)}$-methane line is resolved ($\gamma \cong 5 \ldots 7$ kHz, $\gamma/\omega \cong (5 \ldots 7) \cdot 10^{-11}$), at the stationary one (TE-200) the recoil doublet is resolved ($\gamma \cong 400 \ldots 800$ Hz, $\gamma/\omega_0 \cong (4 \ldots 8) \cdot 10^{-12}$).

2.3   Absolute Frequency Measurements

The reference methane lines with relative widths of the order of $5 \cdot 10^{-12}$ obtained at the stationary laser permit achieving a frequency reproducibility at the level of $10^{-13} - 10^{-14}$, comparable to a primary cesium standard. The accuracy of the transportable laser (TE-30T) is lower, but its attractive feature is a possibility of direct frequency comparison with the cesium standard. The recent progress in the radio-optical frequency chains allows one to multiply the frequency from the microwave to the optical region without losing the accuracy determined by cesium standard. The transportable standards make possible the comparison and verification of complicated radio-optical chains developed in different countries.

For clarifying the metrological possibilities of the synthesized standards, two types of studies were made. The first one was carried out at the radio-optical fre-

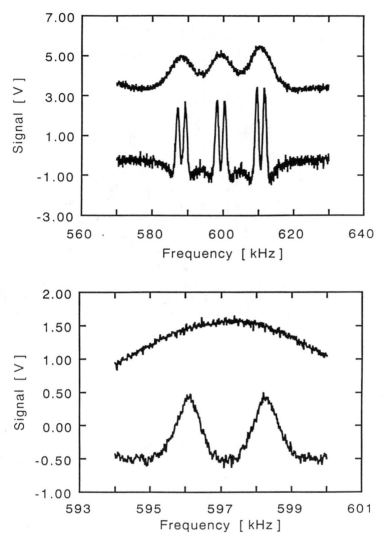

FIGURE 2.3. (a) Magnetic hyperfine structure and recoil doublet (first derivative of saturated dispersion) recorded with the help of TE-30T and TE-200 lasers. (b) More precise record of the central component.

quency chain at the PTB (Braunschweig), where the frequency of the transportable laser was compared directly with a cesium standard. Before and after this absolute frequency measurement, the frequency of the transportable laser was compared to the frequency of the stationary one at the Lebedev Institute. The results of both types of measurements permit claiming that during the whole period of study, the frequency of the transportable laser remained stable with the accuracy $3 \cdot 10^{13}$.

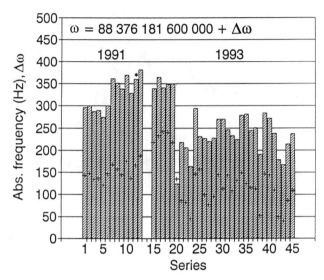

FIGURE 2.4. Results of the absolute frequency measurements. Series 1-6 and 19-34: transportable optical frequency standard was adjusted to optimal conditions (Γ = 6.4 kHz); series 7-18, laser power was double. * denotes analog servo, + denotes ω_0 values calculated according to Eq. (2.2).

Figure 2.4 presents the results of the absolute frequency measurements (AFM) [13]. Thirteen series of AFM with a transportable laser were made during November 1991, and 31 series during November-December 1993. Each series lasted from 0.5 to 4 hours. Only a single parameter—the output power of the telescopic laser (TE-30T)—was changed. The series from 1 to 6 and from 19 to 44 were made under optimal conditions. These conditions correspond to the laser power and methane pressure resulting in a total resonance width (including collisional + optical Stark + transit-time broadening) of about γ = 6.4 kHz. Computer processing of the magnetic hyperfine structure spectra recorded before and after each series of the AFM allowed us to find the parameters of the reference line during the AFM and to compare them with the parameters which were determined at frequency comparisons with the stationary laser "TE-200". This gave the possibility of having a better repeatability of the transportable laser parameters.

The results of short- and medium-term stability measurements are presented in Fig. 2.5. At sample times less than 10 s the Allan variance of the optical frequency standard is better than that of the hydrogen maser which serves as a reference for the chain. At longer times quasi periodical (200-400s) oscillations of the laser frequency (about 50 Hz deviation) were found. These oscillations arise mainly due to accidental acoustic perturbations and the delayed reaction of the slow feedback loop of the computer servo. (At the present time these dynamic errors are strongly suppressed with a new software version.)

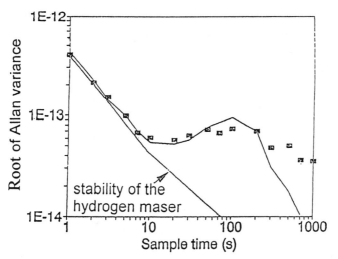

FIGURE 2.5. Allan variance of the laser frequency fluctuation measured on the chain (computer servo is on). Squares denote data of 1991, solid curve denotes data of 1993. Improvement has been achieved due to new version of software.

Taking into account the data of AFM for two years we have determined the frequency reproduced by the laser "TE-30T":

$$\omega_{(laser)} = 88\,376\,181\,600\,260 \pm 90\text{Hz}.$$

The uncertainty of this value was determined as 2σ, where σ is a standard deviation of a frequency value with 31 series of the AFM taken into account. No additional correction of the laser frequencies has been applied.

During absolute frequency measurements carried out at PTB, the laser was stabilized using feedback error, derived by a parabolic approximation. To reduce the relatively high value of the frequency shift, using this method a correction of the stabilized frequency was made before and after the process of frequency measurement. The value of correction was calculated using modeling (see below) of the observed line shape. The corrected values are marked as crosses in Fig. 2.4.

The use of this method results in a frequency uncertainty of 24 Hz (standard deviation) over a period of 1-2 months for the transportable laser. Over a period of two years some frequency drift was observed, which cannot be explained in terms of the model used. However, due to the frequency comparisons with the stationary device this drift did not cause an additional systematic error of unperturbed transition frequency determination.

2.4   Determination of the Unperturbed Transition Frequency ($\lambda = 3.39\,\mu$m)

There are many effects which influence the shape of the resonances shown in Fig. 2.3. Among them are the detuning of line centers of active and absorption media, the transverse gain inhomogeneity, the second-order Doppler shift, the magnetic hyperfine structure of the line, the so-called "crossing" resonances and many others. Almost each of these effects results in asymmetrical distortions of resonance line. For this reason we used the following procedure for line-center determination. The observed signal $S(\omega)$ was approximated by the expression

$$S(\omega) = \sum_{i=1}^{4} A_i [D'(\omega - \omega_0 - \Delta_i \pm \delta, \gamma_i) + \alpha L'(\omega - \omega_0 - \Delta_i \pm \delta, \gamma_i)] \quad (2.2)$$

where i is the index of spectral components (three components of magnetic hyperfine structure plus the nearest group of crossing lines), A_i are amplitudes, L' and D' are the functions of first derivatives of absorption and dispersion, respectively, α is the factor of asymmetry, ω_0, Δ_i, δ are the frequency of the unperturbed CH_4 transition, component shift, and recoil splitting, respectively, and γ_i is the width of the separate component. The analytical formula, Eq. (2.2), was compared with experimental spectra. By the method of least-squares fitting, the parameters γ_i, α, and ω_0 were determined. The values ω_0 determined in experiments with the transportable laser are shown by the crosses in Fig. 2.4. In the case of the stationary laser, instead of the simple formula Eq. (2.2), a more rigorous and complicated one which contains the so-called transit-time integral [14] was used. The frequency of transportable laser serves as a reference in this measurement.

The systematic laser frequency shifts were usually less then 100 Hz in the experiments. Multiple recording of the resonance and averaging of the corresponding laser frequency shifts allowed us to reduce statistical uncertainty below 10 Hz.

Figure 2.6 presents the frequency difference between two lasers, using the stationary laser as a reference. The transportable laser was operating at different output powers. The data show little power dependence of the unperturbed transition frequency, when the position of line center was calculated on the basis of a theoretical model expressed by Eq. (2.2) (curve 1). Curve 2 gives the frequency of the peak of the line, determined by the parabolic approximation. As the top of the parabola coincides approximately with the zero crossing of the derivative, the results reflected by curve 2 give approximately the same value of frequency shift as for lasers with analog servo control using a second derivative of dispersion.

The influence of the He-Ne pressure on the line asymmetry and line-center frequency determination was studied in order to estimate the accuracy of the stationary laser (Fig. 2.7). He-Ne pressure is one of the most critical parameters for this laser, because the pressure induces frequency shift of Ne-line with respect to the CH_4 resonance and changes the line asymmetry. The shape of resonance at different Ne pressure was compared with the theoretical model, which allowed

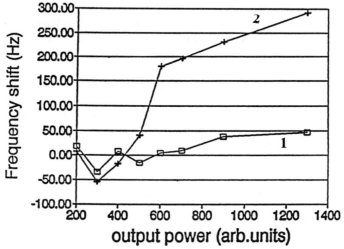

FIGURE 2.6. Power shift of the transportable laser with respect to the stationary laser. Curve 1 denotes the frequency ω_0 evaluated from Eq. (2.2); curve 2 denotes the frequency ω_0 resulting from the parabolic approximation.

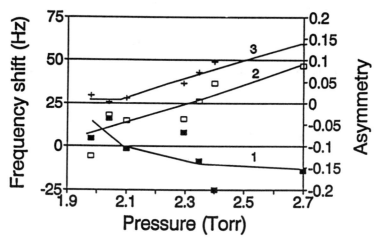

FIGURE 2.7. Frequency shift and line asymmetry of the stationary laser as a function of pressure-induced detuning of the Ne-line. Calculation was performed on the base of a complicated model (curve 1 corresponds to the frequency axis, curve 2 corresponds to the asymmetry axis) and a simple one (curve 3 corresponds to the asymmetry axis). The transportable laser served as the reference.

us to determine the asymmetry coefficient α and the line center position ω_0. The calculations with a simple model given by Eq. (2.2), and the more complicated ones with transit-time integrals have given approximately the same uncertainty in ω_0 determination $| \Delta\omega | = 10$ Hz. Power broadening and cavity-mirrors alignment give smaller uncertainty. The influence of these factors was also studied.

The value ω_0 determined by this procedure does not coincide with the position of the experimental line-shape maximum. This difference was taken into account and allowed us to make the correction to line-center position. The value of the unperturbed transition frequency ω_{UTF} was determined equal to

$$\omega_{UTF} = 88\,376\,181\,600\,106 \pm 55\text{Hz}.$$

The optical frequency chain, developed at PTB [7], demonstrated reliable operation, and the most significant source of the error appeared from the uncertainty of the transportable laser frequency.

2.5   Theoretical Estimates of Methane Standard Accuracy

2.5.1   Second-order Doppler Effect

The shift caused by this effect is one of the main obstacles to the reproducibility of the molecular transition frequency. When collisional or power broadening Γ dominates $\Gamma \gg 1/\tau$, the shift of resonance is equal to

$$\Delta = -\Omega_D = -\frac{1}{2}\omega_0 \frac{v_0^2}{c^2} = -\omega_0 \frac{k_B T}{mc^2}, \tag{2.3}$$

where m is the mass of molecule. In the case of methane at room temperature T the value Ω_D is 150 Hz.

When transit-time broadening is comparable with Γ, which is typical for our experimental conditions, $\Gamma\tau \leq 1$, due to the influence of slow molecules [15], the shift Δ caused by second-order Doppler effect depends on the parameters $\Gamma\tau$ and $\delta\tau$, where δ is the splitting due to the recoil effect.

This dependence was calculated numerically [14]. In the case of the line split by the recoil effect, the position of a line center was determined as a half-sum of the frequencies corresponding to the maxima of recoil components. The dependence of this position on Γ values for different transit-time parameters $1/\tau$ is shown in Fig. 2.8. In the case of unsplit lines, which occur for larger Γ and $1/\tau$ values, the center of dispersion resonance position is shown in Fig. 2.9. Note that for $\Gamma \geq 4$ kHz values, the decreasing of the shift mirrors the influence of hyperfine structure of $F_2^{(2)}$ methane line, which also was taken into account in the numerical calculation [14]. For small Γ and $1/\tau$ values (Fig. 2.8) this influence is not important. In the case of the stationary laser TE-200 the transit-time parameter is approximately equal to $1/\tau \approx 1000$ Hz with an uncertainty ± 200 Hz which reflects the uncertainty of the laser beam radius determination. The comparison of the calculated shape of resonance (with $1/\tau = 1000$ Hz) with the experimental one (Fig. 2.3) permits determining the homogeneous linewidth $\Gamma = (350 \pm 20)$ Hz. From Fig. 2.8 we see that the shift Δ of the line due to the second-order Doppler effect does not exceed $\Delta \leq 10$ Hz. In the case of the transportable laser the transit-time parameter

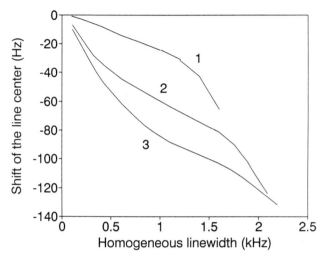

FIGURE 2.8. The shift of the methane dispersion line center due to second-order Doppler effect ($\Omega_D = 150$ Hz) for different transit-time τ values: curve 1 has $1/\tau = 1100$ Hz, curve 2 has $1/\tau = 440$ Hz, curve 3 has $1/\tau = 275$ Hz.

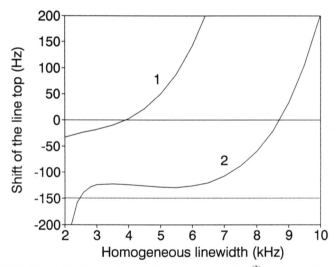

FIGURE 2.9. The shift of central hyperfine component of $F_2^{(2)}$-methane dispersion line ($\lambda = 3.39\,\mu m$) due to the second-order Doppler effect for different transit-time τ value: curve 1 has $1/\tau = 6$ kHz, curve 2 has $1/\tau = 1100$ Hz.

is approximately equal to $1/\tau \approx 6$ kHz. The comparison of the calculated line shape with the experimental one shown in Fig. 2.3 gives the homogeneous line width value $\Gamma \approx 3.5$ kHz. As seen from Fig. 2.9, due to the compensation of shifts caused by the second-order Doppler effect and by the influence of the magnetic hyperfine structure of the line, the total shift as in the case of a stationary laser does not exceed the value $\Delta \leq 10$ Hz.

2.5.2   Detuning of the Active and Absorption Line Centers

The shift of the dispersion resonance caused by this detuning [16] is given by:

$$\Delta = \frac{1}{2} \frac{\Gamma}{\gamma_+} (\omega_0 - \omega_+), \tag{2.4}$$

where ω_+ and γ_+ are the line center and a homogeneous width of the active medium, respectively. The detuning also gives the asymmetry of resonance which can be used as a signal for controlling the value $(\omega_0 - \omega_+)$ in the experimental conditions. This permits restriction of the shift to a value $|\Delta| \leq 10^{-2}\Gamma$ [16]. This means that in the case of the stationary laser, $|\Delta| \leq 4$ Hz and for the transportable laser, $|\Delta| \leq 40$ Hz.

2.5.3   Transverse Inhomogeneity of the Gain

The transverse inhomogeneity of the gain leads to the shift of dispersion resonance [16]

$$\Delta = \chi G_+ \Gamma, \quad G_+ = \frac{g_+}{8\pi} \frac{\lambda R}{R_+^2}, \quad G \ll 1, \tag{2.5}$$

which depends on the dimensionless increment of a single pass through amplifying medium g_+, a curvature radius R of the output mirror, and on the transverse dimension of active medium R_+. The coefficient χ is of the order of unity and changes its value from $\chi = 1/3$ when $\Gamma \tau \gg 1$, to the value $\chi \cong -1/2$ when $\Gamma \tau \ll 1$. This means that under typical experimental conditions ($g_+ = 0.5$, $R = 400$ cm, $R_+ = 0.3$ cm, $\lambda = 3.39 \cdot 10^{-4}$ cm, $|\chi| \approx 1/2$) $|\Delta \leq 10^{-2}\Gamma|$. This large value of the shift shows that transverse inhomogeneity of the gain medium can be one of the main sources of the standard's uncertainty. In the case of a stationary laser this gives $|\Delta| \leq 4$ Hz and for a transportable laser $|\Delta| \leq 40$ Hz. These values are comparable with the shifts caused by the detuning of the active and absorption line centers.

Among the other effects, the influence of the transverse field profile and the influence of collisions are worth mentioning. The shift caused by the transverse field profile is very similar, but to some degree smaller, than the shift caused by transverse inhomogeneity of the gain.

The upper limit of the collisional shift can be estimated. In the case of methane transition this shift is much smaller than the collisional width of the line. That is why it is very difficult to measure it, but the experiment has shown [17] that this shift is at least 100 times smaller than the collisional width. This means that the collisional shift Δ_c is smaller than $15 \cdot 10^4$ Hz/Torr. As a result $\Delta_c < 1.5$ Hz (p $\leq 5 \cdot 10^{-5}$ Torr) in the case of the stationary laser, and $\Delta_c < 10$ Hz (p $\leq 5 \cdot 10^{-5}$ Torr) in the case of the transportable one.

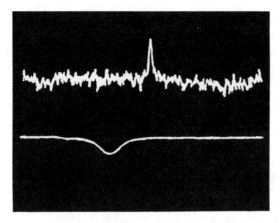

FIGURE 2.10. Doppler-free saturated absorption resonance of the A1(1) component of the R(4) methane line ($\lambda = 3.25\,\mu m$, full-width half-maximum of $\sim$ 300 kHz, upper trace) and a resonance of a confocal scanning interferometer (free spectral range $\sim$ 37.5 MHz, finesse $\sim$ 10, lower trace).

Summarizing all the reasons of line shift we can claim the following: In the case of the stationary laser TE-200 the main reason of reproduced frequency detuning from the center of transition is the second-order Doppler effect which gives the shift $\Delta \leq 10$ Hz. In the case of transportable laser, the main reasons of the shifts are the effects of the active and absorption line center detuning, and the influence of transverse inhomogeneity of the gain medium. Both of them give the comparable shifts $\Delta \sim 10^{-2}\Gamma \cong 40$ Hz.

2.6   Future Possibilities

It seems promising to use for stabilization the methane transitions from initial vibration-rotational molecular states with small rotational quantum numbers. At cryogenic temperatures the intensity of these lines becomes at least two orders larger due to the increase of the population in the initial states. The increase of reference line intensity will permit us to reduce the homogeneous linewidth Γ value which permits us to reduce all systematic shifts of the resonance.

For the nonlinear spectroscopy of these lines, the tunable color-center laser with RbCl:Li crystal operating in the range $\lambda = 3.2 - 3.3\,\mu m$ with an intracavity methane cell, was constructed [18]. The Doppler free A1(1) component of the R(4) methane line (ν_3) is shown in Fig. 2.10. The fifty-fold increase in the intensity of the R(2) methane line was observed after cooling of the methane cell to liquid nitrogen temperature.

At present for the optical pumping of the color-center laser, we use a rather cumbersome and technically complicated krypton ion laser. The development of powerful laser diodes in the visible range of spectra ($\lambda = 640 - 670$ nm) permits us to use such laser diodes for optical pumping instead of a krypton ion laser. We hope that in this way a compact and high precision methane-based optical frequency standard with the accuracy $\sim 10^{-15}$ can be developed. This work has been started now at the Lebedev Institute in collaboration with the Phillips Laboratory (Albuquerque, New Mexico).

References

[1] J. P. Gordon, H. J. Zeiger, and C. H. Townes, Phys. Rev. **95**, 282 (1954).

[2] N. G. Basov and A. M. Prokhorov, JETP **27**, 431 (1954).

[3] D. Kleppner, H. M. Goldenberg, and N. F. Ramsey, Applied Opt **1**, 55 (1962).

[4] C. H. Townes and A. L. Schawlow, *Microwave Spectroscopy* , McGraw-Hill (1955).

[5] V. S. Letokhov, Pis'ma Zh. Eksp. Teor. Fiz **6**, 597 (1967).

[6] K. M. Evenson *et al.*, Appl. Phys. Lett. **22**, 192 (1973).

[7] G. Kramer, B. Lipphardt, C. O. Weiss, Proc. of the IEEE Frequency Control Symposium, p. 39 (1992).

[8] V. S. Letokhov and V. P. Chebotaev, *Non-linear Laser Spectroscopy*, (Springer-Verlag, Berlin, 1977).

[9] W. Demtroder, *Laser Spectroscopy*, (Springer-Verlag, Berlin, 1982).

[10] A. L. Schawlow and C. H. Townes, Phys. Rev. **112**, 1940 (1958).

[11] N. G. Basov, M. A. Gubin, V. V. Nikitin, and E. D. Protsenko, Kvantovaya Elektronika **11**, 1084 (1984).

[12] M. A. Gubin, D. A. Tyurikov, E. V. Koval'chuk, A. S. Shelkovnikov, in the Digest of the Conference on Precision Electrical Measurements (Paris), p. 38 (1992).

[13] M. A. Gubin, D. A. Tyurikov, A. S. Shelkovnikov, E. V. Koval'chuk, J. Quantum Electronics, in press (1995).

[14] D. D. Krylova and A. E. Kiselev, Sov. J. Quantum Electronics **19**, 952 (1989).

[15] S. G. Rautian and A. M. Shalagin, JETP **58**, 962 (1970).

[16] V. A. Alekseev, M. A. Gubin, and E. D. Protsenko, Laser Physics **1**, 221 (1991).

[17] R. L. Barger and J. L. Hall, Phys. Rev. Lett. **22**, 4 (1969).

[18] A. N. Kireev, E. V. Koval'chuk, D. A. Tyurikov *et al.*, Quantum Electronics **24**, 839 (1994).

3

Mid-infrared Lines as Astrophysical Diagnostics: Two Decades of Problems and Promise

Sara C. Beck

3.1   Youthful Enthusiasm

I arrived in Berkeley in August 1976 with an introduction to Charles Townes from my undergraduate advisor, 70 dollars, and the idea that 0.9 μm was pretty far into the infrared. Forty-eight hours later I was at Lick Observatory with John Lacy and Fred Baas and a large and unwieldy spectrometer, looking for an ionic species ([NeII]) I had never heard of, at a wavelength (12.8 μm) that seemed unimaginably long. And in complete disregard for all my preconceptions the [NeII] came booming in so strongly from every HII region we looked at that we could see it on the voltmeter. That first venture into the middle-infrared convinced me that it was the spectral region of the future, and although there have been ups and downs in the last two decades I still think so.

In the late '70s at Berkeley we had John Lacy's extraordinary spectrometer and the whole mid-infrared sky was open to us with very little competition—none within a factor of 10 of our spectral resolution. We could be fairly confident of detecting the big three ionic lines ([NeII] 12.8 μm, [ArIII] 8.99 μm, and [SIV] 10.5 μm) in just about any HII region or planetary nebula we looked at. We were not always so sure what the results meant, however. In hindsight, that was a period when our observational capabilities often outstripped the theoretical and laboratory background available. This has been a common and recurring situation throughout astronomical infrared spectroscopy. Some of the projects on which we had beautiful data turned out to have limited value because the atomic parameters or the astrophysical models were lacking. For example, we did a big study of the elemental abundances in planetary nebulae, only to have the sulfur collision strengths changed by large factors a few years later. Another big study of the infrared ionic lines in HII regions tried to use the infrared line ratios to deduce the temperatures of the ionizing stars, which from the simple nebular models available at the time should have been straightforward. The stellar temperatures, however, consistently came out too low from other evidence. (This was the first sign of a fundamental problem which bothered me for years and which I now believe has been resolved; it is discussed in Sect. 3.) Because we suspected that the models were not adequate we tended to use the infrared ionic lines to measure gas motions with high resolution rather more than we tried to use the line strengths as diagnostics (although we were also of course genuinely interested in the kinematics of infrared galaxies, the Galactic Center, and other objects).

Besides the infrared ionic lines, we did molecular spectroscopy of young stars (with Tom Geballe) and of planetary atmospheres (with Alan Tokunaga), and at first I was impressed at how well molecular constants were known in contrast to the situation with ions. I was rapidly disillusioned when I started my thesis work on the 12.3 μm line of H_2 and found that if I used the existing line constants the Orion Molecular Cloud (OMC) suddenly acquired a large blue-shift. This was before H_2 became a cottage industry among astronomers and before the laboratory spectroscopists turned their attention to it; for about a year the best constants known were those we found by forcing the OMC velocity to agree with other observations.

3.2   General Confusion

In my student years I actually enjoyed the feeling that we, the astronomical spectroscopists, were way ahead of the laboratory spectroscopists. But as my interest in the astrophysics of the astronomical sources increased it became frustrating that the middle-infrared lines gave contradictory results when used to find quantities such as the stellar temperature and the nebular density. So it was a refreshing change when after my Ph.D. I worked with Steve Beckwith at shorter wavelengths (2–4 μm, still a far cry from my original idea of the infrared) and on a species (shocked H_2 and hydrogen recombination lines) whose spectroscopy and diagnostic uses seemed well understood. (I was, again, naive: my colleague Amiel Sternberg has shown that our approach to H_2 was oversimplified. But in the sources we looked at, it worked satisfactorily.)

The success of the Infrared Astronomical Satellite (IRAS) was a tremendous gift to every infrared astronomer. To me the most exciting IRAS results were the great lists of infrared fluxes from galaxies. I had always loved galaxies above all other astronomical sources, but working before IRAS there were only a few galaxies known to be good infrared targets. IRAS not only made it possible to select the best galactic samples, its measurements of total infrared luminosity could be combined with ground-based spectroscopy to find the populations of young stars and get some long-sought understanding of star formation processes. In the late 1980s, while working with Jean Turner and Paul Ho, I used the infrared Brackett lines, radio Very Large Array (VLA) maps and IRAS fluxes to find the range of stellar masses and stellar ages in many nearby starbursts.

All went well with our multi-wavelength studies of starburst galaxies until we tried to include the middle-infrared ionic lines. In more than one galaxy the Brackett lines, the IRAS fluxes, and the radio continuum would agree that the starburst was very young and contained stars as massive as $60M_\odot$. Such massive stars are hot and in all the models should produce generous amounts of the [SIV] 10.5μm fine structure line, the diagnostic of high-excitation ionization. But in none of the spiral galaxy starbursts could [SIV] be seen, while [NeII], the diagnostic of low excitation and comparatively cool stars, was very strong. The only starburst galaxies in which [SIV] could be seen at all were the star-forming blue compact dwarf galaxies such as NGC 5253 and II Zw 40. But the strong [SIV] and the lack of [NeII] in those galaxies gave stellar temperatures as high as 60,000K, which were hard to believe, and contributed to the general feeling that we had fallen off the edge of the models. (Details of the arguments on the contradictions in starburst galaxies may be found in [1].)

The problem of the missing [SIV] in starburst galaxies bothered me to the extent that in the early 1990s I was completely frustrated with both the middle-infrared ionic lines and the near-infrared Brackett line spectroscopy about which I had felt so enthusiastic. It was annoying to have the beautiful IRAS fluxes and not be able to trust the spectroscopy that should explain them, and worse than annoying to think that more beautiful spectra would soon be coming (we hope) from Space Infrared Telescope Facility (SIRTF) and Infrared Space Observatory (ISO), and that we would not be able to understand them. But I could not even think of where to start looking for the cause of the problem.

3.3   Maybe We Know What We're Doing After All

The first hint of where we should go with the middle-infrared lines came at the 1992 "Nearest Active Galaxies" meeting. I had given a talk that basically asked how we are expected to reconcile the $20M_\odot$ upper mass limit that we invariably find from the [NeII]/[SIV] line ratios in spiral galaxy starbursts with the much higher masses found from other diagnostics. Reinhard Genzel pointed out that the problem was also seen in our own Galactic Center. (The reader may with some justice say that considering the intense interest of the Townes group in the Galactic Center while I was a student, I should have seen this myself years ago; all I can offer in my defense is that I was even then focussed on problems of star formation regions to the occasional exclusion of other phenomena.) Joe Shields then showed me models he had done for the Galactic Center that covered ranges of nebular density and ionization parameter far beyond what the classic nebular models included, and which showed several ways of suppressing the [SIV] line in a hot radiation field and giving a falsely cool stellar temperature.

I began to think seriously about the inadequacy of the nebular models most infrared astronomers use. In the roughly twenty years that middle infrared spectroscopy had been in astronomical use, there had been major changes in our picture of "a typical HII region". An example is the persistence of the assumption that density in an HII region is less than around 10^3 cm^{-3}, while we now know from radio and millimeter-wave observations of molecules that the nebular density in compact HII regions and spiral galaxy nuclei is typically 10^4 cm^{-3} with regions of 10^5 cm^{-3}. Another weakness is the assumption that HII regions are spherical and uniform; high-resolution maps in the radio and infrared always find shells, clumps, filaments and other structures which can affect the ionization process. It looked to be the time to do a new set of nebular models. Luckily, it was also time for a sabbatical which I spent as a Visiting Fellow at the Joint Institute for Laboratory Astrophysics (JILA). JILA was probably the best place in the world to work on this problem; it has excellent laboratory spectroscopists who could keep me current on atomic and ionic parameters and when I was there it also had Ralph Sutherland, an expert at photoionization and radiative transfer codes, who made it possible to bring my ideas of what a realistic model nebula should be into reality, or at least into computer reality.

The grid of model HII regions we produced to study the behavior of the infrared ionic lines incoporated the most recent atomic parameters from Opacity Project data, which especially for sulfur differed considerably from the old ones in use, and included densities of 10^2 to 10^6, filling factors of 1, 10 and 100 metallicities from twice to 0.01 solar, and dust. The emitted spectra were calculated with MAPPINGS II, which handles the ionization fractions explicitly. Results are presented in [2]. The major result from the first examination of the model grid is that the [SIV] 10.5 μm line is a very unreliable diagnostic of stellar temperature because it is easily suppressed by nebular effects. If the density is higher than a few 10^3, the ionization parameter can be lower than $\log(Q) = 8.0$; if the nebula is highly clumped, and above all if the metallicity is solar or above, the [SIV] line can be significantly damped compared to what a star of the same temperature could produce in other conditions. We found that it is perfectly possible to have exciting stars of 45,000K in a compact or ultra-compact HII region and not have detectable [SIV]. On the other hand, we also found that in low-metallicity environments the [SIV] line becomes very strong and the [NeII] line weakens, even if the exciting stars are of only moderate temperature.

I will illustrate these findings with the infrared spectra of two very different galaxies, the low-metallicity ($Z \approx 0.15\,Z_\odot$) star-forming dwarf NGC 5253 and the high metallicity (about twice solar) starburst M82. Spectra of both galaxies were obtained with IRSHELL, the University of Texas cryogenic infrared spectrometer built by John Lacy. In M82 [NeII] is the strongest line, [SIV] is barely detectable, and [ArIII] is weak. If solar abundance models are used, the stellar temperature is found to be extremely cool, around 33,000K. This leads to an upper mass cutoff of only 20$M_\odot$, the value usually found for disk galaxy starbursts from this method and one which implies that the Initial Mass Function in such starbursts is quite different from that seen in other galaxies [3]. But if the new model grid with metallicity twice solar is used instead, the stellar temperatures are found to be 40,000 ± 2,500K, giving a more reasonable upper mass cutoff of 30-35$M_\odot$. In contrast the [SIV] line in NGC 5253 is very strong, [NeII] is not detected and [ArIII] is weak. The first analysis of this type of spectrum [4] used solar metallicity and could not fit this spectrum with stars of less than 55,000K temperature, but a model grid of suitable low metallicity found that the exciting stars are of temperature 42,500 ± 2,500K and 35-40$M_\odot$ [5]. So in both cases we see that the model grid has to be carefully chosen to suit the source, or the stellar temperatures and masses found can be badly misleading.

While I think that most astronomers would agree that the corrected stellar populations for the above galaxies are preferable to the original estimates, they would also agree that spectral line diagnostics that are not useful unless one is already sure of the metal abundance (which is not usually the easiest quantity to find, especially in highly obscured objects where optical lines cannot be detected) are not ideal diagnostics. Are there better infrared diagnostics of stellar temperature and nebular parameters? The ratio of the 6.99 μm [ArII] line to the 8.99 μm [ArIII] line is an excellent diagnostic of stellar temperature for temperatures lower than around 40,000K; it is only negligibly affected by density, metal abundance, and ionization parameter. The only problem is that the Earth's atmosphere only rarely gets to be as transmissive as 50% at the [ArII] wavelength. The [ArIII] to [NeII] ratio is a very good diagnostic of stellar temperature over the whole OB star range; it is slightly more affected by density and metallicity than is the pure argon ratio, but is still far better than any ratio involving [SIV]. The [ArIII] line is much weaker than [NeII], so astronomers wishing to use this line ratio as a diagnostic must be prepared to go to signals as weak as 1% of the [NeII] flux. Finally, the 18 μm line of [SIII], which although it is often considered an airplane altitude (and above) line can in fact be observed routinely from a dry site such as Mauna Kea with the high resolution of IRSHELL, makes a good density diagnostic in ratio with [NeII]. (Full details and usable model grids are in [2].) To summarize, the [ArIII], [NeII], and [SIII] lines together make an acceptable diagnostic grid for stellar temperature and density, although they are best used in cases where the metal abundance is at least approximately known. If [ArII] can be detected as well the determination of stellar temperature becomes even better and is immune to the metal abundance.

There are now several possible directions to take. The first and easiest, which will probably be well underway by the time this appears, is to review the stellar types and masses found from the infrared lines in the literature. There are many galaxies, of which M83 is the best example, where the Brackett line flux and the infrared luminosity indicate a very high upper mass cutoff, but [SIV] is not detected. These contradictions should be quickly resolvable with the new model grids. The second, which is also underway, is to check and refine the theoretical findings with ground-based observations of HII regions and galaxies. Time has already been allocated at the Infrared Telescope Facility (IRTF) on Mauna Kea for high resolution observations of [ArIII], [NeII], and [SIII] in compact and ultra-compact HII regions. By comparing the infrared results to the density, ionization parameter and structure seen in high-resolution radio maps we can check the validity of the models in realistic settings. We also intend to search for [ArII], which may be the single most useful line for stellar temperature determinations and which, if the atmospheric models can be trusted, should be observable under some circumstances from Mauna Kea. After understanding the behaviour of the infrared lines in relatively simple HII regions we hope to apply them to new observations of star forming regions in galaxies. Finally, Ralph Sutherland and I are part of an ISO team headed by Allan Willis and Peter Conti which has been allotted time to observe these infrared ionic lines in Wolf-Rayet and other emission line galaxies. We are very enthusiastic about the prospect of not only being able to see all these lines and others, with high spatial and spectral resolution, but of being able to interpret the data with some confidence that our diagnostics are reliable.

3.4   Conclusion

This essay has been, by intention, highly personal and even idiosyncratic. I have not attempted to give a fair description of what other people were doing or of how the field as a whole has developed and I hope that the many fine scientists and interesting problems which are not mentioned here do not feel slighted. I have tried to show how one infrared astrophysical spectroscopist saw some frustrating problems in the field, which arose because our observational skills had gotten ahead of our theoretical underpinning. I believe (and hope) that these contradictions have now been resolved and look forward to forthcoming ground and ISO observations with new confidence in our ability to understand obscured star formation regions.

References

[1]   S. C. Beck, in *The Nearest Active Galaxies*, Beckman, Colina, and Netzer (eds.), CSIC Madrid, p.169 (1993).

[2]   S. C. Beck, and R. S. Sutherland, submitted to AJ (1995).

[3]   J. M. Achtermann, and J. H. Lacy, ApJ. **439**, 163 (1995).

[4]   D. K. Aitken, P. F. Roche, M. C. Allen, and M. M. Phillips, MNRAS **199**, 31p (1982).

[5]   S. C. Beck, J. T. Turner, P. T. P. Ho, J. H. Lacy, and D. M. Kelly, submitted to ApJ (1995).

4

The Laser Stabilitron

William R. Bennett, Jr.
and Vieniamin P. Chebotayev

The late Vieniamin Pavlovich Chebotayev was a visiting professor in my laboratory at Yale University, on leave from the Institute for Laser Physics, Academy City, Novosibirsk, Russia, when we started this paper. Although his sudden death during that stay prevented completion of the work in the form we originally had envisioned, I am confident that he would have approved publishing this paper in a volume honoring Charles Townes. Both he and I benefitted greatly from Townes' early work on masers and lasers. In addition, Chebotayev was corecipient of the C. H. Townes Prize of the Optical Society of America in 1984 for his work on saturated absorption spectroscopy—W.R.B.

4.1   Background

The reduction of amplitude noise in continuous-wave (cw) lasers is obviously desirable for many reasons. In pure physics, very low-noise lasers would be an asset to the detection of gravitational waves in large-scale interferometers [1]. In the realm of applied physics, reduced noise levels in applications ranging from metrology to retinal coherence tomography could result in improved spatial resolution. Theoretical interest has also developed recently in light sources with noise levels below the standard quantum limit through the production of "squeezed" light states arising in the nonlinear interaction of radiation with matter [2] and a decrease in noise level below the quantum limit has been reported [3].

We have shown that it should be possible to obtain a large decrease in the amplitude noise from a laser by use of a coupled system involving a nonlinear saturable absorber [4]. This system uses three mirrors with which two interactive cavities are produced: one contains an amplifier and the other an absorber. Because of strong nonlinear effects, there are two substantially different regions of operation: One of these is characterized by hysteresis effects and regions of instability. The other region should not only be stable, but should also permit strong suppression of fluctuations in the laser intensity. In that case, the laser output should depend very little on changes in gain or loss in the system. This type of device, which we have called a "Laser Stabilitron," could be produced with almost any of the commonly known cw lasers.

It has been shown previously that saturated absorption inside a single laser cavity produces a bistable regime with hysteresis effects [5]. There, positive feedback between the optical field and the amount of absorption results. Increasing the field reduces the absorption, thereby decreasing the loss and increasing the intensity. Hysteresis phenomena appear in such a system when the change in saturated absorption exceeds the change in saturated gain, and it has been shown theoretically that a large increase in quantum fluctuations can occur in a laser with such an intra-cavity saturable absorber [6]. It has also been found that propagation of a strong beam through a nonlinear absorbing medium can exhibit regions of instability in which intensity fluctuations are amplified [7]. Furthermore, when a laser beam is transmitted through a Fabry-Perot interferometer filled with an absorber, bistability and differential amplification of noise fluctuations can occur [8]. However, when the incident and output beam from such an absorber-filled cavity are combined interferometrically, and the system is operated near the turning points in the bistable region, photon noise can be reduced below the shot noise at nonzero frequency [9]. This noise reduction is associated with a temporal redistribution of the photons inside the cavity. Unfortunately, the optimum "squeezing" condition in the passive case occurs right at the turning points where the system is on the edge of instability. However, shown below, one can decrease these amplitude fluctuations by using *negative* feedback between the field and the absorber inside the cavity. The optimum stabilization point in this new system appears to be equivalent to the optimum squeezing condition for photon noise reduction in the passive, absorber-filled Fabry-Perot discussed by Reynaud *et al.* [9].

4.2   Method of Intensity Stabilization

A schematic diagram of our three-mirror laser system is shown in Fig. 4.1. The amplifying and absorbing media are separated by a high-reflecting mirror (M_2 in Fig. 4.1(a)) which divides the system into two coupled resonators. The first cavity contains the amplifying medium and the second one contains the saturable absorber. If the resonant frequencies of the two cavities are different, the coupling between the cavities is negligible. In that limit, oscillation in this system may be considered as that in a conventional laser with a cavity formed by mirrors M_1 and M_2. However, when the frequencies of the two resonators coincide, appreciable coupling occurs. Then, saturation of the absorber by a strong field leads to an increase in output intensity through mirror M_3. The strong negative feedback mechanism between the field intensity in cavity 1 and the absorption in cavity 2 works as follows: Increasing the gain and intensity in cavity 1 leads to increased field intensity in cavity 2 and, correspondingly, to a decrease in absorption. That, in turn, leads to increased transmission from cavity 1 to 2—hence, an increase in the loss of the laser and a decrease of the laser intensity in cavity 1. In this system,

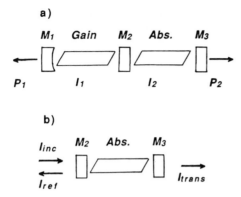

FIGURE 4.1. (a). Schematic diagram of the three-mirror laser system. (b). Effective laser mirror (composed of cavity 2 in Fig. 4.1 (a). The effective transmission $T_{\text{eff}} = I_{\text{trans}}/I_{\text{inc}}$ and the effective reflectance $R_{\text{eff}} = I_{\text{ref}}/I_{\text{inc}}$ are used to determine the laser oscillation threshold.

the intensity I_1 in cavity 1 (hence, the power output P_1 to the left in the Fig. 4.1(a)) is stabilized, whereas the intensity I_2 in cavity 2 (hence, the power output P_2 to the right in Fig. 4.1(a)) is strongly variable.

For simplicity, consider the case where all the mirrors have the same transmission coefficient T. Here the total loss involves the transmission of mirror 1 and of mirror 3, the internal dissipative losses, and the absorption in cavity 2. Due to the difference between the field intensities in cavities 1 and 2, the laser powers P_1 emitted to the left from cavity 1 and P_2 emitted to the right from cavity 2 in Fig. 4.1(a) are given by

$$P_1 = I_1 T \quad \text{and} \quad P_2 = I_2 T \tag{4.1}$$

where I_1 and I_2 are the field intensities in the first and second cavities.

To determine the laser oscillation conditions, it is convenient to consider the second cavity as a mirror with an effective transmission coefficient, T_{eff}, and an effective reflection coefficient, R_{eff}. (See Fig. 4.1(b).) By summing the infinite series for the transmitted and reflected field amplitudes by the second (Fabry-Perot) cavity at resonance, it is seen that

$$T_{\text{eff}} \approx \left(\frac{T}{T + f + A_s} \right)^2 \quad \text{and} \quad R_{\text{eff}} \approx \left(\frac{f + A_s}{T + f + A_s} \right)^2 \tag{4.2}$$

where the total fractional energy loss per pass in cavity 2 during oscillation is given by $(T + f + A_s)$, A_s is the saturated absorption, f is the dissipative loss from sources such as scattering and it has been assumed that $f, A_s \ll 1$. Note that $T_{\text{eff}} + R_{\text{eff}} \neq 1$ because of the loss in cavity 2.

Clearly, from Eqs. (4.1) and (4.2), the power out of the absorptive end of the laser is $P_2 = I_2 T = I_1 T_{\text{eff}}$. (See Fig. 4.1(b).) Hence,

$$I_2 = I_1 \frac{T}{(T + f + A_s)^2} . \tag{4.3}$$

In practice, the system requires "hard excitation" in the sense that the gain must be turned up high enough to exceed the threshold,

$$2G > 1 - R_{\text{eff}} + T + 2f = 1 - \left(\frac{f + A}{T + f + A}\right)^2 + T + 2f , \qquad (4.4)$$

involving the *unsaturated* gain and loss, for oscillation to occur. However, after oscillation starts, G and A are reduced to their saturated values and the condition for cw laser oscillation becomes

$$2G_s = L_{\text{eff}} = 1 - \left(\frac{f + A_s}{T + f + A_s}\right)^2 + T + 2f \qquad (4.5)$$

where L_{eff} is the effective loss in the laser (cavity 1) *after* oscillation has reached steady-state, G_s is the saturated gain per pass, and the fractional dissipative loss per pass f is assumed to be the same in each cavity. Equation (4.5) is valid for small gain and states that the round-trip saturated gain, $2G_s$, equals the effective loss, L_{eff}, at steady-state. The latter, of course, includes the saturated absorption, A_s, in cavity 2.

We shall assume homogeneous saturation in both media, hence

$$G_s = \frac{G}{1 + g_1 I_1} \quad \text{and} \quad A_s = \frac{A}{1 + g_2 I_2} .$$

Equation (4.5) then takes the form,

$$2G_s = L_{\text{eff}} ,$$

where

$$2G_s = \frac{2G}{1 + g_1 I_1} ,$$

$$L_{\text{eff}} = 1 - \left(\frac{f + A_s}{T + f + A_s}\right)^2 + T + 2f , \qquad (4.6)$$

and I_1 and I_2 are related by Eq. (4.3).

Equation (4.6), along with Eqs. (4.1) and (4.3), permit determining the output laser intensities P_1 and P_2. However, the equations are very nonlinear and the analytic solution of Eq. (4.6) is complicated [10]. Eliminating I_1 in Eq. (4.6) permits expressing G directly in terms of the intensity I_2:

$$G = \frac{1}{2} \left[1 + \frac{g_1 I_2}{T} (A_s + T + f)^2\right]$$
$$\times \left[1 - \left(\frac{f + A_s}{A_s + T + f}\right)^2 + T + 2f\right] . \qquad (4.7)$$

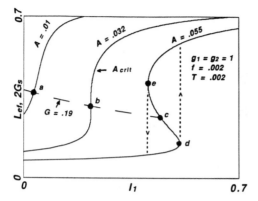

FIGURE 4.2. Behavior of the effective loss L_{eff}, as a function of I_1 for different values of A (solid curves) and graphical solution for the laser oscillation condition for representative parameters. Note that there is a critical value of the unsaturated absorption A_{crit} above which L_{eff} is a triple-valued function of the intensity I_1. The desired oscillation condition for low-noise operation is obtained by adjusting the gain so that $2G_s$ (dashed curve) intersects L_{eff} at the point b on A_{crit}. As illustrated by the vertical dashed lines on the curve for $A = 0.055$, strong hysteresis effects would occur for $A > A_{crit}$ as I_1 is tuned through the triple-valued region. In regions such as the point c between d and e where $\partial L_{eff}/\partial I_1 < 0$, oscillation would be unstable.

Graphic solutions to Eq. (4.6) will help to explain the behavior and the main properties of the proposed three-mirror system. As noted, steady-state laser oscillation corresponds to the requirement that $2G_s(I_1) = L_{eff}(I_1)$. Although the saturated gain has a simple, monotonic dependence on I_1, the effective loss L_{eff} has much more complex behavior. As shown in Fig. 4.2, there are two distinctly different regions: For small values of A, the effective loss is also monotonic in its dependence on intensity I_1, and in that region the system would have noise properties similar to a conventional laser. For large values of the absorption parameter, L_{eff} has a triple-valued dependence on the intensity. As indicated by the dashed vertical lines at the right in Fig. 4.2, strong hysteresis effects would occur if the intensity I_1 were tuned back and forth through the triple-valued region; however, the vertical jump upwards at point d would not occur without a substantial increase in the unsaturated gain. Because $\partial L_{eff}/\partial I_1 < 0$, oscillation between the points e and d would be completely unstable.

These two regions are separated by a critical value, $A_{crit} = 8(T + f)$, for the absorption for which there is one point labelled b in Fig. 4.2 where $\partial I_1/\partial L_{eff} = \partial^2 I_1/\partial L_{eff}^2 = 0$. Hence, by adjusting the parameters so that the saturated gain curve ($2G_s$) intersects the effective loss at point b on A_{crit}, very stable low-noise oscillation should be obtained. The actual intensities in the two cavities for this condition can be determined from Eq. (4.7) and are shown as a function of G in Fig. 4.3 for the same conditions assumed in Fig. 4.2. Note that the intensity in the left cavity I_1 goes through a point where $\partial I_1/\partial G = \partial^2 I_1/\partial G^2 = 0$, whereas that in the right cavity I_2 increases strongly with G. The desired condition for oscillation

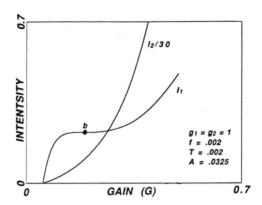

FIGURE 4.3. The intensities I_1 and I_2 versus gain for the optimum value of the absorption A_{crit} for the parameters used in Fig. 4.2. Note the plateau in the laser intensity I_1 at point b, corresponding to the optimum operating conditions in Fig. 4.2. Note that I_1 is independent of the gain over this plateau and the output of the laser (to the left in Fig. 4.1 (a)) should be nearly independent of fluctuations in the gain.

corresponds to the saturated gain curve (dashed curve in Fig. 4.2) intersecting the effective loss at point b on the critical absorption curve and leads to oscillation at point b in the middle of the plateau on the curve for I_1 vs. G in Fig. 4.3. Although the intensity (I_2) in the absorption cavity will depend strongly on variations in the gain G, the intensity I_1 in the gain cavity will be nearly independent of fluctuations in the gain at operating point b. Thus, the output P_1 to the left in the apparatus shown in Fig. 4.1(a) should be unusually noise free. The particular conditions assumed in Figs. 4.2 and 4.3 are intended only for illustrative purposes. Other choices in the laser parameters might well provide more optimum results.

4.3   Coupled Field Equations and Photon Noise

It should be emphasized that we are dealing with the case of weak coupling between the two cavities for which the middle-mirror transmission coefficient satisfies, $T \ll A_s \ll 1$. In order to demonstrate the photon noise reduction properties of the system, it is necessary to consider the coupled field equations for the two cavities in this limit. Specifically, we shall modify some of the earlier work by Lamb and his students to address this problem.

Some related problems in the strong-coupling limit have been investigated previously by others. Three-mirror laser cavities were considered long ago as a method of mode suppression by several authors [11]. As shown by Perel and Rogova [12] for the empty three-mirror cavity, and by Spencer and Lamb [13] in the case of two coupled lasers, there is a splitting of the two separate cavity resonant frequencies in the strong-coupling case when the separate resonances are tuned in close coincidence. As in the case of strongly-coupled resonant AC circuits [14], symmetric and anti-symmetric composite modes of the system develop whose resonant frequencies repel each other as the independent cavity frequencies are tuned to exact coincidence. The minimum splitting of the resonant frequencies goes up by an amount which increases nonlinearly and monotonically with the coupling coefficient. The antisymmetric mode has the higher frequency, and, for near co-incidence of the cavity frequencies, at least one or the other of these modes will be stable (depending on the cavity detuning) if the gain is above threshold. In the strong coupling case, the effective mirror reflectance coefficient, R_{eff} in Fig. 4.1(b), becomes complex and contains a small phase shift which changes the frequency of the laser. The laser (cavity 1) sees a reactive load, and the system can exhibit additional instabilities to those shown in Fig. 4.2.

In order to avoid these problems, we impose the following limiting conditions on the system:

$$
\begin{array}{ll}
\text{a)} & L_1 \ll A + L_2 \; (Q_1 \gg Q_2) \\
\text{b)} & T \ll L_1 \; \text{(small coupling)} \\
\text{c)} & \text{Length}_1 \gg \text{Length}_2 \,.
\end{array}
\tag{4.8}
$$

Under these conditions, the laser (cavity 1) sees a resistive load and the effective mirror reflectance method used in Eqs.(4.2) - (4.7) is valid. We note that conditions (4.9) are readily satisfied by the systems suggested above.

For the present conditions, the field equations developed by Spencer and Lamb take the form:

$$
\ddot{E}_1 + \frac{\Omega_1}{Q_1}\dot{E}_1 + \Omega_1^2 E_1 = 4\pi\omega^2 P_1 - 2g_1\Omega_1(E_1 - E_2)
$$

$$
\ddot{E}_2 + \frac{\Omega_2}{Q_2}\dot{E}_2 + \Omega_2^2 E_1 = 4\pi\omega^2 P_2 - 2g_2\Omega_2(E_2 - E_1)
\tag{4.9}
$$

where,

$$
\mathcal{E}_1 = \text{Re}\left\{E_1(t)e^{-i\omega t}\right\}, \quad \mathcal{E}_2 = \text{Re}\left\{E_2(t)e^{-i\omega t}\right\},
$$

and,

$$
\mathcal{P}_1 = \text{Re}\left\{P_1(t)e^{-i\omega t}\right\}, \quad \mathcal{P}_2 = \text{Re}\left\{P_2(t)e^{-i\omega t}\right\}.
$$

Here, Q_1 and Q_2 are the quality factors and Ω_1 and Ω_2 are the resonant frequencies of the two cavities, $\mathcal{P}_1$ and $\mathcal{P}_2$ are the polarizations in the two cavities, and the coupling coefficients are given by $G_1 = \sqrt{T}c/2\text{Length}_1$ and $G_2 = \sqrt{T}c/2\text{Length}_2$), where T is the intensity transmission coefficient of the middle mirror.

In the slowly-varying amplitude and phase approximation used in the Lamb theory, $\mathcal{E}_1$ and $\mathcal{E}_2$ satisfy

$$
\dot{\mathcal{E}}_1 + \left[\frac{\omega}{2Q_1} + i(\Omega_1 + G_1) - i\omega\right]\mathcal{E}_1 - iG_1\mathcal{E}_2 = i2\pi\mathcal{P}_1
$$

$$
\dot{\mathcal{E}}_2 + \left[\frac{\omega}{2Q_2} + i(\Omega_2 + G_2) - i\omega\right]\mathcal{E}_2 - iG_2\mathcal{E}_1 = i2\pi\mathcal{P}_2 \,.
\tag{4.10}
$$

The polarizations $\mathcal{P}_1$ and $\mathcal{P}_2$ may be written in terms of the induced and stochastic parts as

$$\mathcal{P}_1 = \mathcal{P}_1^i + \mathcal{P}_1^{st} \quad \text{and} \quad \mathcal{P}_2 = \mathcal{P}_2^i + \mathcal{P}_2^{st} . \tag{4.11}$$

The stochastic parts of the polarization in the two cavities are independent statistically, hence we may consider the effects of $\mathcal{P}_1^{st}$ and $\mathcal{P}_2^{st}$ separately. Because we take $Q_1 \gg Q_2$ (i.e., the loss in the absorber cavity is much greater than that in the gain cavity), we need only consider the effects of $\mathcal{P}_1^{st}$. The induced parts of the polarization in the gain and absorber media may be written

$$\mathcal{P}_1^i = i\chi_1\mathcal{E}_1 , \, \chi_1 = \chi_1^0 - \chi_1^1|\mathcal{E}|^2 , \text{ and } \mathcal{P}_2^i = -i\chi_2\mathcal{E}_2 \tag{4.12}$$

where χ_1 and χ_2 are the susceptibilities of the gain and absorber media.

If we measure the field in units of $(8\pi h\nu)^{1/2}/V_1$ where V_1 is the volume of cavity 1, $|\mathcal{E}_1|^2$ is simply the number of photons in the cavity. (We are considering the case where $Length_1 \gg Length_2$ and $\mathcal{E}_2 < \mathcal{E}_1$.) For small gain saturation, it follows from Eqs. (4.9) - (4.12) that

$$\dot{\mathcal{E}}_1 - \left[G_t - \beta|\mathcal{E}_1|^2 - i(\Omega_1 + G_1 + i\omega)\right]\mathcal{E}_1 - iG_1\mathcal{E}_2 = f_1(t)$$

and

$$\dot{\mathcal{E}}_2 + [\Gamma_2 + i(\Omega_2 + G_2 - i\omega)]\mathcal{E}_2 - iG_2\mathcal{E}_1 = 0 \tag{4.13}$$

where $G_t = 2\pi\omega\chi_1^0 = \omega/2Q$ is the difference between the linear gain G and losses per unit time in cavity 1, $\beta = (8\pi h\nu/V_1) \cdot 2\pi\omega\chi_1^1$ is the gain saturation parameter,

$$f(t) = i2\pi\omega\mathcal{P}_1^{st}(t)\sqrt{\frac{V}{8\pi h\nu}} , \quad \text{and} \quad \Gamma_2 = \frac{A_t}{(1 + \kappa|\mathcal{E}_2|)^2} + \frac{\omega}{2Q}$$

is the full loss in cavity 2 (which includes both saturated absorption and dissipative loss), where A is the absorption coefficient, and where

$$\kappa|\mathcal{E}_2|^2 = \frac{4\omega V_2 d^2}{\hbar\gamma_1\gamma_2}|\mathcal{E}_2|^2$$

is a dimensionless saturation parameter for the absorption medium, where d is the dipole moment of the transition, γ_1 and γ_2 are the decay rates for the lower and upper levels. The time-dependent gain, absorption and losses are connected with laser parameters we used before by the relations,

$$G_t = \frac{(G - L_1)c}{2Length_1} , \quad A_t = A\frac{c}{2Length_2} , \quad \frac{\omega}{2Q_j} = \frac{L_jc}{2Length_j} , \tag{4.14}$$

where $j = 1, 2$.

It is apparent that the second cavity only has a significant effect when $\omega \approx \Omega_2$. We consider the resonant case where the laser frequency is close to Ω_1 and Ω_2. In the approximations Eq. (4.9), it is apparent that field $\mathcal{E}_2$ closely follows field $\mathcal{E}_1$.

This arises because the time-delay effects in cavity 2 are negligible and the time derivative in the second Eq. (4.13) is small. At steady state, field $\mathcal{E}_2$ and intensity I_2 are related to $\mathcal{E}_1$ and I_1 by

$$\mathcal{E}_2 = \mathcal{E}_1 \frac{iG_2}{\frac{A_t}{1+\kappa_2 I_2} + \frac{\omega}{2Q_2} + i(\Omega_2 + G_2 - \omega)}$$

$$I_2 = I_1 \frac{G_2^2}{\left(\frac{A_t}{1+\kappa_2 I_2} + \frac{\omega}{2Q_2}\right)^2 + (\Omega_2 + G_2 - \omega)^2} . \qquad (4.15)$$

At resonance, these equations reduce to

$$\mathcal{E}_2 = \mathcal{E}_1 \frac{iG_2}{\frac{A_t}{1+\kappa_2 I_2} + \frac{\omega}{2Q_2}}$$

and

$$I_2 = I_1 \left(\frac{G_2}{\frac{A_t}{1+\kappa_2 I_2} + \frac{\omega}{2Q_2}}\right)^2 . \qquad (4.16)$$

Using Eqs. (4.13) and (4.16) with Eq. (4.14), we obtain the Langevin equation in standard form (at resonance):

$$\dot{\mathcal{E}}_1 - \left(G_t - \beta I_2 - \frac{c}{2Length_1} \frac{T}{\frac{A}{1+\kappa_2 I_2} + L_2}\right)\mathcal{E}_1 = f(t) . \qquad (4.17)$$

The steady-state solution to Eq. (4.17) when $f(t) = 0$ yields the same result we had obtained earlier in Eq. (4.3) and relates the intensities I_1 and I_2 in terms of the laser parameters:

$$G_t - \beta I_1 - \frac{c}{2Length_1} \frac{T}{\frac{A}{1+\kappa I_2} + L_2} = 0 . \qquad (4.18)$$

The term $L_{eff} = (c/2Length_1)T/[(A/(1+\kappa_2 I_2) + L_2]$ is the additional nonlinear loss which transforms from cavity 2 to cavity 1. Near the critical region, the dependence of L_{eff} on intensity I_1 rises very steeply. Figures 4.4(a) and 4.4(b) illustrate this dependence and are analogous to the results in Figs. 4.2 and 4.3 obtained from the simpler effective reflectance model.

The function $f(t)$, describes the noise from spontaneous emission. The spectral density of the intensity noise coincides with the profile of the spontaneous emission line. The width of this line is much broader than the cavity width, ω/Q. In this case, the noise is δ-correlated:

$$\langle f(t)f(t')\rangle = J\delta(t - t') . \qquad (4.19)$$

Below threshold for laser oscillation, the coefficient J is the rate of spontaneous photon emission in the cavity mode. At small-gain saturation, the probability of

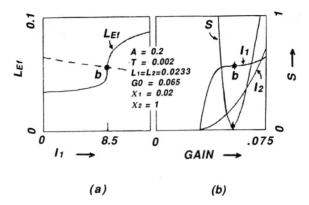

(a)   *(b)*

FIGURE 4.4. (a) and (b). Solutions to the coupled field equations for properties of the system equivalent to those shown in Figs. 4.2 and 4.3. The quantity S shown in (b) is the sub-Poissonian coefficient for the particular parameters assumed (see Eq. (4.25)). Note that for the optimum operating point b, $S \ll 1$ implying that the intensity fluctuations in the gain cavity are substantially below the quantum noise limit. (Also, see Fig. 4.5.)

spontaneous emission is the same. This means simply that J is proportional to the number of emitting atoms.

Substituting I_2 from Eq. (4.16) into Eq. (4.17) yields the solution for $\mathcal{E}_1$. However, the exact form for $I_2 = f(I_1)$ is much too complicated for practical purposes. In the classical limit ($I_1 \gg 1$) the fluctuations are very small in comparison with the intensity. For that reason, we may linearize Eq. (4.17) near the steady-state region:

$$\dot{\mathcal{E}}_1^t + \left[\beta \left(|\mathcal{E}_1^t|^2 - |\mathcal{E}_1^0|^2 \right) + \kappa \left(|\mathcal{E}_1^t|^2 - |\mathcal{E}_1^0|^2 \right) \right] \mathcal{E}_1^t = f(t) , \qquad (4.20)$$

where, defining

$$K \equiv \frac{\partial L_{\mathrm{eff}}}{\partial I_2} \frac{\partial I_2}{\partial I_1} = \frac{c}{2 Length_1} \frac{\kappa T^2 A}{\alpha^3 \gamma^2} \frac{1}{\alpha - \frac{2A\kappa I_2^0}{\gamma^2}}$$

and

$$\alpha = \frac{A}{1 + \kappa I_2^0} + L_2 \quad \text{and} \quad \gamma = 1 + \kappa I_2^0 .$$

$\mathcal{E}_1^0$ and I_2^0 are the solutions of Eq. (4.16).

The Fokker-Planck equation corresponding to Eq. (4.19) has the form

$$\frac{\partial W}{\partial \tau} = \frac{2\partial\{[(\beta + \kappa)\bar{n}_1 - (\beta + \kappa)n_1]n_1 W\}}{\partial n} + \frac{J\partial \left[n_1 \frac{\partial W}{\partial n_1} \right]}{\partial n_1} . \qquad (4.21)$$

At steady-state ($\partial W/\partial t = 0$),

$$W(n_1, t) = W(n_1) = C \exp \left[\frac{-(n_1 - \bar{n})^2}{2\sigma} \right] \qquad (4.22)$$

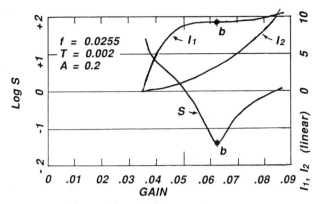

FIGURE 4.5. Log(S), where S is the sub-Poissonian coefficient in Eq. (4.25), is shown as a function of gain in the vicinity of the optimum operating point b for conditions similar to those in Figs. 4.4 (a) and (b). The intensities I_1, I_2 are shown plotted on a linear scale. The results again imply that a substantial reduction below the standard quantum noise limit can be achieved.

where

$$\sigma = \frac{J}{2(\beta + \kappa)} , \quad C = \frac{\sqrt{2/\pi}}{\sigma[1 + F(z)]} , \quad z = \frac{4(\beta + \kappa)}{\sqrt{2J/(\beta + \kappa)}} \gg 1 .$$

The noise characteristics of the stabilitron are determined by the coefficient K. As an example, let us compare the noise in the stabilitron with the noise in a usual laser at the same intensity level. In this case $K = 0$ and the losses in the laser should be given by

$$\frac{\omega}{2Q'_1} = L_{\text{eff}} = G' - \beta \bar{n}_1 .$$

Points b on Figs. 4.4(b) and 4.5 correspond to this region. At this point, the dispersion σ_1 and $\bar{n}_1$ are given by:

$$\sigma_1^2 = \frac{J}{2\beta} , \text{ and } \bar{n}_1 = \frac{G'}{\beta} . \tag{4.23}$$

The ratio of probability of spontaneous emission to stimulated emission (per atom) into the cavity mode is equal to $1/n$, where n is the number of photons in the cavity. In the steady-state condition the rate of stimulated emission R is given by $R = 2L_{\text{eff}}n$. Using Eq. (4.22), we obtain

$$\sigma_1^2 = \frac{G'}{G' - L}\bar{n}_1 . \tag{4.24}$$

Equation (4.23) agrees with the quantum noise calculations of Scully and Lamb [15]. In the stabilitron mode, the characteristics of spontaneous emission are the same as in the previous case when the same values of L_{eff} are used. Then we have,

$$\sigma_1^2 = S\bar{n}_1$$

where

$$S = \frac{G}{G' - L_{\text{eff}}} \frac{\beta}{\beta + \kappa} \qquad (4.25)$$

is a sub-Poissonian coefficient. This depends both on the parameters of the laser cavity and the passive interferometer. As shown by the logarthmic plot in Fig. 4.5, near the bistable region in the second cavity, S can be much smaller than 1. This implies a substantial reduction in noise below the standard quantum limit.

One can find the fluctuations in the second cavity using the relation

$$W(n_1)dn_1 = W(I_2)dI_2 \qquad (4.26)$$

where $W(I_2)$ is the probability distribution for intensity I_2. The fluctuations in the second cavity will be much greater than those in the first cavity. The decrease of fluctuations in the first cavity is compensated by the effect of dynamic amplification in the bistable interferometer.

4.4   Possible Systems for Realizing a Stabilitron

In principle, this method could be used to stabilize the power of almost any cw laser. There are of course many well-known examples of amplifier-absorber pairs where the frequencies may be closely matched, e.g., He-Ne/CH$_4$ at 3.39 μm, He-N$_2$-CO$_2$/SF$_6$ at 10.6 μm, He-Ne/I$_2$ at 0.633 μm, and Ar$^+$/I$_2$ at 0.5145 μm. In these cases, frequency stabilization can also be achieved by locking the laser cavity to the center of the saturable absorber line. In principle, one could always use the same medium and transition for both the amplifier and the absorber by suitably choosing the conditions of operation; that is, almost any amplifier can be operated in an absorber mode. However, one does need a large absorption coefficient to permit making the absorber cavity short compared to the gain cavity. For that reason, a diode laser medium might prove the most viable for dual use as both amplifier and absorber.

Regardless of the choice of medium, the method requires that the laser frequency should closely correspond to the resonant frequency of cavity 2, and that requirement provides the main experimental difficulty. Using the notation of Fig. 4.1, initial adjustment is obtained by varying the length of cavity 2 (the absorber) so that P_2 is a maximum and by varying the length of cavity 1 (the gain) so that P_1 is a minimum. Once found, the laser cavity could then be phase-locked to the absorber cavity at the optimum condition.

References

[1] See, e.g., R. P. W. Drever, in *Gravitational Radiation* NATO Advanced Physics Institute, Les Houches (edited by N. Deruelle and T. Piran, North Holland Publishing, 1983); p.321; D. Shoemaker, R. Schilling, W. Winkler, K. Maischberber, and A. Rüdiger, Class. Quantum Gravity **6,** 1781 (1989).

[2] E. Giacobino, T. Debuisschert, A Heidemann, J. Mertz, and S. Reynaud, in the *Proceedings of NICLOS–Bretton Woods, N.H., June 19-23, 1989* (Academic Press, New York, 1989), p. 180.

[3] H. J. Kimble, in *Atomic Physics II,* edited by S. Haroche, J. C. Gay, and G. Gimbera (World Scientific Press, New York, 1989), p. 467.

[4] Preliminary reports of this work were given by W. R. Bennett, Jr., and V. P. Chebotayev in Bull. Amer. Phys. Soc. **36,** No. 10, p. 2741 (1991); Abstract DB 9; Appl. Phys. B **54,** 552-555 (1992); and U.S. Patent No. 5,251,229 (Oct. 5, 1993).

[5] V. N. Lisitsyn and V. P. Chebotayev, Zh. Eksp. i Teor. Fiz. **54,** 419 (1968). [Trans. in Sov. Phys. JETP **27,** 227 (1968)]; JETP Letters **7,** 3 (1968).

[6] A. P. Kazantsev and G. I. Suredutovich, in *Progress in Quantum Electronics,* edited by J. H. Sanders and S. Stenholm (Pergamon Press, Oxford, 1975).

[7] S. L. McCall, Phys. Rev. A **9,** 1515 (1974); also see, B. R. Mollow, Phys. Rev. A **7,** 1319 (1973).

[8] H. M. Gibbs, S. L. McCall, and T. N. C. Venkatesan, Phys. Rev. Lett. **36,** 1135 (1976).

[9] S. Reynaud, C. Fabre, E. Giacobino, and A. Heidmann, Phys. Rev A **40,** 1440 (1989).

[10] In order to solve Eq. (4.6), it is easiest to treat T_{eff} as the independent variable and compute: I_1 as $F(T_{\text{eff}})$; I_2 as $f(I_1, T_{\text{eff}})$; A_s as $f(I_2)$; R_{eff} as $f(A_s, T_{\text{eff}})$; and then find $L_{\text{eff}} = 1 - R_{\text{eff}} + T + 2f$ as a function of I_1.

[11] See, for example, D. A. Kleinman and P. P. Kisliuk, Bell System Tech. J. **41,** 453 (1962); N. Kumagai, M. Matsuhara, and H. Mori, IEEE Trans. on Quant. Elect. **QE-1,** 85 (1965).

[12] V. I. Perel and L. V. Rogova, Optics and Spectroscopy **25,** 401 (1968).

[13] M. B. Spencer and W. E. Lamb, Jr., Phys. Rev. A **5,** 893 (1972).

[14] See, for example, S. A. Schelkunoff, *Electromagnetic Fields* (Blaisdel Publishing Co., London, 1963), Chapter 8.

[15] M. O. Scully and W. E. Lamb, Jr., Phys. Rev. **159,** 208 (1967).

5

Self-Regulated Star Formation in Molecular Clouds

Frank Bertoldi
and Christopher F. McKee

5.1 Introduction

It is now known that molecular clouds are the sites of star formation in the Galaxy. Forty years ago, however, it was not clear that molecules were even an important constituent of the Galaxy. In 1955, Townes [1] presented a paper at IAU Symposium 4 in which he listed a number of molecules that might have radio transitions. Thirteen years after this conference, Townes and his collaborators [2] discovered the first polyatomic molecule in interstellar space, NH_3. Now, with more powerful radio telescopes, ammonia is used to trace regions in which low mass stars are forming. Many of the molecules Townes listed in his 1955 talk have been discovered, along with an even larger number of other molecules. From observations of these molecules, particularly CO, we have learned an enormous amount about the conditions under which molecules can exist in the interstellar medium. Here we consider how these conditions lead to the formation of low-mass stars, and how this star formation in turn governs the structure of the molecular clouds in which most molecules are embedded. A fuller account of this work will be given elsewhere [3].

Recent infrared surveys of nearby molecular clouds reveal that most of the young stars are located in compact clusters associated with massive, dense molecular clumps. In fact, it appears that the formation of compact clusters of stars is the dominant mode of star formation [4–9]. Recent radio observations of the Orion B molecular cloud [4], the Rosette molecular cloud [10], Ophiuchus [11, 12], and the Taurus-Auriga complex [13] show that the molecular clumps in which star formation is occurring are among the most massive substructures within these molecular clouds. When the clump mass is estimated with tracers such as ^{13}CO, not high-density tracers like CS or NH_3 that reveal only the densest substructures, one finds that the star formation rate per unit gas mass is highest in the most massive clumps [14]. Although these massive clumps account for an appreciable fraction of the molecular mass in the clouds, most of the mass, which is located in the remaining clumps, appears to show almost no signs of star formation.

We propose a global model for low-mass star formation in Giant Molecular Clouds (GMCs) that accounts for the fact that the most massive clumps in the

GMCs are the dominant sites for star formation. This model is an extension of the theory of photoionization-regulated star formation [15], which made no distinction among the clumps in GMCs. In addition, the model discussed here allows for density variations within the clump, for the effect of clumping on the penetration of the FUV (far-ultraviolet) ionizing radiation, and, in calculating the ionization of the clump, for the rapid dissociative recombination of the H_3^+ ion [16].

We argue that, like GMCs as a whole, massive clumps cannot be supported by static magnetic fields alone. As a result, the massive clumps would contract in the time it takes for the motions supporting them to dissipate unless there is a source of energy that maintains the motions. Since the massive clumps cannot be reformed through collisional agglomeration in a dissipation time, and therefore must have lifetimes exceeding the dissipation time, there must be such an energy source. Wave damping is too strong to permit this energy to be supplied from outside the clump; the source must be internal. We show that energy injection from newly formed stars can support the cloud against collapse; the resulting star formation rate is consistent with observations of massive star-forming clumps and with the overall star formation rate in the Galaxy. We can calculate the rate of low-mass star formation expected under the assumption that it is determined by the rate at which the weakly ionized gas can diffuse through the magnetic field (ambipolar diffusion). The timescale for ambipolar diffusion is directly proportional to the ionization, which is determined by FUV photoionization throughout much of the clump. When the photoionization-regulated star formation rate is equated with the rate required to balance the clump's dissipative energy loss, the result is a prediction of the star formation rate *and* the mean density of massive clumps in GMCs. Since most of the star formation in GMCs occurs in their massive clumps, this theory has implications for the structure and evolution of GMCs as a whole. We also discuss the stability of the equilibrium clumps, and suggest that bursts of massive star formation may be the consequence of strong compressive perturbations of such clumps.

5.2   Structure and Stability of Molecular Clouds

Giant molecular clouds (GMCs) are large condensations of dense, cold molecular gas that are embedded in more widely distributed, tenuous, mostly atomic gas. The dense molecular gas in a given GMC appears to be confined in smaller sub-structures, clumps of diverse shapes and small volume filling factor. For molecular clouds in the solar neighborhood, a typical clump identified in ^{13}CO position-velocity maps has a mean density $n_H \sim 10^3$ cm^{-3} (500 H_2 molecules cm^{-3}), while the most massive, star–forming clumps have $n_H \simeq 3000$ cm^{-3} [17]. By comparison, the mean density of a GMC is of order 50 cm^{-3}, so the clumps occupy only a small fraction of the volume of the GMC [18, 14]. There is no clear evidence that the quoted densities increase in the inner Galaxy [19, 20]. The most massive clumps often contain high-density "cores" that are traced by NH_3 or CS observations. These cores appear to be the sites of star formation, since most of them are associated with strong IR point sources, HH objects, or outflows [21].

5.2.1 Gravitational Stability

Most molecular clouds with masses in excess of 10^3 $M_\odot$ are strongly self-gravitating [22–24]. This is true for GMCs with $M \sim 10^5 - 10^6$ $M_\odot$ (such as the Rosette), smaller isolated clouds with 10^4 $M_\odot$ (such as Ophiuchus), and the largest clumps of mass $M \sim 10^3$ $M_\odot$ within the more massive GMCs. An observable measure for the degree of gravitational binding of a cloud is the virial parameter [17],

$$\alpha \equiv \frac{5\sigma^2 R}{GM} \simeq \frac{2T}{|W|} , \qquad (5.1)$$

where σ is the cloud's one-dimensional velocity dispersion, R its mean radius, M its mass, T the kinetic energy, and W the gravitational binding energy. When $\alpha \simeq 1$, the cloud is strongly self-gravitating, whereas for $\alpha \gg 1$, its gravity is unimportant and the cloud is pressure-confined by the surrounding medium. Within a given GMC, the individual clumps follow $\alpha \propto M^{-2/3}$, and only the most massive clumps are strongly self-gravitating [17], i.e., only they have $\alpha \simeq 1$.

If a cloud is threaded by a magnetic flux Φ, then in the absence of any internal kinetic energy, it can be in a stable equilibrium provided its mass is less than the magnetic critical mass [25, 26],

$$M_\Phi \equiv 0.12 \, \Phi/G^{1/2} . \qquad (5.2)$$

On the other hand, if the cloud is supported entirely by thermal pressure, then it will be in a stable equilibrium as long as its mass is less than the Jeans mass,

$$M_J = 1.182 \, \frac{\sigma^4}{(G^3 P_0)^{1/2}} , \qquad (5.3)$$

where P_0 is the surface pressure. With both thermal and magnetic pressure, the cloud can be stable up to the critical mass M_{cr}, which is approximately [15]

$$M_{cr} \simeq M_J + M_\Phi . \qquad (5.4)$$

Clouds with $M > M_\Phi$ are termed *magnetically supercritical:* for such a cloud, the static magnetic field by itself cannot prevent gravitational collapse. On the other hand, gravitational collapse is impossible for *magnetically subcritical* clouds, which have $M < M_\Phi$. Note that if the cloud is centrally condensed, this criterion must be applied throughout the cloud, not just at its surface; it is possible for a magnetically subcritical cloud to have a magnetically supercritical core.

Measurements of magnetic field strengths and turbulent velocity dispersions in GMCs and their clumps indicate that the energy density in turbulent motions is about equal to that of the magnetic field, implying that the turbulence is approximately Alfvén-like [27]. Since a GMC as a whole as well as the massive clumps are strongly self-gravitating, their respective gravitational energies are also about equal to their magnetic or kinetic energies. It follows that $M \simeq M_{cr} \simeq 2M_J \simeq 2M_\Phi$, which implies that such clouds are both Jeans supercritical ($M > M_J$) and magnetically supercritical ($M > M_\Phi$): neither the magnetic field nor the kinetic pressure alone could support such a cloud against gravitational collapse – the cloud is supported by the *combined* pressure of the magnetic field *and* the kinetic motions [15]. In analyzing the physical state of the clumps in three molecular clouds (Ophiuchus, Orion B, and Rosette) with the assumption that the most massive clump in a GMC has a mass close to the critical mass, i.e., $M_{max} \simeq M_{cr}$, we previously [17] estimated the magnetic critical mass of the massive clumps as $M_\Phi/M_{cr} \simeq 0.5$, and confirmed that the most massive clumps are indeed magnetically supercritical.

5.2.2   The Dissipation Problem

The fact that GMCs and massive clumps are magnetically supercritical has a crucial implication: if the turbulent support in the GMC or clump were to diminish due to dissipation, it would contract and at some point undergo gravitational collapse. Thus, dissipation of the turbulent energy is a crucial process in determining the stability of a molecular cloud.

The dynamical timescale of a self-gravitating cloud is the gravitational free-fall time,

$$t_{ff} = \left(\frac{3\pi}{32G\rho} \right)^{1/2} = \frac{4.35 \times 10^7}{n_H^{1/2}} \text{ yr} , \qquad (5.5)$$

which is about 6×10^6 yr for a typical GMC with a mean density $n_H \simeq 50 \text{ cm}^{-3}$. Estimates of the characteristic lifetime of GMCs, t_{GMC}, are significantly greater than t_{ff}: Blitz & Shu [18] argued that the total lifetime is of order a few times 10^7 yr based on the time it takes for a cloud to produce an OB association and for the OB association to destroy the cloud that formed it. McKee [15] made a similar argument and concluded that GMCs with active star formation have $t_{GMC} \simeq 3 - 5 \times 10^7$ yr. Scoville & Sanders [24] suggested that the lifetimes must exceed 10^8 year based on the presence of GMCs between spiral arms. As we have just argued, magnetic fields alone cannot support GMCs against gravitational collapse. How then can they survive for so many free-fall times?

The observed linewidths of mass tracer molecules such as ^{13}CO show that the gas in molecular clouds is in supersonic "turbulent" motion. Inside the clumps, the complex spatial and velocity structure of the gas can be understood as resulting from the nonlinear interactions among hydromagnetic waves that are subject to nonlinear steepening, damping through ohmic dissipation, and loss due to transmission across the clump surface. In the absence of energy injection, GMCs as a whole and the massive clumps within them would contract in the time it takes to dissipate this motion, t_{diss}. Approximate analytic and numerical estimates of the dissipation rate of turbulent motions in clumpy molecular clouds indicate that the dominant damping processes scale such that the dissipation time is approximately proportional to the dynamical time of the cloud, t_{ff}. Let $t_{diss} \equiv \eta t_{ff}$ be the timescale for the cloud's loss of turbulent energy. Elmegreen [28] estimated that $\eta \sim 3 - 10$ for clumpy, magnetized clouds, and Carlberg & Pudritz [29] found $\eta \sim 4$ in numerical simulations of such clouds. McKee [15] adopted $\eta = 5$ as a typical value for the dissipation of energy in the GMC as a whole. Massive clumps should have a somewhat smaller value of η than GMCs because the MHD waves dissipate more rapidly in the weakly ionized clump gas and because of the loss of energy due to wave transmission through the clump surface. With a mean density $n_H \simeq 3000$ cm^{-3} and assuming $\eta \sim 3$, the massive clumps' dissipation time is $t_{diss} \simeq 2.4$ Myr. Blitz & Shu [18] have estimated that the formation time of a clump, as set by the collision time, is $t_{SFC} \simeq 10$ Myr, so that the ratio of the formation time to the dissipation time for star-forming clumps is large, $t_{SFC}/t_{diss} \simeq 10/2.4 \simeq 4$.

We conclude that massive clumps cannot form fast enough to explain their presence in relatively old ($\gg 10$ Myr) GMCs if the clumps lose their kinetic energy in a dissipation time. The turbulent energy must therefore be replenished constantly. There appear to be two possible sources: the large-scale kinetic energy of the GMC, i.e. the bulk kinetic energy of the clumps and the energy stored in the distorted magnetic field [30], or the outflows of the young stars that form in the massive clumps [31]. In a forthcoming paper [3], we demonstrate that wave damping is sufficiently strong that the first mechanism cannot work. We therefore turn our attention to stellar energy injection.

5.3   Self-Regulated Star Formation

Based on an original suggestion by Norman and Silk [31], McKee [15] demonstrated that the energy injection from newly formed low-mass stars can counteract the dissipation in star-forming GMCs and thereby support them against gravitational collapse for a time long compared with their turbulent dissipation time. Here we apply the basic idea of this model to the self-gravitating, massive clumps within a GMC. We shall determine the conditions under which the dissipation of the turbulent energy within a massive clump is just offset by the energy injected by young low-mass stars that form in the interiors of such clumps.

5.3.1   Energy Gain: Low-Mass Star Formation

Very young low-mass stars inject large amounts of energy and momentum into their surroundings (see the review by Fukui et al. [32]). During the formation of a star, while it is still accreting mass from the dense gas core out of which it is forming, a powerful, collimated wind (a jet) is driven off the star-disk system. The mass of the wind m_w is a significant fraction of the system's mass, and it has a velocity v_w of several hundred km s^{-1}. The duration of the outflow is of order 10^5 years, and the total energy can be as large as 10^{47} erg, comparable to the typical energy of a molecular clump. Thus outflows must play a significant role in the kinetic energy balance of star-forming clumps.

When a jet from a newly formed star shocks and piles up against the surrounding dense molecular gas, it deposits much of its momentum, $m_w v_w$, into the natal clump. Most of the kinetic energy however is dissipated in radiative shocks, and only a fraction v_{rms}/v_w contributes to the turbulence in the clump, provided the outflow remains trapped in the clump. The energy injected into a clump through stellar outflows per unit cloud mass turned into stars is then $\epsilon_{in} \approx m_w v_w v_{rms}/2 m_*$, where m_* is the mass of of the star that is formed. The typical total momentum of an outflow can be estimated from theoretical considerations, which suggest that about 1/3 of the final mass of a star is ejected at typical velocities of 200 km s^{-1} [33], resulting in a momentum of 70 km s^{-1} per solar mass turned into stars. Observations suggest a somewhat smaller value: from the distribution of observed outflow momenta [34, 35] we infer a value closer to 40 km s^{-1} per solar mass.

5.3.2   Equilibrium Star Formation Rate

The equation governing the total kinetic energy per unit mass of a given clump can be written $d(T/M)/dt = Gain - Loss$, where $T = M v_{rms}^2/2$ is the total kinetic energy of the clump. We make the "isothermal" assumption that the energy per unit mass is constant throughout the clump and that it does not change as a result of mass gain or loss. Thus, the cloud energy equation is [15]

$$\frac{d}{dt}\left(\frac{T}{M}\right) = \frac{\epsilon_{in}}{t_*} - \frac{v_{rms}^2}{2\eta t_{ff}} , \tag{5.6}$$

where t_* is the star formation time. In equilibrium, the dissipative energy loss is offset by the gain due to winds from young stars [$d(T/M)/dt = 0$], which implies

$$\frac{\epsilon_{in}}{t_{*eq}} = \frac{v_{rms}^2}{2\eta t_{ff}} . \tag{5.7}$$

In equilibrium, the star formation timescale, t_{*eq}, which is the time for the clump to be converted into stars if star formation continued at the instantaneous rate, is then

$$t_{*eq} = \frac{\eta t_{ff}\, m_w v_w}{m_* v_{rms}} . \tag{5.8}$$

The mean velocity dispersion of the clump, $v_{rms} = \sqrt{3}\,\sigma$, can be expressed in terms of the dimensionless virial parameter α (Eq. (5.1)). In terms of the mean visual extinction of the clump,

$$\bar{A}_V = \frac{M/(\mu_H \pi R^2)}{2 \times 10^{21}\mathrm{cm}^{-2}} \ \mathrm{mag} , \tag{5.9}$$

we can rewrite the equilibrium star formation timescale as

$$t_{*eq} = 470\,\phi_1\,\bar{A}_V^{-1}\ \mathrm{Myr} , \tag{5.10}$$

where

$$\phi_1 \equiv \left(\frac{\eta}{3}\right) \left(\frac{m_w v_w/m_*}{40 \text{ km s}^{-1}}\right) \left(\frac{\alpha}{1.4}\right)^{-1/2} \tag{5.11}$$

is a dimensionless parameter that is of order unity in typical clumps. According to Eq. (5.10), the star formation time in a self-gravitating clump is therefore mainly a function of the clump's mean extinction, or, equivalently, its mean column density. If we apply values that are typical for observed star-forming, massive clumps ($\phi_1 = 1, M = 10^3 \, M_\odot, \bar{A}_V \approx 5$) the equilibrium star formation timescale turns out to be of order $t_* \simeq 100$ Myr, corresponding to a star formation rate of 10 $M_\odot$/Myr in such a clump, a value not unusual for the star-forming clumps such as those in Orion B.

5.3.3   Photoionization-regulated Star Formation

Quiescent low-mass star formation is governed by the gradual gravity-driven slippage of neutral gas through the magnetic field (see the review by Shu *et al.* [36]). Small, gravitationally bound substructures, termed "cores", with typical masses of 1–$10 \, M_\odot$, thereby contract until they become gravitationally unstable and collapse to form one or more low-mass stars. The processes governing the formation of such star-forming cores out of a turbulent, massive clump are not well understood. To estimate the star formation timescale in a turbulent clump we shall assume that, on average, the growth rate of small condensations that evolve into star-forming cores is given by the average rate of ambipolar diffusion in the clump. The clump as a whole does not contract significantly via ambipolar diffusion, since the average fractional ionization of the clump remains high due to the ionization of Mg and Fe, elements that are photoionized throughout most of the clump.

The cores form preferentially in the "darkest" and densest part of the clump, toward the center where the gas is sufficiently shielded from the diffuse FUV radiation. The gas ionization closer to the surface is high enough that the gas is strongly coupled to the magnetic field, and effectively no ambipolar diffusion occurs. Thus, the rate of star formation in a clump is strongly determined by the extent to which photoionization inhibits ambipolar diffusion in the dense gas.

To estimate the rate of ambipolar diffusion, we adopt the procedure of McKee [15]. Nakano's [37] calculations show that the time for the central region of a flattened cloud to undergo quasi-static contraction via ambipolar diffusion to the point of collapse is about $8 \times 10^{13} \, x_e$ yr, where x_e is the fractional ionization in the cloud. In a somewhat flattened cloud similar to those studied by Mouschovias [38], about half the cloud undergoes ambipolar diffusion in that time. Therefore, the time for the entire cloud to lose its flux and collapse is about twice this, or

$$t_{AD} \simeq 1.6 \times 10^{14} \, x_e \text{ yr} . \tag{5.12}$$

The star formation rate in a clump can now be estimated by the mass-averaged ambipolar diffusion rate

$$\frac{M}{t_{*AD}} = \int_M \frac{dM}{t_{AD}(x_e)} \,, \tag{5.13}$$

which depends on the level of fractional ionization in the clump.

3.3.1. Ionization Structure

There are two mayor sources of ionization in a dense molecular cloud: photoionization of chemically nonreactive metals like Mg or Fe by the diffuse FUV background, and cosmic ray ionization of H_2. In the outer layers of a clump, FUV photoionization leads to a degree of ionization $x_{e\pi} = 6 \times 10^{-5} G_0 n_3^{-1} \exp(-1.6 A_{Vs})$, where n_3 is the local density of H nuclei in units of 10^3 cm^{-3}, G_0 is the ratio of the mean intensity of the incident FUV radiation field to that in the solar vicinity, and $1.6 A_{Vs}$ is an effective FUV optical depth to the surface of the clump. On the other hand, in the absence of FUV radiation, cosmic rays would produce an ionization level $x_{ecr} \simeq \sqrt{2} \times 10^{-7} \zeta_{-17} n_3^{-1/2}$. Taking into account both sources of ionization, the fractional ionization becomes [15, 3]

$$x_e = \frac{1}{2} \left[x_{e\pi} + (x_{e\pi}^2 + 4x_{ecr}^2)^{1/2} \right] \,. \tag{5.14}$$

The critical value of A_{Vs} at which UV and CR ionization are equally important is given by $A_{cr} = 3.8 - 0.3 \ln n_3 \approx 4$ mag. Since a significant fraction of the molecular mass in GMCs is in clumps that have average extinctions smaller than 4 mag, most molecular gas in GMCs is dominated by FUV photoionization [15].

5.3.4   Self-regulated Equilibrium States

Equation (5.10) gives the star formation timescale, $t_{*eq} \propto \bar{A}_V^{-1}$, that would be required for a self-gravitating clump to offset its loss of turbulent energy through the energy input from the outflows of young, embedded stars. Equation (5.13) provides an estimate for the star formation rate in a clump from its average rate of ambipolar diffusion, which depends on the ionization. The degree of ionization depends on the extinction and on the density n_H; the latter can be eliminated in favor of the mass and mean extinction since $\bar{A}_V \propto n_H R \propto n_H^{2/3} (n_H R^3)^{1/3} \propto (n_H^2 M)^{1/3}$. Equating the two timescales for a given clump mass,

$$t_{*eq}(\bar{A}_V) = t_{*AD}(\bar{A}_V, M) \,, \tag{5.15}$$

we can solve for an equilibrium extinction, $\bar{A}_{eq}$, at which the clump forms stars via ambipolar diffusion at a rate such that the dissipative loss of turbulent energy is exactly offset by the stellar outflows.

To numerically evaluate the ionization structure and thereby the star formation rate in a clump of given mass and size, we assume that the clump has a uniform,

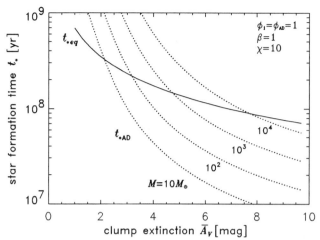

FIGURE 5.1. Determining the self-regulated equilibria from equating the equilibrium star formation time, t_{*eq}, (*solid line:* Eq. (5.10) and the photoionization-regulated (ambipolar-diffusion limited) star formation time, t_{*AD} (Eq. (5.13)), which is plotted for four different clump masses *(dotted lines)*. The intersection points of the solid with the dashed lines correspond to equilibrium clump states for which star formation balances the dissipation of turbulent energy.

spherical density distribution of the form

$$n(r) = \chi\, n_0 \left[1 + \left(\frac{r}{R}\right)^2 \left(\chi^{2/\beta} - 1\right)\right]^{-\beta/2}, \qquad (5.16)$$

where $\chi \equiv n(0)/n_0$ is the central-to-surface density ratio. Observations [39] indicate that $\beta \approx 1 - 2$. We adopt $\chi \simeq 10$, which is comparable to the density ratio in an isothermal sphere at the brink of collapse ($\chi = 14$). The dotted lines in Fig. 5.1 are solutions to the integral in Eq. (5.13) for different clump masses and $\chi = 10$, $\beta = 1$. The dotted line shows t_{*eq} as a function of the mean clump extinction $\bar{A}_V$ for $\phi_1 = 1$. The intersections of the dotted line with the solid lines represent the clump extinction at which star formation balances dissipation, i.e., where $t_{*eq} = t_{*AD}$.

If the clump were at a size such that its mean extinction were below the equilibrium value $\bar{A}_{eq}$, then the star formation rate would be too low to balance dissipation, and $d(T/M)/dt < 0$. Since the clump is assumed to be strongly self-gravitating, it would gradually contract as it lost turbulent support, while the star formation rate would increase because of the increasing density and opacity, until the clump reaches its equilibrium extinction, and $\bar{A}_V = \bar{A}_{eq}$. Conversely, if the clump were smaller than its equilibrium size and $\bar{A}_V > \bar{A}_{eq}$, star formation would produce more energy than the clump could dissipate, and as a result, the clump would expand while its turbulent energy increased and it expanded to its equilibrium size. Thus the equilibria are attractive, stable states [15].

Figure 5.2 shows how the clumps' equilibrium star formation timescale, extinction, and density depend on the clump mass for two degrees of central

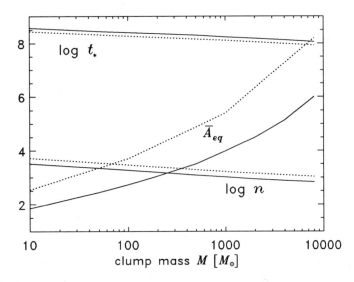

FIGURE 5.2. Self-regulated star formation in centrally condensed clumps. The equilibrium star formation time, mean clump extinction, and mean clump density resulting from a solution of Eq. (5.15), i.e., the intersection points of Fig. 5.1, as a function of the clump mass, for two degrees of central condensation. *Dotted lines:* mildly condensed, $\beta = 1$, $\chi = 5$; *solid lines:* strongly condensed, $\beta = 1.5$, $\chi = 10$.

concentration. Since smaller clumps tend to be less gravitationally bound [17], and gravity is the main cause of the central condensation, smaller clumps may be less centrally condensed.

The dependence of the solutions on parameters such as ϕ_1, for which we have adopted "typical" values, will be discussed in detail elsewhere [3]. Note that a higher FUV or CR flux increases the equilibrium star formation rate in the clumps: enhanced ionizing radiation causes a decrease in the star formation rate and therefore a contraction of the clump. As a result, its turbulent dissipation rate increases, so that in equilibrium, a higher star formation rate is required to offset the dissipative loss. A sudden strong increase in the incident FUV radiation intensity could in fact cause a surge – delayed by one ambipolar time – of star formation activity in otherwise rather quiescent clumps.

5.4   Discussion

Despite uncertainties in the numerical values for some of the parameters entering the theory, we find that under typical conditions, the equilibrium states for self-gravitating clumps that are derived from a balance between dissipation and star formation resemble the observed star-forming clumps. We find star formation timescales in the clumps that are of order 100 Myr, well above the free-fall or dissipation time of a typical massive clump.

Because of unreliable pre-main sequence stellar evolution models and a lack of understanding concerning the dynamical evolution of young, embedded clusters, it has proven difficult to determine the ages of such clusters. Recent modeling of the observed near-infrared (NIR) luminosity function of the embedded cluster IC348 [40] has yielded a cluster age of order 5–7 Myr, as well as strong evidence that the star formation rate has been constant over this period. A NIR survey of parts of the Orion A cloud [41] indicates that stars have been forming there for more than 7 Myr, with the younger stars being more strongly clustered than the older ones. Note that if an embedded cluster is slowly evaporating, the beginning of star formation would lie even further back than the age derived from its present population.

To compare the number of stars observed in embedded clusters with those predicted by our equilibrium model, we need to set an age at which the clumps began to form stars. The beginning of star formation could have been set by its collisional mass growth towards a strongly self-gravitating state, or by an increase in the pressure of its environment, through, for example, the overall contraction of its GMC. Star formation could also have increased from a low to a high level through an increase in the incident UV radiation field.

In Orion, star formation at the currently observed level has probably commenced some 10 Myr ago. The most massive clumps in Orion B have a current gas mass of about 300–500 $M_\odot$ as derived from ^{13}CO measurements (Bally and Bertoldi, unpublished). In 10 Myr of star formation, they would be expected to have bred about (500 $M_\odot$/100 Myr) $\times$ 10 Myr = 50 $M_\odot$ into $\sim$ 100 stars for a mean stellar mass of 0.5 $M_\odot$. Indeed, these clumps contain between 30 and 300 stars each, consistent with this prediction. Differences in gravitational binding or UV illumination could in part be responsible for the variation in the stellar content. Also, the nearby OB associations appear to have affected the Orion B molecular cloud, and this in turn may have altered the the conditions under under which star formation is occurring in the four massive clumps.

A more accurate way to probe the current rate of star formation may be a count of active outflow sources, dense cores, or deeply embedded young stellar objects. If outflows are active for a time of order 10^5 yr, we would expect to find about one of them in each of the Orion B clumps. Similarly, only a few dense cores visible in NH_3 or CS emission lines should be present, consistent with the CS observations [43].

The number of outflows ($\sim$ 10) found by several surveys [44–46] in $\sim$ 2 $\times$ $10^4 M_\odot$ of molecular gas in the Orion A and Mon OB1 clouds agrees with that expected from the equilibrium star formation rate if we assume that about 30% of the molecular mass in these clouds is in massive, star-forming clumps.

The global star formation timescale for star-forming molecular clouds in the Galaxy was estimated [15] as $\sim$ 300 Myr. Our derived star formation rates can account for this average star formation rate if we assume that about 30% of the molecular gas in Galactic GMCs that are actively forming stars is in equilibrium star-forming clumps. This fraction is consistent with that observed in the Orion B, Ophiuchus, and Rosette molecular clouds.

Our prediction of the mean extinction and density of the star-forming clumps (Fig. 5.2) agrees with those observed in the Rosette and in Orion, where the observed mean density of the most massive clumps is around 1000 and 2000 cm^{-3}, respectively [17]. The clumps comprising the Ophiuchus core, however, have higher densities, which could be due to a stronger incident radiation field, or a lower value of the average outflow momentum for stars in this cloud.

Our results suggest that quiescent star forming clumps should be stable: with a typical star formation timescale of order 100 Myr, the number of stars forming in a dissipation time is $\simeq 60 \, M/(10^3 \, M_\odot)$, and is large enough that statistical variations in the rate of star formation will not lead to strong perturbations away from the equilibrium state. However, the equilibrium could be fatally disturbed through a sudden increase in surface pressure that leads to a compression and an enhanced rate of turbulent dissipation [47]. If the dissipation time in the compressed state is shorter than the typical formation time of a star-forming core, then the clump could become unstable before a sufficient number of stars can form in the clump that could offset the enhanced energy dissipation. The clump could then collapse unhindered by its magnetic field because it is magnetically supercritical, possibly leading to the synchronous formation of several massive stars. The instability of magnetically supercritical clouds has often been cited as a possible mechanism for massive star formation [36].

5.5   Summary

We propose that magnetically supercritical clumps in molecular clouds can be gravitationally stable over timescales of 10^8 yr, much longer than their dynamical or their dissipation time. The onset of massive star formation will eventually disperse the clumps and the GMC; this will happen in a time that is significantly shorter than the star formation timescale. The constant dissipation of the turbulent energy that secures the gravitational stability of the clumps is offset by the steady injection of energy through the winds of newly formed stars. The self-regulated balance between dissipation and star formation is due to the fact that star formation can only proceed in the fraction of the clumps that is shielded from the diffuse FUV radiation. McKee [15] has shown that a similar mechanism can explain the moderate star formation rates in GMCs as well as their observed mean extinctions.

Since the time for stars to form out of the turbulent density enhancements is comparable to the clump's dissipation time, the clump equilibria may not be as stable as that of the GMC as a whole: if a clump is suddenly perturbed to a higher density, it may take too long for the star formation rate to adjust to a higher equilibrium value, so the clump may continue to collapse and overshoot—a burst of star formation can follow that is likely to destroy the clump. This may be similar to the idea for starbursts in irregular galaxies: small galaxies have episodic star formation, whereas in large ones like the Milky Way it averages out.

5.6   Acknowledgments

One of us (Chris McKee) wishes to express his appreciation to Charlie Townes for the essential role he has played in the development of astrophysics at Berkeley, and for the sound advice he has provided over the years. The research of Frank Bertoldi is supported by the Deutsche Forschungsgemeinschaft. The research of Chris McKee is supported by NSF grant AST 92-21289, and, for research on star formation, by a NASA theory grant to the Center for Star Formation Studies.

References

[1]   C. H. Townes, in IAU Symposium 4, *Radio Astronomy*, ed. H. C. van de Hulst (London: Cambridge University Press), p. 92 (1957).

[2]   A. C. Cheung, D. M. Rank, C. H. Townes, D. D. Thornton, and W. J. Welch, Phys. Rev. Lett. **21**, 1701 (1968).

[3]   F. Bertoldi and C. F. McKee, in preparation (1995).

[4]   C. J. Lada and E. A. Lada, in *The Formation and Evolution of Star Clusters*, ed. Janes (ASP), p. 3 (1991).

[5]   E. A. Lada, N. J. I. Evans, D. L. DePoy, and I. Gatley, ApJ **371**, 171 (1991).

[6]   E. A. Lada, ApJ **393**, L25 (1992).

[7]   J. M. Carpenter, R. L. Snell, F. P. Schloerb, and M. F. Skrutskie, ApJ **407**, 657 (1993).

[8]   E. A. Lada, K. M. Strom, and P. C. Myers, in *Protostars and Planets III*, eds. E. Levy and J. Lunine (Arizona), p. 245 (1993).

[9]   H. Zinnecker, M. J. McCaughrean, and B. A. Wilking, in *Protostars and Planets III*, eds. E. Levy and J. Lunine (Arizona), p. 429 (1993).

[10]   D. L. Block, T. R. Geballe, and J. E. Dyson, A&A **273**, L41 (1993).

[11]   R. B. Loren, ApJ **338**, 902 (1989).

[12]   R. B. Loren, ApJ **338**, 925 (1989).

[13]   M. Gomez, L. Hartmann, S. J. Kenyon, and R. Hewett, AJ **105**, 1927 (1993).

[14]   L. Blitz, in *The Physics of Star Formation and Early Stellar Evolution*, eds. C. J. Lada and N. D. Kylafis (Dordrecht: Kluwer), p. 3 (1991).

[15]   C. F. McKee, ApJ **345**, 782 (1989).

[16]   G. Sundström *et al.*, Science **263**, 785 (1994).

[17]   F. Bertoldi and C. F. McKee, ApJ **395** 140 (1992).

[18]   L. Blitz and F. H. Shu, ApJ **238**, 148 (1980).

[19]   H. S. Liszt, ApJ **411**, 720 (1993).

[20]   D. B. Sanders, N. Z. Scoville, R. P. J. Tilanus, Z. Wang, and S. Zhou, in *Back to the Galaxy*, ed. S. S. Holt and F. Verter (New York: AIP), p. 311 (1993).

[21]   P. C. Myers, in *Molecular Clouds and Star Formation*, ed. C. Yuan (Singapore: World Press) (1995).

[22]   R. B. Larson, MNRAS **194**, 809 (1981).

[23]   P. M. Solomon, A. R. Rivolo, J. W. Barrett, and A. Yahil, ApJ **319**, 730 (1987).

[24]   N. Z. Scoville and D. B. Sanders, in *Interstellar Processes*, eds. D. Hollenbach and H. Thronson (Dordrecht: Reidel), p. 21 (1987).

[25]   T. Mouschovias and L. Spitzer, ApJ **210**, 326 (1976).

[26]   K. Tomisaka, S. Ikeuchin, and T. Nakamura, ApJ **335**, 239 (1988).

[27]   P. C. Myers and A. A. Goodman, ApJ **326**, L27 (1988).

[28]   B. G. Elmegreen, ApJ **299**, 196 (1985).

[29] R. G. Carlberg and R. E. Pudritz, MNRAS **247**, 353 (1990).

[30] J. L. Puget, in *Fragmentation of Molecular Clouds and Star Formation*, ed. E. Falgarone *et al.* (Dordrecht: Kluwer), p. 75 (1991).

[31] C. A. Norman and J. Silk, ApJ **238**, 158 (1980).

[32] Y. Fukui, T. Iwata, A. Mizuno, J. Bally, and A. P. Lane, in *Protostars and Planets III*, eds. E. Levy and J. Lunine (University of Arizona Press), p. 603 (1993).

[33] F. H. Shu, S. Lizano, S. P. Ruden, and J. Najita, ApJ **328**, L19 (1988).

[34] S. Cabrit and C. Bertout, A&A **261**, 274 (1992).

[35] L. Chernin, private communication.

[36] F. H. Shu, F. C. Adams, and S. Lizano, ARA&A **25**, 23 (1987).

[37] T. Nakano, Fund. Cosmic Phys. **9**, 139 (1984).

[38] T. Mouschovias, ApJ **207**, 141 (1976).

[39] J. P. Williams and L. Blitz, ApJ, submitted (1995).

[40] E. A. Lada and C. J. Lada, AJ **109**, 1682 (1995).

[41] K. M. Strom, S. E. Strom and K. M. Merrill, ApJ **412**, 233 (1993).

[42] C. F. Prosser, J. R. Stauffer, L. Hartmann, D. R. Soderblom, B. F. Jones, M. W. Werner, and M. J. McCaughrean, ApJ **421**, 517 (1994).

[43] E. A. Lada, J. Bally and A. A. Stark, ApJ **368**, 432 (1991).

[44] Y. Fukui, K. Sugitani, T. Tabaka, A. Mizuno, H. Ogawa, and K. Kawabata, AJ **311**, L85 (1986).

[45] M. Margulis, C. J. Lada, and R. L. Snell, ApJ **333**, 316 (1988).

[46] J. A. Morgan, F. P. Schloerb, R. L. Snell, and J. Bally, ApJ **376**, 618 (1991).

[47] F. Bertoldi, C. F. McKee, and R. I. Klein, in *Massive Stars: Their Lives in the Interstellar Medium*, eds. J. Cassinelli and E. Churchwell (ASP Conf. Ser. 35), p. 129 (1993).

6

Long-baseline Interferometric Imaging at 11 Microns with 30 Milliarcsecond Resolution

Manfred Bester
and William C. Danchi

6.1   Introduction

Breakthroughs in fundamental astrophysics are often the result of major advances in the instrumentation available for observations of interesting phenomena. Historically instrumental advances provide major improvements in one or more capabilities, typically resolution, either spectral or spatial, or sensitivity. An important example is that of the discovery of the cosmic microwave background, which depended on the availability of sensitive (low-noise maser) receivers [34]. Another example is the discovery of interstellar molecules more complex than OH. The vast majority of these molecules were discovered from their millimeter- or submillimeter-wave transitions. Observations of these transitions depended on heterodyne mixer receivers that provided both high sensitivity and very high spectral resolution [10].

Improved spatial resolution has been particularly important to the evolution of radio astronomy because of the intrinsically low spatial resolution of a conventional reflecting antenna. At one centimeter wavelength, a 15 m antenna has a diffraction-limited beam size of 2.3 arcminutes. This can be compared to the seeing-limited resolution of a typical telescope (without adaptive optics) which is approximately 1 arcsecond, and hence more than two orders of magnitude better. Thus if the wavelength of observation is fixed, the only way to improve the spatial resolution is to increase the effective size of the antenna aperture. A major technical advance came from the realization that a very large aperture could be synthesized out of individual small ones by interferometric combination of the signals from pairs of these small apertures. This is possible because the signals from the individual pairs are just Fourier components of the surface brightness distribution. This technique brought the spatial resolution available to radio astronomy to the level where it has equaled or exceeded that in optical or infrared astronomy [41]. Sir Martin Ryle [38] was given a Nobel Prize in physics partly for his contributions to the development of these techniques. Radio interferometers are now in daily use around the world investigating fundamental problems in all areas of astronomy and astrophysics.

Long-baseline interferometry at optical and infrared wavelengths is at a much earlier stage of development than radio interferometry. One important reason is that refractivity fluctuations of the atmosphere have more than a proportionally larger effect at optical wavelengths than at radio wavelengths. For example at 500 nm, root-mean-square (rms) fluctuations range from ~ 5-10 μm for a baseline of 10 m, which is a variation of 20 wavelengths. This can be compared to fluctuations at centimeter radio wavelengths, such as have been measured at the VLA. In this case they are in the range of 2-10 mm for 20 km baselines, which is typically a fraction of a wavelength. Thus, interferometers operating at optical and infrared wavelengths have to cope with an atmosphere that fluctuates much more wildly than it does in the radio region. Atmospheric fluctuations and their implications for interferometry are discussed in a review by Masson [30].

Despite the difficulties caused by a strongly fluctuating atmosphere, considerable progress has been made in the past decade in long-baseline interferometry in the optical and infrared. A number of two-element single-baseline interferometers have been built and quite a number of significant observational results have been obtained. The right hand part of Fig. 6.1 displays the resolution and wavelength coverage of some of these instruments. The instruments that have operated or are currently operating at optical wavelengths include the Mark III Interferometer (USA, [39]), GI2T (France, [27]), Sydney University Stellar Interferometer (SUSI) (Australia, [15]). The Infrared/Optical Telescope Array (IOTA) interferometer operates at 2.2 μm (USA, [8]) and was preceded by the Infrared Michelson Array (IRMA) (USA, [21]). In France the I2T interferometer was also successfully used at 2.2 μm [16]. The Infrared Spatial Interferometer (ISI) (USA, [5]) is unique among currently operating US instruments both in spectral coverage, i.e., mid-infrared (9-12 μm), as well as having been designed, built, and operated by a university research group.

A number of instruments are currently under construction or in the planning stages. For example, the Cambridge (UK) group is building the Cambridge Optical Aperture Synthesis Telescope (COAST), a 4-element interferometer for the optical and near-infrared [2]. Three of four telescopes have been constructed and phase closure has been demonstrated [2]. Four elements will allow for amplitude closure as well as phase closure [33]. The Navy Prototype Optical Interferometer (NPOI) consists of an Imaging Array (IA) and Astrometric Array (AA) built as collaboration between the Naval Research Laboratory and the U.S. Naval Observatory. The Imaging Array (IA) is a 6-element optical interferometer under construction by the Naval Research Laboratory [1]. The Astrometric Interferometer developed by the Naval Observatory is a 4-element optical interferometer [26]. As part of NASA's Mission to Planet Earth, the Jet Propulsion Laboratory is constructing a special two-element interferometer for the near-infrared, called ASEPS0, which is designed for very high precision narrow-angle astrometry to address the problem of planet detection [11]. The Center for High Angular Resolution Astronomy (CHARA) array is under construction by Georgia State University (USA) for the optical and near infrared [32], and is expected to have 5 telescopes. Currently in the

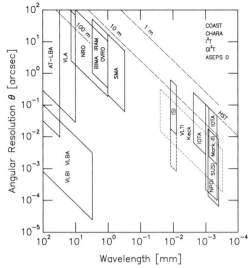

FIGURE 6.1. Comparison of angular resolution and wavelength coverage of radio, infrared, and visible wavelength long baseline interferometers.

planning stage is the Very Large Telescope Interferometer (VLTI) of the European Southern Observatory (ESO), which is expected to operate in the infrared [29].

6.2   The Infrared Spatial Interferometer

The Infrared Spatial Interferometer, installed at Mount Wilson Observatory since 1988, is a two-telescope interferometer. It operates very much like a radio interferometer, but at a much higher frequency of 27,000 GHz, or a wavelength of 11 microns. The ISI telescopes each consist of a 65-inch (1.65 m) parabolic mirror and an 80-inch (2 m) flat mirror, and are mounted in standard-size semi-trailers as seen in Fig. 6.2. The parabolic mirror is fixed while the flat mirror acts as a siderostat, tracking stars or other objects. All associated optics, electronics and computer control systems, which are stationary, are contained in the trailers as well. This unique design, known as Pfund-type optics, makes the telescopes very compact and light weight, and therefore easy to transport in order to reconfigure the baselines. Similar techniques are employed in radio interferometry where telescopes often are moved along railroad tracks.

In order to cover a wide range of angular resolutions, the instrument employs a technique known as Earth rotation aperture synthesis [37]. This means that for a fixed separation of the telescopes on the ground the rotation of the Earth changes the apparent separation between them, as seen from the vantage point of the object being observed. Therefore, a whole range of angular resolutions can be covered in one single night. By moving the telescopes from time to time with respect to each other, various ranges of angular resolution can be selected.

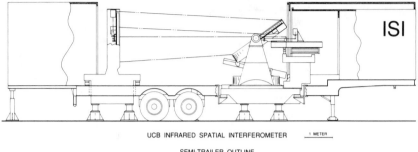

UCB INFRARED SPATIAL INTERFEROMETER   1 METER

SEMI-TRAILER OUTLINE

FIGURE 6.2. Schematic of one of the Pfund telescopes of the ISI. An 80-inch flat mirror on an alt-azimuth mount acts as a coelostat and sends radiation from an astronomical source to a 65-inch parabola shown on the left. The parabola focuses radiation through a hole in the flat mirror to the detection system. Each mirror mount stands on a 3-legged kinematic mount independent of the trailer when the telescope is not moved. Dashed lines represent HeNe laser interferometer beams used to monitor distances.

The available baselines presently extend from 4 to 35 m, giving angular resolutions between 250 and 30 milliarcsec, respectively. Longer baselines up to 85 m will be available in the near future. A third telescope has been funded and is under construction. This will allow two-dimensional imaging and faster sampling of the visibility data.

The infrared detection system employs heterodyne techniques. The infrared signal from the star is down-converted to microwave frequencies using a HgCdTe photodiode and a CO_2 laser local oscillator on each telescope. The signals are then amplified and combined electronically. The delay line, lobe rotator, and correlator, all use conventional radio technology. The detection system is constantly improved and upgraded in order to further increase both sensitivity and precision of the visibility measurements being made.

Other recent technical developments are in the areas of first-order adaptive optics and spectroscopy. At present time, the ISI telescopes are being equipped with near-infrared guiding cameras and a tip/tilt system for fast autoguiding. A large improvement in the precision of measured stellar visibilities is expected from this upgrade. Also under construction is a filter bank system for spectroscopy that will allow investigation of the formation of molecules, such as ammonia, around late-type stars. Visibilities will be measured with high spatial and spectral resolution at the same time, much like in radio interferometry.

6.3 Highlights of Recent Results from the ISI

The ISI is being intensively used for the observations of circumstellar material around late-type stars [3, 12, 13], stellar diameters [14], refractivity fluctuations of the atmosphere [4, 5], and wide-angle astrometry [44, 45]. A number of important results have been published, and with continued improvements to the ISI system, more and better quality data are being produced all the time. In this section we provide some highlights of the present observational program. Observations and detailed analyses of data on thirteen late-type stars are reported in [13]. These results are part of a larger survey of about 30 such stars. The observations of circumstellar material are discussed in relation to three prototypical stars, α Orionis, IRC +10216, and o Ceti, and terms of the general relationship to the problems of dust formation, mass loss, and stellar pulsations.

In the observations discussed here, four baselines were used. Extensive measurements of visibility were made first with the shortest baseline, 3.9 m E–W (position angle (PA), 90°). This provided an effective resolution for the interferometer of about 0.″25, better than that provided by the largest conventional telescopes (4 m class) available. That is, two point sources of equal intensity separated by 0.″25 would produce a visibility very near zero, whereas the FWHM of the diffraction pattern of a 4 m telescope is approximately twice as large. This baseline was used until 1989 November 2, and was followed by a baseline of 13.1 m (PA 113°) until 1992 October 21, when the baseline was changed to 9.6 m (PA 123°). In 1993 October the telescopes were moved to the 32 m baseline (PA 118°). Most recently, the telescopes were moved back to the 3.9 m E-W baseline in 1994 August.

The complex visibility of a source is the Fourier transform of its brightness distribution. For a single baseline interferometer, such as the ISI, the measured visibility is in fact a single track in the $u - v$ plane [41]. If the source has reflection symmetry, then the visibility is real and its phase is constant. In the analyses published thus far, spherical symmetry is assumed so that both the fringe phase and the baseline orientation can be ignored. In this case a curve of visibility as a function of spatial frequency contains all the information. Spherically symmetric radiative transfer models of the dust distribution surrounding the stars are computed and visibility curves calculated from these models are compared with the observed ones. A χ^2 fitting procedure is used to optimize parameters of the model from the data. The models also provide source spectra, which are compared with available spectra so as to further refine the parameters.

The radiative transfer model constructed by [46] was adapted for the dust envelopes surrounding late-type stars for these calculations. In their approach, the equations of radiative equilibrium are solved iteratively to find the radial temperature distribution. As discussed in [46], the variable Eddington factor method was used to perform the radiative transfer. The calculation spans wavelengths from the Lyman edge to 1000 μm and naturally computes the specific intensity as a function of impact parameter[1] and computes the spectrum. The dust is assumed to be fully condensed at the inner radius of the dust shell R_0. The physical and optical characteristics of the dust, including the composition and grain size distribution, are assumed to have constant radial dependence. The grain size distribution is assumed to be that of the interstellar medium [31] although more study of its

---

[1]The impact parameter is the angular distance from the center of the star.

applicability to the circumstellar environment is needed. The minimum radius is 0.005 μm and the maximum is 0.25 μm. The grain composition can be varied from pure silicates, graphites, or amorphous carbon (two phases), to mixtures of any two of the four possibilities. The two amorphous carbon phases available in our radiative transfer computations are the amorphous carbon (AC) and benzene (BE) phases. The AC phase is defined by Rouleau and Martin [36] to represent soot produced by an arc between two amorphous carbon electrodes in an argon atmosphere. The BE phase represents soot produced from benzene burned in the atmosphere at room temperature and pressure. The complex dielectric constants of the silicate (i.e., olivine or $(Mg, Fe)_2SiO_4$) and graphite grains were taken from Draine and Lee [19, 20] while those for amorphous carbon were from [36]. Efficiencies for scattering and absorption were calculated (based on the complex indices of refraction) for dielectric spheres based on the work of Toon & Ackerman [43].

The essentially fixed inputs to trial models were the stellar effective temperature, the angular radius of the star, the dust composition, outer radius, and primary beam response. To fit a visibility curve, an inner radius was chosen and the density was adjusted to minimize the χ^2. A new inner radius was chosen and the density again adjusted to minimize the χ^2 for this choice of inner radius. This process was repeated until a global minimum χ^2 was reached.

6.3.1   Broad Conclusions on Dust Shell Characteristics

An important conclusion from the present work is that the stars can be separated into distinct groups on the basis of the inner radii of their dust shells, as can be seen from Fig. 6.3. Roughly half of the stars have inner radii many stellar radii away from their photospheres. These stars include the M supergiants α Ori, α Sco, α Her, the F supergiant IRC +10420, as well as the S stars χ Cyg and W Aql. This indicates a sporadic production of dust, and for the M supergiants, indicates a mean interval of $\gtrsim$ 30 years. For the S stars the present data indicate a somewhat shorter interval between episodes of dust emission, perhaps as short as 20 years. The Mira variables R Leo, o Cet, IK Tau, and the carbon star IRC +10216 have dust shells with inner radii less than about 6 stellar radii, indicating an almost continuous production of dust. The supergiants VX Sgr, VY CMa, and NML Cyg and the symbiotic star R Aqr also have inner radii of their dust shells close to their photospheres, indicating a nearly continuous formation of dust.

Important correlations exist between the characteristics of the dust shell and the presence of maser activity in these stars (Table 2, [13]). Stars having inner radii close to their photospheres are very likely to have all three masers, SiO, H_2O, and OH. The S stars W Aql and χ Cyg lack both OH and H_2O masers and yet have SiO masers, which links the SiO masers closely to the pulsational activity of the stars, but less closely to the production of dust. The lack of dust close to the photospheres of these two stars is also an indication of the importance of the chemistry of the atmospheres to the production of dust, since S stars have carbon-oxygen ratios very close to unity. The other stars are either oxygen- or carbon-rich

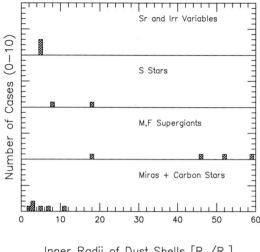

Inner Radii of Dust Shells $[R_{in}/R_*]$

FIGURE 6.3. Summary of inner radii of dust shells determined from visibility data and modeling. Note the wide variation of inner radii which is dependent on spectral type.

Mira variables. A more thorough discussion of these issues is presented below and in Danchi *et al.* [13].

The physical conditions of the material at the inner boundary of the dust shell as well as mass loss rates and total masses of the dust and gas shells (within a specified angular radius) are presented in Table 10 and 11 of Danchi *et al.* [13]. The densities of hydrogen at the inner radii are such that dust, if nucleated at densities $\approx 10^{11}$ cm^{-3}, would be formed at a distance of ≈ 0.8 R_* from the photosphere, in rough agreement with the result of Draine [18]. The mass-loss rates listed in Table 11 of [13] obtained from the visibility models can be compared to rates estimated by fitting the shapes of collisionally excited rotational lines of CO and other molecules. Our own estimates of mass-loss rates use outflow velocities obtained from these molecular lines. The mass-loss rates are in approximate agreement over 4 orders of magnitude in the rates (Fig. 29, [13]). This indicates that the velocities of the gas and dust must be very similar in value, and are consistent with the expectation that radiation pressure of dust grains drives the molecular outflows. Also it means that the mass-loss rates may not have changed substantially during approximately the past one hundred years.

Thus considerable progress has been made in the analysis of visibility data from the ISI using radiative transfer models of the dust shells. This analysis has lead to an improvement in our understanding of the location of the dust formation region surrounding these stars and its relationship to the pulsational and maser activity occurring in these stars.

6.3.2 *Masers*

Several conclusions have been made concerning the relative location of the inner radii of the dust shells, and the OH, H_2O, and SiO masers [13]. The masers farthest from the photospheres of the stars, the OH masers, for the most part are located well outside the inner radius of the dust shells for stars in their sample. An example is IK Tau, for which the OH maser shell is at an angular radius of $\sim 1''65 \pm 0''14$ or 33 R_0 [7]. The symbol R_0 denotes the angular distance of the inner radius of the dust shell from the center of the star. The OH masers for the supergiant VY CMa are at a distance greater than 42 R_0. On the other hand the OH maser shell surrounding U Ori is at only 2.6 R_0, which is rather closer to the inner radius than the others. In general the OH shells are ≥ 30 R_0 away from the star, or equivalently ≥ 100 R_*, however there is certainly considerable variability.

Compared to the OH masers, the H_2O masers lie much closer to the inner radii of the dust shells, but are generally outside of it. For example IK Tau has two H_2O maser shells, one with radius 93 mas, the other at 218 mas [28]. The inner radius of the dust shell is 50 mas, hence the closest H_2O masers are at 1.9 R_0 from the inner radius. Examination of other stars shows that generally the H_2O masers lie within a few R_0 away from the inner radius of the dust shell, i.e. ≤ 4 R_0. This is in the region where dust is accelerated by radiation pressure. It has for some time been recognized that intense H_2O masers require a substantial non-equilibrium input of power to operate. Near the inner radii of dust shells either shock waves or intense radiation from newly formed dust could serve this purpose. Although it is not clear that shock waves can be generated by radiative acceleration of dust in this region [42], shock wave excitation of H_2O masers has been modeled by Hollenbach *et al.* [25] in densities of $10^6 - 10^8$ cm^{-3}, which is the range of densities found for the H_2O maser regions based on the dust models ([13]).

The SiO masers are less well understood than the OH and H_2O masers. Elitzur [22] concludes that for a collisional pump mechanism, the hydrogen density is probably in the range of $10^9 - 10^{10}$ cm^{-3}. Given a critical density for nucleation of dust of $\sim 10^{11}$ cm^{-3} from homogeneous nucleation theory [18], it would seem that SiO masers should lie just outside the location where dust is formed. This is contrary to most expectations, since once the dust is formed the gas phase SiO should be depleted since it should be incorporated into the grains. Recent observations of the dust and SiO masers help to clarify and resolve this apparent conflict. For VX Sgr, Greenhill *et al.* [23, 24] found that the SiO masers were located in a ring at roughly 1.3 R_* from the center of the star, which is well inside the inner radius of the dust shell, 4.6 R_*. The hydrogen density at the inner radius is 4×10^7 cm^{-3}. If the densities at the maser region and at the inner radius of the dust shell are connected by an exponential density law, the corresponding scale height is 1 R_*, with an uncertainty of about 0.3 R_*, because of intrinsic uncertainties in the contributing densities ([24]). This is approximately the scale height expected from the pulsation-driven shock wave models of the atmospheres of Mira variables. For piston amplitudes of 2-4 km s^{-1} these models give atmospheric scale heights in the range of 0.6 to 0.7 R_* [6], which is close to the observational result.

If homogeneous nucleation theory is not applicable to the formation of dust [9], then the conflict can be resolved.

Recently Diamond *et al.* [17] measured the locations of SiO masers around TX Cam and U Her and found they were at $\sim$ 2-4 R_*. However, the inner radii of the dust shells for these stars are not known, nor are there any infrared or optical diameters available for these stars. Diamond *et al.* [17] suggest the diameters used could be at least a factor of 2 too small. This implies that the SiO masers could be located at $\leq 2\ R_*$, which is approximately consistent with the results of Greenhill *et al.* [23, 24]. Improvements to the interpretation of the results for TX Cam and U Her will be possible when direct measurements of the dust shells and diameters become available.

6.4   Direct Inversion of Visibility Data

6.4.1   Methodology

At the present time visibility data are available from the ISI covering more than a factor of ten range in spatial frequencies, from a minimum of about $1.4 \times 10^5\ \mathrm{rad}^{-1}$ using the 4 m baseline to $28 \times 10^5\ \mathrm{rad}^{-1}$ using the 32 m baseline. For the purposes of this discussion we neglect the fact that the observations actually lie on tracks in the two-dimensional $u - v$ (Fourier) plane. Instead we use only the distance of points on the tracks to the origin, i.e., a one-dimensional set of points. The visibility data are reasonably well sampled along stretches corresponding to the baselines used but they also have significant gaps in between. An example is Fig. 6.4(a), which shows data from 1988 to 1992 for α Orionis, an M supergiant star. Here it is easy to see that the data are in two groups, one corresponding to the 4 m baseline, the other to the 10 and/or 13 m baselines. Figure 6.5(a) shows similar data from 1993 and 1994, one set with the 4 m baseline, the other with the 32 m baseline. It is also clear that visibilities for the 4 m baseline are higher in 1994 than in the earlier data. In these figures the solid lines are interpolations of the visibility data as explained below.

In the case of α Orionis, the earlier data were consistent with dust located in a thin spherical shell at a radial distance of $1.''0$ from the center of the star [13]. This result can be understood intuitively by noting that the visibility is relatively constant from about $2.0 \times 10^5\ \mathrm{rad}^{-1}$ and beyond, while for spatial frequencies below that, the visibility changes from unity to this approximately constant value of 0.6. This is because the visibility rapidly decreases as the dust is resolved, while it decreases much more slowly as the star itself is resolved. The thickness of the shell is less than $0.''2$, and is determined by fitting the peak in the low spatial frequency data near $2.0 \times 10^5\ \mathrm{rad}^{-1}$ [13]. For α Ori, the star itself contributes about 60% of the total 11 μm radiation. The recent data show that the star (or dust close to the star) contributes a larger fraction of the 11 μm radiation than previously. The newer data can be modeled with the same distribution of dust as previous data, but with additional dust added close to the star.

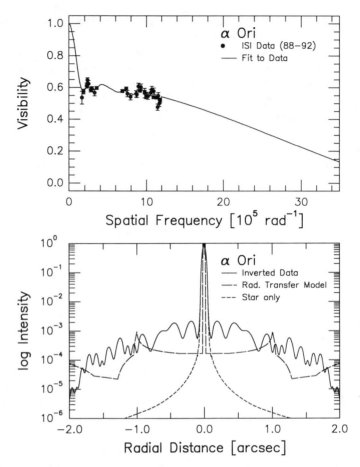

FIGURE 6.4. The upper panel (a) displays visibility data (filled circles) for α Orionis from 1988 to 1992 with the resampled visibility curve superposed (solid line). The lower panel (b) displays the intensity distribution derived from the visibility curve from the upper panel (solid line), the intensity distribution from the radiative transfer model (long-dashed line), and the result of the inversion for a visibility curve of a star only (dashed line).

As the program stars are observed over longer periods of time, significant changes in the visibility curves of several of them have been noted, and sometimes it is difficult to fit the data very well with simple spherical radiative-transfer models. However it is possible to derive intensity distributions from visibility data in a manner that is independent of radiative transfer models. The remainder of this paper is devoted to initial results with a simple direct inversion technique that is close in spirit to the techniques used in radio astronomy to recover images from poorly sampled visibility data [40, 41]. These images are compared with those obtained from radiative transfer models having similar values of χ^2 for the fit to the visibility data.

The method for inverting the visibility data is as follows. A smooth curve with uniform sampling in spatial frequency is obtained from the visibility data by employing interpolation and extrapolation techniques. Cubic splines are used to interpolate in the region from the visibility data point with the smallest spatial frequency to the one with the largest spatial frequency [35]. The maximum spatial frequency is usually 1.024×10^7 rad^{-1} and the spatial frequencies are sampled at an interval of 2.0×10^5 rad^{-1}. The region between the maximum spatial frequency data point and the maximum spatial frequency of the resampled curve is filled in by the visibility of a star having a uniform disk of diameter determined either by measurements in the literature, or from the total luminosity and effective temperature of the star (cf. [13]). The region between the data point with the smallest spatial frequency and zero spatial frequency is filled in two steps. First, from zero spatial frequency to one-half the smallest spatial frequency data point is filled in using the visibility of a Gaussian of full-width at half-maximum (FWHM) of $3''.0$. This is the measured primary beam size from 5 point beam maps, which includes residual telescope guiding errors. (Only a slow ~ 1 Hz bandwidth autoguider has been used so far on the ISI.) The remaining region is filled in by a cubic spline interpolation. After the resampled visibility curve is created it is multiplied by a Hanning window, and is then re-sampled onto a two-dimensional grid having the same spatial frequency sampling as the one-dimensional curve. A two-dimensional Fast Fourier Transform (FFT) is applied to create the image. A slice through the center of this image is the radial intensity distribution which can be compared to the results of the radiative transfer calculation. In all of the cases studied here the χ^2 of the fits of the visibility curves from the radiative transfer models and those from the resampled visibility curves are essentially the same, indicating a similar goodness-of-fit.

6.4.2 Results

We now compare the results from the direct inversion approach with those from the radiative transfer models. The results are compared for the prototypical stars α Orionis, o Ceti, and IRC +10216 as mentioned previously. Figure 6.4(a) displays data from 1988-1992 previously published in [13] and a model fit to the data using the methodology discussed above. The normalized χ^2 for this fit is approximately 2 which indicates a reasonably good fit to the data. This is about equal to the χ^2 of the previously published radiative transfer model [13]. The intensity distribution derived from the radiative transfer model is shown in Fig. 6.4(b). The direct inversion intensity distribution is also shown. For comparison the intensity distribution (also by direct inversion) for a star of the same size as used with the actual visibility data is displayed. Both of these curves have been smoothed slightly using a $0''.05$ width boxcar average. Clearly the directly inverted intensities agree well with the radiative transfer model both in the general shape and magnitude. The major difference between the results is that the directly inverted intensity distribution is less smooth, i.e. it shows moderate amplitude quasi-sinusoidal variations. However the published radiative transfer model visibilities do not fit the quasi-sinusoidal vari-

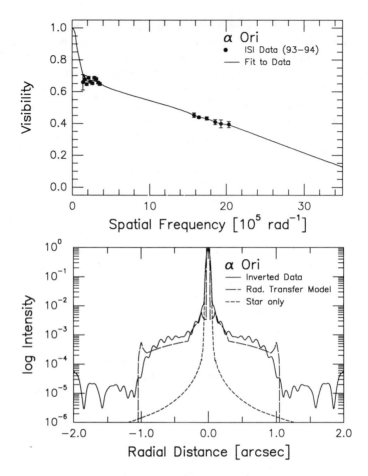

FIGURE 6.5. The upper panel (a) displays visibility data (filled circles) for α Orionis from 1993 to 1994 with the resampled visibility curve superposed (solid line). The lower panel (b) displays the intensity distribution derived from the visibility curve from the upper panel (solid line), the intensity distribution from the radiative transfer model (long-dashed line), and the result of the inversion for a visibility curve of a star only (dashed line). Note the emission from dust near the star that was not observed in the earlier data.

ations in the visibility data as closely as this model. In both models it is clear that there is little dust close to the star since the intensity distribution is still relatively flat from 1″0 inward. This is consistent with the model of a thin shell at a distance of approximately one arcsecond from the star.

Figure 6.5(a) displays more recent visibility data (1993-1994) and the visibility model used for the direct inversion process. A comparison of the best fit radiative transfer model intensity, directly inverted intensity, and stellar intensity are shown in Fig. 6.5(b). In this case the agreement is quite good, in both cases it is clear there is more emission close to the star than was previously observed. This means that

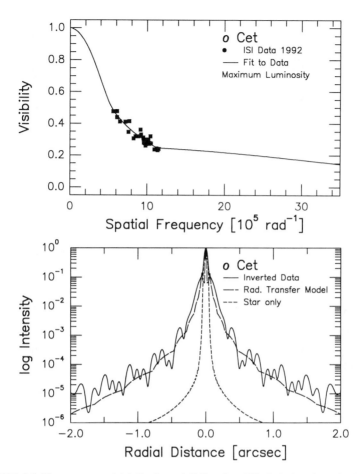

FIGURE 6.6. The upper panel (a) displays visibility data (filled circles) for *o* Ceti in 1992 with the resampled visibility curve superposed (solid line). The lower panel (b) displays the intensity distribution derived from the visibility curve from the upper panel (solid line), the intensity distribution from the radiative transfer model (long-dashed line), and the result of the inversion for a visibility curve of a star only (dashed line).

some hot dust close to the star may have been produced after the 1992 visibility data were taken. Further work is in progress estimating the physical conditions of this material close to the star.

The results for α Orionis can be compared with those for *o* Ceti, an oxygen rich long period variable star. Figure 6.6(a) displays previously published data and a model visibility curve used for inversion. As before, the fit of this model curve to the data is comparable to that in the previously published radiative transfer model. The intensity distribution for the direct inversion, radiative transfer model, and star only, are shown in Fig. 6.6(b). For this star the dust is very close to the star and both intensity distributions are quite consistent with each other. In this case

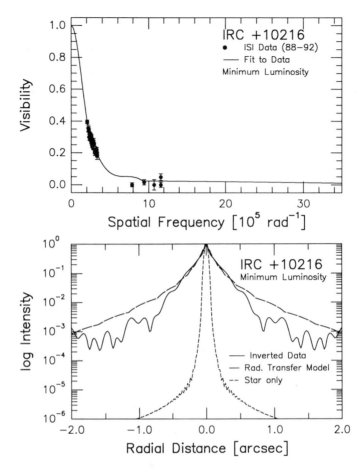

FIGURE 6.7. The upper panel (a) displays visibility data (filled circles) for IRC +10216 from 1988 to 1992 for the minimum luminosity phase with the resampled visibility curve superposed (solid line). The lower panel (b) displays the intensity distribution derived from the visibility curve from the upper panel (solid line), the intensity distribution from the radiative transfer model (long-dashed line), and the result of the inversion for a visibility curve of a star only (dashed line).

the intensity variations in the directly inverted distribution may be due in part to the sharp change in slope observed in the model fit of Fig. 6.6(a). The scalloped appearance in the radiative transfer model is due to the use of a simple linear interpolation between the points calculated from the models.

Figure 6.7(a) displays previously published visibility data and a model fit used for direct inversion for the minimum luminosity phase of IRC +10216. Figure 6.7(b) displays the intensity distribution from the direct inversion of the visibility data, the radiative transfer model, and star only. In this case the published radiative transfer model never did fit the data particularly well, and indeed

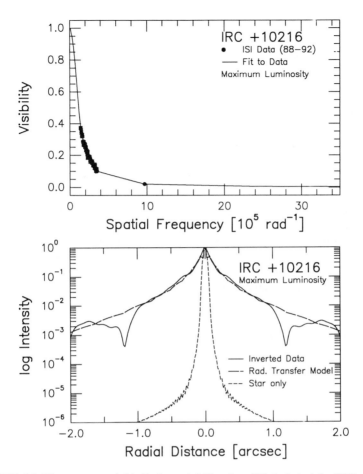

FIGURE 6.8. The upper panel (a) displays visibility data (filled circles) for IRC +10216 from 1988 to 1992 for the maximum luminosity phase with the resampled visibility curve superposed (solid line). The lower panel (b) displays the intensity distribution derived from the visibility curve from the upper panel (solid line), the intensity distribution from the radiative transfer model (long-dashed line), and the result of the inversion for a visibility curve of a star only (dashed line).

tended to lie below the actual data, indicating too high a luminosity for this model compared to the data (Fig. 6.1(b), [13]). The fit of the model for direct inversion is much closer to the data, and produced an intensity distribution that lies below the radiative transfer model intensities for distances greater than about 0″.5 from the star. Given the differences in the visibility models the differences observed in the intensity distribution are not surprising. For comparison Fig. 6.8(a) and (b) show the result of the same calculations for IRC +10216 at its maximum. In this case both the model for direct inversion and the radiative transfer model visibility distributions agree well with the data. The intensity distributions of Fig. 6.8(b) agree well with each other except for the dip in the direct inversion visibility, which is probably due to the derivative discontinuity observed in the visibility curve for the direct inversion model.

6.5   Conclusions

In the past few years long baseline interferometry at infrared and optical wavelengths has become a powerful tool for studies of mass loss phenomena occuring close to the photospheres of AGB stars, including pulsation, surface features, dust formation, and stellar masers. Probably the most important next step for infrared and optical interferometry is to produce high-quality maps, and for this step, the construction of instruments with three or more telescopes is critical. As a first step in this direction we have developed simple methods for directly inverting the visibility data to produce a radial intensity distributions. These intensity distributions agree well with those produced by radiative transfer calculations which fit the visibility data with similar quality.

6.6   Acknowledgments

Long baseline interferometry at the University of California, Berkeley is supported in part by the Office of the Chief of Naval Research (N00014-89-J-1583, N00014-93-0775), and by the National Science Foundation (AST-9315485, AST-9321289, and AST-9221105). We are pleased that much of the work reported here has been performed in collaboration with Professor C. H. Townes over the course of the past decade. We also acknowledge the contributions of the following people to this work: Drs. C. G. Degiacomi, L. J. Greenhill, S. T. Lowe, and R. N. Treuhaft.

References

[1] J. T. Armstrong, Proc. SPIE **2200**, 62 (1994).

[2] J. E. Baldwin *et al.*, Proc. SPIE **2200**, 231 (1994).

[3] M. Bester, W. C. Danchi, C. G. Degiacomi, Townes, C.H., and T. R. Geballe, ApJ **367**, L27 (1991).

[4] M. Bester, W. C. Danchi, C. G. Degiacomi, L. J. Greenhill, and C. H. Townes, ApJ **392**, 357 (1992).

[5] M. Bester, W. C. Danchi, C. G. Degiacomi, and P. R. Bratt, Proc. SPIE **2200**, 274 (1994).

[6] G. H. Bowen, ApJ **329**, 299 (1988).

[7] P. F. Bowers, K. J. Johnston, and P. de Vegt, ApJ **330**, 339 (1989).

[8] N. P. Carleton *et al.*, Proc. SPIE **2200**, 152 (1994).

[9] I. Cherchneff and A. G. G. M. Tielens, in *Circumstellar Media in Late Stages of Evolution,* R. Clegg and I. Stevens (eds.), (Cambridge: University Press), in press (1994).

[10] A. L. Cheung, D. M. Rank, C. H. Townes, D. D. Thornton, and W. J. Welch, Phys. Rev. Lett. **21**, 1701 (1968).

[11] M. M. Colavita *et al.*, Proc. SPIE **2200**, 89 (1994).

[12] W. C. Danchi, M. Bester, C. G. Degiacomi, P. R. McCullough, and C. H. Townes, ApJ **359**, L59 (1990).

[13] W. C. Danchi, M. Bester, C. G. Degiacomi, L. J. Greenhill, and C. H. Townes, AJ **107**, 1469 (1994).

[14] W. C. Danchi, M. Bester, C. G. Degiacomi, L. J. Greenhill, and C. H. Townes, Proc. SPIE **2200**, 286 (1994).

[15] J. Davis, W. J. Tango, A. J. Booth, R. A. Minard, S. M. Owens, and R. R. Shobrook, Proc. SPIE **2200**, 231 (1994).

[16] G. P. Di Benedetto and R. Foy, in *High Resolution Imaging by Interferometry II*, ESO Conference and Workshop Proceedings **29**, 691 (1992).

[17] P. J. Diamond, A. J. Kemball, W. Junor, A. Zensus, J. Benson, and A. Dhawan, ApJ **430**, L61 (1994).

[18] B. T. Draine, *Physical Processes in Red Giants*, I. Iben and A. Renzini eds., (Reidel: Dordrecht), 317 (1981).

[19] B. T. Draine and T. Lee, ApJ **285**, 89 (1984).

[20] B. T. Draine and T. Lee, ApJ **314**, 485 (1987).

[21] H. M. Dyck, J. A. Benson, S. T. Ridgway, and D. T. Dixon, AJ **104**, 1982 (1992).

[22] M. Elitzur, *Astrophysical Masers* (Kluwer: New York) (1992).

[23] L. J. Greenhill, J. M. Moran, D. C. Backer, M. Bester, and W. C. Danchi, Proc. SPIE **2200**, 304 (1994).

[24] L. J. Greenhill, J. M. Moran, D. C. Backer, W. C. Danchi, and M. Bester, ApJ, in press (1995).

[25] D. H. Hollenbach, M. Elitzur, and C. F. McKee, *Astrophysical Masers*, A. W. Clegg and G. E. Nedoluha, eds. (Springer: New York), 159 (1993).

[26] D. J. Hutter, Proc. SPIE **2200**, 81 (1994).

[27] A. Labeyrie *et al.*, A&A **162**, 359 (1986).

[28] A. P. Lane, K. J. Johnston, P. F. Bowers, J. H. Spencer, and P. J. Diamond, ApJ **323**, 756 (1987).

[29] O. von der Lühe, D. Ferrand, B. Koehler, Z. Neng-hong, T. Reinheimer, Proc. SPIE **2200**, 168 (1994).

[30] C. Masson, *Very High Angular Resolution Imaging*, J. G. Robertson, and W. J. Tango, eds., IAU Symposium 158, (Kluwer: Dordrecht), 1 (1994).

[31] J. S. Mathis, W. Rumpl, and K. H. Nordsieck, ApJ **217**, 425 (1977).

[32] H. A. McAlister *et al.*, Proc. SPIE **2200**, 129 (1994).

[33] T. J. Pearson and A. C. S. Readhead, ARA&A **22**, 97 (1984).

[34] A. A. Penzias and R. W. Wilson, ApJ **142**, 419 (1965).

[35] W. Press, S. Teukolsky, R. Flannery, and W. Vetterling, *Numerical Recipes in C*, (Cambridge University Press: New York 1986).

[36] F. Rouleau and P. G. Martin, ApJ **377**, 526 (1991).

[37] M. Ryle, Nature **194**, 517 (1962).

[38] M. Ryle, *Les Prixes Nobel* (1973).

[39] M. Shao *et al.*, A&A **193**, 357 (1988).

[40] R. A. Sramek and F. R. Schwab, *Synthesis Imaging*, R. A. Perley, F. R. Schwab, and A. H. Bridle, (eds.), National Radio Astronomy Observatory, Green Bank, Ch. 5 (1985).

[41] A. R. Thompson, J. H. Moran, and G. W. Swenson, *Interferometry and Synthesis in Radio Astronomy* (John Wiley & Sons, New York 1986), Ch. 2.

[42] A. G. G. M. Tielens, ApJ **271,**, 702 (1983).

[43] O. B. Toon and T. P. Ackerman, Appl. Optics **20,** 3657 (1981).

[44] R. N. Treuhaft, S. T. Lowe, M. Bester, W. C. Danchi, and C. H. Townes, in *Amplitude and Intensity Interferometry II*, J. B. Breckinridge, eds., Proc SPIE **2200**, 316 (1994).

[45] R. N. Treuhaft, S. T. Lowe, M. Bester, W. C. Danchi, and C. H. Townes, ApJ, in press (1995).

[46] M. G. Wolfire and J. P. Cassinelli, ApJ **310,** 207 (1986).

7

Ammonia in the Giant Planets

Albert L. Betz

7.1   Introduction

The largest reservoirs of ammonia in the solar system are the atmospheres of the gas giant planets Jupiter and Saturn. The planets are composed of primarily hydrogen and helium, but trace elements are present in roughly solar abundance and are generally found in the form of hydrides. Of these the most important is ammonia. It is the opacity of ammonia which governs the radiative transfer of the atmosphere up to the tropopause, and it is ammonia which dictates the continuum temperature we measure in the region of the visible cloud tops. In the region of convective flow below the tropopause of Jupiter, the fractional abundance of NH_3 relative to H_2 is about 2×10^{-4}, similar to the solar abundance of nitrogen. For these reasons, and the fact that ammonia has long been of particular interest to Charles Townes, the following commentary is presented. The review is not intended to be an exhaustive summary on the subject, but rather a short compendium emphasizing the historical development of the field, and the progress achieved through advances in instrumentation. The contributions of Charles Townes and his students and associates in the development and application of infrared spectroscopy to the studies of planetary atmospheres are of course noted. For studies of ammonia in the interstellar medium, the reader is referred to the extensive review of Ho and Townes [1].

7.2   The Upper Atmospheres of Jupiter and Saturn

Despite its enormous abundance ammonia was not always easy to detect. It has been known to be a component in the atmosphere of Jupiter since 1932, when Wildt first detected the $5\nu_1$ band in absorption at 6450 Å [2]. At the time photographic media were the only available detectors. The weakness of the $5\nu_1$ overtone band, however, prevented a corresponding detection in Saturn until 1966 [3]. The fundamental and low overtone bands are of course far stronger, but infrared technology was poorly developed before WWII. The stronger $3\nu_1$ band at 1.1 microns was detected in Jupiter in 1947 soon after the first Cashman PbS detectors became available to the astronomical community [4]. The much stronger ν_2 fundamental band at 10 microns was detected in Jupiter in 1972 by Aitken and Jones who used a relatively new Ga:Ge bolometer with a grating spectrometer [5]. These low resolution band spectra were adequate to identify ammonia as a significant species in the upper atmosphere of Jupiter and to quantify its abundance approximately. The measurements also showed that for Jupiter the atmospheric temperature was about 135 K at the visible cloud tops, which is 30 K higher than the temperature expected from solar heating. The fact that Jupiter radiated 1.6 times more energy than it received from the Sun (and Saturn a factor of 2.7) revealed that the two planets had internal sources of energy, presumably associated with their formation and subsequent gravitational contraction.

As wavelength coverage expanded to the infrared, new discoveries were made. The first indication of a temperature inversion in the stratosphere of Jupiter was the observation of Gillett, Low, and Stein in 1969 which showed a higher than expected flux in the 8 micron region, presumably from the 7.7 μm ν_4 band of methane [6]. At about this time the first infrared spectrometers with the ability to detect individual lines and line manifolds became available. Observations of emission features in the 12-13 micron region from C_2H_6 and C_2H_2 by Ridgway in 1974 confirmed the existence of warmer gas above the temperature minimum at the tropopause [7]. Using a high resolution Fourier Transform Spectrometer (FTS—the new name for a Michelson interferometer), they measured peak brightness temperatures near 145 K for unresolved lines in the ν_9 fundamental of C_2H_6 near 822 cm^{-1}. This minimum stratospheric temperature was 10 K higher than the infrared continuum, and more than 30 K higher than the temperature of cooler tropospheric gas producing the absorption cores of ammonia lines. Much as ozone in the Earth's upper atmosphere absorbs solar UV and warms the stratosphere, so too does methane absorb solar UV and warm the stratospheres of Jupiter and Saturn. Ethane and acetylene are expected by-products of the UV-initiated photochemistry of methane. Ammonia, on the other hand, is easily photodissociated by solar UV and is not expected to exist in the upper stratosphere. Without stable UV absorbers in the stratosphere, one would expect an isothermal temperature above the tropopause. Below the tropopause the temperature is controlled by convective rather than radiative effects and is expected to increase adiabatically with penetration into the atmosphere.

The existence of a thermal inversion in the upper atmosphere of Jupiter led to the speculation that warm ammonia might exist above the temperature minimum at the tropopause [8]. The stratospheric mixing ratio would be governed by the saturation abundance at the 115 K "freeze out" temperature of the tropopause, and could be in the range of 10^{-7} to 10^{-8}. This much gas should be detectable with the techniques available at the time, because it would lead to the appearance of an emission core in the center of each broad absorption line, with the emission much more discernible in the longer wavelength infrared and far-infrared transitions.

This conjecture led to a number of observational efforts to detect line core emission. All proved to be negative. Some of the first measurements were those of Lacy *et al.* in 1975, who used a Fabry-Perot spectrometer to observe selected lines in the ammonia ν_2 band at 10 μm [9]. Spectra taken with resolutions between 0.5 and 0.15 cm^{-1} resolved the line profiles completely and indicated that no core emission was present. The absorption lines were observed to be saturation broadened many times the pressure broadened linewidth, but a pressure of about

0.5 atm at the 145 K level below the tropopause was nevertheless indicated. The observations also showed that the 135 K continuum is primarily formed by the wings of NH_3 lines, and that the temperature minimum indicated by the absorption cores is about 118 K. Subsequent observations by Tokunaga and coworkers [10] of the same spectral region pinned down the partial pressure of tropospheric ammonia to be half the saturated vapor pressure.

Other observations of NH_3 rotation-inversion lines in the far-infrared supported the conclusion that the stratospheric abundance of ammonia must be very small ($<10^{-9}$). In 1977 Greenberg *et al.* reported far-infrared observations of the $J = 6 \rightarrow$ 5 rotation-inversion line manifold at 120 cm^{-1} [11]. The instrument was a Fabry-Perot spectrometer with a spectral resolution of 1.5 cm^{-1} that was flown aboard NASA's Kuiper Airborne Observatory (KAO). The observations showed the line to be completely in absorption with no evidence of core re-emission. Although UV photodissociation plays a significant role in limiting the NH_3 abundance in the upper stratosphere, a low temperature (<115 K) at the tropopause leads to a significant "freeze-out" of ammonia there and, as a consequence, a low mixing ratio for NH_3 in the lower stratosphere.

Other observations supporting the non-existence of emission cores in the FIR were reported in 1978: Furniss *et al.* used a balloon-borne Michelson interferometer with 2 cm^{-1} resolution to observe Jupiter's spectrum between 70 and 200 cm^{-1} [12], and Baluteau *et al.* used a Michelson interferometer with 0.2 cm^{-1} resolution to observe NH_3 FIR lines near 120 cm^{-1} aboard the KAO [13]. The only report of an infrared emission feature from ammonia was that of Kostiuk *et al.* in 1977, who observed possible transient line emission of a nonthermal nature at 970 cm^{-1} from the polar region of Jupiter [14]. Their instrument was a laser heterodyne spectrometer with a resolution of 5 MHz. However, subsequent observations in 1984 by Betz and Goldhaber using a similar CO_2 laser heterodyne spectrometer failed to detect any such emission from Jupiter [15].

7.3   The Collision of Comet Shoemaker-Levy 9 with Jupiter

The fact that ammonia is not a permanent constituent of the stratosphere of Jupiter makes it an ideal probe for transient effects, such as the collision of a large comet. As everyone is no doubt aware by now, multiple fragments of Comet Shoemaker-Levy 9 (SL-9) impacted the atmosphere of Jupiter in July, 1994. One of the expected consequences from these impacts was the explosive ejection of plumes of tropospheric material high into the stratosphere. As the material fell back into the stratosphere and cooled, it would serve as a temporary probe of ambient conditions. The best instrument for measurements of the narrow line profiles that were expected is a laser heterodyne spectrometer, and the best telescope is any one you could get your hands on for the entire duration of the event.

On Mt. Wilson, a team of astronomers and physicists led by Charles H. Townes (CHT) had constructed a long-baseline infrared interferometer for stellar observations [16]. Each 1.6 m telescope is instrumented with a laser heterodyne receiver, which could be tuned to observe selected lines of ammonia near 930 cm^{-1}. A team of observers, mostly ex-students and postdocs of CHT from the University of California, Berkeley and the University of Colorado, Boulder, waited for the first impact observable from Mt. Wilson on July 18, 1994. Line emission from the ν_2 band was soon detected from the large G impact site, and subsequently from many others over the following days. Overall we detected emission from the E, G, H, K, L, and Q1 sites (the larger impacts). Figure 7.1 shows emission from two lines at the G impact site on July 21, 1994, 3 days after our initial detection.

The linewidths within the first few days after an impact were about 170 MHz (1.7 km s^{-1}) which confirmed that the emission was stratospheric. True linewidths are substantially less than those observed because of the finite 1.5 arcsec beam size, and Jupiter's rotation during a 4-minute integration period. After appropriate corrections, the true linewidths (FWHM) are close to 50 MHz, which corresponds to pressure altitudes higher than the 2 mbar level. The lack of broad Lorentzian wings on the observed profiles, despite the fact that the aQ(2,2) and aQ(6,6) lines are optically thick (see below), provides further evidence that the emission comes from the upper regions of the stratosphere. This is what would be expected if the material is falling down from a collapsing high altitude plume, rather than upwelling from below, or being uniformly mixed in the stratosphere.

Besides the strong aQ(2,2) and aQ(6,6) lines, the weaker aQ(9,8) transition was also detected at the G impact site. The peak intensities indicate that the two stronger lines are optically thick, with an excitation temperature close to 200 K. At this temperature, the 10.7 μm line emission is on the Wien side of the blackbody curve, and thus quite sensitive to temperature, as illustrated in Fig. 7.1, a 15% uncertainty for the intensity calibration corresponds to a 2.5% uncertainty for stratospheric gas temperature.

The column density of stratospheric ammonia derived from these observations is 3.5×10^{16} cm^{-2} (0.001 atm cm). The lack of equivalently strong emission from water vapor, which is expected to be much more abundant than ammonia in cometary ice, leads us to conclude that the observed ammonia is indeed an plume of explosively ejected tropospheric material. This conclusion helps establish minimum penetration depths for the impact fragments producing detectable NH$_3$ emission. Continued monitoring of the line emission showed that ammonia persisted in the stratosphere for a period of weeks. The lifetime of stratospheric ammonia is limited by solar UV radiation, and this observation will help in refining photochemical models of the upper atmosphere. Many more details, probably of interest only to the specialist, are available from reference [17].

The success of these observations is primarily due to the fact that, during and after the impacts, we had continued access to a telescope big enough to achieve a small beam size matched to the diameters of the collapsed impact plumes. We did not have to apply to a national observatory and take our chances with a scheduling committee for a time-critical observation. As a result, our ammonia line database

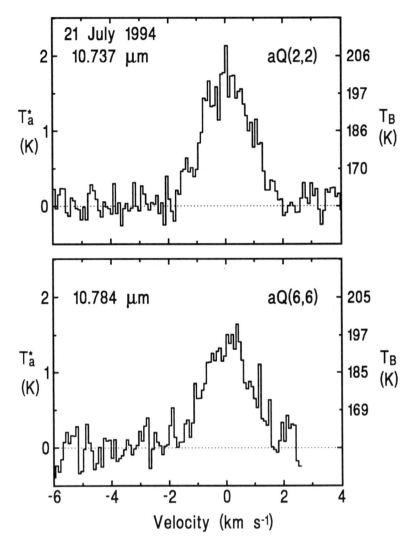

FIGURE 7.1. Emission lines from stratospheric ammonia at the G fragment site on Jupiter 4 days after impact. The noise level can be ascertained from the scatter in the data points between −4 and −6 km s^{-1}.

of this "once in a lifetime" event is probably the best available, and should be of considerable interest to detailed modelers of the upper atmosphere of Jupiter. For this and many other kind considerations, we thank Professor Townes, whose persistence is always an inspiration, and whose determination to build the ISI telescopes made it all possible.

7.4   Acknowledgments

I thank R.T. Boreiko for assistance in the preparation of this manuscript, and acknowledge support from NASA Grant NAGW-3196.

References

[1] P. T. P. Ho and C. H. Townes, Ann. Rev. Astron. Astrophys. **21**, 239 (1983).

[2] R. Wildt, Veröff. U.-Sternw. Göttingen **22**, 171 (1932).

[3] L. P. Giver and H. Spinrad, Icarus **5**, 586 (1966).

[4] G. P. Kuiper in *The Atmospheres of the Earth and Planets*, ed. G. Kuiper, The University of Chicago Press, Ch. 12, pp. 306-405 (1952).

[5] D. K. Aitken and B. Jones, Nature **240**, 230 (1972).

[6] F. C. Gillett, F. J. Low, and W. A. Stein, ApJ **157**, 925 (1969).

[7] S. T. Ridgway, ApJ **187**, L41 (1974).

[8] Th. Encrenaz, D. Gautier, L. Vapillon, and J. P. Verdet, A&A **11**, 431 (1971).

[9] J. H. Lacy, A. I. Larrabee, E. R. Wollman, T. R. Geballe, C. H. Townes, J. D. Bregman, and D. M. Rank, ApJ **198**, L145 (1975).

[10] A. T. Tokunaga, R. F. Knacke, S. T. Ridgway, and L. Wallace, ApJ **232**, 603 (1979).

[11] L. Greenberg, R. McLaren, L. Stoller, and C. H. Townes, *Proc. Symp. Recent Results Infrared Astrophys.*, ed. P. Dyal (NASA Tech. Memo. X-73190), 9 (1977).

[12] I. Furniss, R. E. Jennings, and K. J. King, Icarus **35**, 74 (1978).

[13] J. P. Baluteau, A. Marten, E. Bussoletti, M. Anderegg, A. F. M. Moorwood, J. E. Beckman, and N. Coron, A&A **64**, 61 (1978).

[14] T. Kostiuk, M. J. Mumma, J. J. Hillman, D. Buhl, L. W. Brown, J. L. Faris, and D. L. Spears, IR Phys. **17**, 431 (1977).

[15] A. L. Betz and D. M. Goldhaber, BAAS **16**, 649 (1984).

[16] M. Bester, C. G. Degiacomi, W. C. Danchi, L. J. Greenhill, and C. H. Townes, in *Robotic Telescopes in the 1990s*, ASP Conference Series, Vol. 34, ed. A. Filippenko, pp. 213-222 (1992).

[17] A. L. Betz, R. T. Boreiko, M. Bester, W. C. Danchi, and D. D. Hale (manuscript in preparation).

8

Collision Broadening and Radio-frequency Spectroscopy

Brebis Bleaney

8.1 Prehistory

Spectroscopy naturally began at optical wavelengths, then spread to the ultraviolet and infrared. Gas discharge tubes at low pressures reduced collision broadening; narrow lines and interferometric techniques made possible high-resolution spectrometry, including hyperfine structure. Wolfgang Pauli's interpretation of this as due to nuclear spin [1], was soon followed by a measurement in Oxford of the hyperfine structure of caesium by Derek Jackson [2]. From this, he estimated a value for the nuclear magnetic moment of ^{133}Cs, though the nuclear spin was not known. (In later years, he remarked: "Sommerfeld did it by wave mechanics; I did it by arithmetic.") Discharge tubes with cathodes cooled in liquid air were used to reduce Doppler broadening.

In 1939, the outbreak of World War II in Europe turned the attention of most physicists in Britain from fundamental research to radar. From 1940 to 1945 wavelengths were reduced from 10 cm to 3.15 cm and ultimately to 1.25 cm; that is, frequencies near 3 GHz (S-band), 10 GHz (X-band) and 24 GHz (K-band). It was soon realised that absorption in the atmosphere might be significant, since an attenuation of only a few tenths of a decibel per kilometre can give a marked loss of echo over round-trip distances of 50 to 100 km. In 1942, the problem was referred to John Hasbrouck Van Vleck at Harvard University, who identified two sources of absorption: a weak rotational transition of the water molecule, and magnetic dipole transitions of the oxygen molecule. Reports written by him for the Radiation Laboratory at M.I.T. were published in 1945 [3,4].

8.2 Emission Spectroscopy

Absorption is accompanied by emission, and an ingenious radiometer was invented by Robert Dicke [5] to detect emission from the atmosphere. To separate it from other sources of noise, the amplitude of the incoming radiation was modulated at 30 Hz by a rotating shutter, followed by synchronous detection. The thermal radiation power is proportional to the absolute temperature, and the sensitivity was equivalent to an absolute temperature of 0.4 K. Emission from water vapour and oxygen in the atmosphere was measured [6], and also from the sun [7]. Later, emission from interstellar hydrogen atoms was detected at Harvard University [8] and in the Netherlands [9]. In general, however, absorption rather than emission formed the basis of microwave spectroscopy.

8.3   Laboratory Measurements of Microwave Absorption

To determine the absorption caused by water vapour at atmospheric pressure, a copper cavity was built at Columbia University in the form of an 8 foot cube, maintained at 45°C to increase the vapour pressure of water. Power was injected from pulsed magnetrons, and the energy density was monitored by strings of thermocouples suspended inside. Such a cavity is essentially the same as the "hohlraum" postulated in the previous century in developing the theory of thermal radiation; measurements [10] and theory [11] were developed side by side. The quality factor Q was close to 10^6; its decrease when water vapour was admitted provided a measure of the absorption at wavelengths of 1.0, 1.25 and 1.50 cm. Calibration was obtained from the decrease in Q when a window was opened to allow a known amount of radiation to escape.

At wavelengths below 1 centimetre, a more important source of atmospheric absorption is the oxygen molecule. The ground state of O_2 is an electronic triplet, $S = 1$, split by dipolar interaction between the two electron spins, and perturbed by molecular rotation. Resonant transitions lie mainly at wavelengths 5 to 6 mm [12–14] with one line at 2.5 mm wavelength [15].

8.4   The Inversion Spectrum of Ammonia

The NH_3 molecule is a flat pyramid, formed by an equilateral triangle of hydrogen atoms with the nitrogen atom at the peak. There are two equivalent configurations, between which the potential barrier is rather low, allowing tunnelling between them, as illustrated in Fig. 8.1.

The molecule has an electric dipole moment of about 1.34 Debye units (10^{-18} esu), and since it turns "inside out" some 10^{10} times per second, this produces a strong absorption band. This "inversion spectrum" was first detected by Cleeton and Williams [16] near 1.3 cm wavelength. As a source they used split-anode magnetrons, but otherwise their technique was similar to that used for infra-red spectroscopy. The radiation was collimated by mirrors, and wavelengths were determined by a diffraction grating. A cloth bag containing ammonia gas at atmospheric pressure was interposed in the path of the radiation, and the drop in transmission was detected using a crystal of iron pyrites with a phosphor bronze "cat's-whisker."

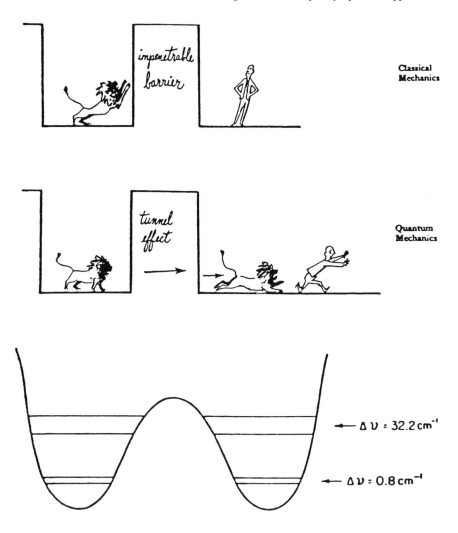

FIGURE 8.1. Tunnelling through a potential barrier (above) and the two potential minima of the ammonia molecule (below). (As used by J.H. Van Vleck to illustrate his last publication, the Julian E. Mack Lecture at his Alma Mater, the University of Wisconsin, in 1979.)

At Oxford, on behalf of the Board of Admiralty, reflex klystrons were developed during World War II as local oscillators for radar receivers. In 1944, permission was obtained to study collision broadening in the ammonia spectrum. A resonant cavity, constructed as a wavemeter, and used for measurements of the dielectric constants of six non-polar liquids [17], was then applied to the ammonia spectrum. From the change in its value of Q between full and empty, an absolute measure of absorption is obtained. At pressures of about 10 torr and below, structure is resolved

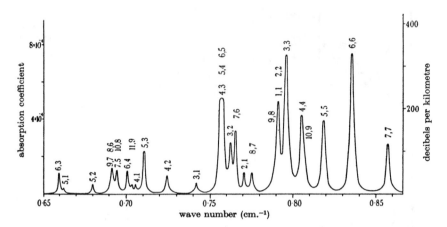

FIGURE 8.2. The inversion spectrum of ammonia gas at a pressure of 1.2 torr. Each line arises from a different rotational state, identified by the quantum numbers, J, K; the spread is due to centrifugal distortion. The line widths, about 10^{-3} cm^{-1}, are wholly due to collision broadening (from Bleaney and Penrose [18]).

in the inversion spectrum, as the tunnelling frequency in the various rotational states varies appreciably through centrifugal distortion, as shown in Fig. 8.2 [18]. The frequencies were fitted to a theoretical formula of Hsi-Yin Sheng, Barker and Dennison [19].

8.5   Further Studies of Collision Broadening at Oxford

At pressures up to about 10 torr, the intensities in the ammonia spectrum fit well with theory [20]. Broadening is mainly due to dipole-dipole interaction; the line widths [21] vary with the rotational state and correspond to collision diameters of 0.9 to 1.4 nm; at higher pressures the rotational structure is not resolved, and the line breadth ceases to vary linearly with pressure, presumably owing to multiple collisions. In his thesis for the degree of Doctor of Philosophy at Oxford University, J. H. N. Loubser extended this work up to pressures of 6 atmospheres. The collision diameter is further reduced, and above 2 atmospheres the tunnelling frequency appears to approach zero, so that the spectrum is similar to that of Debye non-resonant absorption in liquids [22]. In another study, collision cross-sections with six non-polar gasses were found to be markedly smaller than those for self-broadening from other ammonia molecules [23]. Experiments were also made on the inversion spectra of CH_3Cl and CH_3Br [24].

8.6   High Resolution Microwave Spectroscopy

As the pressure is reduced, a feature of collision broadening is that a single line becomes narrower with no loss of intensity at the centre, so long as collisions are the main source of line width. The Doppler effect becomes significant only at pressures of about 10^{-3} torr, and hyperfine structure from the nuclear electric quadrupole moment of ^{14}N in NH_3 was resolved in the ammonia inversion spectrum [25, 26]. Combined with Stark modulation from electric fields at kilohertz frequencies to eliminate "flicker effect" noise in crystal detectors, microwave spectroscopy advanced on a broad front. For all except the lightest molecules, low rotational transitions occur at microwave frequencies; an early example was the linear molecule OCS, for which the $J = 1$ to 2 transition is close to $24\,GHz$ [27]. From the Stark effect, its electric dipole moment was determined [28] as $0.709(3)$ Debye units. By using isotopic substitution, bond lengths were determined, together with other features such as hyperfine structure, mainly from nuclear electric quadrupole coupling.

8.7   The Switch to Electron Paramagnetic Resonance

A resonant cavity is less useful at the lowest pressures, because broadening from power saturation [20] obscures the finer detail. This, together with competition from rapid advances in microwave gas spectroscopy in the U.S.A., made it desirable for Penrose and myself to switch research to another field.

The first experiments in electron paramagnetic resonance were made by Zavoisky [30] at Kazan in the U.S.S.R. in 1944–5, while ferromagnetic resonance was discovered at Oxford [31] by J. H. E. Griffiths. Paramagnetic resonance experiments began at room temperature; of particular interest, novel effects of spin-spin interaction were discovered in $CuSO_4 \cdot 5\,H_2O$ [32]; the theory was discussed by Maurice Pryce [33]. In the same year a measurement at room temperature determined the crystal field splitting of the electron spin quartet of trivalent chromium in potassium chrome alum [34]. Soon afterwards this splitting was found to vary rapidly as the temperature was reduced [35]. Neither this nor the exchange effects in copper sulphate could have been detected by classical thermal experiments at ordinary temperatures.

Many paramagnetic compounds have very rapid spin-lattice relaxation, requiring measurements at low temperatures. A preliminary survey of the field was published in 1948 [36]. Experiments on single crystals of paramagnetic salts were of interest because they were related to other cryomagnetic research, a main field of research at Oxford. For example, following the discovery of hyperfine structure [37], an important suggestion [38] led to the first successful experiment on nuclear alignment [39].

8.8   Atomic and Molecular Beams

The ultimate method of avoiding collision broadening makes use of a beam of atoms or molecules, for other types of experiment. In the first of these, Otto Stern [40] passed a beam of silver atoms through a magnetic field gradient, and demonstrated the existence of spatial quantization: atoms with "up" spins were deviated in the opposite direction from those with "down" spins. In optical spectra, Doppler broadening was minimized by use of an atomic beam, through which light from atoms in a discharge tube was transmitted transversely. Absorption of the resonance radiation produced narrow lines, and effective temperatures of a few degrees absolute were achieved by Jackson and Kuhn [41], making possible measurements of the hyperfine structure of light atoms such as Li, Na and K.

Ingenious deflection experiments with atomic beams were invented as a method of determining nuclear spins and moments [42]. In 1936 magnetic resonance at radio-frequencies was suggested [43], and applied to nuclear moments at Columbia University by I. I. Rabi and his colleagues (see N. F. Ramsey, [44]). Proposals for using stimulated emission from a system with inverted population were put forward by several authors [45], and soon afterwards were realized experimentally [46]. A beam of ammonia molecules was passed through an electric field gradient to select those in the upper state of the inversion doublet; on entering a resonant cavity, they produce "microwave amplification by the stimulated emission of radiation," given the acronym "maser."

8.9   Conclusion

A conference to mark the Jubilee of the Zeeman effect was held in Amsterdam in September 1946. The papers were concerned with its applications in astronomy and optical spectroscopy, except for the last: "Spectroscopy at radio frequencies" [47]. In this, it was suggested that a strong line in the ammonia inversion spectrum could be used for an absolute frequency standard with a precision of "one or two parts in 10^7," comparable with existing quartz clocks. In reply to a question, the recent investigation of atomic hyperfine structure of the ^{133}Cs atom [48] was quoted. This has proved to be more reliable than the ammonia spectrum, and has become the ultimate frequency standard; the second is now defined as

$$9192531770 \text{ cycles}$$

of the hyperfine structure of the ^{133}Cs atom.

8.10   Postscript

This article is mainly a personal account of how microwave research began at Oxford, with my research students and colleagues. Sadly, R. P. Penrose contracted a brain tumour and died at Leiden on April 28, 1949, shortly after he had discovered hyperfine structure in the electron paramagnetic resonance spectrum of a dilute copper compound [37]. A theoretical treatment was given by Abragam and Pryce [37]. J. H. E. Griffiths was later President of Magdalen College, Oxford (1968-79), and died on 28 August 1981. J. H. N. Loubser, in a post-doctoral year at Columbia University (1948-49), extended microwave gas spectroscopy into the region of millimetre waves [49–51]. After his return to South Africa, he had a distinguished academic career as Head of the Physics Department at the University of the Orange Free State, Bloemfontein (1952-63), and then at Witwatersrand (1962-82). His main field of research became electron paramagnetic resonance, particularly on point defects in diamonds. After a long illness, he died on 13 May 1994; he is survived by his widow, Ena, three children, and seven grandchildren.

The 80th birthday of C. H. Townes occurs during a period of important anniversaries. This year is the centenary of X-rays, discovered on November 8, 1895 by Wilhelm Konrad Röntgen in Wurzburg, Germany, while investigating the effects of cathode rays in gas discharge tubes. Within a few months X-rays were being used as a powerful new tool in medical diagnosis. The first experiments on nuclear magnetic resonance in condensed matter were carried out by Felix Bloch at Stanford University and Edward Purcell at Harvard University in 1945.

References

[1]   W. Pauli, Naturwiss. **12**, 741 (1924).
[2]   D. A. Jackson, Proc. Roy. Soc. Lond. **121**, 432 (1928).
[3]   J. H. Van Vleck and V. W. Weisskopf, Rev. Mod. Phys. **17**, 227 (1946).
[4]   J. H. Van Vleck, Phys. Rev. **71**, 413 and **72**, 513 (1947).
[5]   R. H. Dicke, Rev. Sci. Inst. **17**, 269 (1946).
[6]   R. H. Dicke, R. Beringer, R. L. Kyhl, and A. B. Vane, Phys. Rev. **70**, 340 (1946).
[7]   R. H. Dicke and R. Beringer, Astrophys. J. **103**, 375 (1946).
[8]   H. I. Ewen and E. M. Purcell, Phys. Rev. **83**, 881 (1951).
[9]   C. A. Muller and J. H. Oort, Nature **168**, 351 (1952).
[10] G. E. Becker and S. H. Autler, Phys. Rev. **70**, 300 (1946).
[11] W. E. Lamb, Jr., Phys. Rev. **70**, 308 (1946).
[12] R. Beringer, Phys. Rev. **70**, 53 (1946).
[13] H. R. L. Lamont, Proc. Phys. Soc. Lond. **61**, 562 (1948); Phys. Rev. **74**, 353 (1948).
[14] M. W. P. Strandberg, C. Y. Meng, and J. G. Ingersoll, Phys. Rev. **75**, 1524 (1949).
[15] R. S. Andersen, C. M. Johnson, and W. Gordy, Phys. Rev. **83**, 1061 (1951).
[16] C. E. Cleeton and N. H. Williams, Phys. Rev. **45**, 234 (1934).
[17] B. Bleaney, J. H. N. Loubser, and R. P. Penrose, Proc. Phys. Soc. Lond. **A59**, 185 and 194 (1947).
[18] B. Bleaney and R. P. Penrose, Nature **137**, 339 (1946); Phys. Rev. **70**, 775 (1946); Proc. Roy. Soc. Lond. **A189**, 358 (1947).
[19] Hsi-Yin Shen, E. F. Barker, and R. M. Dennison, Phys. Rev. **60**, 786 (1941).
[20] J. H. Van Vleck and V. W. Weisskopf, Rev. Mod. Phys. **17**, 227 (1945).
[21] B. Bleaney and R. P. Penrose, Proc. Phys. Soc. Lond. **59**, 418 (1947).
[22] B. Bleaney and J. H. N. Loubser, Nature **161**, 522 (1948).
[23] B. Bleaney and R. P. Penrose, Proc. Phys. Soc. Lond. **60**, 540 (1948).

[24] B. Bleaney and J. H. N. Loubser, Proc. Phys. Soc. Lond. **63A**, 483 (1950).

[25] W. E. Good, Phys. Rev. **70**, 109A (1946).

[26] B. P. Dailey, R. H. Kyhl, M. P. Strandberg, J. H. Van Vleck, and E. Bright Wilson, Jr., Phys. Rev. **70**, 984L (1946).

[27] T. W. Dakin, W. E. Good, and D. K. Coles, Phys. Rev. **71**, 640L (1947).

[28] M. W. P. Strandberg, T. Wentink, and R. L. Kyhl, Phys. Rev. **75**, 270 (1949).

[29] B. Bleaney and R. P. Penrose, Proc. Phys. Soc. Lond. **60**, 83 (1948).

[30] E. Zavoisky, Fiz. Zh. **9**, 211 and 245 (1945).

[31] J. H. E. Griffiths, Nature **158**, 670 (1946).

[32] D. M. S. Bagguley and J. H. E. Griffiths, Nature **162**, 538 (1948).

[33] M. H. L. Pryce, Nature **162**, 539 (1948).

[34] D. M. S. Bagguley and J. H. E. Griffiths, Nature **160**, 532 (1947).

[35] B. Bleaney and R. P. Penrose, Proc. Phys. Soc. Lond. **60**, 395 (1948).

[36] D. M. S. Bagguley, B. Bleaney, J. H. E. Griffiths, R. P. Penrose, and B. I. Plumpton, Proc. Phys. Soc. Lond. **A61**, 542 and 551 (1948).

[37] R. P. Penrose and C. J. Gorter; Nature **163**, 992 (1949); A. Abragam and M. H. L. Pryce, Nature **163**, 993 (1949).

[38] B. Bleaney, Phil. Mag. **43**, 441 (1951).

[39] B. Bleaney, J. M. Daniels, M. A. Grace, H. Halban, N. Kurti, and F. N. H. Robinson, Phys. Rev. **85**, 668 (1952).

[40] O. Stern, Z. Phys. **7**, 249 (1921).

[41] D. A. Jackson and H. G. Kuhn, Proc. Roy. Soc. Lond. **A148**, 335 (1935).

[42] G. Breit and I. I. Rabi, Phys. Rev. **38**, 2082 (1931).

[43] C. J. Gorter, Physica **3**, 503 and 995 (1936).

[44] N. F. Ramsey, Molecular Beams, The Clarendon Press, Oxford (1956).

[45] J. Weber, IRE Trans. Prof. Group on Electron Devices **3** (1953) and N. G. Basov and A. M. Prokhorov, J. Exptl. Theor. Phys. (USSR) **95**, 282 (1954).

[46] J. P. Gordon, H. J. Zeiger, and C. H. Townes, Phys. Rev. **95**, 282 (1954).

[47] B. Bleaney, Physica **12**, 595 (1946).

[48] Arthur Roberts, Yardley Beers, and A. G. Hill, Phys. Rev. **70**, 112 (1946).

[49] J. H. N. Loubser and C. H. Townes, Phys. Rev. **76**, 178A (1949).

[50] J. H. N. Loubser and J. A. Klein, Phys. Rev. **78**, 348A (1950).

[51] J. A. Klein, J. H. N. Loubser, A. H. Nethercot, and C. H. Townes, Rev. Sci. Inst. **23**, 78 (1952).

Fuller accounts of the subjects outlined above are given in two books with the title "Microwave Spectroscopy" by W. Gordy, W. V. Smith and R. F. Trabarulo, John Wiley and Sons, Inc., New York (1953), and by C. H. Townes and A. L. Schawlow, McGraw-Hill Book Company, Inc., New York, Toronto and London (1955).

See also "Electron Paramagnetic resonance of Transition Ions" by A. Abragam & B. Bleaney, Clarendon Press, Oxford (1970), Dover Publications, New York (1986).

9

Meeting Charles H. Townes

Nicolaas Bloembergen

Being in the eighth decade of life myself, I shall not attempt to honor Charles Townes on his 80th birthday with a scientific paper. That would hardly do justice to the undisputed leader who originated and stimulated the development of masers and lasers, quantum electronics and quantum optics. It appears more appropriate for me to reminisce about some of our early encounters.

I believe I met Charles Townes for the first time at the January 1947 meeting of the American Physical Society at Columbia University in New York City. I had presented a ten-minute paper on nuclear magnetic relaxation on behalf of Purcell's group at Harvard and William Nierenberg had made a ten-minute presentation on behalf of the Columbia group. I. I. Rabi invited Purcell, Ramsey, Townes, Nierenberg and me to his office in the Pupin laboratory. The discussion was about the future applications of radio- and microwave electronic techniques to spectroscopy. As a foreign graduate student, I just listened to what the professors had to say.

I remember hearing Charles Townes at some meetings of the American Physical Society, but a brief conversation at the 1953 International Physics Conference in Kyoto, Japan, stands out. We had independently been traveling for days to get there. Transportation in DC-4's of the Military Air Transport Service required intermediate stops on the West Coast, in Hawaii and Wake Island, or in Alaska and the Aleutians. In addition, there were days of waiting at military air force bases, before civilians like us got space on a stand-by basis, as military personnel actively involved in the Korean War had priority. Charles arrived several days late and had to leave early because of his duties as Chairman of the Physics Department at Columbia University. In a brief conversation, and obviously not in the best of spirits, he told me he was thinking about switching from microwave spectroscopy to high energy physics. Fortunately for low energy physics this did not come to pass. The successful operation of the ammonia beam maser by Townes and his students J. P. Gordon and H. J. Zeiger at Columbia kept his interests focused on masers and optical masers during the next decade.

Townes' ideas were essential for the rapid development of quantum electronics that began to take shape between 1955 and 1960. The field grew so fast that more specialized get-togethers than could be provided by the American Physical Society were required. Townes organized the first conference in Quantum Electronics in 1959, in High View, New York. Although the basic paper on optical masers by

Schawlow and Townes had already been published, the proceedings of that first conference show that nearly all contributions were still concerned with microwave physics and devices.

One meeting in those exciting times stands out in my memory: the meeting when Charles and I had a chance to introduce our wives. It was the 1959 annual banquet of the Institute of Radio Engineers in New York City. On this occasion, we shared the Morris C. Liebman Memorial Award for our work on masers. Before dinner, my wife Deli admired the ruby pendant that Frances Townes was wearing. She explained that Charles had crafted it for her in commemoration of the maser. Talking about the eventful day in our hotel room after the banquet, Deli mentioned the ruby pendant and said, "When are you going to make something nice for me?" I answered, "My maser works with cyanide." Our Harvard-built maser did not use a ruby, but the active chromium ions were embedded in a solution-grown crystal of potassium cobalti-cyanide.

The 1959 Quantum Electronics Conference was the birth of the ICQE series, International Conferences on Quantum Electronics, which continues to flourish to this day. The second conference, organized by Jay Singer in Berkeley in 1961, showed about a fifty-fifty distribution of papers between optics and microwaves. I was program chairman of the third conference held in Paris in February 1963. At that time, the papers on lasers and nonlinear optics eclipsed the microwave contributions. In that same year, Charles Townes chaired the first E. Fermi School on Lake Como, Varenna, Italy, which was devoted to quantum electronics. Frances and Charles, Deli and I enjoyed mountain hikes and other outings together. We treasure a snapshot taken on a Sunday morning. Ben Lax, the secretary of the Fermi School, Deli and I had been swimming in the lake. When the Townes's returned from church, the bikini-clad ladies proposed a joint picture for contrast. The Townes's agreed to the picture only after they had changed to more casual attire. The contrast in the picture was, however, provided by Willis Lamb in an elegant dark blue silk suit. He explained that he had not gone to church, but that his casual suit was unusable. Thinking that it was a wash-and-dry outfit, he had washed it the night before. It turned out not to be "wash-and-dry". His sense of humor made the snapshot memorable.

During the years Townes was provost of the Massachusetts Institute of Technology, our social contacts were not limited to scientific conferences, but included cocktail and dinner parties in our homes and in the homes of colleagues. Charles was fortunate to have Frances as a wife, as she provided emotional and intellectual support, and was furthermore a superb hostess. Charles and I had some vigorous discussions about physics, about stimulated Raman scattering in particular. We also met regularly at laser advisory committee meetings in Washington, D.C.

Townes left Cambridge to join the faculty at U.C. Berkeley in 1966, and his scientific interests shifted to infrared astronomy. At an age when many scientists slow down, Charles did not hesitate to explore new territory.

We continued to meet at least once a year at various scientific gatherings. I recall in particular the "Thousand and One Nights" conference that Ali Javan had organized in Esfahan, Iran, in the summer of 1971. The prime minister, Hoveida, opened the conference and hosted a luncheon in the Shah Abbas Hotel. He was a charming host, engaging in a lively conversation with Deli and Frances. Charles and I saw him again a week later at a tea reception in the Shah's palace in Teheran. By contrast, Hoveida did not utter one sentence in the presence of the Shah, who conversed with Townes and a few other invited foreign scientists.

The energy that Frances and Charles displayed during that last day in Iran was impressive. They had gone on an early morning mountain and bird watching hike north of the city and they departed that same evening on a midnight flight to the Soviet Union. Another quantum electronics conference in Iran is unlikely to be scheduled during our lifetimes. We remember that Hoveida was executed a year later, after the Shah's demise.

In 1984, Charles showed me his laboratory on the Berkeley campus when I was visiting there as the Hitchcock lecturer. I was much impressed by the range of his activities and by his unabating scientific curiosity. His presence continues to add luster to the not infrequent events at which mutual friends and colleagues in the fields of quantum electronics and laser spectroscopy are honored. His 80th birthday is the appropriate time to pay our respects to him, who started it all. We wish him, his wife, his children and grandchildren many more enjoyable years together.

10

Population Inversion and Superluminality

Raymond Y. Chiao

I dedicate this essay to my teacher, colleague, and friend, Charles H. Townes.

10.1   Introduction: The Ammonia Maser Revisited

For this Festschrift in honor of Professor Charles Hard Townes on the occasion of his 80th birthday, it seems appropriate to begin by thinking once again about ammonia, one of his favorite objects of study. The ammonia molecule, upon the inversion of its populations, was the active constituent of the first maser [1]. It was also the constituent of the first molecular cloud, which was discovered by Professor Townes and his coworkers in interstellar space. Here we shall return to ammonia in a laboratory setting, and discuss the possibility that population inversion in a medium such as ammonia gas can lead to further surprises, such as the recently predicted phenomena of superluminality [2-9] and parelectricity [6-9].

Superluminality can occur when wave packets are tuned to a transparent spectral window outside the gain-line profile of a medium with inverted populations. It turns out that in such a medium, these wave packets can travel faster than the vacuum speed of light c. This kind of propagation can even be dispersionless, so that an arbitrary wave form—but one which is restricted in its bandwidth—can propagate faster than c without any significant change in its shape or amplitude. Surprisingly, this is not forbidden by Einstein causality.

Parelectricity can occur when the zero-frequency electric susceptibility of a medium becomes negative. This can happen when there is an inversion of population. The existence of such a parelectric medium implies the possibility of stable electrostatic configurations of charges placed in its vicinity, but surprisingly this is not forbidden by Earnshaw's theorem. A strongly parelectric medium can lead in principle to the levitation of charges above it, and to the possibility of a purely electrostatic trap for charged particles which is fundamentally different from the Paul and the Penning traps [9].

However, condensed-matter parelectric media, which appear difficult to produce at the present time, seem to be necessary for the experimental realization of this new kind of trap. Hence, we shall not discuss in detail the phenomenon of parelectricity here. Instead, we shall concentrate on the phenomenon of superluminality, which does appear practicable to be realized experimentally in an optically pumped rubidium vapor cell [8, 10].

In 1954 the ammonia molecular-beam maser was first demonstrated by Gordon, Zeiger, and Townes [1]. This historic, pioneering achievement pointed out the importance of population inversion for the amplification, and hence the generation, of coherent electromagnetic radiation, and led inexorably to the development of the laser [11]. At the heart of this experiment was the ammonia molecule, NH_3. This simple molecule has a pyramidal structure with a nitrogen atom at the apex, and a triangle of hydrogen atoms at the base. Because of tunneling, the nitrogen atom can be either above or below the plane of the hydrogens. This leads to a splitting (approximately 24 GHz) of the $(J, K \neq 0)$ symmetric-top rotational levels of the molecule into doublets, in which the lower energy state $|+\rangle$ is the symmetric linear combination of the two possible configurations of the molecule arising from tunneling, and the upper state $|-\rangle$ is the antisymmetric linear combination of these two possibilities [12]. Upon the application of a direct current (DC) electric field, the two energy eigenvalues associated with a doublet repel away from each other (see Fig. 10.1). The energy level shifts (i.e., the DC quadratic Stark effect) can be readily calculated by means of second order perturbation theory, which yields for the upper and lower states, respectively,

$$\Delta W_- = +\frac{1}{2}|\alpha|\mathcal{E}^2 > 0$$
$$\Delta W_+ = -\frac{1}{2}|\alpha|\mathcal{E}^2 < 0 \quad , \tag{10.1}$$

where $\mathcal{E}$ is the DC electric field strength, and $|\alpha|$ is a positive constant.

Now the change in energy of a small sphere of polarizability α placed in an applied DC electric field $\mathcal{E}$ is given by

$$\Delta W = -\int_0^{\mathcal{E}} \mathbf{p} \cdot d\mathbf{E} = -\frac{1}{2}\alpha\mathcal{E}^2 \, , \tag{10.2}$$

where the induced dipole moment $\mathbf{p}$ is related to the inducing electric field $\mathbf{E}$ by $\mathbf{p} = \alpha\mathbf{E}$. Hence the polarizability of molecules in the lower state $|+\rangle$

$$\alpha_+ = +|\alpha| > 0 \tag{10.3}$$

is *positive*, implying that ammonia molecules in the lower state of the doublet are high-field seekers, whereas the polarizability of molecules in the upper state $|-\rangle$

$$\alpha_- = -|\alpha| < 0 \tag{10.4}$$

is *negative*, implying that ammonia molecules in the upper state are low-field seekers (see Fig. 10.1). One immediate consequence of Eq. (10.4) is that the DC susceptibility, $\chi(0) = -N|\alpha|$, of ammonia gas with all its molecules in the upper

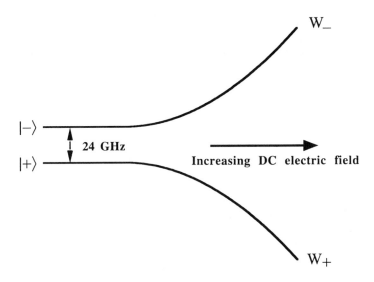

FIGURE 10.1. The DC quadratic Stark effect in the doublets of ammonia.

state, is negative. Thus this medium is parelectric. (*Parelectric* media should not be confused with *paraelectric* media, which are ferroelectrics just above their Curie points.)

Gordon, Zeiger, and Townes utilized the low-field seeking property of the upper states of the ammonia doublets to construct an upper-state selector. The upper states in a molecular beam of ammonia were focused by means of a quadrupole (later, a hexapole) electrostatic lens along its central axis, where the inhomogeneous electric field vanishes; the lower states were defocussed away from the central axis by this electrostatic lens. Thus an inverted population was created in the central beam, which allowed the direct observation of the process of stimulated emission as first conceived by Einstein [13]. The central beam constituted a medium with a negative temperature [14] and therefore negative absorption [15], or gain [1]. When combined with feedback inside a microwave resonator, this inverted medium led to oscillation [1]. Such a medium was in a metastable, but not thermodynamic, equilibrium, since the lifetime of the upper state due to spontaneous emission was very long. The *imaginary* part of the linear susceptibility of this inverted medium underwent a sign reversal with respect to that for an uninverted one, and this led to "*m*icrowave *a*mplification by *s*timulated *e*mission of *r*adiation," or, in the acronym first introduced by Professor Townes, the "*maser*." The *real* part of the linear susceptibility of the population-inverted medium, which also undergoes a sign reversal, leads to further surprises.

Let us now introduce superluminality by noting some simple consequences of the sign reversal of the polarizability of molecules prepared in the upper state. The polarizability is the effective volume of a molecule. In the upper state this volume is negative. A gas of upper-state-selected molecules will then have a negative susceptibility near DC. Therefore the low-frequency dielectric constant of this gas will be less than unity. Since the index of refraction is the square root of the dielectric constant, the index will also be less than unity at low frequencies. Also, since there is little dispersion in the refractive index of dielectric media near DC, any wave packet or wave form whose bandwidth is restricted to low frequencies will travel with little dispersion through the inverted ammonia gas *faster than light travels in the vacuum.* We shall elaborate further on this below.

10.2   Historical Review of Some Faster-than-Light Phenomena

Einstein's famous 1905 paper on special relativity [16] implied that no signal, i.e., information, could be communicated faster than light. Otherwise, some inertial observers would observe that an effect would precede its cause, which would violate the principle of causality. However, in 1935 Einstein, Podolsky, and Rosen [17] showed that in quantum mechanics two distant particles which have flown apart from each other in an entangled state should exhibit upon detection *correlations,* which can be interpreted in terms of faster-than-light influences [18-20]. Nevertheless, because of the *acausal* randomness of quantum events, i.e., that there exists at a fundamental level no physical cause (such as local hidden variables) behind the uncontrollable randomness observed in quantum phenomena, it is fundamentally impossible to utilize these influences for faster-than-light communication. Here we shall not discuss any further these Einstein-Podolsky-Rosen effects, which are multiparticle quantum effects, but shall concentrate instead on single-particle propagation phenomena which can possess apparent speeds exceeding c. (We note in passing that this year, 1995, marks the 90th anniversary of Einstein's 1905 papers and the 60th anniversary of his 1935 paper with Rosen and Podolsky, as well as, on a sadder note, the 40th anniversary of Einstein's passing.)

Group velocities greater than c have already been pointed out in several optical and non-optical contexts. In 1907, Sommerfeld noted that optical media with anomalous dispersion near an absorption line could exhibit group velocities which exceed c [21]. Much later in a 1968 theoretical paper [22] Bludman and Ruderman pointed out in a non-optical context the possibility that the speed of sound in ultradense matter (e.g., neutron star matter) may exceed the speed of light c in the vacuum without any violation of Einstein causality. Shortly afterwards in 1969, Aharonov, Komar, and Susskind also noted the possibility of superluminal group velocities which do not violate Einstein causality, in the context of classical nonlinear field theories which involve unstable configurations [23]. As a specific example, they pointed out that a 1D periodic lattice of *inverted* pendula coupled by springs to their nearest neighbors can exhibit modes of excitation with a tachyonic dispersion relation in which the group velocity always exceeds c. However, the front velocity (i.e., the velocity at which discontinuous disturbances propagate, such as the retarded Green's function) in the corresponding field theory is exactly c. Thus, Lorentz invariance and Einstein causality are not violated. More recently

Scharnhorst and Barton have performed a quantum electrodynamics calculation of the speed of light between Casimir plates, and found light signals faster than c, or amplified by the Casimir vacuum [24-26]. Also recently, Hegerfeldt has raised the question of faster-than-c signals in Fermi's two-atom problem [27, 28].

In an optical context, the possibility that population-inverted two-level media might exhibit group velocities greater than c was apparently first pointed out in 1966 by Basov et al. [29, 30]. They experimentally demonstrated faster-than-c propagation of laser pulses whose bandwidth lay *inside* the gain-line profile of an amplifier. They also calculated stationary solutions which propagated faster than c. This phenomenon arose from the fact that the amplification of the earlier parts of the pulse depleted the population inversion of the medium (i.e., the gain became saturated), so that the later parts of the pulse experienced less gain than its forward portion. The peak of the laser pulse was thereby displaced forward relative to the peak of a pulse propagating at the vacuum speed of light, i.e., its group velocity was greater than c. This pulse advancement was due to a *pulse reshaping process*, an idea which will reappear in all the following cases of optical faster-than-c phenomena. Here the specific mechanism for the pulse reshaping process was the saturation of the gain. Basov et al. noted that relativistic causality was not violated because a finite pulse could not get ahead of a front of zero intensity, which travels at c (i.e., at the front velocity; see below). Icsevgi and Lamb [31] later analyzed the problem of causality in the work of Basov et al. and pointed out that the apparent violation of relativistic causality arose from from their use of unphysical stationary solutions of infinite duration. Propagation of laser pulses with group velocities greater than c which result from gain saturation is now called *superluminous* propagation [32, 33].

In 1993 we pointed out that propagation of wave packets whose spectral content lies *outside* the gain-line profile can also lead to group velocities exceeding c, which we called *superluminal* propagation [2-9]. This effect arises from the fact that the real part of the linear susceptibility of an inverted two-level medium suffers a sign change relative to that of a normal, uninverted medium. We show theoretically below that faster-than-c group velocities result. However, to the best of our knowledge, no experiments have yet demonstrated this effect, and we describe here an experiment which we are presently performing. Just as in the superluminous case, the pulse peak cannot overtake the front, and no violation of Einstein causality occurs. Superluminous and superluminal propagation can both be thought of as the result of pulse reshaping processes, but the underlying physical mechanisms are quite different. Superluminal propagation outside the gain-line profile involves only *virtual* transitions, and therefore no spontaneous emission noise is added to the transmitted pulse, whereas superluminous propagation depends on a *real* amplification of this pulse, with the unavoidable addition of such noise. Also, superluminous propagation is intrinsically *nonlinear*, so that only certain pulse amplitudes and shapes can propagate without distortion. By contrast, the superluminal propagation we consider below is *linear*. Therefore, wave forms with arbitrary shapes and amplitudes can propagate faster than c without appreciable distortion, provided only that the bandwidth of these wave forms are restricted to a spectral region of little dispersion.

In 1970 Garrett and McCumber [34] considered a different optical context in which faster-than-c group velocities could arise, which was briefly considered earlier by Sommerfeld [21]. They calculated that Gaussian wave packets propagating in the region of anomalous dispersion near an absorption line can in fact propagate at faster-than-c, infinite, or negative group velocities. Within certain realistic approximations they showed that an incident Gaussian wave packet will be reshaped by the absorption process (in which later parts of the wave packet are absorbed to a greater extent than the earlier parts) in just such a way as to produce a smaller, but undistorted, Gaussian wave packet at the exit face of the medium. The peak of the wave packet thus appears to have moved at an abnormal group velocity inside the medium.

In 1982 Chu and Wong verified experimentally that this abnormal behavior actually occurred for weak picosecond laser pulses propagating near the center of the bound A-exciton line of a GaP:N sample [35]. This abnormal result has been confirmed in a millimeter-wave propagation experiment [36]. Hence it was demonstrated that the group velocity, even when it exceeds c or becomes infinite or negative, possesses a definite physical meaning and is in fact measurable in all cases, thus contradicting statements in standard textbooks such as Born and Wolf's [37]:

> "...in regions of anomalous dispersion the group velocity may exceed the speed of light or become negative, and in such cases it has no longer any appreciable physical significance."

In 1993 we observed experimentally that a faster-than-c effect also occurs in quantum tunneling [19, 38, 39]. We studied the tunneling of single-photon wave packets,[1] and showed that these wave packets tunneled with an effective speed of $1.7c \pm 0.2c$ through a multilayer dielectric mirror whose frequency of maximum reflection (or minimum transmission) coincided with the center frequency of the wave packet. This mirror was effectively a tunnel barrier, because the amplitude of the wave packet decayed exponentially within the dielectric layers of the mirror. Again, the transmitted single-photon wave packet was Gaussian in shape, and although much smaller in amplitude (most of the incident wave packet was reflected), it had essentially the same shape and width as the incident Gaussian wavepacket.[2] We observed that the peaks of these single-photon wave packets appeared on the far side of the tunnel barrier *earlier* than the peaks of control wave packets, which had propagated through air instead of the barrier. The particle-like aspect of tunneling was manifested in our experiment by the fact that each "click"

---

[1] We use the word "photon" throughout this paper to denote a single quantum of excitation, i.e., a $n = 1$ Fock state, of the electromagnetic field, whether in a standing wave, or in a traveling wave packet.

[2] This results from the lack of dispersion in the mirror near its minimum transmission.

of a single-photon detector used in coincidence detection registered the arrival of a single, individual-photon wave packet which had tunneled through the barrier. In 1994 our result was confirmed at the classical-field level by experiments involving femtosecond laser pulses propagating also through multilayer dielectric mirrors [40], where these femtosecond pulses were detected by means of a second harmonic correlator similar to the one used earlier by Chu and Wong [35].

Our recent work answered a long-standing, controversy-provoking question first raised in a 1932 paper by MacColl [41]: How long does it take for a particle to traverse a tunnel barrier? Our experimental result is consistent with the theoretical calculations of Eisenbud [42] and Wigner [43] for the tunneling time based on the method of stationary phase. These calculations led to predictions that the peaks of wave packets will appear on the far side of a nearly opaque barrier with an effective speed faster than c. Again, it is possible to understand this result in terms of a pulse reshaping process, in which the later parts of the wave packet is causally attenuated (by reflection) more than the earlier parts. Thus there is no violation of Einstein causality. Faster-than-c tunneling of classical electromagnetic wave packets was also recently observed in experiments involving microwaves propagating through wave guides beyond cutoff [44], but the authors interpreted their result as a possibly genuine violation of Einstein causality, in the sense that information was indeed being communicated faster than c in their experiments.

These last few kinds of faster-than-c effects, however, are less dramatic than the phenomena which we propose to observe here, since they are accompanied by large attenuations. This is to be contrasted with the unattenuated, i.e., *transparent*, superluminal wave packet propagation to be discussed below. (There has been a recent experiment demonstrating a $\chi^{(3)}$-induced faster-than-c pulse propagation in a transparent medium [45], but again the propagation process in this case is a nonlinear one; also, the authors, like those of [44], have called causality into question.)

10.3   Theory of Wave Packet Propagation in Transparent, Population-inverted Media

The refractive index of a completely inverted two-level medium can be obtained from a Lorentz model in which the sign of the oscillator strength of the transition is reversed due to the population inversion [2]

$$n(\omega) = \left(1 - \frac{\omega_p^2}{\omega_0^2 - \omega^2 - i\gamma\omega}\right)^{1/2}, \qquad (10.5)$$

where γ is a (small) phenomenological linewidth, ω_0 is the resonance frequency of the medium, and ω_p is the effective plasma frequency,

$$\omega_p = (4\pi|f|Ne^2/m)^{1/2}, \qquad (10.6)$$

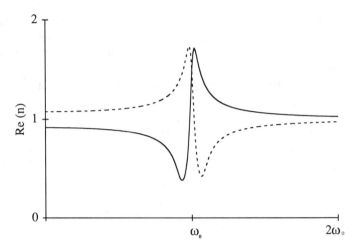

FIGURE 10.2. Real part of the refractive index versus frequency for a completely inverted two-level atomic or molecular medium (solid line), compared with that for the same medium with completely uninverted populations (dashed line).

where $|f| = 2m\omega_0 |\langle -|x|+\rangle|^2/\hbar$ is the absolute value of the oscillator strength of the transition, N is the number density of atoms or molecules, which are all in the upper state, e is the electron charge, and m is the electron mass. Let us consider the typical situation in which the inequalitites $\gamma \ll \omega_p \ll \omega_0$ are obeyed. A plot of the real part of Eq. (10.5) is shown in Fig. 10.2. Note that the minus sign in front of the second term under the square root in Eq. (10.5) arises from population inversion: It differs from the positive sign for a normal medium obtained from the usual Lorentz model [46]. The physical meaning of this second term is that it represents the complex, frequency-dependent susceptibility of the medium (apart from a constant of 4π). This complex susceptibility reverses sign upon an inversion of population. As a result of this sign change, the index of refraction near zero frequency is less than unity[3]

$$n(0) = (1 - \omega_p^2/\omega_0^2)^{1/2} < 1 . \tag{10.7}$$

From Eq. (10.5) it also follows that the slope $d[\text{Re}\,n(\omega)]/d\omega$ approaches zero as $\omega \to 0$. Since the resulting group velocity dispersion vanishes near zero frequency, the medium is dispersionless near DC (see Fig. 10.2).

Now consider a classical, finite-bandwidth wave packet, whose carrier frequency $\omega_c/2\pi$ (say around 1 GHz) and spectrum lie far below the resonance frequency $\omega_0/2\pi = 24\,\text{GHz}$ of ammonia molecules. Let this wave packet be incident upon a medium consisting of a gas of these molecules prepared entirely in the upper state.

---

[3] Note that this implies the unusual phenomenon of total *external* reflection of low-frequency plane waves incident beyond a certain critical angle upon this medium from the vacuum. This is in contrast to the total external reflection of high-frequency radiation, i.e., X-rays, from normal media noted by Einstein. However, the usual Cerenkov effect does not occur in this inverted medium.

The amplitude of this wave packet will be chosen sufficiently small so that only the *linear* response of the medium to this weak perturbation need be considered.

The fact that $n(0) < 1$ means that the phase velocity

$$v_p(0) = c/n(0) > c \tag{10.8}$$

is greater than the vacuum speed of light c. It is well known that the phase velocity can exceed c without any violation of relativity. More surprisingly, here as zero frequency is approached, the group velocity

$$
\begin{aligned}
v_g(0) &= \left(\frac{d\operatorname{Re} k(\omega)}{d\omega}\right)^{-1}_{\omega_c \to 0} \\
&= c\left[\operatorname{Re} n(\omega) + \omega\frac{d\operatorname{Re} n}{d\omega}\right]^{-1}_{\omega_c \to 0} \\
&= \frac{c}{n(0)} = v_p(0) > c
\end{aligned}
\tag{10.9}
$$

is equal to the phase velocity, and is therefore also superluminal: *The group velocity also exceeds the vacuum speed of light.* Still more surprisingly, in contrast to a medium in its ground state, since this inverted ammonia medium can temporarily loan part of its stored energy to the forward tail of the wave packet, the energy velocity [47], which is the velocity of the energy transported by the wave packet, is also superluminal near zero frequency:

$$v_E(0) \equiv \frac{\langle S \rangle}{\langle u \rangle} = \frac{c}{\sqrt{\epsilon(0)}} = \frac{c}{n(0)} = v_p(0) > c , \tag{10.10}$$

where $\langle S \rangle$ is the time-averaged Poynting vector, $\langle u \rangle$ is the time-averaged energy density, and $\epsilon(0)$ is the zero-frequency dielectric constant. Even still more surprisingly, the "signal" velocity of Sommerfeld and Brillouin [47], which they arbitrarily defined as the propagation velocity of the point of half-maximum wave amplitude, is the same here as the group velocity, since there is little change in the shape of the wave packet during its propagation. Note that dispersion is negligible in this large, transparent spectral window stretching from DC to the low-frequency side of resonance. These velocities are all equal to each other because this medium with inverted populations is transparent and dispersionless near zero frequency, as it is for any dielectric medium near $\omega = 0$.

10.4 The Kramers-Kronig Relations Necessitate Superluminality

A more general way to demonstrate these results is to start from the Kramers-Kronig relations for the complex electric susceptibility $\chi(\omega)$, from which one can derive the zero-frequency sum rule (section 62 of [48])

$$\chi(0) = \frac{2}{\pi} \int_0^\infty \frac{\operatorname{Im} \chi(\omega)}{\omega} d\omega , \qquad (10.11)$$

where $\chi(0)$ is the (real) zero-frequency susceptibility. In the case of the ammonia gas with inverted populations, the imaginary part of the susceptibility

$$\operatorname{Im} \chi(\omega) < 0 \qquad (10.12)$$

is negative in the range of frequencies near the strong, low-frequency resonance at 24 GHz which gives the dominant contribution to the integral in Eq. (10.11). From this it follows that the DC electric susceptibility $\chi(0)$ is negative, or equivalently, that the DC dielectric constant

$$\epsilon(0) = 1 + 4\pi \chi(0) < 1 \qquad (10.13)$$

is less than unity. Thus the medium is parelectric. It also follows that the index of refraction near zero frequency

$$n(0) = \epsilon(0)^{1/2} = (1 + 4\pi \chi(0))^{1/2} < 1 \qquad (10.14)$$

is also less than unity. Since the medium is dispersionless near DC (a fact that also follows from the Kramers-Kronig relations; see section 64 of [48]), the phase, group, energy, and "signal" velocities all exceed the vacuum speed of light. Hence superluminality and parelectricity are closely connected phenomena. However, since we have now proved that both parelectricity and superluminality follow from the Kramers-Kronig relations, these results do not depend on the validity of any specific model such as the Lorentz model or the two-level model. Also, since causality was used to derive these relations, causality cannot be violated by these conclusions. In general the Kramers-Kronig relations imply that any medium with sufficient gain (with or without population inversion) gives rise to superluminal group velocities in transparent spectral windows separate from the region with gain [2-4].

In electrical engineering, Bode's law, which relates the gain of a linear amplifier system to its phase shift, is equivalent to the Kramers-Kronig relations. Hence in principle superluminality and parelectricity can also occur in transistor networks with spatially distributed gain.

10.5   Considerations of Energy and of Superposition

For a medium in its ground state, there is loss at all frequencies ($\operatorname{Im} \chi(\omega) > 0$ for all ω), and therefore the susceptibility $\chi(0)$ must always be positive (see Eq. (10.11)). If one were to insert this medium between a pair of charged capacitor plates, energy must flow from the electric field into the medium, since all atoms are initially in their ground states. (This can also easily be seen from the Lorentz model, in which energy must always flow from the field into the spring upon the application of the DC electric field.) As a consequence of the fact that energy must be initially transferred from the wave to the medium in order to polarize it, the phase shift of the forward scattering amplitude from each atom near zero frequency must have a sign such as to retard the transmitted wave packet. Thus the index of refraction must be greater than unity, and the wave propagation speeds must be subluminal. However, when the medium possesses an inverted population, energy is stored in it. It now becomes possible to transfer initially this energy from the medium into the wave.[4] The sign of the susceptibility of the medium $\chi(0)$ is now reversed. The sign of the phase shift of the forward scattering amplitude from each atom is also reversed. As a result, the transmitted wave packet is *advanced* rather than *retarded*, and the index of refraction is now less than unity with little dispersion, leading to superluminal propagation [6].

By the superposition principle, or more specifically, by Fourier's theorem, any finite-bandwidth wave packet or wave form should travel faster than c through this linear inverted medium without appreciable change in shape or amplitude, provided only that its bandwidth lies sufficiently far below resonance [2]. This *dispersionless* superluminal propagation can be thought of as a causal wave form reshaping process in which the earlier parts, i.e., the weak forward tails, undergo virtual amplification by the inverted medium, followed by virtual absorption of the later parts of the incident wave form.[5] An *advanced* wave form is thus produced (versus a *retarded* one produced in an uninverted medium), which faithfully reproduces the entire incident wave form, no matter how complex, e.g., Beethoven's ninth symphony. Energy is temporarily loaned by the medium to the wave, so that the medium is merely a catalyst for this process. No spontaneous emission noise is added to the transmitted superluminal wave form [2, 4, 6], since only virtual emissions occur.

---

[4]Under certain circumstances, the inverted medium will be unstable to superradiance, to superfluorescence, or to amplified spontaneous emission. However, these effects are insignificant here, since the lifetime of the inverted ammonia molecules due to spontaneous emission is extremely long (due to the ω_0^3 dependence of the Einstein A coefficient). Moreover, spontaneous emission can be suppressed for example by placing the gas inside a transmission line with a stopband (e.g., a stripline with a periodically corrugated upper conductor) centered at 24 GHz. Such a system still permits superluminal propagation at low frequencies such as at 1 GHz; the negative polarizability of the upper states is insignificantly changed by the presence of the stopband.

[5]It should be noted that this wave form reshaping process is by nature no different from that in a normal, uninverted medium, except for the reversal of the order of virtual absorption and virtual emission (i.e., the retarded wave form, just like the advanced wave form, should also be thought of as being generated by pulse reshaping). Hence this pulse-reshaping process is already present in the ordinary theory of the refractive index, and antedates that of Basov *et al.* [29, 30].

10.6   Einstein Causality, and Sommerfeld and Brillouin's Wave Velocities

Einstein causality is not violated by these conclusions, because there is no information contained in the peaks of the wave packet or wave form which is not already present in its weak forward tails. New information is communicated only when there is an unexpected change, such as a discontinuity, whose arrival time cannot be inferred from the past behavior of the wave. A simple example of such a discontinuity is that of a step-modulated sine wave, which Sommerfeld and Brillouin used in their study of precursors [47]. Their wave form had a sharp front of zero intensity, which could not be overtaken by the later parts of the wave. Such discontinuous wave forms, in contrast to the smooth, finite-bandwidth wave packets or wave forms which we consider here, contain components at infinite frequency, so that it is the infinite-frequency index of refraction $n(\infty)$ which determines the propagation speed of the discontinuity. Closely related to this is the fact that the microscopic Green's function, which results from a delta-function disturbance of an atom in the medium, must propagate in such a way that its influence on nearby atoms is causally retarded by the speed $c/n(\infty)$. This is true whether the atoms are inverted or not. Moreover, the extremely-high-frequency behavior of the index of refraction of any medium, whether inverted or not, is dominated by the response of the nearly free electrons in the atoms. The inertia of these electrons always causes a retarded response to the passage of a wave form with a sharp front. Asymptotically, the index approaches unity [46],[6] so that $n(\infty) = 1$. Thus the propagation speed of discontinuities, i.e., Sommerfeld and Brillouin's front velocity, is exactly c. Hence Einstein causality cannot be violated under any circumstances in these media.

It is helpful at this point to review Sommerfeld and Brillouin's definitions of various wave velocities [47]. For an absorptive, dispersive medium, they found it necessary to demarcate wave velocities into five different kinds:

(1) the *phase* velocity, at which the zero-crossings of the carrier wave would move,

(2) the *group* velocity, at which the peak of a wave packet would move,

(3) the *energy* velocity, at which energy would be transported by the wave,

(4) the *"signal"* velocity, at which the half-maximum wave amplitude would move, and

(5) the *front* velocity, at which the first appearance of the discontinuity would move.

---

[6] We assume here that the vacuum is not modified by QED effects such as those reviewed in "Speed of Light in Nontrivial Vacua" by J. I. Latorre, P. Pascual, and R. Tarrach (Nucl. Phys. **B437**, 60 (1995)).

These five velocities differed from each other in the region of anomalous dispersion near an absorption line. The first two kinds of wave velocities, the phase and group velocities, might be faster than c (in fact they might become infinite, or even negative), but the next two kinds, the energy and "signal" velocities, were less than c under all the circumstances they considered. The last kind, the front velocity, was equal to the vacuum speed of light c. However, their "signal" velocity should not be confused with the information velocity in special relativity, which must be strictly less than or equal to c. We have shown that in the case of the transparent medium with inverted populations, due to its dispersionless nature, all the first four kinds of wave velocities (the phase, group, energy, and "signal" velocities) when extended to the low-frequency, finite-bandwidth wave forms considered here, are greater than c, and that only the fifth (the front velocity) is not. We have identified the front velocity as the information velocity of special relativity. Thus there is no violation of Einstein causality.

Although relativistic causality is not violated, the above conclusions still can lead to some surprising consequences. For all practical purposes, all detectors, which must have a finite threshold of detection, when placed on the far side of the medium, would be triggered earlier than if the medium were replaced by the vacuum. At the fundamental, quantum level, the "click" of a single-photon detector, which requires a quantum of energy to trigger it, would also be triggered earlier. For example, consider a two-photon light source, e.g., continuous-wave (cw) spontaneous parametric down-conversion, in which a conjugate pair of photons is simultaneously emitted at a slight angle with respect to each other [19]. We have already used this nonclassical light source to confirm experimentally that single-photon wave packets propagate through a (normal) medium at the group velocity [49]. Now let two single-photon counters be placed at equal distances from this source so as to detect the photon pair in coincidence. Let us then insert an (inverted) medium with a group velocity greater than c in the path of one of these photons. According to the Born interpretation of the single-photon wave packet [50], which we have confirmed experimentally [49], the peak of the wave packet is the point of highest probability for the detection of the single photon contained within the wave packet. Hence the counter following the population-inverted medium will most probably click *earlier* than the counter which registers the arrival of the conjugate photon whose path lies entirely in the vacuum. Thus the photon in the medium can apparently travel faster than c.

In the solutions of the nonrelativistic Schrödinger equation, the group velocity of a wave packet in the correspondence principle limit is identified with the particle velocity (e.g., the electron velocity). If one were to extend this identification to the present case where there exist situations in which the group velocity exceeds c, would one then draw the conclusion that it is possible for the particle (e.g., the electron) to travel faster than c? Or if the particle were dressed by a medium to become a quasiparticle, could the quasiparticle travel faster than c when its group velocity exceeds c? Of course the use of nonrelativistic quantum mechanics is precluded here, since we are dealing with the ultrarelativistic limit. However, the identification of the group velocity with the particle velocity holds true also for the solutions of relativistic wave equations (e.g., the Dirac equation). One possible answer to these questions in the case of the electron starts from the fact that it is a fermion. Stimulated emission of electrons, which would result in more than one electron occupying the same single-polarization plane-wave state, is fundamentally impossible due to the Pauli exclusion principle. Hence the amplification of electron waves, in contrast to amplification of electromagnetic waves, is fundamentally impossible, and thus by the Kramers-Kronig argument, group velocities exceeding c arising from gain is fundamentally impossible for electrons. However, this conclusion does not apply to bosons,[7] as the case of the photon demonstrates.

10.7   An Experiment in an Optically Pumped Rubidium Vapor Cell

Superluminal propagation occurs not only in the spectral window near DC (although pedagogically this is the easiest case to understand), but also in transparent windows next to a gain line where resonant enhancement can give rise to large effects. In fact, much-faster-than-c, infinite, or negative group velocities occur within sidebands of the order of the effective plasma frequency on either side of this line. Computer simulations show that the meaning of a negative group velocity is that the peak of the transmitted wave packet leaves the exit face of the sample cell *before* the peak of the incident wave packet enters the entrance face of this cell [5].[8] The typical bandwidth of these abnormal, highly superluminal effects is given by the effective plasma frequency, Eq. (10.6), which can be more conveniently expressed in terms of the gain-bandwidth product as

$$\omega_p = \sqrt{G_0 \gamma c} \,, \tag{10.15}$$

where G_0 is the peak gain and γ is the bandwidth of the gain-line profile. Dispersionless, highly superluminal propagation can be produced in the spectral region between a pair of gain lines (produced for example by a doublet of Raman pump lines), whose spacing is on the order of the effective plasma frequency: There exists a point of zero group-velocity dispersion between these lines [4].

We are performing an experiment to demonstrate these striking phenomena using a resonantly-enhanced stimulated Raman effect in an optically pumped rubidium vapor cell. Pulses propagating in transparent spectral regions with band-

---

[7] The Bose-Einstein statistics of atoms like ^{4}He can also lead to superluminal effects such as those suggested to occur in superfluid helium [6]. For example, in low-energy, coherent forward scattering off a drop of superfluid helium, an incident helium atom may possibly be transmitted superluminally through the drop via its Bose condensate.

[8] The early tail of the incident wave packet, before the peak enters the cell, generates two counterpropagating wave packets at the exit face, one of which leaves the sample, and the other of which moves backwards through the sample and annihilates with the incident wave packet at the entrance face. This is a *macroscopic* virtual process.

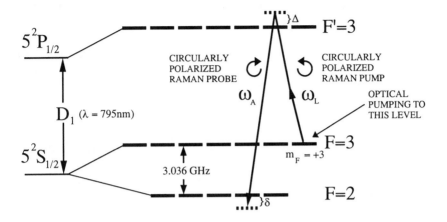

FIGURE 10.3. Stimulated Raman effect in optically pumped ^{85}Rb vapor (not to scale).

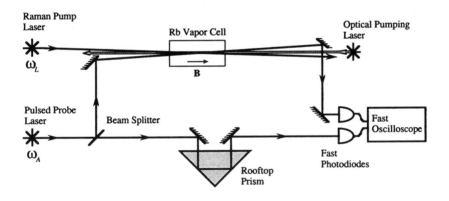

FIGURE 10.4. Experimental setup to observe superluminal pulse propagation.

widths on the order of hundreds of megahertz adjacent to the Raman gain line should exhibit highly superluminal, indeed, negative group velocities.

The Raman transition which we are using in this experiment involves the hyperfine splitting of $\Delta E = 3.036$ GHz in the ground state of ^{85}Rb; see Fig. 10.3. The atoms are prepared in the maximum angular momentum state of the ground-state hyperfine manifold (the $m_F = +3$ state) by optically pumping with an on-resonance left-circularly polarized diode laser beam propagating opposite to the magnetic field direction shown in Fig. 10.4. The optical-pumping laser can be tuned on the

D_1 resonance simultaneously for both the $F = 2 \rightarrow F' = 3$ and the $F = 3 \rightarrow F' = 3$ transitions, by tuning to halfway between these transitions, and then producing side bands by modulating the diode injection current at 1.518 GHz. (To prevent coherent population trapping, we add as a buffer a noble gas around 10 torr pressure into the vapor cell.) In this way the vapor cell is made transparent to left-circularly polarized light at frequencies near the D_1 transition. We use isotopically pure ^{85}Rb in the vapor cell to remove the unwanted nearby absorption lines of ^{87}Rb.

To produce Raman gain [51] we let a left-circularly polarized Raman pump laser beam (a cw titanium sapphire laser of intensity 1.2 kW/cm^2 and a linewidth of 50 MHz), which is detuned by Δ (approximately a Doppler width) from the $F = 3 \rightarrow F' = 3$ transition frequency, pass through the vapor cell in the same direction as that of the magnetic field; see Figs. 10.3 and 10.4. Doppler-free Raman gain will then be produced for a probe beam injected into a direction close to that of the pump beam. The use of this Doppler-free configuration eliminates the necessity for the use of atomic beams. Population inversion in the hyperfine levels resulting from optical pumping leads to Raman gain in the anti-Stokes transition (rather than the Stokes transition, as would be the case without population inversion) for a circular polarization opposite to that of the Raman pump. The Raman gain can then be measured by means of a weak right-circularly polarized probe beam from a diode laser by scanning its frequency. This gain (approximately 1 per centimeter) will be maximum at an anti-Stokes frequency which is detuned by Δ from the $F = 2 \rightarrow F' = 3$ transition frequency. We are using a vapor cell with a number density on the order of 10^{11} atoms/cm^3 in this experiment.

Observation of superluminal pulse propagation may be done with 0.8 ns pulses whose central frequency is detuned from the Raman gain maximum by $\delta \approx \omega_p/2$. The weak probe pulses may be produced by a commercial electrooptic modulator modulating a cw diode laser beam. Pulse observations may be performed using direct detection by means of high-speed photodiodes and a fast oscilloscope. Precise measurement using computer averaging will be required to detect the pulse advancement and to look for any distortion in the pulse shape.

Calculations [8] show that for a detuning $\delta/2\pi = 170$ MHz and for the other parameters given above, the group velocity is around $-c$. Thus for a 10 cm cell the pulse arrives about 0.7 ns ahead of a pulse which travels through the same distance of vacuum. Since the maximum bandwidth of the pulse is about 200 MHz, its duration is about 0.8 ns. Hence the shift in arrival time of the pulse is comparable to its width, and we hope to be able to satisfy Rayleigh's criterion of resolution for distinguishing the superluminally transmitted pulse from the pulse which has propagated through air.

10.8   Concluding Personal Remarks

It was a great privilege to have been one of the two graduate students of Professor Townes (Elsa Garmire being the other one) during his tenure as the provost at

M.I.T. (1961-1967), which in retrospect seems to be a such a brief period of time. However, for me it was a crucial period during which I learned firsthand from a great physicist what physics, and more generally, what science is all about. The lessons which I learned then have lasted a lifetime; his influence on my thinking continues to this day, as this paper attests. His valuable physical insights in many discussions over the years, including the ones on the topic of this paper, will always be treasured. Thus for many years and in different ways, he and I have shared the same desire to come to the full knowledge of the Truth (ΕΙΣ ΕΠΙΓΝΩΣΙΝ ΑΛΗΘΕΙΑΣ ΕΛΘΕΙΝ).

10.9   Acknowledgments

I thank E. L. Bolda, J. Bowie, J. Boyce, J. C. Garrison, J. D. Jackson, M. W. Mitchell, K. Scharnhorst, and A. M. Steinberg for helpful discussions, and G. W. McKinney for help in preparing this manuscript. This work was supported by the Office of Naval Research under ONR Grant No. N00014-90-J-1259.

References

[1] J. P. Gordon, H. Zeiger, and C. H. Townes, Phys. Rev. **95**, 282L (1954).

[2] R. Y. Chiao, Phys. Rev. A **48**, R34 (1993).

[3] E. L. Bolda, R. Y. Chiao, and J. C. Garrison, Phys. Rev. A **48**, 3890 (1993).

[4] A. M. Steinberg and R. Y. Chiao, Phys. Rev. A **49**, 2071 (1994).

[5] E. L. Bolda, J. C. Garrison, and R. Y. Chiao, Phys. Rev. A **49**, 2938 (1994).

[6] R. Y. Chiao, J. Boyce, and J. C. Garrison, in the proceedings of the conference, *Fundamental Problems in Quantum Theory* in honor of Professor John Archibald Wheeler, edited by D. M. Greenberger and A. Zeilinger, Ann. N. Y. Acad. of Sci. **755**, 400 (1995).

[7] R. Y. Chiao, J. Boyce, and M. W. Mitchell, Applied Physics B **60**, 259 (1995).

[8] R. Y. Chiao, E. Bolda, J. Bowie, J. Boyce, J. C. Garrison, and M. W. Mitchell, Quantum and Semiclassical Optics **7**, 279 (1995).

[9] R. Y. Chiao and J. Boyce, Phys. Rev. Lett. **73**, 3383 (1994).

[10] A. Kastler, J. Phys. Radium **11**, 225 (1950).

[11] A. L. Schawlow and C. H. Townes, Phys. Rev. **112**, 1940 (1958).

[12] C. H. Townes and A. L. Schawlow, *Microwave Spectroscopy* (McGraw-Hill, New York, 1955), p. 300-313.

[13] A. Einstein, Phys. Z. **18**, 121 (1917).

[14] E. M. Purcell and R. V. Pound, Phys. Rev. **81**, 279 (1951).

[15] W. E. Lamb and R. C. Retherford, Phys. Rev. **79**, 549 (1950).

[16] A. Einstein, Annalen der Physik **17**, 841 (1905).

[17] A. Einstein, B. Podolsky, and N. Rosen, Phys. Rev. **47**, 777 (1935).

[18] R. Y. Chiao, P. G. Kwiat, and A. M. Steinberg, Advances in Atomic, Molecular, and Optical Physics **34**, 35 (1994).

[19] R. Y. Chiao, P. G. Kwiat, and A. M. Steinberg, Quantum and Semiclassical Optics **7**, 259 (1995).

[20] A. M. Steinberg, P. G. Kwiat, and R. Y. Chiao, "Quantum Optical Tests of the Foundations of Physics," in *The AIP Handbook of Atomic, Molecular, and Optical Physics*, edited by G. W. F. Drake, to be published by the American Institute of Physics.

[21] A. Sommerfeld, Phys. Z. **8**, 841 (1907).

[22] S.A. Bludman and M.A. Ruderman, Phys. Rev. **170**, 1176 (1968).

[23] Y. Aharonov, A. Komar, and L. Susskind, Phys. Rev. **182**, 1400 (1969).

[24] K. Scharnhorst, Phys. Lett. B **236**, 354 (1990).

[25] G. Barton, Phys. Lett. B **237**, 559 (1990).

[26] G. Barton and K. Scharnhorst, J. Phys. A **26**, 2037 (1993).

[27] G. C. Hegerfeldt, Phys. Rev. Lett. **72**, 596 (1994).

[28] D. Buchholz and J. Yngvason, Phys. Rev. Lett. **73**, 613 (1994).

[29] N. G. Basov, R. V. Ambartsumyan, V. S. Zuev, P. G. Kryukov, and V. S. Letokhov, Sov. Phys. Doklady **10**, 1039 (1966).

[30] N. G. Basov, R. V. Ambartsumyan, V. S. Zuev, P. G. Kryukov, and V. S. Letokhov, Sov. Phys. JETP **23**, 16 (1966).

[31] A. Icsevgi and W. E. Lamb, Phys. Rev. **185**, 517 (1969).

[32] E. Picholle, C. Montes, C. Leycuras, O. Legrand, and J. Botineau, Phys. Rev. Lett. **66**, 1454 (1991).

[33] D. L. Fisher and T. Tajima, Phys. Rev. Lett. **71**, 4338 (1993).

[34] C. G. B. Garrett and D. E. McCumber, Phys. Rev. A **1**, 305 (1970).

[35] S. Chu and S. Wong, Phys. Rev. Lett. **48**, 738 (1982); Phys. Rev. Lett. **49**, 1293 (1982).

[36] B. Segard and B. Macke, Phys. Lett. A **109**, 213 (1985).

[37] M. Born and E. Wolf, *Principles of Optics* (Pergamon Press, Oxford, 1975), p. 23.

[38] A. M. Steinberg, P. G. Kwiat, and R. Y. Chiao, Phys. Rev. Lett. **72**, 708 (1993).

[39] A. M. Steinberg and R. Y. Chiao, Phys. Rev. A **51**, 3525 (1995).

[40] Ch. Spielmann, R. Szipoecs, A. Stingl, and F. Krausz, Phys. Rev. Lett. **73**, 2308 (1994).

[41] L. A. MacColl, Phys. Rev. **40**, 621 (1932).

[42] L. Eisenbud, Ph. D. thesis, Princeton University (unpublished, 1948).

[43] E. P. Wigner, Phys. Rev. **98**, 145 (1955).

[44] A. Enders and G. Nimtz, J. Phys. (France) I **3**, 1089 (1993).

[45] Y. L. de Silva, Y. Silberberg, and J. P. Heritage, Opt. Lett. **18**, 580 (1993).

[46] J. D. Jackson, *Classical Electrodynamics* (Wiley & Sons, New York, 1975), p.285, p. 315.

[47] L. Brillouin, *Wave Propagation and Group Velocity* (Academic Press, New York, 1960).

[48] L. D. Landau and E. M. Lifshitz, *Electrodynamics of Continuous Media* (Pergamon Press, Oxford, 1960).

[49] A. M. Steinberg, P. G. Kwiat, and R. Y. Chiao, Phys. Rev. Lett. **68**, 2421 (1992).

[50] I. H. Deutsch and J. C. Garrison, Phys. Rev. A **43**, 2498 (1991).

[51] A. C. Tam, Phys. Rev. A **19**, 1971 (1979).

11

The Autler-Townes Effect Revisited

Claude N. Cohen-Tannoudji

11.1 Introduction

In 1955, Autler and Townes showed that a microwave transition of the OCS molecule can split into two components when one of the two levels involved in the transition is coupled to a third one by a strong resonant microwave field [1]. The corresponding doublet is called the Autler-Townes doublet, or the dynamic Stark splitting. Such an effect is in fact quite general. Using a dressed-atom approach, I would like to show in this paper that the basic features of the Autler-Townes effect show themselves in several new research fields, such as high resolution optical spectroscopy, cavity quantum electrodynamics and laser cooling. This will be a way for me to express my admiration and gratitude to Charles Townes: my admiration for the fundamental concepts that he has so fruitfully introduced in so many different branches of physics; my gratitude for all that I have learned from his writings and his lectures.

11.2 Dressed-atom Approach to the Autler-Townes Effect

The first theoretical treatment of the Autler-Townes effect used a semiclassical theory where the strong resonant microwave field is described as a c-number sinusoidal field. Such a description is perfectly valid and can be rigorously justified when the driving field is in a coherent state [2]. Quantizing the driving field, however, provides interesting physical insights. Although not essential, such a quantum treatment dealing with the total coupled system "atom + driving photons," also called "dressed atom," has the advantage of correlating all the observable phenomena with the properties of the energy diagram of a time-independent Hamiltonian. In particular, the Autler-Townes effect is associated with a level anticrossing in this energy diagram [3]. Furthermore, in certain new domains, like cavity quantum electrodynamics, the quantization of the field becomes essential. We shall therefore introduce here the Autler-Townes effect from a dressed-atom point of view. We shall restrict ourselves to the essential points which will be useful for the discussions presented in the next sections. More details about the dressed-atom approach may be found elsewhere [4].

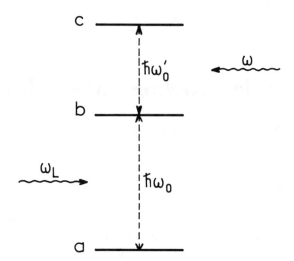

FIGURE 11.1. Three-level atom. The intense quasi-resonant field ω_L drives the transition $a \longleftrightarrow b$. The weak field ω probes the transition $b \longleftrightarrow c$.

We consider a three-level atom a, b, c (Fig. 11.1), with two allowed transitions $a \longleftrightarrow b$ and $b \longleftrightarrow c$, the energy splittings being $\hbar\omega_0$ and $\hbar\omega_0'$, respectively. An intense quasi-resonant field, with frequency ω_L, drives the transition $a \longleftrightarrow b$. The detuning

$$\delta = \omega_L - \omega_0 \tag{11.1}$$

between ω_L and ω_0 is very small compared to ω_0 and ω_0'. The difference between ω_0 and ω_0' is sufficiently large to allow one to consider the field ω_L as being completely nonresonant for the transition $b \longleftrightarrow c$. A very weak field with frequency ω probes the transition $b \longleftrightarrow c$. The problem is to understand how the absorption of the probe field is modified when the transition $a \longleftrightarrow b$ is driven by the field ω_L.

The Hamiltonian H of the atom dressed by the field ω_L may be written

$$H = H_A + H_L + V_{AL} \tag{11.2}$$

where H_A is the atomic Hamiltonian, H_L the Hamiltonian of the field mode L corresponding to the driving field, and V_{AL} the atom-field coupling. The left part of Fig. 11.2 represents a few eigenstates of $H_A + H_L$ (uncoupled states), labeled by two quantum numbers, one (a, b or c) for the atom, and one (N) for the number of photons ω_L. The two states $|a, N + 1\rangle$ (atom in a in the presence of $N + 1$ photons ω_L) and $|b, N\rangle$ (atom in b in the presence of N photons) are separated by a distance $E_a + (N+1)\hbar\omega_L - E_b - N\hbar\omega_L = \hbar(\omega_L - \omega_0) = \hbar\delta$ which is very small compared to the distance $\hbar\omega_0'$ between $|c, N\rangle$ and $|b, N\rangle$. At resonance ($\delta = 0$), the two states $|a, N + 1\rangle$ and $|b, N\rangle$ are degenerate. The interaction Hamiltonian V_{AL} couples these two states: the atom in the state $|a\rangle$ can absorb one photon ω_L and go to the state $|b\rangle$. The corresponding matrix element is written

$$\langle b, N | V_{AL} | a, N + 1 \rangle = \hbar\Omega_1/2 \tag{11.3}$$

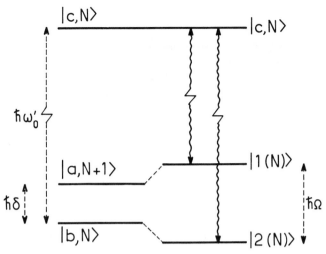

FIGURE 11.2. A few uncoupled states (left part) and corresponding perturbed or dressed states (right part) of the atom + photons ω_L system. The dotted lines indicate the energy splittings between states. The wavy lines give the allowed transitions between dressed states which can be probed by the weak field ω.

where Ω_1 is the so-called "Rabi frequency" which is proportional to the field amplitude and to the transition dipole moment between a and b. Strictly speaking, Ω_1 is proportional to $\sqrt{N+1}$ and thus depends on N. We shall neglect this dependence here, assuming that the field ω_L is initially excited in a coherent state with a Poisson distribution for N having a width ΔN much smaller than the mean value $\langle N \rangle$ of N. As a result of the coupling (3), the two uncoupled states $|a, N+1\rangle$ and $|b, N\rangle$ repel each other and are transformed into two perturbed or dressed states $|1(N)\rangle$ and $|2(N)\rangle$, which are given by two orthogonal linear combinations of $|a, N+1\rangle$ and $|b, N\rangle$ whose energies are separated by a distance $\hbar\Omega$ with

$$\Omega = \sqrt{\Omega_1^2 + \delta^2} \tag{11.4}$$

(see right part of Fig. 11.2). Because we assume that the field ω_L is completely off resonance for the transition $b \longleftrightarrow c$, we neglect any effect of V_{AL} on $|c, N\rangle$.

The transitions with frequencies close to ω_0' which are probed by the weak field ω are those which reduce to the transition $|c, N\rangle \longleftrightarrow |b, N\rangle$ in the limit $\Omega_1 \longrightarrow 0$. Because the two dressed states $|1(N)\rangle$ and $|2(N)\rangle$ contain both admixtures of $|b, N\rangle$, we see that the two transitions $|c, N\rangle \longleftrightarrow |1(N)\rangle$ and $|c, N\rangle \longleftrightarrow |2(N)\rangle$, represented by the wavy lines of Fig. 11.2, are both allowed for the probe field. The absorption spectrum of this probe field, which reduces to a single line of frequency ω_0' in the absence of V_{AL} becomes a doublet when the $a \longleftrightarrow b$ transition is driven by the field ω_L. This is the Autler-Townes doublet.

In order to understand how the frequencies of the two lines of the doublet vary with the detuning δ of the driving field, it is convenient to introduce an energy-level diagram giving the energies of the dressed states of Fig. 11.2 versus $\hbar\omega_L$ (see

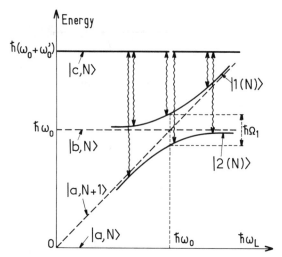

FIGURE 11.3. Energies of the dressed states of Fig. 11.2 (solid lines) versus $\hbar\omega_L$. The dashed lines represent the energies of the uncoupled states. Whereas the two uncoupled states $|a, N + 1\rangle$ and $|b, N\rangle$ cross for $\hbar\omega_L = \hbar\omega_0$, the two dressed states $|1(N)\rangle$ and $|2(N)\rangle$ form an "anticrossing." The wavy lines represent, for different values of $\hbar\omega_L$, the two components of the Autler-Townes doublet.

Fig. 11.3). The uncoupled state $|a, N\rangle$ is chosen as the energy origin. The energies of the uncoupled states $|b, N\rangle$ and $|c, N\rangle$ are then represented by horizontal dashed lines with ordinates equal to $\hbar\omega_0$ and $\hbar\left(\omega_0 + \omega_0'\right)$, respectively. The energy of $|a, N + 1\rangle$ is represented by a dashed straight line with slope 1 passing through the origin and intersecting the horizontal line associated with $|b, N\rangle$ at $\hbar\omega_L = \hbar\omega_0$. When the effect of V_{AL} is taken into account, we get the two dressed states $|1(N)\rangle$ and $|2(N)\rangle$, represented by the solid lines of Fig. 11.3, which form the two branches of a hyperbola having the above-mentioned dashed lines as asymptotes. The minimum distance between the two branches of the hyperbola occurs for $\hbar\omega_L = \hbar\omega_0$ and is equal to $\hbar\Omega_1$. Finally, the state $|c, N\rangle$ is (in the neighbourhood of $\hbar\omega_L = \hbar\omega_0$) unaffected by the coupling V_{AL} and remains represented by a horizontal line.

The effect of the coupling V_{AL} is thus to transform the crossing between $|a, N + 1\rangle$ and $|b, N\rangle$, which occurs at $\hbar\omega_L = \hbar\omega_0$, into an "anticrossing". One clearly sees in Fig. 11.3 how the two components of the Autler-Townes doublet, represented by the wavy arrows, vary with the detuning. At resonance $(\omega_L = \omega_0)$, one gets two lines with frequencies $\omega_0' \pm (\Omega_1/2)$. Off resonance $(|\delta| \gg \Omega_1)$, one of the two lines of the doublet has a frequency close to ω_0', the other line a frequency close to $\omega_0 + \omega_0' - \omega_L$. By evaluating the admixture of $|b, N\rangle$ in each dressed state, one can determine the intensities of the two components of the doublet. One finds that they are equal at resonance, whereas the line with a frequency close to ω_0' becomes the most intense off-resonance. Note that, near the asymptotes of Fig. 11.3, i.e. for $|\delta| \gg \Omega_1$, the distance between each dressed state and its corresponding asymptote is nothing but the ac Stark shift of level a or b due to its coupling with the field ω_L which is then nonresonant. If ω_L lies in the optical domain, such a shift is also called "light shift," and it has been observed in optical pumping experiments [5].

In the previous discussion, we have neglected the width γ of the levels due to various damping processes such as spontaneous emission and collisions. We have supposed implicitly that the Rabi frequency Ω_1 is sufficiently large compared to γ. Such a situation corresponds to a "strong coupling" regime where the two lines of the doublet are clearly resolved, even at resonance.

11.3   The Autler-Townes Effect in the Optical Domain

The spectacular development of laser sources in the early seventies stimulated several experimental and theoretical studies dealing with the behaviour of atoms submitted to intense monochromatic optical fields. The light intensity became high enough to achieve Rabi frequencies larger than the width of the levels. The tunability of laser sources also allowed the frequency of the laser light to be continuously varied across the atomic frequency. Finally, spontaneous emission of radiation, which is negligible in the microwave and RF domains, becomes important in the optical domain and gives rise to new types of signals, such as the intensity or the spectral distribution of the fluorescence light, which can be used for probing the consequences of the laser-atom interaction. Note also that the Doppler effect is important for optical lines and must be taken into account for moving atoms. We briefly review in this section a few experiments showing how the Autler-Townes effect manifests itself in the optical domain.

11.3.1   Case of Two Optical Transitions Sharing a Common Level

Such a case corresponds to a straightforward extension of the experiments performed in the microwave domain. One intense laser field ω_L drives an optical transition $a \longleftrightarrow b$. A second weak laser field probes a second transition $b \longleftrightarrow c$ sharing a common level b with the first transition. The three levels a, b, c can form a "cascade" configuration as in Fig. 11.1, or a V−configuration, or a Λ-configuration.

A few experiments used an atomic beam, perpendicular to both the driving and probe laser fields [6], in which case the Doppler effect for both transitions vanishes. The detection signal is then the intensity I_F of the fluorescence light emitted from level c (or from excited levels populated from c), which is proportional to the population of level c, and also to the absorption of the probe field ω. A clear splitting of the curve giving I_F versus ω, for a fixed resonant value of ω_L, was observed, giving evidence for an Autler-Townes splitting of the optical line $b \longleftrightarrow c$. It was also checked that the frequency splitting between the two components of the doublet was proportional to $\sqrt{I_L}$, where I_L is the intensity of the driving field ω_L, as expected for a Rabi frequency which is proportional to the laser electric field.

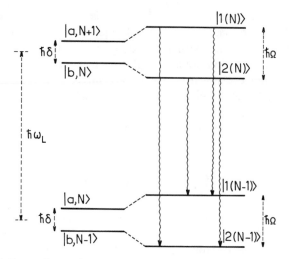

FIGURE 11.4. Dressed-atom interpretation of the Mollow fluorescence triplet. The left part of the figure represents two adjacent manifolds of uncoupled states; the right part, the corresponding Autler-Townes doublets of dressed states. The wavy arrows give the allowed transitions between dressed states whose frequencies give the centers of the various components of the fluorescence spectrum.

A few experiments have also been performed with atomic vapors [7]. The Doppler width of optical lines is generally much larger than the natural width of the atomic states. As a consequence of the large Doppler shifts which modify ω_L and ω for a moving atom, the Autler-Townes splitting of the $b \longleftrightarrow c$ line generally washes out when the average is taken over atomic velocities. It may be shown, however, that narrow Doppler-free structures can survive, under certain circumstances, the washing out of the Autler-Townes effect due to Doppler averaging, and give rise to an observable Autler-Townes splitting of the $b \longleftrightarrow c$ line [8]. For example, for the cascade configuration of Fig. 11.1, the driving field ω_L and the probe field must propagate in opposite directions. Simple graphic constructions, using energy-level diagrams analogous to the one of Fig. 11.3, have been proposed for interpreting these narrow Doppler-free structures [9]. They lead to the idea that the Doppler effect can be compensated by velocity-dependent light shifts [10]. Experiments have been performed to demonstrate this idea [11].

11.3.2   Single Optical Transition—The Mollow Triplet

Suppose now that the probe field ω is switched off and that we have a single laser beam with frequency ω_L driving the $a \longleftrightarrow b$ transition. For detecting the perturbation of the $a \longleftrightarrow b$ transition by the field ω_L, we use the fluorescence light spontaneously emitted by the atom from the excited state b. More precisely, we consider the spectral distribution of this fluorescence light. Using optical Bloch equations, Mollow [12] has calculated the correlation function of the atomic dipole moment and shown that, at high intensity or large detuning ($|\delta|$ or $\Omega_1 \gg$ natural width Γ), the fluorescence spectrum consists of a triplet. Several groups have observed experimentally such a triplet [13].

We will not discuss here the various features of the Mollow fluorescence triplet. We just want to point out that it can be related, as the Autler-Townes effect, to the existence of doublets of dressed states [14]. The left part of Fig. 11.4 represents two adjacent manifolds of uncoupled states, analogous to the one of Fig. 11.2: $\{|a, N + 1\rangle, |b, N\rangle\}$ and $\{|a, N\rangle, |b, N - 1\rangle\}$. Since $|a, N + 1\rangle$ and $|a, N\rangle$ differ by one laser photon, as well as $|b, N\rangle$ and $|b, N - 1\rangle$, the distance between the two manifolds is $\hbar\omega_L$. When the coupling V_{AL} is taken into account, one gets the two doublets of dressed states $\{|1(N)\rangle, |2(N)\rangle\}$ and $\{|1(N - 1)\rangle, |2(N - 1)\rangle\}$ represented in the right part of Fig. 11.4. In each doublet the splitting is $\hbar\Omega$, and the distance between the two doublets is $\hbar\omega_L$. The allowed spontaneous transitions between dressed states correspond to pairs of levels between which the atomic dipole moment operator **d** has a nonzero matrix element. In the uncoupled basis, **d**, which cannot change the quantum number N, couples only $|b, N\rangle$ and $|a, N\rangle$. The two dressed states $|1(N)\rangle$ and $|2(N)\rangle$ are both "contaminated" by $|b, N\rangle$. Similarly, the two dressed states $|1(N - 1)\rangle$ and $|b(N - 1)\rangle$ are both contaminated by $|a, N\rangle$. It follows that the four transitions connecting $|1(N)\rangle$ and $|2(N)\rangle$ to $|1(N - 1)\rangle$ and $|2(N - 1)\rangle$ are allowed (wavy lines of Fig. 11.4). One immediately understands in this way why the fluorescence spectrum consists of three lines with frequencies $\omega_L + \Omega$, $\omega_L - \Omega$ and ω_L associated with the transitions $|1(N)\rangle \longrightarrow |2(N - 1)\rangle$, $|2(N)\rangle \longrightarrow |1(N - 1)\rangle$ and $|i(N)\rangle \longrightarrow |i(N - 1)\rangle$ (with $i = 1$, 2), respectively. We suppose here that Ω is large compared to the natural width Γ, so that the three lines are well resolved. According to (4), such a condition is equivalent to $\Omega_1 \gg \Gamma$ or $|\delta| \gg \Gamma$.

11.4 The Autler-Townes Effect in Cavity Quantum Electrodynamics

In the previous discussion, the number N of photons is not a well-defined quantity. The field is quantized in a fictitious box, having a volume V which can be arbitrarily large. Only the energy density N/V at the position of the atom is relevant to the experiment being analyzed.

During the last few years, spectacular progress has been made allowing one to study the behavior of atoms put in resonant cavities having a very high finesse. In such (real) cavities, the number N of photons has a definite meaning. For example, the cavity can be empty ($N = 0$) and one can then study how the spontaneous emission rates and the Lamb shift are modified by the boundary conditions imposed by the cavity walls. By introducing excited atoms in the cavity, one can produce a maser or a laser action with a very small number of atoms and a very small number of photons. For a survey of the field, we refer the reader to recent reviews [15]. We will focus here on an effect usually called "vacuum Rabi splitting" in cavity

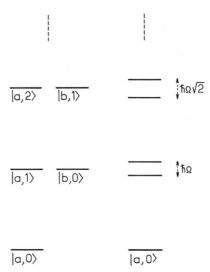

FIGURE 11.5. Uncoupled states (left part) and coupled or dressed states (right part) of the atom-cavity mode system.

quantum electrodynamics and which is in fact quite similar to the Autler-Townes effect. We will use the so-called Jaynes-Cummings model [16]. This model is quite similar to the dressed-atom model used in the previous sections, except that the two-level atom is now coupled to a single mode of an actual cavity, so that N can take very small values, and it is no longer possible to neglect the variations with N of the Rabi frequency.

The left part of Fig. 11.5 represents a few uncoupled states of the system formed by a two-level atom $a - b$ with frequency ω_0 and a single mode field of the cavity with frequency ω_L. The state $|a, 0\rangle$ (atom in the lower state a with 0 photon in the cavity) is nondegenerate and has the lowest energy. We suppose that the cavity is resonant ($\omega_L = \omega_0$), so that the two states $|a, 1\rangle$ and $|b, 0\rangle$ are degenerate, as well as $|a, 2\rangle$ and $|b, 1\rangle$. When the atom-field coupling V_{AL} is taken into account, one finds that level $|a, 0\rangle$ is not perturbed (in the rotating wave approximation). Levels $|a, 1\rangle$ and $|b, 0\rangle$ are coupled by V_{AL}, as well as $|a, 2\rangle$ and $|b, 1\rangle$. One thus gets the series of doublets of dressed states represented in the right part of Fig. 11.5. Since $\langle a, N | V_{AL} | b, N - 1 \rangle$ varies as $\sqrt{N}$, the splittings of the first two doublets are equal to $\hbar\Omega$ and $\hbar\Omega\sqrt{2}$, respectively. The frequency Ω can be considered as the Rabi frequency corresponding to the field associated with a single photon in the cavity. We suppose here that we are in a strong coupling regime: the damping rate of the field in the cavity and the damping rate of the atom are slow enough compared to Ω that the two dressed states of each doublet are well resolved.

Suppose now that a very weak probe field with frequency ω close to ω_L is sent through the cavity, and that one measures the transmission $T(\omega)$ of the cavity versus ω. Initially, no photon is present in the cavity which contains a single atom in the lower state a, so that the total system is in the state $|a, 0\rangle$. The probe field can enter into the cavity only if its frequency ω coincides with an eigenfrequency of the cavity. The frequencies of the new modes of the cavity in the presence of the atom-field coupling are the frequencies of the two transitions connecting $|a, 0\rangle$ to the first two dressed states of Fig. 11.5. One thus expects the transmission spectrum $T(\omega)$ to exhibit two peaks at $\omega = \omega_L \pm (\Omega/2)$. In particular, if the two peaks are well resolved, $T(\omega)$ can become negligible in the middle, i.e., when $\omega = \omega_L$. This means that a resonant probe field $\omega = \omega_L$ which can enter into the empty cavity is reflected from the cavity if the cavity contains a single atom. All the previous results can be easily extended to the case where the cavity contains n identical atoms instead of 1. If the coupling is symmetric, i.e., if all atoms are identically coupled to the field, one can show that the first doublet above the ground state has a splitting $\hbar\Omega\sqrt{n}$ instead of $\hbar\Omega$.

Recent experiments [17] have reached the strong coupling regime and have allowed the observation of a doublet in the transmission spectrum of a probe field in conditions where the cavity contains a very small number of atoms, down to 1. This shows that the physical mechanism responsible for the Autler-Townes effect (transitions involving doublets of dressed states) can give rise to observable phenomena even in the limit of a single atom coupled to a single photon.

11.5 Doublets of Dressed States with a Position-dependent Rabi Frequency

We come back to an atom in free space, which we suppose to interact with a laser electric field varying in space. The Rabi frequency $\Omega_1(\mathbf{r})$ characterizing the atom-field coupling for an atom at $\mathbf{r}$ thus depends on $\mathbf{r}$. By considering, as in the previous sections, doublets of dressed states, which now have position-dependent splittings and position-dependent widths, we want to show in this section that it is possible to get physical insights into important features of the radiative forces which govern atomic motion in laser light. This will be another example of the close connections which exist between the Autler-Townes effect and new research fields, such as laser cooling and trapping.

11.5.1 Gradient (or Dipole) Forces

The splitting $\hbar\Omega$ between the two dressed states $|1(N)\rangle$ and $|2(N)\rangle$ of Fig. 11.2 increases when the Rabi frequency Ω_1 increases (see Eq. (11.4)). Figure 11.6 gives the variations of the energies of these two dressed states when the position of the atom is varied across a laser beam with a finite waist. Inside the laser beam, Ω_1 is large and the splitting between the dressed states is large. Outside the laser beam, this splitting tends to a constant value $\hbar\delta$, equal to the energy separation between the uncoupled states $|a, N+1\rangle$ and $|b, N\rangle$, and the dressed states tend to the uncoupled states. For a positive detuning $(\omega_L > \omega_0)$, $|a, N+1\rangle$ is above $|b, N\rangle$, so that $|1(N)\rangle$ and $|2(N)\rangle$ tend to $|a, N+1\rangle$ and $|b, N\rangle$, respectively, outside

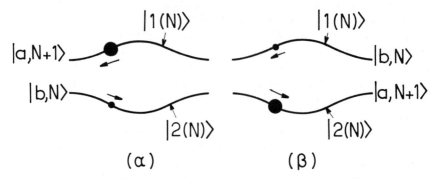

FIGURE 11.6. Energies of the two dressed states $|1(N)\rangle$ and $|2(N)\rangle$ versus the position of the atom across a laser beam. The splitting between the dressed states is maximum in the center of the laser beam. Outside the laser beam the dressed states tend to the uncoupled states $|a, N + 1\rangle$ and $|b, 1\rangle$. Figures 6α and 6β correspond to $\delta > 0$ $(\omega_L > \omega_0)$ and $\delta < 0$ $(\omega_L < \omega_0)$, respectively. The size of each filled circle is proportional to the probability of occupation of the corresponding dressed state.

the laser beam (Fig. 11.6α). For $\omega_L < \omega_0$, the previous conclusions are reversed (Fig. 11.6β).

Consider now an atom, initially at rest, at a certain position in the laser beam and suppose that it is in a certain dressed state. Because the dressed state energy is position-dependent, the atom experiences a gradient force whose direction is indicated by the arrows of Fig. 11.6. If the atom is in the other dressed state, the gradient force has the opposite sign. To get the mean gradient force experienced by the atom, one must average the gradient forces associated with each dressed state using the probability of occupation Π_1 or Π_2 of these states which are indicated by the size of the filled circles of Fig. 11.6. For $\delta > 0$ (Fig. 11.6α), the dressed state $|2(N)\rangle$, which tends to $|b, N\rangle$ outside the laser beam, is more contaminated by the unstable state $|b, N\rangle$ than $|1(N)\rangle$ is. It is therefore less populated than $|1(N)\rangle$, because it has a shorter lifetime: $\Pi_1 > \Pi_2$. The sign of the mean gradient force is thus determined by the sign of the gradient force associated with $|1(N)\rangle$ and this explains why the atom is expelled from the high intensity regions when $\omega_L > \omega_0$. For $\omega_L < \omega_0$ (Fig. 11.6β) these conclusions are reversed and the atom is attracted towards the high intensity regions. This explains how an atom can be trapped in a focal zone of a red-detuned laser beam. For $\omega_L = \omega_0$, the two dressed states contain equal admixtures of $|b, N\rangle$, they are equally populated, and the mean gradient force vanishes.

The previous picture also gives a simple interpretation of the fluctuations of gradient forces. Consider the radiative transitions induced by spontaneous emission between doublets of dressed states (see the wavy lines of Fig. 11.4). They are at the origin of a radiative cascade of the dressed atom falling down its ladder of dressed states. For example, the dressed atom can jump from $|1(N)\rangle$ to $|2(N-1)\rangle$, then from $|2(N-1)\rangle$ to $|1(N-2)\rangle$, and so on. The probabilities of occupation Π_1 and Π_2 of the two dressed states, introduced above, are nothing but the proportions of time spent in dressed states of type 1 and 2 during such a radiative cascade. Every time the dressed atom jumps from a dressed state of type 1 to a dressed state of type 2, or vice versa, the sign of the instantaneous gradient force changes. We thus arrive at the picture of a radiative force oscillating back and forth in a random way between two opposite values. Such an analysis can be made quantitative and provides correct results for the mean value of dipole forces and for the momentum diffusion coefficient associated with their fluctuations [18].

11.5.2 High Intensity Sisyphus Effect

We now consider an atom moving with velocity v in an inhomogeneous laser beam, for example, in an intense laser standing wave, and we shall try to understand the velocity dependence of the mean force. Velocity damping forces are of course important for laser cooling. Here also, we will consider doublets of dressed states. We will put the emphasis on the correlations which exist, in a standing wave, between the modulations of the dressed-state energies and the modulations of the spontaneous departure rates from these dressed states.

The Rabi frequency associated with a laser standing wave along the z-axis can be written

$$\Omega_1(z) = \Omega_1 \cos kz . \tag{11.5}$$

Its variation with z is represented in Fig. 11.7α. The modulus of $\Omega_1(z)$ is maximum at the antinodes A, at $z = 0$, $\lambda/2$, λ..., where λ is the laser wavelength, and vanishes at the nodes N, at $z = \lambda/4$, $3\lambda/4$... The dashed lines of Fig. 11.7β give the energies of the uncoupled states $|a, N+1\rangle$ and $|b, N\rangle$, which are independent of z. We suppose here $\delta > 0$ ($\omega_L > \omega_0$), so that $|a, N+1\rangle$ is above $|b, N\rangle$. The splitting between the two dressed states $|1(N)\rangle$ and $|2(N)\rangle$, represented by the full lines of Fig. 11.7β, is $\hbar\Omega(z)$, where, according to (4) and (5)

$$\Omega(z) = \left[\delta^2 + \Omega_1^2 \cos^2 kz\right]^{1/2} . \tag{11.6}$$

This splitting is an oscillating function of z. It is maximum at the antinodes A, and minimum and equal to $\hbar\delta$ at the nodes N. Each dressed state is also represented in Fig. 11.7β with a thickness proportional to its radiative width, i.e., to the departure rate from this state due to spontaneous emission. At the nodes N, $|1(N)\rangle$ reduces to $|a, N+1\rangle$ which is radiatively stable and has no width (we suppose that a is the atomic ground state), whereas $|2(N)\rangle$ reduces to $|b, N\rangle$ which has a width Γ equal to the natural width of the atomic excited state b. When one goes from a node N to an antinode A, the contamination of $|1(N)\rangle$ by $|b, N\rangle$ increases progressively, so that the width of $|1(N)\rangle$ increases from zero to a maximum value at the antinode. Conversely, the admixture of $|b, N\rangle$ in $|2(N)\rangle$ decreases,

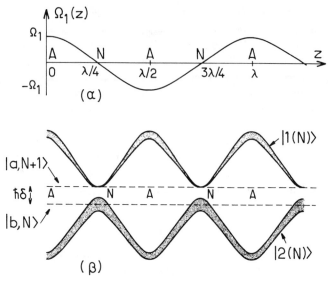

FIGURE 11.7. α: Variations of the Rabi frequency $\Omega_1(z)$ along the axis of a standing wave with wavelength λ. The nodes and antinodes are labelled by N and A, respectively. β: Variations with z of the energies of the uncoupled states $|a, N + 1\rangle$ and $|b, N\rangle$ (dashed lines) and of the dressed states $|1(N)\rangle$ and $|2(N)\rangle$ (full lines). The thickness of the full lines is proportional to the radiative width of the corresponding dressed state. We suppose that $\delta > 0\,(\omega_L > \omega_0)$, so that $|a, N + 1\rangle$ is above $|b, N\rangle$.

so that the width of $|2(N)\rangle$ decreases from Γ at the node to a minimum value at the antinode. The radiative widths of the dressed states are therefore spatially modulated as their energies are. The important point here is the correlation which exists between these two modulations and which clearly appears in Fig. 11.7β. When one moves along each dressed state, one sees a succession of potential hills, centered at A for $|1(N)\rangle$, and N for $|2(N)\rangle$, and potential valleys, centered at N for $|1(N)\rangle$, and A for $|2(N)\rangle$. In either case, the width of this dressed state is maximum at the tops of the potential hills.

Suppose now that the atom is moving along the standing wave with a velocity v such that it can travel over a distance of the order of λ in a time Γ^{-1}. The atom can thus remain in a dressed state and go from a node to an antinode, or vice versa, before leaving this dressed state by spontaneous emission. Figure 11.8 represents a few steps of the radiative cascade of the moving atom. The full lines represent the dressed state occupied by the atom between two successive spontaneous emission processes represented by the wavy lines. According to the previous discussion, the atom has the largest chance to leave a dressed state, be it of type 1 or 2, at the top of a hill. If the atom jumps to the other type of dressed state (from 1 to 2, or from 2 to 1), it arrives in the bottom of a valley from where it must climb to the top of a hill of this new dressed state before leaving it, and so on. We therefore have an atomic realization of the Sisyphus myth, because the atom is running up the potential hills more frequently than down. Such an argument can be made more

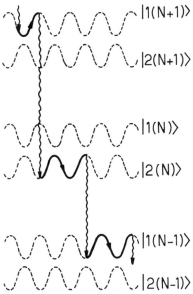

$|1(N+1)\rangle$

$|2(N+1)\rangle$

$|1(N)\rangle$

$|2(N)\rangle$

$|1(N-1)\rangle$

$|2(N-1)\rangle$

FIGURE 11.8. Radiative cascade for an atom moving along a blue detuned ($\omega_L > \omega_0$) intense laser standing wave. The full lines represent the dressed state occupied by the atom between two successive spontaneous transitions (wavy lines). Because of the correlations which exist between the modulations of the energies of the dressed states and the departure rates from the dressed states, the dressed atom is running up potential hills more frequently than down, as did Sisyphus in Greek mythology.

quantitative [18], and it can explain physically all the features of the velocity dependence of the radiative forces in a high-intensity laser standing wave, a problem which has been extensively studied by several authors [19]. Note that we obtain here a velocity damping force for $\omega_L > \omega_0$, whereas the usual Doppler cooling force in a weak standing wave [20] appears for $\omega_L < \omega_0$. Experimental demonstration of the change of sign of the velocity-dependent force at high intensity has been achieved [21]. Evidence has also been obtained for a channeling of atoms in the nodes of a blue-detuned standing wave [22].

Note also that similar effects occur for an atom having several Zeeman sublevels in the ground state and moving in a low-intensity laser configuration exhibiting a spatially modulated polarization. The light shifts of two ground-state sublevels can then be spatially modulated as well as the optical pumping rates from one Zeeman sublevel to the other, the two modulations being correlated in such a way that they give rise to a low-intensity Sisyphus effect [23]. Such a laser cooling mechanism turns out to be two orders of magnitude more efficient than the usual Doppler cooling mechanism (see the review paper [24] and references therein).

Although other physical problems may be analyzed along the same lines, I will stop here, with the hope that I have given an idea of the generality of the basic ingredients of the Autler-Townes effect.

References

[1] S. H. Autler and C. H. Townes, Phys. Rev. **100**, 703 (1955). See also C. H. Townes and A. L. Schawlow, *Microwave Spectroscopy* (Dover, New York, 1975), ¤ 10-9.

[2] B. R. Mollow, Phys. Rev. A **12**, 1919 (1975).

[3] C. Cohen-Tannoudji and S. Haroche, in *Polarisation, Matière et Rayonnement, Livre jubilaire en l'honneur d'Alfred Kastler*, edited by Société Française de Physique (Presses Universitaires de France, Paris, 1969), p. 191.

[4] C. Cohen-Tannoudji, J. Dupont-Roc, and G. Grynberg, *Atom-Photon Interactions - Basic Processes and Applications* (Wiley, New-York, 1992), chapter VI.

[5] C. Cohen-Tannoudji, C. R. Acad. Sci. **252**, 394 (1961). See also C. Cohen-Tannoudji and J. Dupont-Roc, Phys. Rev. A **5**, 968 (1972).

[6] J.L. Picqué and J. Pinard, J. Phys. B **9**, L77 (1976). J. E. Bjorkholm and P.F. Liao, Opt. Commun. **21**, 132 (1977).

[7] A. Shabert, R. Keil, and P. E. Toschek, Appl. Phys. **6**, 181 (1975). C. Delsart and J.C. Keller, J. Phys. B **9**, 2769 (1976). P. Cahuzac and R. Vetter, Phys. Rev. A **14**, 270 (1976).

[8] T. W. Hänsch and P. E. Toschek, Z. Phys. **236**, 213 (1970). B.J. Feldman and M.S. Feld, Phys. Rev. A **1**, 1375 (1970). I.M. Beterov and V.P. Chebotayev, *Progress in Quantum Electronics*, Vol. 3, ed. by J. H. Sanders and S. Stenholm (Pergamon Press, 1974). S. Feneuille and M. G. Schweighofer, J. de Physique **36**, 781 (1975).

[9] C. Cohen-Tannoudji, Metrologia (Springer Verlag) **13**, 161 (1977).

[10] C. Cohen-Tannoudji, F. Hoffbeck, S. Reynaud, Opt. Commun. **27**, 71 (1978).

[11] S. Reynaud, M. Himbert, J. Dupont-Roc, H. Stroke, and C. Cohen-Tannoudji, Phys. Rev. Lett. **42**, 756 (1979).

[12] B.R. Mollow, Phys. Rev. **188**, 1969 (1969).

[13] F. Schuda, C. R. Stroud, and M. Hercher, J. Phys. B **7**, L198 (1974). F. Y. Wu, R. E. Grove and S. Ezekiel, Phys. Rev. Lett. **35**, 1426 (1975). W. Hartig, W. Rasmussen, R. Schieder and H. Walther, Z. Phys. A **278**, 205 (1976).

[14] C. Cohen-Tannoudji, in "Laser Spectroscopy," p. 324, ed. by S. Haroche, J. C. Pébay-Peyroula, T. W. Hänsch, and S. E. Harris (Springer-Verlag, Berlin, 1975).

[15] S. Haroche, in *Fundamental Systems in Quantum Optics, Les Houches Summer School Session 53*, ed. by J. Dalibard, J. M. Raimond and J. Zinn-Justin (North-Holland, Amsterdam, 1992). See also the various articles published in *Cavity Q.E.D., Advances in Atomic, Molecular and Optical Physics*, Suppl. 2, ed. by P. Berman (Academic Press, New-York, 1994).

[16] E.T. Jaynes and F. W. Cummings, Proc. I.E.E.E. **51**, 89 (1963).

[17] Y. Zhu, D. J. Gauthier, S. E. Morin, Q. Wu, H. J. Carmichael, and T. W. Mossberg, Phys. Rev. Lett. **64**, 2499 (1990). F. Bernardot, P. Nussenzveig, M. Brune, J. M. Raimond, and S. Haroche, Europhys. Lett. **17**, 33 (1992). R. J. Thomson, G. Rempe, and H. J. Kimble, Phys. Rev. Lett. **68**, 1132 (1992).

[18] J. Dalibard, C. Cohen-Tannoudji, J.O.S.A. **B2**, 1707 (1985).

[19] See for example A. P. Kazantsev, Sov. Phys. J.E.T.P. **39**, 784 (1974).

[20] T.W. Hänsch and A. L. Schawlow, Opt. Commun. **13**, 68 (1975).

[21] A. Aspect, J. Dalibard, A. Heidmann, C. Salomon, and C. Cohen-Tannoudji, Phys. Rev. Lett. **57**, 1688 (1986).

[22] C. Salomon, J. Dalibard, A. Aspect, H. Metcalf, and C. Cohen-Tannoudji, Phys. Rev. Lett. **59**, 1659 (1987).

[23] J. Dalibard and C. Cohen-Tannoudji, J.O.S.A. **B6**, 2023 (1989).

[24] C. Cohen-Tannoudji and W. D. Phillips, Physics Today **43**, 33 (1990).

12

Parity Nonconservation in Atoms and Searches for Permanent Electric Dipole Moments

Eugene D. Commins

12.1 Introduction

In this paper, I would like to discuss briefly two separate but related subjects: parity nonconservation in atoms, and the search for permanent electric dipole moments. Both are of considerable interest in elementary particle physics, but in each case the methods of investigation are those of low-energy physics: optical pumping, spectroscopy, atomic beams, and magnetic resonance. The laser, that wonderful invention of Charles Townes and others, plays a crucial role in most of the experiments to be described. Those who work in this field therefore owe Townes a considerable debt, as do so many other researchers in diverse branches of science.

12.2 Parity Nonconservation in Atoms

12.2.1 General Background

Although the possibility that parity nonconservation (PNC) might occur in atoms was first suggested by Zel'dovitch [1] in 1959, the first detailed analysis of the phenomenon was presented by Bouchiat and Bouchiat [2] shortly after the discovery of weak neutral currents. The first observation of PNC, by optical rotation in bismuth, was announced by Barkov and Zolotorev [3] at Novosibirsk in 1978; almost simultaneously our group reported the first observation of PNC by Stark interference in thallium [4]. Subsequently, additional optical rotation measurements were done in bismuth [5-7] and lead [8], and more refined Stark interference experiments were carried out with cesium [9, 10] and thallium [11].

By the early 1980's reasonable qualitative agreement between experimental results and the predictions of the standard model had already been established. However, PNC experiments are still actively pursued to make *quantitative* tests of the standard model where radiative corrections are important. One now seeks to discover new physics beyond the standard model.

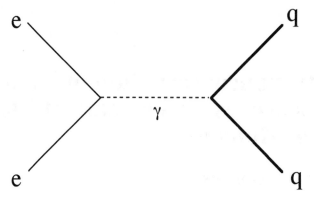

FIGURE 12.1. Interaction of an electron and a quark by single photon exchange.

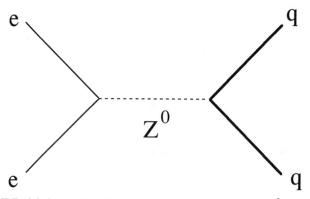

FIGURE 12.2. Interaction of an electron and a quark by single Z^0 exchange.

We begin a brief analysis of the phenomenon with something very familiar: the ordinary Coulomb interaction between an atomic electron and the nucleus (see Fig. 12.1). The Coulomb Hamiltonian H_0 is of course inversion-symmetric, and its eigenstates have definite parity: even or odd. However, we have known for two decades or so that there is another way to couple an atomic electron to the nucleus: by exchange of a massive neutral intermediate vector boson Z^0 (see Fig. 12.2). This yields an additional contribution H' to the atomic Hamiltonian. H' has three important characteristics. First, it is weak, (proportional to the Fermi weak coupling constant G_F), hence very feeble compared to H_0. Second, because the Z^0 is very massive (about 100 times as massive as the proton), H' is to a very good approximation of zero range. This means that Z^0 exchange can only occur when the atomic electron is at the nucleus. And third, H' essentially *violates parity*. Since the electronic and nucleonic neutral weak currents each contain vector and axial vector components:

$$J_e = V_e + A_e, \quad J_N = V_N + A_N$$

H' contains both scalar and pseudoscalar parts:

$$
\begin{aligned}
H' & \approx (V_e + A_e) \cdot (V_N + A_N) \\
& = (V_e V_N + A_e A_N) + (A_e V_N + V_e A_N) \quad\quad (12.1) \\
& = H_s + H_P.
\end{aligned}
$$

The scalar arises from the $V_e V_N$, $A_e A_N$ terms; its contribution is negligible and may be ignored. However, the pseudoscalar H_P is important. When it is combined with the Coulomb Hamiltonian H_0, we obtain a total Hamiltonian $H = H_0 + H'$ that is not inversion symmetric, and this has significant observable consequences.

Let us first examine the $A_e V_N$ term in H_P. To simplify matters we shall take the non-relativistic limit for the nucleons and the electron. (This approximation is always valid for the nucleons. For the electron it is not necessarily so; but to compensate we can add a relativistic correction later.) Then according to the standard electroweak model of Glashow, Weinberg, and Salam, it can be shown that we obtain for the $A_e V_N$ portion the following effective Hamiltonian:

$$
H_1 = \frac{G_F}{2\sqrt{2}} \frac{\boldsymbol{\sigma} \cdot \mathbf{p}}{m_e} Q_W \delta^3(\mathbf{r}). \quad\quad (12.2)
$$

Here $\boldsymbol{\sigma}$ and $\mathbf{p}$ refer to the electron spin and momentum, respectively; $\boldsymbol{\sigma} \cdot \mathbf{p}$ is the simplest time-reversal-invariant pseudoscalar we can form from the electron's variables. Also,

$$
Q_W = -N + Z(1 - 4\sin^2 \theta_W) \quad\quad (12.3)
$$

is the so-called "weak charge", where Z and N are the atomic number and neutron number of the nucleus, respectively; and ϑ_W is the Weinberg (weak mixing) angle. The presence of Q_W in (12.2) reflects the fact that in H_1 the nucleons *add coherently*. Finally the delta function in the electron position $\mathbf{r}$ expresses the zero-range nature of Z^0 exchange.

The small but significant changes induced in atoms by Z^0 exchange may be treated by using first order perturbation theory. Thus we need to employ matrix elements of H_1 between eigenstates of H_0. Such matrix elements are non-zero only for electron orbitals with finite value and/or gradient at the nucleus; hence in almost all practical circumstances the pseudoscalar H_1 connects only $s_{1/2}$ and $p_{1/2}$ orbitals. These matrix elements are approximately proportional to Z^3 (since a factor $\approx Z$ arises from each of the contributions: Q_W, $\boldsymbol{\sigma} \cdot \mathbf{p}$, and the **delta** function). This of course means that atomic parity nonconservation (PNC) effects are most pronounced in heavy atoms, (where relativistic corrections are important and contribute an additional increase with Z). This is why experiments have been carried out with cesium, thallium, lead, and bismuth ($Z = 55, 81, 82, 83$ respectively), and why other experiments are contemplated or in progress with Ba^+, Dy, Yb, Fr, and Ra^+ ($Z = 56, 66, 70, 87$, and 88 respectively).

We turn next to the $A_N V_e$ term in (12.2); the corresponding effective Hamiltonian is called H_2. In the non-relativistic limit for the nucleons, A_N yields a contribution proportional to the nucleon spin. Now, we know that in an even-odd nucleus, the

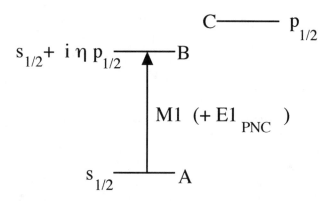

FIGURE 12.3. Energy levels of a hypothetical atom. Mixing of states B, C, by H' causes the AB transition amplitude, nominally M1, to acquire an electric dipole component E1_{PNC}.

spins cancel in pairs, leaving a residual spin due to the last odd nucleon. Hence, while in H_1 the nucleons add coherently, only one nucleon contributes to H_2 in every practical case. Furthermore the vector neutral weak current V_e is proportional to the small quantity $(1 - 4\sin^2\vartheta_W)$. The net result is that matrix elements of H_2 are extremely small compared to those of H_1.

In fact, it turns out that an entirely different effect has exactly the same signature as H_2 and is somewhat larger in magnitude. We refer here to parity nonconservation within the nucleus itself, which comes about because the nucleons experience weak forces (charged *and* neutral), as well as strong and electromagnetic forces. Consider for a moment a nucleus with a spin and magnetic moment. We can associate with this magnetic moment a vector potential, which is ordinarily considered to be a polar vector. However, because of parity violation in nuclear forces, the vector potential acquires an axial component as well; (the so-called nuclear "anapole moment"). When we couple the anapole moment to the electron electromagnetic current in the usual way ("$\mathbf{j} \cdot \mathbf{A}$"), we obtain a contribution that looks just like that of H_2 and must be included with it.

Atomic PNC can also arise in principle from Z^0 exchange between atomic electrons, but the effects are negligibly small in magnitude, at least for present day experiments or those contemplated in the foreseeable future.

12.2.2   PNC Experiments

In the typical experiment, one observes an optical transition between two atomic levels of the same nominal parity (almost invariably an M1 transition). Fig. 12.3 illustrates this for a simplified hypothetical atom, where the two levels of interest, A and B, are both nominally $s_{1/2}$. Suppose that a third level C, nominally $p_{1/2}$, lies close to B. Then state B acquires a small admixture of $p_{1/2}$:

$$\psi_B \Rightarrow s_{1/2} + i\eta p_{1/2} \tag{12.4}$$

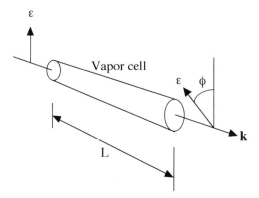

FIGURE 12.4. Schematic diagram of an optical rotation experiment, showing how the plane of polarization of a laser beam is rotated on passage through a vapor cell, when the frequency is tuned near resonance.

where

$$i\eta = \frac{\langle B \,|H'| \,C\rangle}{E_B - E_C}. \tag{12.5}$$

Thus the transition amplitude, nominally M1, acquires in addition a parity nonconserving electric dipole component E1_{PNC}. Interference between M1 and E1_{PNC} gives rise to observable effects such as circular dichroism or optical rotation. In some circumstances one applies an external electric field, which causes B and C to be mixed by Stark effect. The transition amplitude then acquires an E1 amplitude due to Stark effect, in addition to M1 and E1_{PNC}. Interference between E1_{PNC} and E1_{Stark} can then be detected. It is important to emphasize that the magnitude of η depends not only on the matrix element $\langle B \,|H|' \,C\rangle$ but also the energy denominator. If one can find a set of levels where the energy denominator is very small, then $|\eta|$ may be substantial even if the matrix element is also very small. Such a situation occurs in the $2^2S_{1/2}$ and $2^2P_{1/2}$ states of hydrogen, which are separated only by the Lamb shift, 1058 MHz. This motivated several attempts to observe PNC in $n = 2$ hydrogen [12], but unfortunately they failed for a variety of reasons.

Fig. 12.4 is a schematic illustration of a typical optical rotation experiment. Here, one transmits a linearly polarized laser beam through a cell of length L containing the vapor of interest. When the laser beam frequency is tuned close to the transition resonance, one observes a rotation in the plane of polarization. This occurs because the linear polarization can be thought of as a superposition of right and left circular polarization states, and these have slightly different indices of refraction because of the simultaneous presence of M1 and E1_{PNC} amplitudes. At present, the most significant optical rotation results have been achieved using the 1.28 μ transition in ^{205}Tl, (see Fig. 12.5), at Seattle [13], and independently at Oxford [14]. This transition is nominally allowed M1; but the $6^2P_{1/2}$ state is admixed with $7S, 8S,...$ states by H'. The experimental precision achieved by the Seattle group in measurement of the rotation angle ($\approx 10^{-7} rad$) is about 1%, and

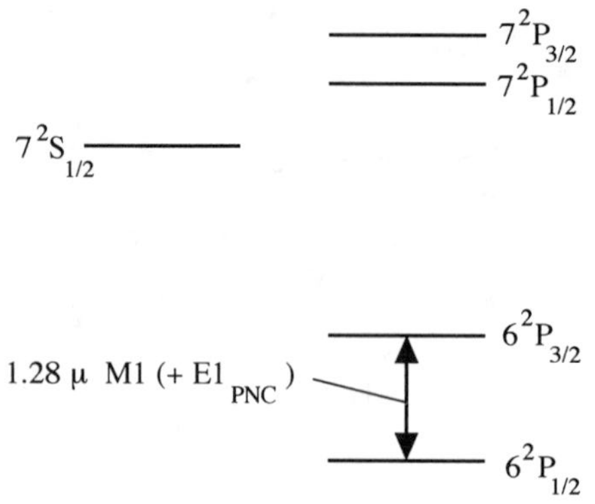

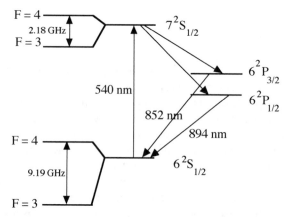

FIGURE 12.5. Low-lying energy levels of atomic thallium. Recent optical rotation experiments on the 1.28 μ transition in thallium at Seattle and Oxford have achieved quite precise results.

FIGURE 12.6. Low-lying energy levels of atomic cesium. The forbidden M1 transition $6S \rightarrow 7S$ (540 nm) has been used in Stark interference experiments at Boulder and at Paris to study PNC.

theoretical calculations [15] of the atomic matrix element of H', which require relativistic many-body techniques of great sophistication, have an uncertainty $\approx$ 3%. The latter measurements yield a determination of Q_W (see eqs. (12.2), (12.3)):

$$Q_W \left({}^{205}_{81}\text{Tl, opt rot} \right) = -114.2 \pm 3.8. \qquad (12.6)$$

At present, the case of greatest interest for Stark interference is the forbidden M1 transition $6S \rightarrow 7S$ in cesium (540 nm) because of precise experimental results from Boulder [16] and highly refined theoretical analysis of cesium atomic wave functions [17]. The relevant cesium levels are illustrated in Fig. 12.6. Because of PNC, the $6S, 7S$ states in Cs are mixed with $P_{1/2}$ states. In addition, when an external electric field is imposed, there is Stark mixing of $6S, 7S$ with neighboring

$P_{1/2,3/2}$ states. The resulting Stark-induced E1 transition amplitudes for stimulated absorption, when the incoming photon $\hat{\varepsilon}$ polarization is parallel (perpendicular) to the applied field E are called αE, (βE) respectively. One observes interference between βE and E1$_{PNC}$ and thereby determines the ratio Im $\{E1_{PNC}/\beta E\}$ by means of the following experimental arrangement. An intense atomic beam of cesium moves in the z direction. An elliptically polarized optical standing wave is generated with a laser and a Fabry-Perot interferometer with wave vectors $\mathbf{k}$, $-\mathbf{k}$ along y. It can be tuned to individual transitions $6S, F, m_F \to 7S, F', m_{F'}$ which are split by imposition of a magnetic field B_z. An electric field $\mathbf{E}$ is applied in the x direction. The intensity of the signal (fluorescence at 852 and 894 nm accompanying decay of the $7S$ state) exhibits a term arising from Stark-PNC interference that is proportional to the pseudoscalar invariant:

$$h(\mathbf{k} \times \mathbf{E}) \cdot \mathbf{B} \tag{12.7}$$

where h is the helicity of the absorbed laser photon. The results of this experiment are as follows:

$$\begin{aligned} \text{Im}\{E1_{PNC}/\beta\} &= -1.639(50)\,\text{mV/cm} \quad (F = 4 \to F = 3) \tag{12.8} \\ &= -1.513\,(48)\,\text{mV/cm} \quad (F = 3 \to F = 4) \tag{12.9} \\ &= -1.576\,(35)\,\text{mV/cm average.} \tag{12.10} \end{aligned}$$

These values may be compared with the theoretical formulae:

$$E1_{PNC}(theo) = -0.905(9) \left[\left(\frac{Q_W}{-N}\right) + A(F', F)K \right] i\,|e|\,a_0 \cdot 10^{-11} \tag{12.11}$$

$$\beta(\text{theo}) = 27.00(20)a_0^3 \tag{12.12}$$

where in Eq. (12.11) $A(F', F)$ are certain numerical coefficients and K is a factor expressing the effects of H_2 and the anapole moment. All of this leads to the following conclusions. From the difference between (12.8) and (12.9), one obtains a result for the anapole moment (albeit with very large uncertainty) that is consistent with theoretical estimates. From the average (12.10) one obtains:

$$Q_W \left(^{133}\text{Cs, Stark interf.}\right) = -71.04\,(1.58)\,(0.88) \tag{12.13}$$

where the first uncertainty is experimental and the second is associated with atomic calculations.

Let us consider how result Eq. (12.13) may be compared with predictions based on the standard model, with radiative corrections included. The main input for such an analysis is the very precise determination of the Z^0 mass at LEP ($M_Z = 91.175 \pm 10.021\text{GeV}/c2$). In addition, the total decay width of Z^0, its partial widths into leptons and into hadrons, the ratio of masses M_W/M_Z and a variety of other measured weak interaction parameters are used in a global analysis [18] to derive the following prediction:

$$Q_W \left(^{133}\text{Cs, theo}\right) = -73.21 \pm 0.085. \tag{12.14}$$

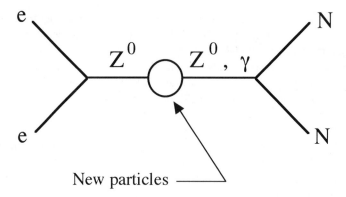

FIGURE 12.7. Single-loop diagrams involving Z exchange and possible new particles can be classified into two types. "S": weak isospin conserving; "T": weak isospin breaking.

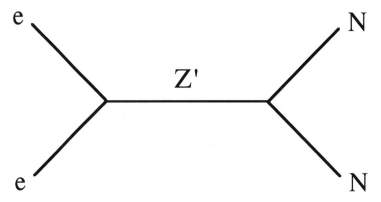

FIGURE 12.8. "Tree" diagram illustrating exchange of additional neutral intermediate vector boson Z'.

There is a slight disagreement between the results of the cesium and thallium PNC experiments on the one hand, and the predictions of the standard model on the other hand. Is this really significant, and if so, what does it mean? Peskin and Takeuchi [19] have analyzed the effect of new physics beyond the standard model. They have shown quite generally that new particles would manifest themselves in lowest order in diagrams of the form illustrated in Fig. 12.7; these may in turn be classified into two categories: "S" (weak isospin conserving) and "T" (weak isospin breaking). In addition tree-like diagrams might occur, with an additional, more massive neutral weak boson or bosons Z' (See Fig. 12.8). Peskin and Takeuchi, and others, have shown that Q_W should be very insensitive to "T" type diagrams, and should depend only on "S" and the new tree-like diagrams.

It is useful to plot various experimental results on a graph, where the axes are labelled by parameters S and T defined by Peskin and Takeuchi and referring to the aforesaid interactions. (See Fig. 12.9). The atomic physics PNC results for cesium and thallium (Eqs. (12.13) and (12.6), respectively) may be shown to imply

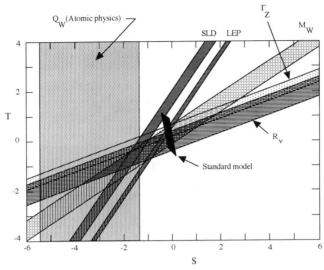

FIGURE 12.9. Allowed regions of the S-T plane from results of LEP, SLC, and atomic PNC experiments. S and T parameters defined in Ref. [19].

a vertical allowed band (independent of T). Also plotted are LEP and SLC results, which depend on both S and T. As can be seen, the present situation is somewhat muddied by the fact that LEP and SLAC results do not agree exactly, (although this 'disagreement' is very minor). The uncertainties in the atomic physics results are still so large that no definite interpretation can yet be made; the one thing we can say for sure is that it would be very helpful to reduce these uncertainties.

The Boulder group is carrying out a sophisticated second-generation experiment, with the hope of reducing the experimental uncertainty in the result (Eq. (12.13)) by a factor of five. However, a corresponding reduction in the theoretical uncertainty of Eq. (12.13) is quite difficult. As for result Eq. (12.6) some improvement in both measurement and theory seems possible. However, it may never be possible to obtain results on a single isotope of a given atom, where the experimental results *and* the atomic theory together are sufficiently precise to make a really critical test.

There is another alternative, however. It may be possible to make PNC measurements on a string of isotopes of a single element. The ratios of PNC results for successive isotopes would be freed from uncertainties in atomic theory, since the atomic wave functions for these isotopes are the same, up to small corrections for nuclear size (and specifically for the neutron spatial distributions in the nuclei). Meanwhile, Q_W depends linearly on neutron number N, and thus in principle it can be extracted from the ratio measurements. Experiments have thus been proposed at Boulder and elsewhere to trap and cool the various isotopes of cesium (all 20 of them radioactive except for ^{133}Cs!) and also the isotopes of francium (all radioactive). At Seattle an experiment has been proposed to trap and cool single ions of the various barium isotopes. Finally, we have found attractive possibilities

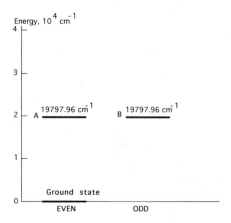

FIGURE 12.10. Partial energy level scheme in atomic dysprosium ($Z = 66$). The ground configuration is $4f^{10}6s^2$. The dominant configurations of A, B are $4f^{10}5d6s$ and $4f^95d^26s$, respectively. The atom actually has a very dense spectrum of bound states, but only three are shown.

in the rare earths dysprosium and ytterbium, and these are being pursued in our laboratory.

Let us describe the latter cases very briefly. In atomic dysprosium (see Fig. 12.10) there is a pair of nearly degenerate excited levels (A, B) of opposite parity and the same total angular momentum $J = 10$ ($E = 19797.96$ cm^{-1} above the ground state, energy separation less than 10^{-2} cm^{-1}). In work done by D. Budker for his Ph. D. thesis, the lifetimes, electric polarizabilities, hyperfine splittings, and isotope shifts of these states have been investigated in detail [20]. Dysprosium has 7 stable isotopes, 4 of which ($A = 161, 162, 163, 164$) have reasonably large natural abundance. For one of them ($A = 163$, nuclear spin $I = 5/2$) it was found by Budker that the hyperfine components $F = 10.5$ of states A and B lie only 3.1 MHz apart! It is useful to compare this to the well-known $2S_{1/2}, 2P_{1/2}$ states in hydrogen, which are virtually degenerate, being separated by the Lamb shift $S = 1058$ MHz, as we mentioned earlier. We see that the $F = 10.5$ states in $A = 163$ dysprosium lie 300 times closer together. By imposing a weak external magnetic field, certain m_F components of A and B may be made to cross, with a minimum effective energy separation determined by the natural lifetime of the shortest lived of the two components. In $n = 2$ hydrogen, the same thing can be done, of course, but there the minimum effective separation is 100 MHz, because of the short lifetime of the 2P state. For dysprosium, the A state lifetime is 5000 times longer than for 2P in hydrogen, and the B state lifetime is much longer still. Hence the minimum effective separation is about 20 kHz. All of this means that the effective energy denominator for PNC mixing of states A, B, at least for these hyperfine components, can be made about 10^8 times smaller than the analogous denominator in cesium. Meanwhile the weak matrix element in dysprosium is about 10^{-3} to 10^{-4} of that in cesium. Consequently, one expects a PNC effect in

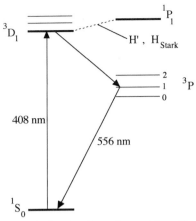

FIGURE 12.11. Low lying energy levels of atomic ytterbium ($Z = 70$). The 408 nm transition is forbidden M1, and can occur only because of spin-orbit mixing and configuration mixing of the ground and $^3D_1$ states with other states. In the presence of an electric field, the transition acquires a Stark-induced E1 amplitude, because of Stark mixing of $^3D_1$ with spin-orbit-induced ^3P component of the nearby $^1P_1$ state. Parity nonconservation induces mixing of $^3D_1$ with the 5d6p component of $^1P_1$. The resulting PNC E1 amplitude for the 408 nm transition is estimated to be more than 100 times as large as that for the 540 nm transition in cesium.

dysprosium as large as 1%; and this is being sought after in an experiment currently underway in our laboratory.

The other rare earth of interest to us is ytterbium ($Z = 70$), which also has 7 stable isotopes ($A = 168, 170, 171, 172, 173, 174, 176$). The level structure of this atom is relatively simple compared to the other lanthanides, because the dominant configuration for various levels of interest consists in each case of a closed 4f shell plus two outer electrons (See Fig. 12.11). Hence, in many ways ytterbium resembles an alkaline earth rather than another lanthanide. David DeMille of our laboratory has shown [21] that with ytterbium a very favorable case exists for a PNC experiment analogous to earlier Stark interference experiments [11, 15]. Here one would employ the forbidden transition

$$\left(6s^2\right)^1 S_0 \rightarrow (5d6s)^3 D_1 \ (408\text{nm}).$$

The $^3D_1$ state is mixed through PNC with the nearby $^1P_1$ state, because the latter is known to have a considerable amount of the configuration $5d6p$, in addition to the dominant configuration $6s6p$. Detailed calculations show that the resulting transition amplitude E1$_{PNC}$ for the 408 nm transition is more than 100 times as large as the analogous amplitude for the 540 nm transition in cesium. In addition, calculations show that in the presence of an external electric field, the Stark amplitude is comparable to that found for the $6P \rightarrow 7P$ transition in thallium [11]. This experiment looks extremely promising, and it is being pursued by us in collaboration with Professor S. Freedman of Berkeley.

TABLE 12.1. Summary of theoretical predictions for the electron EDM, from [22] and references therein.

CP Violation Model	Prediction
Standard model	$|d_e| \leq 10^{-38}$e cm
Super-symmetric models	$|d_e| \leq 3 \cdot 10^{-27}$e cm
Left-right symmetric models	10^{-26} e cm $\geq |d_e| \geq 10^{-28}$e cm
Higgs models	$|d_e| \leq 3 \cdot 10^{-27}$e cm
Lepton flavor-changing models	10^{-26}e cm $\geq |d_e| \geq 10^{-29}$e cm

12.3   Search for Electric Dipole Moments

An elementary particle, nucleus, atom, or molecule cannot possess a truly permanent electric dipole moment (EDM) unless parity (P) and time reversal invariance (T) are both violated. (The so-called "permanent" electric dipole moments of certain polar molecules, familiar to molecular spectroscopists and chemists, are not really permanent at all). Experimental searches for the EDM of the neutron and atomic and molecular EDM's are motivated by the well established fact of CP violation in neutral kaon decay, which is known to be equivalent to T violation. Although this phenomenon was discovered more than 30 years ago and has been investigated in great detail, the basic explanation for it remains unknown, and it is one of the major unsolved problems of modern physics. A variety of theoretical models of CP violation have been proposed, some of which predict finite EDM's for the electron and/or nucleon (see Table 12.3, and [22]). In particular, while in the unadorned standard model the electron EDM is so small that no experiment of the present or foreseeable future could possibly observe it, various extensions of the standard model, and in particular currently popular "grand-unified supersymmetric" models predict values within range of practical experiments.

One searches for an EDM by exposing the neutral system of interest (neutron, atom, molecule) to an external electric field E and looking for an energy shift linear in E. This seems straightforward enough for the free neutron. However, even if the nucleus in a neutral atom were to possess an EDM, it would appear at first that this should not result in such an energy shift, since a neutral atom is not accelerated by an electric field, and thus the average electric field must be zero at the nucleus (complete screening). However, Schiff demonstrated that this conclusion is avoided if we take into account magnetic hyperfine interaction, or more important, the finite size of the nucleus and the possibility that the nuclear EDM spatial distribution is different from its charge distribution [23]. The nucleus

TABLE 12.2. Calculated enhancement factors.

Atom	Atomic No.	State	Enhancement factor
Li	3	$2^2S_{\frac{1}{2}}$	0.004
Na	11	$3^2S_{\frac{1}{2}}$	0.33
K	19	$4^2S_{\frac{1}{2}}$	3.0
Rb	37	$5^2S_{\frac{1}{2}}$	27
Cs	55	$6^2S_{\frac{1}{2}}$	114
Fr	87	$7^2S_{\frac{1}{2}}$	1150
Tl	81	$6^2P_{\frac{1}{2}}$	-585

is then said to possess a "Schiff moment", a quantity with dimensions of charge $\times$ (length)^3.

Screening of the external electric field at a given atomic electron would also appear to occur, since the average force on any such electron in a neutral atom must also vanish. Indeed, if non-relativistic quantum mechanics is assumed, it can be shown that even if the electron were to possess an intrinsic EDM d_e, the atom would exhibit no linear Stark effect to first order in d_e. However, when relativistic effects are taken into account this conclusion is also avoided. In fact, it was first demonstrated by Sandars [24] that in suitable heavy paramagnetic atoms where relativistic effects are large, the ratio of the atomic EDM d_a to d_e (the "enhancement factor") is 10^2 to 10^3 in magnitude (see Table 12.2). Even larger enhancement factors exist in certain paramagnetic molecules [25].

If an atomic or molecular EDM were to exist, it could arise in principle from one or more of the following causes [26]:

a) an intrinsic EDM of the neutron and/or proton;
b) a P,T-odd component of the nucleon-nucleon interaction;
c) an intrinsic EDM of the electron;
d) a P,T odd component of the electron-nucleon interaction.

Although many experiments have been done or are being contemplated, at present the results of just four experiments provide the best limits on these various possibilities. We refer here to two separate experiments on the free neutron, one at St. Petersburg [27] and the other at Grenoble [28]; these are sensitive to case a) above. Next, there is an optical-magnetic resonance experiment at Seattle [29] on the diamagnetic atom ^{199}Hg, which is sensitive mainly to a), b) via the Schiff moment. Finally there is an atomic beam magnetic resonance experiment on ^{205}Tl with laser optical pumping for state selection, performed by our group at Berkeley [30]. It is sensitive mainly to cases c) and d), and provides the best limit on the electron EDM. A summary of the results of these experiments is given in Table 12.3. In Fig. 12.12 we provide a schematic diagram of the Berkeley experiment.

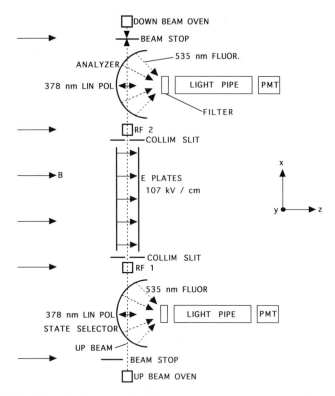

FIGURE 12.12. Schematic diagram of the Berkeley apparatus used to search for the electron electric dipole moment. Two atomic beams of ^{205}Tl are employed, one travelling upward, the other downward. This is done to reduce the possibility of a systematic effect due to precession of the atomic magnetic moment in the motional magnetic field $\mathbf{E} \times \mathbf{v}/c$. State selection and analysis are achieved by optical pumping with 378 nm laser beams travelling in the y direction and polarized linearly along z. The experiment also includes two coherent oscillating fields for magnetic resonance (RF1,2), separated by a space containing an electric field of 107 kV/cm and a length of 1 meter.

I believe that the search for the electron EDM has now entered an extremely interesting stage, because we are now in the range of sensitivity where a rather critical test of supersymmetric grand unified theories may be possible [31]. This has stimulated a number of investigators to propose a variety of new experimental searches for the electron EDM, some of them very imaginative. We ourselves are continuing our experiment, by making a number of basic changes to improve sensitivity, reduce noise, and eliminate the possibility of a class of systematic errors. With these gains we hope in the next several years to reach a level of sensitivity of 1×10^{-28}e cm for the electron EDM.

I hope that this brief review has shown how research in the "old-fashioned" field of atomic physics can still contribute something interesting to problems of current interest in "forefront" elementary particle physics. Of course, hardly any of this

TABLE 12.3. Results of Experimental Searches for Electric Dipole Moments

1. **The free neutron**

$$d_n = (+2.6 \pm 4.2 \pm 1.6) \times 10^{-26} \text{ e cm} \quad \text{(Ref. 27)}$$
$$d_n = (-3 \pm 2 \pm 4) \times 10^{-26} \text{ e cm} \quad \text{(Ref. 28)}$$

2. ^{199}Hg

$$|d(^{199}\text{Hg})| \leq 1.3 \times 10^{-27} \text{ e cm} \quad \text{(Ref. 29)}$$

3. ^{205}Tl

$$d_e = (1.8 \pm 1.2 \pm 1.0) \times 10^{-27} \text{ e cm} \quad \text{(Ref. 30)}$$

would be possible without the laser, an indispensable and miraculous tool invented in large measure by Charles Townes.

12.4   A Brief Personal Note

I had the great good fortune to do my graduate work in the Columbia University Physics Department in the 1950's. It was a truly exceptional place. Its faculty included C. H. Townes, as well as T. D. Lee, C. S. Wu, I. I. Rabi, P. Kusch, H. Foley, N. Kroll, R. Serber, J. Steinberger, L. Lederman, J. Rainwater, and many other luminaries. I was privileged to do some research work for, and become a close friend of, the gifted and imaginative Erwin Hahn (who had recently invented spin echoes, and was then a young adjunct professor at Columbia from IBM's nearby Watson Lab). Later I did my Ph. D. research with Polykarp Kusch and in particular with Bob Novick, (a superb experimentalist, and at the time a young post-doctoral physicist at the Columbia Radiation Lab). I was well-acquainted with numerous outstanding Columbia graduate students including Ali Javan, Bill Bennett, Martin Perl, T. C. Wang, Mel Schwartz, Pat Thaddeus, Eyvind Wichmann, Marcel Bardon, Al Lurio, Herb Lashinsky, and many others. Of course several of those mentioned were in Townes' very active group. The atmosphere was stimulating and lively, graduate students could still read the *Physical Review* and understand most of the articles, and it was still possible to do interesting and important work with modest physical and financial resources. Townes and his group produced their wonderful ammonia maser in 1954; Lamb and Kusch received the Nobel Prize in 1955 for research on the fine structure of atomic hydrogen and the anomalous magnetic moment of the electron; Lee and Yang made their remarkable discovery of parity violation in 1956. My contemporaries and I held Townes in awe for his vast knowledge and scientific abilities. I was even more in awe because on my prelim oral exam, and again on my thesis defense, Townes asked me questions I could not answer. Forty years later I still remember those questions vividly and with some embarrassment!

Charlie Townes came to Berkeley in 1967; I had already arrived there seven years before. Thus it has been my good fortune to be a friend and colleague of Charlie Townes at Berkeley for almost 30 years. I see him frequently in our department, looking the same as ever at age eighty, going energetically and cheerfully about his outstanding and fruitful work in infra-red astronomy. Not many months ago Charlie gave a colloquium on what he and others have learned recently about the center of our galaxy. It was a fascinating lecture, one of the best talks I have heard in a long time. Charlie, with warm affection and great respect I wish you many more happy and productive years! I would also like to extend my most sincere good wishes and kind regards to your devoted wife Frances, who has always shown my wife Ulla and me great kindness.

References

[1] Ya. B. Zel'dovitch, Sov. Phys. JETP **9**, 682 (1959).

[2] M. A. Bouchiat and C. Bouchiat, Phys. Lett. B **48**, 111 (1974); J. Phys. (Paris) 35, 899 (1974); J. Phys. (Paris) **36**, 493 (1975).

[3] L. M. Barkov and M. Zolotorev, JETP Lett. **27**, 357 (1978); JETP Lett. **28**, 503 (1978); Phys. Lett. B **85**, 308 (1979).

[4] R. Conti, P. Bucksbaum, S. Chu, E. Commins, and L. Hunter, Phys. Rev. Lett. **42**, 343 (1979).

[5] J. H. Hollister et al., Phys. Rev. Lett. **46**, 643 (1981).

[6] G. N. Birich et al., Sov. Phys. JETP **60**, 442 (1984).

[7] J. D. Taylor et al., J. Phys. (U.K.) **B20**, 5423 (1987).

[8] T. P. Emmons, J.M. Reeves, and E. N. Fortson, Phys. Rev. Lett. **51**, 2089 (1983).

[9] M. A. Bouchiat, J. Guena, L. Hunter, and L. Pottier, Phys. Lett. B **117**, 358 (1982); B **134**, 463 (1984).

[10] S. L. Gilbert and C. E. Wieman, Phys. Rev. A **34**, 792 (1986).

[11] P. S. Drell and E. D. Commins, Phys. Rev. A **32**, 2196 (1985).

[12] For a review of early PNC experiments, including efforts to observe PNC in the $n = 2$ states of hydrogen, see: E. D. Commins, in Atomic Physics 7, p. 121; eds D. Kleppner and F. Pipkin, Plenum, New York (1981).

[13] P. A. Vetter et al., Phys. Rev. Lett. 1995, to be published.

[14] N. H. Edwards et al., Phys. Rev. Lett. 1995, to be published.

[15] V. A. Dzuba et al., J. Phys. B **20**, 3297(1987).

[16] M. Noecker, B. P. Masterson, and C. E. Wieman, Phys. Rev. Lett. **61**, 310 (1988).

[17] S. A. Blundell, W. R. Johnson, and J. Sapirstein, Phys. Rev. Lett. **65**, 1411 (1990); Phys. Rev. D **45**, 1602 (1992).

[18] P. Langacker and M. Luo, Phys. Rev. D **44**, 817 (1991).

[19] M. E. Peskin and T. Takeuchi, Phys. Rev. D**46**, 381 (1992).

[20] D. Budker, D. DeMille, E. D. Commins, and M. S. Zolotorev, Phys. Rev. A **50**, 132 (1994).

[21] D. DeMille, Phys. Rev. Lett. **74**, 4165 (1995).

[22] W. Bernreuther and M. Suzuki, Rev. Mod. Phys. **63**, 313 (1991).

[23] L. I. Schiff, Phys. Rev. **132**, 2194 (1963).

[24] P. G. H. Sandars, Phys. Lett. **14**, 194 (1965); **22**, 290 (1966).

[25] E. A. Hinds and K. Sangster, AIP Conference Proceedings **270**, 77 (1993) (Time Reversal: The Arthur Rich Memorial Symposium, Ann Arbor, Michigan, Oct. 25-26,1991).

[26] V. M. Khatsymovsky, I. B. Khriplovitch, and A. S. Yelkhovsky, Ann. Phys. **186**, 1 (1988).

[27] I. S. Altarev, *et al.*, Phys. Lett. B **276**, 242 (1992).

[28] K. F. Smith, *et al.*, Phys. Lett. B **234**, 191 (1992).

[29] J. P. Jacobs, W. M. Klipstein, S. Lamoreaux, B. R. Heckel, and E.N. Fortson, Phys. Rev. Lett. **71**, 3782 (1993).

[30] E. D. Commins, S. B. Ross, D. DeMille, and B. C. Regan, Phys. Rev. A **50**, 2960 (1994).

[31] R. Barbieri, L. Hall, and A. Strumia, L. B. L. Report #36381 (1995); Phys. Rev. D., to be published.

13

Stark Dynamics and Saturation in Resonant Coherent Anti-Stokes Raman Spectroscopy

David A. Coppeta
Paul L. Kelley
Patrick J. Harshman
T. Kenneth Gustafson
and Jean-Pierre E. Taran

13.1   Introduction

In this paper, we study the effects of strong pump and Stokes fields in near resonant coherent anti-Stokes scattering. We include the dynamic Stark effect as well as the effect of population redistribution.

Stimulated Raman scattering (SRS) was first observed by Woodbury and Ng in a nitrobenzene Kerr cell used to Q-switch a ruby laser [1]. Along with the expected laser output, fields which were both up-shifted (anti-Stokes) and down-shifted (Stokes) by the Raman frequency were observed. Oscillation on the Raman active modes of several organic liquids was observed [2] soon thereafter. The first theory of SRS proposed by Hellwarth was phenomenological in nature [3]. Using rate equations based on the probabilities of occupation for matter energy levels and detailed balance, relations that describe the Raman gain in terms of measurable parameters were obtained.

The interpretation of SRS as the scattering of the pumping field by coherent vibrations forced by the pump and Stokes fields was first proposed by Garmire, Pandarese, and Townes [4]. Starting with Placzek's polarization theory [5], they obtained expressions for the vibrational coordinate and the associated electric dipole moment. Conditions for the generation of the Stokes and anti-Stokes fields were then found. Experimental verification of the existence of coherent vibrations was performed by scattering a weak probe from the vibrations set up by a pump beam undergoing SRS [6].

Coherent anti-Stokes Raman scattering (CARS) is an important spectroscopic tool for the study of combustion, fluid flow, and plasma physics. Figure 13.1 shows the energy level diagram corresponding to the CARS process.

The input pump field at frequency ω_{p_1} and Rabi frequency Ω_{p_1} interferes with the Stokes field at frequency ω_s and Rabi frequency Ω_s. This interference drives

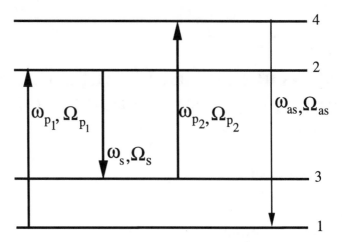

FIGURE 13.1. Energy level diagram corresponding to CARS. The pump fields are denoted by frequency ω_{p_1} and ω_{p_2} and Rabi frequencies Ω_{p_1} and Ω_{p_2}. The Stokes field has frequency ω_s and Rabi frequency Ω_s. The anti-Stokes field has frequency ω_{as} and Rabi frequency Ω_{as}.

the coherent Raman vibration at frequency ω_{31}. The pump, with Rabi frequency Ω_{p_2}, then scatters from the vibrating molecular species to generate the anti-Stokes field at frequency ω_{as}. In analyzing this problem, we only consider the four-wave mixing part of ρ_{41}. When no optical fields are present, only the lowest level is assumed to be occupied.

An extensive review of CARS spectroscopy and applications has been given by Druet and Taran [7]. Increasing attention has been given to the enhancement of CARS signals using strong fields and electronic and Raman resonances with particular attention given to saturation of the Raman transition [8]. Ouellette and Denariez-Roberge [9] have considered the somewhat different situation of one-photon saturation together with dynamical Stark processes produced by the ω_{p_1} field.

The generated anti-Stokes field should be as large as possible for efficient detection. The strength of the pump and/or the Stokes fields cannot be increased indefinitely or the closeness to resonance decreased arbitrarily without encountering new nonlinear effects. When the field Rabi frequencies become large compared to the coherence dephasing and population rates as well as the detuning from resonance, strong field effects limit the generation of the anti-Stokes field. The limiting mechanisms are Stark shifting and level mixing, Autler-Townes splitting [10], and saturation of the electronic and/or the vibrational-rotational (Raman) resonances.

In order to correctly treat these effects, we use the non-perturbative theory of radiative renormalization [11] to treat the pump and Stokes fields in a non-perturbative manner, i.e., they are included to all orders, while the anti-Stokes field is treated perturbatively. The method of radiative renormalization involves substituting the density matrix equation of motion into itself to achieve a form that is bilinear in the field coupling. The first benefit of writing the equation in this form is that when the system eigenstates have definite parity or the resonant couplings dominate in such a way as to produce an equivalent separation, the density matrix elements decouple into two independent sets. This reduces the number of independent coupled equations by a factor of two, making calculations substantially easier. The second benefit is that effects such as Stark shifts and Autler-Townes splittings are incorporated as resonance denominators which are more readily interpreted.

For the case of a four-level system, we have obtained steady-state expressions that allow one to analytically solve for the four-wave mixing component of the polarization, which is independent of the anti-Stokes field, as well as to obtain simplified numerical solutions for the absorption, which is linear in the anti-Stokes field [12]. In this paper, we will consider the four-wave mixing component of ρ_{41} in three limits. The first is the weak transfer limit in which the rate out of the ground state is so slow that no significant population transfer occurs, but the dynamical Stark effect is important. The second limit considered is pure Raman saturation, where population is transferred to state 3 and the dynamical Stark effect remains important. The final limit considered is when full electronic and Raman saturation as well as the dynamical Stark effect occur. In this case, population may be transferred to any level. We show parametric plots of $|\rho_{41}|$ for each case. It should be noted that the intensity of the anti-Stokes field grows with $|\rho_{41}|^2$. In addition to the $|\rho_{41}|$ signal term, the CARS signals are often plotted relative to the simultaneously observed $\chi^{(3)}$ signal from an inert gas. This has the virtue of reducing the effect of shot-to-shot fluctuations and the resulting ratioed signal brings the properties of the material more clearly to the fore. To obtain the ratioed result, the factor $\Omega_{p_1}\Omega_{p_2}\Omega_s^*$ is removed from the expressions shown below.

For all cases considered in this paper, we consider OH "like" molecules as the molecular species under test. For an OH molecule, typical radiative lifetimes range from 60 to 80 ns. At 1 atmosphere, OH vibrational lifetimes are typically 20 ns. In this paper, we take the population decay rates to be uniform and the thermal generation rates to be zero. Thus,

$$\Gamma_1^4 = \Gamma_3^4 = \Gamma_1^2 = \Gamma_3^2 = \Gamma_2^4 = \Gamma_1^3 \equiv -\Gamma, \tag{13.1}$$

and

$$\Gamma_4^1 = \Gamma_4^3 = \Gamma_2^1 = \Gamma_2^3 = \Gamma_4^2 = \Gamma_3^1 = 0, \tag{13.2}$$

where Γ_j^i is the negative of the decay rate from level i to level j [12]. The coherence dephasing rates are calculated using

$$\Gamma_{ij} = \frac{1}{2}\left(\Gamma_i^i + \Gamma_j^j\right) + \Gamma_{ij}', \tag{13.3}$$

where Γ_i^i is the total population decay rate out of state i and Γ_{ij}' is the pure dephasing term for the coherence between i and j. In the present case, we set $\Gamma_{ij}' = 0$ and $\Gamma = 5 \times 10^7 \text{ s}^{-1}$.

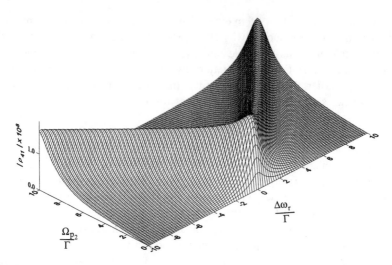

FIGURE 13.2. Autler-Townes splitting of $|\rho_{41}|$ as $\Delta\omega_r$ and $|\Omega_{p_2}|$ are varied. $\Delta\omega_r$ is detuned by varying the Stokes frequency, ω_s. $\Delta\omega_{p_1} = 6 \times 10^5\Gamma$, $\Delta\omega_{p_2} = 0$, $\Omega_{p_1} = \Gamma$, $\Omega_{p_2} = 0.02\Gamma$.

13.2   Weak Ground State Coupling Limit

In this section, we take the coupling out of the ground state via one, two, or three photon processes to be so weak that no significant population transfer occurs. For the single photon rate, this is done by taking Ω_{p_1} to be much less than the larger of the coherence dephasing rate or the detuning from resonance. Similar considerations hold for the two and three photon rates. In this limit, the expression for the four-wave mixing term is

$$\rho_{41} = i \frac{\Omega_{p_1} \Omega_{p_2} \Omega_s^*}{D_{p_1} D_{as} \left[D_r + \frac{|\Omega_s|^2}{D_{p_1}} + \frac{|\Omega_{p_1}|^2}{D_s} + \frac{|\Omega_{p_2}|^2}{D_{as}} \right]}, \tag{13.4}$$

where,

$$D_{p_1} = i(\omega_{21} - \omega_{p_1}) + \Gamma_{21} = i\Delta\omega_{p_1} + \Gamma_{21}, \tag{13.5}$$

$$D_s = i(\omega_{32} - \omega_s) + \Gamma_{32} = i\Delta\omega_s + \Gamma_{32}, \tag{13.6}$$

$$D_r = i(\omega_{31} - \omega_{p_1} + \omega_s) + \Gamma_{31} = i\Delta\omega_r + \Gamma_{31}, \tag{13.7}$$

$$D_{as} = i(\omega_{41} - \omega_{p_1} + \omega_b - \omega_{p_2}) + \Gamma_{41} = i\Delta\omega_{as} + \Gamma_{41}. \tag{13.8}$$

In the first case we consider, the pump field is taken to be detuned from the 1-2 transition by $\Delta\omega_{p_1} = 6 \times 10^5\Gamma$. The Rabi frequencies Ω_s and Ω_{p_1} are 0.02Γ and Γ respectively. The pump field, ω_{p2}, is taken to be resonant with the 3-4 transition. In Fig. 13.2, the four-wave signal term, $|\rho_{41}|$, is shown as the Stokes field detuning from the Raman resonance is varied over the range $-10\Gamma \le \Delta\omega_r \le 10\Gamma$, and the pump Rabi frequency is varied over the range $0 \le \Omega_{p_2} \le 10\Gamma$.

Due to the detunings, the terms $\frac{|\Omega_s|^2}{D_{p_1}}$ and $\frac{|\Omega_{p_1}|^2}{D_s}$ do not significantly contribute to the denominator in Eq. (13.4). As a result, Eq. (13.4) may be approximated as

$$\rho_{41} = i \frac{\Omega_{p_1} \Omega_{p_2} \Omega_s^*}{D_{p_1}\left(D_{as}D_r + |\Omega_{p_2}|^2\right)}. \tag{13.9}$$

Since D_{as} and D_r both depend on $\Delta\omega_r$, Eq. (13.9) may be expanded as a partial fraction decomposition to yield

$$\rho_{41} = \frac{\Omega_{p_1} \Omega_{p_2} \Omega_s^*}{2D_{p_1}\sqrt{|\Omega_{p_2}|^2 - \gamma^2}} \left(\frac{1}{D_r + \gamma - i\sqrt{|\Omega_{p_2}|^2 - \gamma^2}} \right.$$
$$\left. - \frac{1}{D_r + \gamma + i\sqrt{|\Omega_{p_2}|^2 - \gamma^2}} \right), \tag{13.10}$$

where,

$$\gamma = \frac{\Gamma_{41} - \Gamma_{31}}{2}. \tag{13.11}$$

We see from Eq. (13.10) that the strong pump field Ω_{p_2} splits the resonance into two Autler-Townes lines with linewidths $\frac{\Gamma_{41}+\Gamma_{31}}{2}$. When $\Omega_{p_2} \gg \gamma$, the splitting is roughly linear in Ω_{p_2} and the amplitude of the magnitude of resonance peaks become independent of Ω_{p_2}. As a measure of where the single peak may be considered split, we use a Rayleigh-type criterion. We assume the value of $|\Omega_{p_2}|$ required to resolve the peaks occurs where the shift in frequency of the resonance is equal to the HWHM of the resonance. This gives

$$|\Omega_{p_2}| = \sqrt{\frac{\Gamma_{41}^2 + \Gamma_{31}^2}{2}}. \tag{13.12}$$

From Eqs. (13.1) and (13.2) we find $\Gamma_{31} = \Gamma/2$ and $\Gamma_{41} = 3\Gamma/2$ which results in a value of $|\Omega_{p_2}| = 1.12\Gamma$ for the peaks to be resolved. In this case, $\gamma = \Gamma/2$.

As the second case of the weak ground state coupling regime, we take the Stokes field to be exactly resonant with the 2-3 transition so that $\Delta\omega_s = 0$ and the second pump frequency, ω_{p_2}, is exactly resonant with the 3-4 transition. We vary both the Stokes Rabi frequency over the range $0 \le |\Omega_s| \le 20\Gamma$ and the detuning of ω_{p_1} from its simultaneous resonances with the 1-2, 1-3 and 1-4 transitions over the range $-20\Gamma \le \Delta\omega_{p_1}, \Delta\omega_r, \Delta\omega_{as} \le 20\Gamma$. The four wave mixing term $|\rho_{41}|$ is shown in Fig. 13.3. In this case, the pump-field Rabi frequency is held fixed at a value small enough that population transfer may be neglected, $|\Omega_{p_1}| = |\Omega_{p_2}| = 2 \times 10^{-8}\Gamma$. We see in this case that the single resonance splits into three distinct lines as the Stokes Rabi frequency is increased. This case may be understood as follows. Since the pump Rabi frequency is weak compared to the coherence dephasing rates, it may be neglected in the denominator of Eq. (13.4). The four-wave mixing term may then be expressed as

$$\rho_{41} = i \frac{\Omega_{p_1} \Omega_{p_2} \Omega_s^*}{D_{as}\left(D_{p_1} D_r + |\Omega_s|^2\right)}. \tag{13.13}$$

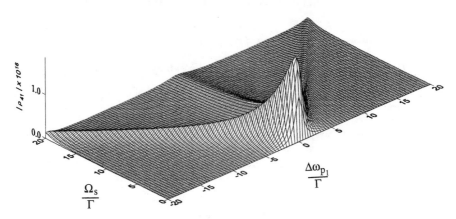

FIGURE 13.3. Triply split resonance in the weak field limit. $|\rho_{41}|$ is plotted as $\Delta\omega_{p_1}$ and Ω_s are varied. The Stokes field is held resonant, $\Delta\omega_s = 0$, and the pump Rabi frequency is held at $\Omega_{p_1} = \Omega_{p_2} = 2 \times 10^{-8}\Gamma$. $\Delta\omega_{p_2} = 0$. The split resonances decay asymptotically to zero.

One can see that the resonance denominator D_{as} is unaffected by $|\Omega_s|$ while the terms in parenthesis in the denominator form the two split resonances. As the Stokes field is increased, the resonances drop off as $|\Omega_s|^{-1}$. One may find the magnitude of the Stokes field at which the peak occurs by setting the derivative of ρ_{41} with respect to $|\Omega_s|$ equal to zero. One finds that a maximum occurs when

$$|\Omega_s|^2 = |D_{p_1} D_r|. \tag{13.14}$$

Since the peak occurs when the detuning is zero, $D_{p_1} = \Gamma_{21}$ and $D_r = \Gamma_{31}$, the values of the assumed decay rates yield $|\Omega_s| = \Gamma/\sqrt{2}$.

13.3   Pure Raman Saturation

In this section, we consider the case where the single photon coupling out of the ground state is weak but the two photon (Raman) coupling is strong. Under these conditions, population is transferred between states 1 and 3. This corresponds to a significant fraction of the molecular species being driven into a coherent vibration. The coupling of the states 1 and 3 can be treated as an effective two-level atom since the pump is far off its single photon resonances. From the two-level atom, we know that the population of level three will saturate at 50% of the total population as the coupling field increases in strength. There is a level of field strength where the vibrational coherence, ρ_{31}, is maximized and the anti-Stokes field reaches its limiting value. If one tunes slightly off the dynamically Stark-shifted Raman resonance, a larger field will be required to saturate the transition. Therefore, as a

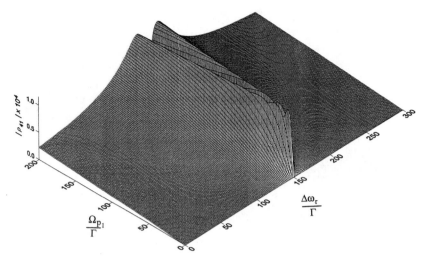

FIGURE 13.4. Pure Raman saturation of $|\rho_{41}|$ as $|\Omega_{p_1}|$ and $\Delta\omega_r$ are varied. $\Delta\omega_r$ is varied by tuning the Stokes field. $\Delta\omega_{p_1} = 6 \times 10^4\Gamma$ and $\Delta\omega_{p_2} = 6 \times 10^5\Gamma$. $\Omega_s = 3000\Gamma$ and $\Omega_{p_2} = 200\Gamma$.

function of field strength, we expect $|\rho_{41}|$ to first saturate on the Raman resonance while stronger fields will be required for off-resonance saturation.

In the case considered, we take the pump and Stokes fields to be so far off resonance with the 1-2, 3-4, and 2-3 transitions, respectively, that there is no significant population in states 2 and 4. If, in addition, the detuning from the Raman resonance is much less than the detuning of the pump from the 1-2 transition and the Stokes field from the 2-3 transition, then we may write

$$\rho_{41} = \frac{\Omega_{p_1}\Omega_{p_2}\Omega_s^*}{\Delta\omega_s \Delta\omega_{as}\left(D_r + \frac{|\Omega_{p_1}|^2}{D_s} + \frac{|\Omega_{p_2}|^2}{D_{as}} + \frac{|\Omega_s|^2}{D_{p_1}}\right)}(\rho_{11} - \rho_{33}). \tag{13.15}$$

The population difference is given in [13] as

$$\rho_{11} - \rho_{33} = \frac{1}{1 + \frac{2|\Omega_{p_1}|^2|\Omega_s|^2}{(\Delta\omega_s)^2|D_r|^2}}. \tag{13.16}$$

In Fig. 13.4, $|\rho_{41}|$ is plotted as a function of detuning of the Stokes field about the Raman Resonance, $0 \le \Delta\omega_r \le 300\Gamma$, for pump Rabi frequencies in the range $0 \le \Omega_{p_1} \le 200\Gamma$. The detunings $\Delta\omega_{p_1}$ and $\Delta\omega_{p_2}$ were taken as $6 \times 10^4\Gamma$ and $6 \times 10^5\Gamma$ respectively. The Stokes Rabi frequency, Ω_s, was held fixed at 3000Γ while the pump Rabi frequency, $|\Omega_{p_2}|$, was held constant at 200Γ.

We see that $|\rho_{41}|$ quickly saturates on line center. Although not resolvable from the plot, evidence of saturation on the Raman transition starts around $|\Omega_{p_1}| = 10\Gamma$ where roughly 30% of the population has moved from the ground state to level three. As Ω_{p_1} increases, the width of the saturated region increases while the peaks in $|\rho_{41}|$ move further from the Raman resonance. It is found that $|\rho_{41}|$ is asymmetric

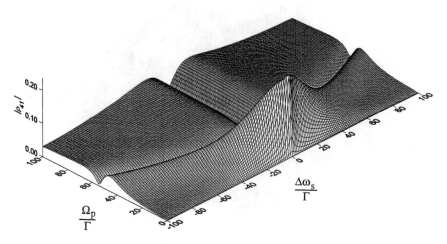

FIGURE 13.5. Electronic and Raman saturation of $|\rho_{41}|$ are shown as $\Delta\omega_s$ and $|\Omega_p|$ are varied. In this case, the pump field is held resonant with both single photon transitions, $\Delta\omega_{p_1} = \Delta\omega_{p_2} = 0$. The Stokes Rabi frequency is fixed at $\Omega_s = 20\Gamma$. Autler-Townes splitting is evident for weak $|\Omega_p|$ and strong $|\Omega_p|$ far off resonance. An extra resonance which has a "hole" in it on Raman Resonance comes in for strong $|\Omega_p|$.

about the Raman resonance. This is due to the fact that there is a Stark shift brought about by the Stokes field which is detuned from the 2-3 transition, as well as a Stark shift due to the detuned pump when Ω_{p_1} is large. These Stark shifts can be seen from the denominator of Eq. (13.15). We note that the saturation value reached is a nonzero constant. The splitting observed is not due to the two-photon Autler-Townes effect but arises because saturation makes the reactive or real part of ρ_{31} larger than the absorptive or imaginary part. The peaks in the real part occur at increasing detuning as $|\Omega_{p_1}\Omega_s|$ is increased.

13.4   Full Resonance

In the previous sections, we have considered cases where: 1) the system population was confined to the ground state so that only the dynamical Stark effect played a role in the higher order nonlinearities and 2) the case where population was allowed to move to state 3 (Raman transition). In this section, we consider the case of full resonance where one, two, and three photon transition rates are important as is the dynamical Stark effect. We expect that the four-wave mixing profiles in this case, will be difficult to interpret due to the fact that population redistribution may occur in a complex manner depending upon the parameters varied and the population decay rates. To simplify the parameterization, we take $\omega_{p_1} = \omega_{p_2} = \omega_p$ and $\Omega_{p_1} = \Omega_{p_2} = \Omega_p$.

In Fig. 13.5, we plot $|\rho_{41}|$ as a function of the Stokes detuning from the 2-3 transition, $-100\Gamma \leq \Delta\omega_s \leq 100\Gamma$, while the pump Rabi frequency is varied over the range $0 \leq \Omega_p \leq 100\Gamma$. The pump field is held resonant with both the 1-2 and 3-4 transitions while the Stokes Rabi frequency is held at $|\Omega_s| = 20\Gamma$. The expressions from which this plot was obtained are too complex to be given here; however, they may be found elsewhere [12].

In the case of weak pump fields, we would expect that the population would remain in the ground state and the terms of importance would lead to an Autler-Townes splitting of the resonance. From the plot we see that the original Lorentzian peak does indeed split for weak fields, and for strong fields far off resonance. The plot also indicates a new resonance which is reminiscent of a field-induced extra resonance. This new resonance is very broad, centered at the point where the Stokes detuning is zero, and has a hole burned at zero detuning (where the peak would have been). The origin of the hole is most likely the equalization of the populations of the four levels. Since $|\rho_{41}|$ is driven by population differences, if they vanish so will $|\rho_{41}|$. When the Stokes field is tuned off the 2-3 transition, the populations will not be equal due to the asymmetry of the coupling.

References

[1] E. J. Woodbury and W. K. Ng, Proc. IRE **50**, 2367 (1962).

[2] G. Eckhardt, R. W. Hellwarth, F. J. McClung, S. E. Schwarz, D. Weiner, and E. J. Woodbury, Phys. Rev. Lett. **9**, 455, (1962).

[3] R. W. Hellwarth, Phys. Rev. **130**, 1850 (1963).

[4] E. Garmire, F. Pandarese, and C. H. Townes, Phys. Rev. Lett. **11**, 160 (1963).

[5] G. Placzek, in *Handbuch der Radiologie*, edited by E. Marx (Akademishe Verlagsgesellschaft, Leipzig, 1934), 2nd ed., pp 209-374.

[6] J. A. Giordmaine and W. Kaiser, Phys. Rev. **144**, 676 (1966).

[7] S. A. J. Druet and J.-P. E. Taran, Prog. Quant. Electr. **7**, pp 1-72 (1981).

[8] M. Péalat, M. Lefebvre, J.-P. E. Taran, and P. L. Kelley, Phys. Rev A **38**, 1948 (1988). This paper contains extensive references to earlier literature on Raman saturation.

[9] F. Ouellette and M. M. Denariez-Roberge, Can. J. Phys. **60**, 877 (1982); **60** 1477 (1982).

[10] S. H. Autler and C. H. Townes, Phys. Rev. **100**, 703 (1955).

[11] O. Blum, P. J. Harshman, T. K. Gustafson, and P. L. Kelley, Phys. Rev. A **47**, 5165 (1993).

[12] D. Coppeta, P. L. Kelley, P. J. Harshman, and T. K. Gustafson (to be published).

[13] P. L. Kelley, P. J. Harshman, O. Blum, and T. K. Gustafson, J. Opt. Soc. Am. B **11**, 2298 (1994).

14

A Raman Study of Fluorinated Ethanes Adsorbed on Zeolite NaX

Michael K. Crawford
David R. Corbin
and Robert J. Smalley

14.1   Introduction

Due to the destructive effect of chlorofluorocarbons (CFCs) upon the ozone layer, the major global producers of CFCs have agreed to cease production by 1996. As a result of this agreement there is a need to find replacements for CFCs in a number of applications. Hydrofluorocarbons (HFCs) offer reasonably effective alternatives in uses such as refrigeration and air conditioning. One of the primary replacements for CFCs such as CFC-12 (CF_2Cl_2) for use in automobile air conditioners is HFC 134a (CF_3CFH_2). HFC 134 (CF_2HCF_2H), an isomer of 134a, is also a potentially useful HFC. A number of routes in the syntheses of these HFCs involve the use of heterogeneous catalysts to improve the efficiency and specificity of the chemical reaction sequences. Since the HFCs interact with the catalyst surface, it is of general interest to understand the nature of these molecule-surface interactions. For this reason we have undertaken this study of the adsorption of HFCs 134, 134a, and 143 (CF_2HCFH_2) on the zeolite NaX ($Na_{86}Al_{86}Si_{106}O_{384}$). Zeolites have found wide application as heterogeneous catalysts and sorbs in many industrial chemical processes, and from our perspective they have the advantages of large surface areas and relatively weak Raman scattering. These features make zeolites ideal hosts for studies of molecule-surface interactions. Furthermore, NaX binds the HFC molecules nondissociatively at room temperature with high enough number densities to produce reasonably intense Raman bands.

The vibrational and rotational spectroscopy of fluorinated ethanes has a long and rich history [1]. Of particular interest is the fact that these molecules possess intramolecular torsional vibrational potentials which exhibit energy barriers for rotation about the C-C bond [2]. The magnitudes of these barriers are on the order of one to several thousand cm^{-1}, leading to strongly hindered intramolecular rotations at room temperature. Some of the earliest work on the subject of hindered rotations was performed using microwave spectroscopy and has been described rather completely in the text by Townes and Schawlow [3].

14.2   Experimental

The data presented here include gas phase infrared and Raman spectra and Raman spectra of HFCs adsorbed on the zeolite surface.

The gas-phase infrared data were obtained with a Bruker 113v Fourier transform spectrometer at a nominal spectral resolution of 0.2 cm^{-1}. A HgCdTe detector operated at 77 K and a Si composite bolometer operated at 4.2 K were used for the 500–5,000 cm^{-1} and 30–650 cm^{-1} regions, respectively. The gas cell had a path length of 24 cm, and typical gas pressures used were between 0.1 and 30 torr.

The gas phase Raman data were obtained with a triple spectrometer equipped with a liquid nitrogen cooled charge-coupled device (CCD) detector. An Ar^{+} laser operating at 5145 Å was the exciting source, and typical powers of 1 – 2 watts at the sample were used. The gas pressures were 400 torr and the Raman light was collected in 180° back-scattering geometry (90° for the polarization measurements).

Spectra of the HFCs adsorbed on the zeolite surface were obtained with the same Raman spectrometer although a microscope was used to focus the excitation light and collect the Raman backscattered radiation. Typical excitation powers were on the order of 150 mW at the sample.

The zeolites were vacuum-dehydrated at 125°C for 24 hours before the HFCs were adsorbed at room temperature. The number of HFC molecules adsorbed per supercage in the zeolite was determined both by direct weighing of the samples before and after HFC adsorption, and by measurements of the intensities of emission of 2.23 MeV gamma rays following neutron capture by the hydrogen atoms of the HFCs at the Cold Neutron Research Facility of the National Institute of Standards and Technology.

14.3   Results and Discussion

14.3.1   Gas Phase

In Fig. 14.1 we show gas phase Raman and infrared data for HFC 134a. Accurate data have been reported for the HFCs 134 and 143 [4, 5], but the literature data which exist for 134a [6] are inconsistent regarding many of the vibrational frequencies and assignments. Since the ultimate goal of our studies is a detailed understanding of bonding geometries and dynamics of the HFCs on surfaces it is necessary to first obtain high-quality gas-phase Raman and infrared data to aid in assigning and modeling the vibrations of the adsorbed molecules, and we will thus briefly describe our results for HFC 134a.

In Table 14.3.1 we list vibrational band positions determined from Raman and infrared measurements for HFC 134a. HFC 134a has C_s point group symmetry, the only symmetry element being a plane of symmetry which includes the C-C axis and bisects the H-C-H bonds. Using group-theoretical analysis one can predict that there should be a total of 18 vibrational modes, with 11 A$'$ modes (totally

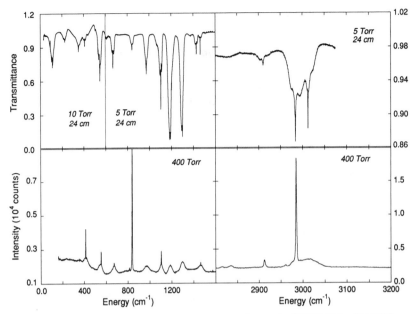

FIGURE 14.1. Infrared (top) and Raman spectra (bottom) of HFC 134a, CF_3CFH_2.

symmetric) and 7 A″ modes (non-totally symmetric). Both A′ and A″ modes are Raman and infrared active. The Raman vibrations of A′ symmetry are expected to be (fully or partially) polarized, while the A″ vibrations are depolarized [1]. In Table 14.3.1 the measured polarizations are also listed for the more intense Raman lines. The total number of fundamentals observed is 18, consistent with the group theoretical prediction. We believe that the assignments given in Table 14.3.1 significantly improve upon the results of the earlier studies [6].

Although determination of the eigenvectors for all the vibrations of HFC 134a is best accomplished by detailed force-field calculations, several vibrations are readily assigned qualitatively. The lowest frequency infrared-active vibration near 110 cm^{-1} is the intramolecular torsion of A″ symmetry, whereas the highest frequency modes near 3000 cm^{-1} are clearly stretching vibrations of the -CH$_2$ group. In particular, the vibration at 3013 cm^{-1} is the asymmetric stretch of A″ symmetry, while the vibration at 2984 cm^{-1} is the symmetric stretch of A′ symmetry. These assignments follow from the line shapes of the Raman bands, since an intense Q branch-type maximum is present at 2984 cm^{-1}, whereas the Raman mode at 3013 cm^{-1} is broad and does not have a sharp central maximum [1]. Furthermore, the 2984 cm^{-1} band is polarized, whereas the 3013 cm^{-1} band is depolarized, as expected [1] for A′ and A″ modes, respectively. Finally, HFC 134a is a nearly symmetric top [6], where the moments of inertia about the b and c axes differ by less than one percent, so in some cases it is possible to identify parallel or perpendicular (or hybrid) bands in the infrared spectrum, and these identifications

TABLE 14.3.1 Infrared and Raman bands of HFC 134a (CF_3CFH_2).

Point group: C_s
Irreducible representations: $11A' + 7A''$

Activity:
Infrared: $11A' + 7A''$
Raman: $11A' + 7A''$

Symmetry	Infrared* (cm^{-1})	Raman* (cm^{-1})	Polarization*
A″	3013 w	3010 m	DP
A′	2984 w	2985 vs	P
A′	1465 vw	1464 w	P
A′	1429 w	—	—
A″	1298 vs	1301 w	DP
A″	1186 vs	1189 w	DP
A′	1106 s	1107 w	P
A″	975 m	975 w	DP
A′	—	874 vw	—
A′	845 w	843 vs	P
A′	—	810 vw	—
A′	666 m	673 m	P
A′	549 m	557 m	P
A″	—	542 w	—
A′	410 vw	415 m	P
A″	351 w	355 vw	—
A′	226 vw	230 vw	—
A″	109 w	113 vw	—

* Band intensities are designated as *vs* (very strong), *s* (strong), *m* (medium), *w* (weak), or *vw* (very weak). The Raman band polarizations are deisgnated as P (polarized) or DP (depolarized).

also aid in making assignments. Symmetry assignments for the vibrations listed in Table 14.3.1 have been made from such considerations.

The remaining infrared bands may be assigned to skeletal vibrations of the molecule [6], with the features near 1400 cm^{-1} the C-H rocking, and wagging vibrations, whereas the lower frequency bands correspond to various C-F stretching, rocking, and wagging motions. In particular, the very strong polarized Raman band at 843 cm^{-1} is probably the C-C stretching vibration of A′ symmetry [6].

In the cases of both HFC 134 and HFC 143 the high-quality gas-phase infrared and Raman spectra have been previously reported [4, 5] so we will not discuss them in detail here. There is, however, a very important structural aspect of both of these molecules, the presence of *distinguishable* molecular conformations which

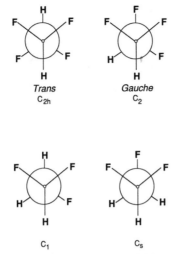

FIGURE 14.2. Conformational diagrams for HFC 134 (top) and HFC 143 (bottom).

differ by internal rotations of the two end groups about the C-C bond, which we will now describe.

In Fig. 14.2 we show schematically the two conformers of HFC 134 and HFC 143. In the former case, these are called the *trans* and *gauche* conformers, whereas in the latter case they are referred to as the C_1 and C_s conformers because of their respective point group symmetries. Since the trans conformer of 134 has a center of symmetry (inversion center), its infrared and Raman modes obey the rule of mutual exclusion. The gauche conformer, on the other hand, does not have a center of symmetry; thus a given vibration may be both infrared and Raman active. These differences allow easy assignment of infrared and Raman bands to each conformer. For HFC 143 neither the C_1 nor C_s conformer has a center of symmetry so the process of making vibrational assignments is not as straightforward [5]. Nevertheless, once the bands are assigned to individual conformers, it is a simple matter to use the intensities of the bands to follow the population of each as a function of temperature, and thereby estimate the difference in energy between them. In this way the trans conformer of 134 has been found to be 1.5 kcal/mole more stable than the gauche conformer, while for 143 the C_1 conformer has been found to be 1.6 kcal/mole more stable than the C_s conformer [5]. It should be noted that similar information may be obtained from electron diffraction [7] and microwave spectroscopy [3, 8].

14.3.2   HFCs adsorbed on NaX

The structure of one supercage of the zeolite NaX is shown in Fig. 14.3. The unit cell consists of eight such units. The HFCs are able to enter the supercage and bind nondissociatively to surface sites. In this section the use of Raman spectroscopy to elucidate the binding mechanism will be illustrated.

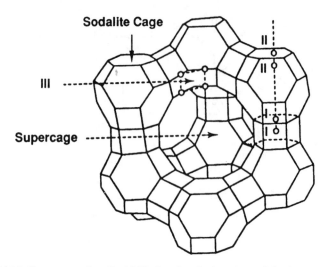

FIGURE 14.3. Supercage of zeolite NaX, showing the locations of the three types (labeled I, II, and III) of Na^+ ions. Data not described in this paper indicate that the HFCs bind in the vicinity of Na^+ ions at the III site.

In Fig. 14.4 the Raman spectra of HFC 134 in the gas phase is compared to the spectra of the molecule bound to NaX. The Raman data for 134 in the NaX matrix suggest that this molecule binds to the surface via the CH groups rather than the CF_2 groups. This conclusion follows from examination of the line broadening of the C-H stretching vibration near 3000 cm^{-1}, which clearly indicates the strong perturbation of the C-H vibration by the surface. A plausible interpretation for this broadening is the formation of hydrogen bonds between the CH_2 group and the oxygen ions located at the surface of the zeolite cavity. The ability of halogenated CH groups to hydrogen bond is well-established [10]. In the case of HFC 134, the presence of the strongly electronegative F atoms bound to the same carbon will tend to increase the acidity of the H atoms and thereby enhance their tendency to form hydrogen bonds.

There is a second interesting observation which can be made from the data in Fig. 14.4. As mentioned above, 134 can exist in two conformations, the trans conformer being about 1.5 kcal/mole more stable than the gauche conformer in the gas phase. In the case of 134 bound to the NaX surface, however, the data of Fig. 14.4 clearly show that vibrations which arise from the gauche conformer are about twice as intense as those due to the trans conformer. Thus some aspect of the molecule-surface interaction must stabilize the gauche conformation.

We postulate that the interaction which is responsible is the hydrogen bond formation between the C-H groups on the molecule and oxygen atoms on the NaX surface described above. HFC 134 has *two* C-H groups, allowing both ends of the molecule to bind to the surface and thereby altering the conformational stability. The gauche form could have both C-H bonds directed toward the surface and this would be favorable for formation of two hydrogen bonds to the surface

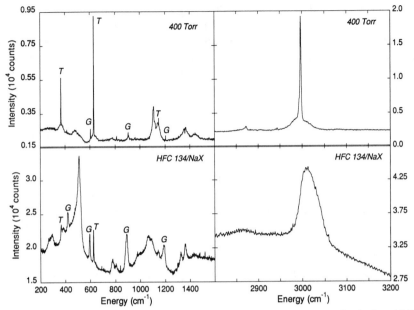

FIGURE 14.4. Raman spectra of HFC 134 in the gas phase (top) and adsorbed on NaX (bottom). The Raman bands of the trans (T) and gauche (G) conformers are labeled.

oxygen atoms. It is worthwhile noting that typical hydrogen bond strengths are 3–5 kcal/mole[10], certainly sufficient to overcome the difference in potential energy (1.5 kcal/mole) between the trans and gauche conformers.

As a second example of the effect of the NaX surface on HFC conformations, we have measured the Raman spectrum of HFC 143, CF_2HCFH_2, adsorbed on NaX. This molecule has the two conformers shown in Fig. 14.2. In the gas phase the C_1 conformer is more stable by 1.6 kcal/mole. We have used the Raman assignments for the two conformers [6] to estimate the relative population of each on the NaX surface, and we find that the C_s conformer has indeed been stabilized by interaction with the surface. This conformer would have all three hydrogen atoms oriented toward the NaX surface (Fig. 14.2), and apparently the additional surface bonding allowed by this molecular geometry is greater than the difference in intramolecular conformational energy. The role of hydrogen bonding is again suggested by the presence of strongly broadened C-H stretching vibrations. Thus, in general, we find that these heterogeneous systems favor the conformers which maximize the number of hydrogen bonds between the adsorbed HFCs and the surface.

Although our data suggest that hydrogen bonding is important in these systems, there are other interactions which may contribute to the stability of a given HFC conformer in the zeolite. For example, dipolar interactions between the zeolite and the HFC, and between the HFCs themselves, will be conformation-dependent since the permanent dipole moments of the HFCs depend upon conformation. HFC 134 offers a good example of this dependence, since the trans conformer has no dipole moment, while the gauche conformer has a large dipole moment (2.8 D). In the context of this Festschrift it is interesting to note in closing that the permanent dipole moments of HFCs were in many cases determined by measurements of first-order Stark splittings in microwave spectra, a technique which Townes and his coworkers were instrumental in developing [3, 11].

14.4   Conclusion

We have described the use of vibrational spectroscopy to study the conformations of adsorbed molecules. The fluorinated ethanes are among the simplest molecules for this purpose. Furthermore, the important technological role of HFCs requires increased understanding of their interactions with the surfaces of solids. Thus we believe that studies of this type will serve as a starting point to understand, on a microscopic level, the molecule-surface interactions which underlie so many scientifically interesting and industrially important chemical processes.

14.5   Acknowledgments

The authors thank their colleagues at DuPont (C. Grey, S. Schwartz, R. Balback) and NIST (T. Udovic, R. Cavanagh, J. Rush, G. Fraser, F. Lovas, and J. Nicol) for their contributions to this work. Michael K. Crawford would also like to thank Charles H. Townes for the opportunity to spend the years 1982–85 as a posdoctoral member of his research group at Berkeley.

References

[1] G. Herzberg, *Molecular Spectra and Molecular Structure II. Infrared and Raman Spectra of Polyatomic Molecules* (Van Nostrand Reinhold Co., New York, 1945).

[2] S. Mizushima, *Structure of Molecules and Internal Rotation*, (Academic Press, New York, 1954).

[3] C. H. Townes and A. L. Schawlow, *Microwave Spectroscopy*, (McGraw-Hill, New York, 1955).

[4] P. Klaboe and J. R. Nielsen, J. Chem. Phys. **32**, 899 (1960).

[5] V. F. Kalasinsky, H. V. Anjara, and T. S. Little, J. Phys. Chem. **86**, 1351 (1982).

[6] J. R. Nielsen and C. J. Halley, J. Mol. Spectrosc. **17**, 341 (1965); W. F. Edgell, T. R. Ri-ethof, and C. Ward, J. Mol. Spectrosc. **11**, 92 (1963); A. Danti and J. L. Wood, J. Chem. Phys. **30**, 582 (1959).

[7] D. E. Brown and B. Beagley, J. Mol. Struct. **38**, 167 (1977).

[8] T. Ogata and Y. Miki, J. Mol. Struct. **140**, 49 (1986); W. L. Meerts and I. Ozier, Phys. Rev. Lett. **41**, 1109 (1978); A. B. Tipton, C. O. Britt, and J. E. Boggs, J. Chem. Phys. **46**, 1606 (1967); N. Solimene and B. P. Dailey, J. Chem. Phys. **22**, 2042 (1954).

[9] D. W. Breck, *Zeolite Molecular Sieves* (Wiley, New York, 1974).

[10] G. C. Pimentel and A. L. McClellan, *The Hydrogen Bond* (W. H. Freeman, San Francisco, 1960).

[11] R. G. Shulman, B. P. Dailey, and C. H. Townes, Phys. Rev. **78**, 145 (1950).

15

Laser Light-scattering Spectroscopy of Supercooled Liquids and the Glass Transition

Herman Z. Cummins

15.1 Introduction

In 1958, as a first-year Ph.D. student at Columbia, I visited Charles Townes to discuss the possibility of pursuing thesis research with him. After hearing that my background was in optical spectroscopy, he described a new light source he was then thinking about, and proposed first building it and then using it as a thesis project. He explained that this new "Optical Maser" (later rechristened as the laser) would revolutionize optical spectroscopy, making it possible to perform optical experiments with the resolution and sensitivity previously available only in radio frequency or microwave spectroscopy.

Later that year, I began working in the Columbia Radiation Laboratory (in 1028 Pupin) on the first version of an optically-pumped alkali vapor laser, the scheme that Townes and Schawlow described in the paper they published that year [1]. Despite an increasingly complicated design which consumed massive amounts of machine- and electronics-shop time and floor space, the alkali-vapor system steadfastly refused to reach the necessary level of optical gain needed to produce oscillation. After the announcement by Ted Maiman of successful lasing action in the ruby laser (soon followed by the He-Ne laser), we abandoned the alkali-vapor scheme and proceeded to work with the successful ruby and He-Ne lasers, moving on to the new spectroscopies for which the laser, as predicted by Townes, was soon shown to be a truly revolutionary light source.

At Columbia, we soon developed light-beating spectroscopy (subsequently known as photon correlation spectroscopy or PCS) and applied it to the study of fluid flow and Brownian motion, exploiting the high intensity and coherence of the laser. Before long, Townes, with his new group at the Massachusetts Institute of Technology (including *inter alia*, Ray Chiao) had begun using lasers for stimulated and nonlinear optical spectroscopy, opening a whole new era for optics and materials science. The rest, as they say, is history.

Many new laser-based spectroscopies have evolved during the intervening 30 years, but lasers have also had an enormous impact on the conventional optical light-scattering spectroscopies of Raman and Brillouin scattering. Although these techniques had been around since the 1920s, it was only with the advent of lasers that they became convenient and sensitive enough to encourage their widespread use as routine analytical tools. In my own work, laser-excited Raman, Brillouin, and PCS light-scattering spectroscopy have been the principal techniques exploited in the study of various classes of phase transitions and critical phenomena, starting with a preliminary study of critical opalescence in a binary fluid mixture begun at Columbia in 1963. Recently, the power of the laser to enhance conventional spectroscopic methods, foreseen by Townes in 1958, has once again played a critical role.

During the past few years, at the City College of New York (CCNY), we have investigated the weak broad "Rayleigh wing" found in the light-scattering spectra of supercooled fluids, as part of a major international effort to elucidate the dynamical processes underlying the still mysterious liquid-glass transition. While this "transition" has been studied almost continuously since Maxwell's time, its origins, and even the question of whether or not it actually is a transition, remain unsettled. The recent renewal of activity in this area was catalyzed by the appearance in 1984 of two papers describing a new theoretical approach to the dynamics of supercooled liquids known as the mode-coupling theory (MCT). W. Götze (the principal architect of this theory) and his coworkers have produced a series of papers predicting many details of the dynamics of supercooled liquids, many of which have been confirmed by experiments, especially by inelastic neutron scattering techniques. (For a review of MCT and its experimental tests, see [2].)

Our experiments at CCNY, as described below, revealed that the light-scattering spectra of supercooled fluids exhibit many of the subtle details predicted by MCT. Furthermore, thanks to the high intensity of the laser, it is possible to investigate this behavior over a much wider range of temperatures than is currently possible with neutron scattering techniques. While the scattering mechanism responsible for the observed light-scattering spectra has not yet been completely explained, the spectra themselves provide compelling evidence for the dynamical liquid-glass transition scenario predicted by MCT.

15.2   Experiments

The experiments described below have been carried out on a number of well-known glass-forming materials [4]. We will limit the discussion to a single substance, the mixed-salt glass-former CKN [40% $Ca(NO_3)_2$, 60% KNO_3] [3]. Above the glass transition temperature ($\sim 60°C$) the viscosity of CKN decreases very rapidly with increasing temperature, much faster than the better-known network glass-forming materials like SiO_2.

In a typical experiment, the material to be studied is purified by multiple vacuum distillation, loaded into well-cleaned cylindrical glass sample tubes, and flame sealed under vacuum. The sample cell is then mounted in a cryostat (or oven), and illuminated with a focused laser beam, typically the 488 nm line of a continuous wave argon ion laser. Scattered light is collected in a near-backscattering geometry with polarization orthogonal to the incident light. This geometry eliminates most of the scattering from longitudinal and transverse acoustic modes which would otherwise overwhelm the weak signal under study.

Scattered light is analyzed with two instruments: a conventional tandem grating spectrometer (Spex 1401) and a tandem Fabry-Perot interferometer (TFPI) (Sandercock) operating in a six-pass configuration. As shown in Fig.15.1, a single Fabry-Perot spectrum covers a frequency range of about one decade.

But by collecting spectra with four different spacings plus a Raman spectrum, the full set of spectra, plotted on a log-log scale, span over four decades in frequency. (The lower limit, ~ 0.2 GHz, is set by the combined strong elastic scattering plus the limit set by the plane Fabry-Perot interferometer). By adjusting the intensities of these spectra relative to one set for which the intensities are carefully controlled, we obtain the set of combined spectra for each temperature shown in Fig. 15.2(a). Note that at lower temperatures there is some leakage of the intense longitudinal-acoustic Brillouin lines due to imperfect polarization selection.

The composite spectra $I(\omega)$ shown in Fig. 15.2(a) are then converted to susceptibility spectra $\chi''(\omega)$ via the fluctuation-dissipation relation $I(\omega) \propto [n(\omega) + 1]\chi''(\omega)$, where $n(\omega)$ is the Bose factor. The resulting set of $\chi''(\omega)$ spectra are shown in Fig. 15.2(b). These susceptibility spectra exhibit several obvious characteristics, independent of any theoretical interpretation. First, there is a high-frequency peak near 2 THz which is nearly temperature-independent. Second, there is a strongly temperature-dependent low frequency peak (the α peak) centered at $\sim$ 3 GHz at $T = 195°$C that moves rapidly towards lower frequency with decreasing temperature and moves out of our available spectral window at $\sim 120°$C. Third, there is a susceptibility minimum (the β minimum) between these two peaks that also moves towards lower frequency with decreasing temperature, although not as rapidly as the α peak. Fourth, for the lowest temperatures, the low-frequency part of the $\chi''(\omega)$ spectrum becomes concave downward rather than upward. Since each of these characteristics is predicted by the MCT, we have carried out extensive comparisons of our data with that theory.

15.3   Comparison with MCT

The mode coupling theory of the liquid-glass transition, developed in the past decade by W. Götze and his coworkers, is based on a Langevin-type generalized oscillator equation of motion for each $\phi_q(t)$, the normalized autocorrelation function of the density-fluctuation mode $\rho_q(t)$:

$$\ddot{\phi}_q(t) + \Omega_q^2 \phi_q(t) + \Omega_q^2 \int_0^t M_q(t - t')\dot{\phi}_q(t')dt' = 0 \qquad (15.1)$$

where the kernel $M_q(t-t')$ in Eq. (15.1) is the correlation function of the fluctuating forces.

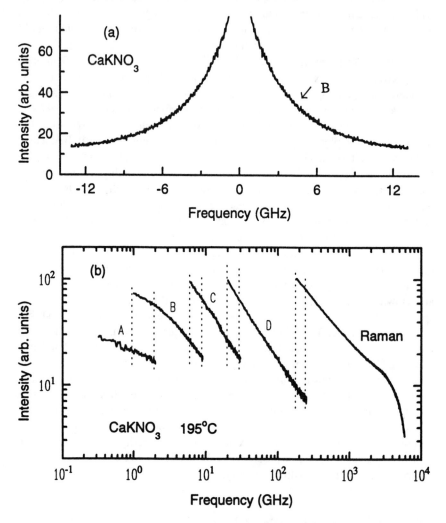

FIGURE 15.1. (a) Depolarized $\theta = 173°$ light scattering spectrum of CKN at 195°C obtained with the TFPI. (b) log-log plot of the Stokes sides of four TFPI spectra and a Raman spectrum, subsequently combined to form a single composite spectrum. Spectrum B in (b) corresponds to the spectrum shown in (a).

The idealized (or simplified) mode-coupling approximation consists of projecting the fluctuating forces onto pairs of density fluctuations, which leads to

$$M_q(t - t') = [\gamma_q \delta(t - t') + m_q(t - t')] \,. \tag{15.2}$$

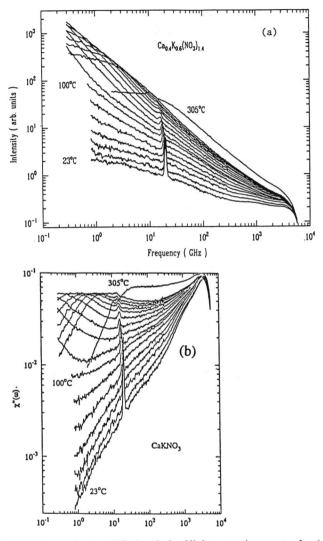

FIGURE 15.2. (a) Composite $\theta = 173°$ depolarized light-scattering spectra $I_{VH}(\omega)$ of CKN for $T = 305, 195, 180$ to 60 in $10°$C steps, 45, and $23°$C. The small peaks near 20 GHz are due to slight leakage of the intense LA Brillouin components. (b) Susceptibility spectra $\chi''(\omega)$ derived from the intensity spectra in (a).

$$m_q(t) = \frac{1}{2} \sum_{q_1, q_2} V^{(2)}(q, q_1, q_2)\phi_{q_1}(t)\phi_{q_2}(t) . \tag{15.3}$$

(The extended mode-coupling theory, which we will not discuss here, includes additional terms in $M_q(t)$ that arise from coupling to currents [2].)

The vertices (coupling constants) $V^{(2)}(q, q_1, q_2)$ in Eq. (15.3) are determined by the equilibrium structure factors S_q and can be calculated numerically if the interatomic potentials are specified. The full dynamics of $\phi_q(t)$ can then be predicted by self-consistent solution of Eqs. (15.1)-(15.3). This analysis has been performed for hard-sphere and Lennard-Jones potentials, allowing the full MCT predictions for a hard-sphere colloidal glass to be tested quantitatively, as shown by van Megen and Underwood [5]. Such analysis shows that the coupling constants increase with decreasing temperature; the increasing nonlinear feedback via the memory integral in Eq. (15.1) eventually blocks the decay of $\phi_q(t)$, leading to structural arrest, i.e., an ergodic to nonergodic transition.

For structural glasses containing nonspherical molecules, the interatomic potentials are complicated and the coupling constants have not yet been evaluated. However, because there is an intrinsic singularity in Eq. (15.1), asymptotic formulae describing the vicinity of the singularity can be found using power series methods. The results of such analyses [2] show that there is an intermediate time (or frequency) regime with universal features where the detailed structure of the coupling constants is unimportant. This β-relaxation region occurs in the $\chi''(\omega)$ spectrum around the minimum between the α-relaxation peak and the microscopic peak. [The microscopic peak results from Ω_q in Eq.(15.1).] In this region, MCT predicts that $\chi''(\omega)$ is given by the factorized equation

$$\chi_q''(\omega) = h_q |\sigma|^{1/2} \hat{\chi}_\pm''(\omega/\omega_\sigma) , \tag{15.4}$$

where $\pm$ refers to $\sigma \gtrless 0$, and all q-dependence is contained in the prefactor h_q.

In Eq.(15.4), σ is the separation parameter, taken as approximately linear in T:

$$\sigma \propto (T_C - T)/T_C \tag{15.5}$$

where T_C is the crossover (or critical) temperature corresponding to the glass transition singularity (GTS).

To a good approximation, $\chi_q''(\omega)$ can be represented by the interpolation equation

$$\chi''(\omega) = \chi_{\min}''[b(\omega/\omega_{\min})^a + a(\omega_{\min}/\omega)^b]/(a+b) . \tag{15.6}$$

The form of the susceptibility master function $\hat{\chi}''(\omega/\omega_\sigma)$ is determined by a single material-dependent parameter, the exponent parameter λ, which in turn determines the two critical exponents a and b:

$$\lambda = \Gamma^2(1-a)/\Gamma(1-2a) = \Gamma^2(1+b)/\Gamma(1+2b) \tag{15.7}$$
$$(0 \leq a \leq 0.4; \ 0 < b \leq 1)$$

The scaling frequency ω_σ in Eq. (15.4) is given by

$$\omega_\sigma = \omega_0 |\sigma|^{1/2a} \tag{15.8}$$

where ω_0 is a microscopic matching frequency. (Note that there are two $\hat{\chi}''$ master functions: $\hat{\chi}_-''$ for $T > T_C$ and $\hat{\chi}_+''$ for $T < T_C$.)

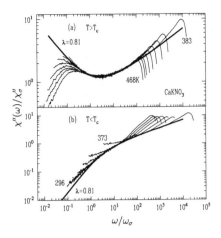

FIGURE 15.3. Scaling of the CKN $\chi''(\omega)$ spectra to obtain optimum overlap for (a) $T > 110°C$ (383 K) and (b) $T < 110°C$. The MCT master curves are also shown, corresponding to $\lambda = 0.81$.

An idealized MCT analysis can be carried out by exploiting the scaling behavior implied by Eq. (15.4), adjusting λ to produce a best fit of Eq. (15.4) or (15.6) to the complete set of experimental data, and finding $\omega_\sigma(T)$ and $\sigma(T)$ from the individual fits.

In Fig. 15.3, we show the CKN $\chi''(\omega)$ data of Fig. 15.2(b) scaled empirically by shifting vertically and horizontally to achieve optimum overlap.

The data for $T > 110°C$ (383 K), which exhibit a minimum, are shown in (a); the data for $T < 110°C$, which exhibit downward concavity (a "knee"), are shown in (b). Note that in both (a) and (b) the scaling region expands as $T \to T_C$.

Having established the scaling region for each $\chi''(\omega)$ spectrum, we can fit the data for $T > T_C$ to Eq. (15.4) or (15.6). Fits of the $\chi''(\omega)$ spectra of Fig. 15.2(b) for $T \geq 110°C$ to Eq. (15.6) gave $\lambda = 0.81 \pm 0.4$, $a = 0.27 \pm 0.02$, $b = 0.46 \pm 0.07$. In Fig. 15.3 we show the two MCT master functions (heavy lines) for $\lambda = 0.81$ superimposed on the scaled CKN $\chi''(\omega)$ susceptibility spectra.

The scaling procedure provides a scaling frequency (ω_e or ω_σ) for each spectrum while the fits to Eq. (15.6) provide ω_{min} and χ''_{min}. Using these results, we can test the central MCT prediction of Eqs. (15.5) and (15.8), that

$$\omega_e^{2a} \propto \omega_{min}^{2a} \propto \omega_\sigma^{2a} \propto |T_C - T| \tag{15.9}$$

$$\chi''_{min} \propto (T - T_C)^{1/2} \qquad (T > T_C). \tag{15.10}$$

In Fig. 15.4(a) we plot ω_e^{2a} vs T (for $T < 110°C$) and ω_{min}^{2a} vs T (for $T > 110°C$). Both plots exhibit the linear T-dependence predicted by Eq. (15.9), with their intercepts determining the crossover temperature $T_C = 105 \pm 5°C$. In 15.4(b) we plot the resulting temperature-dependent scaling times $\tau_\beta^+ = 1/\omega_e$ and $\tau_\beta^- = 1/\omega_{min}$.

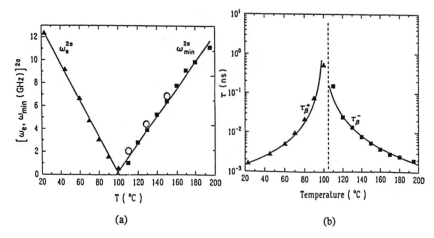

FIGURE 15.4. (a) Scaling frequency ω_e for $T < 110\,°C$ (triangles) and ω_{min} for $T > 110°C$ (squares) from the analysis shown in Fig. 15.3. The open circles are the results of neutron scattering experiments [7]. (b) Scaling times $\tau_\beta^+ = 1/\omega_e$ and $\tau_\beta^- = 1/\omega_{min}$.

The cusp shown in Fig. 15.4(b) is reminiscent of critical slowing down effects usually found in the study of thermodynamic phase transitions. However, the increase in τ_β as $T \rightarrow T_C$ does *not* represent slowing down of a relaxation time in the sense of some $f(t/\tau_\beta)$ such as an exponential decay law. It represents the temperature dependence of the crossover between the short-time t^{-a} critical decay of $\phi_q(t)$ (sometimes called the "fast β-relaxation") and, at longer times, either the von Schweidler decay ($-t^{-b}$ decay of $\phi_q(t)$ for $T > T_C$) or the plateau representing frozen-in density fluctuations (for $T < T_C$).

Note also that the cusp in Fig. 15.4(b) appears to diverge although we could not follow τ_β closer to T_C because of the limited frequency range available. This divergence would occur only if complete structural arrest took place as predicted by the idealized MCT. In the more realistic extended MCT [6], however, where ergodicity-restoring effects are included, there is no divergence of $\tau_\beta^\pm$.

This brief description was intended to demonstrate that laser-based light scattering spectroscopy has made it possible to examine the weak depolarized light scattering spectra of supercooled liquids in sufficient detail to provide a major test of the mode-coupling theory. The cusp shown in Fig. 15.4 is among the most compelling evidence for the applicability of MCT to the liquid-glass transition.

15.4   Acknowledgments

The experiments described in this article were carried out and analyzed by the graduate students and postdoctoral researchers in our group at CCNY, Nongjian Tao, Gen Li, Weimin Du, Joel Hernandez, and Guoqing Shen. We have all benefited from discussions and correspondence with Wolfgang Götze, Matthias Fuchs, and

Robert Pick. Finally, we acknowledge the support provided for this research by the National Science Foundation under Grant No. DMR-9315526.

References

[1] A. Schawlow and C.H. Townes, Phys. Rev. **112**, 1940 (1958).

[2] W. Götze and L. Sjögren, Rep. Prog. Phys. **55**, 241 (1992).

[3] G. Li, W. M. Du, X. K. Chen, H. Z. Cummins, and N. J. Tao, Phys. Rev. A **45**, 3867 (1992).

[4] For a review, see: H. Z. Cummins, G. Li, W. M. Du, and J. Hernandez, Physica A **204**, 169 (1994).

[5] W. van Megen and S. M. Underwood, Phys. Rev. E **47**, 248 (1993).

[6] W. Götze and L. Sjogren, Z. Phys. B **65**, 415 (1987); J. Phys. C **21**, 3407 (1988).

[7] W. Knaak, F. Mezei, and B. Farago, Europhys. Lett. **7**, 529 (1988).

16

The Electronic Emission Spectra of Triatomic Hydrogen: The 6025Å Bands of H_2D and HD_2

Izabel Dabrowski
and Gerhard Herzberg

16.1 Introduction

In the original work on triatomic hydrogen [1–5], transitions were reported which characterized the $n = 2$ and the $n = 3$ Rydberg states, the lowest bound states of the molecule. Only the isotopomers H_3 and D_3 were considered in detail. The spectra of the triatomic molecule were observed in the cathode glow of a hollow cathode discharge, superimposed on the spectra of the diatomic molecule. In fact the lines of the triatomic were distinguishable from those of the diatomic only because they did not occur in the anode glow.

The study of the two other isotopomers, H_2D and HD_2, did not appear promising at that time. Spectra taken with a mixture of isotopes simultaneously produced lines of H_3, D_3, H_2, HD, D_2, as well as a few of H_2D and HD_2.

A new development arose when we began to study the spectra of the rare gas hydrides [6–8] by means of an afterglow tube of a design similar to that of Cossart [9]. We found to our surprise that when argon was used as the exciting gas to which H_2 was added, a very clean spectrum of H_3 was obtained without any superimposed H_2 lines. ArH spectra were also observed but were in general much less intense, and could be eliminated completely by varying the experimental conditions.

Figger *et al.* [10–13] have been able to observe the emission spectra of each separate isotopomer by neutralization of mass-selected ion beams. However, their spectra were not obtained under high resolution, nor did they attempt any rotational analysis of the mixed isotopes.

We considered it therefore to be worthwhile to study the spectra of H_2D and HD_2. The 6025 Å band was chosen since it has the simplest rotational structure and because the lines are fairly sharp for all the isotopomers. New spectra taken on our Bomem Fourier Transform Spectrometer permitted us to improve the accuracy of the earlier measurements on H_3 and D_3, as well as to obtain fairly complete new data on H_2D and HD_2. At the same time more careful measurements of linewidths were possible. The object of the present paper is to report the results of these studies.

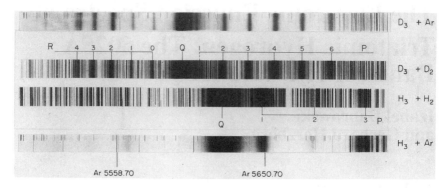

FIGURE 16.1. Photographic emission spectra of D_3 and H_3 near 5600 Å. D_2 and H_2 lines, which partially obscure the D_3 and H_3 spectra in the two middle spectra are completely eliminated in the top and bottom spectra by the addition of argon to the discharge. The broad lines in the top and bottom spectra all belong to the D_3 (H_3) $3p\,^2A_2'' \rightarrow 2s\,^2A_1'$ transition; the sharp lines on the right belong to the $D_3(H_3)$ $3d\,^2E'$, $^2E''$, $^2A_1'' \rightarrow 2p\,^2A_2''$ transition centered at 5800 Å.

16.2   Observed Spectra

The afterglow tube used in these experiments was set up in very much the same way as described by Cossart *et al.* [9]. H_2, D_2, or mixtures of the two in various proportions, were added to the stream of metastable argon atoms produced in the primary discharge. The most favorable conditions for the production of H_3 were those when the gas pressures were Ar:H_2 (2.5:0.2 torr) and the discharge current was 40 mA. If only a trace of hydrogen was used, the H_3 spectrum disappeared, to be replaced by that of ArH. If an excess of H_2 was present, then the H_2 spectrum appeared as well.

Figure 16.1 shows a comparison of the photographic spectra of the hollow cathode discharge and of the afterglow tube for both H_2 and D_2 showing the 5600 Å ($3p\,^2A_2'' \rightarrow 2s\,^2A_1'$) band. The remarkable freedom from H_2 and D_2 lines in the afterglow is very obvious. The sharp lines in the afterglow spectrum are either atomic argon, or, at the red end of the spectrum, sharp lines of H_3 or D_3 belonging to the 5900 Å ($3d\,^2E'$, $^2E''$, $^2A_1'' \rightarrow 2p\,^2A_2''$) band. No further work was done on the 5600 Å and 5900 Å bands; only the 6025 Å ($3s\,^2A_1' \rightarrow 2p\,^2A_2''$) band was measured. As shown in our earlier paper [2], this band consists of very sharp lines for D_3, and even H_3 has some sharp lines.

Figure 16.2 shows a section of the spectrum of the afterglow for the four isotopomers as obtained with the Fourier Transform spectrometer. Figure 16.2b and 16.2c show predominantly H_2D and HD_2 respectively. The differences in the

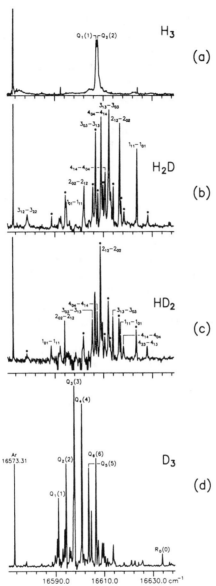

FIGURE 16.2. Observed Q-branches at 16600 cm^{-1} for (a) H_3, (b) H_2D, (c) HD_2, and (d) D_3. The most intense lines are identified. Lines overlapped by lines of other isotopomers are marked by an asterisk (*).

relative intensities are a result of different pressures of H_2:D_2 used in the experiment [2:1 for H_2D and 1:2 for HD_2]. It should be noted that the spectra shown in Figs. 16.2b and 16.2c have been somewhat manipulated. In both cases H_3 and D_3 lines were present in the spectra. They have been subtracted from the spectra. In some cases this subtraction is incomplete, since the relative emission line intensities were not always identical. Overlapping lines are marked by asterisks (*) in the diagram.

The results of these measurements are given in Tables 1, 2, 3, and 4 for H_3, H_2D, HD_2, and D_3 respectively. The atomic argon lines observed in the spectrum were used as standards [14]. The spectra were taken with a resolution of 0.1 cm^{-1}. For sharp, unblended lines, we believe that these measurements are reliable to 0.01 cm^{-1}. Tables 1 and 4 are very similar to Tables III and IV of [2], except that a few lines are added and the estimated accuracy of the wavenumbers is slightly improved. Tables 2 and 3 for H_2D and HD_2 are entirely new.

16.3   Assignments

The assignments of the lines in the 6025 Å bands of H_3 and D_3 given in Tables 1 and 4 are the same as in [2] and need not be further discussed.

The structure of the bands of H_2D and HD_2 is much more complicated. These molecules are asymmetric tops and therefore the assignments are not trivial. They can most easily be established by simulating band structures with approximate A, B, and C values. For all four isotopomers the electronic transition moment is perpendicular to the plane of the molecule, that is, for H_2D and HD_2 we expect C type bands with the difference that in H_2D the a-axis coincides with the axis of symmetry (the C_2-axis) while in HD_2 the a-axis is perpendicular to it. Using approximate A, B, and C values (identical with those of the H_2D^+ core from Foster et al. [15] and of the HD_2^+ core from Foster et al. [16]) there was little difficulty in simulating the bands and in assigning the observed lines as given in Tables 2 and 3. The intensities of the lines predicted by the asymmetric rotor simulations were also found to agree reasonably well with the observed intensities.

In Fig. 16.3 a diagram of the lowest rotational levels of H_2D in the upper and lower vibronic states $^2A_1$ and $^2B_2$ is presented. The overall symmetry species are shown for the levels and the allowed transitions are indicated. According to well-known rules [17], **A** levels combine only with **A** levels, **B** with **B**. Levels with $K_a \neq 0$ have both **A** and **B** species. Because the two identical nuclei are protons, the statistical weights follow Fermi statistics, and therefore the **B** levels have three times the weight of the **A** levels. The resulting intensity alternation is clearly seen in the spectrum in Fig. 16.2b.

For HD_2 the a-axis does not coincide with the symmetry axis (C_2), but the b-axis does. Therefore the symmetry species are different. In addition, because now the identical nuclei are deuterons and they follow Bose statistics, the **A** levels should be strong, the **B** levels weak, with a ratio of 2:1. In Fig. 16.4 the lowest rotational levels, their symmetry species, and the allowed transitions are shown for HD_2. Unfortunately, because of perturbations (see the following discussion) and overlapped lines, the intensity alternations are not as obvious in HD_2 (Fig. 16.2c) as they are in H_2D (Fig. 16.2b).

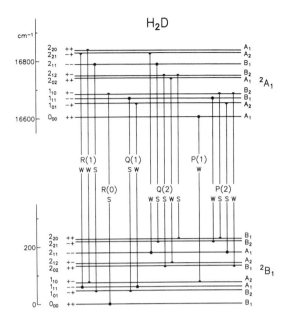

FIGURE 16.3. Rotational levels, quantum numbers, and symmetry species are given for the $^2A_1 - {}^2B_1$ transition observed in H_2D. The transitions are labeled s (strong) or w (weak) according to their expected statistical weight.

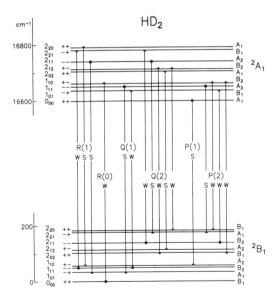

FIGURE 16.4. Rotational levels, quantum numbers, and symmetry species are given for the $^2A_1 - {}^2B_1$ transition observed in HD_2. The transitions are labeled s (strong) or w (weak) according to their expected statistical weight.

Although there was no great difficulty in assigning the lines of both H_2D and HD_2, considerable difficulties arose in fitting the lines to obtain the constants of HD_2. As a result, the constants of HD_2 in Table 16.5 have a very low accuracy. The reason for this is a perturbation which is apparently caused by the interaction with a vibrational level of the $3p\,^2E'$ state. In this state (only for HD_2 but not for H_2D) when v_2 is doubly excited, a near resonance occurs with $3s\,^2A_1'(v_2 = 0)$. We can use the data of Foster $et\ al.$ [16] for HD_2^+ as an approximation for HD_2, where v_2 is observed to be 1968 cm^{-1}. $2v_2$ is then approximately 3940 cm^{-1} so that the total energy above $2p\,^2A_2''$ would be 17000 cm^{-1} which is close to the energy of $3s\,^2A_1'(v_2 = 0)$. For H_2D^+, $2v_2 = 4412$ cm^{-1} [15], and thus no resonance is expected for H_2D. This is the same interaction which causes the perturbation in the higher $\mathbf{N}$, $\mathbf{K}$ levels of D_3. The effect on the spectrum can be seen in Fig. 16.2d where anomalous positions and intensities are observed in the $Q_6(6)$, $Q_7(7)$, and $Q_8(8)$ lines.

Unfortunately, there are not enough data at present to study this perturbation in a more quantitative way. The molecular constants for both H_2D and HD_2 are given in V.

16.4   Predissociation and Line Widths

The first band spectra of H_3 that were discovered, near 5600 Å (see Fig. 16.1) and 7100 Å, consisted of very broad lines. They are broad because the lower state, $2s\,^2A_1'$, is strongly predissociated by the repulsive ground state, $2p\,^2E'$. This predissociation is vibronically allowed (if there were no vibronic interaction it would be forbidden), and it is strong because the Franck-Condon factors are appreciable.

As described in [2], shortly after the discovery of the 5600 Å and 7100 Å bands a further band at 6025 Å was found which consisted of fairly sharp lines. It was assigned as the $3s\,^2A_1' \rightarrow 2p\,^2A_2''$ transition. The upper state could in principle also predissociate into the $2p\,^2E'$ ground state, but since it lies much higher in energy the Franck-Condon factor will be much smaller, so that emission from the $3s\,^2A_1'$ state can take place. The lower state, $2p\,^2A_2''$, cannot predissociate into $2p\,^2E'$ since there is no vibration that could cause vibronic interaction. It is only if ro-vibronic interaction takes place that a weak predissociation will occur. For D_3 this predissociation is so weak that little line broadening can be detected, but for H_3 a slight broadening occurs, which increases rapidly with $N(N+1) - K^2$. This was shown in Fig. 8 of [2]. The new results, which confirm this observation, are shown in Fig. 16.5, which in addition contains the new data for H_2D and HD_2. The numerical values of the line widths are included in Tables 1 to 4. It is gratifying to note that the data for H_2D and HD_2 form lines of intermediate slope between those of H_3 and D_3. In Fig. 16.5 it can be seen that the linewidths converge to ≈ 0.15 cm^{-1} at $\mathbf{N} = \mathbf{K} = \mathbf{0}$. This non-rotating level cannot predissociate ro-vibronically. The observed linewidth is simply due to Doppler broadening.

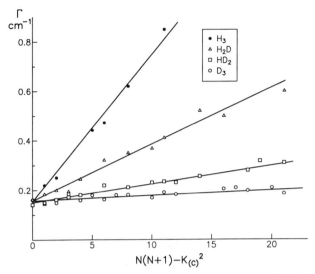

FIGURE 16.5. Observed linewidths for H_3, H_2D, HD_2, and D_3 from Tables 16.5 to 16.5, showing predissociation in the lower state.

It should be emphasized that the observed linewidths are unaffected by any possible predissociation in the upper electronic state since a predissociation lifetime that would have an observable effect on the linewidth would be so short that the emission would be unobservable. Notwithstanding the linewidths, however, there is some indication that predissociation is occurring in the upper state, particularly for the lighter isotopomers. In Fig. 16.6, the relative line intensities in the Q-branches are plotted as a function of N. In all cases $N = K_{(c)}$. Sharp cutoffs in emission due to predissociation in the upper state can be seen in H_3 and HD_2. In H_2D there is no sharp cutoff in the Q-branch spectrum. However, the $4_{14}-4_{04}$ line is considerably weaker than the $3_{13}-3_{03}$ line and no lines with higher energies in the upper state have been observed. In D_3 there is no observed predissociation in the upper state. The irregularities in the Q-branch intensities are presumably caused by the perturbations discussed previously.

16.5   Discussion and Conclusion

The fortuitous observation in the Cossart cell of spectra of H_3 free of overlapping H_2 spectra, has made it possible to study rotationally resolved spectra of the mixed isotopes H_2D and HD_2. Although the new results do not change the previously reported interpretation of the observations of the H_3 and D_3 spectra, they do provide a welcome confirmation and extension of the knowledge of an interesting and unusual molecular system.

The reason for the complete elimination of the H_2 (or D_2) spectrum in the Cossart cell may not be immediately obvious. However the explanation is actually

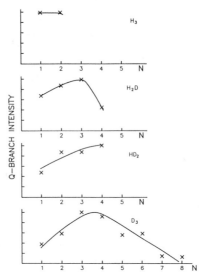

FIGURE 16.6. Observed intensities in the Q-branches of H_3, H_2D, HD_2, and D_3. In all cases $N = K$ (or K_c). The sharp cutoffs indicate predissociation in the upper state.

relatively simple. The excitation in the Cossart cell produces a stream of metastable argon atoms mixed with H_2. The energy of metastable Ar in the ^3P state is very close to the excitation energy of the metastable $c^3\Pi_u$ state of H_2. That is,

$$Ar(^3P) + H_2(X^1\Sigma_g^+) \rightarrow Ar(^1S) + H_2(c^3\Pi_u).$$

In turn, an H_2 molecule in the $c^3\Pi_u$ state, upon colliding with another H_2 molecule in the ground state, has just enough energy to form H_3 in the $n = 3$ Rydberg state

$$H_2(c^3\Pi_u) + H_2(X^1\Sigma_g^+) \rightarrow H_3\,(n = 3) + H(n = 1).$$

The two reactions above describe very satisfactorily the formation mechanism for H_3 or D_3 in the Cossart cell. Although a search was made for higher Rydberg levels of H_3, no transitions with $n > 3$ were observed with these experimental conditions despite the fact that there were no obscuring H_2 lines, confirming the formation mechanism described above.

TABLE 16.1. Observed lines and assignments in the 6025Å region of H_3 : the $3s\,^2A_1' \rightarrow 2p\,^2A_2''$ band

ν_{obs} cm^{-1}	I_{obs} ‡	Γ^a cm^{-1}	assignment
16430.203	0.4	0.48	$P_1(2)$
604.066	0.5	b	$Q_2(3)$
605.851	0.5	b	$Q_1(2)$
607.046	6.7	b	$Q_1(1)$
607.566	6.7	b	$Q_2(2)$
669.615	0.7	1.20	$R_2(4)^c$
694.988	4.9	0.16	$R_0(0)$
700.671	0.8	0.62	$R_2(3)^c$
728.581	0.9	1.00	$R_1(4)^c$
735.700	0.9	0.50	$Q_0(2)^c$
741.566	0.6	0.83	$Q_2(4)^c$
755.372	2.2	1.25	$R_0(4)^c$
768.756	1.2	0.85	$R_1(3)^c+Q_1(3)^c$
781.807	1.5	0.22	$R_1(1)$
797.957	0.7	0.44	$R_1(2)^c$
809.403	0.2	0.65	
834.305	0.4	0.60	
841.737	1.1	0.45	
859.094	0.6	0.42	$Q_1(2)^c$
868.059	0.2	0.25	$R_2(2)$
891.453	0.3	0.46	$R_0(2)^c$

^{a} observed linewidth, half width at half height
^{b} part of close doublet, linewidth could not be measured
^{c} line belongs to the overlapping
$3d\,^2E',\ ^2E'',\ ^2A_1' \rightarrow 2p\,^2A_2''$ transition
‡ arbitrary units

TABLE 16.2. Observed lines and assignments in the 6025Å region of H_2D : the $3s\,^2A_1' \rightarrow 2p\,^2A_2''$ band

ν_{obs} cm^{-1}	I_{obs} ‡	Γ^a cm^{-1}	N'	K_a'	K_c'	N''	K_a''	K_c''
16262.806	0.9	0.52	4	1	4	5	2	4
279.420	0.9	0.36						
282.968	0.6	0.48						
284.842	1.1	0.60	4	1	3	5	2	3
300.854	0.3	0.46						
304.779	0.5	0.37						
308.320	0.3	0.22						
327.577	1.8	0.40	3	1	3	4	2	3
348.552	1.6	0.50	3	1	2	4	2	2
363.950	0.5	0.33						
389.719	2.9	0.35	2	1	2	3	2	2
401.309	0.5	0.26						
404.957	3.7	0.34	2	1	1	3	2	1
411.713	1.5	0.30	2	0	2	3	1	2
447.681	4.1	0.31	1	1	1	2	2	1
454.868	5.0	0.31	1	1	0	2	2	0
476.948	3.0	0.22	1	0	1	2	1	1
522.418	0.4	0.31	2	1	2	2	2	0
535.618	2.4	0.19	0	0	0	1	1	0
549.232	1.0	0.21	1	1	0	2	0	2
564.874	1.6	0.31	2	1	1	2	2	1
578.940	1.8	0.36	3	1	2	3	2	2
591.933	1.2	0.42	4	1	2	4	2	3
594.686	3.4	0.19	1	0	1	1	1	1
602.064	5.7	0.23	2	0	2	2	1	2
606.882*	13.2	0.19	3	0	3	3	1	3
609.182*	15.4	0.18	4	0	4	4	1	4
610.830	7.7	0.23	4	1	4	4	0	4
612.388	16.3	0.23	3	1	3	3	0	3
613.112	4.9	0.22						

^{a} observed linewidth, half width at half height.
* overlapped by a line of HD_2
‡ arbitrary units

TABLE 16.2. Continued

ν_{obs} cm^{-1}	I_{obs} 3	Γ^a cm^{-1}	N'	K_a'	K_c'	N''	K_a''	K_c''
616.766	14.5	0.18	2	1	2	2	0	2
623.841	10.9	0.17	1	1	1	1	0	1
637.275	0.5	0.34	3	2	2	3	1	2
651.785	0.9	0.16	2	2	1	2	1	1
682.658	8.0	0.16	1	1	0	0	0	0
688.414*	0.9	0.21	3	3	1	3	2	1
741.078	7.5	0.18	2	1	1	1	0	1
760.711	0.8	0.33						
761.817	1.8	0.21	2	2	0	1	1	0
768.016	1.0	0.35						
769.444	1.4	0.18	2	2	1	1	1	1
775.889	1.0	0.40						
805.980	3.4	0.21	3	1	2	2	0	2
808.870	1.0	0.30						
821.467	0.8	0.30						
827.454	0.9	0.18	3	2	2	2	1	2
844.346	2.0	0.37	3	3	0	2	2	0
849.681	1.1	0.30	3	3	1	2	2	1
876.714	1.4	0.18	4	1	3	3	0	3
888.140	0.6	0.43	4	3	1	3	2	1
889.682	0.5	0.36						
895.444	0.4	0.30						
901.352	0.6	0.42	4	3	2	3	2	2

^{a} observed linewidth, half width at half height.

* overlapped by a line of HD_2

3 arbitrary units

TABLE 16.3. Observed lines and assignments in the 6025Å region of HD_2: the $3s\,^2A_1' \rightarrow 2p\,^2A_2''$ band

ν_{obs} cm^{-1}	I_{obs} ‡	Γ^a cm^{-1}	N'	K_a'	K_c'	N''	K_a''	K_c''
16282.854	0.6	0.31						
304.808	0.6	0.37						
308.322	0.4	0.22						
326.507	0.8	0.32	3	3	1	4	4	1
335.027	1.1	0.38	4	2	2	5	3	2
335.981	1.3	0.31	4	1	3	5	2	3
341.093	0.5	0.26	4	1	4	5	2	4
345.505	1.4	0.25	4	0	4	5	1	4
367.519*	1.7	0.32						
375.147	0.5	0.28	3	2	1	4	3	1
387.426	1.0	0.26	3	1	3	4	2	3
401.198	0.5	0.25	2	2	1	3	3	1
402.755	0.8	0.26	2	2	0	3	3	0
403.774	1.1	0.23	3	0	3	4	1	3
441.739	3.0	0.20	2	1	1	3	2	1
451.310	1.4	0.24	2	0	2	3	1	2
470.688	1.9	0.19	1	1	1	2	2	1
476.875*	1.5	0.22	1	1	0	2	2	0
500.532	1.4	0.16	1	0	1	2	1	1
510.842	0.3	0.20	3	2	2	3	3	0
544.362	2.9	0.16	0	0	0	1	1	0
560.104	1.2	0.16	2	1	1	2	2	1
588.786	1.8	0.14	1	0	1	1	1	1
594.373	5.1	0.12	2	0	2	2	1	2
598.865	0.8	0.14						
605.705	5.2	0.18	3	0	3	3	1	3
606.750*	8.8	0.24						
607.870	6.2	0.18	4	0	4	4	1	4
609.160*	15.0	0.18	2	1	2	2	0	2
609.926	4.9	0.18						

^{a} observed linewidth, half width at half height
*overlapped by a line of H_2D
‡ arbitrary units

TABLE 16.3. Continued

ν_{obs} cm^{-1}	I_{obs} [3]	Γ^a cm^{-1}	N'	K_a'	K_c'	N''	K_a''	K_c''
614.189	5.5	0.14	3	1	3	3	0	3
617.331	3.0	0.16	1	1	1	1	0	1
618.565	1.7	0.20	4	1	4	4	0	4
628.346	1.6	0.26	4	2	3	4	1	3
639.953	2.0	0.18	3	2	2	3	1	2
661.806	0.9	0.14	1	1	0	0	0	0
663.951	0.9	0.22						
669.195	1.0	0.30						
706.793	3.9	0.14	2	1	1	1	0	1
727.649	1.5	0.17						
730.355	1.5	0.22	2	2	1	1	1	0
755.500	0.5	0.16	3	1	2	2	0	2
772.607	1.0	0.15	3	2	1	2	1	1
782.973	2.4	0.17	3	2	2	2	1	2
793.143	0.6	0.17	4	1	3	3	0	3
804.841	1.5	0.20	3	3	1	2	1	1
806.699	4.2	0.19						
807.779	1.3	0.34	4	2	2	3	1	2
823.056	0.7	0.20						
830.219	1.2	0.18	4	2	3	3	1	3
833.357	1.1	0.25						
838.383	1.7	0.16						
855.319	1.0	0.18						
857.644	0.5	0.26						
861.588	1.0	0.18						
870.912	1.5	0.25						
879.913	1.0	0.17						

^{a} observed linewidth, half width at half height
*overlapped by a line of H_2D
[3] arbitrary units

TABLE 16.4. Observed lines and assignments in the 6025Å region of D_3: the $3s\,^2A_1'\;\rightarrow\;2p\,^2A_2''$ band

ν_{obs} cm^{-1}	I_{obs} ‡	Γ^a cm^{-1}	assignment
16413.890	0.1	0.20	$P_0(4)$
415.372	0.3	0.20	$P_1(4)$
418.856	0.2	0.22	$P_2(4)$
422.142	0.3	0.20	$P_3(4)$
457.486	1.2	0.17	$P_0(3)$
459.014	1.0	0.16	$P_1(3)$
462.351	0.7	0.17	$P_2(3)$
501.578	0.4	0.16	$P_0(2)$
503.079	2.0	0.15	$P_1(2)$
545.793	3.1	0.15	$P_0(1)$
588.783	0.4	0.21	$Q_1(4)$
589.893	0.9	0.17	$Q_1(3)$
590.759	2.2	0.14	$Q_1(2)$
591.286	5.4	0.15	$Q_1(1)$
591.901	0.6	0.19	$Q_2(5)$
592.986	1.5	0.20	$Q_2(4)$
593.795	3.0	0.18	$Q_2(3)$
594.391	8.0	0.16	$Q_2(2)$
595.889	1.6	0.17	$Q_3(5)$
596.400	1.3	0.18	
597.463	5.7	0.19	$Q_3(4)$
597.803	14.1	0.18	$Q_3(3)$
600.931	12.7	0.19	$Q_4(4)+Q_4(5)$
601.666	1.1	0.22	$Q_4(6)$
603.710	7.7	0.18	$Q_5(5)$
604.796	4.4	0.21	$Q_5(7)$
606.694	8.4	0.18	$Q_6(6)$
607.589	2.1	0.22	$Q_5(7)$
609.250	1.9	0.23	$Q_6(7)$
609.863	1.9	0.18	$Q_7(7)$
610.501	0.8	0.20	
611.465	0.6	0.22	
613.788	1.8	0.19	$Q_8(8)$

[a] observed linewidth, half width at half height
[b] line belongs to the overlapping $3d\,^2E',\;^2E'',\;^2A_1'\;\rightarrow\;2p\,^2A_2''$ transition
‡ arbitrary units

TABLE 16.4. Continued.

ν_{obs} cm^{-1}	I_{obs} 3	Γ^a cm^{-1}	assignment
618.197	0.4	0.23	
621.208	0.5	0.20	
622.668	0.5	0.20	
625.753	0.5	0.30	
633.980	1.0	0.16	$R_0(0)$
653.031	0.9	0.18	
674.724	0.3	0.26	
675.807	0.9	0.18	$Q_2(5)^b$
677.275	6.1	0.14	$R_0(1)$
678.950	5.2	0.15	$R_1(1)$
681.020	0.5	0.30	
699.685	1.4	0.18	
713.038	0.4	0.22	
717.608	1.2	0.25	
719.799	0.8	0.18	$R_0(2)$
721.658	4.6	0.15	$R_1(2)$
722.312	0.5	0.22	
725.835	2.8	0.15	$R_2(2)$
727.858	1.0	0.18	
745.118	1.0	0.20	
745.734	1.6	0.18	
747.443	0.5	0.20	
748.377	0.4	0.31	
752.672	0.7	0.20	
755.828	0.5	0.25	
756.844	0.7	0.25	
761.304	4.6	0.19	$R_0(3)$
763.301	3.3	0.17	$R_1(3)$
767.943	2.8	0.18	$R_2(3)$
771.038	0.6	0.18	
773.063	2.3	0.19	$R_3(3)$

[a] observed linewidth, half width at half height
[b] line belongs to the overlapping $3d\,^2E'$, $^2E''$, $^2A_1' \rightarrow 2p\,^2A_2''$ transition
3 arbitrary units

TABLE 16.4. Continued.

ν_{obs} cm^{-1}	I_{obs} [3]	Γ^a cm^{-1}	assignment
782.790	0.5	0.20	
784.169	1.0	0.21	
791.892	1.4	0.19	
793.524	1.3	0.22	
797.088	0.3	0.24	
800.204	0.3	0.30	
801.244	0.4	0.22	$R_0(4)$
802.502	1.1	0.24	
803.549	1.5	0.18	$R_1(4)$
805.155	0.5	0.18	
808.864	1.1	0.19	$R_2(4)$
811.682	0.5	0.26	
813.633	1.6	0.18	$R_3(4)$
814.276	0.3	0.20	
816.892	0.7	0.23	$Q_0(5)^b$
818.110	0.5	0.19	
819.201	0.4	0.20	
819.911	0.9	0.17	$R_4(4)$
832.486	0.4	0.26	
836.765	0.4	0.24	
837.879	1.5	0.20	$R_3(5)^b$
839.037	1.2	0.21	$R_0(5)$
840.418	0.7	0.16	$R_3(4)^b$
841.459	0.8	0.21	$R_1(5)$
845.434	0.2	0.21	$P_2(4)^b$
848.235	0.6	0.27	$R_2(5)$

[a] observed linewidth, half width at half height
[b] line belongs to the overlapping $3d\,^2E'$, $^2E''$, $^2A_1' \rightarrow 2p\,^2A_2''$ transition
[3] arbitrary units

TABLE 16.5. Rotational constants for the $3s\,^2A_1'$ and the $2p\,^2A_2''$ electronic states $(v_2 = 0)$ of H_2D and HD_2.[a]

H_2D

	$3s\,^2A_1'$	$2p\,^2A_2''$	H_2D^{+b}
ν_0	16609.25(19)		
A	43.992(84)	44.125(197)	43.466(8)
B	29.439(50)	29.486(47)	29.138(2)
C	16.954(79)	16.841(83)	16.603(2)
D_K	0.054(10)	0.038^c	0.0376(22)
D_{JK}	0.0^c	0.0^c	0.0025(10)
D_J	0.014(2)	0.012(2)	0.0100(13)

HD_2

	$3s\,^2A_1'$	$2p\,^2A_2''$	H_2D^{+b}
ν_0	16602.51(45)		
A	37.73(24)	36.90(17)	36.243(19)
B	22.17(15)	22.07(10)	21.869(5)
C	13.69(77)	13.58(78)	13.057(3)
D_K	0.027^c	0.027^c	0.027(3)
D_{JK}	0.008^c	0.008^c	0.0082(23)
D_J	0.005^c	0.005^c	0.0047(22)

[a] observed values in cm-1 with standard deviation on last digits in parentheses.
[b] rotational constants for the ground state of the ion: H_2D^+ from [15]; HD_2^+ from [16].
[c] These constants could not be adequately derived. These are constrained values.

References

[1] G. Herzberg, J. Chem. Phys. **70**, 4806 (1979).

[2] I. Dabrowski and G. Herzberg, Can. J. Phys. **58**, 1238 (1980).

[3] G. Herzberg and J. K. G. Watson, Can. J. Phys. **58**, 1250 (1980).

[4] G. Herzberg, H. Lew, J. J. Sloan, and J. K. G. Watson, Can. J. Phys. **59**, 428 (1981).

[5] G. Herzberg, J. T. Hougen, and J. K. G. Watson, Can. J. Phys. **60**, 1261 (1982).

[6] I. Dabrowski, G. Herzberg, B. P. Hurley, R.H. Lipson, M. Vervloet, and D. -C. Wang, Mol. Phys. **63**, 269 (1988).

[7] I. Dabrowski, G. Herzberg, and R. H. Lipson, Mol. Phys. **63**, 289 (1988).

[8] I. Dabrowski, G. DiLonardo, G. Herzberg, J. W. C. Johns, D. A. Sadovskii, and M. Vervloet, J. Chem. Phys. **97**, 7093 (1992).

[9] D. Cossart, C. Cossart-Magos, G. Gandra, and J. M. Robbe, J. Mol. Spectrosc. **109**, 166 (1985).

[10] H. Figger, M.N. Dixit, R. Maier, W. Schrepp, H. Walther, I.R. Peterkin, and J. K. G. Watson, Phys. Rev. Lett. **52**, 906, (1984).

[11] H. Figger, Y. Fukida, W. Ketterle, and H. Walther, Can. J. Phys. **62**, 1274 (1984).

[12] H. Figger, W. Ketterle, and H. Walther, Zeit. für Physik D **13**, 129 (1989).

[13] W. Ketterle, H. Figger, and H. Walther, Zeit. für Physik D **13**, 139 (1989).

[14] V. Kaufman and B. Edlén, Jour. Phys. Chem. Ref. Data **3**, 825 (1974).

[15] S. C. Foster, A. R. W. McKellar, I. R. Peterkin, J. K. G. Watson, F. S. Pan, M. W. Crofton, R. S. Altman, and T. Oka, J. Chem. Phys. **84**, 91 (1986).

[16] S. C. Foster, A. R. W. McKellar, and J. K. G. Watson, J. Chem. Phys. **85**, 664 (1986).

[17] G. Herzberg, Molecular Spectra and Molecular Structure: II. Infrared and Raman Spectra of Polyatomic Molecules, Krieger Publishing Co. Malabar, Florida (1991).

17

Limitations for Frequency-based Absolute Length Measurements

Michael Danos

17.1 Introduction

The advent of the laser has provided for a quantum jump in the accuracy available for the definition of length and for the stability and accuracy of its measurement. In particular, it has allowed the replacement of the independent unit, "the length," by the wavelength of a suitable optical frequency. This frequency, by means of a chain of frequency multipliers, could be related to the time unit, the second, which is by far the most accurate and reproducible of the various physical quantities.

Several problems exist in implementing the absolute length definition, one of them is the diffraction by the apertures of the setup. For example the question of diffraction from a corner cube for particular measurement setups has been considered in the Gaussian beam approximation by Monchalin et al. [1]. The present paper examines the question of the influence of light beam characteristics which are independent of the particular setup; they exist universally and must be addressed in every setup to be used for absolute length determination.

The basic problem lies in that the measurement of the wavelength along the axis of the apparatus, taken to be the z axis, i.e., determines at best k_z. On the other hand, in a single-mode laser it is the frequency which is fixed, and the frequency determines $k = \sqrt{k_x^2 + k_y^2 + k_z^2}$, which is the quantity one is actually interested in. (In fact, as we will see below, the problem is somewhat more involved as one is presented only with certain average quantities.) This way the presence of transverse momenta in the photon beam induces a systematic shift towards longer wavelength in the measured value. Since the transverse wavelength is of the order of the beam diameter, i.e., of the order of millimeters, the relative wavelength shift is of the order 10^{-6} to 10^{-7}, which in terms of metrology is very far from negligible, in particular considering the present-day long-term stability of lasers which is of the order 10^{-12}. In today's technology such an accuracy of 10^{-12} for length measurement seems perfectly feasible; for example, mechanical motion controlled to a small fraction of the atomic spacing has been achieved by Deslattes [2] in the determination of the lattice constant for Si crystals by means of X-ray interferometry. Even though a uniform shift of the unit length is irrelevant for most practical applications, differences in laser characteristics, i.e., beam profile, will show up even in wavelength comparisons as differences in the measured value of the length of a given object—which may be the putative secondary length standard. If one wants to achieve an absolute implementation of the length standard in terms of the definition of the value of the vacuum speed of light, one must know the characteristics of the individual setup very well indeed in order to be able to ascertain, and then to account for these systematic shifts. In order to match the long-term stability of today's frequency-stabilized lasers one would have to know the corrections arising from systematic shifts to an accuracy of 10^{-5} to 10^{-6}, which certainly is not trivial. In this note only the properties of the light beam are considered; the difficulties related to the other components of the experiment are not addressed.

In the present paper we shall not discuss photon correlation measurements. Hence a classical treatment of the light beam is fully adequate; the quantum aspect is here totally irrelevant.

17.2   Basics

We consider the light beam only for the region in space beginning several centimeters after the last member of the laser system, e.g., the last collimator (see below). Then we have to deal strictly with the solutions to the free Maxwell equations. That is, the collimators or other optical components interact with the beam by supplying structures which support convection or displacement currents, or more accurately, both. These currents produce secondary radiation, which depends not only on the geometry but also on the properties of the material making up these structures, i.e., conductivity, index of refraction, etc. In lowest order, assuming infinite conductivity, this is described as diffraction. However, the assumption may not be accurate. The secondary radiation is coherent with the beam radiation. Hence it will interfere with the beam, and the influence on the beam characteristics will be proportional to its amplitude and not to the square of that amplitude. In order to be negligible the intensity of that scattered radiation therefore must be 10^{-24} of the beam intensity. To achieve this low a level of parasitic stray radiation is certainly not feasible, even by placing all structures well out of the way of the beam. The characteristics of this secondary radiation thus must be well understood.

We now collect the underlying formulae. We use the radiation gauge, i.e., we set for the vector potential

$$\nabla \cdot \mathbf{A} = 0. \tag{17.1}$$

Every component of $\mathbf{A}$, hence $\mathbf{A}$ itself, obeys the wave equation (units such that $\hbar = c = 1$)

$$[(\partial_j)^2 - (\partial_t)^2]\mathbf{A} = 0 \tag{17.2}$$

and

$$(\partial_t)^2 \mathbf{A} = -\omega^2 \mathbf{A}, \tag{17.3}$$

where ω is determined by the laser. We shall expand the field of the beam in terms of plane-wave solutions; hence, writing $\omega^2 = k^2$ (setting $c=1$) there holds for each Fourier component

$$k^2 = k_x^2 + k_y^2 + k_z^2.$$ (17.4)

Supposedly one determines k_z by the measurement; however, see below. To obtain the desired length $\lambda = 2\pi/k$ one must know k_x and k_y. Since the light beam has finite transverse dimensions, in addition to k_x, $k_y \neq 0$ the surfaces of constant phase are of necessity non-planar. The beam polarization thus is not strictly transverse, i.e., it does not lie exactly in the $x - y$ plane. So, for example, the component propagating along the x-direction will support polarization along the y- and z-directions. Hence, dropping the time-dependence $\exp(-i\omega t)$, there holds for a given Fourier component,

$$\mathbf{a}(k_x) = e^{ik_x x} \left[\widehat{y} f_y(k_x) + \widehat{z} f_z(k_x) \right],$$ (17.5)

where $\widehat{x}$, $\widehat{y}$ are unit vectors. This way the radiation field is given by

$$
\begin{aligned}
\mathbf{A} = {} & \int dk_x \, dk_y \, e^{i\left[k_x x + k_y y + \sqrt{k^2 - k_x^2 - k_y^2} + \phi(\mathbf{k})\right]} \\
& \times \left\{ \widehat{x} \left[f_x(k_y) + f_x(k_z) \right] + \widehat{y} \left[f_y(k_z) + f_y(k_x) \right] \right. \\
& \left. + \widehat{z} \left[f_z(k_x) + f_z(k_y) \right] \right\},
\end{aligned}
$$ (17.6)

where we have set explicitly

$$k_z^2 = k^2 - k_x^2 - k_y^2.$$ (17.7)

Any non-linear $\mathbf{k}$-dependence of the phase $\phi(\mathbf{k})$ will only decrease the depth of the interferometer "zeroes" (or the visibility of the fringes). It will not introduce systematic errors in the length determinations. It seems likely that such phase non-linearities are small.

Being sufficiently far from the interfering objects only real k_j should be encountered. Violations of this assumption could happen if in the boundary, i.e., in the interfering objects, there exist periodicities with, say, $k_x^2 > k^2$, which is highly unlikely. At any rate, the propagation of such a component in the direction orthogonal to this periodicity will be damped exponentially, and its influence on the radiation field should be totally negligible a few centimeters away from the object. Of course, this assumption must be checked out. (The set of real-k solutions being a complete set when dropping the condition (17.7), the exponentially damped solutions could be expressed in terms of this complete set.)

The finite transverse size of the light beam enforces some properties for the beam. Thus, since we are dealing with a mixture of standing and running waves, the plane-wave relation $E = B$ here does not hold, and the energy flux must be computed from the Poynting vector. The other aspect, already mentioned above, is that the polarization cannot be purely transverse. Even more generally, the propagation along the z-direction is fully determined by the transverse characteristics of the beam. (This is actually simply a consequence of the hyperbolic character of the wave equation.) Thus from the gauge condition, Eq. (17.1), we have

$$k_x \left[f_x(k_y) + f_x(k_z) \right] + k_y \left[f_y(k_z) + f_y(k_x) \right]$$
$$+ k_z \left[f_z(k_x) + f_z(k_y) \right] = 0. \tag{17.8}$$

Since this equation must hold for all allowed k_x and k_y, we obtain two more conditions by setting the derivatives of Eq. (17.8) with respect to k_x and k_y to zero, taking into account Eq. (17.7). Indeed, taking for simplicity the transverse polarization to be purely in the x direction we find

$$k_x \left[f_x(k_y) + f_x(k_z) \right] + k_z \left[f_z(k_x) + f_z(k_y) \right] = 0, \tag{17.9}$$

and for the derivatives

$$f_x(k_y) + f_x(k_z) + k_x \, \partial_{k_x} f_x(k_z) + k_z \, \partial_{k_x} f_z(k_x) = 0, \tag{17.10}$$

and

$$k_x \left[\partial_{k_y} f_x(k_y) + \partial_{k_z} f_x(k_z) \right] \frac{dk_z}{dk_y} = 0, \tag{17.11}$$

with dk_z/dk_x and dk_z/dk_y from Eq. (17.7). Indeed, given $f_z(k_x)$ and $f_x(k_y)$, $f_x(k_z)$ is fully determined. (Of course, in general the polarization is not constant over the beam cross section.) Thus, in principle an accurate knowledge of the transverse beam characteristics would be sufficient for evaluating the required corrections. Still, the required accuracy may be beyond reach for an experimental determination.

17.3   Accuracy Estimate

Without further information one cannot proceed beyond the (exact) relation Eq. (17.6). One can, however, introduce an approximate effective, i.e., the measurable, periodicity along the z direction, say κ_z. Herewith we have

$$\kappa_z = \sqrt{k^2 - \kappa_r^2} \tag{17.12}$$

where z is the *nominal* beam direction, i.e., the axis of the apparatus, while the quantity κ_r is similarly the effective periodicity in the radial directions, obtained by a suitable averaging over the azimuthal angle and the transverse momentum. In full generality we know that

$$\kappa_r^2 > 0; \tag{17.13}$$

consequently

$$\kappa_z^2 < k^2. \tag{17.14}$$

This being an inequality is unavoidable. Furthermore, the effective κ_z arises from the superposition of a continuous distribution of values of k_z, cf. Eq. (17.6). This can be written approximately as

$$\kappa_z^2 \sim \int g(k_r) \sqrt{k^2 - k_r^2} \, dk_r \ . \tag{17.15}$$

Therefore the interference "zeroes" are actually only minima in the intensity; the minima disappear at the distance where $z \, \Delta k_z$ (Δk_z = widths of the distribution) approaches unity. For high accuracy lasers the resulting coherence length turns out to be in the hundreds of meters. More important for length measurements is the decrease of the sharpness of the minima, which, as all other beam characteristics, is determined by the distribution of the transverse momenta.

The difficulty in defining the length is apparent in Eq (17.15): the desired quantity, k^2, is associated with the measured quantity, κ_z, in the form of an integral over the not-directly-accessible weight function $g(k_r)$. And, in order to improve the accuracy of the length definition beyond the immediate scale factor κ_r^2 / k_z^2 this weight function must be known to a very high accuracy, as can be seen from

$$\Delta \kappa_z \ = \ \frac{\kappa_r^2}{k_z^2} \, \Delta \kappa_r \ : \tag{17.16}$$

the uncertainty of κ_r propagates linearly into the uncertainty of κ_z and hence of k_z. This relation, Eq. 17.15, in fact, pinpoints the effect which provides the ultimate limitations for the accuracy with which the unit of length can be implemented.

It seems that it is hopeless to determine the beam characteristics experimentally to the desired accuracy, say to 10^{-5}. Instead, one needs to use an apparatus the characteristics of which can be accurately computed. This, not too surprisingly, points to the choice of a Fabry-Perot interferometer, for example in a Deslattes-type setup [2, 3] for which the internal field configuration is somewhat (roughly by the finesse) decoupled from the laser beam characteristics. The information thus produced will allow the evaluation of the measurements directly in terms of the exact treatment following, e.g., from Eq. (17.6) above.

17.4   Conclusions

In order to approach the precision afforded for absolute length determination by today's laser technology one must obtain a sufficiently precise handle on the electromagnetic field in the setup; most importantly in the measuring Fabry-Perot cavity—if that is the technique employed in the measurement. It seems that this goal can only be reached by a sufficiently exact solution of the Fabry-Perot cavity modes, including the characteristics of the mirrors. These solutions then must be matched to the beam characteristics, to ascertain the relative excitation of the higher modes in the cavity—which, of course have different wave lengths in the z direction. Of course, the "wrong" modes will be off-resonance, but not by much;

furthermore, besides the mismatch to the beam, the intensity suppression factor is only of the order $(\Gamma/\Delta\omega)^2$ which may not render these modes negligible in the present context. The most important effect of these "wrong" modes is to introduce a non-symmetric change in the line shape; this being an interference effect it is proportional not to the ratio of the intensities but to the ratio of the amplitudes, and hence it can be rather substantial.

The principal systematic shift can be evaluated in first order by using an experimentally determined more or less accurate beam profile. This can lead to a precision of the absolute length determination of perhaps 10^{-7} to 10^{-8}. To go beyond that seems to be possible, but certainly not easy.

References

[1] J.-P. Monchalin, M. J. Kelly, J. E. Thomas, N. A. Kurnit, A. Szoke, F. Zernike, P. H. Lee, A. Javan, Applied Optics **20**, 736 (1981).

[2] R. D. Deslattes and E. G. Kessler, IEEE Transact. Instr. Meas. **40**, No. 2, 92 (Apr. 1991).

[3] H. P. Layer, R. D. Deslattes, and W. G. Schweitzer, Applied Optics **15**, 734 (1976).

18

Microcavity Quantum Electrodynamics

Francesco De Martini and Marco Giangrasso

18.1   Introduction

The concept of the microscopic optical cavity, or microcavity introduced by one of us in the domain of quantum optics a few years ago [1, 2, 3], has produced a series of scientific advances in that field on the technological and on the fundamental sides. As far as the first aspect is concerned, efficient nearly-thresholdless solid state microlasers are today available [4]. Furthermore, monolithic arrays of transversely interacting microlasers [3], micro-amplifiers or optical active multiplexers represent a significant step forward in the domain of optoelectronics, and are conceivable in the near future. Among the other applications, the use of the microcavity in optical nonlinear spectroscopy, such as in molecular Raman spectroscopy, has already been investigated successfully [5]. On a fundamental side, the peculiar topology of the field-confinement which is intrinsic to the microcavity structure offers new conceptual perspectives and leads to a series on new results in the context of all quantum interaction processes involving the emission or absorption of a photon, i.e., in any field-confined process of quantum scattering. Any appropriate theoretical account of such a confinement should, of course, be based first on the description of a complete set of cavity modes supporting the field. Accordingly, in Sec. 1 of the present work a detailed theory of the field quantization over such a complete set of modes is given. Then some relevant and sometimes intriguing aspects of quantum electrodynamics (QED) scattering commonly encountered in quantum optics will be analyzed. The first, and most important of these processes, is related with the spontaneous emission in the presence of vacuum-confinement [2]. In this connection we shall analyze in Secs. 2 and 3 the process of transverse coupling between two equal quantum objects, atoms or microlasers, via a superradiant-type process. One may wonder, for instance, how fast the interaction travels between these objects A and B, resonating at a wavelength λ_0 and placed in the symmetry plane of a single Fabry-Perot microcavity, in the transverse direction, i.e., orthogonal to the plane-wave wavevector $\mathbf{k}$ relative to the only longitudinal mode allowed by a microcavity with mirror spacing $d = \lambda_0/2$ [3]. From this point of view, if one can carry out an experiment with sufficient time resolution, i.e., by using a sub-picosecond laser, the active microcavity offers a unique opportunity to clarify important features of quantum electrodynamics in a confined structure. At the same time it offers an ideal, general test of relativistic causality in QED interactions involving atoms. Before venturing in the details of the quantum theory of the interaction process, let us first comment on the issue of relativistic causality. It is well-known that the causality issue within the two-atom interaction is considered to be the very first problem ever analyzed by QED, as it was discussed by a classic paper published by Enrico Fermi back in 1936 [6]. In more recent times the causality claim made in that work was criticized, and the relativistic issue reemerged within the context of the Local Quantum Field Theory (LQFT) [7]. This was the last of a long series of works dedicated by various theoretical physicists in the last two decades to the investigation of the quantum-dynamical aspects of the two-atom model. In fact, this problem is now taken as a paradigmatic example expressing in the form of a gedanken experiment the basic conceptual features of any quantum measurement process, namely, the nonlocality of particle states in LQFT [7], the atomic superfluorescence [8, 9], the role of the vacuum field within the atomic correlations or the interaction by photon-exchange [10, 11], the causality problem in LQFT and, most important, the very role of relativity in quantum mechanics [12, 13]. In spite of such extended theoretical endeavours, today the emerging conceptual picture concerned with causality in quantum mechanics may be considered at least contradictory.

Consequently, we believe that such a fundamental issue should be conclusively investigated experimentally by taking advantage of the modern femtosecond time-resolved laser methods [14]. The experiment we have in mind, and that is actually being done in our laboratory, may be considered the short-pulse version of our early experiment [3] in which two equal atoms A, B, with atomic resonance $\omega_0 = 2\pi c/\lambda_0$, are trapped within a planar, symmetrical Fabry-Perot cavity, with mirror reflectivity $|r_i| \cong 1$, "finesse" $f \gg 1$ and relevant dimension $d = m\lambda_0/2 \equiv m\pi/k_0$ with cavity-order $m = 1, 3, 5$. As shown in [3], this configuration implies a relevant relativistic causality argument we may express by the following question. Assume, according to Fig. 18.1, that A and B are localized on the symmetry plane $z = 0$ of the single-mode cavity, i.e., $m = 1$, at a mutual transverse distance $R = |\mathbf{R}| \leq l_c \cong 2\lambda_0 \sqrt{mf}$, l_c being the cavity transverse coherence length relative to the mode of spontaneous emission having k_0-vector parallel to $\mathbf{z}$ and orthogonal to $\mathbf{R}$. Since the field involved in the inter-atomic coupling does belong to that mode, what is the "transverse speed" at which the interaction is established betweeen the atoms? This argument is close to the famous single-photon, plane-wave, transverse delocalization problem expressed by A. Einstein back in 1917 [15]. As we shall see, the theory provides an interesting answer to this problem. In this connection, and more generally, our present work provides a definite criterion for a correct formulation of the dynamics of the field-atom interaction process that preserves causality in the interaction. In addition to the basic two-atom issue, and for the sake of generality, in Sec. 4 several fundamental QED field properties in microcavity confinement condition are evaluated, namely the field's commutation-relations, the field's propagator, and the field's chronological pairing generally adopted in the Wick's formulation of the S-matrix. The problem of the electron's QED mass-

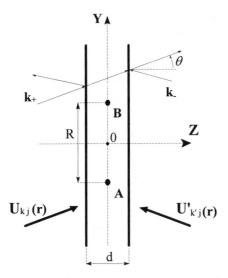

FIGURE 18.1. Two interacting atoms A and B in an optical microcavity.

renormalization is then analyzed in the last section again for the condition of microcavity confinement. The theory shows that the cavity contribution to the renormalized mass can be accessible to experimental detection.

18.2   Mode Structure and Field Quantization in the Microcavity

In order to determine the atomic dynamics in the microcavity we first determine the appropriate spatial modes for quantization of the electromagnetic field [2, 16]. The z-axis is taken normal to the mirrors with its origin in the middle of the cavity (Fig. 18.1). The mirrors are assumed to have infinite extents in the xy plane. Multiple reflections couple together waves of wavevectors:

$$\mathbf{k}_+ = \mathbf{k}(\sin \Theta \cos \Phi, \sin \Theta \sin \Phi, \cos \Theta);$$
$$\mathbf{k}_- = \mathbf{k}(\sin \Theta \cos \Phi, \sin \Theta \sin \Phi, -\cos \Theta) \qquad (18.1)$$

for $(0 \leq \Theta \leq 1/2\pi)$. Four distinct spatial modes can be constructed from contributions with the same two-wave vectors. For each set of polar angles Θ and Φ, there are two transverse polarization directions whose unit vectors are chosen to be:

$$\varepsilon(\mathbf{k}_+, 1) = \varepsilon(\mathbf{k}_-, 1) = (\sin \Phi, -\cos \Phi, 0) \qquad (18.2)$$

$$\varepsilon(\mathbf{k}_+, 2) = (\cos \Theta \cos \Phi, \cos \Theta \sin \Phi, \sin \Theta);$$
$$\varepsilon(\mathbf{k}_-, 2) = (\cos \Theta \cos \Phi, \cos \Theta \sin \Phi, -\sin \Theta) \qquad (18.3)$$

where $\mathbf{k}_+$ and $\mathbf{k}_-$ designate the polarizations of the respective wave-vector contributions. It is convenient to indicate the polarizations in Eqs. (18.2) and (18.3) by an index $j = 1, 2$. The complex reflection and transmission coefficients r_{1j}, t_{1j} and r_{2j}, t_{2j} of the cavity mirrors are generally different for the two polarizations and depend on the polar angle Θ. They are assumed to have the following unitary, lossless properties for all values of Θ:

$$
\begin{aligned}
|r_{1j}|^2 + |t_{1j}|^2 &= |r_{2j}|^2 + |t_{2j}|^2 = 1; \\
r_{1j}^* t_{1j} + r_{1j} t_{1j}^* &= r_{2j}^* t_{2j} + r_{2j} t_{2j}^* = 0
\end{aligned}
\tag{18.4}
$$

where r^* and t^* are the complex conjugates (c.c.) of the parameters r, t. Optical propagation within the mirrors is not important for the present study and we accordingly ignore their internal mode structure. For each pair of coupled wave vectors $\mathbf{k}_+$, $\mathbf{k}_-$ designated by $\mathbf{k}$ for brevity, and for each transverse polarization there are two distinct mode functions corresponding to incoming plane waves of unit amplitude that are incident respectively from the negative and positive z-sides of the cavity. The forms of these functions are obtained by summing the geometric series resulting from the multiple mirror reflections [17]. The two kinds of spatial dependence are given as follows,

	Mode function $U_{\mathbf{k}j}(\mathbf{r})$:	
$\mathbf{k}_+$	$\mathbf{k}_-$	z
$\exp(i\mathbf{k}_+ \cdot \mathbf{r})$	$R_{\mathbf{k}j}\exp(i\mathbf{k}_- \cdot \mathbf{r})$	$-\infty < z < -(1/2)d$
$t_{1j}\exp(i\mathbf{k}_+ \cdot \mathbf{r})/D_j$	$t_{1j}r_{2j}\exp(i\mathbf{k}_- \cdot \mathbf{r} + ikd\cos\Theta/D_j$	$-(1/2)d < z < +(1/2)d$
$T_{\mathbf{k}j}\exp(i\mathbf{k}_+ \cdot \mathbf{r})$	0	$+(1/2)d < z < +\infty$

$$\tag{18.5}$$

	Mode function $U'_{\mathbf{k}j}(\mathbf{r})$:	
$\mathbf{k}_-$	$\mathbf{k}_+$	z
$T'_{\mathbf{k}j}\exp(i\mathbf{k}_- \cdot \mathbf{r})$	0	$-\infty < z < -(1/2)d$
$t_{2j}\exp(i\mathbf{k}_- \cdot \mathbf{r})/D_j$	$t_{2j}r_{1j}\exp(i\mathbf{k}_+ \cdot \mathbf{r} + ikd\cos\Theta/D_j$	$-(1/2)d < z < +(1/2)d$
$\exp(i\mathbf{k}_- \cdot \mathbf{r})$	$R'_{\mathbf{k}j}\exp(i\mathbf{k}_+ \cdot \mathbf{r})$	$+(1/2)d < z < +\infty$

$$\tag{18.6}$$

where the expressions in each row of Eqs. (18.5) and (18.6) represent, as shown in Fig. 18.1, the plane-wave mode functions propagating in the space portions indicated at the right-hand side of each row and excited for the sets $U_{\mathbf{k}j}$, $U'_{\mathbf{k}j}$ by the waves $\exp(i\mathbf{k}_+ \cdot \mathbf{r})$, $\exp(i\mathbf{k}_- \cdot \mathbf{r})$ respectively. In Eqs. (18.5) and (18.6) the various quantities are defined as:

$$
D_j \equiv [1 - r_{1j}r_{2j}\exp(2ikd\cos\Theta)]
\tag{18.7}
$$

$$
R_{\mathbf{k}j} \equiv [r_{1j}\exp(-ikd\cos\Theta) + r_{2j}(t_{1j}^2 - r_{1j}^2)\exp(ikd\cos\Theta)]/D_j
\tag{18.8}
$$

$$
T_{\mathbf{k}j} \equiv T'_{\mathbf{k}j} = t_{1j}t_{2j}/D_j
\tag{18.9}
$$

$$
R'_{\mathbf{k}j} \equiv [r_{2j}\exp(-ikd\cos\Theta) + r_{1j}(t_{2j}^2 - r_{2j}^2)\exp(ikd\cos\Theta)]/D_j.
\tag{18.10}
$$

The last three quantities represent the reflection and transmission coefficients of the cavity as a whole. It is not difficult to show, using Eq.(18.4), that they satisfy,

$$|R_{\mathbf{k}j}| = |R'_{\mathbf{k}j}|; \qquad |R_{\mathbf{k}j}|^2 + |T_{\mathbf{k}j}|^2 = |R'_{\mathbf{k}j}|^2 + |T'_{\mathbf{k}j}|^2 = 1. \qquad (18.11)$$

These properties ensure the normalization and orthogonality of the two modes possessing wavevectors and polarizations, and the general relations are:

$$\int d\mathbf{r} \; \varepsilon(\mathbf{k}, j) \cdot \varepsilon(\mathbf{k}', j') U_{\mathbf{k}j}(\mathbf{r}) U'^{*}_{\mathbf{k}'j'}(\mathbf{r}) \;\; = \;\; 0;$$

$$\int d\mathbf{r} \; \varepsilon(\mathbf{k}, j) \cdot \varepsilon(\mathbf{k}', j') U_{\mathbf{k}j}(\mathbf{r}) U^{*}_{\mathbf{k}'j'}(\mathbf{r}) \;\; = \;\; (2\pi)^3 \delta_{jj'} \delta(\mathbf{k} - \mathbf{k}') \; (18.12)$$

together with the identical normalization integral for the primed mode function Eq. (18.6). The modes Eqs. (18.5) and (18.6) form a complete set of functions for all of space, including the interior of the cavity and the exterior regions on either side: this property will be further verified by our causality results. These modes allow calculations to be made of the spontaneous emission rates and radiated field operators for atoms that are excited in cavities whose mirrors transmit nonzero fractions of the emitted intensity [2, 5]. Thus, in the limiting case of a perfectly-reflecting closed cavity, the travelling-wave mode functions used here reproduce results ordinarily obtained with standing-wave modes, while in the opposite extreme of an absent cavity, the mode functions Eqs. (18.5) and (18.6), taken together produce the usual complete set of plane-waves in infinite free-space. In intermediate conditions the modes form a convenient basis for general calculations, and they are free of the potential limitations inherent in modes restricted to exterior regions of finite extent, or to only one side of the cavity. The field is quantized in terms of a set of mode creation and destruction operators. The operators for the modes with spatial functions $U_{\mathbf{k}j}(\mathbf{r})$ and $U'_{\mathbf{k}j}(\mathbf{r})$ are denoted $\hat{a}^2_{\mathbf{k}j}$, $\hat{a}_{\mathbf{k}j}$ and $\hat{a}^2_{\mathbf{k}j}$, $\hat{a}'_{\mathbf{k}j}$ respectively, where $j = 1, 2$ indicates the mode-polarization. With $\mathbf{k}$ taken to be a continuous variable, the operators satisfy the commutation relations,

$$\left[\hat{a}_{\mathbf{k}j}, \hat{a}^2_{\mathbf{k}'j'} \right] \;\; = \;\; \left[\hat{a}'_{\mathbf{k}j}, \hat{a}'^2_{\mathbf{k}'j'} \right] = \delta_{jj'} \delta(\mathbf{k} - \mathbf{k}');$$

$$\left[\hat{a}_{\mathbf{k}j}, \hat{a}'^2_{\mathbf{k}'j'} \right] \;\; = \;\; \left[\hat{a}'_{\mathbf{k}j}, \hat{a}^2_{\mathbf{k}'j'} \right] = 0. \qquad (18.13)$$

The field quantization now proceeds in the usual way, and we quote only the main results [18]. The Heisenberg electric-field operator is separated into two parts: $\mathbf{E}(\mathbf{r}, t) = \mathbf{E}^+(\mathbf{r}, t) + \mathbf{E}^-(\mathbf{r}, t)$, where:

$$\mathbf{E}^+(\mathbf{r}, t) \;\; = \;\; i \int d\mathbf{k} \sum_j \sqrt{\left(\hbar k c / 16 \pi^3 \, \varepsilon_0 \right)} \; \varepsilon(\mathbf{k}, j)$$

$$\times \left\{ U_{\mathbf{k}j}(\mathbf{r}) \hat{a}_{\mathbf{k}j} + U'_{\mathbf{k}j}(\mathbf{r}) \hat{a}'_{\mathbf{k}j} \right\} \exp(-ickt) \qquad (18.14)$$

and $\mathbf{E}^-(\mathbf{r}, t)$ is given by the Hermitian-conjugate (h.c.) expression. In writing out the field operators explicitly, the polarization vectors, Eqs. (18.2) and (18.3), are those associated with the vectors $\mathbf{k}_+$ or $\mathbf{k}_-$ appropriate to the corresponding terms in the mode-functions (18.5) and (18.6). Because of the way in which wavevector-space is divided into two half-spaces by the cavity, the 3-dimensional integral in Eq. (18.14) is

$$\int d\mathbf{k} \equiv \int d^3k = \int_0^\infty dk \int_0^{\pi/2} d\Theta \int_0^{2\pi} k^2 d\Phi \sin \Theta. \qquad (18.15)$$

The normal-ordered part of the free-field Hamiltonian is,

$$\hat{H}_0 = \int d^3k (\hbar ck) \sum_j \left\{ \hat{a}_{\mathbf{k}j}^\dagger \hat{a}_{\mathbf{k}j} + \hat{a}'^\dagger_{\mathbf{k}j} \hat{a}'_{\mathbf{k}j} \right\}, \quad j = 1, 2. \qquad (18.16)$$

18.3   Two Atom Dynamics: Correlated Spontaneous Emission and Relativistic Causality

The Hamiltonian accounting for the interaction in the cavity of the field with two atomic dipoles, A and B, in the microcavity is now expressed in the form:

$$\begin{aligned}
\hat{H}_I \;=\; & i \int d\mathbf{k} \sum_i \sum_j \sqrt{(\hbar kc/16\pi^3 \, \varepsilon_0)} [\exp(-ickt)/D_j] \\
& \times \{\hat{a}_{\mathbf{k}j}[\, \varepsilon(\mathbf{k}_+, 1)t_{1j} \exp(i\mathbf{k}_+ \cdot \mathbf{r}_i) \\
& + \varepsilon(\mathbf{k}_-, 1)t_{1j}r_{2j} \exp(i\mathbf{k}_- \cdot \mathbf{r}_i + ikd \cos \Theta)] \\
& + \hat{a}'_{\mathbf{k}j}[\, \varepsilon(\mathbf{k}_-, 1)t_{2j} \exp(i\mathbf{k}_- \cdot \mathbf{r}_i) \\
& + \varepsilon(\mathbf{k}_+, 1)t_{2j}r_{1j} \exp(i\mathbf{k}_+ \cdot \mathbf{r}_i + ikd \cos \Theta)]\} \cdot \boldsymbol{\mu}_i + h.c. (18.17)
\end{aligned}$$

where Eqs. (18.5) and (18.6) have been used. The vector-operators $\boldsymbol{\mu}_i \equiv \boldsymbol{\mu} \times (\hat{\pi}_i^+ + \hat{\pi}_i)$, $(i = A, B)$ express the two dipoles, assumed mutually parallel in real space and placed at the coordinates $\mathbf{r}_i$ within the cavity, and that are here represented by equal two-level systems resonating at a common radiation wavelength λ_0. The dipole operators are represented respectively by transition operators $\hat{\pi}_i^+, \hat{\pi}_i$ which satisfy the completeness and commutation relations

$$\hat{\pi}_i^+ \hat{\pi}_i + \hat{\pi}_i \hat{\pi}_i^+ = 1$$

$$\begin{aligned}
\hat{\pi}_i^+ \hat{\pi}_i^+ \;=\; & \hat{\pi}_i \hat{\pi}_i = 0, \quad [\, \hat{\pi}_i^+ \hat{\pi}_i, \, \hat{\pi}_j^+ \,] = \delta_{ij} \hat{\pi}_i^+ \\
[\, \hat{\pi}_i^+, \hat{\pi}_j \,] \;=\; & \delta_{ij}(2\hat{\pi}_i^+ \hat{\pi}_i - 1), \quad [\, \hat{\pi}_i^+ \hat{\pi}_j, \hat{\pi}_j \,] = -\delta_{ij} \hat{\pi}_i
\end{aligned}$$

The *complete* Hamiltonian of the overall dynamical system is

$$\begin{aligned}
\hat{H} \;=\; & (\hbar ck_0)(\hat{\pi}_A^+ \hat{\pi}_A + \hat{\pi}_B^+ \hat{\pi}_B) \\
& + c\hbar \int d\mathbf{k}\, k \sum_j \{\hat{a}_{\mathbf{k}j}^\dagger \hat{a}_{\mathbf{k}j} + \hat{a}'^\dagger_{\mathbf{k}j} \hat{a}'_{\mathbf{k}j}\} + i \int d\mathbf{k} \sum_j \sqrt{(\hbar kc/16\pi^3 \, \varepsilon_0)}
\end{aligned}$$

$$\times [\exp(-ickt)/D_j] \, \mu \cdot \big\{ \{ \hat{a}_{\mathbf{k}j} [\, \varepsilon(\mathbf{k}_+ j) t_j \exp(i\mathbf{k}_+ \cdot \mathbf{r}_A)$$
$$+ \, \varepsilon(\mathbf{k}_- j) t_j r_j \exp(i\mathbf{k}_- \cdot \mathbf{r}_A + i \, kdC)]$$
$$+ \hat{a}'_{\mathbf{k}j} [\, \varepsilon(\mathbf{k}_- j) t_j \exp(i\mathbf{k}_- \cdot \mathbf{r}_A) + \varepsilon(\mathbf{k}_+ j) t_j r_j \exp(i\mathbf{k}_+ \cdot \mathbf{r}_A + i \, kdC)] \}$$
$$\times (\hat{\pi}_A^+ + \hat{\pi}_A)$$
$$+ \{ \hat{a}_{\mathbf{k}j} [\, \varepsilon(\mathbf{k}_+ j) t_j \exp(i\mathbf{k}_+ \cdot \mathbf{r}_B) + \varepsilon(\mathbf{k}_- j) t_j r_j \exp(i\mathbf{k}_- \cdot \mathbf{r}_B + i \, kdC)]$$
$$+ \hat{a}'_{\mathbf{k}j} [\, \varepsilon(\mathbf{k}_- j) t_j \exp(i\mathbf{k}_- \cdot \mathbf{r}_B) + \varepsilon(\mathbf{k}_+ j) t_j r_j \exp(i\mathbf{k}_+ \cdot \mathbf{r}_B + i \, kdC)] \}$$
$$\times (\hat{\pi}_B^+ + \hat{\pi}_B) \, \big\} + h.c \tag{18.18}$$

where $C \equiv \cos \Theta$. We may now simplify the dynamical problem, without loss of generality, by making the following assumptions: (a) the optical parameters of the parallel, plane, equal mirrors placed at a mutual distance d, viz.: $r_{1j} = r_{2j} = -|r_j|, t_{1j} = t_{2j} = i |t_j|$, that make the microcavity are independent of the directions of the $\mathbf{k}$-vectors and of the field's polarizations; (b) the common dipole-vectors μ have directions parallel to the mirror planes. (c) The two dipoles are placed on the y-axis on the cavity symmetry plane $z = 0$, and are separated by a mutual distance $\mathbf{R}$ such that $\mathbf{r}_B = \mathbf{r}_A + \mathbf{R}$ (Fig. 18.1). The two "fictitious-spin" system is assumed to be prepared in the state $|\Psi_{AB} >$. By assumption (c) we may express the confined electromagnetic field in a simple fashion by the following "quasimode" substitutions:

$$\hat{c}_{\mathbf{k}j} \to \{ \hat{a}_{\mathbf{k}j} + \hat{a}'_{\mathbf{k}j} \}; \quad \hat{c}_{\mathbf{k}j}^\dagger \to \{ \hat{a}_{\mathbf{k}j}^\dagger + \hat{a}'^\dagger_{\mathbf{k}j} \}.$$

These new operators satisfy the commutation relations: $[\hat{c}_{\mathbf{k}j}, \hat{c}^\dagger_{\mathbf{k}'j'}] = 2\delta_{jj'}\delta \cdot (\mathbf{k} - \mathbf{k}')$. The interaction Hamiltonian appearing in Eq. (18.18) may now be re-expressed in the simple form:

$$\hat{H}_I = \int d\mathbf{k} \sum_j g_j(C) \times \exp[-i(ckt - \mathbf{k} \cdot \mathbf{r}_A)] \hat{c}_{\mathbf{k}j}$$
$$\times \{ \hat{\pi}_A^+ + \hat{\pi}_B^+ \exp(i\mathbf{k} \cdot \mathbf{R}) + \hat{\pi}_A + \hat{\pi}_B \exp(i\mathbf{k} \cdot \mathbf{R}) \} + h.c. \tag{18.19}$$

and the overall field-atom coupling parameter is now expressed by

$$g_j(C) \equiv [\hbar kc/(16\pi^3 \, \varepsilon_0)]^{1/2} \times G_m(kC)(\mu \cdot \varepsilon_j) \text{ and}$$
$$G_m(kC) \equiv (1 - |r|^2)^{1/2}/[1 + |r| \times \exp(i\pi mC)]. \tag{18.20}$$

We find that for large "finesse" $f = \{ \pi |r| /(1 - |r_j|^2) \} \gg 1$ is $G_m(kC) \approx 2(f/\pi)^{1/2}$ [19]. The symbol m represents the microcavity "length" in terms of the resonant half-wavelength: $m \equiv 2d/\lambda_0 \geq 1 : m$ is the characteristic parameter representing the cavity "mode order": the minimum value $m = 1$ corresponds to the minumum size of a planar Fabry-Perot microcavity that can resonate at λ_0, viz. that does not inhibit the resonant energy exchange between the atoms and the field in confinement condition [1, 2]. The Airy factor is now expressed by $D_j^{-1} \equiv [1 - |r_j|^2 \times \exp(i2\pi mC)]^{-1}$. Suppose first that the vectors μ are parallel to the x-axis. We may now investigate the time behaviour of the upper level $|A, \uparrow\rangle\langle A, \uparrow| = \hat{\pi}_A^+ \hat{\pi}_A$, for atom A, in the Heisenberg representation:

$$\hbar \frac{d\hat{\pi}_A^+ \hat{\pi}_A}{dt} = -i[\hat{\pi}_A^+ \hat{\pi}_A, \hat{H}] = i \int d\mathbf{k} \sum_j g_j(C) \exp(i\mathbf{k} \cdot \mathbf{r}_A) \hat{c}_{\mathbf{k}j}(\hat{\pi}_A^+ - \hat{\pi}_A) + h.c.$$

The time evolution of the operators in the right-hand side is obtained by the formal solution of an analogous equation involving the field operator $\hat{c}_{\mathbf{k}j}$ [18]. Evaluate now the ensemble average of the spin operators over the state of the (vacuum-field+atom) system: $|\{n_{\mathbf{k}j}\}, \Psi_{AB}>$, with $n_{\mathbf{k}j} = 0$. In other words, consider the process of Spontaneous Emission (SpE) of the intercorrelated atoms A or B, placed within the spatial extension of the single, transverse microcavity mode and suitably prepared in a state $|\Psi_{AB}>$. By formal integration of the above equation we obtain the evolution of the $\hat{\pi}^+ \hat{\pi}$ for atom A in second-order approximation, in the Heisenberg picture [18]:

$$\begin{aligned}
\frac{d\langle \hat{\pi}_A^+ \hat{\pi}_A \rangle_t}{dt} &= -\frac{2i}{\hbar^2} \int d\mathbf{k} \sum_j |g_j|^2 \{\langle \hat{\pi}_A^+ \hat{\pi}_A \rangle_0 \, \zeta(k_0 - k) \\
&\quad -\langle \hat{\pi}_A \hat{\pi}_A^+ \rangle_0 \, \zeta(k_0 + k) + [\langle \hat{\pi}_A^+ \hat{\pi}_B \rangle_0 - \langle \hat{\pi}_A \hat{\pi}_B \rangle_0 \exp(-2i\omega_0 t)] \\
&\quad \times [\exp(-i\mathbf{k} \cdot \mathbf{R})\zeta(k_0 - k) - \exp(i\mathbf{k} \cdot \mathbf{R})\zeta(k_0 + k)]\} \\
&\quad +c.c.
\end{aligned} \qquad (18.21)$$

where $\zeta(x) \equiv [1 - \exp(ictx)]/(cx)$, and:

$$\begin{aligned}
\sum_j |g_j|^2 &= \sum_{j=1}^{2} \left(\frac{\hbar c k}{16\pi^3 \varepsilon_0}\right) |G_m(k\, d\, C)|^2 \, \varepsilon_i^{(j)} \, \mu_i \, \varepsilon_h^{(j)} \, \mu_h \\
&= \left(\frac{\hbar c k}{16\pi^3 \varepsilon_0}\right) \mu_i \, \mu_h \, |G_m(kdC)|^2 \sum_{j=1}^{2} \varepsilon_i^{(j)} \, \varepsilon_h^{(h)} \\
&= \left(\frac{\hbar c k}{16\pi^3 \varepsilon_0}\right) \mu_i \, \mu_h \, |G_m(k\, d\, C)|^2 \, (\delta_{ih} - \hat{k}_i \hat{k}_h).
\end{aligned} \qquad (18.22)$$

The sum is calculated on the two transverse modes of the field, and the indexes $i, h = 1, 2, 3$ represent the spatial vector and tensor components, while $j = 1, 2$ represent the polarization components. The Airy parameter may be expressed by its Fourier expansion:

$$|G_m(k\, d\, C)|^2 \approx 1 + 2 \sum_{n=1}^{\infty} (-|r|)^n \cos(n\, k\, d\, C) \qquad (18.23)$$

where $|r| < 1$ is the reflectivity of the two equal mirrors. Equation (18.21) transforms into:

$$\frac{d\langle \hat{\pi}_A^+ \hat{\pi}_A \rangle_t}{dt} = -\frac{i\, \mu_i \, \mu_h c}{8\pi \, \varepsilon_0 \hbar} \int_0^{\infty} dk\, k^3 \int_{\tilde{\Omega}} d\Omega \left[1 + 2 \sum_{n=1}^{\infty} (-|r|)^n \cos(n\, k\, d\, C)\right]$$

$$\times \left\{ \langle \hat{\pi}_A^+ \hat{\pi}_A \rangle_0 \, \zeta(k_0 - k) + \langle \hat{\pi}_A \hat{\pi}_A^+ \rangle_0 \, \zeta(k_0 + k) \right.$$
$$+ \left[\langle \hat{\pi}_A^+ \hat{\pi}_B \rangle_0 - \langle \hat{\pi}_A \hat{\pi}_B \rangle_0 \exp(-2i\omega_0 t) \right]$$
$$\times \left. \left[\exp(-i\mathbf{k} \cdot \mathbf{R}) \zeta(k_0 - k) - \exp(i\mathbf{k} \cdot \mathbf{R}) \zeta(k_0 + k) \right] \right\}$$
$$\times (\delta_{ih} - \hat{k}_i \hat{k}_h) + c.c. \tag{18.24}$$

The result of the angular integration over the solid-angle $\overline{\Omega} : \{0 \le \Phi \le 2\pi; 0 \le \Theta \le \pi/2\}$ is expressed by:

$$
\begin{aligned}
I(R, n) &= \int_{\overline{\Omega}} d\Omega \, (\delta_{ih} - \hat{k}_i \hat{k}_h) \cos(n\, k\, d\, C) \exp(\pm i \mathbf{k} \cdot \mathbf{R}) \\
&= 2\pi \left\{ \left[\delta_{ih} - (\hat{R}_n)_i (\hat{R}_n)_j \right] \frac{\sin(k R_n)}{k R_n} \right. \\
&\quad + \left. \left[\delta_{ih} - 3(\hat{R}_n)_i (\hat{R}_n)_j \right] \left[\frac{\cos(k R_n)}{k^2 R_n^2} - \frac{\sin(k R_n)}{k^3 R_n^3} \right] \right\} \tag{18.25}
\end{aligned}
$$

where $(\hat{R}_n)_i$ is the direction cosine of the vector $\mathbf{R}_n = \mathbf{R} + n\mathbf{d} \equiv (R_x, R_y, n\, d)$. The explicit evaluation of the integral Eq. (18.24) is reported in [19]. Finally Eq. (18.24) can be written in the form showing the contributions to the population transition rate for atom A due respectively to the "free-space", i.e., non-confined, vacuum and to the microcavity-confined vacuum,

$$\frac{d\langle \hat{\pi}_A^+ \hat{\pi}_A \rangle_t}{dt} = I_{VAC} + I_{CAV}$$

The explicit formal expressions of these contributions are:

$$
\begin{aligned}
I_{VAC} &= -\frac{ic}{8\pi^2 \, \varepsilon_0 \hbar_0} \int_0^\infty dk\, k^3 \\
&\quad \times \left\{ \frac{4}{3} |\boldsymbol{\mu}|^2 \{ \langle \hat{\pi}_A^+ \hat{\pi}_A \rangle_0 \, \zeta(k_0 - k) + \langle \hat{\pi}_A \hat{\pi}_A^+ \rangle_0 \, \zeta(k_0 + k) \} \right. \\
&\quad + 2\mu_i \, \mu_h \left[\langle \hat{\pi}_A^+ \hat{\pi}_B \rangle_0 - \langle \pi_A \pi_B \rangle_0 \exp(-2i\omega_0 t) \right] \\
&\quad \times [\zeta(k_0 - k) - \zeta(k_0 + k)] \\
&\quad \times \left\{ (\delta_{ih} - (\hat{R}_i \hat{R}_h)) \frac{\sin(k R)}{k R} + (\delta_{ij} - 3\, \hat{R}_i \hat{R}_h) \right. \\
&\quad \times \left. \left. \left[\frac{\cos(k R)}{k^2 R^2} - \frac{\sin(k R)}{k^3 R^3} \right] \right\} \right\} + c.c. \tag{18.26}
\end{aligned}
$$

$$
\begin{aligned}
I_{CAV} &= -\frac{i\, \mu_i \, \mu_h c}{2\pi^2 \, \varepsilon_0 \hbar} \sum_{n=1}^{+\infty} (-|r|)^n \int_0^\infty dk\, k^3 \\
&\quad \times \{ \langle \hat{\pi}_A^+ \hat{\pi}_A \rangle_0 \, \zeta(k_0 - k) + \langle \hat{\pi}_A \hat{\pi}_A^+ \rangle_0 \, \zeta(k_0 + k) \} \\
&\quad \times \left\{ (\delta_{ij} - \hat{d}_i \hat{d}_h) \frac{\sin(knd)}{knd} \right.
\end{aligned}
$$

$$+ (\delta_{ij} - 3\,\hat{d}_i\hat{d}_h)\left[\frac{\cos(knd)}{k^2 n^2 d^2} - \frac{\sin(knd)}{k^3 n^3 d^3}\right]\Bigg\}$$

$$+ \left[\langle\hat{\pi}_A^+\hat{\pi}_B\rangle_0 - \langle\hat{\pi}_A\hat{\pi}_B\rangle_0\exp(-2i\omega_0 t)\right] \times \left[\zeta(k_0 - k) - \zeta(k_0 + k)\right]$$

$$\times\left\{\left(\delta_{ij} - (\hat{R}_n)_i(\hat{R}_n)_h\right)\frac{\sin(kR_n)}{kR_n}\right.$$

$$\left. + (\delta_{ij} - 3(\hat{R}_n)_i(\hat{R}_n)_h)\left[\frac{\cos(kR_n)}{k^2 R_n^2} - \frac{\sin(kR_n)}{k^3 R_n^3}\right]\right\} + c.c. \quad (18.27)$$

The first integral in Eq. (18.26) does not converge for $k \to +\infty$. We may then set the upper integration limit to a finite value equal to the inverse of the atomic Bohr radius: $k_m = r_0^{-1}$. This value of $|\mathbf{k}|$ establishes the upper limit of the frequency domain that is consistent with the electric-dipole approximation for the field-atom interaction [10, 19]. The explicit evaluation of the contour integrals appearing in Eqs. (18.26) and (18.27) is given in [19]. We limit ourselves to give here the final result for the SpE rate for atom A, including the overall free- and the confined-vacuum contributions:

$$\frac{d\langle\hat{\pi}_A^+\hat{\pi}_A\rangle_t}{dt} = -2\gamma\left\{\langle\hat{\pi}_A^+\hat{\pi}_A\rangle_0\right.$$

$$+\frac{3\,\hat{\mu}_i\,\hat{\mu}_h}{k_0^3}\langle\hat{\pi}_A^+\pi_B\rangle_0 D_{ih}^R[\sin(k_0 R)]\theta(ct - R)$$

$$+\frac{3\,\hat{\mu}_i\,\hat{\mu}_h}{k_0^3}\sum_{n=1}^{+\infty}(-|r|)^n$$

$$\times\left\{\langle\hat{\pi}_A^+\hat{\pi}_A\rangle_0 D_{ih}^{nd}[\sin(k_0 nd)]\theta(ct - nd)\right.$$

$$\left.\left. + \langle\hat{\pi}_A^+\hat{\pi}_B\rangle_0 D_{ih}^{R_n}[\sin(k_0 R_n)]\theta(ct - R_n)\right\}\right\} \quad (18.28)$$

where

$$\gamma = \frac{|\,\mu|^2 k_0^3}{6\pi\,\varepsilon_0\hbar} = \frac{1}{2\tau_{RAD}} = \frac{\Gamma_0}{2} \quad (18.29)$$

being τ_{RAD} the free-space spontaneous-emission lifetime, and the general expression holds:

$$D_{ih}^x[\sin(k_0 x)] = k_0^3\left\{(\delta_{ih} - \hat{x}_i\hat{x}_h)\frac{\sin(k_0 x)}{k_0 x}\right.$$

$$\left. + (\delta_{ih} - 3x_i x_h)\left[\frac{\cos(k_0 x)}{k_0^2 x^2} - \frac{\sin(k_0 x)}{k_0^3 x^3}\right]\right\}. \quad (18.30)$$

Note in Eq. (18.28) the appearance of the self-interaction contribution to the SpE of the atom A: this one is expressed by the first term at the r.h.s. of Eq. (18.28). The effect of the presence of the atom B process takes place through two different channels, i.e., by a direct non-confined vacuum-field correlation, expressed by the second term at the r.h.s., and by a *confined-vacuum* process with intensity proportional to $|r|$. This last contribution is expressed by the series with alternate signs appearing in Eq. (18.28). A most relevant result of the present work consists precisely of the demonstration, expressed formally by Eq. (18.28), that the principle of relativistic causality is indeed verified in the process of spontaneous-emission via atom-atom interactions, a process involving the vacuum-field in any condition of confinement [11, 19]. Formally the effect of relativistic causality is expressed in Eq. (18.28) by the multiplication of all mutual interaction terms by the Heaviside step-functions: $\theta(z) = 1$, for $z > 0$ and $\theta(z) = 0$, $z < 0$, where the argument z is the appropriately retarded time [20]. Note that this result has been obtained in the framework of the nonrelativistic quantum theory, and that no preselection of retarded vs. advanced field propagation solutions has been found necessary. Note however in this respect that the transverse character of the coupling field, i.e., the fundamental relativistic behaviour of the free photon due to its zero mass, has been expressed right at the outset of the quantization procedure [12]. Note the peculiar interference with alternate signs affecting the increasingly retarded interaction contributions to the spontaneous emission rate of atom A due to the large number of images of atoms A and B provided by the reflections within the cavity itself. This interference leads to the peculiar quasi-oscillatory result of a numerical computation of Eq. (18.28) shown in Fig. 18.2. Such behaviour is found to be largely dependent on the cavity finesse f and on the ratio $\gamma = R/l_c$, expressing the degree of coexistence of the atoms within the same transverse cavity-mode. Very importantly, note that the adoption at the origin of the calculations of a Dicke-Hamiltonian, a common choice in this sort of problems, or of any truncated Hamiltonian [8, 9] would have made it impossible to obtain rigorously the relativistically causal result just expressed.

For the sake of completeness, we report also the result of an explicit evaluation of the spontaneous emission rates for different spatial orientation of the mutually parallel dipoles μ_A and μ_B ($|\mu| = |\mu_A| = |\mu_B| = \mu$) placed in the cavity symmetry plane $z = 0$ (see Fig. 18.1).

Case 1: Dipoles parallel to the x-axis.
The explicit evaluation of the tensor parameters appearing in Eqs. (18.28) and (18.29), leads to [19]:

$$\frac{d\langle \hat{\pi}_A^+ \hat{\pi}_A \rangle_t}{dt} = -2\gamma \left\{ \langle \hat{\pi}_A^+ \hat{\pi}_A \rangle_0 + \frac{3}{k_0^3} \langle \hat{\pi}_A^+ \hat{\pi}_B \rangle \right.$$

$$\times \left[\sin(k_0 R) \left(-\frac{1}{R^3} + \frac{k_0^2}{R} \right) \right.$$

$$\left. + \cos(k_0 R) \frac{k_0}{R^2} \right] \theta(ct - R) + \frac{3}{k_0^3} \sum_{n=1}^{+\infty} (-|r|)^n$$

$$\times \left\{ \langle \hat{\pi}_A^+ \hat{\pi}_A \rangle_0 \left[\sin(k_0 nd) \left(-\frac{1}{n^3 d^3} + \frac{k_0^2}{nd} \right) \right. \right.$$

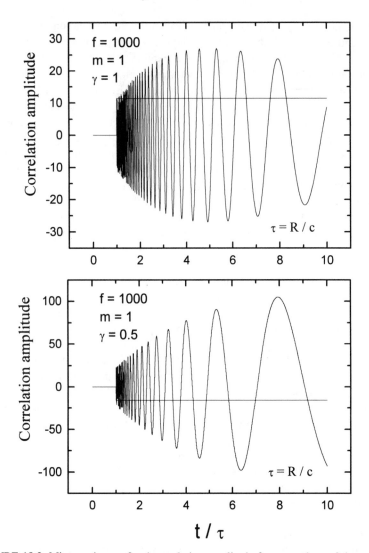

FIGURE 18.2. Microcavity-confined correlation amplitude for two values of the relative interatomic distance $= R/l_c$. The oscillatory behaviour due to the cavity effect is superimposed on the step-function behaviour expected in free-space, i.e., with cavity "finesse" **f**=0.9.

$$+ \cos(k_0 nd)\frac{k_0}{n^2d^2}\Bigg]\theta(ct - nd)$$

$$+ \langle\hat{\pi}_A^+\hat{\pi}_B\rangle_0\Bigg[\sin(k_0 R_n)\left(-\frac{1}{R_n^3} + \frac{k_0^2}{R_n}\right)$$

$$+\cos(k_0 R_n)\frac{k_0}{R_n^2}\Bigg]\theta(ct - R_n)\Bigg\}\Bigg\}. \qquad (18.31)$$

Case 2: Dipoles parallel to the y-axis and aligned on the same axis:

$$
\begin{aligned}
\frac{d\langle \hat{\pi}_A^+ \hat{\pi}_A \rangle_t}{dt} &= -2\gamma \Bigg\{ \langle \hat{\pi}_A^+ \hat{\pi}_A \rangle_0 + \frac{3}{2}\, \hat{\vec{\mu}}_i\, \hat{\vec{\mu}}_h\, \langle \pi_A^+ \hat{\pi}_B \rangle_0 \\
&\quad \times \left\{ 2\left[-\frac{\cos(k_0 R)}{k_0^2 R^2} + \frac{\sin(k_0 R)}{k_0^3 R^3} \right] \right\} \times \theta(ct - R) \\
&\quad +3\, \hat{\vec{\mu}}_i\, \hat{\vec{\mu}}_h\, \sum_{n=1}^{+\infty} (-|r|)^n \\
&\quad \times \Bigg\{ \langle \hat{\pi}_A^+ \hat{\pi}_A \rangle_0 \left[\frac{\sin(k_0 nd)}{k_0 nd} + \frac{\cos(k_0 nd)}{k_0^2 n^2 d^2} - \frac{\sin(k_0 nd)}{k_0^3 n^3 d^3} \right] \\
&\quad \times \theta(ct - nd) \\
&\quad +\langle \hat{\pi}_A^+ \hat{\pi}_B \rangle_0 \Bigg\{ 2\left[-\frac{\cos(k_0 R_n)}{k_0^2 R_n^2} + \frac{\sin(k_0 R_n)}{k_0^3 R_n^3} \right] \\
&\quad + \frac{n^2 d^2}{R_n^2}\left[\frac{\sin(k_0 R_n)}{k_0 R_n} + 3\frac{\cos(k_0 R_n)}{k_0^2 R_n^2} - 3\frac{\sin(k_0 R_n)}{k_0^3 R_n^3} \right] \\
&\quad \times \theta(ct - R_n) \Bigg\} \Bigg\} \Bigg\}.
\end{aligned}
\tag{18.32}
$$

Note in this second case the absence of the long-range interaction term proportional to R^{-1}. At the end of this calculation we may summarize the physical content of the results of the present section of the work by the following relevant prescription: The use of a truncated Hamiltonian (like a Dicke-type Hamiltonian) within a dynamical quantum-scattering theory can be admissible when only a limited range of frequencies over the entire spectrum is actually relevant in the context of the problem at hand. This is the case for all kinds of resonant field-atom interactions where only a small portion of the positive frequency spectrum around the resonance at ω_0 determines the most important physical features of the scattering process. The approximation made in this case can be acceptable only for a strongly resonant process, i.e., where only a restricted frequency range is really important to determine the relevant feature of the resonance. An example is given in the context of a spin-field coupling process by the adoption of the Dicke-Hamiltonian within the calculation of the Rabi frequency at resonance. In this case the results are generally affected by a very small, higher order, Bloch-Siegert frequency-shift [21, 22]. This last contribution is precisely due a more accurate, albeit clumsy, calculation involving the full spectrum of the interacting field. In general many important non-resonant interaction physical processes require an adequate formal account of the entire radiation spectrum. These may be for instance the ones connected with relativistic causality, as we have just seen, with pulse-shape evolution in material media (e.g., in the theory of "Self-Induced-Transparency", or in soliton-pulse propagation in a resonant optical amplifier [23]), and with wide-band photon-detection (as in Glauber's photo-detection theory [24]). In these last cases, the formal elegance intrinsic of a simple and symmetric Hamiltonian formulation of the problem is generally paid by a severe lack of accuracy of the final results. As we have seen in the present work, the causality issue just cannot be resolved by a shorter and more elegant formulation. This may lead, as it did recently, to some misleading epistemological conclusions about the role of special relativity in the context of quantum-mechanics.

18.4   Field Commutation Relations and Field Propagator

The correspondence existing in QED between the phenomenon of spontaneous emission and the commutation property of the field interacting with the atom now suggests comparing the above results, which involve pairs of cavity-confined interacting atoms, with properties involving quadratic combinations of cavity-confined fields evaluated in a quantum relativistic framework [12]. Let us start evaluating the commutator of the four-vector field $A(x)$ evaluated at different space-time points x belonging to the plane $z = 0$ within the microcavity. We may use again the mode structure described by the spatial functions $U_{kj}(r)$, $U'_{kj}(r)$, the corresponding plane-wave field operators $\hat{a}_{kj}$, $\hat{a}'_{kj}$ and integrate the different plane-wave commutators over the full four-momentum space filtered by the cavity boundaries. Formally the **k**-space "filtering" procedure is simply obtained by multiplying by the cavity function Eq. (18.23) the field functional expressions evaluated by standard QED for free-space, before integration over the momentum space. The final result of the integration appropriate for the commutators of the electromagnetic field is [12]:

$$[\hat{A}_\mu(x), \hat{A}_\nu(x')] = 4\pi \, ig_{\mu\nu} \left\{ D_0(x - x') + 2 \sum_{n=1}^{\infty} (-1)^n \, |r|^n \, D_0(x - x')_n \right\}$$

where the invariant $D_0(x - x')_n \equiv (1/2\pi)\delta(R_n^2 - c^2 T^2)$ sign $T = [1/(4\pi R_n)] \times [\delta(R_n - cT) - \delta(R_n + cT)]$ is introduced, as usual. We may also write the equal-time commutator in terms of the canonical field-momentum operator $\hat{p}_\nu = -(1/4\pi)(\partial \hat{A}_\nu/\partial t)$ in the form:

$$[\hat{A}_\mu(x), \hat{p}_\nu(x')]_{t=t'} = ig_{\mu\nu} \left\{ \delta(R) + 2 \sum_{n=1}^{\infty} (-1)^n \, |r|^n \, \delta(R) \right\}$$

where $T = (t - t')$, $\mathbf{R} = (\mathbf{r} - \mathbf{r}')$ and $\mathbf{R}_n$ is a vector having modulus R_n, already used within Eq. (18.25), and connecting one atom with the nth -order "image" of the other through the mirrors. The modern analysis of any quantum scattering process involving photons is based on the evaluation of the corresponding S-operator, usually by a rather clumsy perturbation procedure. In quantum optics we can apply this theory in various contexts, for instance in atomic SpE, nonlinear molecular scattering, nonlinear parametric processes, etc. The explicit evaluation of the terms of the perturbation technique is generally simplified by the application of the

Wick's theorem for the Bose fields. This requires the evaluation of a quantity called field chronological-pairing, or QED field-contraction [12]. The cavity "filtering" procedure, just outlined, may be adopted to extend the standard QED evaluation of this quantity made for free-space to the condition of microcavity confinement. The result of such calculation is found to be:

$$\hat{A}_\mu(x)\hat{A}_\nu(x') \equiv T\hat{A}_\mu(x)\hat{A}_\nu(x') - N\hat{A}_\mu(x)\hat{A}_\nu(x') = \langle T\hat{A}_\mu(x)\hat{A}_\nu(x')\rangle_0$$

$$= -4\pi\, g_{\mu\nu}\left\{ D_c(x-x') + 2\sum_{n=1}^{\infty}(-1)^n\, |r|^n\, D_c(x-x')_n \right\} \quad (18.33)$$

where T and N represent time- and normal-ordering, respectively and $D_c(x)$ is the field's causal function, or propagator [12]

$$D_c(x) = \theta(x)D_+(x) + \theta(-x)D_-(x) \text{ and}$$
$$D_\pm(x) = -[(2\pi x)^{-2} \pm (i/4\pi)\delta(x^2)\text{sign}\,x^0]. \quad (18.34)$$

The expressions Eqs. (18.33) and (18.34) are important since, within the context of any process of quantum scattering, they fully account for the effect of the microcavity field-confinement. Accordingly, the use of these expressions at the outset of the calculations allows a straightforward solution of the overall problem by a further application of the standard theory of QED.

18.5   Electron Mass Renormalization

In addition to the effects on QED scattering, considered in the previous sections, the process of field-vacuum confinement determined by the microcavity structure can affect substantially the strength of the self-interactions of charged elementary particles, e.g., of the electron. According to the standard renormalization program of QED, this may be interpreted as a change of the renormalized mass due to the modification of the vacuum-field distribution around the particle. We may formally evaluate this change by the application of the momentum-space "filtering" procedure outlined in Sec. 4 to the evaluation of the contribution to the one particle irreducible Green's function corresponding to the one-loop QED process shown diagrammatically, for free space and to lowest order, in Fig. 18.3 [12, Chap.7]. The renormalized electron mass is then obtained by the evaluation over the four-momentum space of the integral:

$$\Sigma_{cav}(p) = \frac{ie^2}{(2\pi)^4}\int d^4k\, f(k_3)\gamma_\mu \frac{i(\hat{p}-\hat{k})-m}{(p-k)^2+m^2}\gamma^\mu \frac{1}{k^2} \quad (18.35)$$

where γ_μ represents Dirac matrices, $\hat{p} = \gamma^\mu p\gamma_\mu$, $\hat{k} = \gamma^\mu k\gamma_\mu$, and we use the shorthand notation for the bare electron mass $m = 1/\lambda_c = m_{el}c/\hbar = 2.59 \times 10^{10}$ cm^{-1}. The "filter" function is expressed, according to Eq. (18.23), by the Airy

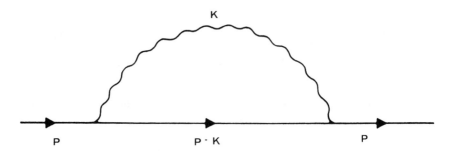

FIGURE 18.3. Free-space electron self-energy Feynman diagram of second order.

function:

$$f(k_3) = 1 + 2 \sum_{n=1}^{+\infty} (-|r|)^n \cos(n \, d \, k_3) \qquad (18.36)$$

The overall renormalized correction to the mass is then given by the two contributions:

$$\Sigma_{cav}(p) = \Sigma_{vac}(p) + 2 \sum_{n=1}^{+\infty} (-|r|)^n I_{cav}(p, n) \qquad (18.37)$$

where the cavity effect is provided by the series, with alternate signs and terms proportional to the reflectivity parameter $|r| \le 1$. Note that, since a Fabry-Perot cavity is a high-pass frequency filter, our present renormalization theory is free from the infrared divergences affecting the standard, free-space theory [12]. The evaluation of Eq. (18.35) is obtained through the integral:

$$I_{cav}(p, n) = \frac{ie^2}{(2\pi)^4} \int d^4k \cos(n \, d \, k_3) \, \gamma_\mu \frac{i(\hat{p} - \hat{k}) - m}{(p - k)^2 + m^2} \gamma^\mu \frac{1}{k^2}. \qquad (18.38)$$

This highly nontrivial integral is solved using the result [25]:

$$\frac{1}{a_1 a_2} = \int_0^1 \frac{dx}{[a_1 x + a_2(1 - x)]^2}.$$

Then

$$
\begin{aligned}
I_{cav}(p, n) &= \frac{ie^2}{(2\pi)^4} \int d^4k \cos(n \, d \, k_3) \int_0^1 dx \, \gamma_\mu \frac{i(\hat{p} - \hat{k}) - m}{[(k - px)^2 + a^2]^2} \gamma^\mu \\
&= -\frac{2ie^2}{(2\pi)^4} \int d^4k \, \cos(n \, d \, k_3) \int_0^1 dx \, \frac{i\hat{p} + 2m - i\hat{k}}{[(k - px)^2 + a^2]^2} \qquad (18.39)
\end{aligned}
$$

The last result is obtained by setting $\gamma_\mu \hat{p} \gamma^\mu = -2\hat{p}$; $\gamma_\mu \gamma^\mu = 4$, $a^2 = (p^2 + m^2)x(1-x) + m^2x^2$ and then $a^2 = m^2x^2$ for the free particle $i\hat{p} = -m$. The integral Eq. (18.39) is evaluated, by a quite complex procedure, for the case of the free particle and with the approximation $ndm \gg 1$. In fact, assuming $d \cong 1000\text{Å} = 10^{-5}$ cm, we have $d\,m \cong 10^5$. The details will be given by a forthcoming comprehensive paper [26]. We provide here only the final result, which is expressed by:

$$I_{cav}(n) = \frac{\alpha m}{2} \left\{ \frac{3}{2} \frac{K_0(ndm)}{\pi} - \frac{1}{\sqrt{\pi}} \Gamma(3/2)[K_1(ndm)L_0(ndm) \right.$$
$$+ K_0(ndm)L_1(ndm)] - K_2(ndm) + \left. \frac{1}{(ndm)^2} \right\}$$
$$- \left(\frac{3}{4}\right)^{3/4} \frac{\alpha}{8nd} \exp\left(-\frac{ndm\sqrt{3}}{2}\right) \tag{18.40}$$

where $K_n(z)$ are the modified Bessel function and $L_n(z)$ the modified Struve function [20]. By the properties of these functions, one can simplify this expression for $(ndm) \gg 1$ and obtain,

$$I_{cav}(n) \cong \frac{\alpha m}{4} \left\{ \frac{1}{2ndm} + \frac{2}{\pi(ndm)^2} + O[1/(ndm)^3] \right\}$$

$\alpha = e^2/4\pi$ being the fine-structure constant. The final expression of the renormalized free-electron mass is expressed by $m_{cav} = m_{el} + \delta m_{cav}$ with

$$\delta m_{cav} = \frac{\alpha m_{el}}{2} \sum_{n=1}^{+\infty} (-|r|)^n \left\{ \frac{1}{2ndm} + \frac{2}{\pi(ndm)^2} + O[1/(ndm)^3] \right\}.$$

The expression of first term of the first series at the right-hand side of the last equation coincides with an approximate result previously reported [27]. The larger relative contribution to the mass due to the renormalization in the presence of confinement is then simply expressed by $(\delta m_{cav}/m_{el}) = -[\alpha |r|/(4dm)] \approx [\alpha \lambda_c/(8\pi d)]$. This is found to be $(\delta m_{cav}/m_{el}) \approx -10^{-8}$ for $d = 1000$ Å. By assuming that the free-particle approximation holds for an electron orbitating on the outer shell of a Rydberg atom placed in a microcavity, the cavity contribution to the relative mass $(\delta m_{cavity}/m_{el}) = (\delta E/E)$ could be measured by a spectroscopic technique through the measurement of the energy E of the atomic level as a function of d. Note nevertheless that, in the above calculation we have assumed that $|r| \approx 1$ through the full frequency spectrum $\mathbf{k}$. A more realistic calculation must account for the presence of a plasma-frequency cutoff gives a smaller value of $(\delta m_{cavity}/m_{el})$. The details will be reported elsewhere [26].

References

[1] F. De Martini, G. Innocenti in *Quantum Optics IV*, J. Harvey, F. Walls eds. (Springer, Berlin, 1986); F. DeMartini, G. Innocenti, G. Jacobovitz, and P. Mataloni, Phys. Rev.

Lett. **59**, 2955 (1987); W. Jhe, A. Anderson, E. A. Hinds, D. Meschede, L. Moi, S. Haroche, Phys. Rev. Lett. **58**, 666 (1987).

[2] F. DeMartini, M. Marrocco, P. Mataloni, L. Crescentini, R. Loudon, Phys. Rev. A **43**, 2480 (1991).

[3] F. DeMartini, M. Marrocco and D. Murra, Phys. Rev. Lett. **65**, 1853 (1990); A. Aiello, F. DeMartini, M. Marrocco and P. Mataloni, submitted for publication.

[4] F. DeMartini and G. R. Jacobovitz, Phys. Rev. Lett. **60**, 1711 (1988); F. DeMartini, F. Cairo, P. Mataloni, F. Verzegnassi, Phys. Rev. A **46**, 4220 (1992); F. De Martini, M. Marrocco, P. Mataloni, D. Murra, and R. Loudon, J. Opt. Soc. Am. B **10**, 360 (1993).

[5] F. Cairo, F. De Martini and D. Murra, Phys. Rev. Lett. **70**, 1413 (1993); F. De Martini, M. Marrocco, C. Pastina and F. Viti, submitted for publication.

[6] E. Fermi, Rev. Mod. Phys. **4**, 87 (1932).

[7] M. I. Shirokov, Sov. J. Nucl. Phys. **4**, 774 (1967) and Sov. Phys. Usp. **21**, 345 (1978); M. H. Rubin, Phys. Rev. D **35**, 3836 (1987); G. C. Hegerfeldt, Phys. Rev. Lett. **72**, 596 (1994); D. Buckholtz and J. Yngvanson, Phys. Rev. Lett. **73**, 613 (1994).

[8] P. W. Milonni and P.L. Knight, Phys. Rev. A **10**, 1096 (1974).

[9] F. T. Arecchi and E. Courtens, Phys. Rev. A **2**, 1730 (1970).

[10] A. Biswas, G. Compagno, G. Palma, R. Passante, and F. Persico, Phys. Rev. A **42**, 4291 (1990).

[11] A. Valentini, Phys. Lett. A **153**, 321 (1991).

[12] C. Itzykson and J. Zuber, *Quantum Field Theory*, Ch. 3 (McGraw Hill, New York) (1980).

[13] L. Maiani and M. Testa, unpublished.

[14] A. Aiello, F. De Martini, M. Giangrasso, P. Mataloni, and D. Murra, *Ultrafast Phenomena IX*, Vol. 60, edited by W. Knox (Springer Verlag, Berlin), (1994).

[15] A. Einstein, Phys. Z. **18**, 121 (1917).

[16] M. Ley and R. Loudon, J. Mod. Opt. **34**, 227 (1987).

[17] M. Born and E. Wolf, *Principles of Optics*, Ch. 7 (Macmillan, New York) (1964).

[18] R. Loudon, *The Quantum Theory of Light* (Oxford University Press, New York) (1983).

[19] F. De Martini and M. Giangrasso, Appl. Phys. B **60**, S49 (1995).

[20] *Handbook of Mathematical Functions*, edited by M. Abramowitz and I. Stegun, (Dover, New York) (1965).

[21] R. H. Dicke, Phys. Rev. **89**, 472 (1953).

[22] A. Abragam, *The Principles of Nuclear Magnetism*, Ch. 2 (Clarendon Press, Oxford) (1961).

[23] S. L. McCall and E.L. Hahn, Phys. Rev. Lett. **18**, 908 (1967) and Phys. Rev. 183, 457 (1969).

[24] R. J. Glauber in *Quantum Optics*, edited by R. J. Glauber (Academic Press, New York) (1969).

[25] J. M. Jauch and F. Rohrlich, *The Theory of Photons and Electrons* (Springer, New York) (1976).

[26] F. De Martini and M. Giangrasso, submitted for publication.

[27] E. A. Power, Proc. Roy. Soc. London A **292**, 424 (1966); R. Golub and G. L. Guttrich, Phys. Lett. A **24**, 224 (1967); G. Barton, J. Phys. **B7**, 2134 (1974); P. W. Milonni, Int. J. Theor. Phys. **22**, 323 (1983).

19

Testimonial for Celebration of Professor Charles Townes' 80th Birthday

Sidney D. Drell

The scientific achievements and inventions of Professor Charles Townes have profoundly influenced the world and the work of today's scientists. In my own case, Charlie has had an exceptional impact, both personal and professional, and I am pleased to have this opportunity to express my appreciation and friendship, and to send warm greetings.

The personal dimension started with a phone call in early 1960 from Charlie, who was then the Executive Vice President of the Institute for Defense Analyses. This call was to invite me to Washington to attend an organizational meeting for a group of "bright, young physicists." The idea was to bring new talent together with fresh ideas to work on problems of national importance for the nation. This meeting was the genesis of the group of scientists called JASON that since then has been active, largely through summer studies, in analyzing technical problems to assist the nation as it faces new challenges.

Coming as it did from Charlie, whom I admired so much as a scientist, this invitation caused me to pause and think about many issues that had been pretty distant from my mind in my life as a physics professor in those days. Charlie presented cogent arguments about the need for a new and younger generation of scientists to address some of the major problems that the nation and the world were facing at that time during the Cold War, as technology took us to space and nuclear weapons continued to grow as a threat to survival. Further encouraged by my colleague and close friend, Pief Panofsky, whose own personal commitment and enormously valuable contributions in this area I admired very much, I attended that first briefing to learn what the JASON idea was all about. The rest, as they say, is history. For better or worse Charlie's phone call, my first personal interaction with him, triggered a major change in my career. Technical issues of national security, arms control, and nuclear safety joined pure theoretical physics as a major component of my scientific life.

About six years later at the beginning of 1966 Charlie and I were both appointed to serve on the President's Science Advisory Committee by President Lyndon Johnson. We served together for four years under both Presidents Johnson and Nixon, and saw a lot of each other. There were the regular monthly meetings in Washington, and, in addition, joint work on several panels formed to address specific problems. It was a wonderful opportunity to share technical, and also broader, insights and inquiries with Charlie. Our warm and broad friendship also developed. More than that, I learned to appreciate how much Charlie was contributing in the nation's service, above and beyond his prodigious achievements in pure science.

At about the same time, early in 1966, I literally "saw" one of Charlie's lasers for the first time. I suffered a detached retina which was repaired in an operation. Subsequently the condition that led to it has been treated successfully by means of periodic "patching up" with a laser by my ophthalmologist. This has preserved my eyesight and prevented the need for any further major operations. Thanks for your laser, Charlie!

Most recently Charlie and I, joined by a third colleague (Dr. John Foster, the former Director of Livermore National Laboratory and Director of Defense, Research and Engineering for the Department of Defense), were tasked by the U.S. Congressional Armed Services Committees to undertake a thorough review of the safety of the U.S. nuclear weapons stockpile. This study occupied a large fraction of 1990. The three of us worked together very closely. Once again I had the pleasure of a close collaboration and stimulating intellectual experience with Charlie.

It is against this extended background of more than 30 years of contact, collaboration, and friendship with Charlie that I am so delighted to have this opportunity to send my very warm personal greetings on the occasion of his 80th birthday. I hope our association will continue for many years to come.

20

Marine Physical Laboratory: A Brief History

Carolyn L. Ebrahimi
and Kenneth M. Watson

20.1 Introduction

It is a pleasure and a privilege to contribute to this volume commemorating Charles Townes' 80th birthday. One of us (KMW) met Charlie Townes shortly after World War II in the context of some research for the Navy. We had both worked on radar development for the Navy during the War. The association with Charlie Townes continued when we were colleagues at the Physics Department at the University of California (UC), Berkeley. Since my (Watson) transfer to the San Diego campus of the University of California, it has been a pleasure to see Charlie during his visits here.

The Marine Physical Laboratory (MPL) had its origin as an acoustics laboratory operated by the University of California for the Navy during World War II. The laboratory continued with Navy support after the war, becoming a part of the Scripps Institution of Oceanography (SIO). The Marine Physical Laboratory is now an administrative unit within the Scripps Institution of Oceanography, which is both an academic department and a research facility of the University of California at San Diego (UCSD). The new campus of what is now UCSD in San Diego was housed at the Scripps Institution of Oceanography (SIO) for a time, until facilities were constructed on the "upper campus." As the new campus evolved, SIO continued to expand its research into phenomena of the sea, atmosphere, and of the earth's crust. The Marine Physical Laboratory evolved from an acoustics laboratory into a significant facility developing and using acoustic instrumentation for geophysical and oceanographic research. It is the history of this laboratory that we will now describe.

20.2 World War II Research and the Creation of MPL

During the postwar years, the need for intense research in underwater sound and the training of marine acousticians led to the formation of the Marine Physical Laboratory. It was a successor to the University of California Division of War Research (UCDWR), which was set up under the University of California by the National Defense Research Council to conduct research and development in subsurface warfare. Dr. Roger Revelle, who as a Naval Reserve Commander was in charge of the oceanographic section of the Navy's Bureau of Ships in Washington, recalled that the formation of the Marine Physical Laboratory had the long-term pledge and interest of support from the Navy Bureau of Ships.

The main purpose of the Division of War Research was threefold: (1) to compile information on sound and all aspects of the ocean: temperature, currents, chemical properties, bottom topography, etc., which might have some bearing on the behavior of underwater sound; (2) the development of offensive and defensive equipment for submarine warfare; and (3) the selection and training of Navy personnel to operate these sensitive instruments.

Development of new submarine-detection instruments, contributions to the knowledge of shallow-water bottom topography, understanding of sound propagation, ambient noise, temperatures, salinity, and currents of the Pacific were wartime contributions of UCDWR to the security of the United States. These demonstrated the value of oceanographic research, and the need for study of scientific problems in underwater physics, which were of immediate value to the Navy and could be continued in higher academic institutions.

In June 1946, UCDWR was dissolved. The Navy Electronics Laboratory absorbed most of the UCDWR employees and the programs that were applicable features of wartime research: development of devices for submarine detection, underwater-sound propagation and training of personnel. Concurrently, by agreement between the Navy and the Regents of the University of California, the Marine Physical Laboratory was formed with an initial scientific staff of five people. It was the consensus of a number of high-ranking Navy officials that the Navy's research objectives "can best be achieved if part of this basic research is carried out by one of the nation's leading universities." In a letter to U.C. President Robert Gordon Sproul from Vice Admiral E. L. Cochrane, Chief of Bureau of Ships, stated:

> "In addition, the wider intellectual interests involved in academic positions would attract more capable personnel to this work and would also militate against the possible stagnation of the program. It appears, moreover, that such a program should be of intrinsic interest to the University....In particular, close liaison should be maintained between such a program and the related oceanographic studies at the Scripps Institution of Oceanography... It is therefore proposed that the University of California enter into a contract with the Navy Department to carry on for the Bureau of Ships, as part of its academic research program, the fundamental studies of the physics and acoustics of underwater sound started by the University of California, Division of War Research."

He further stated:

> "The advantages to the Navy:

(1) Research on fundamental sonar problems, which are vital to the development of the most effective underwater sound gear, would be carried out on a broad basis by scientists of high caliber and national reputation.

(2) Much of the work could be done by graduate students, who could be depended upon to supply a continuous stream of new and ingenious ideas.

(3) A large number of physicists and engineers would be familiarized with and trained in underwater sound problems, and consequently, there would be available a pool of qualified personnel if rapid expansion of Naval research on underwater sound problems should again be necessary in the future.

The advantages to the University of California:

(1) The University would gain in prestige through the conduct of scientific research of fundamental geophysical importance and national scope.

(2) Facilities, including ships, would be available for research projects required of graduate students seeking degrees.

(3) The proposed investigations would furnish a significant field of research for several departments of the University. Some research might also be carried out on shipboard in other fields not of primary interest to the Bureau of Ships."

The Marine Physical Laboratory was opened as a research unit of the University of California with Dr. Carl Eckart as Director. He had worked at UCDWR from 1942. Director Eckart gave exceptional leadership in the first laboratory projects initiated at MPL. He was keenly interested in signal processing. He also pursued studies of sound scattering from rough surfaces, and absorption of sound by liquids. In the Spring of 1948, Eckart also assumed the directorship of Scripps, while simultaneously MPL became a part of SIO. In 1961, the University formalized its policy on Organized Research Units (ORUs) and MPL became an ORU within Scripps.

Dr. Carl Eckart continued as Director of MPL until his sabbatical in 1952, to be replaced by Sir Charles Wright. Wright had been Director of Scientific Research for the British Admiralty during World War II. In 1955 Wright retired from MPL and was replaced by Albert Focke as Director. Focke left in 1958 and became a professor at Harvey Mudd College.

Dr. Fred Spiess (a Berkeley physics graduate student and contemporary of Brueckner) became Director of MPL following Focke. He had been at MPL since

FIGURE 20.1. The past directors of the Marine Physical Laboratory in April 1946. From left are Fred N. Speiss, Charles S. Wright, Carl Eckart, and Alfred B. Focke. Photo courtesy of Elizabeth Shor.

1952. He remained as MPL Director until 1980, when he took the directorship of the University-wide Institute of Marine Resources. During his time as Director of MPL the UCSD campus was created. Both Drs. Eckart and Spiess played significant roles in the development of UCSD. Fred Spiess sometimes wore two hats from 1961 to 1965 when he also served some of that time as Acting Director of SIO and from 1964-1965 as Director of SIO.

In 1981, Kenneth Watson, who was a theoretical physicist at the University of California, Berkeley, became Director of MPL. He remained as Director until his retirement in 1991.

20.3   Research History

A major portion of MPL's historical context has centered around the study of underwater acoustics: reverberation, sound propagation, sonar signal processing, the distribution of sound scatterers throughout the water, and the background noise of the sea. These studies bring other elements of oceanography into focus: signal processing; physical, chemical, and biological subjects; marine geophysics; and Earth's magnetism.

The "naval relevance" of this broad-based research program in MPL's infancy brought to fruition other areas of oceanographic specialization: research platforms and devices, seafloor mapping and navigation systems, Doppler sonars and arrays, and unmanned underwater vehicles. The research platform FLIP has a busy schedule and MPL scientists continue active at-sea research aboard the SIO fleet of ships and ships of other facilities. The Deep Tow system remains in frequent use.

Russell Raitt (a Caltech graduate in physics) was a member of the staff of UCDWR and one of the original staff members of MPL when it was formed. During WWII he worked on problems of sonar use, including discovery of the Deep Scattering Layer, explanation of the "afternoon effect" (degradation of sonar systems during the day due to surface heating of the water), and measurements of seafloor reflectivity. After the war he developed effective methods of doing seismic-refraction work in the deep sea, and over the following decades used these methods to determine the thickness and seismic velocity of the sediments, basement, crust, and mantle below the floor of the Pacific and Indian Oceans, and the anisotropy of mantle velocity.

In 1947, a UCLA graduate student, Victor C. Anderson, came to MPL to do research under Dr. Russell Raitt's guidance. His interest in electronics and signal processing continued during a postdoctoral period with F. V. Hunt at Harvard, where Anderson developed the Delay Line Time Compressor (DELTIC) correlator. Dr. Anderson returned to MPL and later developed the Digital Multibeam Sonar Processor (DIMUS), an extension of DELTIC. The DIMUS system has been widely adopted by the Navy for use in sonar systems.

George Shor joined MPL in 1953. With a background in geology and geophysics, he provided a link to the marine geology community at SIO, working primarily on geological structure of ocean margins and island arcs. He worked with Raitt on exploration of the oceans using seismic refraction and reflection, and on improvement of echosounding systems.

Victor Vacquier, inventor during WWII of the Magnetic Airborne Detector (MAD) joined MPL in 1957 to continue the program of magnetic measurements initiated by Ronald Mason. Vacquier discovered the large offsets of magnetic lineations of the seafloor that became a key to crustal movements, and measured the original magnetic inclination of the rocks of seamounts, thereby determining their horizontal movement between their dates of formation and the present. He carried out extensive work on heat flow both through the ocean floor and in continental areas.

Because of varied interests on the part of the scientific staff, research at MPL has become diverse. The investigation of ambient ocean noise and of acoustic propagation continues as a major effort. Acoustic receiving arrays as large as 3 km in length have been developed by F. H. Fisher, W. M. Hodgkiss, and J. A. Hildebrand. These arrays provide information on the frequency spectrum and levels and sources of ocean sound. A substantial signal-processing effort is required to process the array data. Seismic signals and the propagation of sound in the sediment and rocks of the ocean floor provide information on the structure and composition of the earth's crust. "On-the-bottom" seismometers have been developed and used by L. Dorman, J. Hildebrand and S. Webb. These researchers have applied tomographic analysis with these instruments to study the structure of seamounts.

The use of acoustic data in Arctic research is pursued by M. J. Buckingham and R. Pinkel. Sonars have been developed and deployed by R. Pinkel, J. Smith, and K. Melville to study upper-ocean mixing and currents.

Swath-mapping echo-sounding systems are used by P. Lonsdale and C. de Moustier to study the morphology of the seafloor.

As a part of UCSD, MPL scientists give oceanographic students in the SIO graduate training in ocean research, both at sea and in the laboratory.

20.4   Research Platforms and Devices

The development of oceanographic instrumentation and data processing techniques has been a major part of the work of MPL scientists, who traditionally have also taken their equipment and skills to sea. The revolutionary development in the technology of scientific instrumentation is highlighted by a remark once made by Roger Revelle, referring to pre-World War II oceanography, "We even had a slogan that the best oceanographic instruments should contain less than one vacuum tube."

20.4.1   Deep Tow Instrumentation System

In providing new ways of making observations at sea the MPL has developed many innovative research platforms and devices that support people and instruments in or on the ocean. New vehicle and research platform development evolves from oceanic research needs.

In the early 1960s the Office of Naval Research had an interest in deep-submersible research and development. The Navy had purchased the bathyscaphe Trieste and ONR watched with interest in the development of Aluminaut by Reynolds Aluminum. A small group under the leadership of F. N. Spiess at MPL, realizing the limitations of these vehicles, wanted an instrumentation system that could be towed close to the ocean floor utilizing geophysical remote-sensing techniques (echo sounding, magnetics, gravity). Inherent in the instrumentation design would be the capability to examine the statistical distribution of seafloor slopes (which could be measured) averaging over lateral distances of less than 100 m. With initial funding from the Navy's submarine reconnaissance program (SUB-ROC), an opportunity to develop and build a near-bottom towed echo-sounding device came in 1961.

The resulting Deep Tow system consisted initially of a pressure-proof case for electronics, an up- and a down-looking echosounder, and a long baseline acoustic-transponder navigation system. The Deep Tow unit is towed near the seafloor by an electromechanical cable (about 9 km long) that has a coaxial core over which outgoing and returning signals and power signals are transmitted. This provides the convenience for data storage and display to be done aboard ship, with the sensors several kilometers below.

FIGURE 20.2. Deep Tow Instrumentation System prototype (dated 1961).

Since its inception, development of new system options has made Deep Tow a novel instrumentation system for fine-scale surveying, geologic mapping, and for seafloor search and recovery operations. A proton-precession magnetometer was added to allow near-bottom investigation of lineated magnetic anomalies, which became a cornerstone of the evidence of sea-floor spreading and plate tectonics. With complementary cameras, side-looking sonar (110 kHz), and a bottom-penetrating 4-kHz echo sounder, the Deep Tow system has provided further insight into the nature of the seafloor. The side-looking sonar capability enabled the group to make quantitative measurements of acoustic backscatter, and the fine-scale variability of surface reflectivity. Later a precision (0.001° C) temperature-measurement system, an opening-closing plankton net, water-sampling devices, and a salinity/temperature/depth measurement capability were added. These developments were the outgrowth of interest in properties of the water column close to the seafloor.

With a diversity of funding sources from the Navy Deep Submergence Program, Office of Naval Research and the National Science Foundation, Deep Tow has contributed to a variety of scientific programs. The most valuable contribution has been the fostering of 23 Ph.D. theses, well over 200 refereed publications, and the placement of researchers and educators into higher-education institutions.

20.4.2   Floating Instrument Platform (FLIP)

The research platform FLIP was developed to determine the limitations imposed by the ocean environment on the Navy's Submarine Rocket (SUBROC) program. One of the major questions concerned bearing accuracy obtained acoustically out to convergence-zone ranges. Horizontal temperature/salinity gradients in the ocean could introduce bearing errors in the volume of the ocean, and sloping bottoms could do the same for acoustic paths that interacted with the bottom. What was needed were measurements to determine the extent to which environmental gradients and fluctuations could affect bearing accuracy out to convergent zone ranges.

In the late 1950s F. N. Spiess and F. H. Fisher embarked on experimental work to address these problems with Spiess attacking the effect of bottom topography and Fisher attacking the effect of horizontal gradients and fluctuations on bearing accuracy.

Fisher initially used the research submarine, the USS BAYA (AGSS 318) to make such measurements, piggybacking on research cruises with Frank Hale and Henry Westfall of the U.S. Naval Electronics Laboratory (NEL). After a few cruises, it became clear that the submarine was not a satisfactory platform for this work, even though it had hydrophone booms with a 100-foot span. There was no way to obtain either an optical or electromagnetic-bearing reference in a satisfactory manner at periscope depth or at the desired 300-foot depth needed for separating acoustic multipaths. Even at the 300-foot depth, wave-action effects caused sufficient yawing motion of a submarine to limit measurements of bearing fluctuations in acoustic propagation. Fisher described these problems to Spiess in January, 1960, who responded by mentioning Allyn Vine's idea of upending a submarine to make a stable platform in the ocean. This interchange resulted in the development of the research platform FLIP. Spiess and Fisher determined that it would not be possible to convert a submarine to a vertical platform and that it would be better to start from scratch. To measure bearing accuracy and the fine-scale fluctuations in bearing (phase difference) and amplitude of sound waves due to inhomogeneities in the ocean such as thermal microstructure and internal waves, a manned spar buoy was proposed as an ideal platform to make such measurements.

The 300-foot draft was dictated by the desire to resolve acoustic multipaths of 5-millisecond pulses and the 20-foot diameter was a limitation imposed by the desire to berth it at the B finger of the existing Navy Electronics Laboratory pier. The device was designed to be a manned spar buoy able to work in 30-foot seas. This led to an overall length of 355 feet with a bow section that had an engine room, electronics laboratory, and room for personnel. However, a spar buoy with a constant cross-section and a draft of 300 feet would have a resonant period of 18 seconds, which is too close to the periods of waves encountered at sea. Philip Rudnick of MPL investigated the heave response of spar buoys with reduced diameters and, together with Spiess and Fisher, it was decided to tune FLIP to have a 27-second heave period with a null response at 22 seconds. This was accomplished by having FLIP's diameter taper from 20 feet at a depth of 150 feet to 12.5 feet at a depth of 60 feet. Although FLIP's heave response is a function of frequency, its response to waves is generally about 10% of the surface-wave amplitude.

Extensive model studies with different 1/10 scale model configurations were conducted by Fisher to work out the flipping operation. The idea was to simply flood tanks by opening valves to vent the control tanks. These studies led to the need for a hard tank that could withstand the full differential pressure during the flipping operation in order to prevent plunging that occurred when the tanks were allowed to flood freely. By dividing the hard tank into top and bottom sections and adding ballast in the horizontal keel, it was possible to ensure a safe flipping operation.

On June 22, 1962, the full-scale FLIP was launched at the Gunderson Brothers Engineering Company in Portland, Oregon. On July 23, after completion of out-fitting, it was tested for the first time in the Dabob Bay area of the Hood Canal in Washington state on the Navy tracking range. After successfully completing trials of the flipping operation, it was towed to San Diego to begin its research program.

The flipping operation takes from 20 to 30 minutes; once enough water is in the tanks to get it about 10 degrees down by the stern, it then takes only 90 seconds to go to the vertical. Compressed air stored in ten large air bottles is used to blow out water in FLIP's tanks to bring it back to the horizontal. Over 300 flipping operations have been conducted since FLIP was launched.

Since it has no propulsion of its own, it is always towed to station at speeds of up to 10 or 11 knots, depending on the weather and the size of the tugboat. It has a crew of five or six and can comfortably carry a scientific party of eight. Its engines and galley are mounted in trunnions to permit operation in both the horizontal and vertical position. Electronic equipment is generally mounted in racks that can be rotated before loading on FLIP and bolted to the laboratory decks so that they are in the upright position when FLIP is vertical. With resupply of food and fuel, FLIP has been at sea for nearly two months.

Although designed for only 30-foot waves, FLIP endured 80-foot swells (22-second period) with minimal damage off Hawaii in 1969. Originally designed for drifting operations, FLIP has been placed in moorings routinely for over twenty years. In a tight three-point mooring recently in 4000-meter water, its watch circle was less than 100 meters. Booms to launch sensors and an orientation motor to keep a constant heading are part of its operations equipment. Recently, a 3000-meter long vertical array with 200 hydrophones was deployed from FLIP. Acoustic navigation of the elements to a precision of 2 meters was a feature of the experiment.

Over thirty years of operations have included many deployments in the Pacific as far as Hawaii and one deployment in the Atlantic. While originally intended for acoustic research, it has become a versatile platform for research in upper-ocean physical oceanography, meteorology, geophysics, and biology.

20.4.3 Doppler Sonar Development

Instruments that measure ocean properties at one location—such as velocity, conductivity, and temperature meters—are limited in their ability to determine large-scale phenomena. In the mid-1970's to provide information relating to ocean currents within a substantial volume, R. Pinkel began development of acoustic Doppler profilers. The first effort used an 87.5-kHz 2-beam sonar built by Fred Fisher.

An array of these sonars mounted on FLIP, having a range of 1 to 2 km, has provided detailed information concerning the upper-oceanic internal wavefield,

FIGURE 20.3. (Top) P. Rudnick, F. H. Fisher and F. N. Spiess, designers of FLIP, holding the first model of FLIP made from a Louisville slugger bat by M. Crouch of MPL; (center left to right) FLIP returning to horizontal position during initial trials conducted at Dabob Bay in the Hood Canal, Washington; launching of FLIP from Gunderson Brothers Engineering Company, Portland, Oregon; (bottom) Christening ceremonies of FLIP at Gunderson Brothers Engineering Company, June 22, 1962. Left to right: Captain C. B. Momsen, Jr. (Office of Naval Research), F. N. Spiess (MPL), Warren G. Magnesun (Washington State Senator), and Sally Spiess.

near-surface turbulence, and surface wave properties. Later developments have included electronic beam scanning, sonars in the 100 to 300 kHz range, and sonars to be deployed from ships.

Pinkel has deployed his system under arctic ice to observe flow fields, turbulence and mixing under the ice pack. Jerry Smith has adapted the higher frequency Doppler system to study circulation in the surf zone.

20.5   Acknowledgments

We wish to express our special thanks to F. N. Spiess, F. H. Fisher, and George and Betty Shor for their special contributions to this paper.

21

Searching for the Cause of Alzheimer's Disease

Gerald Ehrenstein

21.1   Introduction

In 1962, when I completed my Ph.D. at Columbia under Charlie's supervision, jobs for physicists were plentiful. It seemed to me like a good opportunity to try something that interested me in a general way, but about which I knew very little. I asked Charlie's advice about whether to take a job in biophysics. He replied that it was worth a try, if I were interested, but that it would be prudent to set a deadline of 2 or 3 years to decide whether to continue or to return to the area of my basic training. I followed that advice, considered several areas of biophysics, and took a job at the National Institutes of Health (NIH) in Bethesda, Maryland in a laboratory started by another physicist, Kenneth Cole. My early work at NIH was on the squid giant axon, which was then the standard preparation for studying the electrical properties of nerves. I found the work quite interesting, and have remained at NIH for 33 years. (By Charlie's standards, I am now at about mid-career.)

Most of my work at NIH has been related to the electrical properties of the ionic channels that are located in the plasma membranes surrounding cells. This was a relatively small field when I started, but is now actively pursued by many laboratories. In 1963 at the first Biophysical Society Annual Meeting that I attended, there were 210 abstracts. By comparison, at the Annual Meeting in 1995, the number of abstracts was about 2700. During this period, the fraction of presentations relating to ion channels has also increased.

Most research on ionic channels is primarily basic research, and the overall approach is quite similar to the approach in basic physics research. There are a number of key questions, such as the molecular basis for the voltage-dependence of channel opening and the molecular basis for the ionic selectivity of the channels. Answers to these questions are developed gradually as more evidence is obtained and as better hypotheses are developed and tested. In general, information relating to the properties of specific channels or to the properties of the membranes in which they are embedded suggests further experiments, and is useful in the quest for answers to the key questions.

For the past two years, I have been working on a new project, whose goal is to determine the cause of Alzheimer's Disease (AD). Before describing some of the scientific issues involved in this project, I would like to point out a difference in approach between this type of project and basic research. In searching for the cause of AD, we are focusing on mechanisms that may be important in the etiology of the disease. At this time, it appears likely that a peptide called beta-amyloid ($A\beta$) plays a crucial role in causing the disease, and much of the effort involves studies of the various effects of this peptide. Because of the specific goal of this effort, however, information relating to effects of $A\beta$ is useful only if it relates to the cause of the disease. Thus, if we were to find an effect of $A\beta$ that is of biochemical or physiological interest, but that turned out to be unrelated to AD, we would not pursue it. Instead, we would reassess the field, and consider alternative hypotheses. At this time, this issue is hypothetical, since we do not yet know enough to rule out any reasonable hypothesis.

In this paper, I will try to summarize the essential facts concerning the cause of AD, describe some of the hypotheses being considered, and indicate the path that I am currently following.

21.2   Important Facts about Alzheimer's Disease

AD is the most common type of senile dementia, and involves the loss of large numbers of brain neurons. A useful source for essential facts concerning AD is a recent issue of the Journal of Neuropathology and Experimental Neurology, in which Roses and Selkoe presented two conflicting views of the primacy of apolipoprotein E (apoE) and $A\beta$. Roses [1] emphasized the importance of genetic factors. In his view, the correlation between the e4 allele of apoE and the probability of AD is much more significant than other genetic correlations. Selkoe [2] emphasized the importance of the correlation between the presence of $A\beta$ plaques and AD. In assessing these two views, Hyman and Terry [3] stated that, despite the areas of disagreement, they agree for the most part on four essential facts. I will summarize these four facts below:

(1) Plaques containing $A\beta$ are present in the brains of all AD patients [4].

(2) $A\beta$ is a fragment of a protein called the amyloid precursor protein (APP), and mutations in APP are associated with a genetic form of AD [5]. This association strongly implicates APP as a causative factor for AD, and $A\beta$ is a reasonable candidate for the mediator of the mutation's effect.

(3) Deposition of $A\beta$ plaques, by itself, is not the proximate cause of AD, since the plaques do not accumulate preferentially in the parts of the brain most implicated by clinical symptoms [6] and since there is a lack of correlation between the degree of amyloid deposition and the severity of the illness [7].

(4) The e4 allele of apoE is a positive risk factor [8] and the e2 allele of apoE is a negative risk factor [9, 10] for AD. It is clear that the e4 allele is not a causative agent however, since about half of individuals who develop AD do not have the e4 allele and some very elderly individuals with that allele do not develop AD.

I will also summarize some additional important facts about AD:

(5) There is evidence that a reduction in acetylcholine concentration is a causative factor for AD. Firstly, the severity of dementia in AD patients correlates with the reduction of acetylcholine in cerebrospinal fluid [11]. Secondly, tacrine [12] and physostigmine [13], drugs that cause an increase in acetylcholine concentration, have been moderately successful in treating AD patients.

(6) Choline, which is a necessary ingredient in the synthesis of acetylcholine, is pumped into neurons that use acetylcholine as a neurotransmitter. These neurons employ a specialized mechanism, called the high affinity choline uptake system (HACU) to pump choline. AD patients have a significant increase in HACU activity in regions of the brain that are particularly affected by AD [14].

(7) Neurofibrillary tangles (NFT), composed of tau protein, are present in the brains of most, but not all, AD patients [4].

21.3   Current Hypotheses about Alzheimer's Disease

Because of the importance of plaques and tangles in assessing the presence of AD, a lot of effort has been focused on learning more about these two features. Indeed, emphasis on one or the other of these aspects has led to an informal classification of scientists working on AD into two groups: BAPtists, who consider beta-amyloid protein to be central to the disease and Taoists, who consider tau protein to be central.

A mainstream view among BAPtists is that $A\beta$ deposition forming senile plaques is necessary, but not sufficient, to cause AD [2]. Since such deposition is not the proximate cause of AD (fact (3) above), this view requires another step between $A\beta$ deposition and AD. So far, however, there is no compelling experimental evidence for such a step [1]. One suggestion, referred to as the amyloid cascade hypothesis [15], proposes that $A\beta$ deposition somehow disrupts calcium homeostasis and increases intraneuronal calcium concentrations. According to this hypothesis, the increased calcium concentration ultimately leads to cell death.

A mainstream view among Taoists is that NFT are a causative factor for AD. One problem with this view is that some AD patients show little or no evidence of NFT [4]. On the other hand, there is evidence that the density of tangles correlates with the severity of AD more than does the density of plaques [7]. At least two mechanisms have been proposed to relate the presence of tangles to AD [1]. One is that the tangles act directly to interfere with neurons and kill them, and another is that the formation of tangles by tau proteins diverts these proteins from the important task of stabilizing neuronal microtubules, thus leading to earlier cell death [16].

An interesting and important question regarding AD is the role of the three alleles of apoE. It is well-established [1] that those born with the e4 allele have the highest probability, those born with the e3 allele have an intermediate probability, and those born with the e2 allele have the least probability of developing AD, but the mechanism is not yet known. Interestingly, apoE immunoreactivity has been shown for both $A\beta$ and NFT [17]. Further experiments have shown that tau protein binds to apoE3, but not to apoE4, leading to the suggestion that interaction with apoE3 (or apoE2) reduces the tendency of tau proteins to form tangles [16]. From this point of view, the increased probability of AD associated with apoE4 results not from the presence of apoE4, but from the absence of apoE3 or apoE2.

Until recently, hypotheses relating to $A\beta$ were all based on effects of the plaques containing $A\beta$. More recently, mechanisms involving $A\beta$ in membranes have been proposed. Fragments of $A\beta$ were found to increase intracellular calcium in cultured cortical neurons [18] and in PC12 tissue-cultured cells [19]. This suggests increased calcium influx. Also, three independent groups at NIH have considered mechanisms for AD involving the entrance of the complete $A\beta$ peptide into the plasma membrane surrounding neuronal cells. Etcheberrigaray et al. have shown that $A\beta$ in fibroblasts prevents the normal functioning of the same potassium channel previously shown to be present in normal fibroblasts, but absent in fibroblasts from patients with AD [20]. Arispe et al. have shown that high concentrations of $A\beta$ in lipid bilayer membranes can form calcium-selective channels, and have proposed that the basic cause of AD is the leakage of calcium into neurons by means of these $A\beta$ channels [21]. Our NIH group examined the effect of $A\beta$ on PC12 tissue-cultured cells, and found that $A\beta$ in the membranes of these cells caused them to become leaky to both choline [22] and calcium [23]. This leakiness appears to be caused by the formation of carriers, rather than channels.

There is a possible connection between the ability of $A\beta$ to cause leakiness of neurons to calcium, whether by means of channels or carriers, and the amyloid cascade hypothesis, since that hypothesis includes a putative increase in intraneuronal calcium concentration. Such an increase could lead to cell death, but the mechanism whereby increased calcium might lead to cell death in AD is not known.

There is also a possible connection between the development of AD and the ability of $A\beta$ to cause leakiness of neurons to choline. That is because choline is required for the production of the neurotransmitter acetylcholine, and a reduction in acetylcholine concentration is a characteristic of AD (fact (5) above). Thus, choline might leak out of cholinergic neurons (neurons using acetylcholine as a neurotransmitter) faster than it is pumped in by the HACU, reducing the intracellular concentration of choline. Since the choline concentration is the rate-limiting step in the production of acetylcholine, the concentration of acetylcholine would also decrease. An interesting hypothesis to explain how a reduced concentration of acetylcholine could lead to AD has been proposed by Wurtman [24]. This hypothesis is based on the fact that cholinergic neurons require choline for the synthesis of both phosphatidylcholine (a membrane phospholipid) and acetylcholine. If choline is in short supply, the neurons give priority to acetylcholine synthesis, resulting in a decrease in phosphatidylcholine synthesis [25]. This, in turn, could change the properties of the plasma membrane or even decrease the membrane area [26]. Cell death might eventually ensue.

A decrease in the concentration of acetylcholine could have another deleterious effect. There are multiple pathways by which APP is processed. One pathway leads to the production of $A\beta$, which is generally thought to be toxic (at least by BAPtists), but there is at least one other pathway that does not seem to produce any toxic products. Recently, it was shown that acetylcholine decreases the production of $A\beta$ [27, 28], presumably by promoting APP processing by a nontoxic alternative pathway. This implies that a decrease in acetylcholine concentration would increase the concentration of $A\beta$. Since an increase in $A\beta$ concentration would further decrease the acetylcholine concentration (because of increased choline leakage), a positive feedback loop would be generated, and this would increase toxicity [29].

The hypotheses described above that are based on the toxicity of $A\beta$ are summarized schematically in Fig. 21.1. There are other hypotheses regarding $A\beta$ not considered in Fig. 21.1 and there are also many others that do not involve $A\beta$ at all. Fig. 21.1 is not meant to be representative of AD research, but rather to illustrate some of the different levels of causality being considered. Also, the manner in which any of the processes shown in Fig. 21.1 could actually cause destruction of brain neurons is not indicated. This issue is as controversial as the issue concerning the earlier steps in AD.

21.4 Future Plans

The hypothesis that seems most compelling to me is that the basic cause of AD is the insertion of $A\beta$ into the plasma membrane surrounding neuronal cells and the consequent increased leakage of choline out of the neurons and increased leakage of calcium into the neurons. Choline leakage could account for the well-known loss of acetylcholine and could also be responsible for a positive feedback (as described above) that would enhance the effect of $A\beta$. There are several possible scenarios whereby these effects could lead to the death of neurons. One possibility is the one offered by Wurtman [24] that decreased acetylcholine would cause a loss of phosphatidylcholine, and that this would compromise the plasma membranes of cholinergic neurons. Another possibility is that acetylcholine is a trophic factor, and that a significant loss of acetylcholine would therefore lead directly to neuronal death [30]. Finally, choline leakage and the consequent reduction in acetylcholine concentration could be responsible for increasing the concentration of $A\beta$, but

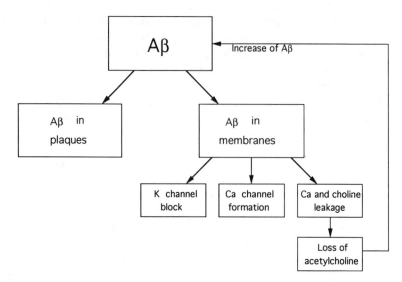

FIGURE 21.1. Schematic diagram of some proposed causes of Alzheimer's Disease that are based on effects of beta-amyloid.

calcium leakage and consequent increase in intracellular calcium concentration could be the proximate cause of neuronal death.

The reason that I find this hypothesis compelling is that it avoids the difficulties associated with basing the cause of AD on the plaques [1, 6, 7] or tangles [4], but is consistent with the fact that Aβ is always present in AD. There are also a number of other facts that are consistent with this hypothesis:

(1) There is a correlation between impaired memory functioning and low concentrations of acetylcholine in the cerebrospinal fluid of AD patients [11].

(2) There is a decreased concentration of choline in postmortem brains [31], and an increased concentration of choline in the cerebrospinal fluid [32] of AD patients. This is consistent with leakage of choline out of brain neurons.

(3) There is an up-regulation of the HACU (fact (6) above). This is consistent with a loss of choline and a consequent activation of a feedback mechanism to increase intracellular choline.

The major limitation of this hypothesis is that the experimental evidence for leakage of choline and calcium is based on experiments in which tissue-cultured cells were incubated with relatively high concentration of Aβ for 1 day. We do not yet know whether similar effects would occur *in vivo*, where Aβ concentrations are much lower, but where incubation times are much longer.

We are now planning experiments to test this hypothesis and also to find possible therapeutic agents. One goal is to examine post-mortem tissue from AD patients to determine whether or not these tissues exhibit significant leakage of choline and calcium. Another goal is to seek competitive inhibitors of the binding of choline to the putative Aβ carriers. Such inhibitors might be able to slow the course of AD. Since AD is a disease of old age, even a modest slowdown might delay the disease enough so that most potential victims of AD would die of natural causes before the disease develops.

References

[1] A. D. Roses and J. Neuropathol, Exp. Neurol. **53**, 429 (1994).

[2] D. J. Selkoe and J. Neuropathol, Exp. Neurol. **53**, 438 (1994).

[3] B. T. Hyman and R. D. Terry, J. Neuropathol. Exp. Neurol. **53**, 427 (1994).

[4] R. D. Terry, *et al.*, J. Neuropathol. Exp. Neurol. **46**, 262 (1987).

[5] A. Goate, *et al.*, Nature **349**, 704 (1991).

[6] S. E. Arnold, B. T. Hyman, J. Flory, A. R. Damasio, and H. G. Van, Cereb. Cortex **1**, 103 (1991).

[7] P. V. Arriagada, J. H. Growdon, W. E. Hedley, and B. T. Hyman, Neurology **42**, 631 (1992).

[8] W. J. Strittmatter, *et al.*, Proc. Natl. Acad. Sci. U.S.A. **90**, 1977 (1993).

[9] E. H. Corder, *et al.*, Nat. Genet. **7**, 180 (1994).

[10] H. L. West, G. W. Rebeck, and B. T. Hyman, Neurosci. Lett. **175**, 46 (1994).

[11] K. L. Davis, *et al.*, Psychopharmacol. Bull. **18**, 193 (1982).

[12] M. J. Knapp, *et al.*, Jama **271**, 985 (1994).

[13] R. C. Mohs, *et al.*, J. Am. Geriatr. Soc. **33**, 749 (1985).

[14] T. A. Slotkin, *et al.*, Proc. Natl. Acad. Sci. U. S. A. **87**, 2452 (1990).

[15] J. A. Hardy, and G. A. Higgins, Science **256**, 184 (1992).

[16] W. J. Strittmatter, *et al.*, Exp. Neurol. **125**, 163 (1994).

[17] Y. Namba, M. Tomonaga, H. Kawasaki, E. Otomo, and K. Ikeda, Brain Res. **541**, 163 (1991).

[18] M. P. Mattson, *et al.*, J. Neurosci. **12**, 376 (1992).

[19] R. Joseph, and E. Han, Biochem. Biophys. Res. Commun. **184**, 1441 (1992).

[20] R. Etcheberrigaray, E. Ito, C. S. Kim, and D. L. Alkon, Science **264**, 276 (1994).

[21] N. Arispe, E. Rojas, and H. B. Pollard, Proc. Natl. Acad. Sci. U.S.A. **90**, 567 (1993).

[22] Z. Galdzicki, R. Fukuyama, K. C. Wadhwani, S. I. Rapoport, and G. Ehrenstein, Brain Res. **646**, 332 (1994).

[23] R. Fukuyama, K. C. Wadhwani, Z. Galdzicki, S. I. Rapoport, and G. Ehrenstein, Brain Res. **667**, 269 (1994).

[24] R. J. Wurtman, Trends Neurosci. **15**, 117 (1992).

[25] M. Ando, M. Iwata, K. Takahama, and Y. Nagata, J. Neurochem. **48**, 1448 (1987).

[26] R. J. Wurtman, *et al.*, Adv. Neurol. **51**, 117 (1990).

[27] A. Y. Hung, *et al.*, J. Biol. Chem. **268**, 22959 (1993).

[28] J. D. Buxbaum, A. A. Ruefli, C. A. Parker, A. M. Cypess, and P. Greengard, Proc. Natl. Acad. Sci. U.S.A. **91**, 4489 (1994).

[29] G. Ehrenstein, and Z. Galdzicki, Biophys. J. **68**, A126 (1995).

[30] S. H. Appel, Ann. Neurol. **10**, 499 (1981).

[31] R. M. Nitsch, *et al.*, Proc. Natl. Acad. Sci. U.S.A. **89**, 1671 (1992).

[32] R. Elble, E. Giacobini, and C. Higgins, Neurobiol. Aging **10**, 45 (1989).

22

Radio and Infrared Spectroscopy of Interstellar Molecules

Neal J. Evans II
and John H. Lacy

22.1   History

A few diatomic interstellar molecules (CH, CN, and CH^+) were detected in 1937-1940 with optical spectroscopy [1, 2, 3]. After a long delay, another diatomic (OH) was discovered at radio wavelengths in 1963 [4]. Despite these early results and the publication of lists of other candidate molecules by Townes [5] and Shklovsky [6], further progress awaited the development of radio telescopes and receivers working at higher frequencies. The discovery of ammonia (NH_3) and H_2O in 1968-1969 by the group at Berkeley [7, 8], followed by the discovery of formaldehyde (H_2CO) [9], opened the floodgates. By 1995, about 100 molecules have been discovered in space, mostly with radio frequency spectroscopy of rotational transitions.

Because infrared vibration-rotation transitions have small dipole moments, infrared spectroscopy of interstellar molecules has lagged behind longer wavelength observations. Near-infrared emission by interstellar H_2 was first detected in 1976 [10] from shocked gas in the Orion Molecular Cloud, where the gas is heated to about 2000 K, sufficient to produce strong emission. Soon thereafter, absorption by CO was detected [11], but no other interstellar molecules were observed until 1989, when C_2H_2 was detected [12], followed shortly by CH_4 [13]. Several additional molecules have been observed in the infrared over the last few years, but it is still much more difficult to observe vibrational, rather than rotational, transitions of interstellar molecules.

22.2   Techniques

Two developments have stimulated the productivity of radio spectroscopy for studies of interstellar molecules. One is the development of low-noise receivers for short wavelength observations. Most of the rotational transitions of simple molecules composed of abundant elements (H, C, O, N) have their lowest rotational transitions around 3 mm, while transitions between higher rotational states extend into the submillimeter region. Early work at Berkeley produced millimeter-wave receivers with noise temperatures around 10,000 K [14], allowing detection only of quite strong emission lines. More recently SIS junctions [15] have achieved noise temperatures only a few times the quantum limit even at wavelengths as short as 1 mm [16], allowing detection of very weak lines. A second development was of very broad band backends with many channels. Discrete-element filter banks with many channels (up to 500) were used for most of the discoveries, but these have largely been replaced with digital correlators or acousto-optic spectrometers with more channels.

The challenge for infrared spectroscopy has been combining high spectral resolution with high sensitivity. Heterodyne receivers easily achieve the desired resolution but are severely limited by the increase of quantum noise with frequency [17]. Consequently, spectrometers with incoherent photon detectors are almost always used for interstellar observations at $\lambda < 300\mu$m. Four types of spectrometers have been used: heterodyne, Fourier transform, Fabry-Perot, and grating, with much of the early work on Fabry-Perot and heterodyne spectrometers being done at Berkeley. With a single detector the three types of incoherent spectrometers have similar sensitivities, depending on the spectral and spatial extent of interest, but detector arrays favor gratings because they can use different pixels to observe many spectral channels simultaneously. The main drawback to using gratings at long wavelengths is the size of the grating required to obtain sufficient resolution to resolve narrow interstellar lines.

22.3 Application to Star Formation and Interstellar Chemistry

Studies of interstellar molecules have found two primary, related applications. One is determining the physical conditions in interstellar molecular clouds, where new stars are formed. The other is the study of the abundances of the molecules, the observational input for the study of interstellar chemistry.

Studies of the distribution of populations over the rotational energy levels can be used to derive the temperature and density in molecular clouds, while the line profiles yield information about the motions in the clouds, via the Doppler effect. These studies have produced a wealth of information about molecular clouds, especially the regions in the clouds that are forming stars [18]. Stars form from gas which is cold ($T \sim 10$ K), molecular, and lightly ionized (fractional ionization less than 10^{-7}). Densities in molecular clouds are very low by terrestrial standards, but much higher than the average interstellar density ($n = 1$ cm^{-3}). Average densities as low as 100 cm^{-3} typify most of the volume of a molecular cloud, but the distribution is far from homogeneous; in regions of star formation, densities ranging up to about 10^7 cm^{-3} have been inferred. Since the average density of a star is about 10^{24} cm^{-3}, most of the stages of compression have not yet been probed. Gravitational collapse is the mechanism for achieving the enormous compression factors needed to turn molecular clouds into stars. For low mass stars (like the Sun), theoretical analysis [19] predicts that collapse initiates from a centrally condensed configuration, with $n(r) \propto r^{-2}$ and begins from near the center. A wave of infall propagates outward, modifying the density and velocity fields inside the infall radius. Recent analysis of line profiles of various interstellar molecules has found that this theoretical model fits the data very well in at least one case [20, 21]. Formation of more massive stars is considerably more complex, partly because they typically form in groups, with many stars condensing in a small space. Generally speaking, massive stars form from gas which is initially denser, more turbulent, and perhaps warmer. The massive stars quickly heat up their surroundings; hence determining the temperature before the first stars form is difficult.

Understanding the composition of molecular clouds has proven to be a challenge. The most common interstellar molecule, H_2, is believed to form on the surfaces of dust grains, sub-micron sized particles of either silicate or carbon material, which comprise about 1% of the mass in interstellar clouds. Most other molecules have been assumed to form via gas-phase reactions, many of which involve molecular ions, ionized by the occasional cosmic ray. Tests of the chemical modeling require determination of the abundances of molecules over a range of physical conditions. Early studies assumed that the molecular abundances were constant along a line of sight through a cloud, thus given by the ratio of the column densities of the species of interest to that of the dominant species (H_2). In most cases, however, the H_2 is not directly measured because it has no radio lines and its infrared lines are very weak under typical interstellar conditions, so CO has been used as a substitute, with an assumed H_2/CO $\sim 10^4$. Only recently has the H_2/CO ratio been directly measured in a molecular cloud [22], with a result somewhat less than 10^4. However, the main complication to the usual approach is the probable variation of abundances along a line of sight. One important reason for such variations is that at the higher densities found in regions of star formation, the timescale for molecules to impact a dust grain becomes shorter than relevant dynamical and chemical timescales. Under these conditions, some molecules freeze onto grain surfaces, producing mantles on the silicate or carbon cores. Indeed, infrared spectroscopy of modest resolution has revealed mantles of H_2O, CO, CH_4 and other ices [23, 24, 13]. Unfortunately, the precise composition of the grain mantles is only poorly known because of the widths of the spectral features, the similarity of the spectra of similar molecules, and the fact that many of the spectral features are in regions of poor atmospheric transmission. The process of depletion onto grains occurs throughout the process of chemical evolution, which takes about 10^6 yr to reach steady state. Consequently, the composition of the grain mantles may reflect abundances from earlier periods in the chemical evolution of the cloud.

The existence of molecules in icy mantles near forming stars suggests interesting links to other subjects. The icy dust grains may survive infall onto the outer parts of disks around forming stars, providing raw materials for the formation of comets and the cores of giant planets. Analysis of interplanetary dust particles, thought to be the fragments of comets that have penetrated the inner solar system, show chemical signatures, such as enhancement of deuterium [25], that are also seen in molecular clouds, suggesting a direct connection between these objects and chemical processes in the molecular cloud that formed our Solar System. Further reactions among the mantle molecules, perhaps driven by radiation from the forming star, can lead to more complex species, such as the amino acids identified in some meteorites. Copious infall of such material early in the history of the Earth may have supplied raw materials for the origin of life [26, 27].

More massive stars become much more luminous and are likely to evaporate the ices from the grain mantles. The molecules released back to the gas phase were deposited on the dust earlier in the cloud's chemical evolution and thus differ from the composition of the gas into which they are released. Thus, an overabundance of "early-time" molecules is a signature of this process. Further chemistry may be driven by this infusion of molecules which are out of equilibrium with their surroundings.

Our recent interest has been in combining radio and infrared techniques to study star formation and the chemistry that occurs as part of the process. In the next section, we describe some of our results in two regions forming relatively massive stars, where molecules stored on grains at early times appear to have been liberated.

22.4   Observations and Interpretation in Orion IRc2 and GL2591

Using an echelle spectrograph operating at 5-14 microns [28] we have detected rovibrational transitions from a number of interstellar molecules. This work is complementary to radio observations for a number of reasons. First, it can detect species without permanent dipole moments, like C_2H_2 [28] and CH_4 [13], which are invisible to radio techniques. Second, because the molecules are seen in absorption, the effective resolution is the size of the emitting background source, usually on the order of an arcsecond. Radio observations have much worse resolution in general but are not restricted to lines of sight toward infrared sources, because the lines are usually seen in emission. In addition, the column densities needed for detection in the radio are usually considerably less than those needed for detection in the infrared, as long as the molecule has a substantial permanent dipole moment. The combination of these two techniques can be very powerful.

We have detected molecules by their infrared absorption toward Orion IRc2 [29] and GL2591 [30], massive, luminous young stars or protostars, which are deeply embedded in clouds of gas and dust. We have detected C_2H_2, HCN, and probably NH_3 in the gas surrounding both sources; OCS and CO were also detected toward Orion IRc2. Combining our results with previous detections of CO toward GL2591 [31] and CH_4 toward IRc2 [13], we were able to compute abundances relative to CO, the most common interstellar molecule after H_2. The abundances of C_2H_2 and HCN (and CH_4 in IRc2) are orders of magnitude greater than predicted by chemical models which have reached steady state abundances. Instead, they are roughly consistent with abundances after about 10^5 yr of chemical evolution, suggesting a very young cloud. However, this explanation would require that the abundances be enhanced throughout the cloud. In GL2591, we have two pieces of evidence against this idea. First, the CO absorption has been analyzed in terms of three temperature components, with T = 38, 200, and 1000 K. The enhanced abundances seem to be restricted to the hotter components. In addition, we have extensive radio observations of GL2591. The abundance of HCN determined from these is depleted, rather than enhanced, compared to chemical models of gas-phase chemistry. If the column density of HCN inferred from the infrared observations filled the beam of the radio observations, the radio lines would be much stronger than observed. These facts indicate that the enhanced abundances are restricted to a small region ($< 2 - 3$ arcseconds or $2000 - 3000$ AU). The absorption from high rotational levels ($J = 21$ and 22) is clearly less than expected from thermal equilibrium (LTE) populations, allowing us to estimate the density in the absorbing gas to be about 3×10^7 cm^{-3}. Thus, the molecules seen in the infrared appear to be restricted to a small, dense, hot region, probably close to the forming star. This situation fits nicely with the picture that the molecules have recently been liberated from grain mantles by processes associated with star formation. The picture which seems to fit the data best is one in which the molecules were frozen onto dust grains early in the cloud evolution, when their abundances were relatively high, stored there, and released as the forming star heated the grains. Shock waves associated with outflow from the newly-formed star may also have played a role, as indicated by the blue-shift of the infrared lines, relative to the mean cloud velocity measured with the radio lines, and the very high temperature (1000 K) of one of the components. Similar conclusions apply to IRc2.

In addition to analyzing the cloud composition, we were able to use the combined data set to study the physical conditions in the clouds. As mentioned above, non-LTE analysis of the HCN populations seen in infrared absorption indicate densities of about 3×10^7 cm^{-3} for GL2591. A similar analysis in Orion IRc2 indicated densities between 3×10^6 and 1×10^7 cm^{-3}. Gas at these densities had already been indicated by radio observations in IRc2, but not in GL2591. In GL2591, our extensive radio data set allowed us to model the surrounding cloud in some detail. Since the maps indicated that spherical symmetry was a reasonable approximation, we employed spherical models, allowing variations in temperature and density with radius from the forming star. The temperature distribution was fixed by analysis of the infrared radiation from dust and the assumption that the dust temperature and gas temperature are well coupled, as is expected at the densities seen in this source. The gas density distribution was modeled as a power law ($n(r) \propto r^{-\alpha}$). Values of α ranging from 1 to 2 were tested, with $\alpha = 1.5$ providing the best fit to the data. In particular, the models were constrained by observations of CS, for which we had data on 5 rotational transitions of the main isotope and 3 transitions of C^{34}S. Constant density models clearly could not fit the data. These results are interesting because theoretical work suggests that stars are likely to form in regions with density gradients. In fact, $\alpha = 1.5$ is predicted for the region inside the infall radius in the Shu picture of low-mass star formation [19]. However, we see no kinematic evidence of collapse, indicating that the picture is more complex for the formation of more massive stars, like GL2591.

22.5   Future Prospects

Future progress in radio observations relies primarily on improving the spatial resolution and sensitivity. With radio receivers approaching the minimum theoretical noise, sensitivity improvements will generally require greater collecting areas. For mapping projects, receiver arrays offer an alternate path to higher sensitivity, but many of the interesting processes operate on small scales. Larger telescopes offer both improved resolution and sensitivity, but the high cost and need for accurate pointing of very large telescopes provide advantages to the use of aperture synthesis. In this technique, many smaller telescopes are used to synthesize a beam which is smaller than that achieved by any plausible single dish, and collecting area can be increased by adding more telescopes. This approach has been used effectively at centimeter wavelengths, for example in the Very Large Array, operated by the National Radio Astronomy Observatory. This approach has been extended to millimeter wavelengths, the prime territory for molecular studies, with arrays in California, Europe, and Japan. All these millimeter arrays are modest in size ($<$ 10 dishes), but there are plans for much larger arrays. For example, the NRAO is developing the design for the Millimeter Array (MMA), which envisions about 40 antennas, each with 8-m diameter. Spatial resolutions would range from 0.1 to 10 arcsec, with the best resolution only possible with excellent atmospheric phase stability. The MMA would allow us to extend radio studies to the regions currently probed only by infrared techniques. In particular, we could begin to study the physical and chemical processes associated with material infalling onto circumstellar disks, and perhaps subsequent processes occurring in those disks, the likely site of planet formation.

   In the case of infrared observations, it is primarily spectral resolution and sensitivity that must be improved. Particularly for observations of compact sources, instruments involving large diffraction gratings and infrared detector arrays seem most promising for the foreseeable future. To resolve interstellar molecular lines at wavelengths near 10 μm will require gratings about 1 m in length, which must be cooled to cryogenic temperatures, a daunting but perhaps not impossible task at infrared wavelengths. Unfortunately, the high background radiation severely limits the sensitivity achievable from a ground-based telescope. The advent of 10-m class telescopes at good infrared sites will help substantially, but ultimately observations will have to be made from cooled telescopes in space. So far only relatively low spectral resolution has been achieved from such telescopes, but the upcoming ISO mission will provide substantial improvements, although not with the sensitivity that could be achieved with a larger instrument optimized for high spectral and spatial resolution. A nearly ideal instrument, capable of orders of magnitude improvement in our capabilities in infrared studies of interstellar molecules, is now within our capabilities, if not yet within our budget.

22.6 Acknowledgments

Our research is supported by NSF grants AST-9317567 and AST-9120546.

References

[1] T. Dunham, Jr., and W. S. Adams, Pub. Astron. Soc. Pac. **49**, 26 (1937).

[2] P. Swings and L. Rosenfeld, Astrophys. J. **86**, 483 (1937).

[3] A. McKellar, Publ. Astron. Soc. Pac. **52**, 307 (1940).

[4] S. Weinreb, A. H. Barrett, M. L. Meeks, and J. C. Henry, Nature **200**, 829 (1963).

[5] C. H. Townes, in Radio Astronomy, IAU Symp. 4, ed. H. C. van de Hulst (Cambridge: Cambridge University), 92 (1957).

[6] I. S. Shklovsky, Astr. Zhur. **26**, 10 (1949).

[7] A. C. Cheung, D. M. Rank, C. H. Townes, D. D. Thornton, and W.J. Welch, Phys. Rev. Letters **21**, 1701 (1968).

[8] A. C. Cheung, D. M. Rank, C. H. Townes, D. D. Thornton, and W.J. Welch, Nature **221**, 626 (1969).

[9] L.E. Snyder, D. Buhl, B. Zuckerman, and P. Palmer, Phys. Rev. Letters **22**, 679 (1969).

[10] T.N. Gautier III, U. Fink, R. R. Treffers, and H. P. Larson, Astrophys. J. **207**, L129 (1976).

[11] D. N. B. Hall, S. G. Kleinmann, S. T. Ridgway, and F. Gillett, Astrophys. J. **223**, L47 (1978).

[12] J. H. Lacy, N.J. Evans II, J. M. Achtermann, D. E. Bruce, J. F. Arens, and J. S. Carr, Astrophys. J. **342**, L43 (1989).

[13] J. H. Lacy, J. S. Carr, N. J. Evans II, F. Baas, J. M. Achtermann, and J. F. Arens, Astrophys. J. **376**, 556 (1991).

[14] P. F. Goldsmith, R. L. Plambeck, and R. Y. Chiao, Astrophys. J. **196**, L39 (1975).

[15] W. C. Danchi and E. C. Sutton, J. of Applied Phys. **60**, 3967 (1986).

[16] J. W. Kooi, M. Chan, B. Bumble, T. G. Phillips, Int. J. IR and MM Waves **15** (1994).

[17] A. L. Betz, Infrared Physics **17**, 427 (1977).

[18] N. J. Evans II, in Frontiers of Stellar Evolution, ed. D. L. Lambert (San Francisco: Astronomical Society of the Pacific), 45 (1991).

[19] F. H. Shu, Astrophys J. **214**, 488 (1977).

[20] S. Zhou, N. J. Evans II, C. Kömpe, and C. M. Walmsley, Astrophys. J. **404**, 232 (1993).

[21] M. Choi, N. J. Evans II, E. Gregersen, and Y. Wang, Astrophys. J. (in press Aug. 1, 1995).

[22] J. H. Lacy, R. Knacke, T. R. Geballe, and A. T. Tokunaga, Astrophys. J. **428**, L69 (1994).

[23] F. C. Gillett and W. J. Forrest, Astrophys. J. **179**, 483 (1973).

[24] J. H. Lacy et al., Astrophys. J. **276**, 533 (1984).

[25] S. J. Clemett, C. R. Maechling, R. N. Zare, P. D. Swan, and R. M. Walker, Science **262**, 721 (1993).

[26] W. F. Huebner, and D. C. Boyce, Origins of Life **21**, 299 (1992).

[27] C. Chyba and C. Sagan, Nature **355**, 125 (1992).

[28] J. H. Lacy, J. M. Achtermann, D. E. Bruce, D. F. Lester, J. F. Arens, M. C. Peck, and S. D. Gaalema, Pub. Astron. Soc. Pac. **101**, 1166 (1989).

[29] N. J. Evans II, J. H. Lacy, and J. S. Carr Astrophys. J. **383**, 674 (1991).

[30] J. S. Carr, N. J. Evans II, J. H. Lacy, and S. Zhou, Astrophys. J., (in press, Sept. 10, 1995).

[31] G. F. Mitchell, C. Curry, J-P. Maillard, and M. Allen, Astrophys. J. **341**, 1020 (1989).

23

Lessons Learned

Elsa Garmire

23.1   Introduction

"Charlie" was my teacher at M.I.T.; but for him I probably would not have a Ph.D. today, much less be on my way to a position as Dean of Engineering at Dartmouth College. Even after 30 years I still find it awkward to say "Charlie". As students, we called him "Dr. Townes"; we were told that when we received our Ph.D.'s we could call him "Charlie". However, because I learned so much from him, and hold him in such regard, he has never moved in my mind from "Dr. Townes" to a first-name basis. It is safe to say that Dr. Townes made me who I am today. From him I learned, not only about physics, but about technical leadership. Also, by his example, I obtained a role model for life.

23.2   Lessons about Physics

At M.I.T. I learned physics from classes; from Dr. Townes I learned about physics: the importance of a finely honed curiosity, the value in simple approaches to technical questions, and the pursuit of quality in all endeavours.

23.3   Scientific Curiosity

We hear much today about "curiosity-driven research," often said in a disparaging manner. Little is said about the many years of learning that precede the properly honed curiosity that leads fruitfully to new understandings about the universe. I learned first-hand from Dr. Townes' example about the power of curiosity as a driver to understand physical phenomena. I was fortunate enough to write one of the first Ph.D. theses in nonlinear optics, with experiments performed on the second commercially sold ruby laser. The field was so new that almost any experiment led to new results. Charlie was curious about all the new phenomena that were being reported, particularly right after Q-switching was invented, enabling nanosecond bursts of intense light. He heard about the colored rings that were emitted from liquids such as carbon disulfide and benzene in an apparent anti-Stokes Raman process. His curiosity about this process led us to understand that the stimulated Raman generation of Stokes radiation caused coherent molecular vibrations that would interact with additional laser light, causing coherent anti-Stokes radiation. The stimulated anti-Stokes emission that we first explained, became known later as a stimulated four-photon process. The same curiosity led Charlie to extend his thinking about these coherent molecular vibrations; solid state physicists would call them optical phonons. He wondered about whether there would be analogous stimulated emission on the acoustic phonon branch. From this came the idea of stimulated Brillouin scattering and a successful experiment carried out by Ray Chiao, with help from Boris Stoicheff, specifically to observe it (in quartz and sapphire). Meanwhile, I was pursuing experiments on stimulated Raman scattering, attempting to understand why the spectral linewidth of the emission was so broad. As part of the study, I used a Fabry-Perot interferometer to measure the laser linewidth. On a Saturday, as I recall, I showed Dr. Townes my spectra. He commented with characteristic curiosity that there was a "fuzziness" about the Fabry-Perot rings and wondered what this was. With typical graduate student impatience I said, "Oh, they are probably because I had the lens out of focus, I am sure they are nothing." Nonetheless, Dr. Townes insisted that I pursue this "fuzziness" by using a Fabry-Perot with larger spacing, in order to provide better resolution. His curiosity about my experimental result led to the unexpected observation of stimulated Brillouin scattering in liquids, and the subsequent amplification of the retro-reflected Stokes component in the ruby laser rod. This provided the foundation for what later turned out to be phase-conjugation in stimulated Brillouin scattering, the phenomenon that unfortunately provided a limit to the power-handling capability of optical fibers in communication systems.

A third demonstration of curiosity was our exploration into nonlinear self-guiding of light. Dr. Townes had seen a photograph of an extremely long damage trail made in glass by Mike Hercher with the University of Rochester's intense laser system. The long trail defied diffraction and Dr. Townes was curious to see if optical nonlinearities in glass could counteract the natural tendency of light to diffract. Ray Chiao and I worked with him to find the "spatial soliton" solution to the nonlinear wave equation and to understand mechanisms of optical nonlinearities in glass. Later investigations showed that the spatial soliton was unstable and self-focussed into filaments. A message I received from this education was to be curious about unexpected results and new phenomena. This has allowed my students and me to investigate integrated optics, optical bistability, and nonlinearities in semiconductors. We have discovered a number of new phenomena, the most recent of which is the role of holes in picosecond carrier transport through quantum wells.

23.4   Simplicity of Approach

Perhaps the most important message I learned from Dr. Townes was to seek the simplest possible explanation for physical phenomena. He had a way of making things seem easy; he described the laser as an electronic oscillator. He used classical equations whenever possible, and had an excellent ability to explain things in classical terms. I was an arrogant graduate student and felt that perhaps "he didn't understand quantum mechanics". One day, in his office, I challenged him to explain a laser phenomenon by quantum mechanics. He protested that lasers were inherently classical devices and that quantum mechanics was not needed. I insisted, thinking gleefully that, at last, I would catch him in something he could not do. Instead, he began at the upper left-hand corner of the board, covered it with intense scribbling of Dirac operators, and by the time he reached the lower right-hand corner of the board, he proved the same thing that could be proved in three lines with classical equations. After that time, I never questioned his ability. I learned that, like Picasso, the ability to explain things in simple lines, comes only after years of more complex trying. It has been my intention to communicate to my own students the importance of the simple approaches that I learned from Dr. Townes. I was pleased when a graduate returned to tell me: "The most important thing that I learned from you that I use in my job is how to do a Taylor series. Everyone at work thinks I am so smart because I do not need a computer to estimate how a system will respond." Thanks to Charlie.

23.5   Quality in All Endeavors

In honor of the 25th anniversary of the laser, the Optical Society of America and several other organizations sponsored "The Laser History Project," an oral history of the science and engineering surrounding the invention of the laser, carried out by Joan Bromberg and published by the M.I.T. Press. As a former student of Dr. Townes during the time he received the Nobel Prize and during his research in nonlinear optics, I was interviewed. One question that I was asked stands out in my mind: "How did Dr. Townes ensure the quality of the research that his students did?" The question amazed me. I could not fathom an answer because in my memory this issue never arose. I cannot remember questioning the quality of work we did—we did the best we possibly could. I looked for the answer to this question through my own interactions with my students, deciding that I would surely treat my students as Dr. Townes treated me. What do I do? I have a litany that my students know by heart: "Are your sure?" "Would you stake your Ph.D. on the correctness of that (because, of course, you are)?" I have no doubt that this is, in essence, what Dr. Townes communicated to us. I am sure that he did not need to exhort us to "quality" with words. This came from our own innate desire to be sure about the physics that we were uncovering; the excitement that we were uncovering the "truths" of physics is what drove us in the first place.

23.6   Lessons about Leadership

As a Ph.D. student at M.I.T., I saw Dr. Townes balance interest in physics with interest in making a difference in a larger arena. Before coming to M.I.T. as Provost, he spent a year at the Institute for Defense Analyses in Washington and was articulate on the importance of scientists giving something back to society. From him I learned the important lesson that technical contributions include, not only published research papers, but contributions to policy-making in the national arena. Whether by his academic leadership, through his being Provost of M.I.T., his industrial leadership, as a member of the Board of General Motors, or his government leadership as a member of innumerable panels on government policy, his voice has been that of an articulate statesman. It was the inspiration of Dr. Townes that led me to become active in the Optical Society of America (including acting as President), to found the Laserium light shows, and to serve on advisory boards of the Air Force and the National Science Foundation. It was his legacy that encouraged my former students to become active also. Finally, it is through his example that I have decided to accept an appointment as Dean of the Thayer School of Engineering, Dartmouth College. Dr. Townes' style of leadership, as a Professor at M.I.T. in conjunction with a busy life as Provost, was to provide his students with the maximum freedom. He articulated a given area of interest, and allowed the students to define in detail how an investigation would be carried out. This meant that his students developed a maximum of self-confidence. This could be seen at conferences (such as the Conference on Quantum Electronics in Puerto Rico in 1965), where Ray Chiao and I sat in the front row and actively asked questions. By contrast, some other Ph.D. students tended to sit demurely near the back. It has been my mission to imbue my own students with that self-confidence and ability to self-define research directions. As a result, two of my former students (his grand-students) have tenured positions (at University of Michigan and Georgia Tech) and are now producing Ph.D. students who are his great-grand-students. I learned from Dr. Townes the importance of respecting the right of every scientist to express his/her own point of view. When I was an impatient graduate student, I listened to a visiting scientist present some obviously faulty logic; I would have broken in with a "You're wrong" and stopped the presentation. Instead, Dr. Townes listened politely, allowed the scientist to finish his argument, and then gently suggested a line of reasoning that led the speaker to see the error in his thinking. I learned from this the importance of allowing students to discover for themselves their misconceptions. The purpose of a teacher is to lead students in thought, not to think for them. One last message I learned by example is the importance of professors being willing to accept students of all types. Brilliant students do not usually need teaching; it is the challenging students that need good teachers. I am proudest of my students who have reached beyond apparent limitations to achieve real scientific productivity.

23.7   Lessons about Life

From Dr. Townes as a role model, I learned several lessons that I have applied to my own life. These include his family (wife and four daughters), his hobbies and energy for things outside of physics (his farm and his orchids), his search for religious truth and moral living. I was the first female Ph.D. student who completed a Ph.D. under Dr. Townes (and the only female Ph.D. student entering Physics at M.I.T. in 1961). I found it easy to switch to him as an advisor. Because he had four daughters and an intelligent, articulate wife, he had no difficulty in relating to me as a person. In fact, I do not believe that he treated me any differently because I was a female. This gave me the self-confidence to proceed in later years against some rather severe odds. Dr. Townes led a balanced life, full of energy to explore new things. Not only was he passionate about physics, he had several other interests. His ability to grow orchids was legendary. He also was intensely interested in his New England farm. Every year he would invite his students and post-docs to the farm; there would be long tramps through the woods and delicious dinners. I remember one particular time when it rained and rained. During a break in the rain, Charlie prevailed on the group to go for a walk: "The rain looks like it is over." Luckily I was eight months pregnant and used this as an excuse to avoid the hike. After they left, the rain began again—in earnest. After about half an hour a bedraggled group trudged back. What a funny sight they were! I remember one eminent Italian physicist whose silk shirt had become transparent and was plastered to his body. (He had thought he was being invited to a "Villa" and came dressed up in a silk shirt. With dismay he discovered Charlie and the others in Levi's.) One M.I.T. student wore tennis shoes that had been washed in the washing machine without full rinsing. At every step of his sodden shoes, detergent bubbles squished out! Putting aside the funny stories, I learned from Charlie's example to balance science, family, and other activities. Dr. Townes was an active seeker for religious truth, and the relation of physics to God interested him immensely. While my search for spiritual truth has led in different directions, the role model of a scientist actively pursuing these questions was important to me. Even more important was the fact that he led by example, showing us a person who carries out the tenets of his religion in a moral life. I think of Dr. Townes as an eminently honest man and one who encourages others to live honestly. I remember when, as a graduate student, I "borrowed" a bottle of laboratory-grade ethanol to spike the punch for a party the graduate students were having. Dr. Townes found out. I do not remember exactly what he said to me, but it apparently taught me a lesson. I remember feeling ashamed; I was not living up to the expectations that he had for me, etc. Never again did I do such a thing. Furthermore, I have talked often to my students about the importance of scientists to be honest in all that they do. Honesty must exude from the pores of all reputable scientists.

23.8   Conclusions

Doing Ph.D. research with Charlie Townes at M.I.T. was a magical time. The laser had just been invented. Nonlinear optics was new. My advisor won the Nobel Prize (the students met him at the airport with banners congratulating him!). Because he was Provost at M.I.T., he took on only two Ph.D. students (myself and Raymond Chiao). Not many students have such golden opportunities. The Townes curiosity about science, formulation of simple explanations, and insistence on quality have lived on in his students and his "grand-students". In fact, the line is being continued by his "great grand-students". His academic leadership style has influenced my own: respect for students and other scientists, offering technical leadership in larger arenas, and willingness to accept a wide variety of students. Finally, through his relations to family, hobbies, spirituality, and morality, his example has inspired me to a fuller life.

24

Infrared Spectroscopy of Jupiter in 1970 and in the 1990s

Thomas R. Geballe

24.1   The Early 1970s

Much of Charles Townes' career in astrophysics has been devoted to studying the interstellar medium, in particular the center of the Milky Way, and molecular clouds in the spiral arms of our galaxy and others. However, since the early days of infrared astronomical spectroscopy, in the late 1960s and early 1970s when some of his graduate students (myself included) were developing spectrometers for use in the 10 μm atmospheric window, he also has had an interest in spectroscopy of planetary atmospheres and, in particular, the atmosphere of Jupiter. Whether this originally was because Jupiter was interesting in itself or because Jupiter's atmosphere contains ammonia, a molecule that Townes understands better than anyone ever has and perhaps ever will, I am not certain.

I was fortunate to join the research group of Prof. Townes in 1969, as a second-year graduate student in the physics department at U.C. Berkeley. At that time most of Townes' students designed and constructed astronomical infrared and millimeter wave spectrometers of various types and learned sufficient amounts of astronomy to be able to publish respectable papers about the remarkable new data that they were obtaining. In the latter endeavor, a few members of the neighboring U.C. astronomy department, who were also participating in the opening up of infrared astronomy, were of help to us physicists. In particular Roger Knacke took considerable interest in our research and provided useful advice when asked. Some recent data of his appear below.

For me the struggles of constructing an infrared spectrometer involved, among other work, the complete manufacture of piezoelectrically scanned Fabry-Perot interferometers and the production of doped photoconductive infrared detectors. Along with these came the continual process of persuading the spectrometer to behave, first in the laboratory and then at the telescope; of finding out what worked and what did not. There was the excitement of taking part in the invention of new modes of observational astronomy and the more sobering emotions encountered in obtaining an empirical understanding of signal/noise. The latter learning process was enhanced by countless discussions with Townes about the reality of wiggles in our spectra. These activities were the best training that an infrared astronomer could have had. Of course, back then I had little idea of the bright future of infrared astronomy or that the experience I was gaining would serve me well for at least the following 25 years. I only knew that I was learning and enjoying.

In those early days of astronomical spectroscopy in the thermal infrared, conditions were considerably different than they are now. We observed at the Coudé focus of the Lick 3-m telescope (the "120-inch", as it was called then). In addition to the intrinsically poor transmission and high emissivity of this location, our spectrometer had numerous optics at ambient temperature, including a focal plane chopper, several curved and flat mirrors, a Fabry-Perot interferometer, and a circular variable filter (CVF). It also employed a single bolometer or a single photoconductor as a detector. I believe that the CVF had been loaned to us by the late Professor Pimentel of the chemistry department at U.C. Berkeley, and was identical to the one that recently had been sent to Mars. The loaned CVF had been shock-tested (until it broke), in order to determine if its twin would survive the stresses of a rocket launch. Fortunately Jim Holtz and Dave Rank had managed to piece it back together and, except for a few wavelengths which corresponded to cracks (and turned out to be useful position/wavelength calibrators), it worked well. It worked even better once we began cooling it to liquid nitrogen temperature.

Nevertheless, ground-breaking science could be done. Jim Holtz, Dave Rank, Townes, and I successfully detected the SIV line at 10.5 μm in a few bright planetary nebulae and subsequently we detected other fine structure lines. To do so with modern instruments requires much less than one second of integration, but back then required many all-night marathons. I recollect numerous occasions of dawn breaking, the last nebula fading from view, and we bleary-eyed graduate students feeling more than ready for shutdown of equipment and ourselves (i.e., bed), when Townes would brightly "suggest" that we observe ammonia in Jupiter. Of course a suggestion from Dr. Townes was not to be refused by any of us, and so we observed ammonia lines in Jupiter, often until 10 a.m., unless (by a lucky circumstance) the weather intervened. I did not and still do not understand how he managed to sleep so little and remain so alert. As I approach his age at that time, I am again aware that this is an unusual ability.

24.2   The Mid-1990s

Now, one quarter of a century later, infrared astronomical spectroscopy has undergone numerous technological advances; it has taught and continues to teach us much about the Jovian atmosphere. In addition to the major constituents (e.g., molecular hydrogen, helium, methane, ammonia), an impressive assembly of exotic molecules has been detected via infrared spectroscopy, including carbon monoxide, water, germane, phosphine, arsine, silane, acetylene, ethane, and H_3^+ (see Noll 1990 for an informative article). Many of the above are disequilibrium molecules, created in deep, unobservable parts of the Jovian atmosphere and transported upwards by convection to where they can be detected spectroscopically, often in concentrations of less than a part per billion, before they are destroyed by chemical reactions. H_3^+ is created in Jupiter's ionosphere by the dissociation of H_2, in part due to solar ultraviolet radiation (which also leads to the production of ethane and acetylene), and in part by the impact of the solar wind at Jupiter's poles. The latter process creates luminous infrared aurorae of H_3^+ emission lines, which currently are the subject of intense study.

24.2.1 H_2 Dimers

A particularly interesting molecule detected in the Jovian atmosphere is the hydrogen dimer, $(H_2)_2$. Although monomeric H_2 has no dipole moment and thus cannot absorb radiation except very weakly via quadrupole transitions, dipole moments are induced during brief, close interactions of H_2 molecules or by H_2 molecules in bound pairs. Two absorption features, due to bound-free transitions in hydrogen dimers, were identified by McKellar (1984) and Frommhold et al. (1984) in far-infrared low resolution spectra of Jupiter and Saturn obtained by Voyager (Hanel et al. 1979). Detection of these pure-rotational absorptions led to the expectation that dimer lines in the fundamental ro-vibrational bands in the near-infrared might be detectable near 2 μm, a wavelength region which is much easier to observe from the ground. These expectations have been borne out during the last few years, first by tentative detections of features at the Canada-France-Hawaii Telescope (CFHT) and the United Kingdom Infrared Telescope (UKIRT), and finally by definitive detections of five spectral features (Sang et al. 1995) at UKIRT. Figure 24.1 shows a spectral image obtained by Sang et al. using the cold grating spectrometer CGS4 at UKIRT.

Two absorption lines of the dimer are seen in this spectrum, on either side of an absorption line due to the monomer. The slit of the spectrometer was placed along the equator of the planet; the tilt of the absorption lines along the slit is a Doppler shift due to the rotation of Jupiter.

Inspection of the figure reveals that the absorption lines of $(H_2)_2$ are considerably broader than the line of H_2. Unfortunately, little is known about the behavior of the line profiles of the dimer as a function of temperature and density. Furthermore, the transition strengths of the lines in the fundamental band of $(H_2)_2$ are not known accurately. When better understood the 2 μm dimer lines may become uniquely revealing probes of the Jovian atmosphere.

24.2.2 Hot CO

One would be remiss, even in a short contribution about Jupiter, not to mention the impact of Comet Shoemaker-Levy 9 in July, 1994. At this stage astronomers are still analyzing the many remarkable images and spectra obtained during this event. The comet is thought to have deposited an amount of energy equal to the amount of solar energy incident on the planet in one month, mainly in about a dozen localized areas at latitude south 45 degrees, briefly producing intense fireballs at these sites.

At UKIRT, several experiments successfully detected chemical products of these events. In one experiment at 3.5 μm, originally designed by Steven Miller and

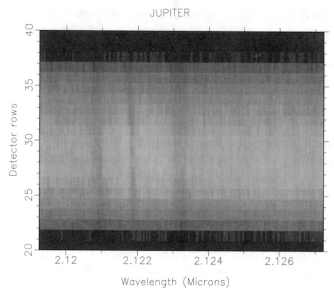

FIGURE 24.1. Spectral image of Jupiter, showing absorption lines of $(H_2)_2$ on either side of a narrower 1-0 S(1) absorption line of H_2. The slit of the spectrometer was positioned along the equator of Jupiter.

Takeshi Oka to study the effects of impacts of fragments C and G on the spectrum of H_3^+ (Miller *et al.* 1995), a set of new lines appeared during impact G and completely overwhelemed the pre-impact spectrum. Many of these lines are now identified as high *J* rotational transitions of methane, but some remain unidentified at present. In a second experiment, Roger Knacke and I obtained a remarkable time sequence of spectra near 2.3 μm of the R impact (Knacke *et al.* 1994). In the post-impact spectra, emission from CO overtone vibrational bands is seen. The entire episode of CO emission in the overtone band lasted no more than ten minutes and is portrayed in Fig. 24.2.

The second spectrum, obtained just after the dramatic rise in intensity (by approximately two orders of magnitude) shows a classic CO 2-0 band profile—a sharp rise at 2.295 μm and a gradual decline to longer wavelengths. The other CO band heads (3-1, 4-2, 5-3) are not as easily distinguishable, due to blending by other emission features (probably due to CH_4), but the CO bands definitely are present.

The subsequent spectrum, obtained four minutes later, is an order of magnitude weaker and shows the 2-0 band emission profile being cut off at longer wavelengths. It is thought that this is due to 0-2 absorption lines from low-lying rotational levels (by CO on the outside of the blast which had already cooled). After another four minutes, the CO 2-0 emission had disappeared leaving the absorption (centered at about 2.31 μm). Within another two minutes this absorption feature had disappeared.

JUPITER   IMPACT R   JULY 21, 1994, 5:30 UT

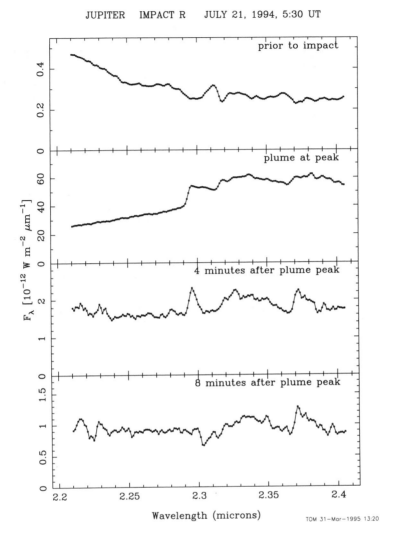

FIGURE 24.2. Sequence of spectra of Jupiter obtained at the site of the R impact of Comet Shoemaker-Levy 9, showing rapid evolution of the overtone bands of CO.

From the measured CO emission intensity and estimated temperature, and after correcting for the field of view of the spectrometer compared to the angular dimensions of the fireball as observed by the Keck telescope, we estimate the total mass of CO observed to be 2×10^{12} g. This value, which assumes optically thin emission, is about one order of magnitude less than the preliminary estimates by Zahnle (private communication) of the CO content of a fireball caused by a 0.5 km diameter impactor. In view of the uncertainties in the observational and theoretical estimates, this appears to be a reasonable preliminary agreement.

24.3   Conclusion

This paper indicates how remarkable technological advances in astronomical infrared spectroscopy during the last quarter century have led to a great increase in our knowledge of the infrared spectrum of Jupiter and the molecular physics revealed by it. Progress in laboratory molecular spectroscopy (spurred on of course by other apects of Professor Townes' research) has been equally remarkable during this period, and our much improved understanding of Jupiter would not be possible without it as well.

References

[1] L. Frommhold, R. Samuelson, and G. Birnbaum, ApJ, **283** L79 (1984).

[2] R. Hanel, B. Conrath, M. Flasar, V. Kunde, P. Lowman, W. Maguire, J. Pearl, J. Pirraglia, R. Samuelson, D. Gautier, P. Gierasch, and S. Kumar, *Science* **204**, 972 (1979).

[3] S. J. Kim, L. M. Trafton, T. R. Geballe, and Z. Slanina, Icarus **113**, 217 (1995).

[4] R. F. Knacke, T. R. Geballe, K. S. Noll, T. Y. Brooke, BAAS **26**, 1589 (1994).

[5] A. R. W. McKellar, Can J Phys **62**, 760 (1984).

[6] S. Miller, N. Achilleos, B. M. Dinelli, H. A. Lam, J. Tennyson, M. -F. Jagod, T. R. Geballe,, L. M. Trafton, R. D. Joseph, G. E. Ballester, R. Baines, T. Y. Brooke, and G. Orton, Geophys. Res. Lett. **22**, 1629 (1995).

[7] K. S. Noll, "The Strange Gases of Jupiter and Saturn," The Planetary Report **10**, 4 (1990).

25

The Galactic Center: Star Formation and Mass Distribution in the Central Parsec

Reinhard Genzel

25.1   Introducing the Phenomena

The nucleus of the Milky Way (adopted distance 8.5 kpc) is one hundred times closer to the Earth than the nearest large external galaxy, M31 (the Andromeda galaxy) and more than a thousand times closer than the nearest active galactic nuclei. Astronomers can therefore study physical processes happening in our own Galactic Center at a level of detail that will never be reached in the more distant, but usually also more spectacular systems. What powers these nuclei and how do they evolve? Do dormant massive black holes reside in their cores? While the black hole model has been the standard paradigm for explaining quasars and other luminous (active) galactic nuclei for about 30 years, there is no definitive proof yet that massive black holes exist. Because of its proximity the Galactic Center thus may be our best chance to test the black hole paradigm and determine the physics of the central core. Charlie Townes has followed this vision for about twenty years.

Until then, investigations of the Galactic Center were severely impeded by the large amount of interstellar dust and gas situated in the Galactic plane between the Sun and the nucleus, and preventing observations in the visible, ultraviolet and soft X-ray bands. In the past two decades, however, our knowledge about the Galactic Center has improved rapidly, as sensitive high resolution imaging and spectroscopy in the radio region, at hard X- and γ-rays and in particular, in the infrared band have become available. In the following, a brief summary of these phenomena as revealed by a number of observations is given, followed by a more detailed account of recent results of the exploration of the central parsec. For a more extensive discussion I refer to a recent review [1].

The nuclear mass is dominated by stars, except probably in the innermost parsec. Infrared observations on a scale of 10^2 to 10^3 pc show that these stars appear to be distributed in a rotating bar [2]. The gravitational torque of this bar may also explain the non-circular motions of interstellar gas clouds found by radio spectroscopy [3, 4]. The non-circular motions in turn may trigger gas infall into the nucleus [3]. X-ray spectroscopy of 6.7 keV Fe K-shell emission has revealed a component of $\sim 10^8$ K intercloud, coronal gas that permeates the central bulge/disk on a scale of about 10^2 pc [5]. The coronal medium cannot be gravitationally bound to the nuclear bulge and probably escapes as a Galactic wind. There is increasing evidence from γ-ray spectroscopy of the 1.8 MeV, ^{26}Al line [6], and from infrared stellar spectrophotometry (see below) that (massive) star formation occurred throughout the Galactic Center region less than 10 million years ago. Hence, the energy input responsible for the coronal gas could have come from supernovae that have been exploding in the central bulge [1, 5]. Also on a scale of 10^2 pc, several variable, spectacular, hard X- and γ-ray sources have been found [7]. They may represent stellar black holes or neutron stars accreting gas from a companion or from nearby dense gas clouds. A broad emission bump at $\sim$500 keV and a twin radio jet seen in one of these sources (the "Great Annihilator" 1E1740.7-2942) may signify the presence of a relativistic electron-positron jet annihilating in the environment of the compact source [8,9]. Recently discovered superluminal motions (apparent motions in the plane of the sky greater than the speed of light) of radio knots in another hard X-ray/γ-ray source near the Center provide direct evidence for the above interpretation of relativistic radio jets in the nuclear region of our Galaxy [10].

Throughout the central few hundred parsecs giant molecular clouds ($10^{5.5}$ – 10^7 $M_\odot$, $N(H_2) \sim 10^{23.5}$ to 10^{24} cm^{-2}) are found whose gas densities ($N(H_2) \sim 10^4$ to 10^6 cm^{-3}) and temperatures (40 to 200 K) are significantly greater than those of the clouds in the Galactic disk [11]. The dust temperature of these clouds, on the other hand, is fairly low indicating that most of them are presently not heated internally by active star formation. The dynamics of this central molecular cloud layer is characterized by large internal random motions and unusual streaming velocities that can be partially explained by the presence of the central bar potential mentioned above. Extended 6.4 keV line emission from iron atoms in low ionization stages [12] and 8.5-22 keV hard X-ray continuum emission [13] appear to be correlated with these very dense molecular clouds. This surprising finding suggests that the molecular clouds act as dense reflectors of a hard X-ray continuum emission source(s) located somewhere in the central 10^2 pc, perhaps analogous to what is observed in some active galactic nuclei. The source(s) of the scattered X-ray emission is (are) yet unknown. It (they) could be identical with the known compact sources discussed above [13], or perhaps with an active source at the very center [12].

Magnetic fields as large as $\sim$1 milligauss appear to permeate the central 50 pc and are aligned approximately perpendicular to the Galactic plane [14, 15]. Where they interact with neutral gas clouds remarkable filaments of nonthermal radio synchrotron emission are seen. The most spectacular set of such filaments, the Radio Arc, located about 13' (34 pc) north of the dynamic center and the central radio source *SgrA*, also contains a set of more than 20 pc long filaments of thermal radio emission that are associated with ionized plasma of temperature about 8000 K. The origin of the widespread ionization, whether caused by collisions of fast moving clouds, by magneto-hydrodynamic effects, or by photoionization, has been the subject of a lively debate for some time [15-19]. The most recent evidence strongly favors photoionization by a number of OB associations [1, 20]. *SgrA* itself can be separated into a thermal source, *SgrA* West (see below), and a non-thermal source, *SgrA* East, emitting synchrotron radiation. *SgrA* East has a distinct shell morphology and appears to be expanding into an accelerating, dense molecular gas [19, 21]. *SgrA* East thus may be direct evidence for a recent explosion (less than 10^5 years ago) in the central 10 parsecs, with an energy of $10^{52\pm1}$ ergs.

The density of the nuclear star cluster increases with decreasing radius R approximately as R^{-2} outside of its core radius of less than a parsec in size [22-24]. Inside this core radius, the stellar density is certainly a few 10^6 and possibly 10^7 times greater [24] than in the solar neighborhood. Infrared spectroscopy has identified individual blue and red supergiants in the core [26-28]. These supergiants are very likely massive stars that have formed in the core within the last few million years. They probably provide a significant and perhaps dominant fraction of the total luminosity of the central few parsecs. A discussion of the present state of our knowledge about this nuclear cluster, its distribution and evolution will be given in the next section.

Also inside about one parsec, the gas and stellar velocities are observed to increase. The first such evidence, shown in Fig. 25.1, came from the work of Wollman [29] and later of Lacy *et al.* [30, 31]. Based on spectroscopy of the 12.8 μm [NeII] ground-state fine structure line, Wollman, Lacy, Geballe, Townes and coworkers found much larger Doppler widths of ionized gas in the central 10″ ($\sim$0.4 pc) than further out. They interpreted these large velocities as signalling a concentration of non-stellar mass in the Galactic Center, possibly caused by a million solar mass black hole at the dynamic center [31, 32]. The picture of the dynamics of the ionized gas that has developed since is largely consistent with these first observations and their interpretation. Recent radial velocity measurements in the [NeII] line (Fig. 25.1) fully confirm the increase of the gas velocities from about 100 km/s at radii of $\geq$ 20″ to $\sim$300 km/s at a few arcseconds from the dynamic center [36]. The ionized gas is arranged in a number of coherent streamers that orbit the Center [33–35]. The orbiting streamers are arranged in the form of a "mini-spiral" that comprises the brightest part of the central thermal radio source *SgrA* (West). While the streamer velocities are largely dominated by the gravitational field, an intense nuclear wind (mostly from massive stars discussed below) probably also affects their motions within about 1 pc of the center [37, 1]. Infrared polarimetry and radio observations of Zeeman splitting indicate that the streamers are permeated by milligauss magnetic fields that are dragged along their orbits [38–41]. The magnetic fields may account for macroscopic viscosity and lead to angular momentum transport in the clumpy, turbulent circumnuclear gas streamers. At the dynamic center is a compact radio source, *SgrA**, which is close to, but not coincident with a bright near-infrared source (IRS16) of blue color [42]. Since its discovery 24 years ago *SgrA** has been the most probable candidate for the central black hole. Surrounding the innermost ionized streamers one finds a system of dense orbiting molecular filaments approximately arranged in form of a circumnuclear 'disk' [43]. The circumnuclear disk is probably fed by gas infall from dense molecular clouds at $\geq$ 10 pc.

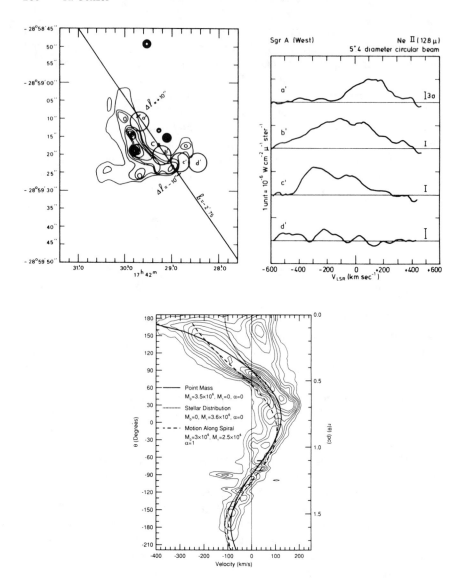

FIGURE 25.1. 12.8 μm [NeII] spectra from the work of Wollman [29] along with the beam positions on a 10 μm map of the SgrA West mini-spiral (upper two insets). Position-velocity diagram of the [NeII] line along the 'western arc', 'northern arm' and 'central bar' of *SgrA* (West) (bottom, from [36]).

Clearly the modern multiwavelength observations of the Galactic Center tell a fascinating story and show that a broad range of phenomena involving a number of physical processes are at work. To help the reader find the way in the jungle of names, Fig. 25.2 gives a "road map" of the central 100 pc.

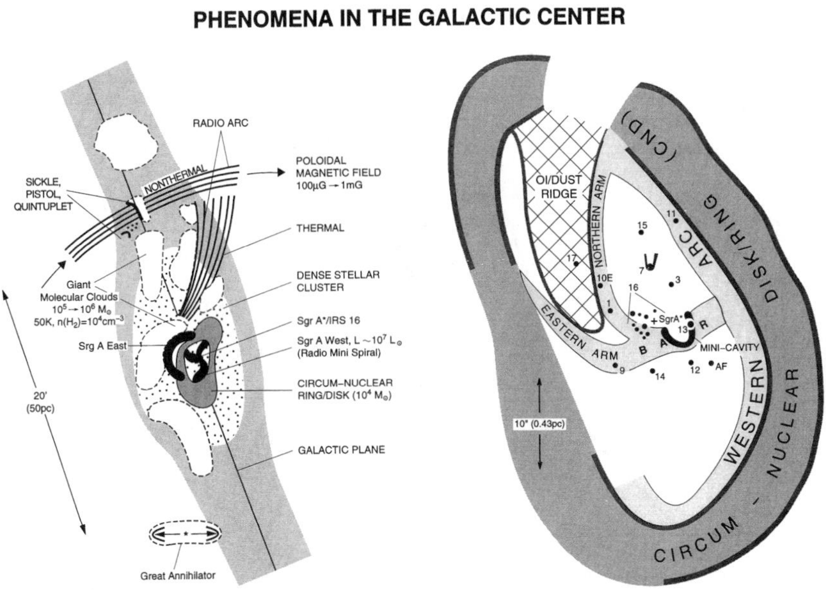

PHENOMENA IN THE GALACTIC CENTER

FIGURE 25.2. Overview of the phenomena (and nomenclature) in the Galactic nucleus (adapted from [1]).

25.2   What Powers the Central Parsec?

Figure 25.3 shows a composite spectral energy distribution of the central few parsecs (adapted from [1]). Strong 10 to 300 μm continuum emission from 50 to 100 K dust grains originates in the circumnuclear disk, and in a cloud ridge associated with the northern arm of the minispiral [43, 44]. The far-infrared emission can be used as a calorimeter for estimating the total short-wavelength (visible and ultraviolet (UV)) luminosity of the central parsec that cannot be accounted for by main sequence and late-type stars in the nuclear stellar cluster. Taking into account the ($\sim$ 50%) fraction of the nuclear stellar radiation not intercepted by the circumnuclear gas, the total UV and visible luminosity of the central parsec has been estimated to lie between 1 and 3 $\times$ 10^7 L$_\odot$) [44,45]. The Lyman continuum flux is about 2 to 3 $\times$ 10^{50} s^{-1} ($\sim$ 1.3 to 1.9 $\times$ 10^6 L$_\odot$) as determined from the thermal radio continuum [31].

Infrared spectroscopy of fine structure lines sampling a wide range of excitation stages of the ionized gas demonstrates that *SgrA* (West) is a low excitation HII region with electron temperature $\sim$ 6000 K and a range of electron densities (10^3 to 10^5 cm^{-3}). This result was first derived by Lacy, Townes and coworkers [31] and implies that the effective temperature of the UV radiation field in the central parsec is only about 30,000 to 35,000 K. The line ratios also suggest a heavy element abundance of about twice that in the Sun. What powers this low excitation HII region?

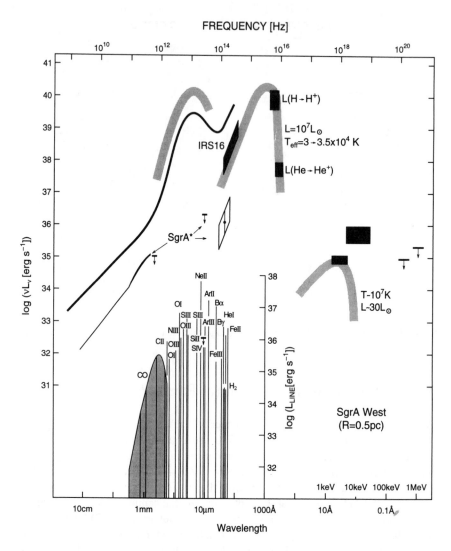

FIGURE 25.3. Schematic spectral energy distribution of the central few parsecs (heavy continuous), spectral line emission (bottom inset), X-ray fluxes and *SgrA** spectrum. Observational constraints on the intrinsically emitted far-infrared spectrum, UV spectrum and X-ray emission are denoted by shaded curves and boxes (adapted from [1]).

At this point we need to turn our attention to a number of recent high resolution near-infrared observations that elucidate the distribution and characteristics of the nuclear stellar cluster. These measurements have become possible in the last half a dozen years through the advent of sensitive, large format infrared detector arrays. In only a few years' time it has become possible to record the first high-quality seeing-limited images ($\sim 1''$, e.g., [46]) and then to improve further—through speckle

The Galactic Center at 2.2μm and a Resolution of 150 mas

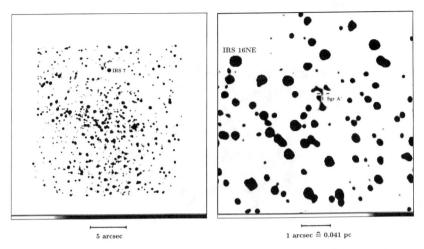

5 arcsec

1 arcsec $\hat{=}$ 0.041 pc

FIGURE 25.4. $0.15''$ resolution 2 μm continuum (K-band) image of the central parsec, obtained in 1994 with the MPE SHARP camera on the ESO NTT [47]. The right inset is a zoom into the central few arcseconds. The position of the compact radio source *SgrA** is denoted by a filled circle, and the bright stars $1''$ to $2.5''$ east/south-east of *SgrA** are stars in the IRS16 complex.

imaging—the resolution to the diffraction limit of 4 m class telescopes ($\sim 0.1''$ at 2 μm [24]). The best current 2 μm image of the central parsec is shown in Fig. 25.4 (adapted from [47]). It was taken with the MPE SHARP camera on the 3.5 m ESO NTT during superb ($\sim 0.3''$) seeing. It resolves the near-infrared emission of the central parsec into about 700 stars with K-band (2.2 μm) magnitudes ≤ 16. The central IRS16 complex located within $1''$ of the compact radio source *SgrA** consists of about two dozen single (or perhaps multiple) stars. From the number distribution of the near-infrared sources it appears that the centroid of the stellar cluster is more likely on *SgrA** than on the IRS16 complex, and that the core radius of the $K \leq 14$, 2 μm stellar number density distribution is about 0.2 pc [24]. If the stars with $K \leq 14$ are representative of the overall mass distribution of the cluster (an assumption that still needs to be proven by spectroscopic identification of the stars), such a small core radius would imply that the stellar density in the core is in excess of $10^{7.5}$ M$_\odot$ pc^{-3}. At these densities stellar mergers and disruption of giant star atmospheres by direct stellar collisions may become significant, with a number of interesting consequences for stellar and cluster evolution [1, 24, 25].

Another important ingredient of the near-infrared story has been the discovery of a 2.06 μm HeI/2.16 μm Br γ near-infrared emission line star (the AF-star, [27, 48]), followed by the discovery [28] of an entire cluster of about 15 such stars in the central parsec centered on the IRS16/IRS13 complex. Recent, less than $1''$ resolution line imaging [47] and imaging spectroscopy with the new MPE 3D-spectrometer (Fig. 25.5, [49]), now show unambiguously that several of the brightest members

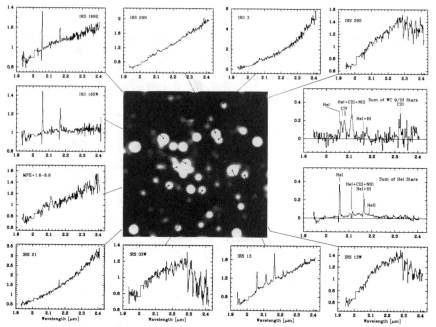

FIGURE 25.5. 3D K-band spectra (resolving power 800 to 1000, spatial resolution 1″) of 10 selected stars in the central 8″ [49], along with a smoothed (0.4″) version of the K-band map of Eckart *et al.* [47] showing the locations of the respective stars. The lower spectrum on the right side is the sum of the appropriately shifted spectra of IRS16C, SW, NE, NW, 34W and 33E. The upper right spectrum is the sum of IRS29N, 6, MPE-1.0-3.5 and MPE+1.6-2.8 which we identify as WC 9 (or Of).

of the IRS16 complex are HeI-stars, as is the nearby bright source IRS 13 (see also [50]). The IRS16 HeI "broad line region" discovered a decade ago by Hall *et al.* ([51], see also [52]) thus is now identified as a group of luminous mass-losing, He-rich stars. Figure 25.5 shows selected 2 μm spectra in the central 8″ × 8″ taken with the 3D spectrometer [49]. A number of the brightest stars in the region show bright HeI/HI emission lines and several have CIII/CIV/NIII emission lines, characteristic of winds and heavy element nucleosynthesis in massive stars. Non-local-thermodynamic-equilibrium stellar atmosphere modeling of the observed emission characteristics of the AF-star [53] confirms and quantifies earlier proposals that the AF-star is similar to WN9/Ofpe stars, a rare class of evolved, luminous blue supergiants related to Wolf-Rayet stars. These stars very likely represent the post-main sequence phase of massive stars (20 to 100 $M_\odot$) before they explode as supernovae. The AF-star has a luminosity of about $10^{5.5}$ $L_\odot$, effective temperature near 20,000 K and main-sequence mass between 25 and 40 $M_\odot$. The surface He/H abundance ratio is near unity and the mass loss rate is 6×10^{-5} $M_\odot$ yr^{-1} at a velocity of 700 km/s [53]. Based on the most recent results from 3D, the brightest HeI stars (IRS16NE, C, SW, IRS 13) are in many respects similar to the AF-star. They have effective temperatures between 20,000 and 30,000 K, and are helium-rich, but are about 5 to 10 times more luminous [49]. Their progenitor O stars likely had masses near 100 $M_\odot$. There is also evidence for classical (WC9 and WN9) Wolf-Rayet stars [49, 54], further strengthening the identification of the HeI stars as evolved massive stars. Combining the contributions from all its members, *the HeI-star cluster can plausibly account for essentially all of the bolometric and Lyman-continuum luminosities of the central parsec* [49]. The HeI-star cluster also provides in excess of 10^{38} erg/s in mechanical wind luminosity which may have a significant impact on the gas dynamics in the central parsec. In fact, the outward acceleration of the winds may explain why a possible central massive black hole is not likely to accrete very much interstellar gas at the present time [1].

The recent infrared stellar spectroscopy of the central parsec can be quantitatively fitted well by a model of a star formation burst between 3 and 7×10^6 years ago in which a few hundred OB stars and perhaps a total of a few thousand stars were formed [49]. This conclusion is in agreement with earlier proposals by Lacy, Townes and Hollenbach [32], Rieke and Lebofsky [55], and Allen and Sanders [56]. In this model the HeI stars are the most massive cluster members that in the meantime have evolved off the main sequence, and the central parsec is now in the late, wind-dominated phase of the burst. A similar object is the famous R136 star cluster that powers the 30 Doradus nebula in the Large Magellanic Cloud. The starburst model accounts naturally for the low excitation of the SgrA (West) HII region. Although there is also evidence for some very young, embedded OB stars, the present star formation activity appears to be significantly less than during the peak of the burst. The present gas density in the central parsec is too low for gravitational collapse of gas clouds to stars in the presence of the strong tidal forces [58]. One may speculate that the burst was perhaps triggered by a large influx of dense gas less than 10 million years ago.

Based on earlier theoretical work, Eckart *et al.* [24] have proposed sequential merging as an alternative to the starburst scenario, a possibility whose likelihood strongly depends on the density of the nuclear cluster [1]. A recent Fokker-Planck calculation of an evolving Galactic Center type, dense cluster with merging, shows, however, that merging can only account for fewer than ten 20 $M_\odot$ stars and none above 30 $M_\odot$ [57]. The basic reason is that in the calculations a sufficiently dense stellar core (density $10^{7.5}$ $M_\odot$ pc^{-3} or greater) cannot be maintained for a long enough time to build up very many massive stars. Morris [58] has suggested that the HeI stars are not classical blue supergiants at all but transitory objects that have been created in collisions between (~ 10 $M_\odot$) stellar black holes and solar-mass red giants. Both accounts of the HeI stars just cited are very specific to the high-density environment of the central parsec. However, a number of stars similar to the SgrA HeI stars have now been found in several clusters 2 to 13′ away from the central, high density SgrA region [20]. In the case of the Morris scenario [58] one probably would also expect a much larger X-ray emission than is observed. These facts, the requirement of having to account for 100 $M_\odot$ stars, and the presence of heavy-element nucleosynthesis products discussed above, in my opinion now strongly favor the star formation model over the other scenarios.

The fact that there are only about four red supergiants ($L \geq 10^4$ $L_\odot$) when compared to the number of blue supergiants suggests that there was relatively little star formation before 10 million years ago unless red supergiants have been preferentially destroyed by collisions in the dense core. By comparison, there is a much greater number of late-type stars with luminosities 10^3 to 10^4 $L_\odot$, both inside [49] and outside [58, 59] the central parsec. These medium luminosity stars are likely asymptotic giant-branch stars of moderate mass (2 to 7 $M_\odot$). They may signify another starburst episode that happened more than 10^8 years ago.

25.3   Is *SgrA** a Massive Black Hole?

The next key issue that I want to discuss is the evidence for a central massive black hole. Ever since the original discovery of the nonthermal compact radio source *SgrA** at the core of the nuclear star cluster [60-62], that source has been the primary black hole candidate, analogous to compact nuclear radio sources in other nearby normal galaxies [63, 64].

In fact, ever more detailed radio observations have confirmed the unique nature of *SgrA** in the Galaxy. Recent very long baseline interferometry (VLBI) observations at 7 mm show its size to be less than a few AU [42, 65]. Its proper motion relative to a background quasar is now known to be less than about 38 km/s [42], at least 6 times smaller than the (2D-) velocity dispersion of the stars. It is therefore a fairly safe conclusion that *SgrA** must have a mass in excess of about 150 $M_\odot$. The radio observations [42] have also provided compelling evidence for intrinsic flux density variability on time scales of months/years. The source shows a mm/submm excess above the flat cm-spectral energy distribution [66; Fig. 25.3] probably indicative of the presence of a very compact ($\sim 10^{12}$cm) radio core of stellar dimensions.

Yet observations at shorter wavelengths indicate nothing particularly impressive toward the radio position *SgrA**. The high resolution maps of Eckart *et al.* [47; Fig. 25.4] for the first time show that *SgrA** is located near the centroid of a T-shaped concentration of ≈ 10 compact 2 μm sources (=*SgrA*(IR)*). These sources are likely stars, and one might speculate whether they signal a central stellar cusp around *SgrA**. Based on the identification of three stars on the radio continuum maps (IRS13E, 1W and 7) and a cross-correlation of nebular Br γ and thermal radio continuum emission, the registration of *SgrA*(IR)* on the 2 μm map in Fig. 25.4 is now accurate to about $\pm 0.03''$. Because of the high local source density the identification of *SgrA** with the closest near-infrared source (K $\sim$15.2 mag) is uncertain. Speckle polarimetry at 2 μm shows that the polarizations of these sources, with the exception of one knot $1''$ southeast of the nominal position of *SgrA**, are small and consistent with the dichroic absorption of magnetically aligned dust grains in the interstellar medium along the line of sight to the Galactic center [47]. *SgrA*(IR)* also does not show intrinsic variability on scales of minutes or years, or significant line emission [47]. The possible infrared counterpart of *SgrA** in Fig. 25.4 has an absolute K-magnitude of about -2.5, similar to an early B main sequence star ($\sim 10^{4.2} L_\odot$) or an early K giant ($L \sim 10^{2.2} L_\odot$).

(Variable) hard X-ray emission is commonly considered a key signature of black holes. However, in contrast to the fairly bright infrared emission, the X-ray luminosity of *SgrA* (West) is fairly low (Fig. 25.3). The ≤ 2.5 keV X-ray luminosity (corrected for interstellar extinction) is only a few $L_\odot$ [67, 68], and the hard X-ray luminosity (2.5 to 100 keV) is less than a few hundred $L_\odot$ [69, 70, 8]. The upper limit originates from the recent finding by the Japanese X-ray satellite ASCA that there is a compact hard X-ray source about $1'$ southwest of *SgrA** that may have contributed significantly to some of the lower resolution hard X-ray observations just mentioned [12]. The inevitable conclusion is that *SgrA** presently contributes only a small fraction of the bolometric luminosity of the central parsec, with a conservative upper limit of perhaps a few 10^5 $L_\odot$ [1, 71]. There are currently three possible constraints on any past activity of *SgrA**. The first two are based on the scattering of hard X-ray emission from *SgrA**. Considering the extended 8.5 to 22 keV emission measured with the ART-P telescope on the Russian satellite GRANAT, Sunyaev, Markevitch, and Pavlinsky [13], on the one hand, conclude that "if there is a supermassive black hole in the Galactic nucleus, it has not emitted at the level of its Eddington luminosity for even a day over the past 400 years." A somewhat more conservative calculation [1] shows that the average 8-22 keV luminosity of *SgrA** was not substantially greater than $10^{5.2}$ $L_\odot$ during the past 400 years. On the other hand, extrapolating from the extended 6.4 keV Fe-line emission, the hard X-ray luminosity of *SgrA** may have been at least as large as $10^{5.4}$ $L_\odot$ [12]. The third possible constraint comes from the thermal energy of the 10^8 K gas that ASCA observes to be associated with *SgrA* which may have originated from a recent energetic explosion ($\geq 10^{52\pm1}$ ergs; see above) from the nucleus. Yet despite the uncertainties, none of these estimates comes even close to the Eddington luminosity of a 10^6 $M_\odot$ black hole ($\sim 10^{10}$ $L_\odot$).

The evidence for a central mass concentration in the Galactic Center thus is based entirely on the gas and stellar dynamics. As mentioned in the introduction, evidence for a central mass concentration based on gas dynamics (Fig. 25.1) had already been growing in the 1980s. However, interstellar gas reacts to forces other than gravity, such as friction, pressure disturbances (shocks), magnetic fields and radiation pressure. Therefore, a definite statement on the mass distribution requires measurements of stellar velocities which have been emerging in the past half a dozen years. The current status of the inferred (dynamic) mass as function of radius is shown in Fig. 25.6. It includes the most recent determinations by Krabbe *et al.* [49] of stellar velocity dispersions derived from radial velocity measurements for about 35 individual stars between $1''$ and $12''$ from *SgrA**. The various mass estimates derived from the gas and stellar dynamics are compared to the mass distribution derived from the stellar light and a constant mass to light ratio (M/L $\sim M_\odot/L_\odot$, dashed line) for a compromise core radius of 0.5 pc. Compared to the situation more than a decade ago, when the first velocity measurements of ionized gas clouds became available [31, 32], the evidence for a central dark mass has become fairly compelling. The stellar velocity dispersion of the 35 innermost stars

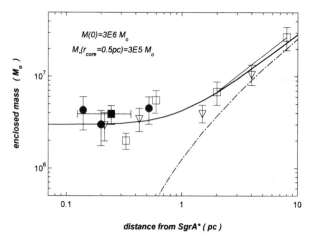

FIGURE 25.6. Enclosed mass as a function of (true) distance from *SgrA**. The new mass estimates derived from early- and late-type star velocity dispersions [49] are marked by filled circles and a filled rectangle, with corresponding 1σ error bars. Shown as open squares (stellar velocity data) and open triangles (gas dynamics) are several mass estimates taken from the literature [36, 77-82]. The dashed curve represents the mass distribution of an isothermal stellar cluster, with an assumed 'compromise' core radius of 0.5 pc [1] and a mass of 3×10^5 $M_\odot$ within the core radius. The continuous curve is a model containing a 3×10^6 $M_\odot$ central mass plus the isothermal cluster.

(153 ± 18 km/s; [49]) is 4.6σ above the velocity dispersion of the stellar cluster at > 1 pc from the center (70 ± 15 km/s [78, 80, 81]). The latter should be the velocity dispersion of the stars also in the innermost core if there were no central dark mass. The fact that the velocity dispersions of the (massive) early type HeI/HI emission line stars and of the intermediate mass late-type stars agree further strengthens the case for a dominant central mass. If one combines the stellar and gas dynamics shown in Fig. 25.6 the case for a dark central mass concentration of 2 to 4 $\times$ 10^6 $M_\odot$ is now quite convincing. The only remaining caveat is the possibility that stellar orbits with large radial anisotropies dominate. This could also result in an increase of velocity dispersion near pericenter. An experiment is now well underway to measure the proper motions of stars between 0.3" (0.012 pc) and 10" (0.4 pc) from *SgrA** from repeated high-resolution near-infrared imaging with the SHARP camera on the ESO NTT. Two or three significant proper motions have already been detected and it is expected that ~ 10 to 20 motions may be available in a few years. This experiment should give a clear-cut answer on the anisotropy of the stellar orbits and the reality of the dark central mass.

Given the present evidence the central mass has a density of at least $10^{8.5}$ $M_\odot$ pc^{-3} and its ratio of mass to luminosity is 10 to 100 times greater than that of the average star in the surrounding stellar cluster. As the dark mass is at least as concentrated as the distribution of the massive HeI/HI stars it cannot be a cluster of solar mass, dark stellar remnants, such as neutron stars. The dark mass concentration could be a very compact cluster of stellar mass (~ 10 $M_\odot$) black holes [58] although theoretical calculations suggest that such a cluster will rapidly form a central massive black hole [85,86]. Thus the most likely configuration is probably a single massive black hole.

If *SgrA** is indeed a million solar mass black hole, the riddle is why it is presently so inactive. It is very interesting that the Galactic Center shares this 'luminosity deficiency' or 'blackness' problem with essentially all nearby nuclei for which there is substantial evidence for dark central masses [72], including the presently most convincing case, the "mega" H_2O maser source NGC4258 [73]. It is possible that the tidal disruption and accretion of stars by the hole (happening in the Galactic Center at a rate of a few $10^{-4}\,yr^{-1}$ [32]) occurs very efficiently albeit at low-duty-cycle [74]. Accretion of interstellar gas streamers by the hole may be prevented by the need to overcome the angular momentum problem, coupled with the outward force of the stellar winds as discussed above. Finally, the wind-gas itself may be accreted largely spherically, with very low radiation efficiency [75]. Nevertheless, current models of black hole accretion have to be stretched to be compatible with *SgrA** being an underfed million-solar-mass black hole [76].

25.4   Acknowledgments

Having had the opportunity to work with Charlie Townes and enjoying his support and friendship ever since we first met 16 years ago has been a key experience in my life. As can be seen from this contribution, my research group and I are still following the trail on which Charlie set us long ago.

References

[1] R. Genzel, D. Hollenbach and C. H. Townes, Rep. Progr. Phys. **57**, 417 (1994).

[2] L. Blitz and D. N. Spergel, ApJ **379**, 631 (1991).

[3] J. J. Binney, O. E. Gerhard, A. A. Stark, J. Bally. and K. A. Uchida., MNRAS **252**, 210 (1991)

[4] H. S. Liszt and W. B. Burton, ApJ **236**, 779 (1980).

[5] K. Koyama *et al.*, Nature **339**, 603(1989).

[6] R. Diehl *et al.*, A&A (Suppl.) **97**, 181 (1993).

[7] G. K. Skinner, A&A (Suppl.) **97**, 149 (1993).

[8] R. Sunyaev *et al.*, ApJ **383**, L49 (1991).

[9] I. F. Mirabel, L. F. Rodriguez, B. Cordier, J. Paul, and F. Lebrun, Nature **358**, 215 (1992).

[10] I. F. Mirabel and L. Rodriguez, Nature **371**, 46 (1994).

[11] R. Güsten, The Center of the Galaxy, p. 89, Kluwer (1989).

[12] K. Koyama, preprint (1994).

[13] R. A. Sunyaev, M. Markevitch, and M. Pavlinsky, ApJ **407**, 606 (1993).

[14] M. Morris, Galactic and Extragalactic Magnetic Fields, R. Beck, P. Kronberg, and R. Wielebinski (eds.), 361, Kluwer (1990).

[15] Y. Sofue, Nuclei of Normal Galaxies: Lessons from the Galactic Center, p. 43, Kluwer (1994).

[16] E. Serabyn and R. Güsten, A&A **184**, 133 (1987).

[17] J. Heyvaerts, C. Norman, and R. Pudritz, ApJ **330**, 718 (1988).

[18] M. Morris and F. Yusef-Zadeh, ApJ **343**, 703 (1989).

[19] R. Genzel *et al.*, ApJ **356**, 160 (1990).

[20] A. S. Cotera, E. F. Erickson, D. A. Allen, S. W. J. Colgan, J. P. Simpson, and M. G. Burton, in [84], 217 (1994).

[21] P. G. Mezger *et al.*, 1989, A&A **209**, 337.

[22] D. A. Allen, Nuclei of Normal Galaxies: Lessons from the Galactic Center, p. 293, Kluwer (1994).

[23] G. Rieke and M. Rieke, in [84], 283 (1994).

[24] A. Eckart, R. Genzel, R. Hofmann, B. J. Sams, and L. E. Tacconi-Garman, ApJ **407**, L77 (1993)

[25] E. S. Phinney, The Center of the Galaxy, p. 543, Kluwer (1989).

[26] M. J. Lebofsky, G. H. Rieke, and A. Tokunaga, ApJ **263**, 736 (1982).

[27] D. A. Allen, A. R. Hyland, and D. J. Hillier, MNRAS **244**, 706 (1990).

[28] A. Krabbe, R. Genzel, S. Drapatz, and V. Rotaciuc, ApJ **382**, L19 (1991).

[29] E. Wollman, Ph.D. Thesis, Univ. of California, Berkeley (1976).

[30] J. H. Lacy, F. Baas, C. H. Townes, and T. R. Geballe, ApJ **227**, L17 (1979).

[31] J. H. Lacy, C. H. Townes, T. R. Geballe, and D. J. Hollenbach, ApJ **241**, 132 (1980).

[32] J. H. Lacy, C. H. Townes, and D. J. Hollenbach, ApJ **262**, 120 (1982).

[33] K. Y. Lo and M. J. Claussen, Nature **306**, 647 (1983).

[34] E. Serabyn and J. Lacy, ApJ **293**, 445 (1985).

[35] E. Serabyn, J. Lacy, C. H. Townes, and R. Bharat, ApJ **326**, 171 (1988).

[36] J. H. Lacy, J. M. Achtermann, and E. Serabyn, ApJ **380**, L71 (1991).

[37] T. R. Geballe *et al.*, ApJ **284**, 118 (1984)

[38] D. K. Aitken *et al.*, ApJ **380**, 419 (1991).

[39] R. H. Hildebrand *et al.*, ApJ **417**, 565 (1993).

[40] U. Schwarz and J. Lasenby, in Galactic and Extragalactic Magnetic Fields, R. Beck, P. Kronberg, and R. Wielebinski (eds.), p.383, Kluwer (1990)

[41] N. E. B.Killeen, K. Y. Lo, and R. Crutcher, ApJ **385**, 585 (1992).

[42] D. Backer, in [84], 403 (1994).

[43] J. Jackson *et al.*, ApJ **402**, 173 (1993).

[44] J. A. Davidson, M. W. Werner, X. Wu, D. F. Lester, P. M. Harvey, M. Joy, and M. Morris, ApJ **387**, 189 (1992).

[45] E. E. Becklin, I. Gatley, and M. W. Werner, ApJ **258**, 134 (1982).

[46] W. J. Forrest, J. L. Pipher, and W. A. Stein, ApJ **301**, L49 (1986)

[47] A. Eckart, R. Genzel, R. Hofmann, B. Sams, and L. E. Tacconi-Garman, ApJ **445**, L23 (1995).

[48] W. J. Forrest, M. A. Shure, J. L. Pipher, and C. A. Woodward, in The Galactic Center, D. Backer (ed.), AIP Conference Proceedings **155**, 153 (1987).

[49] A. Krabbe *et al.*, ApJ Lett. **447**, L95 (1995).

[50] R. D. Blum, D. L. dePoy, and K. Sellgren, ApJ **441**, 603 (1995).

[51] D. N. B. Hall, S. G. Kleinmann, and N. Z. Scoville, ApJ **262**, L53 (1982).

[52] T. R. Geballe *et al.*, ApJ **320**, 562 (1987).

[53] F. Najarro *et al.*, A&A **285**, 573 (1994).

[54] R. D. Blum, K. Sellgren, and D. L. DePoy, ApJ in press (1995).

[55] G. H. Rieke and M. J. Lebofsky, in The Galactic Center, G. R. Riegler and R. D. Blandford (eds.), AIP Conference Series **83**, 194 (1982).

[56] D. A. Allen and R. H. Sanders, Nature **319**, 191 (1986).

[57] H. M. Lee, in [84], 335 (1994).

[58] M. Morris, ApJ **408**, 496 (1993).

[59] J. W. Haller and M. J. Rieke, in [83], 487 (1989).

[60] R. D. Ekers and D. Lynden-Bell, Ap. Lett. **9**, 189 (1971).

[61] D. Downes and A. Martin, Nature **233**, 112 (1971).

[62] B. Balick and R. L. Brown, ApJ **194**, 265 (1974).

[63] D. Lynden-Bell and M. Rees, MNRAS **152**, 461 (1971).

[64] K. Y. Lo, in [83], 527 (1989).

[65] T. P. Krichbaum, C. J. Schalinski, A. Witzel, K. J. Standke, D. A. Graham, and J. A. Zensus, in [84], 411 (1994)

[66] R. Zylka, P. G. Mezger, and J. Lesch, A&A **261**, 119 (1992).

[67] M. G. Watson, R. Willingale, J. E. Grindlay, and P. Hertz, ApJ **250**, 142 (1981)

[68] P. Predehl and J. Trümper, A&A **290**, L29 (1994).

[69] G. K. Skinner *et al.*, Nature **330**, 544 (1987).

[70] Goldwurm *et al.*, Nature **371**, 5889 (1994).

[71] P. G. Mezger, in [84], 415 (1994).

[72] J. Kormendy, in [84], 379 (1994).

[73] M. Miyoshi, J. M. Moran, J. Hernstein, L. Greenhill, N. Nakai, P. Diamond, and M. Inoue, Nature **373**, 127 (1995).

[74] M. Rees, Nature **333**, 523 (1988).

[75] F. Melia, ApJ **387**, L25 (1992).

[76] L. Ozernoy and R. Genzel, in The Galaxy, L. Blitz (ed.) Kluwer, in press (1995).

[77] T. M. Herbst, S. V. W. Beckwith, W. J. Forrest, and J. L. Pipher, AJ **105**, 956 (1993).

[78] K. Sellgren, M. T. McGinn, E. Becklin, and D. N. B. Hall, ApJ **359**, 112 (1990).

[79] J. W. Haller, M. J. Rieke, and G. H. Rieke, BAAS **24**, 1178 (1992).

[80] M. Lindqvist, H. Habing, and A. Winnberg, A&A **259**, 118 (1992).

[81] G. H. Rieke and M. J. Rieke, ApJ **330**, L33 (1988).

[82] R. Güsten *et al.* ApJ **318**, 124 (1987).

[83] M. Morris (ed.), The Center of the Galaxy (Dordrecht: Kluwer) (1989).

[84] R. Genzel and A. I. Harris, Nuclei of Normal Galaxies: Lessons from the Galactic Center, Kluwer (1994).

[85] H. M. Lee, MNRAS, in press (1995).

[86] M. J. Rees, Ann. Rev. A&A **22**, 471 (1984).

26

Microwave Spectroscopy, the Maser, and Radio Astronomy: Charles Townes at Columbia

Joseph A. Giordmaine[1]

Charles Townes' graduate students at Columbia in the mid-1950s were unusually fortunate.

The scientific applications of the new field of microwave spectroscopy of gases had been remarkably rich during the previous years, many of the most important directions had been pioneered by Charlie himself, the new experimental techniques were well-established at the Columbia Radiation Laboratory, a large and stimulating group of graduate students and postdoctoral research scientists was in place, and the wide range of results of the intense activity in the field had just been organized by Charlie (with Arthur Schawlow) in the comprehensive monograph "Microwave Spectroscopy."

The scope of Charlie's own contributions had extended from basic issues of microwave spectrometer design, sensitivity, mechanisms of broadening, power saturation, lineshapes and splittings to a wealth of molecular spectroscopy results. These covered molecular rotational frequencies, moments of inertia, spin-rotation coupling mechanisms, microwave attenuation processes in the atmosphere, magnetic dipole and quadrupole interactions between nuclei and electrons in atoms and molecules, nuclear spins, magnetic and quadrupole moments, isotopic mass ratios, and inversion frequencies in molecules with hindered motions. Not limited in his interests to molecular spectroscopy, Charlie made some of the first observations of paramagnetic resonance in free radicals, and showed how nuclear resonance frequency shifts observed in radio-frequency spectroscopy of metals arose from the paramagnetism of conduction electrons in the neighborhood of the nucleus, for example. During all of this work, Charlie continued to explore new techniques, from magnetron harmonics to Cerenkov radiation, in a quest to move to yet higher frequencies.

---

[1] Based on remarks at the Birthday Symposium in Honor of Charles Hard Townes, Conference on Lasers and Electro-optics, and Quantum Electronics and Laser Science Conference, Baltimore, Maryland, May 22, 1995.

This rich output of basic research, coupled to the technological goal of generating higher frequencies, culminated in the conception of the ammonia maser and, by 1954, its successful demonstration. Although it is sometimes pointed out that stimulated emission devices could in principle have been demonstrated many years earlier, the invention in fact occurred only in the fertile mix of molecular spectroscopy, molecular beam experimental techniques, and microwave radar technology. The basic quantum mechanical principle of population inversion, known indeed for many years, was not enough. The years between conception and first oscillations of the maser undoubtedly required at least as much determination and confidence in the outcome as understanding of the principles!

In the mid-1950s, Charlie's interest in taking advantage of the maser to open up new research areas led to expansion of the work at Columbia into fields such as solid-state paramagnetic resonance, maser amplifiers for radio astronomy, high-precision tests of special relativity, and stimulated emission in the infrared, culminating in turn in the invention of the laser.

During this period Charlie's style of physics attracted a group of graduate students marked by an unusual degree of camaraderie and mutual support. Among them were Isaac Abella, Lee Alsop, Alan Barrett, Dave Carter, Herman Cummins, George Dousmanis, Jim Gordon, Fred Johnson, Paul Kisliuk, Harold Lecar, Herb Lashinsky, Jim Lotspeich, Frank Nash, Arno Penzias, Elsa Rosenwasser, Bill Rose, Mirek Stevenson, and Pat Thaddeus. Also on hand were outstanding senior visiting scientists and research staff, each of whom brought a unique expertise that was quickly shared with the students. Among them were Bill Bennett, Brebis Bleaney, Oliver Heavens, Ali Javan, and Ned Nethercot.

Charlie maintained a remarkably close degree of contact with students—often including a daily "journal club" and "research status report." The intensity of Charlie's approach to research was always in evidence. One graduate student recalls observing for the first time and reporting to Charlie on a Friday a novel paramagnetic relaxation phenomenon. It was remarkable to see an explanation and an almost completed manuscript by Monday!

During this period, Charlie's work began to be recognized by awards. These events took on a more personal meaning for his graduate students who discovered that Charlie shared cash awards with his students!

Acceptance of substantial risk was a normal research practice for Charlie. Two of his students were assigned the thesis project of building a practical maser amplifier for radio astronomy, to take advantage of the inherently low maser noise levels. Charlie assigned this topic as a thesis project after his demonstration, while on sabbatical in France, that pulsed gain was possible in a 2-level solid state system, and the introduction by Nico Bloembergen of the more powerful 3-level scheme. At the time, however, there existed no suitable material for a practical amplifier. Not until shortly into the project, however, had scientists at the University of Michigan introduced ruby, a highly suitable amplifier material. The risk had been well-taken. The project led to the first application of a maser amplifier and, for example, new data on the surprisingly high temperature of Venus, heated by the greenhouse effect to temperatures close to the melting point of lead.

What were the methods, the scientific culture and the values communicated by Charlie to his students? They were many, and included:

- Respect for students and colleagues at all levels, caring about students' personal welfare (often literally *in loco parentis*), and constructive acceptance of mistakes. On one occasion, at a critical moment in a project, the author of this contribution made a particularly time-consuming error. Charlie's only reaction was a gentle "Joe, sometimes you have to think about things a little bit." That mistake was never made a second time!

- The value of getting an early lead in a research field through use of new technology. This reverse technology transfer from development to research has in the past given industrial labs an important advantage in doing basic research, an advantage being sadly lost today as many companies cut back on fundamental work.

- The necessity of risk-taking, accepting the failures that go along with risk, and recognizing that successes and surprises with a big payoff come from only a small fraction of projects.

- The understanding that surprises and unpredictability are a measure of first-rate research. One recalls that the maser was at first thought of as a mm-wave generator rather than as a stepping-stone to the laser.

- The importance and extreme difficulty of interdisciplinary work. One recalls the combination of electronics and spectroscopy, physics and engineering, feedback principles and quantum mechanics, biology and statistics involved in Charlie's maser work. In his new book, "Making Waves," Charlie recalls the initial difficulty that even John von Neumann experienced in accepting the new experimental result that the output of the maser was highly coherent.

- The importance of socializing in science, but especially in interdisciplinary research.

- The impact on science and society of open-minded, far-sighted funding agencies, exemplified by the sponsorship of Charlie's work at Columbia by the Joint Services, the Army Signal Corp., the Office of Naval Research (ONR), and the Air Force Office of Scientific Research (AFOSR).

How fortunate we were indeed!

27

The Role of Radioactive ^{14}C and ^{26}Al in the Ionization of Circumstellar Envelopes

Alfred E. Glassgold

27.1   Introduction

The circumstellar envelopes of evolved stars are another of the many scientific interests of C. H. Townes. When I became interested in the topic about ten years ago, pioneering studies at infrared and radio wavelengths had already demonstrated the potential of millimeter and submillimeter spectroscopy for understanding this interesting part of stellar evolution [1]. A good part of the scientific and technical basis for this situation had been developed by Townes and his collaborators, as descibed in an early review [2]. After introducing the subject, I will discuss the ionization of circumstellar envelopes in Sec. 27.2 and the role of live radioactivity in Sec. 27.3.

When stars of about the same (or somewhat larger) mass as the Sun run out of nuclear fuel, they end the red giant phase of their lives at the tip of the *asymptotic giant branch* (AGB). At this stage, they are cool (photospheric temperatures $\approx$ 2500 K) and luminous ($L \approx 10^4 L_\odot$), but their most remarkable property is that they shed mass at a prodigous rate in the form of a slow stellar wind, with a typical speed of 10 km s^{-1}. With mass-loss rates in the range $10^{-5} - 10^{-4}$ $M_\odot$ yr^{-1}, they can lose several $M_\odot$ in a characteristic time of fifty thousand years. As a result, the star can reduce its mass enough to end up below the Chandrasekhar limit, and thus have a good chance of becoming a planetary nebula and then a white dwarf, rather than a supernova. An AGB wind also carries dredged-up products of stellar nucleosynthesis that can tell us about nuclear burning inside dying stars, before the ejected material is diluted by the interstellar medium.

The winds of AGB are cool by stellar standards. An important piece of evidence is that dust particles, with typical condensation temperatures of 1500 K, are observed to form quite close to AGB stars. (Much of what we we know about where dust forms comes from Townes' pioneering efforts in developing infrared stellar interferometry [3].) Support for low wind temperatures also comes from the fluxes of molecular lines that cool the wind, e.g., CO. Deductions on the temperature distribution of the best studied case, IRC + 10216, a nearby carbon star, indicate that the temperature at 1 arcsecond (3×10^{15} cm if the star is 200 pc distant) is about 300 K and that it reaches 10 K at 30 arcsecond (10^{17} cm) [4].[1]

The chemical properties of AGB winds are interesting because, on the one hand, they contain information about nucleosynthesis inside the star and, on the other, determine the way in which stars alter the interstellar medium. A good example is circumstellar dust grains, which form the growth centers for interstellar dust. There is an important distinction between the chemical properties of O-rich and C-rich winds based on the special properties of the CO molecule and the condensation processes in the two environments. CO is better bound than any other circumstellar molecule so that, in the near-LTE (local thermodynamic equilibrium) stellar atmosphere, where such molecules originate, it takes up all the O in a C-rich environment, leaving C as the dominant chemically active species after H. Just the opposite happens in an O-rich environment, i.e., CO takes up all the C, leaving O as the dominant chemical species. Thus the gas in C-rich envelopes will be dominated by hydrocarbons, and the dust by carbonaceous forms such as amorphous carbon. O-rich envelopes are dominated by O-bearing species like H_2O and OH and by silicaceous dust. Because of the special chemical properties of carbon, the C-rich environments are more interesting, and we shall concentrate on them.

The richness of circumstellar carbon chemistry is shown by IRC + 10216, where almost 50 molecules have been detected (not counting isotopic variants), as shown in Table 27.1. Though most have been detected with millimeter-wave rotational lines, a number of interesting species have been found in the near infra-red [5] and at submillimeter wavelengths [6]. The propensity toward radicals and chain molecules is probably the result of an active photochemistry [7]. Interstellar UV photons penetrate the dusty outer envelope, and dissociate hydrocarbons that enter the envelope from the inside. Among the main dissociation products are CN and C_2H radicals, which are capable of fast chemical reactions at the low temperatures of the outer envelope, and synthesize a large fraction of the observed species. The tell-tale signature of photoproduction is the distribution of the synthesized molecules in shells. The theory predicts typical shell angular radii for IRC + 10216 of around 15 arcseconds, and sizes of this order have been found for many molecules [8, 9]. A notable aspect of Table 27.1 is the presence of only one ion, HCO^+, at quite low abundance: $x(HCO^+) \approx 4 \times 10^{-10}$ (relative to the density of hydrogen nuclei). Theoretically, the most abundant ions are C^+ in the outer envelope and $C_2H_2^+$ and $C_2H_3^+$ at intermediate distances. Detection of the former at 158 μm awaits suitable space-borne instrumentation; the most favorable line for the detection of $C_2H_3^+$ is in a difficult part of the 350 GHz atmospheric window.

---

[1]The densities span a wide range from those characteristic of stellar atmospheres to those in the interstellar medium. For a steady, spherical wind with terminal velocity V_6 (units 10^6 cm s^{-1} = 10 km s^{-1}) and mass-loss rate $\dot{M}_{-5}$ (units 10^{-5} $M_\odot$ yr^{-1}), the density of hydrogen nuclei (mainly in the form of H_2 for cool winds) is $n_H = Cr^{-2}$, where $C = 3 \times 10^{37}$cm$^{-1}\dot{M}_{-5}/V_6$.

TABLE 27.1. Species Detected in IRC + 10216[1]

10^{-3}	CO							
10^{-4}								
	C		C_2H_2					
10^{-5}					HCN			
			C_2H		CN			
		CH_4	C_4H					
10^{-6}			C_3		HC_3N	CS		
					C_3N	SiS		
	SiH_4				HC_5N	SiC_2		AlCl
10^{-7}	NH_3		C_5		HC_7N	SiO		AlF
			C_6H		HNC	SiN		NaCN
			C_3H_2		HC_9N	SiC		MgNC
10^{-8}			C_3H	H_2C_4	$HC_{11}N$		C_2S	CP
			C_5H	HC_2N	CH_3CN		C_3S	NaCl
				H_2C_3				KCl
10^{-9}						SiC_4	H_2S	
	HCO^+							
10^{-10}								

1. Abundances relative to H_2 (somewhat arbitrary as well as uncertain).

27.2 Ionization

The ionization of a circumstellar envelope is important because it determines (1) whether the wind interacts with the stellar magnetic field (with sufficient ionization, the wind can drag the field along with it); (2) the rate of ionic processes such as electronic recombination and ion-molecule reactions; and (3) the level of thermal bremsstrahlung or synchrotron radiation. Information on the ionization of thick (dust enshrouded) envelopes is limited because only HCO^+ has been detected so far in these winds. The situation in optically visible envelopes is quite different; e.g., numerous lines of ionized atoms have been detected in the optical-UV spectrum of α Ori [10].

Before discussing the significance of HCO^+, we note that a circumstellar envelope can be ionized by either internal or external sources. Unless the wind is hot enough for thermal (collisional) ionization to occur (not uncommon *after* the stars have left the AGB), the internal sources are stellar UV radiation, X-rays and cosmic rays, and radioactivity. Standard indicators suggest that AGB stars are not sources of X-rays or cosmic rays [11], and those undergoing heavy mass loss can block any UV radiation that they emit. Thus the main ionizing agents for AGB stars with massive circumstellar envelopes are radioactive elements dredged up from the interior of the stars and external interstellar UV photoionization and cosmic rays.

Galactic cosmic rays were invoked early on as a way of initiating molecular synthesis in the outer envelopes of relatively quiescent, approximately spherical, circumstellar envelopes such as IRC + 10216. We now believe that the rate at which cosmic rays induce ion-molecule reactions is too small to account for the large number of complex molecules in that envelope. As mentioned above, the present favored method involves CN and C_2H neutral radical reactions [12, 13]. Thus the question is, to what extent do cosmic-ray induced, ion-molecule reactions play any role at all? There are indeed species, like HNC and $C_3H_3^+$, that are produced predominantly by cosmic rays [12], but more direct evidence is desired. It is possible to show that, in C-rich circumstellar envelopes, the molecular ions H_3^+, HCO^+, and $C_2H_3^+$, are signatures of high-energy ionization processes, as opposed to UV photoionization [14], and that HCO^+ is the easiest detectable ion at this time. Lucas & Guélin have in fact detected the $J = 1 \rightarrow 0$ line of HCO^+ in IRC + 10216 [15]. Their data implies an effective cosmic ray ionization rate in this envelope of $\zeta_{CR} \approx 2.5 \times 10^{-18}$ s^{-1}.

How does this value compare with other values of ζ_{CR}? Spitzer and Tomasko [16] calculated this rate using the fluxes of cosmic rays measured at Earth and correcting for modulation by the solar wind. On integrating the demodulated cosmic ray flux all the way down to 1 MeV/nucleon, they obtained $\zeta_{CR} = 6.8 \times 10^{-18}$ s^{-1} per H atom. Because of the uncertainties in the low-energy part of the spectrum, this result has often been considered a lower limit. Values determined from interstellar chemistry are roughly 5×10^{-17} s^{-1} per H atom [17]. Instruments on Pioneer 10 and 11 and the Voyager missions have measured the cosmic ray spectrum down to 60 MeV/nucleon out to Neptune [18], and we now have a better solution of the solar modulation problem [19]. But when the cosmic ray ionization rate is recalculated, the results are not much different than the original result [16]. Thus the rate for cosmic ray energies $E > 1$ MeV is $\zeta_{CR} = 1.8 \times 10^{-17}$ s^{-1} (about 2.5 larger than [16]) and the rate for $E > 1$ GeV is $\zeta_{CR} = 8 \times 10^{-19}$ s^{-1}. With this information, we can interpret the much smaller value determined for IRC + 10216 as the effect of cosmic rays with $E > 100$ MeV; the lower-energy cosmic rays are probably excluded by the wind, along the lines first discussed by Parker [20] in analogy with the solar wind. This conclusion is speculative because the problem of the exclusion of cosmic rays by AGB winds has not yet been solved. Among the physical issues that arise are the strength of the magnetic field of the wind, the cosmic ray diffusion coefficient in the wind, and the possible regeneration of cosmic rays in the terminating wind shock. Moreover, we still have to consider the other internal ionization source, radioactivity.

27.3   Radioactive Winds

The idea that radioactive nuclei might be dredged up to the atmospheres and into the winds of evolved stars is an old one. In 1952 Merrill [21] reported the detection of Tc (an element with no stable isotopes) in the atmospheres of S stars, thereby directly confirming nucleosynthesis *and* dredgeup in red giants. The presence of ^{26}AlH was searched unsuccessfully for in the optical spectra of stars with Tc as early as 1970 [22]. Cowan and Rose [23] estimated the ^{14}C/^{13}C ratio produced by double-shell burning in red giants, and suggested that ^{14}C might be detected with near-infrared or millimeter wave spectroscopy. Such observations were already underway in 1976 and upper limits were obtained on the ^{14}C/^{12}C ratio in IRC + 10216 of 10^{-4} (CO, 4.6 μm [24]) and 0.04 (HCN, 89 GHz [25]). Many searches have been been made for ^{14}C in the succeeding years, all with negative results until 1994, when Wright [26] reported a possible detection of the $J = 1-0$ line of ^{14}CO at 105.871 GHz. (It should be mentioned in passing that precision measurements of the the ground rotational frequency of the six most common CO isotopes were made in Townes' laboratory in 1958 [27].) The most accurate limits obtained in the near-infrared for C stars are in the range $3 \times 10^{-5} - 2 \times 10^{-4}$ [28]; Wright's upper limit ($\approx 10^{-4}$) is in the middle of this range. Radioactive ^{26}AlF may also have been detected in IRC + 10216 by Guélin *et al.* [29].

Although attention has been given to the actual detection of radioactive nuclides, it is of interest to consider their effect on the physical properties of AGB winds. We first considered the decay of ^{26}Al [14], which nucleosynthesis theory predicts is synthesized during either the thermal-pulse stage of double-shell burning for low-mass AGB stars (1-3 $M_\odot$) or hot-bottom, convective envelope burning for heavier stars ($> 5\ M_\odot$) e.g., [30, 31]. The expected photospheric abundance ratio of radioactive ^{26}Al to stable ^{27}Al isotope is $\mathcal{R} \equiv {}^{26}$Al/^{27}Al $\sim 2 \times 10^{-2}$. Isotope enhancements this large should have observable consequences.

It is informative to compare the effects of the β-decays of ^{26}Al and ^{14}C on an AGB wind. The decay of ^{26}Al is accompanied by the emission of γ rays, with 82% producing a positron (^{26}Al $\rightarrow {}^{26}$Mg$^* + $ e$^+ + \nu$) and 18% proceeding by electron capture (e$^- + {}^{26}$Al $\rightarrow {}^{26}$Mg$^* + \overline{\nu}$). The latter channel gives only a few Auger electrons and can be neglected, whereas the positron with mean energy $E_\beta(^{26}$Al) = 0.66 MeV can produce about 20,000 ionizations. AGB winds are too thin to stop the γ rays (they interact only with the Compton cross-section), and they escape. The decay of ^{14}C is simple, (^{14}C $\rightarrow {}^{14}$N + e$^- + \overline{\nu}$), with 100% going to the ground state of ^{14}N with $E_\beta(^{14}$C) = 0.047 MeV. Thus the two β decays behave similarly in the sense that in both cases it is a β-particle which heats and ionizes the wind; the physics of the stopping of electrons and positrons is the same for energies above 5 keV. There is a difference in the depletion of the parent nuclei. Whereas about half of the C is in circumstellar dust particles, most of the Al is probably depleted from the gas phase. A qualitative analysis of the escape of ^{26}Al positrons from dust grains indicates, however, that most of them escape and then get stopped as they excite and ionize the circumstellar gas.

In order to assess the effects of these radioactive decays on a circumstellar envelope, we estimate the ionization rate ζ per H nucleus [14]. If $\mathcal{R}$ is the abundance ratio of a radioactive to a stable nuclide, the latter with absolute abundance x (relative to H nuclei), and if V is the expansion speed of the wind,

$$\zeta = b_\beta \frac{1}{\tau} \mathcal{R} x \frac{E_\beta}{e_{ion}} \exp(-r/V\tau),$$

where b_β is the branching for the emission of a β-particle (0.82 for ^{26}Al and 1.00 for ^{14}C), τ is the mean lifetime (1.0×10^6 yr for ^{26}Al and 8100 yr for ^{14}C), and $e_{ion} = 37$ eV is the mean energy to create an ion pair. For the cases of interest here, $r < V\tau$, and the rates are :

$$\zeta_{26} \approx 1.4 \times 10^{-15} \mathcal{R}_{26} \, s^{-1} \qquad \zeta_{14} \approx 9.0 \times 10^{-13} \mathcal{R}_{14} \, s^{-1},$$

where we have used $x(^{27}\text{Al}) = 3 \times 10^{-6}$ and $x(^{12}\text{C}) = 4 \times 10^{-4}$. Just to pin down the order of magnitudes, we evaluate these formula using the upper limits given by recent radio observations of IRC + 10216: $\mathcal{R}^{26} = 0.04$ [29] and $\mathcal{R}_{14} = 10^{-4}$ [26], and obtain $\zeta_{26} = 6 \times 10^{-17} \, s^{-1}$ and $\zeta_{14} = 9 \times 10^{-17} \, s^{-1}$. These numbers are larger than the ionization rate determined from the HCO^+ abundance by factors of 20 and 35, respectively. We conclude that the ionization of AGB winds similar to IRC + 10216 is likely to be due in part to *in situ* radioactive decays.

There is an interesting connection here with meteoritics where $\mathcal{R}_{26}$ has been measured in presolar grains with AGB signatures [32, 33]. For SiC grains in Murchison with about the same $^{12}C/^{13}C$ ratio as IRC + 10216, the $^{26}Al/^{27}Al$ ratios are in the range $10^{-4} - 10^{-2}$; the upper part of this range is not much different than the values discussed above.

27.4   Acknowledgments

It is a pleasure to dedicate this contribution to C. H. Townes in honor of his 80th birthday. I first met Charlie when I was a member of the Berkeley Physics Department, and had the privilege of visiting his laboratory at Columbia University. At that time, I also met his students, Herman Cummins and Patrick Thaddeus, who later became good personal friends and colleagues in New York City. When I became actively interested in astrophysics, Bill Langer and I spent the summer of 1972 visiting Charlie's group in Berkeley. Not only did we enjoy Charlie's hospitality, but we met many of his graduate students who have since become well-known figures in far-infrared and millimeter-wave astronomy. Mike Werner comes particularly to mind through his help on research related to the subject of this paper, the heating of molecular clouds by cosmic rays. I am grateful for the numerous contacts I have had with Charlie Townes for over three decades and wish him many more years of fruitful research.

This work has been supported in part by a grant from NASA's Infrared-Radio Branch. I would to thank Javier Igea for re-calculating the cosmic ray ionzation rate and Glenn Ciolek for discussions on cosmic ray exclusion from winds.

References

[1] B. Zuckerman, Ann. Rev. Ast. Astrophysics **18**, 263 (1980).

[2] D. M. Rank and C. H. Townes, and W.J. Welch, Science **174**, 1083 (1971).

[3] W. C. Danchi, M. Bester, C. G. Degiacomi, L. J. Greenhill and C. H. Townes, AJ **107**, (1994).

[4] J. Kwan and R. A. Linke, ApJ **254**, 587 (1982).

[5] J.J Keady and S. T. Ridgway, ApJ **406**, 199 (1993).

[6] J. Keene, K. Young, T. G. Phillips, T. H. Büttgenbach, and J. E. Carlstrom, ApJ **415**, L131 (1994).

[7] I. Cherchneff and A. E. Glassgold, ApJ **419**, L41 (1993).

[8] R. Lucas, in *Astronomy with Millimeter and Sub-millimeter Wave Interferometry*, eds. M. Ishaguro and W. J. Welch (PASP, 1994), p. 135.

[9] J. H. Bieging and M. Tafalla, AJ, **105**, 576, (1994).

[10] L. Goldberg, QJRAS **20**, 361 (1979).

[11] T. Montmerle, in *Genesis & Propagation of Cosmic Rays*, eds. M. M. Shapiro and J.P. Wefel (Reidel, Dordrecht, 1988), p.105.

[12] I. Cherchneff, A. E. Glassgold, and G. A. Mamon, ApJ **410**, 188 (1993).

[13] T. J Millar and E. Herbst, A&A **288**, 561 (1994).

[14] A. E. Glassgold, ApJ **438**, L111 (1995).

[15] R. Lucas and M. Guélin, in *Submillimeter Astronomy*, eds. G. D. Watt and A. S. Webster (Kluwer, Dordrecht, 1990), p. 97.

[16] L. Spitzer and M. G. Tomasko, ApJ **152**, 971 (1968).

[17] S. Lepp, in *Astrochemistry of Cosmic Phenomena*, ed. P. Singh (Kluwer, Dordrecht, 1992) p. 471.

[18] S. S. Suess, Rev. Geophys. **26**, 97 (1990).

[19] J. A. Le Roux and V. S. Ptuskin, ApJ **438**, 427 (1995).

[20] E.N. Parker, *Interplanetary Dynamical Processes*, (Interscience, New York, 1963).

[21] P. Merrill, ApJ **116**, 21 (1952).

[22] D. Branch and B. F. Perry, PASP **82**, 1060 (1970).

[23] J. J. Cowan and W. K. Rose, ApJ **212**, 149 (1977).

[24] T. G. Barnes, R. Beer, K. H. Hinkle, and D. L. Lambert, ApJ **213**, 71 (1977).

[25] E. N. Rodriguez Kuiper, T. B. H. Kuiper, B. Zuckerman, and R. K. Kaker, ApJ **214**, 394 (1977).

[26] G. A. Wright, ApJ **436**, L157 (1994).

[27] B. Rosenblum, A. H. Nethercot, and C. H. Townes, Phys. Rev. **109**, 400 (1958).

[28] M. J. Harris and D. L. Lambert, ApJ **318**, 868 (1987).

[29] M. Guélin, M. Forestini, P. Valiron, L. M. Ziurys, M. A. Anderson, J. Cernicharo, and C. Kahane, A&A **297**, 183, 1995.

[30] M. Forestini, G. Paulus, and M. Arnould, A&A **252**, 597 (1991).

[31] G. J. Wasserburg, M. Busso, R. Gallino and C. M. Raiteri, ApJ **424**, 412 (1994).

[32] E. Anders and E. Zinner, Meteorites **28**, 490 (1993).

[33] U. Ott, Nature **365**, 25 (1993).

28

The Clumpy Structure of Molecular Clouds

Paul F. Goldsmith

The structure of molecular clouds has been a puzzle since this dense phase of the interstellar medium was first recognized in the 1960's. One of the very early indications of a "problem" with molecular cloud structure was the highly superthermal $\left(\delta v \gg \frac{\sqrt{kT}}{m_{mol}}\right)$ and even supersonic $\left(\delta v > \frac{\sqrt{kT}}{\langle\mu\rangle}\right)$ line widths found to characterize the emission from almost all types of molecular regions. This is very different, for example, from ionized HII regions, for which line widths reflect the temperature of the gas. The recognition that carbon monoxide and ammonia should be good molecular cloud thermometers [6, 32] allowed astronomers to establish that the temperature of these regions is between approximately 10 and 100 K [39, 14]. The line widths observed in Giant Molecular Clouds (GMC's) are typically an order of magnitude greater than the speed of sound in these regions. Only in the most quiescent cores of dark clouds (without embedded heating sources) does one find molecular line widths that are consistent with the gas temperature [2, 11, 12, 22].

Most models of clouds with highly supersonic gas motions also imply rapid dissipation of energy. These motions apparently require a continuous and large input of energy whose source is not evident. Magnetic fields and associated motions have been one avenue of escaping from this dilemma [1, 43], but while there is evidence that magnetic fields are present, it has been extremely difficult to conclude that Alfvén-like or other types of non-dissipative motion are the general explanation for the observed line widths. Another approach to resolving this dilemma has been to postulate that molecular cloud material is actually highly structured, consisting of relatively small clumps moving around within the cloud with supersonic relative velocities. The clumps have approximately thermal internal motions so that they do not suffer excessive dissipation. The superposition of the thermal line widths of the clumps with the much greater clump-to-clump velocity dispersion can explain the line widths observed in GMC's [20].

The chemical composition of molecular clouds also appears to indicate that they have a very clumpy structure. In the interiors of dense molecular clouds with visual extinctions greater than a few magnitudes and densities greater than a few times 10^3 cm^{-3}, most gas phase carbon is predicted by standard models to be in the form of carbon monoxide [9]. It has thus been a considerable surprise to find that within dark clouds [10] and GMC's [19, 13, 33] the abundance of neutral, atomic carbon (CI) is $\cong 0.2$ times that of CO over very large regions. There have been a variety of explanations proposed, but probably the most successful has been that the very irregular structure of the clouds allows the diffusion of UV photons, which can destroy CO, deep within the cloud [5]. Thus, while each dense clump may be almost totally molecular in its central region, surrounding outer layers, forming an "onion skin" strucure, may have a relatively large CI abundance. A critical point of this model is that the atomic material is distributed throughout the volume of the general molecular cloud, allowing the spatial distribution and line profiles of CI to be relatively well correlated with those of molecular tracers, e.g., ^{13}CO. Clumpiness has also been suggested to be responsible for the existence of ionized carbon over large regions of molecular clouds, and thus to explain the widespread CII 158 μm emission seen in quite a few clouds with nearby sources of ionization [18]. Models of clumpy photodissociation regions have been relatively successful at modeling the both the atomic and molecular emission [26, 27]. The general parameters are that there are clumps with densities between 10^6 and 10^7 cm^{-3}, immersed in an interclump medium of density few times 10^3 cm^{-3}. The Orion Bar, a region severely affected by flux of ultraviolet radiation from the young stars in the Trapezium cluster, has been modeled by an inhomogeneous density structure with 10% of the material in clumps having $n_{H_2} = 10^6$ cm^{-3} and 90% in an interclump medium having density 3×10^4 cm^{-3} [17]. It is likely that these are only characteristic values for what is a distribution of clump and interclump densities.

The fact that the interclump medium in Giant Molecular Clouds can act as a conduit for UV photons has several interesting consequences. One of these is that each clump would have molecular material just inside its ionized and atomic outer layers subject to enhanced heating, and this material would likely be warmer than the central parts of the clump. When we use carbon monoxide as a cloud ther-mometer, an important question is: where in the cloud along the line of sight from the observer does the emission become optically thick? Different rotation tran-sitions of this molecule have different absorption coefficients, but for conditions expected to characterize the clumps, the absorption coefficient increases as one goes up the rotational ladder through the millimeter region and into the submil-limeter. Thus, the higher-J transitions become optically thick sooner, and trace the emission from closer to the surfaces of the clumps than do the lower lines. The observation that the measured temperature increases with increasing J, while not itself absolutely requiring that there be clumps, is consistent with their presence, and strongly supports the idea that the clumps are externally heated [37]. This is also consistent with measurements of gas temperatures using CH$_3$CCH (also established as a good cloud thermometer) that the very dense gas is considerably cooler than the lower-density material probed by CO [3].

A highly-clumped structure for molecular clouds would obviously have direct observational consequences, but the details depend on the scale size of the in-homogeneities. Observations with increasing sensitivity during the 25 years that molecular clouds have been studied in detail have revealed that they certainly are *not* uniform on angular size scales larger than 10 arc-seconds [29, 41, 35, 15, 21, 24, 42, 22]. Clouds are irregular and have considerable internal structure, but what has been observed *directly* does not appear to be sufficient for the clumpiness we need to invoke here. A significant constraint is that if the individual clumps have only thermal line widths, there must be *many* of them in the beam of the telescope used to observe them, in order to produce the extremely smooth line profiles that are observed [31, 36]. Thus, we have to consider the possibility of small clumps that are extremely difficult to observe directly (in all but possibly the nearest clouds) due to their small dimensions.

While awaiting improvements in sensitivity and angular resolution of both single-antenna and interferometric radio astronomical systems used for study of molecular clouds, we might consider what avenues of investigation may contribute to defining the presence of these constituent clumps. In fact, there are several lines of argument that strongly support the existence of unresolved clumps, in addition to those of chemistry, energetics, and line smoothness already discussed.

The first of these can be traced back to early observations of ammonia in dense clouds [25]. This study revealed the NH_3 excitation temperature to be significantly larger than the observed brightness temperature, which is expected for optically thin emission. However, the ratio of the NH_3 hyperfine components, which directly reflects the optical depth, indicated that the line is, in fact, optically thick. The only plausible resolution of this dilemma is to assume that the emission does not fill the antenna beam. The observed ammonia emission can be interpreted as arising from unresolved clumps filling a relatively small fraction (≈ 0.2) of the beam area.

A very different set of observations leading to a similar conclusion consists of efforts to employ multiple transitions of a single molecular species, which have different spontaneous decay rates, to determine the collision rate and thus the gas density. This has been carried out for quite a few species, including $C^{18}O$, $C^{34}S$, and HC_3N, to indicate some of the optically thin rigid rotor molecules which have been utilized. There are two suspicious parts to the general conclusions that emerge [34, 28]. First, the maps of regions look surprisingly similar irrespective of which transition or which species is observed [38]. If, for example, clouds were smoothly centrally condensed, then the emission would be expected to be concentrated in increasingly smaller regions as the A-coefficient of the transition observed increased. This is observed not to be the case. Rather, the boundary of a region such as the Extended Ridge in the Orion Molecular Cloud is almost independent of the tracer being employed. The actual gas densities determined from line ratios are also not at all consistent, but depend only on the characteristic density (roughly $A/\langle \sigma v \rangle$) of the transitions observed. Multiple transition studies of the optically thin carbon monoxide isotope $C^{18}O$ (which has a low dipole moment and hence low spontaneous decay rates) yield densities in the range of 1–3×10^4 cm^{-3} [16], while observations of molecules such as HC_3N with much larger spontaneous decay rates indicate $n(H_2) \geq 10^6$ cm^{-3} [4]. Thus, it appears that there is a large range of densities within any telescope beam.

Another line of argument supporting the presence of a substantial range of densities is the virial theorem. While the validity of the virial theorem for describing molecular cloud structure is not entirely established, most studies using independent methods of mass determination do suggest that giant molecular clouds are not far from virial equilibrium, in the sense that the gravitational potential energy from the gas mass is approximately equal to the kinetic energy implied by the observed line widths [23, 30]. We might ask: what is the density obtained from assuming that the central parts of GMC's, such as the Orion Ridge, are in virial equilibrium? The general answer is that the density is close to the lower-density molecular excitation result, which in turn implies that the fraction of volume filled by material at the higher densities is relatively small. This picture is reinforced by the result from maps made of clouds, which suggest that the H_2 density derived from the high-density tracers does not depend significantly on the position within the cloud, but remains essentially constant, while the line intensities diminish from the center towards the edge of the cloud. Results of this type led Snell *et al.* [34] and Mundy *et al.* [28] to postulate that the cores of three Giant Molecular Clouds consist of clumps having a hydrogen density $0.5-1 \times 10^5$ cm^{-3} embedded in a much less dense interstellar medium. As one maps across the surface of the cloud, the characteristic density of the transition being observed is always present, but the filling factor of this material varies. This of course fits in with the idea that there are numerous clumps, and that it is the clump filling factor that diminishes as one moves from the center of the cloud to its edge.

The picture of clouds with dense clumps embedded in a lower density interclump medium is not new, and it is certainly plausible that the regions between the clumps is not entirely empty. An argument against a "raisins in pudding" structure is the observational result that the line profiles and velocities of the high-density and low-density optically thin tracers are very nearly identical. It would be surprising if an interclump medium and clumps two to three orders of magnitude denser would have the same distribution of gas velocities, although this cannot be entirely ruled out. A possible alternative is that each individual clump has a high density core and lower density envelope [8]; this naturally leads to similar velocity distributions for material at different densities. Such a model for clumps with density gradients has been advanced as an explanation for apparent differences in clump properties derived from different tracers, in the M17SW molecular cloud [40]. With a reasonable selection of parameters, it is possible to have the mass dominated by material towards the low end of the density range, although whether such loosely-bound clumps could survive is certainly an open question. In this regard, the possibility that clump motion could be modified by magnetic fields [7] certainly deserves further investigation.

Interest in the issue of the structure of dense molecular clouds has, in fact, increased as its importance for determining the rate of star formation as well as the properties of newly-formed stars has become apparent. What are the prospects for unraveling this long-standing riddle? We are on the threshold of major instrumental advances which include the development of interferometric systems that have a good chance of directly revealing the clumpy structure of clouds. In addition, the development of focal plane array receivers for single-antenna millimeter and submillimeter systems allows really high sensitivity mapping of extensive regions of clouds in a wide variety of tracers. Finally, the expansion of observational capability through the electromagnetic spectrum is now permitting the multiple transition studies alluded to above, to be carried out for a variety of molecular species, including those for which the emission is optically thin and hence relatively weak. After twenty five years or so during which considerable effort has been devoted to determining just "what molecular clouds are like," it may be that the observational data, which until relatively recently has been very hard to obtain and in consequence quite limited, will now be readily available, and we can hope that it will lead to a real answer.

Given the fact that this is a contribution to the Festschrift for Charles Townes, it seems appropriate to mention briefly how his work has influenced this topic and how his influence as an advisor and scientist has affected my involvement with it. In a general sense, Charlie's prescience concerning the existence of complex molecules in interstellar space set the stage for the study of the molecular interstellar medium. His move to Berkeley was followed by the detection of NH_3 and H_2O, and studies of the former, in particular, have played a critical role in studies of the clumpiness of dense clouds. At the time that I started my research as a graduate student in the Townes Group, other "veteran" students such as Al Cheung, Mike Chui, and Neal Evans were already immersed in this problem to some degree, and so it may be natural that I have been concerned with it ever since.

I think it is most important to convey that as a Research Advisor, Professor Townes really encouraged students by his confidence that they could make a significant contribution to their chosen area of specialization. I certainly have tried to follow his example in dealing with graduate students whom I have supervised. In addition, by means of one-on-one discussions as well as group meetings, he helped me develop my own physical intuition and willingness to apply it to astrophysical problems. This has proven to be very valuable in the confusing effort to unravel the mystery of molecular cloud structure, and I would not want to miss this opportunity to thank Charlie for his guidance and inspiration. I also wish to thank my colleagues E. Bergin, R. Dickman, W. Langer, R. Snell, J. Tauber, and T. Xie for discussions of ideas presented in this paper.

References

[1] J. Arons and C. E. Max, ApJ **196**, L77 (1975).

[2] P. J. Benson and P. C. Myers, ApJ Suppl. **71**, 89 (1989).

[3] E. A. Bergin, P. F. Goldsmith, R. L. Snell, and H. Ungerechts, ApJ **431**, 674 (1994).

[4] E. A. Bergin, P. F. Goldsmith, R. L. Snell, and F. P. Schloerb, in preparation (1995).

[5] P. Boissé, A&A **228**, 483 (1990).

[6] A. C. Cheung, D. M. Rank, C. H. Townes, D. D. Thornton, and W. J. Welch, Nature **221**, 626 (1969).

[7] B. G. Elmegreen, *Nearby Molecular Clouds*, G. Serra (ed). Springer, 52 (1985).

[8] E. Falgarone and J. L. Puget, A&A **142**, 157 (1985).

[9] D. R. Flower, J. Le Bourlot, G. Pineau des Forêts and E. Roueff, A&A **282**, 225 (1994).

[10] M. A. Frerking, J. Keene, G. A. Blake, and T. G. Phillips, ApJ **334**, 311 (1989).

[11] G. A. Fuller and P. C. Myers, ApJ **384**, 523 (1992).

[12] G. A. Fuller and P. C. Myers, ApJ **418**, 273 (1993).

[13] R. Genzel, A. I. Harris, D. T. Jaffe, and J. Stutzki, ApJ **332**, 1049 (1988).

[14] P. F. Goldsmith *Molecular Clouds in the Milky Way and External Galaxies*, R. L. Dickman, R. L. Snell, and J. S. Young (eds), Springer, 1 (1988).

[15] P. F. Goldsmith, M. Margulis, R. L. Snell, and Y. Fukui, ApJ **385**, 522 (1991).

[16] P. F. Goldsmith, E. A. Bergin, and D. C. Lis, in preparation (1995).

[17] M. R. Hogerheijde, D. J. Jansen, and E. F. van Dishoeck, A&A **294**, 792 (1995).

[18] J. E. Howe, D. T. Jaffe, R. Genzel, and G. J. Stacey, ApJ **373**, 158 (1991).

[19] J. Keene, G. A. Blake, T. G. Phillips, P. J. Huggins, and C. A. Beichmann, 1985, ApJ **299**, 967 (1985).

[20] J. Kwan and D. B. Sanders, ApJ **309**, 783 (1986).

[21] W. D. Langer, R. W. Wilson, and C. H. Anderson, ApJ **408**, L45 (1993).

[22] W. D. Langer, T. Velusamy, T. B. H. Kuiper, S. Levin, E. Olsen, and V. Migenes, ApJ, in press 1995.

[23] R. B. Larson, MNRAS **194**, 809 (1981).

[24] A. P. Marscher, E. M. Moore, and T. M. Bania, ApJ **419**, L101(1993).

[25] C. H. Mayer, J. A. Waak, A. C. Cheung, and M. F. Chui, ApJ **182**, L65 (1973).

[26] M. Meixner, M. R. Haas, A. G. G. M. Tielens, E. F. Erickson, and M. Werner, ApJ **390**, 499 (1992).

[27] M. Meixner and A. G. G. M. Tielens, ApJ **405**, 216 (1993).

[28] L. G. Mundy, R. L. Snell, N. J. Evans II, P. F. Goldsmith, and J. Bally, ApJ **306**, 670 (1986).

[29] L. G. Mundy, T. J. Cornwell, C. R. Masson, N. Z. Scoville, L. B. Bååth, and L. E. B. Johansson, ApJ **325**, 382 (1988).

[30] P. C. Myers and A. A. Goodman, ApJ **329**, 392 (1988).

[31] A. A. Penzias, *Atomic and Molecular Physics and the Interstellar Matter*, Les Houches Session XXVI 1974, R. Balian (ed.), **375** (1975).

[32] A. A. Penzias, P. M. Solomon, K. B. Jefferts, and R. W. Wilson, ApJ **174**, L43 (1972).

[33] R. Plume, D. T. Jaffe, and J. Keene, ApJ **425**, L49 (1994).

[34] R. L. Snell, L. G. Mundy, P. F. Goldsmith, N. J. Evans II, and N. R. Erickson, ApJ **276**, 625 (1984).

[35] J. Stutzki and R. Güsten, ApJ **356**, 513 (1990).

[36] J. A. Tauber, P. F. Goldsmith, and R. L. Dickman, ApJ **375**, 635 (1991).

[37] J. A. Tauber and P. F. Goldsmith, ApJ **356**, L63 (1992).

[38] H. Ungerechts, E. A. Bergin, J. Carpenter, P. F. Goldsmith, W. M. Irvine, A. Lovell, D. McGonagle, F. P. Schloerb, and R. L. Snell, *The Astrochemistry of Cosmic Phenomena*, IAU Symposium 150, P. D. Singh (ed.), 271 (1992).

[39] C. M. Walmsley and H. Ungerechts, A&A **122**, 164 (1983).

[40] Y. Wang, D. T. Jaffe, N. J. Evans II, M. Hayashi, K. Tatematsu, and S. Zhou, ApJ **419**, 707 (1993).

[41] T. L. Wilson and K. J. Johnston, ApJ **340**, 894 (1989).

[42] J. J. Wiseman and P. T. P. Ho, *Clouds, Cores, and Low Mass Stars*, D. P. Clemens and R. Barvainis (eds.). Astron. Soc. Pacific **396** (1994).

[43] E. G. Zweibel and K. Josafatsson, K., ApJ **270**, 511 (1983).

29

Spontaneous Emission Noise in Quantum Electronics

James P. Gordon

On this auspicious occasion I thought it might be a good idea to revisit the question of spontaneous emission noise in quantum electronics and photonics. Low noise was one of the important attributes originally ascribed to masers, and so it was important to understand the effects of spontaneous emission noise in maser amplifiers and oscillators. One of the first papers to discuss the problem in detail was the well-known 1957 paper by Shimoda, Takahashi, and Townes, entitled "Fluctuations in Amplification of Quanta with Application to Maser Amplifiers" [1]. This paper, which calculated energy fluctuations, is still referenced in works on photonic communications. Another important early discussion, also in 1957, was by R. V. Pound in a paper entitled "Spontaneous Emission and the Noise Figure of Maser Amplifiers" [2]. This paper extended the Nyquist-Johnson formulation of thermal noise in electric circuits to encompass the case of negative resistances and negative temperatures such as are encountered in maser amplifiers. It demonstrated that all such noise can be identified as spontaneous emission noise. Many other papers have been written on the subject, but the fundamental nature of spontaneous emission noise and how it can most easily be treated to solve real-world problems does not seem even now to be well understood by many people. The challenge is to discuss the subject in simple and yet rigorous ways. This turned out to be more than I could manage properly before the aforementioned auspicious occasion was past, but here is a beginning. I claim little originality here, perhaps just a fresh look. My discussion does not involve any of the details of the matter-radiation interaction, but they are not really needed. The only things that are required to establish the necessary fundamentals of this problem are the fluctuation-dissipation theorem, and Planck's law of radiative thermal equilibrium.

I will follow the track of Pound, since it is based on a discussion of fields rather than energies, and is more versatile. According to the well-known law of Nyquist-Johnson, at low frequencies ($h\nu \ll k_B T$) one must add a random Gaussian noise voltage source $V_n(t)$ in series with any resistor (or an equivalent noise current source in parallel) in order to preserve the thermal equilibrium state in electrical circuits. If R is the value of the resistance, then the mean-square magnitude of the noise voltage in any frequency band $\delta\nu$ is given by $\langle V_n^2(t) \rangle = 4R k_B T \delta\nu$, where

k_B is Boltzmann's constant and T is the absolute temperature of the system. The angular brackets indicate a statistical average.

Translated to the angular frequency domain by the Fourier transform,

$$V_n(t) = \frac{1}{\sqrt{2\pi}} \int_0^\infty d\omega (\tilde{V}_n(\omega)e^{-i\omega t} + \tilde{V}_n^*(\omega)e^{i\omega t}) \tag{29.1}$$

one finds that the appropriate correlation function for the angular frequency components of the noise is

$$\langle \tilde{V}_n^*(\omega)\tilde{V}_n(\omega')\rangle = 2Rk_BT\delta(\omega - \omega') \tag{29.2}$$

where $\delta(\ )$ is the Dirac delta function of its argument. By considering parallel and series circuit combinations of resistances and reactances, it is straightforward to show that the appropriate generalization of this law is to put in series with any impedance Z at temperature T a noise voltage source of the same form as Eq. (29.2), with R replaced by $R(\omega)$, the frequency-dependent real part of the impedance Z. Pure reactances, since they do not dissipate energy, have no attendant noise sources.

If one identifies thermal noise with spontaneous emission from the many degrees of freedom of a resistor, one can translate the Nyquist-Johnson law into a consistent quantum theory. In the quantum theory a "positive" frequency voltage component $\tilde{V}_n(\omega)$ becomes a lowering operator for those thermal excitations in the resistor which interact with the electromagnetic field at frequency ω. Likewise a "negative" frequency component $\tilde{V}_n^*(\omega)$ translates to the Hermitian conjugate raising operator and is relabeled $\tilde{V}_n^\dagger(\omega)$. Since a lowering operator can operate only on excited particles, and a raising operator only on those not excited, one finds that Eq. (29.2) is replaced by a pair of relations of the form

$$\langle \tilde{V}_n^\dagger(\omega)\langle\tilde{V}_n(\omega')\rangle = 2\rho\hbar\omega N_2(\omega)\delta(\omega - \omega') , \tag{29.3}$$

$$\langle \tilde{V}_n(\omega)\tilde{V}_n^\dagger(\omega')\rangle = 2\rho\hbar\omega N_1(\omega)\delta(\omega - \omega') , \tag{29.4}$$

where ρ is a proportionality constant, and $N_1(\omega)$ and $N_2(\omega)$ represent respectively the number of available upward and downward transitions at frequency ω. Equation (29.1) still holds, with the Hermitian-conjugation symbol replacing the complex-conjugation symbol. The corresponding resistance R is proportional to the difference $N_1(\omega) - N_2(\omega)$, and so we arrive at the quantum form of the Nyquist-Johnson noise voltage by identifying R as

$$R = \rho(N_1(\omega) - N_2(\omega)) . \tag{29.5}$$

For the case of thermal equilibrium at temperature T Boltzmann's law says that the ratio of $N_1(\omega)$ to $N_2(\omega)$ is $\exp(\hbar\omega/k_BT)$. In this case Eqs. (29.3-29.5) reduce to

$$\langle \tilde{V}_n^\dagger(\omega)\tilde{V}_n(\omega')\rangle = 2R\hbar\omega\bar{n}\delta(\omega - \omega') , \tag{29.6}$$

$$\langle \tilde{V}_n(\omega)\tilde{V}_n^\dagger(\omega')\rangle = 2R\hbar\omega(\bar{n} + 1)\delta(\omega - \omega') , \tag{29.7}$$

where $\bar{n}$ is the mean modal occupation number ($\bar{n} = 1/(\exp(\hbar\omega/kT) - 1)$).

Equations (29.3-29.7) form a basis from which to deal with most problems in quantum electronics and photonics. In particular, Eqs. (29.3-29.5) are valid not only for equilibrium conditions but also for cases in which the dynamics of N_1 and N_2 are governed adequately by rate equations. Thus they allow for the power saturation of loss ($R > 0$) or gain ($R < 0$) and hence apply to oscillators as well as to linear gain or loss elements. They maintain the proper quantum-mechanical thermal-equilibrium state and preserve the proper commutation relations for the dynamical variables of any circuit, e.g., voltages, currents, and charges.

In the quantum theory, all of the classical dynamical variables of a circuit become quantum operators. In the Heisenberg picture, however, they remain as dynamical variables with equations of motion which mimic the classical ones. For example, the equations for a closed, series-resonant circuit consisting of a resistor (R), capacitor (C), and inductor (L) are $dQ/dt = I$, and $LdI/dt + RI + Q/C = V_n(t)$. Here Q is the charge on the capacitor and I is the current in the circuit. These equations represent a damped simple harmonic oscillator. In their quantum-mechanical embodiment, thermal equilibrium, with its correct quantum-mechanical form consistent with Planck's blackbody radiation law, is maintained by the noise source connected to the dissipative element of the circuit. Furthermore, the correct commutator of Q and I is maintained. Similarly, one can write equations for any electrical or optical system that can be represented by a wire circuit, for example a resonant circuit or a transmission line with both gain and loss elements in it. The noise fields originating from the different gain and loss elements are of course uncorrelated.

For many purposes it is useful to have a classical-noise theory which closely approximates the quantum theory. Those interested in communications, for example, then have a large body of literature which is directly applicable to whatever problem they are addressing. The simplest and most versatile way to do this is to average the orderings of Eqs. (29.3-29.7), yielding what I shall call the Wigner picture, since it is intimately connected with Wigner's quasi-classical representation of quantum systems. This has classical Gaussian noise sources given by

$$\langle \tilde{V}_n^*(\omega) \tilde{V}_n(\omega') \rangle = \rho\hbar\omega(N_1 + N_2)\delta(\omega - \omega') \tag{29.8}$$

$$\langle \tilde{V}_n^*(\omega) \tilde{V}_n(\omega') \rangle = R\hbar\omega(2\bar{n} + 1)\delta(\omega - \omega') \tag{29.9}$$

where R and ρ are related by Eq. (29.5) and the second form applies to the case of thermal equilibrium at temperature T. In the Wigner picture, the low-temperature limit ($k_B T \ll h\nu$) of the energy of a weakly-damped oscillator resonant at frequency ν is its zero-point energy $h\nu/2$. Likewise the low temperature limit of the power flow in a transmission line is the zero-point power ($h\nu/2)\delta\nu$, which is to be treated as a real Gaussian noise field. Note that the complete quantal noise voltage can be regarded as the sum of the classical Wigner noise voltage, and a quantum component which depends only on the value of the resistance, and not otherwise on its temperature. In many cases this quantum noise component can be treated as a perturbation.

Time and space do not permit here more than a few remarks concerning the properties of the Wigner picture. For linear systems such as maser amplifiers, parametric amplifiers, attenuators, directional couplers, etc., the solutions of a problem in the Wigner picture can be translated back to the quantum picture simply by reinstating the quantum commutation relations for the positive and negative frequency components of the field variables, if that turns out to be needed. In general, if the classical noise field in the Wigner picture is much greater than the zero-point field at the location where some measurement is to be made, then the effects of the additional quantum noise field can be mostly ignored.

Finally, let me change the subject, and comment a bit on Charlie Townes. Recently I was asked how he was as my professor, and I replied that he was very good indeed. He was usually available, with good ideas, and the broad picture, but he left plenty of room for contributions from his charges. He was even wrong once in a while, which naturally made a student's day. He has had, of course, a great impact on my professional career. I want to thank him for persevering in the research that led to the maser, in the face of criticism from some of his (at the time) more famous peers. As it turned out, it was not a waste of his student's time after all.

References

[1] K. Shimoda, H. Takahashi and C. H. Townes, J. Phys. Soc. Japan, **12,** 686 (1957).
[2] R. V. Pound, Ann. Phys. **1,** 24 (1957).

Possibility of Infrared Coronal Line Laser Emission in Seyfert Nuclei

Matthew A. Greenhouse
Howard A. Smith
and Uri Feldman

30.1   Introduction

Energetic emitting regions in astronomical sources have traditionally been studied via x-ray, UV, and optical emission lines of highly ionized intermediate mass elements. Such lines are often referred to as "coronal lines" since the ions, when produced by collisional ionization, reach maximum abundance at electron temperatures of $\sim 10^5 - 10^6$ K typical of the sun's upper atmosphere. However, optical and UV coronal lines are also observed in a wide variety of Galactic and extragalactic sources including the Galactic interstellar medium [2], nova shells [3, 4], supernova remnants [5, 6, 7], galaxies and QSOs [8, 9, 10].

Infrared forbidden lines typically result from fine structure transitions within the ground electron configuration and are excited predominantly by electron impact. Although these relatively low energy transitions are easily excited, the coronal ionization states are only produced in energetic environments by collisional ionization in a kinetically hot gas, [11, 12] or by photoionization in power-law continuum radiation fields [12, 13, 14]. In the latter case, coronal emission lines can be produced in relatively low kinetic temperature ($T_e \sim 10^4$ K) or low density ($n_e < 10^5$) plasmas due to their low excitation temperature. As a result of their relatively high critical density ($n_e > 10^6$ cm^{-3}) for collisional de-excitation, coronal fine structure lines are also important coolants of higher density collisionally ionized plasmas. As a consequence, both ionization mechanisms can be important in heterogeneous astronomical sources such as Active Galactic Nuclei (AGN) or rapidly evolving sources such as young nova shells.

Infrared observations of a total solar eclipse [15] and of nova V1500 Cyg [16] resulted in the discovery of several infrared coronal lines. As a result of subsequent advances in near-infrared array spectrometers, infrared coronal lines are providing a new window for observation of energetic emitting regions in heavily dust-obscured sources such as Seyfert nuclei [17, 18, 19], and new opportunities for model constraints on physical conditions in these sources [20]. Unlike their UV and optical counterparts, infrared coronal lines can be primary coolants of collisionally ionized plasmas with $10^4 < T_e (K) < 10^6$ which produce little or no optical or shorter wavelength coronal line emission [1, 21]. In addition, they provide a means to probe heavily dust-obscured emitting regions which are often inaccessible to optical or UV line studies. The wide range of critical densities spanned by infrared coronal lines ($5 < \log n_e$ (cm^{-3}) < 10) combined with their reduced extinction make them ideal probes of Seyfert broad line region cloud kinematics [22, 23]. Finally, they are useful for abundance studies of white dwarf interiors via their novae ejecta [21, 24, 25, 26], planetary nebulae [27], and are modeled to be bright behind bow shocks of cometary compact HII regions [28], in galaxy cluster cooling flows [29], and infrared bright merging galaxies [30]. As a result, infrared coronal lines are providing a new and rich opportunity to study gas-phase chemical abundances, kinematics, and spatial morphology of emitting regions in a wide variety of astronomical sources other than active galactic nuclei.

The majority of infrared coronal lines occur at wavelengths coincident with telluric absorption features, so that only a few are accessible to ground-based spectrometers. Although near-infrared coronal lines can dominate the K-band (2.2 μm) spectrum of active galaxies (e.g., NGC 1068—Moorwood and Oliva [18]), the range of red shifts accessible to ground-based studies is also limited. However, the Infrared Space Observatory (ISO) scheduled for launch during October 1995 will present the first comprehensive opportunity to test model predictions of infrared coronal line emission relevant to a wide range of astrophysical problems.[1]

30.2   Stimulated Emission in Seyfert Nuclei

30.2.1   Population Inversions and Gain Lengths

New detailed-balance calculations presented by Greenhouse et al. [1] reveal that population inversions commonly exist within the ground and first excited terms of high ions. The possibility of infrared laser emission due to several low ionization states ($\chi < 100$ eV) within the isoelectronic sequences listed in Table 1 was first considered by Smith [33] under planetary nebulae temperature and density conditions with ionization balance determined using the Saha equation. Here we assess the possibility of infrared coronal line laser emission due to higher ($\chi > 100$ eV) ionization states at higher densities with collisional ionization balance determined using models [34, 35] which take non-thermodynamic equilibrium processes into account [36].

Light of intrinsic intensity I_0 passing through a uniform medium of length l propagates as

$$I(l) = I_0 e^{-\alpha l} + \frac{j_0}{\alpha}(1 - e^{-\alpha l}), \qquad (30.1)$$

where j_0 is the volume emission coefficient. The coefficient α for a Doppler-broadened line due to a transition $j \to i$ in ion X^{+p}, is maximum at the line center

---

[1]A description of ISO and its instruments is presented in the proceedings of the Les Houches Summer School on *Infrared Astronomy with ISO* [32].

TABLE 30.1. Potential Infrared ($\lambda > 1$ μm) Coronal Line Laser Transitions Within The Ground $2s^2 2p^k$, $3s^2 3p^k$, and Excited $2s2p$, $3s3p$ Configurations[‡]

1	2	3	4	5	6	7	8
Species	Transition $i \leftarrow j$	λ (μm)	A_{ji} (s^{-1})	χ (eV)	log T_{max} (K)	log (E_j/k) (K)	$-\log(A_{ji}n_j/n_e)\|_{<n_{crit}}$ (photon s^{-1} ion^{-1})
C I Isoelectronic Sequence ($2s^2 2p^2$)							
Ne V	$^3P_0 - ^3P_1$	m 24.28(2)	$1.27e-3$	126.21	5.5	2.775	7.9
Na VI	$^3P_1 - ^3P_2$	c 8.61(9)	$2.11e-2$	172.15	5.7	3.427	8.3
Na VI	$^3P_0 - ^3P_1$	c 14.3(3)	$6.14e-3$	"	"	3.002	7.9
Mg VII	$^3P_0 - ^3P_1$	c 9.03(9)	$2.44e-2$	224.95	5.8	3.202	8.1
Al VIII	$^3P_0 - ^3P_1$	c 5.85(10)	$8.96e-2$	284.59	6.0	3.391	8.2
N I Isoelectronic Sequence ($2s^2 2p^3$)							
S X	$^2D_{3/2} - ^2D_{5/2}$	c 8.676(11)	$1.58e-2$	447.1	6.2	6.534	11.0
Ar XII	$^2D_{3/2} - ^2D_{5/2}$	c 2.97(6)	$3.77e-1$	618	6.3		10.7
Ca XIV	$^2D_{3/2} - ^2D_{5/2}$	c 1.3070(40)	$4.19e+0$	817	6.5	5.213	10.7
O I Isoelectronic Sequence ($2s^2 2p^4$)							
Mg V	$^3P_1 - ^3P_0$	c 13.54(5)	$2.17e-2$	141.27	5.4	5.614	9.2
Al VI	$^3P_1 - ^3P_0$	c 9.116(6)	$7.10e-2$	190.47	5.7	3.741	9.3
Si VII	$^3P_1 - ^3P_0$	c 6.515(18)	$1.94e-1$	246.49	5.8	3.903	9.4
S IX	$^3P_1 - ^3P_0$	c 3.75(3)	$1.01e+0$	379.1	6.0	4.185	9.5
Ar XI	$^3P_1 - ^3P_0$	c 2.60(5)	$3.00e+0$	539.0	6.2		9.7
Ca XIII	$^3P_1 - ^3P_0$	c 2.258(15)	$4.46e+0$	726	6.4	4.619	9.9
Si I Isoelectronic Sequence ($3s^2 3p^2$)							
Ca VII	$^3P_0 - ^3P_1$	c 6.154(8)	$7.67e-2$	127.7	5.7	3.369	7.5
P I Isoelectronic Sequence ($3s^3 3p^3$)							
Fe XII	$^2D_{3/2} - ^2D_{5/2}$	c 2.2170(3)†	$8.68e-1$	331	6.2	4.821	9.6
S I Isoelectronic Sequence ($3s^2 3p^4$)							
Fe XI	$^3P_1 - ^3P_0$	c 6.082(19)	$2.23e-1$	290.4	6.1	4.314	9.7

† a shorter wavelength of 2.205 μm is listed in Fuhr et al. [50].
‡ see Greenhouse et al. [1] for data sources and further explanation.
Column Designations
1 – ion
2 – transition
3 – calculated or measured wavelength denoted as c or m respectively, the standard error of the last digit follows in parenthesis ( ), calculated wavelengths < 5 μm are given in air, otherwise in vacuum
4 – spontaneous transition rate A_{ji}
5 – ionization potential
6 – ion characteristic temperature $T_{max} \equiv$ the equilibrium temperature of maximum ion concentration $n(X^{+i})/n(X_{tot})$ in a collisionally ionized plasma
7 – transition excitation temperature E_j/k where E_j is the energy of the upper level of the transition and k is Boltzmann's constant
8 – intrinsic photon rate $\log(A_{ji}n_j/n_e)\|_{<n_{crit}}$ photons s^{-1} ion^{-1}, where n_j is the relative population of the level j

wavelength λ_{ij} and can be written as [37]

$$
\begin{aligned}
\alpha_{max} &= \frac{\lambda_{ij}^3}{8\pi} A_{ji} \left(\frac{m}{2\pi k T_e}\right)^{1/2} \left(\frac{g_j}{g_i} - \frac{n_j}{n_i}\right) N(X_i^{+p}) \\
&= \frac{\lambda_{ij}^3}{8\pi} A_{ji} \left(\frac{m}{2\pi k T_e}\right)^{1/2} \left(\frac{g_j}{g_i} - \frac{n_j}{n_i}\right) \\
&\quad \times \left(\frac{n(X_i^{+p})}{n(X_{tot}^{+p})} \frac{n(X_{tot}^{+p})}{n(X)_{tot}} \frac{n(X)_{tot}}{n(H)} \frac{n(H)}{n_e} n_e\right),
\end{aligned}
\tag{30.2}
$$

TABLE 30.2. Coronal Line Laser-gain Lengths and Column Densities

1	2	3	4	5	6		
Species	λ	$-log\left[\frac{n(X^{+P})}{n(X_{tot})}\right]$	$12+log\left[\frac{n(X)}{n(H)}\right]$	l_6	$log[n_6]$	l_9	$log[n_9]$
	μm			pc	cm^{-2}	pc	cm^{-2}
Ne V	24.28	0.21	8.09	a	...	$1.7e-3$	24.72
Na VI	8.61	0.31	6.33	a	...	$4.0e-2$	26.09
Na VI	14.3	"	"	a	...	$1.7e-1$	26.72
Mg V	13.54	0.22	7.58	$1.2e+0$	24.57	a	...
Mg VII	9.03	0.30	"	a	...	$3.8e-3$	25.07
Al VI	9.12	0.28	6.47	$7.1e+1$	26.34	a	...
Al VIII	5.85	0.38	"	a	...	$4.5e-2$	26.14
Si VII	6.52	0.30	7.55	$1.9e+1$	25.77	$1.6e-2$	25.69
S IX	3.75	0.35	7.21	$3.4e+2$	27.02	$1.9e-3$	24.77
S X	8.68	0.35	"	$9.8e+2$	27.48	$5.6e-3$	25.24
Ar XI	2.60	0.39	6.56	$7.7e+3$	28.38	$1.3e-2$	25.60
Ar XII	2.97	0.38	6.80	$5.3e+4$	29.21	$7.4e-2$	26.36
Ca VII	6.15	0.35	6.36	a	...	$1.0e-3$	24.49
Ca XIII	2.26	0.41	"	$3.3e+4$	29.01	$4.2e-2$	26.11
Ca XIV	1.31	0.41	"	$1.3e+6$	30.60	$1.3e+0$	27.49
Fe XI	6.08	0.57	7.67	a	...	$3.3e-3$	25.01
Fe XII	2.22²	0.53	"	$7.9e+2$	27.39	$3.9e-3$	25.08

1 – see Table 30.1 for transition
2 – see Table 30.1 for more precise wavelengths
3 – collisional ionization equilibrium at $T_e = T_{max}$ (see Table 30.1)
4 – solar photospheric abundance (Grevesse and Anders 1989)
5 - 8 – column length l and corresponding column density n for a laser gain of e evaluated at $n_e = 10^6$ and 10^9 cm^{-3} and collisional ionization equilibrium
² see note Table 30.1
a – no significant inversion exists at this density

where $N(X_i^{+P})$ is the number of ions in the state i, m is the atomic mass of X, A_{ji} is the spontaneous transition probability, $n(X_i^{+P})/n(X_{tot}^{+P})$ is the relative population of the level i in the ion X^{+P}, $n(X^{+P})/n(X)_{tot}$ is the relative abundance of the ionization state p of element X, $n(X)_{tot}/n(H)$ is the *gas-phase* chemical abundance of X, and n_e is the electron density. If $g_j/g_i - n_j/n_i < 0$, then $\alpha < 0$ and $I(l)$ is an increasing function. Under these circumstances, α is a gain coefficient.

In Table 30.2, we list all transitions treated in Greenhouse *et al.* [1] for which Eq. (30.2) yields $\alpha < 0$. In Figs. 30.1 and 30.2, we plot values of α vs. n_e for two typical cases. Similar plots for all of the transitions listed in Table 30.2 can be found in Greenhouse *et al.* [1]. We assume collisional ionization equilibrium at $T_e = T_{max}$, $n(H)/n_e = 1$, and parameters listed in Tables 30.1 and 30.2. Laser gain lengths $l \equiv \alpha^{-1}$ and corresponding column densities evaluated at $n_e = 10^6$ and 10^9 cm^{-3} are listed in Table 30.2 for hot ($T_e = T_{max}$) collisionally ionized regions. We find that under these conditions, several of the bright transitions listed in Table 30.2 have relatively short gain lengths, $\sim 10^{-3}$ pc at $n_e = 10^9$ cm^{-3} and typically less than 1 kpc at $n_e = 10^6$ cm^{-3} corresponding to column densities as low as $\sim 10^{24-25}$ cm^{-2}. This gain column density is larger than the $\sim 10^{23}$ cm^{-2} column assumed by "standard" broad line regions (BLR) models [38], but are roughly equivalent to those occasionally invoked to explain other observed AGN lines [39].

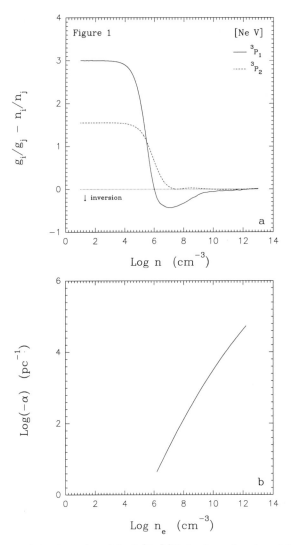

FIGURE 30.1. Population inversion of the $2s^2 2p^2\ ^3P_1$ level as a function of density [a] and corresponding laser gain length [b] for the $^3P_0 - \ ^3P_1$ transition of Ne V.

Laser amplification is suppressed by absorption when $|\alpha|/\kappa \leq 1$, where κ is the absorption coefficient [33]. Free-free absorption is a potentially important source of opacity in the physical environment considered here. The free-free absorption coefficient can be written as

$$\kappa_{ff} = 1.98 \times 10^{-23} Z^2 g_{ff} \lambda^2 n_e^2 T_e^{-3/2} \frac{n(X^{+p})}{n(X_{tot})} \frac{n(X)}{n(H)} \frac{n(H)}{n_e} \quad (\text{cm}^{-1}), \qquad (30.3)$$

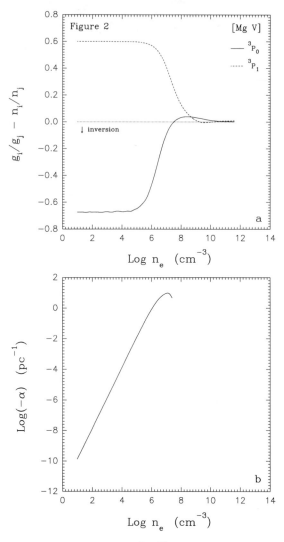

FIGURE 30.2. Population inversion of the $2s^2 2p^4 \, ^3P_0$ level as a function of density [a] and corresponding laser gain length [b] for the $^3P_1 - {}^3P_0$ transition of Mg V.

where g_{ff} is the free-free Gaunt factor. In cases where $|\alpha|/\kappa_{ff} > 1$, net laser amplification can occur. We find that for T_e as low as 10^4, $|\alpha|/\kappa_{ff} \gg 1$ for all transitions listed in Table 30.2.

Relatively little is known about gas phase abundances of these elements in Seyfert nuclei. However, available observations [40] suggest that they are not depleted. Gain lengths were calculated using solar photospheric abundances (Table 30.2, column [4]) which we take as a reasonable lower limit to their gas phase abundance in Seyfert nuclei. Although we have assumed conditions of pure colli-

TABLE 30.3. Photoionization Equilibrium of Selected Coronal Ions[2]

$$-log\left[\frac{n(X^{+p})}{n(X_{tot})}\right]_{phot}$$

Species	U = 0.01	U = 0.10	U = 1.00
Ne V	0.34	0.48	1.42
Mg V	0.61	1.99	3.55
Mg VII	1.01	0.35	1.13
Al VI	0.50	1.05	2.32
Al VIII	1.91	0.39	0.56
Si VII	0.85	0.60	1.47
S IX	2.82	0.54	0.55
S X	4.42	1.12	0.55
Ar XI	5.20	1.72	0.66
Ca VII	0.61	0.92	2.06
Ca XIII	...	4.06	1.41
Fe XI	4.35	1.20	0.60
Fe XII	...	1.97	0.58

[2] Results from [42] plane parallel slab, spectral index $L_v \propto v^{-1.5}$, ionization parameter U $\equiv$ ionizing photon flux$/cn_e$, $n_e = 10^6$ cm^{-3}, and $T_e \sim 10^4$ K, see Sect. 2

sional ionization, ion abundances similar to those listed in column [3] of Table 30.2 can also be produced by pure photoionization at relatively low kinetic temperatures. To illustrate this point, photoionization equilibria of ions listed in Table 30.2 are given in Table 30.3. Ionization fractions listed in Table 30.3 were calculated using the photoionization model CLOUDY [41] using ionization parameters in the range of $-2 < \log U < 0$, power law index $\beta = -1.5$, $n_e = 10^6$ cm^{-3}, and $T_e \sim 10^4$ K [42].[2] Although the relative contribution of photo- and collisional ionization in AGN coronal line emitting regions (CLR) is controversial (Sect. 30.1), the results listed in Table 30.3 and the excitation temperatures listed in Table 30.1 illustrate that the prospects for laser amplification in infrared coronal lines can be largely independent of the relative balance between these ionization mechanisms provided that $T_e > E_j/k$ for the relevant levels (see Table 30.1).

The location of the coronal line emitting region observed in Seyfert nuclei (see Sect. 30.1) is poorly understood insofar as its position within broad line region (BLR) and narrow line region (NLR) structure of the nucleus is concerned. A correlation between line width and both ionization potential and transition critical density has been observed in Seyfert nuclei, e.g. [45, 46, 22], and is predicted by some models, e.g. [23]. Coronal lines observed to date in Seyfert nuclei (primarily optical lines) are typically broader than NLR forbidden lines, but do not achieve line widths characteristic of BLR permitted lines. This trend suggests that their emitting

[2] A general discussion of photoionization in AGN can be found in Ferland and Shields [43] and Netzer [44] and references in Sect. 1.

region occupies an intermediate zone between the BLR and NLR. Alternatively, the emission could be produced in a low density coronal temperature intercloud confining medium described by two-phase BLR models, e.g. [47].

If the CLR occupies a BLR/NLR warm ($T_e > 10^4$ K) intermediate zone characterized by an electron density of $n_e \sim 10^{6-9}$ cm^{-3}, then the results shown in Table 30.2 suggest that infrared coronal line lasers could be produced there. As a consequence, observations designed to reveal laser amplification of these transitions can constrain the spatial extent or volume filling factor of Seyfert CLRs.

30.2.2 Observational Tests

The population inversions $g_j/g_i - n_j/n_i$ we calculate can be constrained by observation of sources such as young novae where the lines in Table 30.2 are produced, but column densities required for laser amplification are not achieved [21, 48]. The specific intensity I_{ij} due to a transition $j \rightarrow i$ in ion X^{+p} is

$$I_{ij} = A_{ji} \frac{hc}{\lambda_{ij}} \frac{n(X_j^{+p})}{n(X_{tot}^{+p})} \frac{n(X_{tot}^{+p})}{n(X_{tot})} \frac{n(X_{tot})}{n(H)} \frac{l n_e}{4\pi} \text{ erg s}^{-1}\text{cm}^{-2}\text{sr}^{-1}, \quad (30.4)$$

where hc/λ_{ij} is the photon energy, and l is the column length through the emitting region. Values for λ and A are listed in Table 30.1, level populations $n_j \equiv n(X_j^{+p})/n(X_{tot}^{+p})$ and $n(X^{+p})/n(X_{tot})$ are given in Greenhouse et al. [1]. Using Eq. (30.4), we can write the intrinsic intensity ratio of I_{ij} in X^{+p} to I_{kl} in Y^{+q} as

$$\frac{I_{ij}}{I_{kl}} = \frac{\lambda_{kl}}{\lambda_{ij}} \frac{A_{ji}}{A_{lk}} \frac{n(X^{+p})}{n(Y^{+q})} \frac{n_j}{n_l}. \quad (30.5)$$

In a case where both lines are emitted by the same ion, one can see that the relative populations n_j and n_l are directly measured. This test is not possible for the C I sequence ions listed in Table 30.1 since the inverted $^3P_1$ level is the first fine structure level above the ground state. The N I and P I sequence ions in Table 30.1 have an inverted $^2D_{5/2}$ level so that one could observe the ratio I($^2D_{3/2} - {}^2D_{5/2}$)/I($^2D_{3/2} - {}^4S_{3/2}$). However, this ratio can be severely effected by selective extinction of the UV line. The O I and S I sequence elements have inverted $^3P_0$ level so that one could observe the fine structure ratio I($^3P_0 - {}^3P_1$)/I($^3P_1 - {}^3P_2$). The wavelength difference is typically large, but not so large that the effect of selective extinction could not be accurately assessed.

Fine structure lines due to transitions from inverted and non-inverted levels in different ionization states of the same element can be used to reveal pronounced laser amplification. In this case, if one assumes that the model level populations are correct, then one can use Eq. (30.5) to look for an observed ionization pattern $n(X^{+p})/n(X^{+q})$ that is inconsistent with the observed ionization equilibrium of the ions involved. For example, one could observe the ratio [Ne V] 24.28 μm/[Ne VI] 7.64 μm to measure $n(Ne^{+4})/n(Ne^{+5})$ which yields an ionization temperature. If the [Ne V]24.28 μm line was significantly amplified, the resulting temperature would be inconsistently low relative to that derived from [Ne V] 14.32 μm/[Ne VI] 7.64 μm.

The spatial extent of Seyfert CLRs could, in principal, be constrained by ground based speckle interferometry of the [Al VI]3.661 μm line in very nearby Seyfert nuclei using techniques similar to those employed by Greenhouse et al. [21] to resolve the coronal line emitting region of Nova QU Vul. For example, a spatial resolution of roughly 20 pc was achieved on NGC 1068 by McCarthy et al. [49] using speckle interferometry on the nuclear continuum at 2.26 μm. If the CLR occupies a physically intermediate location between the BLR and NLR, then it may be possible to resolve it, or significantly constrain its location by employing speckle techniques in near-infrared coronal lines.

30.3   Conclusions

Roughly half of the bright ($\log[A_{ji}n_j/n_e]|_{<n_{crit}} > -10$ photons s^{-1} ion^{-1}) transitions treated by Greenhouse et al. [1] arise from levels with inverted populations (Table 30.2). The calculation of laser gain lengths for these transitions reveal that if column densities $\sim 10^{24-25}$ cm^{-3} are achieved in Seyfert CLRs or quasars, then significant infrared coronal line laser amplification could occur there. We suggest that observations designed to reveal coronal line laser emission in AGN can provide new information about the structure and evolution of Seyfert nuclei. Observations planed as part of the ISO Central Science Program will produce the first complete 2-200 μm spectra of a relatively large sample of AGN, and may reveal the first infrared coronal line lasers in nature.

References

[1] M. A. Greenhouse, U. Feldman, H. A. Smith, M. Klapisch, A. K. Bhatia, and A. Bar-Shalom, ApJ Supp. **88**, 23 (1993).

[2] B. D. Savage, in *Interstellar Processes*, eds. D. J. Hollenbach and H. A. Thronson (Dordrecht, Reidel, 1987) p. 123.

[3] C. Payne-Gaposchkin, *The Galactic Novae* (New York, Dover, 1957).

[4] J. S. Gallagher and S. Starrfield, ARAP **16**, 171 (1978).

[5] R. G. Teske and R. Petre, ApJ **318**, 370 (1987).

[6] K. F. Fischbach, C. R. Canizares, T. H. Markert, and J. M. Coyne, in *Supernova Remnants and the Interstellar Medium*, R. S. Roger and T. L. Landecker, eds. (Cambridge, New York, 1988), p. 153.

[7] G. E. Brown, Zeitschrift für Physik C **38**, 291 (1988).

[8] D. E. Osterbrock, ApJ. **246**, 696 (1981).

[9] R. Fosbury and A. Sansom, MNRAS **204**, 1231 (1983).

[10] D. E. Osterbrock and W. G. Mathews, ARA&A **24**, 171 (1986).

[11] S. Oke and W. Sargent, ApJ **151**, 807 (1968).

304   M. A. Greenhouse, H. A. Smith, and U. Feldman

[12] H. Nussbaumer and D. E. Osterbrock, ApJ **161**, 811 (1970).
[13] S. A. Grandi, ApJ **221**, 501 (1978).
[14] K. T. Korista and G. J. Ferland, ApJ **343**, 678 (1989).
[15] G. Münch, G. Neugebauer, and D. McCammon, ApJ **149**, 681 (1967).
[16] G. L. Grasdalen and R. R. Joyce, Nature **259**, 187 (1976).
[17] E. Oliva and A. F. M. Moorwood, ApJ **348**, L5 (1990).
[18] A. F. M. Moorwood and E. Oliva, Messenger **63**, 57 (1991).
[19] A. Marconi, A. Moorwood, M. Salvati, and E. Oliva, A&A **291**, 18 (1995).
[20] L. Spinoglio and M. A. Malkan, ApJ **399**, 504 (1992).
[21] M. A. Greenhouse, G. L. Grasdalen, C. E. Woodward, J. Benson, R. D. Gehrz, E. Rosenthal, and M. F. Skrutskie, ApJ **352**, 307 (1990).
[22] A. V. Filippenko and W. L. W. Sargent, ApJ **324**, 134 (1988).
[23] N. Scoville and C. Norman, ApJ **332**, 163 (1988).
[24] M. A. Greenhouse, G. L. Grasdalen, T. L. Hayward, R. D. Gehrz, and T. J. Jones, ApJ **95**, 172 (1988).
[25] M. A. Greenhouse, in *Chemistry in Space*, J. M. Greenberg and V. Pirronello, eds. (Dordrecht, Kluwer, 1989), p. 417.
[26] R. D. Gehrz, C. E. Woodward, M. A. Greenhouse, S. Starrfield, D. H. Wooden, F. C. Witteborn, S. A. Sandford, L. J. Allamandola, J.D. Bregman, and M. Klapisch, ApJ **421**, 762 (1994).
[27] M. C. B. Ashley and A. R. Hyland, ApJ **331**, 532 (1988).
[28] D. Van Buren, M. Mac Low, D. O. S. Wood, and E. Churchwell, ApJ **353**, 570 (1990).
[29] C. L. Sarazin and C. M. Graney, ApJ **375**, 532 (1991).
[30] M. Harwit, J. R. Houck, B. T. Soifer, and G. G. C. Palumbo, ApJ **315**, 28 (1987).
[31] A. F. M. Moorwood, in *Infrared Astronomy with ISO*, T. Encrenaz, ed. (Dordrecht, Kluwer, 1992).
[32] T. Encrenaz and M. F. Kessler, eds., *Infrared Astronomy with ISO*, (New York, Nova Science Pub., 1992).
[33] H. A. Smith, ApJ, **158**, 371 (1969).
[34] M. Landini and B. C. Monsignori Fossi, A&A Supp. **7**, 291 (1972).
[35] M. J. Shull and Van Steenberg, ApJ Supp. **48**, 95 (1982).
[36] R. A. McCray, in *Spectroscopy of Astrophysical Plasmas*, A. Dalgarno and D. Layzer, eds. (Cambridge Univ. Press, Cambridge, 1987).
[37] G. R. Fowles, *Introduction to Modern Optics* (Holt, Rinehart, and Winston Inc., New York, 1968).
[38] K. Davidson and H. Netzer, Rev. Mod. Phys. **51**, 715 (1979).
[39] G. J. Ferland and S. E. Persson, ApJ **347**, 656 (1979).
[40] M. C. Gaskell, G. A. Shields, and E. J. Wampler, ApJ **249**, 443 (1981).
[41] G. J. Ferland, OSU Astronomy Department Internal Rept. 89-001 (1989).
[42] L. Spinoglio, *private communication*.
[43] G. J. Ferland and G. A. Shields, in *Astrophysics of Active Galaxies and Quasi-Stellar Objects*, J. S. Miller, ed. (Univ. Science Books, Mill Valley, 1985).
[44] H. Netzer, in *Active Galactic Nuclei*, T. J. L. Courvoisier and M. Mayor, eds. (Springer-Verlag, New York, 1990).
[45] M. M. de Robertis and D. E. Osterbrock, ApJ **286**, 171 (1984).
[46] M. M. De Robertis and D. E. Osterbrock, ApJ **301**, 98 (1986).
[47] J. H. Krolik, C. F. McKee, and C. B. Tarter, ApJ **249**, 422 (1981).
[48] B. Benjamin a nd H. C. Dinerstein, ApJ **100**, 1588 (1990).
[49] D. W. McCarthy, F. J. Low, S. G. Kleinmann, and F. C. Gillett, ApJ **257**, L7 (1982).

[50] J. R. Fuhr, G. A. Martin, and W. L. Wiese, W. L. 1988, J. Phys. Chem. Ref. Data **17**, Suppl. 4 (1988).

[51] N. Grevesse and E. Anders, E. 1989, in *Cosmic Abundances of Matter*, C. J. Waddington, ed. (New York, AIP Conference Proceedings, 1989), p. 1.

31

Concepts of Nuclear Magnetic Resonance in Quantum Optics

Erwin L. Hahn

It was my good fortune and privilege to have known Charlie Townes at Columbia since I was a research physicist (1952-55) at IBM Watson Computing Laboratory. As I remember, he and Art Schawlow had just completed publication of the definitive book "Microwave Spectroscopy," and about the same time Charlie embarked upon his pioneering ammonia maser experiment. This experiment set the stage for of the laser era which followed. What sticks in my mind from that time was how Polykarp Kusch (who functioned as a sort of Mother-Hen advisor-consultant to Watson Lab physicists) groused in his own inimitable style of stentorian verbiage that the ammonia experiment was a crazy idea and that it would never work. I gathered also from the grapevine that Rabi had similar thoughts. However, thanks to the farseeing scientific judgement and persistence of Charlie, we have him to thank for the ground work he and his colleagues provided that gave birth to laser physics and quantum electronics.

Laser physics has provided a revival and rebirth of optical spectroscopy, and has stimulated new investigations of fundamental mechanisms in other branches of resonance physics. In honor of Charlie Townes on his 80th birthday I am pleased to present a review below about nuclear magnetic resonance as it relates to some concepts of laser physics, and some history about it.

31.1   Introduction

The purpose of this review[1] is to outline a few nuclear magnetic resonance (NMR) concepts and relationships that carry over into the realm of quantum and laser optics. It is natural that the semiclassical physics of resonance exitation in two-quantum-level systems should be common to both quantum optics and NMR. Solutions of the Bloch-Maxwell equations serve as a model for analogues of NMR experiments that occur in quantum optics. The resemblances to NMR of effects

---

[1]This review, to appear in the *Encyclopedia of Nuclear Magnetic Resonance*, is reproduced here with kind permission of the publisher, John Wiley and Sons, Ltd.

in quantum optics cannot have a strict one-to-one relationship in every property, but they do exist in varying degrees. There is no intent here to present a comprehensive review of these analogues, but rather to focus upon principles of coherent macroscopic emission of dipole radiation. Although a number of NMR conditions and effects may be conceived in the domain of quantum optics, they are not always easy nor routine to demonstrate because the required experimental parameters in quantum optics are often more difficult to implement.

There are obvious differences between NMR and quantum optics. The effects of spontaneous emission and wave propagation are absent or negligible in NMR, but dominate in quantum optics. For this reason there are numerous quantum optical effects that have no counterpart in NMR. Conversely, there are unique conditions and experiments in NMR that do not have analogues in quantum optical systems. For example, there is no simple direct optical analogue of a NMR dipolar reservoir, as distinguished from the Zeeman reservoir. The validity of an optical electric "dipolar reservoir" temperature in a coupled optical electronic two-level system in the rotating frame of reference so far has not been demonstrated. In solids, strong exchange electric interactions of optical dipoles among themselves, or with the host lattice, prevent the distinction between an optical electric dipolar and a "two-level internal splitting" (analogous to Zeeman splitting) reservoir. In NMR the weakly interacting nuclear moment permits the definition of isolated spin reservoirs to a high degree. Therefore, decoupling of spin-spin magnetic dipole-dipole interactions from externally controlled spin Zeeman interactions can be achieved over long relaxation times. In quantum optics, short lifetimes of excited states are associated with strongly interacting internal atomic electric degrees of freedom. Transitions are often broadened or rapidly transformed by complex mechanisms difficult to analyze, not amenable to the predictions of relatively simple Hamiltonians so useful in NMR. Pulsed coherent laser experimental techniques have not yet reached the stage of sophisticated high resolution pulsed NMR spectroscopy, utilizing elaborate pulse sequences and double resonance methods, made possible by the persistence of coherent nuclear precession over relaxation times ranging widely from microseconds to many seconds.

As NMR is a predecessor of quantum optics, the molecular beam magnetic resonance method should be recognized as the predecessor of NMR. After space quantization of the spin was confirmed by the Stern-Gerlach molecular beam deflection experiment [1], a series of non-resonant type of molecular beam deflection experiments followed. Spins were measured directly, and magnetic moments were measured indirectly by combining laboriously acquired beam deflection data with theory known at the time. Of special interest was the effect of the apparent rotating magnetic field seen by an observer co-moving with the magnetic moments in a molecular beam as it passes through a static inhomogeneous field. For those molecules in the Maxwell spectrum that have just the right velocity, the static inhomogeneous field appears to rotate in partial synchronism with the Larmor frequency of spin precession determined by the main magnetic field. Therefore the synchronism acts as a partial resonance and flips the spin. Rabi published the theory for this effect [2] and in principle defined a magnetic resonance. The application of radio-wave resonance excitation was not carried out by Rabi's Columbia group immediately after Rabi's paper was published. C. J. Gorter visited Columbia in 1937 and suggested to Rabi that it would be more efficient to apply a rotating radio-frequency field produced by an oscillator driven inductance rather than rely on field inhomogeneity. This suggestion stimulated Rabi and his collaborators in 1937 to apply radio-frequency magnetic excitation to a beam of LiCl, and the magnetic resonance spectrum of the Li nucleus was obtained [3] as the first coherent spectroscopic measurement in the history of physics, followed by a series of many ground-breaking resonance experiments. Although line-broadening effects, such as time-of-flight and field inhomogeneity effects initially did reduce the precision of molecular beam resonance spectroscopy, this first demonstration showed that spin resonance electromagnetic energy could be concentrated into a single coherent frequency by an oscillator, and led to the development of precision magnetic resonance and laser spectroscopy as it exists today.

For a number of years after the Bloch and Purcell groups [4, 5] initiated the science of NMR, it was not anticipated that the dynamic Bloch equations of NMR could describe equally well the transient and steady-state behaviour of optical electric-dipole resonance. Measurements of nuclear spins and magnetic moments were an early preoccupation. Studies were and still are focused upon lineshape, relaxation times and high-resolution NMR spectra where the spins serve as probes of molecules, local fields and fluctuations of the medium. Because the applied field energies and intensities usually far exceed total spin coupling energies and NMR signal intensities, applied rf fields are taken as constant. Therefore the direct role of cavity coupling to the spins in the resonance equations is not a major consideration, although this coupling is necessary in order to observe NMR signals in the first place. In order to take into account the alteration of applied electric fields and free precession signals because of radiation damping caused by coupling with the cavity or the Poynting flux in free space in quantum optics, radiation damping, however, must be considered of primary importance.

Tuned circuit cavity-spin coupling effects show up during free precession of the macroscopic nuclear magnetization in terms of coherent radiation damping [6]. Slight effects of radiation damping of free induction signals are evident in magnetic resonance imaging, where magnetic fields are very homogeneous and relaxation times are long. Radiation damping in NMR is essentially "coherent overdamping." The rate at which the cavity resistivity dissipates power absorbed from the precessing magnetization usually far exceeds the rate at which power is emitted by the magnetization as it precesses about a reaction field H_1 in the rotating frame of reference. Radiation damping occurs only over the signal coherence lifetime T_2 (including the effect of inhomogeneous broadening as well as natural broadening), usually much shorter than the radiation damping time τ_r of the magnetization. Therefore the magnetization tips back by a slight amount toward alignment with the polarizing magnetic field.

The radiation damping rate [6] is given by $1/\tau_r = 2\pi Q\gamma M_0\xi$. Suppose cavity losses become less and less, and the emission from the two-level system becomes greater and greater. Let any or all of the terms in the product of Q (cavity figure of merit), M_0 (magnetization), γ (gyromagnetic ratio), and the filling factor ξ be made large enough so that instead of the usual condition $T_2 \ll \tau_r$, the approximate equality $\tau_r \approx T_2$, holds. This is the threshold condition that would make possible spontaneous NMR oscillations, applicable as well to the laser when translated into parameters of macroscopic electric-dipole radiation. The final requirement is to maintain an inverted population of the two-level system, and the maser or laser is therefore invented with benefit of hindsight. The parameters expressed by $1/\tau_r$ reappear upon manipulation of optical Bloch -Maxwell equations, outlined later. The above conditions for laser oscillations indeed became the basis for the maser-laser principle, finally realized [7] by Townes, and by Basov and Prokhorov in 1954.

In 1954 Dicke [8] stressed the idea of radiation coherence from an ensemble of phased electric dipole moments, applicable to the prediction of the dipole pattern of radiation from two-level systems other than magnetic spins. Formally, the incoherent spontaneous emission rate, proportional to N dipoles, can be distinguished from the coherent emission rate, proportional to N^2 by inspecting the rules of angular momentum that apply to collective Bose states. Dicke pointed out that coherent superpositions of atomic two-level states attained by radiation-driven initial states could be prepared, just as in pulsed NMR, and that these states give rise to coherent dipole radiation. Following this work a semantic trend developed in the literature in which the term "superradiance" was applied to almost every process of coherent radiation from phased dipole arrays, particularly in coherent laser optics. Perhaps this happened because the direct application of optical Bloch equations was not yet made, and it appeared as though the generation of coherent optical radiation was a unique phenomenon. However it is standard knowledge from Maxwell's equations that coherent dipole source terms emit power in proportion to N^2, which "is" superradiance, and is absolutely necessary, for example, to detect ordinary NMR signals. It would seem reasonable not to rename an old Maxwell phenomenon just because "superradiance" rolls off the tongue so impressively. It is more apropos to apply the term to those special effects in which the initial incoherent spontaneous emission of power, proportional to N, finally evolves into the coherent emission condition [8, 9] where the power becomes proportional to N^2.

From the above, the obvious conclusion is that all NMR phenomena as we know them involve an enormous number of N dipoles prepared in phased arrays, where $N^{1/2} \ll N$. Likewise many optical analogues of NMR are governed by this condition. There is no need therefore to supplement the semi-classical approach with quantum field theory in the majority of experiments involving macroscopic behaviour. It may be desirable in certain circumstances to deal with subtle aspects of fluctuations and spontaneous emission. Instead, spontaneous emission is introduced as a phenomenological damping term, just as the spin-lattice relaxation time T_1 appears in Bloch's equations.

In 1957 Feynman, Vernon and Hellwarth [10] showed that electric-dipole transitions in a two-level system could be formulated by equations which were the analogue of the Bloch torque equations, originating from the density matrix. Although the discussion in their paper was restricted to a particular maser problem, they utilized the semi-classical method in which the atoms are quantized, but the radiation and applied fields become classical, self-consistent solutions of the Maxwell and Schrödinger equations. The works of Jaynes [11] and Lamb [12] extended the semi-classical method to quantum maser and laser analysis. A number of treatments and reviews have appeared with the express purpose of clarifying and justifying the assumptions of the semiclassical approach. In a monograph by Bloembergen [13] on nonlinear optics, the density matrix formulation commonly used in NMR is spelled out so that it may be applied more universally to optical systems, extending as well to multilevel systems. Allen and Eberly [14] treat a number of topics in greater depth, some of which are briefly outlined in this essay, particularly with regard to radiation damping as it pertains to photon echoes [15] and self-induced transparency [16]. The book by Shen [17] covers modern topics in quantum optics, including transient phenomena related to NMR.

The discussion of NMR foundations in quantum optics which follows will not attempt to hew the line of a notation different from NMR introduced in the original papers and monographs referred to above. A one-to-one notation conventional in NMR will be carried over into a presentation of the optical Bloch-Maxwell equations. From them certain parametric and phenomenological relationships between NMR and quantum optics will be pointed out.

31.2   The Resonance Equations

The optical picture of NMR absorption was first introduced by the Harvard group [6]. The Bloch equations [5] of the Stanford group at first emphasized the nuclear induction dynamics of macroscopic nuclear polarizations, but of course these two approaches lead to the same interpretatation of NMR. The rate equation for population difference $n = N_1 - N_2$ between the number of spins N_1 in the ground state and the number N_2 in the excited state is given by

$$\frac{dn}{dt} = [(n_0 - n)/T_1] - 2nR_{12} . \tag{31.1}$$

In the high temperature approximation $n_0/N \ll 1$ for spin $I = \frac{1}{2}$, the equilibrium propulation difference is given by $n_0 = N\hbar\omega_0/2kT$, where $N = N_1 + N_2$ is the total number of spins, $\omega_0 = \gamma H_0$, $\gamma = 2p/\hbar$ is the gyromagnetic ratio, and H_0 is the applied polarizing magnetic field. Note that the coherence cannot be described by Eq. (31.1). The rate [18] of rf absorption, expressed from standard perturbation theory,

$$R_{12} = \frac{\pi}{2}\gamma^2 H_1^2 g(\delta) , \tag{31.2}$$

tends to equalize the populations, while the spin-lattice relaxation rate T_1^{-1} tends to restore a population difference toward thermal equalibrium. The off-resonance parameter $\delta = \omega_0 - \omega$ appears in the normalized (assumed to be Lorentzian) line-shape function

$$g(\delta) = \left[\frac{T_2}{\pi}\right] \frac{1}{1 + \delta^2 T_2^2} , \tag{31.3}$$

where ω_0 is the Larmor frequency and ω is the applied frequency of the rotating RF field modulus H_1.

Equation (31.1) can now be modified so that it speaks the language of optics, useful later in devising the optical Bloch-Maxwell equations. The energy of the two-level ensemble is first defined as $W = -n\hbar\omega_0/2$, so that Eq. (31.1) becomes

$$\dot{W} = \frac{(W_0 - W)}{T_1} - 2W R_{12} . \tag{31.4}$$

Substitution of $g(\delta) = T_2/\pi$ from Eq. (31.3) into Eq. (31.2), for $\delta = 0$, gives for the last term in Eq. (31.4)

$$2W R_{12} = \omega_0 H_1 v . \tag{31.5}$$

The absorption component of the magnetization $v = np\gamma H_1 T_2$ can be identified as a slowly-varying solution at resonance for the linear, unsaturated driven magnetization of the Bloch equation in the rotating frame of reference, transverse to the applied rotating field H_1. The slowly varying longitudinal magnetization $N_z = np$ is directed along the polarizing field H_0. In Bloch's equations the v and W are interpreted as time dependent, valid both on and off resonance. This may be generally adopted, and therefore the restricted on-resonance quasi-steady state interpretations of v and W in Eq. (31.4) may be dropped. Instead Eq. (31.4) becomes one of the coherence-dependent Bloch equations:

$$\dot{W} = -\omega_0 H_1 v + \frac{(W_0 - W)}{T_1} , \tag{31.6}$$

valid for any δ, expressed in terms of the energy

$$W = -M_z H_0 = \frac{-M_z \omega_0}{\gamma}, \text{ and } W_0 = -M_0 H_0 = -n_0 p .$$

31.3   Optical Bloch-Maxwell Equations

At this point a jump in the interpretation of the Bloch equations can be made from the NMR of a two-level system to the resonance description of a two-level optical system. Now the polarizations representing absorption (v) and dispersion (u) are to be interpreted as electric-source term which are functions of time t and distance z in Maxwell's equations of propagaion. A pseudo-electric polarization may be defined as $P_z = -(\kappa/\omega_0)W$, with P_z replacing M_z, "gyro-electric" ratio $2p_0/\hbar = \kappa$ replacing γ, and electric field modulus ε replacing H_1. The polarizations v and u

are defined in the frame of reference rotating at the applied optical frequency ω of a plane wave

$$E(z, t) = 2\varepsilon(z, t)\cos[\omega t - kz + \phi(z, t)] . \tag{31.7}$$

In the slowly-varying wave approximation $\partial \varepsilon/\partial z \ll k\varepsilon$; $\partial \varepsilon/\partial t \ll \omega \varepsilon$, the terms $\varepsilon(z, t)$, u, v, W and phase shift $\phi(z, t)$ appear as slowly varying parameters.

The optical analogue of the NMR torque equation in the rotating frame reads

$$\begin{aligned}
\frac{d\mathbf{P}}{dt} = \; & \mathbf{P} \times [\mathbf{u}_0 \kappa \varepsilon(z, t) + \mathbf{z}_0(\delta\omega + \dot{\phi}(z, t))] \\
& - \frac{(\mathbf{u}_0 u + \mathbf{v}_0 v)}{T_2} - \frac{\mathbf{z}_0(W - W_0)}{T_1} ,
\end{aligned} \tag{31.8}$$

where $\mathbf{P} = \mathbf{u}_0 u + \mathbf{v}_0 v + \mathbf{z}_0 P_z$ is the total macroscopic electric polarization vector, and $\mathbf{u}_0$, $\mathbf{v}_0$, $\mathbf{z}_0$ are the unit vectors of axes in the rotating frame. The effective frequency and propagation termes are given by $\omega_e = \omega + \dot{\phi}(z, t)$ and $k_e = k + \partial\phi(z, t)/\partial z$ respectively. Instead of the pseudo-polarization component P_z, the energy $W = -n\hbar\omega_0/2$ is often the appropriate physical term to be used in the optical Bloch equations. In absence of damping, the conservation condition

$$N^2 p^2 (u^2 + v^2) + \frac{\kappa^2}{\omega_0^2} W^2 = \frac{\kappa^2}{\omega_0^2} W_0^2$$

applies; and with damping, the optical Bloch equations are written out completely as

$$\dot{u} = v(\delta\omega + \dot{\phi}) - u/T_2 \tag{31.9}$$

$$\dot{v} = -u(\delta\omega + \dot{\phi}) + \frac{\kappa^2}{\omega_0}\varepsilon W - v/T_2 \tag{31.10}$$

$$\dot{W} = -v\varepsilon\omega_0 + (W_0 - W)/T_1 , \tag{31.11}$$

The relaxation time T_1 here is assumed to be the inverse spontaneous emission rate. The upper state of the optical two-level system is empty and therefore the ground state population is $n_0 = N$ at the equilibrium condition $\hbar\omega \gg kT$, in contradiction to the high temperature approximation $n_0/N \ll 1$ applied in Eq. (31.1). However the form of Eq. (31.11) remains valid because the rate of change of population difference due to spontaneous emission in Eq. (31.4) can be written

$$\frac{dn}{dt} = 2R_{spon}N_2 = R_{spon}(N - n) ,$$

so that the identity

$$(W_0 - W)/T_1 = -R_{spon}(N - n)\hbar\omega_0$$

may be retained in Eq. (31.11), where $W_0 = -N\hbar\omega_0/2$, $W = -n\hbar\omega_0/2$, and $1/T_1 = R_{spon}$. The Eqs. (31.9)-(31.11) are time dependent at a given position z, and include the effect of frequency modulation in the term $\dot{\phi}(z, t)$. They are then connected with Maxwell's wave equation

$$\frac{\partial^2 E(z, t)}{\partial z^2} = \frac{\eta^2}{c^2}\frac{\partial^2 E(z, t)}{\partial t^2} + \frac{4\pi}{c^2}\frac{\partial^2 P_{u,v}(z, t)}{c^2 \partial t^2} \qquad (31.12)$$

by assuming the slowly-varying wave approximation expressed earlier, with the definition $P_{u,v} = (u + iv)\exp\{i[\omega t - kz - \phi(z, t)]\}$. Therefore combining Eqs. (31.9)-(31.11) with Eq. (31.12) yields

$$\frac{\partial \varepsilon(z, t')}{\partial z} = \frac{-2\pi\omega}{\eta c}\int_{-\infty}^{\infty} v\, g(\delta)d\delta \qquad (31.13)$$

and

$$\varepsilon\frac{\partial \phi(z, t')}{\partial z} = \frac{2\pi\omega}{\eta c}\int_{-\infty}^{\infty} u\, g(\delta)d\delta\ , \qquad (31.14)$$

in the retarded frame of reference, where the retarded time $t' = t - \eta z/c$. Equations (31.9)-(31.14) may apply to either sign of circular polarization (positive as shown) of the rotating ε field modulus, satisfying the transition selection rule $\Delta m = \pm 1$. Oppositely rotating magnetizations excited by linearly-polarized field $E_x(z, t)$, expressed by Eq. (31.7), involves the selection rule $\Delta m = 0$, and the magnetizations sum to produce linearly polarized magnetization along the x axis in the laboratory frame as follows:

$$P_x = P_{u,v}(+) + P_{u,v}(-) = u \cos \psi(z, t) - v \sin \phi(z, t)\ , \qquad (31.15)$$

where $\psi(z, t) = \omega t - kz - \phi(z, t)$. The Bloch-Maxwell equations above can be generalized to cases where the two-level system is degenerate [16], and not necessarily characterized by one dipole matrix element. In that case u and v are replaced by sums $\sum_i u_i$, $\sum_i v_i$ corresponding to sums over the i matrix elements.

If Eq. (31.12) is applied without the slowly varying wave approximation, extra smaller terms of the type $\ddot{u}$, $\ddot{v}$, $\omega\dot{u}$, $\omega\dot{v}$ would appear in Eq. (31.13) and Eq. (31.14) along with $\omega^2 u$ and $\omega^2 v$ that dominate. The size of the smaller higher-order terms is of the same order of magnitude as the off-resonance response contribution by the dipole matrix elements in the atom involving transitions to quantum levels other than the two-level system at or on resonance. Although other levels are not included, they must contribute in a formal sense, as dictated by the sum rule for optical response. Unless other levels are taken into account, it is quite possible to arrive at unrealistic conclusions concerning optical pulse-propagation effects using extremely sharp and powerful laser pulses.

31.4   Radiation Damping

The mechanism of radiation damping is essential for the detection of free precession signals in NMR and quantum optics. The coupling of precessing spin magnetization to the cavity (LCR tuned circuit) in NMR allows the spin system to radiate coherently into a cavity volume V_c with a single mode confined to a bandwidth $\Delta v = v/Q$, where the wavelength λ is much greater than the cavity and sample dimensions. The density of states is therefore $\rho_c = (V_c \Delta v)^{-1}$, whereas with no cavity present, the magnetic dipole system would radiate incoherently because of vacuum field fluctuations H_{vac} at an infinitesimally small Einstein emission rate, given by

$$\frac{1}{\tau_E} = \frac{2\pi}{\hbar^2} \langle \mathbf{p} \cdot \mathbf{H}_{vac} \rangle \rho_F = \frac{32\pi^3 p_0^2}{\hbar \lambda^3} , \tag{31.16}$$

where the free space density of states is $\rho_F = 8\pi v^2/c^3$. The connection between incoherent spontaneous emission rate of the dipole moment in free space to that in the NMR cavity is made by replacing the density of states ρ_E by $\rho_c = (V_c \Delta v)^{-1}$ in Eq. (31.16), namely,

$$\frac{1}{\tau_r} = \frac{N V_s}{\tau_E} \frac{\rho_c}{\rho_E} = 2\pi M_0 \gamma Q \xi . \tag{31.17}$$

The NMR coherent emission rate in a cavity is enhanced by factors of $N V_s$ coherent spins and the cavity density of states ρ_c much larger than ρ_E. In the LCR circuit, spin energy is dissipated under the condition $\gamma H_1 \gg R/L$, where $Q = \omega/\Delta v$, and the sample filling factor is $\xi = V_s/V_c$. Power emitted by the spins in terms of the voltage V_{ind} is given by

$$M_0 H_0 V_s \frac{d(\cos\theta)}{dt} = \frac{V_{ind}^2}{2R} = \frac{(4\pi M_0 \omega_0 N_t \xi \sin\theta)^2}{2R} , \tag{31.18}$$

with N_t as the number of turns in the inductance $L = 4\pi(N/l)^2 V_c$ of length l, volume $V_c = Al$, and $\Delta v = R/L$. From Eq. (31.18) the radiation damping equation is expressed as

$$\begin{aligned} \gamma H_1(t) &= -2\pi M_0 \gamma Q \xi \sin\theta = \dot{\theta} \\ &= -(1/\tau_r) \sin\theta = -(1/\tau_r) \operatorname{sech}[(t + t_0)/\tau_r]. \end{aligned} \tag{31.19}$$

Following a θ_0 pulse at $t = t_0$, the magnetic moment M_0 precesses about the cavity reaction field $H_1(t)$ in the frame of reference rotating at frequency ω, and is tipped toward the external magnetic field direction at a later time to a smaller angle $\theta(t)$. If the spins remain in phase for a short time $T_2 \ll \tau_r$ after an applied $\theta_0 = \pi/2$ pulse $(t = -t_0)$, the spin energy $W_s = -M_0 H_0 \Delta\theta$ is given up to the cavity after a small decrease of angle $\Delta\theta = \gamma H_1 T_2 \approx T_2/\tau_r$. Therefore the emitted spin precession power immediately after the $\pi/2$ pulse is

$$P_{max} = \frac{M_0 H_0 \Delta\theta}{T_2} V_s = \frac{M_0 H_0}{\tau_r} V_s = 2\pi Q M_0^2 \omega_0 \xi V_s . \tag{31.20}$$

31.5   NMR Maser and Optical Laser Threshold Condition

Equation (31.19) can be adapted to a steady state interpretation of a NMR maser by postulating that $M_0 \rightarrow -M_0$, and that the spin population is maintained in a steady-state inversion. The field H_1 is now time independent instead of a transient, and the steady state condition at resonance for the transverse magnetization in Eq. (31.19) requires the conversion

$$-M_0 \sin \theta \rightarrow v = \frac{-M_0 \gamma H_1 T_2}{1 + (\gamma H_1)^2 T_1 T_2} , \qquad (31.21)$$

obtained from the steady-state solution of Bloch's equations which includes the saturation denominator. Substitution of Eq. (31.21) in place of $M_0 \sin\theta$ in the first equation of Eq. (31.19), left side, yields the relation $H_1 = v/\gamma \tau_r M_0$, and therefore

$$(\gamma H_1)^2 T_1 T_2 = \sqrt{-1 + T_2/\tau_r} . \qquad (31.22)$$

This very simple macroscopic relation shows that the NMR "maser" field H_1 can only be sustained as a real quantity when the condition $T_2/\tau_r \geq 1$ applies at threshold. Obviously Eq. (31.22) cannot take into account statistical properties at threshold due to noise, nor does Eq. (31.22) properly take into account the gain in H_1^2 when the condition $T_2/\tau > 1$ is obeyed. In a qualitative sense this threshold condition defines a phase transformation boundary between the incoherent spontaneous emission phase and coherent emission phase.

In terms of optical parameters the relation Eq. (31.22) satisfies the threshold condition as well for laser action. For later reference, the optical Bloch equation (31.11) is now written using optical parameters:

$$W = -\omega \kappa \tau_{r0} P_0 \varepsilon^2 - (W - W_0)/T_1 , \qquad (31.23)$$

where

$$\varepsilon = \frac{P_0 \sin \theta}{P_0 \kappa \tau_{r0}} = v/P_0 \kappa \tau_{r0} , \qquad (31.24)$$

and the optical damping time τ_{r0}, now governed by feedback in an optical cavity, replaces τ_r. The definition of ε in Eq. (31.24) is analogous to $H_1 = v/\gamma \tau_r M_0$ substituted in Eq. (31.22) in the case of NMR. Note that ε can only be interpreted as a reaction field in Eq. (31.23), and not an applied field. In the weak perturbation limit, the saturation denominator in v is absent, and any dependence of $\varepsilon = v/P_0 \kappa \tau_{r0}$ on T_1 is included only by its effect on T_2. In the next section the damping time τ_{r0} will pertain to optical coherent emission in the absence of cavity feedback.

31.6   Optical Free-Induction Decay and the Photon Echo

In terms of the transverse polarization v, a sample of radiating electric dipoles is prepared in cooperative phase by a traveling-wave short pulse, or two pulses in succession to obtain a photon echo, and thereafter emits an optical free-induction decay (FID) signal. The sample length l is much shorter than the linear absorption length

$$\alpha^{-1} = (8\pi N p_0^2 \omega_0 T_2 / \hbar \eta c)^{-1} . \tag{31.25}$$

This restriction avoids the possibility of nonlinear propagation effects such as self-induced transparancy [16], in which case the applied pulse itself is modified in shape, amplitude and velocity as it transmits through the sample. (The absorption coefficient α is obtained from Eq. (31.11) and Eq. (31.13) in the homogeneous line limit by combining relations $(\eta c/4\pi)\partial \varepsilon^2/\partial z = \dot{W}$ and $v = N p_0 \kappa \varepsilon T_2$, and dropping the last term in Eq. (31.11) in the weak perturbation limit that $W \approx W_0$. Here the field ε is interpreted as an applied field.) The reaction field ε in Eq. (31.23) is now interpreted as the electric FID signal radiated by the polarization v that remains in phase for a coherence time t_p, shorter than T_1. Because the cooperative emission rate is enhanced over the spontaneous rate term $(W - W_0)/T_1$ in Eq. (31.23), this term is dropped. The conservation relation

$$\frac{nc}{4\pi} \varepsilon^2 = \dot{W} l \approx \varepsilon^2 \omega_0 \kappa \tau_{r0} P_0 l , \tag{31.26}$$

now defines the flux per unit area passing out of a sample of length l. The flux represents a power loss analogous to the radiation damping ohmic loss in the NMR case of radiation damping. Coherent radiation is directed along l, and the emission rate $1/\tau_{r0}$ is defined from the identity Eq. (31.26) as

$$\frac{1}{\tau_{r0}} = 2\pi P_0 \kappa \omega_0 \left(\frac{2l}{\eta c} \right) = \frac{1}{\eta \tau_E} N V_{coh} . \tag{31.27}$$

This rate depends on the free space spontaneous emission rate $1/\tau_E$ given by Eq. (31.16). There is no cavity feedback but there is emission enhancement by a factor of the number of phased dipoles contained in a coherence volume defined by $V_{coh} = \lambda^2 l/2\pi$. In the case of NMR the wavelength is large compared to the sample dimesions, and therefore the coherence volume is given by the entire sample volume $V_s = \xi V_c$ as it appears in Eq. (31.17). In the optical FID case the wavelength is much less than sample dimensions, and therefore the radiation rate is determined by a much smaller coherence volume V_{coh} instead of V_s.

Before Hartmann and his collaborators [15] obtained photon echos there was a period of time when the coherent FID emission rate was predicted to be much too large. This resulted because the sample volume V_s was used in place of the coherence volume V_{coh} in Eq. (31.27). The duration of the optical FID pulse burst of power or "optical bomb" [8] was expected to be as short as 10^{-13} sec or shorter, a number intimidating enough to make experimentalists believe that the FID would be just about impossible to observe. The much longer observed times of photon echo bursts extend from about 10^{-8} sec to 10^{-6} sec.

Laser pulse coherent transient experiments often reauire pulses shorter than spontaneous emission times $\approx 10^{-8.-9}$ seconds, so that the initial tipping angle $\theta_0 \approx \theta(t) \ll 1$ must be small if laser output power is limited. Also the optical line is usually inhomogeneously Stark, or Doppler broadened, and only a small fraction ($\approx T_2^*/t_p$) of the N atoms in the line is excited spectrally, where t_p is the pulse width. Therefore the number of dipoles actually excited is given by $n \approx N\theta(t)T_2^*/t_p$, from an optical spectral width of about $1/t_p$ within the optical line. The emitted electric field in Eq. (31.23) appears in the form of a transient FID or a photon echo [15] lasting about t_p seconds. Using Eq. (31.24), Eq. (31.26) may be rewritten to express the FID pulse power emitted by a volume Al of the sample:

$$\dot{W}Al = \frac{v^2}{\hbar P_0 \kappa \tau_{r0}} Al .$$

(31.28)

The number of photons $N_{ph,A}$ emitted per unit area by an electric-dipole FID lasting t_w seconds from n dipoles in coherent superposition is given by

$$N_{ph,A} = lt_w \dot{N}_{ph} = \frac{lt_w \dot{W}}{\hbar\omega_0} = \frac{v^2}{\hbar P_0 \kappa \tau_{r0}} \frac{lt_w}{\hbar\omega_0}$$

(31.29)

$$= \frac{8\pi^2 p_0^2 l^2 n^2 t_w}{\eta\hbar\lambda} ;$$

$$N_{ph,A} \approx \frac{8\pi p_0^2 l^2 N^2}{\eta\hbar\lambda} \frac{T_2^{*2}\theta^2}{t_w} .$$

(31.30)

Note that the coherence volume V_{coh} in Eq. (31.27) appears naturally in the application of the optical Bloch equation and does not have to be introduced explicitly.

31.7   Reaction Field Role During Linear Propagation

The concept of the reaction field is connected with stimulated absorption and emission during steady-state propagation. Consider linear absorption (no saturation denominator in v) through a two-level medium where the propagating beam intensity is given by $\varepsilon^2(z) = \varepsilon^2(z = 0)e^{-\alpha z}$. The field amplitude reduction over a short distance $\Delta z = l$ at position z is given by $\Delta\varepsilon \approx -\varepsilon\alpha z/2$. Substitution of α from Eq. (31.25) and $\varepsilon = v/P_0\kappa T_2$ into this expression for $\Delta\varepsilon$ shows that it is identical to the reaction field given by Eq. (31.24), where $1/\tau_{r0}$ is given by Eq. (31.27). If the incoming driving field ε that induces the steady-state polariztion v is cut off abruptly, then v would radiate an initial FID field $\Delta\varepsilon$ identical to the reaction field that existed before ε was cut off. The $\Delta\varepsilon$ fields add in phase over the entire sample to provide a total optical FID signal at the exit end of the sample. Similarly, if the two-level system is maintained in the inverted state, acting as a traveling wave amplifier, $\Delta\varepsilon$ would represent a gain in field amplitude, with the FID signal shifted in phase by $180°$ following the abrupt cut-off.

31.8   Two-pulse Photon Echo Phase-matching Condition

The two-pulse photon echo is obtained in practice after a first θ_1, k_1 pulse, propagating in the z direction, is followed by a second θ_2, k_2 pulse applied in a slightly different direction by a small angle ξ. The k propagation directions now play an important role. It will be seen that the photon echo signal [15] is emitted and detected in a third k_{echo} direction, deviating by small angle 2ξ with respect to z. This property conveniently avoids saturation of the optical detector by the applied laser pulses along the k_1 and k_2 directions. In a given volume $A\Delta z$ at z the rotating electric laser modulus ε is shifted in phase angle by $\vartheta = (k_1 - k_2)z$ relative to its alignment in the rotating frame during the first pulse. In the NMR case for a two-pulse echo, if the θ_2 pulse is applied at a RF phase shift φ with respect to the phase of the θ_1 pulse, the phase shift of the echo signal is 2ϑ relative to the echo which occurs when both pulses are in phase [19].

The photon echo is now described only with regard to its z dependence as it propagates. Integration must be carried out over all the signals emitted from given volume elements $A\Delta z$ extending from $z = 0$ to $z = L$, the length of the sample. Phase dependencies $k_1 z, k_{echo}(L - z), 2\varphi = 2(k_1 - k_2)z$, account respectively for the phased array set up along z to L, the phase change of the echo emitted from a volume element $A\Delta z$, and the phase shift caused by the shifted direction of the second pulse. The result for $\xi \ll 1$ is

$$E_{echo}(L) \propto \int_0^L e^{2i(k_1-k_2)z - ik_1 z - ik_{echo}(L-z)} dz$$

$$= \frac{\sin[k_{echo} - (2k_2 - k_1)]L/2}{[k_{echo} - (2k_2 - k_1)]L/2}. \qquad (31.31)$$

Since the magnitudes of the two pulse k vectors ($|k_1| = |k_2|$) are the same in the usual experiment, the phase matching condition above cannot be satisfied exactly, given that the direction of k_2 is chosen at a small angle ξ with respect to the z direction of k_1. If ξ were not small, the phase factor in the integrand in Eq. (31.31) should be written as $\exp\{i[\mathbf{k}_{echo} - (2\mathbf{k}_2 - \mathbf{k}_1)] \cdot \mathbf{r}$, which in principle would require integration over all dipoles at points over $\mathbf{r}$ in the sample. Since we ignore diffraction from a strip of dipoles along $\mathbf{r} = \mathbf{z}$, then it is seen for $\xi \ll 1$ that $[\mathbf{k}_{echo} - (2\mathbf{k}_2 - \mathbf{k}_1)] \cdot \mathbf{r} \approx [k_1 - |2\mathbf{k}_2 - \mathbf{k}_1|]z \cos 2\xi$, where the maximum photon echo amplitude is emitted at an angle $\approx 2\xi$ with respect to the $\mathbf{r} = \mathbf{z}$ direction when $\mathbf{k}_{echo}$ is parallel to $2\mathbf{k}_2 - \mathbf{k}_1$. However, because $|\mathbf{k}_1| = |\mathbf{k}_2|$, the propagation vector triangle cannot form an exact closed triangle, and the intensity of the photon echo is therefore reduced by a very small factor. The important point is that the photon echo is emitted in the $\mathbf{k}_{echo}$ direction at an angle $\approx 2\xi$ with respect to the k_1 direction along z in order to avoid the blanking effect of the applied laser pulses.

There is a connection between the various types of photon echo production in the time domain and steady-state nonlinear optical effects [17] such as four-wave mixing and phase conjugation. Propagating phonon backward-wave acoustic echoes [20] are identically phase-conjugated echoes. The phase-conjugated echo represents an apparent reversal in time because δ is effectively changed in sign in the precession phase δt. In addition the sign of the k vector in the propagation phase kz is reversed, and the echo travels in the reverse z direction instead of the usual forward direction of applied pulses. These effects occur because pulse echoes are caused by nonlinear tranformations imposed upon the Bloch vectors during applied pulses. As laser pulse sequences are conceived to merge with one another into the steady state, considerations of nonlinearity [17] persist but in a different form.

31.9   Principal Effects Common to NMR and Quantum Optics

A number of the most important NMR effects and concepts that appear in quantum optics can be listed. However it is sometimes difficult to make clear one-to-one relationships between the two fields. The basic physics of the Bloch-Maxwell equations, together with general application of the density matrix to more than two quantum levels, encompasses a large number of phenomena in both fields. Many items are not listed that involve spontaneous emission, certain propagation effects, and some multilevel Raman processes, all having only remote connections to NMR. Parametric conversion effects are also omitted. No doubt there are some topics overlooked that deserve listing; or vice-versa, some items appearing may not deserve recognition in the opinion of some people. Rather than give references for each item, the reader is referred to standard reviews in the literature. For examples, books by Shen [17], Yariv [21], Sargent, Scully and Lamb [22], and Allen and Eberly [14] treat many of the modern topics in quantum optics.

Linear and nonlinear absorption, dispersion; saturation, hole burning, spectral and spatial diffusion

- Homogeneous, inhomogeneous broadening

- Radiation damping; coherent reaction field dynamics

- Rabi flop; precession about the effective field in the rotating frame (optical nutation)

- Adiabatic following and inversion

- Photon echo, rotary photon echo, tri-level echoes, transfer of coherence, single pulse free precession (FID)

- Stimulated echo pulse holographic memory storage in rare earth crystals

- Coherent quantum beats; stimulated heterodyne coherent Raman beats, coherent Raman scattering

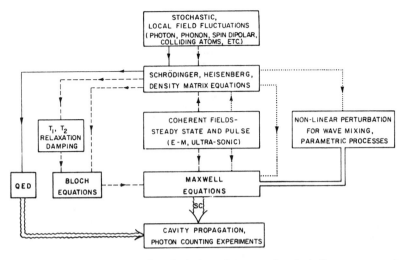

FIGURE 31.1. Various routes of analysis in quantum optics, including quantum electrodynamics (QED). The semi-classical (SC) method connects Maxwell equations to various polarization source terms. The Bloch-Maxwell equations are designated by coupling Maxwell equations to optical Bloch equations adapted from NMR.

- Self-induced transparency

- Two photon transition phenomena; AC Stark effect; multiple quantum transitions

- Electric rotary saturation

- Double resonance

- Two-level laser theory, operation

- Optical four-wave mixing, stimulated echo, backward wave echo

- Optical phase conjugation

- Fourier transform spectoscopy

- Optical pulse chirping, frequency modulation

- Relaxation dynamics, motional narrowing, dipole-dipole coupling

- Autler-Townes effect; dressing of atomic levels, fast driven precession averaging

- Holography, grating scattering

- Nuclear polarization by optical pumping, spin exchange

31.10   Conclusion

This brief review covers only a few fundamental concepts of spontaneous coherent dipole radiation that connect NMR to quantum optics. The block diagram of Fig. (31.1) suggests how the Bloch-Maxwell equations tie in with various routes of analysis that cover a broad number of topics in the literature. The figure shows the path of a perturbation approach often carried out without the use Bloch equations, by taking the .... → route. Instead of using the Bloch equations directly on a par with Maxwell's equations, it is this route often used by insertion of perturbation terms into Maxwell's equations directly. However one discovers that by taking the .... → route, all orders of perturbation in principle are included simultaneously from these equations when they are directly coupled to Bloch's equations. Although full theoretical rigor is not always satisfied by the use of the Bloch-Maxwell equations, they are helpful in providing phenomenological predictions. Fig. (31.1) includes diagrammatically the role of other degrees of freedom and mechanisms involved in quantum optics, and indicates how experiments analyzed by quantum electrodynamics (QED) take a different route.

31.11   Acknowledgment

The author thanks the National Science Foundation for support.

References

[1] O. Stern, Z. Phys. **7**, 249 (1921); W. Gerlach and O. Stern, Ann. Phys. Leipzig, **74**, 673 (1924).

[2] I. I. Rabi, Phys. Rev. **51**, 652 (1937).

[3] I. I. Rabi, J. R Zacharias, S. Millman, and P. Kusch, Phys. Rev. **53**, 318 (1938).

[4] N. Bloembergen, E. M. Purcell, and R. V. Pound, Phys. Rev. **73**, 679 (1948).

[5] F. Bloch, W. W. Hansen, and M. E. Packard, Phys. Rev. **69**, 474 (1946); F. Bloch, Phys. Rev. **70**, 460 (1946).

[6] N. Bloembergen and R. V. Pound, Phys. Rev. **95**, 8 (1954).

[7] J. P. Gordon, H. J. Zeiger, and C. H. Townes, Phys. Rev. **95**, 282 (1954); N. G. Basov and A. M. Prokhorov, JETP (USSR) **27**, 43 (1954).

[8] R. H. Dicke, Phys. Rev. **93**, 99 (1954); *Quantum Electronics,* Proc. Third Intl. Conf. Ed. by P. Grivet and N. Bloembergen, (Columbia Univ. Press, New York, 1964), p. 35.

[9] M. S. Feld and J. C. MacGillivary, *Coherent Nonlinear Optics,* edited by M. S. Feld and V. S. Letokhov (Springer-Verlag, Berlin, 1980).

[10] R. P. Feynman, F. L. Vernon, and R. W. Hellwarth, J. Appl. Phys. **28**, 49 (1957).

[11] E. T. Jaynes and F. W. Cummings, Proc. IEEE **51**, 89 (1963).

[12] W. E. Lamb, Phys. Rev. A **134**, 1429 (1964).

[13] N. Bloembergen, *Nonlinear Optics* (W.A. Benjamin Inc., New York, 1965, 3rd Printing, 1976).

[14] L. Allen and J. H. Eberly, *Optical Resonance and Two-level Atoms* (John Wiley and Sons Inc., New York, 1975).

[15] I. D. Abella, N. A. Kurnit, and S. R. Hartmann, Phys. Rev. **141**, 391 (1966).

[16] S. L. McCall and E. L. Hahn, Phys. Rev. A **2**, 861 (1970).

[17] Y. R. Shen, *The Principles of Nonlinear Optics* (John Wiley and Sons Inc., New York 1984).

[18] A. Abragam, *The Principles of Magnetic Resonance* (Clarendon Press, Oxford, 1986) p.28.

[19] E. L. Hahn, Phys. Rev. **80**, 580 (1950).

[20] N. S. Shiren and T. G. Kazyaka, Phys. Rev. Lett. **28**, 1304 (1972).

[21] A. Yariv, *Quantum Electronics* (John Wiley and Sons Inc., New York, 1975).

[22] M. Sargent III, M. O. Scully, and W. E. Lamb, *Laser Physics* (Addison Wesley Publishing Co., Reading, 1974).

32

Magnetic Resonance Imaging with Laser-polarized Noble Gases

William Happer

There is hardly a corner of science or engineering that has not benefitted, often in the most improbable ways, by the fruits of Charles H. Townes's work on microwave spectroscopy, masers and lasers. In this note, I will describe a promising new addition to the capabilities of magnetic resonance imaging that would have been unthinkable without the lasers that resulted from the pioneering work of Townes and Schawlow [1].

Nuclear magnetic resonance (NMR), which was discovered in 1946 by E. M. Purcell, H. C. Torrey and R. V. Pound [2], and independently by F. Bloch, W. W. Hansen and M. Packard [3], makes use of the small voltages induced in a pickup coil by the changing magnetic flux from nuclear spins, which are precessing about a magnetic field. In NMR experiments, the signal voltage is proportional to the static magnetization density of the nuclei, which can be written as

$$M = N\mu_I P, \qquad (32.1)$$

where N is the number density (spins/cm^3), μ_I is the magnetic moment of a nucleus of spin I, and the polarization P measures how nearly the spins are lined up with each other. For nuclei with $I = 1/2$ in thermal equilibrium at a temperature T and in a magnetic field B, the polarization P is

$$P = \tanh\left(\frac{\mu_I B}{kT}\right). \qquad (32.2)$$

Because the nuclear moment μ_I is so small, the nuclear polarization (32.2) is very small at room temperature, even for very large magnetic fields. For example, for a proton in a magnetic field of 1 T, the polarization is $P = 3.4 \times 10^{-6}$. However, the number density N of (32.1) is often so large (for example $N = 6.7 \times 10^{22}$ cm^{-3} in water) that this weak thermal polarization is adequate to produce good signals.

Among the many applications of NMR, magnetic resonance imaging (MRI) is perhaps the most striking to the average person. In 1973 Paul Lauterbur [4] pointed out that it should be possible to use magnetic resonance of an extended body in an inhomogeneous magnetic field to produce images of the distribution of nuclei in

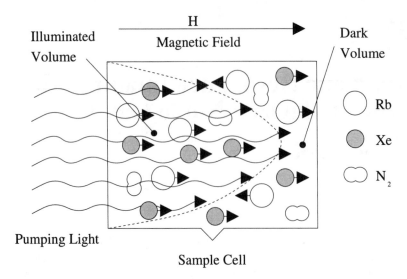

H

Magnetic Field

Illuminated Volume

Dark Volume

○ Rb

◉ Xe

〇 N₂

Pumping Light

Sample Cell

FIGURE 32.1. A series of six magnetic resonance images of a section through the lung of a guinea pig, taken with the same volume of laser-polarized ^{3}He gas. The time interval between images is 8 seconds. There is little loss of signal, either due to the small tip angles of the imaging sequence or to the intrinsic relaxation of the nuclear polarization of the ^{3}He gas in the lung.

a three-dimensional object. Rather than struggling to eliminate all magnetic field inhomogeneities, one applies carefully controlled inhomogeneities to map spatial information onto the free induction decay transients and spin echos of classical NMR. Very beautiful diagnostic images of the brain, the spine, and many other internal organs are routinely obtained with MRI. However, one organ, the lung, has stubbornly resisted all attempts to produce good images—with MRI, x-rays or any other modality. The reason is that the lung is designed to be filled with air. Consequently, it has a very low proton density and a complicated three dimensional structure of tissue and gas spaces, with magnetic susceptibility gradients that produce artifacts in the weak MRI images that can be obtained. There is also little to scatter x-rays in the lung except for the calcified scars of tuberculosis.

Our group at Princeton, in collaboration with research groups at the State University of New York at Stony Brook [5] and at the Center for In Vivo Microscopy at Duke University [6] has recently begun to use the laser-polarized gases ^{129}Xe and ^{3}He to make magnetic resonance images of lungs. For example, Fig. 32.1 shows an image of a guinea pig's lung, after it has been filled with laser-polarized ^{3}He gas. The nuclear spin polarization of the gas was some tens of percent, about 10^5 times greater than that of the protons in the animal's tissue. Even though the number density of ^{3}He nuclei in the gas was about 10^3 times smaller than the number density of protons is tissue, the single-shot gas signal was much better than the proton signals because of the high spin polarization of the ^{3}He nuclei.

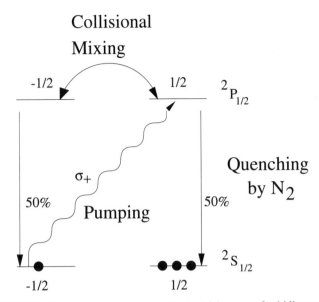

FIGURE 32.2. Optical pumping of an optically thick vapor of rubidium atoms.

The data of Fig. 32.1 were taken in a fruitful collaboration between our group at the Princeton Physics Department, and a group from the Center for In Vivo Microscopy, headed by Professor G. Alan Johnson at Duke University.

The high nuclear polarization of the ^{3}He nuclei used to make the image of Fig. 32.1 has its origins in photons of light, and spin-exchange collisions. In 1948, Albert Kastler [7] showed that it is possible to transfer much of the spin polarization of circularly polarized photons to the electron and nuclear spins of atoms. Kastler called this process *optical pumping*. The great majority of optical pumping work is done with the alkali-metal vapors, mercury vapor or metastable helium atoms [8]. These systems have several features in common. All of the atomic species have spherically symmetric pumped states, ^{2}S states for the alkali-metal atoms, a ^{1}S state for mercury atoms, and a ^{3}S state for metastable helium atoms. Such electronic states are exceptionally resistant to electron or nuclear spin depolarization during collisions. Furthermore, all of these systems can be pumped with near-visible light, and they have adequate atomic number densities at convenient temperatures. Containing the gases and vapors presents manageable materials problems. A typical arrangement for optically pumping rubidium vapor is sketched in Fig. 32.2.

Not shown is the source of the static magnetic field H, which is often only a few Gauss produced by Helmholtz coils and designed to prevent the spin depolarization by stray 60 Hz fields or the ambient field of the earth. A simple oven is used to keep the cell at a constant temperature, usually greater than 80°C and less than 150°C. This ensures that the saturated vapor pressure of a few droplets of Rb metal in the cell is adequate. The number density of Rb atoms is typically 10^{11} to 10^{14} atoms cm^{-3}. More intense pumping light is needed to handle the higher number densities. The sample cell also contains the noble gas which is to be polarized

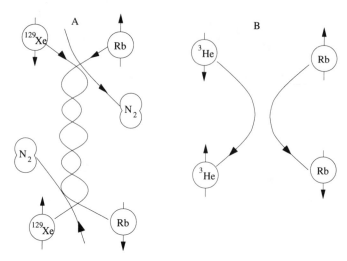

FIGURE 32.3. Optical and collisional transitions involved in the optical pumping of an alkali-metal atom with D_1 light at high gas pressure.

by spin-exchange and a chemically inert quenching gas, normally nitrogen, to prevent reradiation of light from the excited atoms. Since the reradiated light would be nearly unpolarized and can be absorbed by the Rb atoms, it would optically *depump* the atoms. The damage done by the reradiated photons becomes more severe at high Rb vapor densities where the reradiated photon can be scattered several times and depolarize several atoms before escaping from the cell.

The detailed atomic physics of optical pumping for the arrangement of Fig. 32.2 is shown in Fig. 32.3. We consider an imaginary alkali-metal atom for which the nuclear spin is zero. All real alkali-metal atoms have non-zero nuclear spins, but the basic mechanisms of optical pumping and spin exchange are qualitatively similar. Circularly polarized D_1 resonance light is incident on the atoms. The light can excite the atoms from the $^2S_{1/2}$ ground state into the $^2P_{1/2}$ first excited state. In both the ground state and the excited state, the electronic angular momentum of the atom can point up or down. This is represented by two Zeeman sublevels, which are split by their respective Larmor frequencies when the atom is in an external magnetic field. Since the atom must absorb both the energy and the spin angular momentum of the photon, transitions are only possible from the spin-down ground-state sublevel to the spin-up excited-state sublevel. The noble gas and the quenching gas collisionally transfer atoms between the sublevels of the excited state. When a quenching collision finally deexcites the atom, both ground state sublevels will be repopulated with nearly equal probability. Since the atom had $-1/2$ units of spin angular momentum before absorbing the photon and 0 units of spin angular momentum after being quenched from the excited state, each absorbed photon deposits $1/2$ a unit of spin angular momentum in the vapor.

Even though the alkali-metal vapor is often many optical depths thick, intense circularly polarized laser light can still illuminate most of the cell [9], as indicated

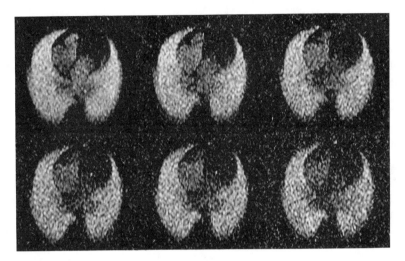

FIGURE 32.4. Spin-exchange polarization of the nuclei of noble gases by three-body or binary collisions with optically pumped alkali-metal atoms.

in Fig. 32.2. In the illuminated part of the cell most of the alkali-metal atoms are pumped into the nonabsorbing $+1/2$ sublevel of the ground state and the spin polarization is nearly 100%. In the dark volume where no light penetrates, the alkali-metal spin polarization is nearly zero, but some small spin polarization may be maintained by spin exchange with the polarized nuclei of the noble gas. Because of its long relaxation time, the spin polarization of the noble gas is nearly the same throughout the cell. The boundary between the illuminated and dark volumes of the cell is about one optical depth thick. In a well-designed system, the gas composition, the cell temperature and the laser intensity are matched to ensure that most of the cell is illuminated.

A key step forward was the discovery by T. R. Carver, M. A. Bouchiat and C. M. Varnum [10] in 1963 that a substantial fraction of the electron spin angular momentum of optically pumped alkali-metal atoms could be transferred to the nucleus of ^{3}He by spin exchange collisions. In 1978 B. C. Grover [11] and his colleagues [12] at Litton Industries reported the extension of the spin exchange method from ^{3}He to isotopes of xenon and krypton. Extensive studies of the physics of spin exchange between the electrons of alkali-metal atoms and noble gas nuclei were subsequently carried out by our group at Princeton University [13].

As indicated in Fig. 32.4, spin exchange from the alkali-metal atomic electron to the nucleus of the noble gas can occur either in a three-body collision (Fig. 32.4A), for which a third body carries away the binding energy of the van der Waals molecule which is formed, or in a simple binary collision (Fig. 32.4B). The main observable difference between the two types of collisions is that the relaxation and spin transfer caused by the van der Waals molecules, while extremely efficient, can be suppressed with external magnetic fields of a few hundred to a few thousand Gauss. The relaxation due to binary collisions is hardly affected by magnetic fields which are readily available in the laboratory [13]. The relaxation due to van der Waal molecules also depends nonlinearly on the gas pressures in the cell [13]. The three-body collisions can account for much of the spin exchange for the heavier noble gases, Kr, Xe or Rn, but binary collisions are the dominant spin exchange mechanism for the light gases He and Ne, which have particularly weak van der Waals attraction for alkali-metal atoms.

The transfer of angular momentum from the electron spin S of the alkali-metal atom to the nuclear spin I of a noble gas atom during a collision is mediated by the Fermi contact interaction [14]

$$\alpha S \cdot I. \tag{32.3}$$

Bernheim [15] first pointed out that a large fraction of the spin angular momentum S of the alkali-metal atoms is lost to the rotational angular momentum N of the alkali-metal atom and its collisional partner through the spin-rotation interaction

$$\gamma S \cdot N. \tag{32.4}$$

The fraction η of angular momentum which is transferred to the nucleus of the noble gas is

$$\eta = \frac{\alpha^2}{\frac{4}{3}(\gamma N)^2 + \alpha^2}. \tag{32.5}$$

For example [13], for the pair ^{129}Xe Rb, the root mean squared values of the coupling coefficients are $\alpha/h = 38$ MHz and $\gamma N/h = 121$ MHz so the efficiency of spin exchange would be about 4%, that is, $1/\eta = 25$ photons are needed to produce each fully polarized ^{129}Xe nuclear spin.

For the lighter noble gases ^{3}He and ^{21}Ne the spin-exchange probability per collision is relatively small, and it is advantageous to operate with as high a number density of alkali-metal atoms as possible to speed up the spin exchange. At high densities, there are frequent spin exchange collisions between alkali-metal atoms A and B

$$A(\uparrow) + B(\downarrow) \rightarrow A(\downarrow) + B(\uparrow). \tag{32.6}$$

The up and down arrows indicate the orientations of the electron spins. The cross section for electron-electron spin exchange collisions [16] are on the order of 10^{-14} cm^{-2}, so the exchange rates can be on the order of 10^4 sec^{-1} at alkali-metal atomic number densities of 10^{14} cm^{-3}. Electron-electron exchange is much faster than any other spin relaxation rate, but since such collisions conserve the total spin of the vapor they do not interfere with the polarization of noble gas nuclei.

Unfortunately, a small fraction of the collisions between alkali-metal atoms transfer spin angular to the translational angular momentum of the vapor [17] in spin-dependent collisions of the form

$$A(\uparrow) + B(\downarrow) \rightarrow A(\downarrow) + B(\downarrow). \tag{32.7}$$

The angular momentum losses are most pronounced for vapors of the heavy alkali metal cesium, less for rubidium, and still less for potassium [18]. For the important case of ^{3}He gas, in which the electron-nucleus spin exchange interaction (32.3) and the spin-rotation interaction (32.4) are both very small, it is the spin-destroying collisions between alkali-metal atoms which sets the practical upper limit on the alkali-metal vapor pressure and on the speed with which the ^{3}He can be polarized.

Although Kastler and his contemporaries did the first optical pumping experiments with a few hundred microwatts of useful resonance light from lamps, it is now straightforward to get watts of useful laser light for optical pumping experiments. A watt of 7947 Å Rb resonance radiation is 4×10^{18} photons per second. Since one unit of angular momentum $\hbar$ is added to the electron spin of the vapor of alkali-metal atoms for every two photons absorbed, 4×10^{18} fully polarized electron spins can be produced per second. In favorable cases, about 10% of the electron spin polarization can be transferred by spin exchange to nuclear spin polarization of noble gases in the same container with the alkali-metal vapor. The spin can be accumulated for a time period on the order of T_1, the longitudinal spin relaxation time of the noble gases. T_1 can be hours or longer for spin-1/2 species like ^{129}Xe or ^{3}He, so some 10^{21} highly polarized spins can be produced. The exceptionally long values of T_1 for these gases are essential to the success of gas imaging, but they are the least understood part of the physics and most in need of further basic research. With more powerful tunable lasers, for example, diode arrays with tens or even hundreds of watts of power, it would be possible to produce many grams of solid ^{129}Xe with nearly complete nuclear polarization.

Like so many other useful new technologies, most especially the laser of Townes and Schawlow, magnetic resonance imaging of lungs resulted from years of curiosity-driven, basic research. As we are about to move into the twenty-first century, let us hope that society will continue to find ways to support basic research for all the benefits it brings, intellectual, moral and practical.

References

[1]   A. L. Schawlow and C. H. Townes, Phys. Rev. **112**, 1940 (1958).

[2]   E. M. Purcell, H. C. Torrey and R. V. Pound, Phys. Rev. **69**, 37 (1946).

[3]   F. Bloch, W. W. Hansen and M. Packard, Phys. Rev. **70**, 474 (1946).

[4]   P. C. Lauterbur, Nature **242**, 190 (1973).

[5]   M. S. Albert, G. D. Cates, B. Driehuys, W. Happer, B. Saam, C. S. Springer, Jr. and A. Wishnia, Nature **370**, 199 (1994).

[6]   H. Middleton, R. D. Black, B. Saam, G. D. Cates, G. P. Cofer, R. Guenther, W. Happer, L. W. Hedlund, G. A. Johnson, K. Juvan and J. Swartz, Magnetic Resonance in Medicine **33**, 271 (1995).

[7]   A. Kastler, J. Phys. Radium **11**, 225 (1950).

[8]   G. K. Walters, F. D. Colgrove and L. D. Schearer, Phys. Rev. Letters **8**, 439 (1962).

[9]   A. C. Tam and W. Happer, Phys. Rev. Letters **43**, 519 (1979).

[10] M. A. Bouchiat, T. R. Carver and C. M. Varnum, Phys. Rev. Letters **5**, 373 (1960).

[11] B. C. Grover, Phys. Rev. Letters **40**, 391 (1978).

[12] C. H. Volk, T. M. Kwon, and J. G. Mark, Phys. Rev. A **21**, 1549 (1980).

[13] N. D. Bhaskar, W. Happer and T. McClelland, Phys. Rev. **49**, 25 (1982); N. D. Bhaskar, W. Happer, M. Larsson and X. Zeng, Phys. Rev. Letters **50**, 105 (1983); W. Happer,

E. Miron, S. Schaefer, D. Schreiber, W. A. van Wijngaarden and X. Zeng, Phys. Rev. 3092 (1984); X. Zeng, Z. Wu, T. Call, E. Miron, D. Schreiber and W. Happer, Phys. Rev. **31**, 260 (1985); S. R. Schaefer, G. D. Cates, Ting-Ray Chien, D. Gonatas, W. Happer and T G. Walker, Phys. Rev. **39**, 5613 (1989).

[14] E. Fermi and E. Segrè, Z. Physik **82**, 729 (1933).

[15] R. A. Bernheim, J. Chem. Phys. **36**, 135 (1962).

[16] W. Happer, Rev. Mod. Phys. **44**, 169 (1972).

[17] N. D. Bhaskar, J. Pietras, J. Camparo and W. Happer, Phys. Rev. Letters, **44**, 930 (1980)

[18] R. L. Knize, Phys. Rev. A **40**, 6219 (1989).

33

Deterministic Order-Chaos Transition of Two Ions in a Paul Trap

John A. Hoffnagle
and Richard G. Brewer

It is a pleasure and an honor to contribute to this Festschrift for Charles H. Townes. His inspired vision has continued to influence us over the years and we say happy birthday to him with our gratitude. This paper is condensed from a review talk delivered at the 91st Nobel Symposium on Trapped Charged Particles and Related Fundamental Physics, Lysekil, Sweden, 19 – 26 August 1994.

33.1   Introduction

In addition to their many applications in spectroscopy and quantum electronics, ion traps offer unique possibilities to study the nonlinear dynamics of simple systems. This was first realized in a beautiful, early demonstration of electrodynamic confinement of charged aluminum particles [1], which were observed to form regular, "crystalline" arrays. Increasing the trap voltage induced "melting," i.e., the arrays abruptly disintegrated into shapeless clouds; when the trap voltage was reduced again, the particles recrystallized. Since this first observation of Coulomb crystals in the Paul trap, the use of laser cooling [2] has allowed experiments to be carried out with individual atomic ions and the theory of deterministic chaos has provided a framework in which to understand irregular motion such as the ion cloud. Cold ions were made to crystallize in Paul traps [3] and transitions between regular and chaotic motion were observed upon varying the trap voltage [4, 5], in clear analogy with Ref. [1].

Here we summarize experimental and theoretical work on the transition between ordered and chaotic motion in the simplest nontrivial case of two ions. This system has the Coulomb nonlinearity which leads to chaos, yet is simple enough to enable detailed calculations of the particles' motion. It is very elementary—two classical charged particles in a periodic potential—yet it exhibits a rich variety of nonlinear behavior, which has been the subject of some controversy. The existence of an order-chaos transition at a well-defined value of the trap voltage was disputed [6] until it was appreciated that the stability of the motion on long-time scales is crucial. Experiments showed that the crystallization of the two-ion system is a manifestation of transient chaos and the observed lifetime scaling suggested a boundary crisis mechanism [7].

In recent years, trapped-ion dynamics has inspired considerable theoretical work [8]. Our basic assumptions are twofold: (1) although laser cooling can, under certain conditions, be an important element of the dynamics, and even induce transitions between ordered and chaotic motion [5], it is not essential for order-chaos transitions, as the seminal experiment of [1] clearly shows; and (2) it is necessary to take into account the full time dependence of the oscillatory trap potential to describe properly the periodic orbits which are at the heart of the nonlinear dynamics. This leads us to explain our experiments [4, 7] with a time-dependent, deterministic model having only one nonlinearity, i.e., that due to the Coulomb force. Laser cooling, which in general can be both nonlinear and non-deterministic, can be adequately modeled, for the present purposes, by linear dissipation of the form $d\mathbf{v}/dt = -\Gamma\mathbf{v}$. A more extensive justification of this model, as well as theoretical and computational details, can be found in [9].

Trapped-ion dynamics also has intriguing similarities to celestial mechanics. The interionic Coulomb force is of the same form, except for its sign, as the gravitational attraction of massive bodies, and the solutions of the trapped-ion equations of motion show some analogies with celestial motion. For instance, the solar system is rife with examples of frequency-locking, in which orbital periods of planets or their satellites acquire rational ratios [10]; we have shown the existence of an infinite family of stable, two-ion orbits with frequencies that are an integral multiple of the trap frequency [11], and observed several of these frequency-locked orbits in a Paul trap for microspheres [12]. In this context, the role of dissipation is especially interesting. The contraction of phase-space volumes and evolution of trajectories to attractors is essential for the observation of frequency-locking in both traps and the solar system. Yet celestial mechanics is a classic example of the success of Hamiltonian (dissipationless) dynamics. The reason is that the fundamental structure of phase-space, e.g., the presence of stable, periodic orbits, does not depend on dissipation. Rather the effect of weak energy dissipation, acting over very long periods of time, is that the system evolves toward attractors which have their origin in the stable periodic solutions of the Hamiltonian equations. A very similar situation applies to trapped ions, for which the motion observed on long time scales is ultimately related to phase-space structures that are independent of dissipation.

33.2   Equations of Motion

In an ideal Paul trap [1, 13] hyperboloidal electrodes generate an oscillatory quadrupole potential with the time-dependent amplitude, $V_{DC} + V_{AC}\cos(\Omega t)$.

The classical motion of two identical particles of charge e and mass m in this potential is most conveniently described in terms of dimensionless electric potentials, $a = -8eV_{DC}/mr_0^2\Omega^2$ and $q = 4eV_{AC}/mr_0^2\Omega^2$, and scaled units of time, $[t] = 2/\Omega$, and length, $[\ell] = (2e^2/m\Omega^2)^{1/3}$. The motion separates into a center-of-mass part, which can be solved immediately in terms of Mathieu functions, and a relative part, described by two coupled equations for the radial and axial coordinates,

$$\ddot{r} + \Gamma\dot{r} + \left(-a/2 + q\cos 2t - r_{12}^{-3}\right)r = 0$$

$$\ddot{z} + \Gamma\dot{z} + \left(a - 2q\cos 2t - r_{12}^{-3}\right)z = 0.$$

(33.1)

Here dots denote differentiation with respect to t and the term proportional to Γ is the linear energy dissipation discussed above. We consider the special case $V_{DC} = 0$, for which there is one control parameter q. Then there always exists a stable, regular solution, denoted as the crystal because it is the two-ion analog of the regular, quasistatic arrays seen in multi-ion systems.

33.3   Experiments on Two Ions in a Paul Trap

In the first experiments to produce ion crystals [3] it was noted that the ions could form either ordered, crystalline structures or a diffuse cloud, and computations showed that the latter was an example of chaotic motion [4, 5]. The first experimental evidence for a *reproducible* transition between ordered and chaotic motion was seen in the melting of two-ion crystals [4] under conditions where the laser cooling parameters were constant. At low q, a pair of Ba^+ ions was crystallized, and the trapping voltage was slowly increased to $q \approx 0.9$ then reduced to the starting value. During this time the scattered fluorescence from the cooling laser was observed in 1 s intervals. In principle, this experiment was very similar to the melting point observations of [1]. The number of melting events versus the q-value at which melting occurred is plotted in Fig. 33.1 for one set of 29 voltage cycles, from which it is clear that melting tends to occur near $q = 0.87$. A puzzling aspect of this result [4] is that numerical solutions of Eqs. (33.1) indicate that the crystal is linearly stable for all q when $a = 0$. We therefore concluded that a perturbation of the particles' motion was required to initiate melting.

A subsequent experiment [7] showed that the perturbations responsible for melting in [4] were collisions with residual gas molecules. A single collision with a H_2 molecule disturbs the crystal enough to induce chaotic motion. The same experiment discovered that the key to the q-dependence of the order-chaos transition is the stability of the chaotic cloud over long periods of time. Although a sufficient perturbation of the crystal can induce chaotic motion at any q, the resulting cloud is only stable when q is greater than a critical value q_c. For $q < q_c$ one observes transient chaos [14], i.e., motion which is chaotic for a long time compared to the natural time scale (the trap period) but ultimately reverts to regular motion.

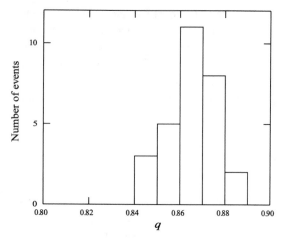

FIGURE 33.1. Onset of melting when q is slowly raised. The histogram shows the number of observed melting events vs. the value of q at which the crystal melted.

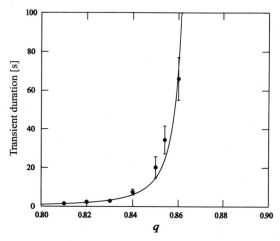

FIGURE 33.2. Measured average transient lifetime as a function of q for two trapped barium ions. The error bars are statistical only. The curve shows a fit to a power law with $q_c^{(expt)} = 0.87$ and $\gamma^{(expt)} = 2.3$.

Although the duration of an individual chaotic transient is unpredictable, the average transient lifetime has a smooth dependence on q, which was measured by perturbing the crystal and observing the ensuing chaotic transient for up to 3 minutes, corresponding to $t = 2 \times 10^9$ in the dimensionless units of Eqs. (33.1). The average chaotic lifetimes, restricted to $q \leq 0.86$ to avoid distortion of the observed lifetime due to the limited observation time, are shown in Fig. 33.2, from which one sees that the transient lifetime diverges at $q = 0.87$. The theoretical lifetime dependence will be discussed in more detail below.

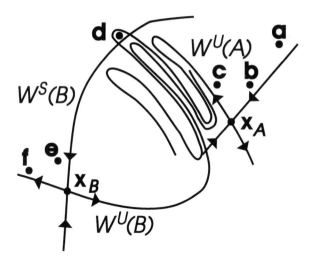

FIGURE 33.3. Schematic illustration of a heteroclinic boundary crisis for $q < q_c$ (after [15]). The curved lines represent the invariant manifolds of the fixed points $\mathbf{x}_A$ and $\mathbf{x}_B$ of a stroboscopic mapping; the points **a** – **f** depict a trajectory leading from the neighborhood of the chaotic repeller into the basin of attraction of the crystal.

33.4 Boundary Crisis

A mechanism for transient chaos which proves to be applicable to trapped-ion transitions is that of boundary crisis, elaborated by Grebogi, Ott, Yorke, and coworkers [15]. Because of the dissipation in Eqs. (33.1), phase-space volumes contract with time and trajectories ultimately tend to attractors. One regular attractor that always exists is the crystal. If there is stationary chaos, there is also a chaotic attractor built around unstable periodic orbits of saddle-like character [14]. Such orbits have both stable and unstable invariant manifolds (designated W^S and W^U), sets of points in phase space that are invariant under the equations of motion and tend to the periodic orbit in the limits $t \to +\infty$ and $t \to -\infty$, respectively. A heteroclinic boundary crisis occurs at a critical value q_c when the unstable manifold of an orbit on the attractor (orbit A) is tangent to the stable manifold of an unstable periodic orbit B on the boundary of the basin of attraction of the chaotic attractor.

It can be proved that the mediating orbits A and B must have the same period [15]. The manifolds change smoothly as q is varied, and at q_c one has a transition between two geometrically distinct situations: for $q < q_c$ $W^U(A)$ and $W^S(B)$ intersect, while for $q > q_c$ they do not. Figure 33.3 schematically illustrates the geometry for a stroboscopically sampled 2-dimensional phase space. The periodic orbits A and B are represented by fixed points $\mathbf{x}_A$ and $\mathbf{x}_B$ in the stroboscopic view, the basin boundary is $W^S(B)$, and when q decreases through q_c the segment of $W^U(A)$ that loops around the point **d**, which had been part of the chaotic attractor, protrudes into the basin of attraction of the crystal. This "leak" destroys the chaotic attractor, which becomes a fractal object called a *repeller* [14]. Orbits near the repeller may, however, remain in its vicinity for a long time before leaving, and during this time a chaotic transient is observed. The points **a** – **f** in Fig. 33.3 represent the end of a chaotic transient, showing how a point **a** initially in the chaotic region of phase space approaches x_A and is mapped, following $W^U(A)$, to **d** on the other side of the basin boundary. The trajectory then briefly visits the vicinity of x_B, and ultimately disappears in the direction of the crystal.

The boundary crisis scenario provides a precise geometrical picture for chaotic transients, in which trajectories spend a long time sampling the phase space in the neighborhood of the repeller before eventually "leaking" past the unstable periodic orbits A and B into the basin of attraction of the crystal. It also has the more practical consequence that the lengths of the individual transients are exponentially distributed with an average lifetime $T(q)$ that scales with the control parameter as a power-law [15]:

$$T(q) \propto (q_c - q)^{-\gamma} . \tag{33.2}$$

Aside from a constant of proportionality, the critical exponent γ and the parameter q_c completely describe the divergence of chaotic lifetimes near a crisis. Equation (33.2) provides a good description of the experimental data, as can be seen from the solid curve in Fig. 33.2, which is the result of a least squares fit yielding

$$q_c^{(expt)} = 0.87(1) , \tag{33.3}$$

$$\gamma^{(expt)} = 2.3(6) . \tag{33.4}$$

The scaling of $T(q)$ with q led to the suggestion that a boundary crisis could be responsible for the two-ion order-chaos transition [7]. More specific evidence of a boundary crisis is provided by numerical solutions of Eqs. (33.1).

33.5   Numerical Studies of Transient Chaos

It is straightforward to show, by choosing random initial conditions and numerically integrating Eqs. (33.1), that long, chaotic transients are indeed typical solutions. The dependence of the average transient lifetime on q, calculated by numerical simulations with a large number of different initial conditions, qualitatively reproduces the lifetime scaling seen in the ion trapping experiment. Integrating 450 simulated transients for 5×10^6 trap cycles, with $\Gamma = 3 \times 10^{-3}$ and q ranging from 0.84 to 0.88, we found average lifetimes distributed according to Eq. (33.2) with

$$q_c^{(sim)} = 0.89(1) , \tag{33.5}$$

$$\gamma^{(sim)} = 1.3(7) . \tag{33.6}$$

The critical exponent is consistent, within the large uncertainty, with the exponent derived from the trapped-ion observations, Eq. (33.4). This approach, however, is plagued by a time-scale problem that generally afflicts calculations of transient chaos, and is also well-known in numerical studies of celestial dynamics [10]. The trapping potential imposes a natural time-scale of order unity in dimensionless units, and it is necessary to take an even smaller time-step to integrate accurately the equations of motion. However, crystallization takes place on an enormously larger time scale—the data of Fig. 33.2 extend to $t \sim 10^9$. Consequently, the trajectory calculation for even a single chaotic transient requires considerable computation, and extracting the critical parameters by simulating data sets and fitting the q-dependence of the transient lifetime to Eq. (33.2) is a difficult task.

More insight into the details of the crisis is obtained from analysis of the simulated trajectories in the last few trap cycles before crystallization. As pointed out by Grebogi et al. [15], for a brief time just before a chaotic transient terminates, its trajectory may approximate the mediating orbits of the crisis. Referring again to the schematic diagram in Fig. 33.3, one sees that as the trajectory makes its way from the chaotic to the regular regions of phase space, there are some points (**b** and **c**) which lie in the neighborhood of $\mathbf{x}_A$, while others (**e** and **f**) are close to $\mathbf{x}_B$. Expressed in the language of differential equations rather than discrete mappings, Fig. 33.3 suggests that short segments of the trajectory can provide a glimpse of the underlying periodic orbits A and B, despite their instability.

This approach was applied to long chaotic transients, obtained by numerical integration of Eqs. (33.1), and allows identification of the orbits responsible for the crisis [9]. Close examination of a computed trajectory just before condensation shows how the mediating orbits "guide" the trajectory from the chaotic repeller into the basin of attraction of the crystal as in Fig. 33.3.

Having found the unstable, periodic orbits that mediate the boundary crisis, we can build a theoretical model of the order-chaos transition upon the geometry of the mediating orbits' invariant manifolds, rather than numerical simulations. For instance, the critical parameter q_c can be found from the intersection of $W^U(A)$ and $W^S(B)$, rather than by fitting transient lifetimes to a power law. This approach not only provides a more fundamental picture of the order-chaos transition than simulations alone, it also has the practical advantage of circumventing the time-scale problem referred to above. In our case the invariant manifolds of A and B are elaborately folded hypersurfaces in a four-dimensional phase space, and finding their intersection is not trivial. However, it is relatively easy to generate rays on the invariant manifolds which originate in the neighborhoods of A and B, where the motion can be linearized. For small q it was possible to construct two rays belonging to $W^U(A)$ and $W^S(B)$ and which intersect, thus establishing the fundamental premise of the boundary crisis scenario [9].

As the parameters of Eqs. (33.1) are varied, the fixed points and their manifolds move smoothly in phase space. The critical parameter can therefore be estimated by repeating the intersection search just described, raising q in small steps until the procedure fails to locate an intersection. The result

$$q_c^{(th)} = 0.869 \,, \qquad (33.7)$$

is in excellent agreement with the experimental value, Eq. (33.3). These calculations were carried out with $\Gamma = 1 \times 10^{-3}$, but the results are insensitive to the exact value of the damping parameter, provided that it is small. This was tested by holding q constant, $q = 0.863$, and varying the damping parameter over the range $\Gamma = 0$—1×10^{-3}. A manifold intersection was always found at nearly the same location in phase space. This is not surprising in light of previous studies of the stable, frequency-locked attractors [11], which showed that the orbits for finite damping transformed smoothly as $\Gamma \to 0$ to periodic solutions of the Hamiltonian equations of motion.

It has been shown [15] that for crises in two-dimensional mappings, the critical exponent γ in Eq. (33.2) can be calculated from the linearized motion near A. With an additional assumption that the points originating near the repeller which escape to the crystal lie in a band of some characteristic width around $W^U(A)$, the argument of [15] can be adapted to the phase-space geometry of our two-ion problem [9]. In the limit $\Gamma \ll 1$, we obtain

$$\gamma^{(th)} \approx 2 \, , \tag{33.8}$$

in good agreement with the fit to the experimental lifetime data, Eq. (33.4). Except for an overall scale factor, we thus have a complete, deterministic description of the lifetime divergence of chaotic transients, based on the underlying geometry of the boundary crisis.

33.6   Conclusions

We have reviewed the experimental evidence for a reproducible order-chaos transition at $q = 0.87$ for two ions in a Paul trap and shown that exact solutions of the equations of motion support the interpretation of [7] that this transition is due to a boundary crisis. The unstable periodic orbits that are central to this scenario were identified and points of intersection of the invariant manifolds $W^U(A)$ and $W^S(B)$ were located. This fulfills a central postulate of boundary crisis theory, and permits the critical parameter q_c to be predicted. A derivation of the critical exponent γ completes a theoretical description of the experimentally observed lifetime scaling. Experimental observations summarized by Eqs. (33.3) and (33.4), are in excellent agreement with the theoretical results given by Eqs. (33.7) and (33.8) which are derived entirely from exact solutions of the deterministic Eqs. (33.1), without any stochastic terms or additional nonlinearity.

References

[1] R. F. Wuerker, H. Shelton, and R. V. Langmuir, J. Appl. Phys. **30**, 342 (1959).
[2] W. Neuhauser, M. Hohenstatt, P. Toschek, and H. Dehmelt, Phys. Rev. Lett. **41**, 233 (1978).
[3] F. Diedrich, E. Peik, J. M. Chen, W. Quint, and H. Walther, Phys. Rev. Lett. **59**, 2931 (1987); D. J. Wineland, J. C. Bergquist, W. M. Itano, J. J. Bollinger, and C. H. Manney, Phys. Rev. Lett. **59**, 2935 (1987).

[4] J. Hoffnagle, R. G. DeVoe, L. Reyna, and R. G. Brewer, Phys. Rev. Lett. **61**, 255 (1988).

[5] R. Blümel *et al.*, Nature (London) **334**, 309 (1988).

[6] R. Blümel, C. Kappler, W. Quint, and H. Walther, Phys. Rev. A **40**, 808 (1989).

[7] R. G. Brewer, J. Hoffnagle, and R.G. DeVoe, Phys. Rev. Lett. **65**, 2619 (1990).

[8] J. Hoffnagle and R. G. Brewer, in the Proceedings of the 91st Nobel Symposium on Trapped Charged Particles and Related Fundamental Physics, Lysekil, Sweden, 19 – 26 August 1994, in press.

[9] J. Hoffnagle and R. G. Brewer, Phys. Rev. A **50**, 4157 (1994).

[10] J. Wisdom, Icarus **72**, 241 (1987).

[11] J. Hoffnagle and R. G. Brewer, Science **265**, 213 (1994).

[12] J. Hoffnagle and R. G. Brewer, Phys. Rev. Lett. **71**, 1828 (1993).

[13] E. Fischer, Z. Physik **156**, 1 (1959).

[14] T. Tél, in *Directions in Chaos, Vol. 3*, edited by Hao Bai-Lin (World Scientific, Singapore, 1990).

[15] C. Grebogi, E. Ott, and J. A. Yorke, Phys. Rev. Lett. **48**, 1507 (1982); Physica **7D**, 181 (1983); Phys. Rev. Lett. **57**, 1284 (1986); C. Grebogi, E. Ott, F. Romeiras, and J. A. Yorke, Phys. Rev. A **36**, 5365 (1987).

34

Infrared Emission and H_2O Masers around Massive Black Holes in Galactic Nuclei

David J. Hollenbach
and Philip R. Maloney

Professor Charles Townes has made numerous important contributions to the fields of physics and astrophysics. We thought this paper might interest him because it brings together three of his interests: masers, infrared astronomy, and massive black holes in the centers of galaxies.

34.1   Introduction

Many galaxies exhibit evidence for nonstellar sources of energy in their centers [1]. The most extreme examples of these active galactic nuclei (AGN)—quasars— reach extraordinary luminosities, exceeding 10^{47} erg s^{-1} (about 10^{13} times the luminosity of the Sun). Time-variability arguments indicate that this energy must arise in very compact regions ($\lesssim 10^{16}$ cm). The generally accepted mechanism for powering AGN is accretion of gas onto a massive black hole: a central black hole of mass $10^6 - 10^9$ solar masses ($M_\odot$) is surrounded by an orbiting accretion disk which may extend $\sim 0.1 - 1$ pc (1 pc is about 3 light-years) from the black hole. Gaseous material spirals through this disk onto the black hole at rates $\dot{M} \sim 10^{-4} - 1$ $M_\odot$ yr^{-1}, and a fraction $\epsilon \sim 0.1$ of the gravitational rest-mass energy is converted to radiation [2].

The resultant luminosities are enormous. Even the relatively common and nearby Seyfert galaxies—AGN in disk galaxies—typically radiate $L \sim 10^{44}$ erg s^{-1}. The AGN generally radiate a power-law spectrum extending far into X-ray wavelengths [3], and a significant fraction of their total photon luminosities are radiated between $1 - 10$ keV.

Most of the luminosity is produced near the black hole. There is considerable evidence, especially for the lower-luminosity AGN (Seyfert galaxies) that there is a distribution of obscuring gas occupying a considerable solid angle around the nucleus [4]; this gas can even be optically thick to hard X-rays [5, 6, 7]. We

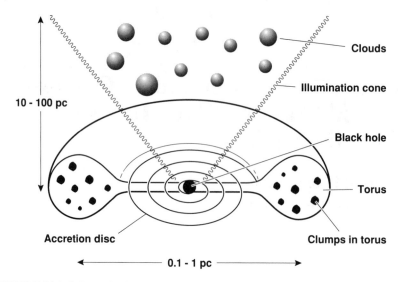

FIGURE 34.1. Schematic diagram (not to scale) of the black hole/torus model of AGN.

shall refer to this obscuring material as a "torus", even though its precise geometry and nature is still uncertain [8, 9]. For example, the torus may consist of orbiting clumps of gas and dust. The torus attenuates all but the highest energy X-rays so that photons mostly emerge in the polar directions, which we call the "illumination cone". Any gas clouds in the illumination cones are subjected to the strong intrinsic flux from the AGN [10]. Observations show the plane of the torus is not generally aligned with the plane of the host galaxy, so that clouds in the galactic disk may fall within the AGN illumination cone [4]. The spatial resolution of existing (non-interferometric) telescopes is such that spatially resolved emitting regions in the illumination cones of nearby Seyferts are $\sim 10 - 100$ pc from the AGN [10]. Figure 34.1 provides a schematic diagram of this model.

Although this paradigm is appealing, there is still much to be proven and there are many mysteries to be solved. For example, the mechanisms which feed material from the clouds to the torus or disk are uncertain, and a definitive proof of the existence of the hypothesized massive central black hole has eluded us. In addition, some Seyfert nuclei exhibit extremely luminous H_2O maser emission, called "megamasers" because they are roughly 10^6 times as luminous as typical interstellar H_2O masers in our own Milky Way Galaxy [11–13]. The origin and excitation mechanism for the megamasers have been controversial [12, 14]. In this paper we discuss the infrared spectrum produced by X-ray-illuminated neutral gas and show how theoretical models compared to the observations can diagnose the mass, dynamics, temperature and morphology of the neutral gas around the black hole. We also discuss the origin of H_2O megamasers and the case for a massive black hole in the Seyfert galaxy NGC 4258.

34.2 The Physical Model

A large number of physical and chemical processes are important in dense, neutral gas illuminated by X-rays. In this section we outline the important physics and describe the physical model. We will model both the clouds in the illumination cone at $\sim 10 - 100$ pc and the opaque torus or disk thought to exist at $\sim 0.1 - 1$ pc. The density of the gas in the former is thought to lie in the range $10^3 - 10^6$ (hydrogen nuclei) cm^{-3} [1]; we adopt $n = 10^5$ cm^{-3} as our standard density. The density of the inner torus is thought to be much higher, and we assume densities of order $n \sim 10^{8-9}$ cm^{-3}, or thermal pressures of order $nT \sim 10^{11}$ cm^{-3} K in the torus [15]. We assume standard cosmic abundances for the elements other than hydrogen [16].

The dense neutral gas is illuminated by the incident X-ray flux from the AGN, which we take to be a power law in the X-ray regime,

$$F_X = F_0 E^{-\alpha}, \tag{34.1}$$

where $\alpha \sim 0.5 - 1$ and Eq. (34.1) holds for photon energies $E < E_{max} \sim 100$ keV. There is considerable observational uncertainty [3] as to whether the X-ray spectrum cuts off above $E \sim 100$ keV, but this has very little effect on the models (see below).

The X-ray photoelectric absorption cross-section per hydrogen nucleus of the gas and dust in the clouds is roughly a power law with energy,

$$\sigma_X(E) \sim \sigma_0 E^{-\gamma}, \tag{34.2}$$

where $\gamma \sim 3$ below 0.5 keV and ~ 2.7 above 0.5 keV [16]. The cross-section is dominated by helium for $E < 0.5$ keV, and by heavier elements such as carbon, oxygen, and iron at higher photon energies. Above $E \sim 10$ keV Compton scattering dominates over photoelectric absorption; the cross-section declines above 100 keV due to Klein-Nishina corrections, so that photons with $E > 100$ keV are unimportant [17].

The equilibrium chemical state and temperature of a parcel of dense gas illuminated by X-rays depends largely on the ratio H_X/n, where H_X is the X-ray energy absorbed per unit time per hydrogen nucleus. The chemical destruction rates (via ionization) and heating rates per unit volume are proportional to nH_X, whereas the chemical formation rates and cooling rates are generally proportional to n^2 times a rate coefficient. The value of H_X at a column depth N into the neutral gas is found by integrating the attenuated incident flux at N times the cross-section (Eq. (34.2)) over photon energy. H_X is dominated by X-ray photons with energies E_1 such that $\tau(E_1) = N\sigma(E_1) \sim 1$. The lower energy photons ($\tau >> 1$) have already been absorbed at smaller columns, and the higher energy photons are transmitted with relatively little absorption. Assuming equal energy per decade in hard X-rays ($\alpha = 1$, as is approximately true for AGN) and a total $1 - 100$ keV X-ray luminosity of $L_X = 10^{44} L_{44}$ erg s^{-1},

$$H_X \sim 7 \times 10^{-22} L_{44} r_{100}^{-2} N_{22}^{-1} \text{erg s}^{-1}, \tag{34.3}$$

where $10^{22}N_{22}$ cm^{-2} is the total hydrogen column density and $100r_{100}$ pc is the distance of the neutral gas from the AGN. Note the decrease in H_X with N caused by the attenuation of the lower energy X-rays. For clouds at ~ 100 pc, the typical column densities are likely $N_{22} \sim 1 - 10$ and the typical absorbed X-ray energy $E_1 \sim 2$ keV. For the torus, $N_{22} \sim 100$ and $E_1 \sim 10$ keV.

Although H_X is the rate at which X-ray energy is absorbed per hydrogen, it is not necessarily the gas *heating* rate per hydrogen, since (at low ionization fractions) much of the absorbed energy goes into ionization and the collisional excitation of atoms and molecules [18–20]. However, the heating rate is proportional to H_X, and, in the neutral gas which we consider, the heating rate is about $0.3 - 0.4H_X$ [21].

The total ionization rate per hydrogen ζ_T is also proportional to H_X. The absorbed X-rays produce primary photoelectrons with energies of order of the photon energy E_1 (i.e., $1 - 10$ keV). The primary electron produces about 30 secondary ionizations per keV of primary energy if the ionization fraction of the neutral gas is low ($< 1\%$), as is typically the case. In this limit, the ionization rate becomes

$$\zeta_T \sim 1.4 \times 10^{-11} L_{44} r_{100}^{-2} N_{22}^{-1} \text{ s}^{-1} . \tag{34.4}$$

For clouds within ~ 100 pc and for the nearer but more opaque torus, both the heating rates and the ionization rates are at least 10^6 times larger than the cosmic ray rates in solar neighborhood molecular clouds [22], and therefore have profound effects on the chemistry and temperature of these regions.

The chemistry adopted in our model includes the major reactions of the ions, atoms and simple molecules formed by the elements H, C and O [21]. Most notable are the species H_2, C, C$^+$, CO, O, OH and H_2O, since these species generally dominate the cooling of the gas; their radiative transitions therefore provide the best observational diagnostics of the physical conditions. Essential ingredients in understanding the chemistry in X-ray illuminated regions are the neutral-neutral reactions $H_2 + O \rightarrow OH + H$ and $H_2 + OH \rightarrow H_2O + H$. These reactions are not important at low temperatures ($T < 300$ K) because of activation energy barriers, but produce copious quantities of water in gas which has been heated to $T > 300$ K [23].

34.3   Results: Infrared Emission from Clouds

To date, two species have been the primary diagnostics for dense neutral clouds near AGN: H_2 and Fe$^+$. The 2 μm vibrational transitions of H_2, e.g., $v = 1 - 0$ S(1) and $v = 2 - 1$ S(1), and the [FeII] 1.26 and 1.64 μm transitions have been observed in a number of Seyfert galaxies [24]. Fe$^+$ is thought to trace predominantly neutral hydrogen gas since iron is generally ionized to Fe^{++} or higher ionization states in regions of fully ionized hydrogen. Recent observations show that the H_2 emission is extended on 100 pc scales (e.g., in NGC 1068 [25]), and that the Fe$^+$ emission is often more extended than the H_2, but also shows larger velocity linewidths, suggesting that much of the Fe$^+$ emission arises closer to the AGN, where the

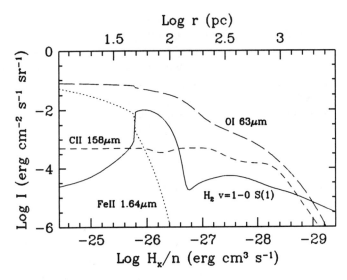

FIGURE 34.2. The infrared intensity of four transitions emerging from the standard (see text) X-ray-illuminated cloud. The distance r of the standard cloud from the standard AGN is given on the top.

orbital velocities are higher [25, 26]. Shocks have often been invoked to explain the H_2 and [FeII] emission near AGN [24], but we propose that X-ray heated clouds produce the emission.

Figure 34.2 shows the results of our model for clouds of density $n = 10^5$ cm^{-3} and column density $N = 10^{22}$ cm^{-2} exposed to a central source of $1 - 100$ keV luminosity $L = 10^{44}$ erg s^{-1}. We plot the model intensities of H_2 and [FeII] 1.64 μm, and of the potentially observable fine-structure lines [CII] 158 μm and [OI] 63 μm. The figure shows that the clouds close to the source are bright in [FeII], while the H_2 intensity peaks at $\sim 50 - 100$ pc from the source, in agreement with the observations. At relatively high values of H_X/n, nearer the AGN, the H_2 abundance is very low $(<< 10^{-3})$, and the atomic gas is heated to 10^4 K, leading to collisional excitation of the [FeII] lines (whose upper states lie $\sim 10,000$ K above the ground state). At intermediate values of H_X/n, the H_2 abundance rises to $\sim 10^{-2}$, the gas temperature $T \sim 1000 - 3000$ K, and collisional excitation of the transitions of H_2 at 2 μm provides the dominant cooling. The gas is now too cool to appreciably excite the [FeII] 1.64 μm line. At larger distances from the AGN, hence at lower values of H_X/n, the gas becomes fully molecular but the temperatures are too low ($T < 1000$ K) to excite either H_2 or [FeII]. The *observed* H_2 1-0 S(1) and [FeII] line intensities in NGC 1068, the nearest and therefore best-studied luminous Seyfert, averaged over 1″ beams (~ 100 pc at the distance of NGC 1068), peak at $\sim 2 \times 10^{-3}$ erg cm^{-2} s^{-1} sr^{-1} but are more typically $0.25 - 0.3$ times that value [25]. The predicted line intensities of clouds in our models indicate that such clouds need only fill $\sim 5\%$ of the beam in order to produce the observed intensities.

Figure 34.2 also shows the model predictions for [CII] 158 μm and [OI] 63 μm intensities from the clouds. If the beam filling factors are similar to those for H_2 and [FeII], then the predicted fluxes should be sufficient for observation with the Infrared Space Observatory (ISO) or the Kuiper Airborne Observatory (KAO). The higher spatial resolution of the planned Stratospheric Observatory for Infrared Astronomy (SOFIA) would be extremely useful in discriminating the [CII] and [OI] fluxes produced by X-ray illuminated clouds from the fluxes produced by clouds in star-forming regions far from the AGN.

34.4 Results: H_2O Megamasers and the Massive Black Hole in NGC 4258

Extremely powerful 22 GHz water maser emission has been discovered over the past ten years in a number of extragalactic sources. Of the 13 known megamasers with apparent (isotropic) maser luminosities larger than 35 solar luminosities ($L_\odot$), all are in galaxies with AGN [13]. Two of the best studied are NGC 1068 and the low-luminosity AGN NGC 4258. Interferometer observations have shown that the maser emission comes from regions $\lesssim$ 1 pc from the AGN [12]. A recent spectacular VLBI (Very Long Baseline Interferometry) observation of NGC 4258 shows [27] that the masers arise in a thin Keplerian ($v \propto r^{-1/2}$) disk which extends $0.13 - 0.25$ pc from the AGN, orbiting a central mass of 3.6×10^7 $M_\odot$. The volume-averaged mass density in the region interior to 0.13 pc is 4×10^9 $M_\odot$ pc^{-3}, more than 40 times that of any other black hole candidate previously observed [27]. It is unlikely that a star cluster could have such a high density; for example, if the stars were solar-type stars, they would collide with each other in less than 10^8 years, which is much less than the age ($\sim 10^{10}$ years) of the host galaxy. Furthermore, the observed rotation curve (within the errors) is so close to being perfectly Keplerian (i.e., as expected for a point mass) that any finite-size mass distribution (such as a star cluster) would have to have an extraordinarily high central density. This observation therefore provides the strongest dynamical evidence to date for a massive black hole in the center of a galaxy.

We have applied our model for X-ray illuminated gas to the dense tori or disks hypothesized to exist around AGN, and, in collaboration with David Neufeld (Johns Hopkins University), have modeled the water maser emission [28]. We find the rather robust conclusion that a thin zone of H_2O maser emission is produced inside the tori, over a wide-range of gas pressures. At a critical pressure which depends on the incident X-ray flux and the total column density to the X-ray source, the gas undergoes a phase transition, from warm ($T \sim 8000$ K) and atomic, to molecular at $T \sim 400$ K, with nearly all oxygen not in CO incorporated into H_2O. The maser transition is excited by collisions followed by radiative decay to the upper state of the masing transition. High gas temperatures are required since the upper level lies ~ 600 K above the ground state of the H_2O molecule. The population of the masing levels become inverted because neither the radiation field or collisions are sufficient to produce local thermal equilibrium (LTE) conditions. With increasing depth beyond the transition region the H_2O rotational transitions become so optically thick that the level populations are driven to LTE and the maser inversion is quenched. Our model maser luminosities are $10^{2\pm0.5}$ $L_\odot$ per pc^2 of illuminated area, sufficient to explain the H_2O megamasers [13].

The maser models can also be used to derive important parameters of the accretion disk in NGC 4258. In standard accretion disk theory [29], the pressure within the accretion disk is related to the mass accretion rate. By comparing the disk pressure with the critical pressure for producing a molecular/masing region, Neufeld and Maloney [30] have shown that the mass accretion rate in NGC 4258 is $\dot{M} \sim 10^{-4}$ $M_\odot$ yr^{-1}, and that the observed luminosity of the AGN requires that the efficiency $\epsilon \sim 0.1$, is in remarkable agreement with theoretical expectations [2, 29].

34.5   Conclusion

Although optical lines from ionized gas have long been used to probe the environment around AGN, we show how the neutral gas, which likely makes up the bulk of the material, can be probed with infrared and radio (masing) transitions. We have provided a self-consistent explanation for the origin of the H_2 $v = 1 - 0$ S(1), [FeII] 1.64 μm, and H_2O 22 GHz maser emission which has been observed in the last decade, and make predictions on [CII] 158 μm and [OI] 63 μm emission from the neutral clouds which lie within a few hundred pc from the AGN. We have also shown how our models can be used in conjunction with observations of water masers in AGN to probe the physical conditions in accretion disks around massive black holes.

34.6   Acknowledgments

We are grateful to David Neufeld for allowing us to discuss our collaborative work. DJH was supported by NASA RTOP 399-20-10. PRM was supported by NASA grant NAGW-766 to the University of Colorado.

References

[1] H. Netzer, in Active Galactic Nuclei, ed. T.J.-L. Courvoisier and M. Mayor (Berlin, Springer-Verlag, 1990) p. 57.

[2] R. D. Blandford, in Courvoisier and Mayor, *op. cit.*, p. 161

[3] R. F. Mushotzky, C. Done and K. A. Pounds, ARAA **31**, 717 (1993).

[4] R. Antonucci, ARAA **31**, 473 (1993).

[5] H. Awaki, K. Koyama, H. Inoue, and J. P. Halpern, PASJ **43**, 195 (1991).

[6] J. S. Mulchaey, R. F. Mushotzky, and K. A. Weaver, ApJ (Letters) **390**, L67 (1992).

[7] K. Iwasawa *et al.*, ApJ **409**, 155 (1993).

[8] J. S. Miller and R. W. Goodrich, ApJ **355**, 456 (1990).

[9] A. Lawrence, M.N.R.A.S. **252**, 586 (1991).

[10] I. N. Evans, Z. Tsvetanov, G. A. Kriss, H. C. Ford, S. Caganoff, and A. P. Koratkar, ApJ **417**, 82 (1993).

[11] M. J. Claussen, G. M. Heiligman, and K.-Y. Lo, Nature **278**, 34 (1984).

[12] M. J. Claussen and K.-Y. Lo, ApJ **308**, 592 (1986).

[13] J. A. Braatz, A. S. Wilson, and C. Henkel, ApJ (Letters) **437**, L99 (1994).

[14] A. D. Haschick and W. A. Baan, ApJ (Letters) **355**, L23 (1990).

[15] J. H. Krolik and M. C. Begelman, ApJ **329**, 702 (1988).

[16] R. Morrison and D. McCammon, ApJ **270**, 119 (1983).

[17] J. H. Krolik and S. Lepp, ApJ **347**, 179 (1989).

[18] J. M. Shull and M. E. van Steenberg, ApJ **298**, 268 (1985).

[19] Y. Xu and R. McCray, ApJ **375**, 190 (1991).

[20] G. M. Voit, ApJ **377**, 158 (1991).

[21] P. R. Maloney, D. J. Hollenbach, and A.G.G.M. Tielens, ApJ, submitted (1995).

[22] J. H. Black, in *Interstellar Processes*, ed. D. J. Hollenbach and H. A. Thronson (Dordrecht, Reidel, 1987), p. 731.

[23] D. A. Neufeld and A. Dalgarno, ApJ **340**, 869 (1989).

[24] A. F. M. Moorwood and E. Oliva, A&A **203**, 278 (1988).

[25] M. Blietz *et al.*, ApJ **421**, 92 (1994).

[26] E. Oliva and A. F. M. Moorwood, ApJ (Letters) **348**, L5 (1990).

[27] M. Miyoshi, J. Moran, J. Herrnstein, L. Greenhill, N. Nakai, P. Diamond, and M. Inoue, Nature **373**, 127 (1995).

[28] D. A. Neufeld, P. R. Maloney, and S. Conger, ApJ (Letters) **436**, L127 (1994).

[29] J. Frank, A. King, and D. Raine, Accretion Power in Astrophysics (Cambridge, Cambridge University Press, 1992) p. 62.

[30] D. A. Neufeld and P. R. Maloney, ApJ (Letters), in press (1995).

35

Knowing Charlie: In the 1950s and Since

Ali Javan

It is hard for me to believe that I first met Charlie forty-five years ago, in the spring of 1950. It was in his office on the tenth floor of the Pupin Building at Columbia University. In my recollection of that first meeting with him, it seems as if it occurred just yesterday. And just as fresh are my remembrances of him throughout the years I have known him since: first as my mentor and thesis advisor, and then as a colleague and dear friend. I had just come to the United States some months earlier, and had enrolled at Columbia to do graduate work in physics with a passion; my English was still rusty. When I went to see him in his office, I asked a close friend to come along to act as a translator in the event that I needed help—other than my native Persian (Iranian), I speak French as my second language. As it turned out Charlie insisted in conversing in French. Although his French was just as shaky as was my English, we did fine. I was in search of the right kinds of physics to put my hands on, although I had yet to matriculate as a graduate student. He spoke of his work with microwaves and molecules; I still remember his words. He asked that I look him up after my graduate entrance was completed—at that time I had only applied for admission to Columbia. And I did look him up, in the Spring of the following year.

Five years had passed since the end of the war, an end that had marked a new beginning in everything. Physics had branched off in directions towards both ends of the energy spectrum, from upwards of hundreds of millions of electron-volts in high-energy physics to downwards of milli- and micro-electron-volts in the fine details of atomic and molecular energy spectra. With an affection for molecules, Charlie had chosen the latter. Atomic and molecular optical spectra had already done their precious things for physics, yielding discoveries in quantum physics through the study of the interaction of matter and radiation. Now it was time for radio and microwave spectra to show what new physics was held in store for us.

In the few short years during the war, microwave technology had come from mere infancy to maturity with true perfection—all as a result of the war effort on radar. To quote I. I. Rabi: "Without the bomb we could have still won the war; without radar we would have lost it." The entire research-and-development work on radar during the war was overseen by the National Defense Research

Committee (NDRC). The bulk of that work was done by the university people and engineers at the M.I.T. Radiation Laboratory for Electronics (as well as at other universities and in industry). Within weeks after the war's end, the NDRC commissioned a six-month project for a select group of people to write up, in book format, the entire technical work on radar—on what had been until then a highly classified project. That monumental undertaking of about 21,000 printed pages in twenty-six volumes, the M.I.T. Radiation Laboratory Series, was first published by McGraw-Hill in 1948. For decades afterwards, well through the 1970s, that entire twenty-six volumes served as the electronics bible for so many of us. Boxes full of K-and X-band klystrons, magnetrons, radio-frequency (RF) detection electronics, microwave plumbing, etc., were laying around on the 10th and the 11th floors of the Pupin Building at Columbia University where I did my graduate research work.

With each discovery, the excitement in physics went through peaks so familiar to all of us. Radio frequency (RF) and microwave techniques took over physics like thunder right after the war, and one can well imagine the excitement: nuclear magnetic resonance discoveries, atomic beam experiments in Rabi's style, the discovery of the anomalous electron g-value, the Lamb-Retherford measurements of the hydrogen $2S_{1/2} - 2P_{1/2}$ QED shift, the observation of the hydrogen 21-centimeter waves from outer space, all took place in a few short years (all in the 1940s) right after the war—as did Charlie's work at Bell Telephone Laboratories on the first observation of high-resolution microwave spectra of molecules in absorption. With these on the center-stage of physics, optics faded into the past.

Microwaves provided the capability of making precision measurements at scales unheard of in optics. Knowing this, everyone had his eyes on making an atomic microwave clock with an accuracy better than one part in 10^8, the accuracy limit last reached in the 1930s by the pendulum clock. The wartime's crystal-controlled oscillator clocks could hardly hold an accuracy anywhere close to that limit. The possibility of a microwave atomic (or molecular) clock challenged everyone's imagination.

While still at the Bell Telephone Laboratories, rumor had it that Charlie had given a talk at a conference in Brookhaven in 1946 on his work on the high-resolution K-band absorption spectrum of NH_3 molecules, proposing it as a candidate for a molecular clock reference-servo. At the conference Rabi severely questioned the idea—he had his eyes on a clock servo using the X-band cesium atomic beam hyperfine transition, which had just then been observed by Roberts and others at M.I.T. In a short while after that meeting, when Rabi returned to Columbia, he extended an offer to Charlie to join the Columbia physics faculty: Charlie accepted. Incidentally, a few years later Harold Lyons at the National Bureau of Standards (NBS) (now the National Institute of Science and Technology (NIST)) made an NH_3 clock with an accuracy better than one part in 10^8. There is a nice photograph of him sitting in front of the clock's control panel with a big round wall-clock in the top of the panel with a K-band waveguide wrapped around it—a waveguide full of the NH_3 molecules that provided the K-band absorption resonance for the clock reference-servo—Charlie's first clock idea before his maser. The Brookhaven meeting story, whether true or not, was what was told by people around the Columbia Radiation Laboratory in the early 1950s when I did my graduate research. Charlie can authenticate it.

I did my dissertation research on the microwave spectra of a series of molecules. My research at the time extended considerably beyond what I did for my thesis—I was all over the scene in physics. I had a wonderful time in Charlie's laboratory, a story all by itself. I did the bulk of my thesis research experiments on the 10th floor of the Pupin Building, and finished and graduated in 1954.

Also in 1954, for the first time in history, with his ammonia beam maser, Charlie extracted monochromatic radiation from a molecule by the stimulated emission effect in his successful experiment with Gordon and Zeiger, on the 11th floor of the Pupin Building. The rest is history.

Charlie did so very much for physics then, and has done so very much for physics since. We can be grateful to him for all his creative work: first in microwave spectroscopy, then in maser and laser research, in optics, in astrophysics, and in everything else he has done for physics—for what he has done for us all.

I gave a talk some time ago at the Conference on Lasers and Electro-Optics (CLEO) in Baltimore on the 35th birthday celebration of my work on the gas laser, a celebration organized by the Optical Society of America. My talk was titled: "Birth of the Gas Laser: How Did It Come About?" I always looked for and did the next thing beyond what had been thought of and discovered. That was how my kind of gas laser came about: a different kind from the optically-pumped laser, a different invention, a continuous-wave (cw) laser: a monochromatic laser. I told the story in my talk. The Optical Society has a video tape of it.

Accurate clocks have always challenged many of us. Within days following the moment that that monochromatic beam of light emanated for the first time from my helium-neon gas laser—at 4:20 PM on December 12, 1960—it had snowed heavily that day—I began thinking of clocks operating at optical frequencies. For me the challenge extended far beyond generating an accurate optical-resonance reference for the clock's servo loop. Once stabilized with the servo the clock frequency had to be transferred to the RF region to be referenced against the RF time standard. A clock is a clock only if its frequency can be measured referenced to the time interval defined by the standard second: the clock's absolute frequency has to be measurable. This part of the challenge for the microwave atomic or molecular clocks had long been well-resolved as a result of the war effort on radar. This effort had already provided the frequency up-conversion and down-conversion technology needed for the absolute frequency measurement of microwaves, but such techniques were unheard of in optics. I gave my first talk on the operation of my helium-neon gas laser as a last-minute post-deadline talk at an American Physical Society meeting at the New Yorker Hotel in New York City on January 29, 1961. At a luncheon after my talk with Charlie, Ed Purcell, Nico Bloembergen, and Norman Ramsey, thinking out loud I said, "Wouldn't it be nice to measure the frequency of this thing?" (referring to my He-Ne 1.15 μm laser—at 3×10^{14} Hz). Everyone laughed—kind of nervously, now that I recollect. We had lunch around a round table at the New Yorker Hotel.

My own early work on stable gas lasers and sub-Doppler laser spectroscopy set the pace for high-resolution laser spectroscopy, the search for accurate high-Q sub-Doppler optical resonances for optical-clock servo-loops, an area to which so many people have since contributed so creatively. But I take pride as well in my work on introducing the technology that made possible the absolute frequency measurement of clocks operating at optical frequencies. Without this technology there would have been no clocks operating at optical frequencies; there would have been no laser measurements of the speed of light, which have since unified the time and length standards.

The night before the day I gave my talk at the CLEO conference, we celebrated Charlie's 80th Birthday. Charlie, you have earned and deserved every recognition that has come your way. You have earned your place in history. I have witnessed how you did your work. You did it graciously. I will always treasure your friendship.

36

The SH Radical: Laboratory Detection of its $J = \frac{3}{2} \leftarrow \frac{1}{2}$ Rotational Transition

Egbert Klisch
Thomas Klaus
Sergei P. Belov
Gispert Winnewisser
and Eric Herbst

36.1   Introduction

The terahertz region has remained one of the last frontiers of the electromagnetic spectrum to be fully exploited by spectroscopic investigations because of the various associated technological difficulties. High-resolution, broad-band scanning spectroscopy in the terahertz region has recently been achieved for the first time with microwave accuracy in the Cologne spectroscopy laboratory [1]. The successful frequency and phase stabilisation [2] of continuously tunable backward wave oscillators (BWOs) produced by the ISTOK Research and Production Company, Fryazino, Moscow region, was essential for this technological breakthrough. The unique feature of these Russian-produced BWOs is the ease with which they can be tuned over wide-frequency ranges, typically about 200 GHz at 1 THz, as has been amply demonstrated in an exemplary way by the recording of the rotational spectra of HSSH [1], the dimer of SH [4], and its isotopomers HSSD and DSSD [1]. So far, high resolution spectroscopy in the terahertz region has been achieved by tunable far-IR laser sideband techniques [5], but with known limitations in continuous broadband scanning capability.

The terahertz frequency region holds great promise both for laboratory and interstellar research alike. In the laboratory, this new terahertz technique of broadband scanning spectroscopy with microwave accuracy opens the doors for solving a variety of long-standing spectroscopic problems, essentially untouched so far. Among these are:

1. The measurement of the pure rotational spectra of light hydrides, molecular radicals, and ions, such as:
   SH, OH, CH, ..., H_2S, H_2O, H_2C, ..., and many others.

2. The high-precision recording of higher rotational transitions of molecules with intermediate masses, such as:
SO_2, HCN, HCCCN, CH_3CN, CH_3OH, ..., and many more. It includes measurements of their various isotopomeric forms as well as measurements in vibrationally excited states, particulary for those of astrophysical interest.

3. The measurement of low-frequency bending motions of species such as carbon chain molecules, which are of particular astrophysical importance. It also includes the direct observation of low-lying bending vibrations of weakly bound van der Waals complexes. Both areas have hardly been investigated at all by spectroscopic techniques.

The broadband scanning technique based on BWOs, particularly when combined with frequency multiplication, is well-suited to laboratory spectroscopy probably up to 3 THz. It may also provide a useful alternative to the notorious local oscillator problem encountered in all heterodyne high-frequency receivers for use in radiotelescopes. The first laboratory results of frequency tripling, using a French-made BWO as pump source operating at a fundamental frequency near at 220 GHz, produced an output power of about 300 μW. It has been used for measuring the $^rQ_3$ branch of HSSD at 656 GHz [6].

In the present study we report the laboratory detection of the $J = \frac{3}{2} \leftarrow \frac{1}{2}$ pure rotational transition of the SH radical in its $^2\Pi_{1/2}$ electronic state. The search for the existence of the SH radical has a long history, and dates back more than 30 years to our early attempts at Duke University in Prof. Walter Gordy's microwave and millimeter wave laboratory. At that time we aimed at detecting the pure Λ-type spectrum of SH. The search was triggered by the discovery of the Λ-type spectrum of the OH radical, and its appropriate theoretical interpretation by Dousmanis, Sanders, and Townes [7]. With the recording of the first absorption lines produced by the H_2S discharge, we thought that the carrier of the observed transitions was SH [3], but it turned out instead that we had discovered the pure rotational spectrum of its dimer, disulfane or HSSH [3, 4]. Because of both the limited sensitivity of the millimeter wave spectrometer at that time and the low yield of the discharge arrangement, we now know that the search was premature and thus ill-fated.

36.2   Experimental

An essential component of the Cologne terahertz spectrometer is the high frequency BWO supplied by the ISTOK Research and Production Company, Fryazino (Moscow region). Although these tubes have been operating in Russia for many years, e.g. in the Nizhnii Novgorod laboratories of the Institute of Applied Physics [8], it was the first successful frequency and phase stabilisation achieved in 1994 in the Cologne laboratories which marked the beginning of high-precision broadband spectroscopy in the terahertz region [9]. Other essential components in the Cologne terahertz spectrometer are: a newly designed multiplier-mixer with low noise HEMT amplifier circuitry, together with two precision-tunable millimeter wave synthesizers, covering the frequency regions 78 to 118 GHz and 118 to 178 GHz. These two sweeper units are also based on frequency and phase-stabilized BWOs and were supplied by the Institute of Electronic Measurement, KVARZ, Nizhnii Novgorod. The power output of the high-frequency BWO OB-82 is split two-fold: a small fraction is used to drive the mixer system fed by the KVARZ frequency synthesizer of the 78-118 GHz region, whereas the major portion of the BWO radiation, typically 1-3 mW, is focused through a free-space absorption cell and then detected with the He-cooled InSb hot-electron bolometer.

The free-space absorption cell is equipped with a DC-discharge unit. The entire absorption cell can be cooled by circulating liquid nitrogen through a copper jacket attached to the cell walls. The SH rotational spectrum was detected by discharging H_2S together with H_2 and He. Simultaneously the dimer disulfane, HSSH, was observed as monitored through its $^rQ_3$ branch. It is interesting to note that in a discharge of pure H_2S an additional radical, HSS, has been recently detected by Yamamoto and Saito [10] and Ashworth et al. [11]. Thus the discharge of H_2S leads to a number of reaction products, and although three have been detected now, i.e., SH, SSH, and HSSH, at present, little is known about their branching ratios and how their relative abundance might depend on the detailed experimental conditions of the discharge.

36.3   Theory and Analysis

Our new measurements of the $J = \frac{3}{2} \leftarrow \frac{1}{2}$ transition were used together with the Λ-type doubling data of Meerts and Dymanus [12, 13] and the recent far-IR-Fourier-Transform pure rotational transitions of Morino and Kawaguchi [14] to determine the appropriate molecular parameters of the effective Hamiltonian

$$H_{eff} = H_{SO} + H_{SR} + H_{Rot} + H_{CD} + H_{\Lambda D} + H_{Hfs}, \qquad (36.1)$$

where the contributions refer to spin-orbit- and spin-rotation interactions, rotation and its centrifugal distortions, Λ-type doubling and hyperfine interactions, respectively.

Earlier predictions based on this Hamiltonian made prior to our new observations were within 3 MHz of the measured transition frequencies, and are in agreement with the recent laser magnetic resonance (LMR) measurements and predictions (Table VII) by Ashworth and Brown [15]. For further references on the earlier work on SH, the reader is referred to [14, 15].

The calculation of the first pure rotational transition of SH ($X^2\Pi$) including hyperfine structure is based on Hund's case (a) coupling scheme. Basis functions corresponding to this coupling scheme of a definite parity are presented by Zare *et al.* [16] as:

$$| n^2\Pi_{|\Omega|}vJ\pm\rangle = \frac{1}{\sqrt{2}}[| nvJ\Omega S\Lambda\Sigma\rangle$$
$$\pm(-1)^{J-S} | nvJ(-\Omega)S(-\Lambda)(-\Sigma)\rangle], \quad (36.2)$$

where n and v represent different electronic and vibrational states and will be neglected in the following discussion, while Λ, Σ and Ω are, respectively, the projections of **L**, **S** and **J** on the molecular axis which are appropriate to Hund's case (a) representation. To consider the effect of the nuclear spin $I = \frac{1}{2}$ of the ^{1}H atom in the coupling scheme $\mathbf{F} = \mathbf{J} + \mathbf{I}$, where **F** is the total angular momentum of the radical, the parity basis functions above have to be coupled with the nuclear spin functions.

Through the calculation of the matrix-elements $\langle {}^2\Pi_{|\Omega'|}J'IF\pm \, | \, \hat{H} \, |\, {}^2\Pi_{|\Omega|}JIF\pm\rangle$, the Hamiltonian representation

$$\begin{pmatrix} \langle {}^2\Pi_{1/2}J'\frac{1}{2}F\pm | \hat{H} |{}^2\Pi_{1/2}J\frac{1}{2}F\pm\rangle & \langle {}^2\Pi_{3/2}J'\frac{1}{2}F\pm | \hat{H} |{}^2\Pi_{1/2}J\frac{1}{2}F\pm\rangle \\ \langle {}^2\Pi_{1/2}J'\frac{1}{2}F\pm | \hat{H} |{}^2\Pi_{3/2}J\frac{1}{2}F\pm\rangle & \langle {}^2\Pi_{3/2}J'\frac{1}{2}F\pm | \hat{H} |{}^2\Pi_{3/2}J\frac{1}{2}F\pm\rangle \end{pmatrix}$$
$$(36.3)$$

is obtained. The individual matrix elements are given in the Appendix. Terms representing all six contributions to H_{eff} have been included, with centrifugal distortion contributions up to J^{10}. The Hamiltonian matrix is numerically diagonalized for the least squares fitting procedure and appropriate frequency predictions.

The newly-observed transitions are listed in Table 1 along with relative intensities and residuals (observed-calculated frequencies). An energy level diagram pertaining to the $^2\Pi_{1/2}$ state is shown in Fig. 36.1, including the transitions with their relative statistical weights as given in Table 1. Since the $^2\Pi_{1/2}$ state is about 375 cm^{-1} above the $^2\Pi_{3/2}$ as indicated in Fig. 36.1, SH has an inverted $^2\Pi$ ground state, as does OH.

The recorded SH spectrum is shown in Fig. 36.2. Both, the $e \leftarrow e$ and the $f \leftarrow f$ transitions split each into three hyperfine components, the relative intensities of which can be calculated from the known relationship

$$I_{FF'} = \frac{(2F+1)(2F'+1)}{2I+1} \left\{ \begin{matrix} J' & F' & I \\ F & J & 1 \end{matrix} \right\}^2 \quad (36.4)$$

as 5:2:1. The hyperfine structure of the two e-e transitions $F = 2 \leftarrow 1, F = 1 \leftarrow 1$ could not be resolved experimentally, because the difference in energy between the two upper hyperfine components $F = 2$ and $F = 1$ is only 1.250 MHz, which is within the Doppler width of around 1.4 MHz. For the higher J transitions the observed hyperfine structure of those transitions which carry most intensity collapses and one observes the pure Λ-doublet structure of the rotational spectrum of SH.

By applying a simultaneous nonlinear least-squares fit of our new data together with the data of Meerts and Dymanus [12, 13] and Morino and Kawaguchi [14], we obtained the molecular constants summarized in Table 2. It should be noticed that

TABLE 36.1. Measured transition frequencies of the $J = \frac{3}{2} \leftarrow \frac{1}{2}$ transition in the $X^2\Pi_{\frac{1}{2}}$ state.

J	F	Ω	Sym.	←	J'	F'	Ω'	Sym.'	ν [MHz]
1.5	1.0	0.5	e		0.5	1.0	0.5	e	866946.734(1260)
1.5	2.0	0.5	e		0.5	1.0	0.5	e	866946.734(1260)
1.5	1.0	0.5	e		0.5	0.0	0.5	e	866960.317(80)
1.5	1.0	0.5	f		0.5	1.0	0.5	f	875236.537(100)
1.5	2.0	0.5	f		0.5	1.0	0.5	f	875267.136(60)
1.5	1.0	0.5	f		0.5	0.0	0.5	f	875288.347(80)

J	F	Ω	Sym.	←	J'	F'	Ω'	rel. Int.	obs. − calc. [MHz]
1.5	1.0	0.5	e		0.5	1.0	0.5	0.125	+ 1.597
1.5	2.0	0.5	e		0.5	1.0	0.5	0.625	+ 0.342
1.5	1.0	0.5	e		0.5	0.0	0.5	0.250	− 0.038
1.5	1.0	0.5	f		0.5	1.0	0.5	0.125	− 0.177
1.5	2.0	0.5	f		0.5	1.0	0.5	0.625	+ 0.040
1.5	1.0	0.5	f		0.5	0.0	0.5	0.250	− 0.058

Numbers in parantheses denote the estimated uncertainty on measured frequencies in unit of the last quoted digit. Uncertainties are 1σ.

we had to include the centrifugal distortion constants up to the M term, in order to reproduce the experimental data within their uncertainties. The Λ-doubling parameters are the same as those used by Morino and Kawaguchi. The reported hyperfine parameters include data from the $^2\Pi_{1/2}$ state for the first time. The hyperfine constants for the $^2\Pi_{3/2}$ state were reported previously by Meerts and Dymanus [12] from observation of pure Λ-doubling transitions and by Ashworth and Brown [15] from the far-IR LMR spectrum. The two sets of hyperfine constants agree very well with each other (see Table 36.3).

36.4   Interstellar Implications

Although SH has not yet been detected in space, the abundance of this radical in assorted interstellar sources has been considered in a variety of models [17–20]. In quiescent, dense interstellar molecular clouds, characterized by a temperature $T \approx 10-50\,\mathrm{K}$ and a gas density $n \approx 10^4\,\mathrm{cm}^3\,\mathrm{s}^{-1}$ where the gas is overwhelmingly H_2, the chemistry is dominated by exothermic ion-molecule and ion-electron dis-

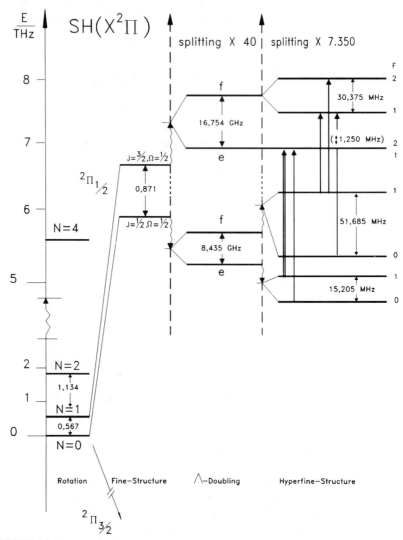

FIGURE 36.1. Energy level diagram of SH. The 'pure' rotational ladder has been included on the left (Hund's case (b)). The effect of fine structure, Λ-doubling, and hyperfine structure is displayed for the $^2\Pi_{1/2}$ state which is energetically about 375 cm^{-1} above the $^2\Pi_{3/2}$ state, which is indicated. The observed $J = \frac{3}{2} \leftarrow \frac{1}{2}$ transitions are indicated by bold arrows.

sociative recombination reactions [21]. In these sources, the calculated fractional abundance of SH is very low ($\approx 10^{-11}$), and probably undetectable because the 'natural' exothermic ion-molecule reactions leading to the precursor ion H_2S^+

$$S^+ + H_2 \rightarrow SH^+ + H ,$$
$$SH^+ + H_2 \rightarrow H_2S^+ + H , \tag{36.5}$$

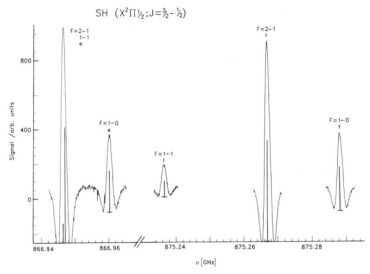

FIGURE 36.2. Observed line pattern of the $J = \frac{3}{2} \leftarrow \frac{1}{2}$ transition. The sticks indicate the relative intensities of the hyperfine components.

TABLE 36.2. Electronic ground state rotational parameters of SH.

Parameter	This work	Morino and Kawaguchi	Unit
A	− 11.297 139[z]	− 11.297 139	THz
γ	− 4.573 1 (44)	− 4.573 0 (66)	GHz
D_γ	461 (15)	456 (25)	kHz
B	283.587 61 (12)	283.616 89 (24)	GHz
D	14.494 1 (31)	14.509 1 (25)	MHz
H	190 (47)	438.3 (78)	Hz
L	− 1.57 (31)	−	Hz
M	− 3.36 (73)	−	mHz
p	9.008 38 (39)	9.007 048 (57)	GHz
D_p	− 0.946 (14)	− 1.067 (36)	MHz
H_p	− 465(65)	− 440 (170)	Hz
q	−286.163 7 (64)	− 284.518 9 (93)	MHz
D_q	56.66 (19)	56.60 (51)	kHz
H_q	− 13.5 (22)	− 12.8 (48)	Hz

[z] fixed to the value given by Morino and Kawaguchi ref. [14]

are endothermic by ≈ 1 eV [20]. Exothermic reactions analogous to these produce a sizeable abundance of the radical OH [21]. In place of these endothermic reactions, only two very slow radiative association reactions which require an electronic spin flip can produce the precursor ions H_2S^+ and H_3S^+, which lead to the SH radical

TABLE 36.3. Hyperfine parameters of SH.

Parameter	This work	Meerts and Dymanus	Unit
a	33.62 (31)	32.58 (7)[2]	MHz
b	− 63.417 (35)	− 63.44 (4)[2]	MHz
c	30.32 (62)	32.44 (14)[2]	MHz
d	27.36 (20)	27.36(12)[2]	MHz

[2] from Meerts and Dymanus ref. [12]

via the dissociative recombination

$$S^+ + H_2 \rightarrow H_2S^+ + h\nu,$$
$$SH^+ + H_2 \rightarrow H_3S^+ + h\nu,$$
$$H_2S^+ + e \rightarrow SH + H,$$
$$H_3S^+ + e \rightarrow SH + H_2 \tag{36.6}$$

Under very high temperature conditions, the fractional abundance of SH is expected to increase because of the above endothermic ion-molecule synthesis, and an endothermic neutral-neutral synthesis

$$S + H_2 \rightarrow SH + H, \tag{36.7}$$

produce SH efficiently. Very high temperatures (thousands of degrees) in the interstellar medium are associated with shock waves that occur in regions of active star formation. Other sulfur-containing species have been detected at elevated abundance in shocked sources [20]. Two types of shocks can occur—the J-type and the C-type—depending on the magnitude of the magnetic field in the vicinity of the shock [20]. In J-type shocks, the influence of the magnetic field is minimal, and the shock produces a swift rise in temperature followed by a more leisurely decline. In C-type, or magnetohydrodynamic two-fluid shocks, the magnetic field tends to moderate the rise in temperature but is associated with a streaming between ions and neutral species so that the importance of endothermic ion-molecule reactions is enhanced. For J-type shocks [18, 19], model calculations show that the fractional SH abundance can reach values as high as 10^{-7} within 0.1 yr of the shock with a fairly rapid decline thereafter.

Another type of region in which an elevated abundance of SH might be found is known as a Hot Core. Hot Cores are warm but quiescent regions in the vicinity of star-forming areas, where the rising temperatures drive off saturated molecules from the surfaces of dust particles [21]. One such saturated molecule is H_2S; this species is subsequently depleted via the reaction

$$H_2S + H \rightarrow SH + H_2 , \tag{36.8}$$

leading to the transitory production of the SH radical, which is itself depleted by reaction with atomic hydrogen to form $S + H_2$. Detailed calculations for the abundance of SH that can be expected are in progress [22].

36.5   Acknowledgments

This work was supported in part by the Deutsche Forschungsgemeinschaft (DFG) via Grant SFB 301 and additional funding by the Science Ministry of the Land Nordrhein-Westfalen and in part by NASA. G. W. and E. H. thank the Max Planck Gesellschaft and Alexander von Humboldt Stiftung for the Max Planck Research Award. The work of S. P. B. at Cologne was made possible by the DFG through grants aimed to support Eastern and Central European Countries and the republics of the former Soviet Union.

Appendix

The matrix elements of the effective Hamiltonian, including all the various contributions, are listed below:

$$\langle {}^2\Pi_{1/2} J' \tfrac{1}{2} F\pm \mid \hat{H} \mid {}^2\Pi_{1/2} J \tfrac{1}{2} F\pm \rangle$$

$$
\begin{aligned}
= \; \delta_{J',J} \Big\{ &-\frac{1}{2} A - (\gamma + D_\gamma J(J+1)) + B(J^2 + J + \frac{1}{4}) - D(J^4 + 2J^3 \\
&+\frac{5}{2}J^2 + \frac{3}{2}J - \frac{11}{16}) + H(J^6 + 3J^5 + \frac{27}{4}J^4 + \frac{17}{2}J^3 + \frac{7}{16}J^2 - \frac{53}{16}J \\
&+\frac{61}{64}) - L(J^8 + 4J^7 + 13J^6 + 25J^5 + \frac{111}{8}J^4 - \frac{37}{4}J^3 - \frac{55}{16}J^2 + \frac{75}{16}J \\
&-\frac{311}{256}) + M(J^{10} + 5J^9 + \frac{85}{4}J^8 + 55J^7 + \frac{233}{4}J^6 + \frac{23}{8}J^5 - \frac{573}{32}J^4 \\
&+\frac{127}{16}J^3 + \frac{1309}{256}J^2 - \frac{1547}{256}J + \frac{1561}{1024}) \mp [p + D_p J(J+1) \\
&+H_p J^2 (J+1)^2] \frac{1}{2}(-1)^{J-\frac{1}{2}}(J+\frac{1}{2}) \mp [q + D_q J(J+1) \\
&+H_q J^2 (J+1)^2](-1)^{J-\frac{1}{2}} (J+\frac{1}{2}) \Big\} + (-1)^{J'+F+\frac{1}{2}} \\
&\times \sqrt{\frac{3}{2}(2J+1)(2J'+1)} \begin{Bmatrix} F & J' & \frac{1}{2} \\ 1 & \frac{1}{2} & J \end{Bmatrix} \Big\{ (-1)^{J'-\frac{1}{2}} \\
&\times \begin{pmatrix} J' & 1 & J \\ -\frac{1}{2} & 0 & \frac{1}{2} \end{pmatrix} [a - \frac{b+c}{2}] \pm d\frac{1}{\sqrt{2}} \begin{pmatrix} J' & 1 & J \\ \frac{1}{2} & -1 & \frac{1}{2} \end{pmatrix} \Big\}
\end{aligned}
$$

$$\langle {}^2\Pi_{3/2} J' \tfrac{1}{2} F\pm \mid \hat{H} \mid {}^2\Pi_{3/2} J \tfrac{1}{2} F\pm \rangle$$

$$
\begin{aligned}
= \; \delta_{J',J} \Big\{ &\frac{1}{2} A + B(J^2 + J - \frac{7}{4}) \cdots D(J^4 + 2J^3 \\
&-\frac{3}{2}J^2 - \frac{5}{2}J + \frac{37}{16}) + H(J^6 + 3J^5 + \frac{27}{4}J^4 + \frac{17}{2}J^3 + \frac{7}{16}J^2 - \frac{53}{16}J
\end{aligned}
$$

$$+ \frac{61}{64}) - L(J^8 + 4J^7 + 5J^6 + J^5 - \frac{1}{8}J^4 + \frac{11}{4}J^3 - \frac{47}{16}J^2 - \frac{77}{16}J$$

$$+ \frac{937}{256}) + M(J^{10} + 5J^9 + \frac{45}{4}J^8 + 15J^7 + \frac{65}{8}J^6 - \frac{57}{8}J^5 - \frac{367}{32}J^4$$

$$- \frac{57}{16}J^3 + \frac{1661}{256}J^2 + \frac{1557}{256}J - \frac{4687}{1024})\Big\} + \sqrt{\frac{3}{2}(2J+1)(2J'+1)}$$

$$\times \begin{Bmatrix} F & J' & \frac{1}{2} \\ 1 & \frac{1}{2} & J \end{Bmatrix} (-1)^{J'-\frac{3}{2}} \begin{pmatrix} J' & 1 & J \\ -\frac{3}{2} & 0 & \frac{3}{2} \end{pmatrix} [a + \frac{b+c}{2}]$$

$$\langle {}^2\Pi_{1/2} J' \tfrac{1}{2} F \pm | \, \hat{H} \, |{}^2 \Pi_{3/2} J \tfrac{1}{2} F \pm \rangle$$

$$= \delta_{J',J}\Big\{ (\gamma + D_\gamma J(J+1))\frac{1}{2}\sqrt{J^2 + J - \frac{3}{4}} - B\sqrt{J^2 + J - \frac{3}{4}} + D$$

$$\times 2\sqrt[3]{J^2 + J - \frac{3}{4}} - H \times 3\sqrt{J^2 + J - \frac{3}{4}}(J^4 + 2J^3 - \frac{1}{6}J^2 - \frac{7}{6}J$$

$$+ \frac{31}{48}) + L \times 4\sqrt{J^2 + J - \frac{3}{4}} \times (J^6 + 3J^5 + \frac{7}{4}J^4 - \frac{3}{2}J^3 - \frac{1}{16}J^2$$

$$+ \frac{19}{16}J - \frac{39}{64}) - M \times 5\sqrt{J^2 + J - \frac{3}{4}}(J^8 + 4J^7 + 5J^6 + J^5 - \frac{37}{40}J^4$$

$$+ \frac{23}{20}J^3 - \frac{11}{80}J^2 - \frac{97}{80}J + \frac{781}{1280}) \pm \frac{1}{2}[q + D_q J(J+1) + H_q J^2(J+1)^2]$$

$$\times (-1)^{J-\frac{1}{2}}\sqrt{J^2 + J - \frac{3}{4}}(J + \frac{1}{2})\Big\} - b \times \frac{1}{2}\sqrt{\frac{3}{2}(2J+1)(2J'+1)}$$

$$\times (-1)^{J'+F+\frac{1}{2}} \begin{Bmatrix} F & J' & \frac{1}{2} \\ 1 & \frac{1}{2} & J \end{Bmatrix} (-1)^{J'-\frac{3}{2}} \begin{pmatrix} J' & 1 & J \\ -\frac{3}{2} & 1 & \frac{1}{2} \end{pmatrix}$$

References

[1] G. Winnewisser, Vib. Spectrosc. **8**, 241 (1995).

[2] S. P. Belov, M. Liedtke, Th. Klaus, R. Schieder, A. H. Saleck, J. Behrend, K. M. T. Yamada, G. Winnewisser, and A. F. Krupnov, J. Molec. Spectrosc. **166**, 489 (1994).

[3] G. Winnewisser, J. Molec. Spectrosc. **168**, 227 (1994).

[4] G. Winnewisser, M. Winnewisser, and W. Gordy, J. Chem. Phys. **49**, 3465 (1968), and Bull. Am. Phys. Soc. **11**, 312 (1966).

[5] K. M. Evenson, *Faraday Discuss. Chem. Soc.* **71**, 7 (1981).

[6] W. Etzenbach, A. H. Saleck, M. Liedtke, and G. Winnewisser, Can. J. Phys. **72**, 1315 (1994).

[7] G. C. Dousmanis, T. M. Sanders, and C. H. Townes, Phys. Rev. **100**, 1735 (1955).

[8] A. F. Krupnov, in *Modern Aspects of Microwave Spectroscopy*, ed. G. W. Chantry, (Academic Press, London, 1979), p. 217.

[10] S. Yamamoto and S. Saito, Can. J. Phys. **72**, 954 (1994).

[11] S. H. Ashworth, K. M. Evenson, and J. M. Brown, J. Mol. Spectrosc. **172**, 282 (1995).

[12] W. L. Meerts and A. Dymanus, ApJ **187**, 445 (1974).

[13] W. L. Meerts and A. Dymanus, Can. J. Phys. **53**, 2123 (1975).

[14] I. Morino and K. Kawaguchi, J. Mol. Spectrosc. (in press) (1995).

[15] S. H. Ashworth and J. M. Brown, J. Mol. Spectrosc. **153**, 41 (1992).

[16] R. N. Zare, A. L. Schmeltekopf, W. J. Harrop and D. L. Albritton, J. Mol. Spectrosc. **46**, 37 (1973).

[17] T. J. Millar and E. Herbst, A&A **231**, 466, (1990).

[18] G. F. Mitchell, ApJ **287**, 665, (1984).

[19] T. M. Leen and M. M. Graff, ApJ **325**, 411 (1988).

[20] G. Pineau des Forêts, E. Roueff, P. Schilke, and D. R. Flower, MNRAS **262**, 915 (1993).

[21] E. Herbst, Ann. Rev. Phys. Chem. **46**, 27 (1995).

[22] S. B. Charnley (private communication).

37

Charlie Townes at Brookhaven

Walter D. Knight

37.1   Fortunate Meeting

Charles Townes went to Columbia University in the late 1940s, and spent considerable time at Brookhaven National Laboratory (BNL) in the summer of 1949. I was spending my second summer at BNL, having essentially completed a nuclear magnetic resonance (NMR) spectrometer in the previous twelve months, and began preliminary experiments in the summer of 1949. Charles participated actively in discussions and seminars at the Lab, and we saw him often in the Nuclear Moments Laboratory, whose program also included a project on microwave spectroscopy of molecules.

Early that summer we obtained the first significant results in the NMR project, and Charles Townes was there to help us understand them.

37.2   How to Find a Thesis

I received the M.A. degree in physics at Duke University in 1943, and completed course requirements for the Ph.D. Then there was an interruption during 1944-45 when I joined the Navy, saw small parts of the world on radar screens, and learned a lot about electronics. Subsequently I accepted a position as an Assistant Professor at Trinity College in Hartford, Connecticut to work under Professor F. W. Constant, who had been one of my teachers and M.A. thesis supervisor at Duke. In seeking a way to complete the doctorate, I consulted Norman Ramsey who was head of the Physics Department at Brookhaven National Laboratory. He pointed out that molecular-beam magnetic resonance (MBMR) had been the preferred method for measuring nuclear moments [1] because of the high accuracy obtainable. However, the new methods based on the nuclear magnetic resonance techniques of Bloch and Purcell had the combined advantages of accuracy, simplicity, and sensitivity for bulk solid or liquid materials.

My conversations with Ramsey resulted in a summer (1948) position for me in the newly established Nuclear Moments Laboratory at Brookhaven. Bill Cohen, who like Ramsey had been a student of Rabi, was running the nuclear moments program. Bill suggested that I visit Bob Pound's lab at Harvard to learn about his marginal oscillator-detector, which was tunable over the wide-frequency ranges needed to search for nuclear magnetic resonances. At the time, interest in nuclear moments was stimulated by the recently developed nuclear shell model. The problem was interesting, and I decided to work along these lines, hoping that a specific thesis topic would emerge.

37.3   Searching for a Resonance

Bill Cohen designed and oversaw the construction of a suitable electromagnet, and I constructed a marginal oscillator based on Pound's circuit, with generous assistance from Bob on details. Thanks to plenty of training in electronics in the Navy radar program, the construction was easy, and we found a proton resonance on the first try. However, it was difficult to maintain stable oscillation and sensitivity over wide frequency ranges, and so I decided to try a simpler circuit based on the cathode-coupled oscillator. Later, I explained the new oscillator to Pound, who liked it and put his student, George Watkins, to work analyzing it and constructing an improved version, which came to be known in the trade as the "Pound Box" [2]. The summer was soon over, and I returned to teaching at Trinity College. During the college year I spent many weekends traveling to Brookhaven improving the NMR system, returning to Hartford in time to teach Monday morning classes. The following summer (1949) I resumed the search for nuclear magnetic resonances at Brookhaven.

The first search was for resonances in antimony, which has two isotopes—Sb-121 and Sb-123 having different nuclear spins and magnetic moments—hopefully, we could bag two birds in one search! The first run was on a sample of antimony trichloride.

A good signal appeared, to the delight and excitement of all in the lab, but unfortunately the observed resonance occurred nowhere near the frequency expected from the spectroscopic magnetic moments. Could it be an impurity?... We dumped the sample and ran the empty test tube ... the signal persisted ... what is in the pyrex glass? ... Boron! Sure enough, the observed frequency was close to that expected for boron. For a double-check, we removed the test tube entirely which eliminated the boron signal. Then a previously unseen signal appeared near the frequency which Pound had reported for Cu-63 signals originating within the rf skin depth of the copper resonance coil ... but our observed frequency was about 20 kHz higher than what is expected from the standard tables of nuclear moments. Were there any other known moments close to copper? ... no ... Was there an error in the tables? ... no ... In order to be sure of the effect we constructed a sample mixture of dry powders of copper metal and cuprous chloride. The resulting experimental resonance profile showed two well-resolved peaks, with the resonance for the pure metal appearing approximately 20 kHz above the resonance for the non-metallic salt, around a 10 MHz average frequency. The measured relative shift was 0.23%.

Later that day Sam Goudsmit, who was Director of Brookhaven at the time, came by the lab to see the new results. I asked his advice whether I should pursue the search for new magnetic moments, or continue studying the shift. He replied that naturally new physics should be more productive than another moment in the tables, and we returned to the experiments, this time to determine what was shifting and why. In a flurry of activity in the next few weeks we investigated several other metals, including sodium, aluminum, and liquid gallium, each with its unique shift, and determined that isotopic pairs such as Cu-63 and Cu-65 possessed equal shifts, fractional shifts, which were independent of the resonance field.

37.4   A Theory to Match the Experiment

Charlie Townes had moved to Columbia, and was spending time at Brookhaven that summer. He mentioned some of his recent work on hyperfine structure in molecules, and suggested that the NMR shift in metals might result from a hyperfine interaction between the magnetically polarized metal conduction electrons and the magnetic moments of atomic nuclei. The hyperfine field should augment the applied field and shift the resonance to a higher frequency. He suggested I try to estimate the magnitude of the effect in terms of the Pauli electronic paramagnetism and the contact term of s-electron hyperfine interaction. I studied up on hyperfine structure, located Goudsmit's formula for the contact interaction, and got a result which agreed quite well with the experimental shift for sodium. In further discussions, Charlie took me to the article of Sommerfeld and Bethe in the *Handbuch der Physik*, and explained how the atomic hyperfine structure differs from the corresponding average for electrons at the Fermi surface of a metal. The average contact hyperfine field at the nucleus for an electron at the Fermi surface is [4–6]

$$\Delta B = \left(\frac{8\pi}{3} \right) \left[\chi_p v_0 B \right] P_F, \tag{37.1}$$

where χ_p is the Pauli susceptibility per unit volume, v_0 is the atomic volume, and B is the applied magnetic field, P_F is the probability density for the s-electron at the nucleus, averaged over electrons at the Fermi surface. The relative frequency shift $K = \Delta f / f = \Delta B / B$, and the above equation may be used in several ways, depending on what quantity is to be found. For example, in the metals where χ_p can be measured directly, P_F may be calculated from Eq. (37.1).

Having established that all of the other metals studied show comparable shifts of a few tenths percent, we were surprised to find no measurable effect in beryllium. Charlie arranged a consultation with Conyers Herring [5], who had recently published an article with Hill on the electronic structure of beryllium. Herring believed that the hyperfine field, Eq. (37.1), is low because of the strong p-character (low P_F) of the wave function and also an abnormally low density of states (low χ_p) at the Fermi level.

In the early investigations we studied the metals Li, Be, Na, Al, V, Cu, liquid Ba, and non-metallic white phosphorous. The shift in white P compared to the non-metallic phosphorous trichloride was approximately 0.1%. Ramsey [1] recognized that the observed non-metal (chemical) shifts were caused mainly by Van Vleck orbital paramagnetism, which produces a smaller hyperfine field than does the contact interaction.

37.5   The Nucleus as a Non-invasive Structural Probe

While I was completing my thesis during the spring of 1950, I made several trips to New York to consult with Charlie Townes, who continued to be my informal adviser, providing me with strong support. Charlie told me that he had given a talk ca. 1949 with a title something like "The Nucleus as a Probe of Structures." Understanding that "structures" include molecular and crystalline structures and electronic structures as well, the probe idea soon became the main theme of NMR experiments. The NMR and quadrupole coupling energies are in most cases negligible in comparison with atomic and molecular binding energies. Consequently, the resonance experiments can be used to measure local internal fields in materials, without disturbing the basic structure of the system under study.

The NMR line shift K in metals, named "Knight Shift" by Bloembergen [6], is a phenomenon unique to the metallic state since it depends on conduction of electrons, and it enables the practical analytical technique of probing systems endoscopically. The Knight shift has been used in many studies, including melting of metals, metal-insulator transitions, normal-superconducting transitions, and phase transitions and magnetic couplings in metal alloy systems.

37.6   Pleasant Memories

I remember the scientific climate at BNL as being enormously stimulating for everyone there. On one memorable occasion, Charlie's seminar talk was suddenly interrupted by vigorous objections from Gregory Breit. During an extended discussion Charlie managed skillfully to calm Gregory down, and to educate the rest of us on the point in question. For me, Charlie's presence was a uniquely valuable experience. His guidance partly compensated for the fact that I had not had a formal thesis supervisor. I surely would have profited from a more prolonged and concentrated dose of Charlie's style of physics. As it was, the experience was one of the most exciting and productive periods of my life. I am happy to say that he has continued his interest and support of my activities during the years we have spent as colleagues in Berkeley.

References

[1] N. F. Ramsey, *Nuclear Moments*, Wiley, New York (1953).

[2] R. V. Pound and W. D. Knight, Rev. Sci. Instr., **96**, 219 (1950).

[3] W. D. Knight, Phys. Rev. **76**, 1259 (1950).

[4] W. D. Knight, Solid State Physics **2**, 93 (1956).

[5] C. H. Townes, C. Herring, and W. D. Knight, Phys. Rev. **77**, 851 (1950).

[6] T. J. Rowland, *Progress in Metal Physics*, Vol. 9, Pergamon (1961).

38

Classical Theory of Measurement: A Big Step Towards the Quantum Theory of Measurement

Willis E. Lamb, Jr. and Heidi Fearn

This paper is written to celebrate the 80th birthday of Professor Charles Townes. I have had the distinct pleasure of knowing Charlie for almost 50 years. As far as I know, he has not had a need to know much about the Quantum Theory of Measurement. However, the invention of the laser has certainly done much to keep the subject of the theory of quantum measurements far more active than it would otherwise have been. Charlie went from Furman University to Caltech, to Bell Telephone Labs (BTL), to Columbia, to M.I.T. and to the University of California at Berkeley. I followed a bit of that path in reverse order, from U.C. Berkeley to Columbia, to Stanford, to Oxford, to Yale and to Arizona. Some people may think that Charlie and I each did our best work at Columbia, but we may think otherwise. In any case, I am very grateful to U.C. Berkeley, where I spent eight years from 1930, and had the opportunity to learn the rudiments of quantum mechanics from Professors William Williams, Victor Lenzen, and especially, Robert Oppenheimer. I am happy that the move to Berkeley has been so good for Charlie's research, and wish him every success and happiness for the future years.

38.1 Introduction

One of us, in previous years, has published several articles [1–8] the quantum theory of measurement. These papers have not been much quoted, perhaps because they have been difficult to understand. After a review of the history of the subject, the present paper outlines a purely classical approach to the measurement problem in nonrelativistic classical mechanics. We regard this as a very simple and trivial problem. In fact, it is so simple and trivial that no one has treated it until now. The remarkable fact is that, once one learns how to treat this completely classical problem, a literal translation of the calculation into the quantum domain provides a very fine model for the quantum theory of measurement. The papers by Lamb [2]–

[5] mentioned above, were in essential agreement with our present formulation. The translation from the classical to the quantum theory of measurement will be given in a later paper.

It is now just about three score and ten years from the founding of Quantum Mechanics in the years 1924-1926. There is no doubt that this subject gives a highly successful description of many phenomena of modern physics. On the other hand, the state of understanding of quantum mechanics has never been very good.

Consider the following citations, given with their Nobel Prizes, to the six founders of the subject:

- Louis de Broglie (1929): "Discovery of the wave nature of the electron."

- Werner Heisenberg (1932): "Creation of quantum mechanics."

- Erwin Schrödinger and Paul Dirac (shared prize in 1933): "Discovery of new forms of atomic theory,"

- Wolfgang Pauli (1945): "Discovery of the Exclusion Principle."

- Max Born (1954): "Statistical interpretation of the wave function."

We have a great deal of admiration for the insightful contributions of all of the six founders of quantum mechanics. However, to a greater or lesser degree, none of them ever had a real understanding of the meaning of the subject. The first of them was the least successful.

For a little more completeness, we also list the six later prize winners whose non-prize work has involved interpretation of, and (or) changes to, quantum mechanics, and begin the list with a person who did not win a prize, but could easily have done so, for his influence on quantum mechanics and many other branches of science:

- John von Neumann

- Willis Lamb (1955): "Fine structure of hydrogen."

- Eugene Wigner (1963): "Discovery and application of fundamental symmetry principles in quantum mechanics."

- Julian Schwinger and Richard Feynman (shared prize in 1965): "Fundamental work in quantum electrodynamics."

- Murray Gell–Mann (1969): "Classification of elementary particles."

- Steven Weinberg (1979): "Theory of unified weak and electromagnetic interactions."

38.1.1   De Broglie

De Broglie [9] was overwhelmed by the conflicting admixture of wave and particle aspects of his 1924 Ph.D. thesis, and never developed a satisfactory synthesis of them. Under influence of Albert Einstein, David Bohm tried to revive de Broglie's theory in 1950. In our opinion, his effort can only be characterized as a disappointing failure. One has only to compare the importance and numbers of problems which can be solved by the two theories. Nevertheless, some lingering support for de Broglie's views can still be found today [10].

Heisenberg (1)

Heisenberg is most widely acclaimed for his invention of matrix mechanics [11] in 1925, and for his formulation of the uncertainty relation [14] in 1927. This takes the form of $\Delta x \Delta p \geq \hbar/2$ and correctly sets limits on the possible simultaneous knowledge of the coordinate x and momentum p of a particle. It is not generally appreciated that Heisenberg made absolutely no use of quantum mechanical theory in this work. Rather, the discussion is qualitative and based on no more physics than an intuitive consideration of Compton scattering of gamma rays by an electron combined with the elementary physical optics of a lens. Nevertheless, Heisenberg was the first one to see that measurement of a quantum mechanical dynamical variable changed the system under consideration.

Schrödinger (1)

The Schrödinger [15] wave equation of 1926 made it possible for quantum mechanics to be used by a much wider scientific community than would otherwise have been the case. In 1929, Howard Robertson [18], using quantum mechanical theory, and in 1930, Schrödinger [17] gave independent derivations of a relation involving Δx and Δp which looks just like Heisenberg's 1927 relation. However, the physical meaning is entirely different. The new relation follows, in a few lines of proof, from the Schwartz inequality of mathematics. It involves a product of the separate measures of scatter of x and p, and has nothing in common (except dimensional correctness) to do with the modeling of simultaneous measurements of x and p attempted in Heisenberg's 1927 work.

Heisenberg (2)

In his Chicago lectures of 1929, Heisenberg [16] discussed the problem of measuring the distribution of electronic momentum values in an atom. To do this, he assumed that he could suddenly turn off the nuclear Coulomb field in the atom. Time of flight techniques applied to the liberated electrons would give their velocity distribution. Of course, we think that this would be rather hard to do in practice. No one could just reach into the atom, and suddenly turn off the Coulomb field of the nucleus. Neither could one suddenly remove the nucleus very quickly from the atom. However, rather similar things can be done: For example, the atom could be bombarded with a high energy nuclear particle that would quickly knock out the atomic nucleus. Coincidence techniques could be used to select the processes desired for study.

Schrödinger (2)

After his wonderful papers of the twenties, Schrödinger went off into very unfortunate attempts to deal with the meaning of quantum mechanics. The paper [19] on the "cat paradox" caused a lot of confusion which still lingers in physics today. Except for a few papers like [8], the important points have largely been missed. The Schrödinger [20] 1944 book *What is Life?*, was charmingly written, and has provided insight and stimulation for molecular biologists from Max Delbruch to Kary Mullis [21]. Unfortunately, the book was full of much manifest nonsense about quantum mechanics. There is no doubt that quantum mechanics is needed to understand the structure and properties of atoms and molecules, and also phenomena like electron transfer in, and between, macromolecules. For other things, something like classical mechanics would seem more appropriate. Of course, the complexity of biological systems is so great that a higher degree of understanding of the fundamental nature of nonrelativistic quantum mechanics may not be of much use in biophysics.

Born

Although Born, very belatedly, received his Nobel Prize in 1954 for the 1926 introduction of the probability interpretation of quantum mechanics, it was probably overlooked by most people that Dirac had already brought probability into his 1925-26 papers [12] in the *Proceedings of the Royal Society*. However, Born's paper [22] was more timely, and had a greater influence in the early days of quantum mechanics. Furthermore, Born deserved much credit for his help to Heisenberg on matrix theory in 1925, a paper of that year with Pascual Jordan, and for the attempts he made in the Born–Einstein Letters [23], to explain quantum mechanics to Einstein.

Dirac

In his 1930 book, Dirac [24] stated that the measurement of an observable for a system would give as a result one of the eigenvalues of that observable. It would also change the state of the system. He did not use quantum mechanical theory to calculate what that change would be in any particular problem, but simply made a postulate that, if a certain result was obtained for one measurement, the same result would be obtained if the same measurement on the system were immediately repeated. He gave no indication how either measurement might be made.

Von Neumann

In Chapter III of his 1932 book, von Neumann [25] made a postulate, equivalent to Dirac's, that measurement of a dynamical variable would reduce, or collapse, the wave function into an eigenstate of the measured observable. As in Dirac's work, von Neumann gave no indication of what might bring about such a change in the wave function. Von Neumann's book ends with a long and complicated Chapter VI. On the last two pages of that chapter, he gave a simple dynamical model for a position measurement. He did not provide any application of this model. The book simply ends with some exercises left to the reader. It should be noted that there is nothing about wave function collapse in Chapter VI. Perhaps it would have been better if von Neumann had interchanged a completed Chapter VI and a rewritten Chapter III.

Arthurs and Kelly

We are not aware of any subsequent reference to von Neumann's meter model before the appearance of a short communication in the *Bell System Technical Journal* of 1965 in which Arthurs and Kelly [26] extended the model to momentum measurements, and discussed the problem of simultaneous measurement of position and measurement. For this, Arthurs and Kelly used two meters at the same time: one of them as suggested by von Neumann for position measurement, and one devised by Arthurs and Kelly for momentum measurement. As far as we know, only one later publication before 1986 by Mario Bunge [27] in 1967 had any significant mention of the von Neumann meter, or of the Arthurs and Kelly work. One of us learned of the 1965 work from Professor Melvin Lax of City University of New York (CUNY) and Bell Laboratories sometime after 1983. We learned of Bunge's book only in May of 1995.

The von Neumann meter consisted of a dynamical system described by a coordinate X and a momentum P. The system to be studied was a particle with coordinate x and momentum p. The measurement process assumed the introduction, for a very brief time τ of an interaction between system and meter. The potential energy of this interaction took the form $(\theta/\tau)A(x, P)$ inside the time range τ, and zero outside. Here the strength of the interaction is described by the dimensionless factor theta. The action $A(x, P)$ was taken to be the simple product of the coordinate x for the system and the momentum operator P for the meter.

Just before the system and meter begin their interaction, the combined system–meter wave function was a simple unentangled product $u(x)v(X)$ of the system wave function $u(x)$ and the meter wave function $v(X)$. The time evolution of the combined system–meter wave function has to be calculated from the time-dependent Schrödinger wave equation for the combined system and meter.

The Hamiltonian for this calculation is approximated by neglect of kinetic energies and other terms which do not contain the very small pulse duration time τ in the denominator. The wave function $u(x)v(X)$ after time τ is changed into a wave function $u(x)v(X - \theta x)$. (In our notation, von Neumann took $\theta = 1$.) The Von

Neumann discussion stopped at this point. Arthurs and Kelly supplied a plausible recipe whereby measurements of the X–meter's coordinate X and of the P–meter's momentum P could give information about the x coordinate and p momentum of the system. We would like to propose a simple test for the merit (or lack of it) in a paper or a book written after 1965 on the quantum theory of measurement. Did it [2–7] (or did it not [1, 10, 30, 34]) make adequate reference to the paper of Arthurs and Kelley?

Pauli

Not knowing of the von Neumann meter, Wolfgang Pauli [29], in his 1933 *Handbuch der Physik* discussion of measurement theory, made a meter out of a modified Stern-Gerlach apparatus. The main purpose of this meter was to study atoms described by a non-stationary state wave function, and to determine the probabilities that the various stationary states were found in the use of the apparatus. This Pauli meter model was also used in a 1969 article in *Physics Today* by Lamb [1], who at that time had no knowledge of the von Neumann meter or of its use by Arthurs and Kelly.

Wigner

Beginning about 1960, Eugene Wigner [35] was active in quantum measurement theory. His dissatisfaction with the results drove him into the absurd view that the consciousness of the observer had to play an important part in the theory. He felt that there had to be nonlinear modifications of the Schrödinger equation. He also had conversations with an unnamed friend about the measurement problem.

Schwinger

Julian Schwinger thought deeply about the foundations of quantum mechanics. In the summer of 1993, he gave three lectures at Ulm on a deduction of the principles of that subject from elementary phenomena in molecular beam experiments. Possibly his ideas for this work may appear in a posthumous publication.

Feynman

In general, Richard Feynman did not show a great amount of interest in the quantum theory of measurement. However, his work on quantum electrodynamics led him to the path integral formulation of quantum mechanics. Some people [30], [31] seem to feel that path integrals must play an important part in a discussion of measurement theory. See some remarks by Lamb [3] on this matter.

Weinberg

Steven Weinberg [32] considered some possible nonlinear modifications of quantum mechanics.

Gell-Mann

Murray Gell-Mann [33] has been exploring extensions of quantum mechanics as applied to the Universe. After Newton's work on classical mechanics, it seemed to many people that the theory of a few body problem could solve the problems of the Universe. Perhaps quantum mechanics will do better, but classical mechanics has the advantage that there is no problem of measurement in this earlier discipline.

Nature of Classical Mechanics

Quantum mechanics is a generalization [7] of classical mechanics. It is therefore desirable to consider some relevant features of classical mechanics. Some of these are to be kept or discarded, and some are changed by the transition to quantum mechanics.

Newtonian mechanics starts with the notion of a particle. Various concepts like position, velocity, acceleration, mass and force are developed. Newton's second law of motion $F = ma$ is obtained. Other dynamical concepts like momentum, kinetic energy, work, impulse, potential energy, angular velocity, torque, angular momentum, and action are defined. The theory has been extended to apply to systems of particles, rigid bodies, deformable bodies, and fluids. Some problems involve friction and other dissipative forces. Constrained systems may involve rolling or sliding bodies, and gear- and ratchet-systems, although these features are probably not important for atomic and subatomic physics. In some problems, a Lagrangian or Hamiltonian reformulation of the $F = ma$ equations might be used.

It is a feature of classical mechanics that there has to be a well-defined dynamical system of interest. Forces internal to the system have to be well specified, or there can be no dynamics. No attention is paid to the rest of the universe, except that certain forces may be applied from outside the system. The forces have to be known, or at least assumed to be known. Otherwise there is no well-defined dynamical problem.

In conventional classical mechanics, there is a complete separation between any problem of mechanics and the problem of measurement. The dynamical system is completely undisturbed by a measurement process. In this paper, we will have to take steps to introduce a little dynamics into the measurement processes. In order to later be able to make a connection with quantum mechanics, we want to limit the discussion of classical mechanics to very simple problems. Only systems with one degree of freedom will be considered. Neither friction nor coupling to an "environment" will be allowed.

For pedagogical, and other reasons, we would like to consider that in classical mechanics the force $F(x, t)$ acting on a particle of mass m at position x can be an arbitrary function of t and the instantaneous value of x at time t. The Hamiltonian function is of the form $H(x, p, t) = T(p) + V(x, t)$ where the $T(p)$ is the usual kinetic energy written in terms of momentum p instead of velocity v and where the potential energy function $V(x, t)$ is an arbitrary function of x and t. This often leads to interesting problems that help to illuminate the subject of classical mechanics. We feel that the same kind of thought processes can help to resolve the mysteries of measurement in quantum mechanics. This method of "Designer Hamiltonians" was applied to the state-preparation problem, and to various kinds of measurement in Lamb's 1969 paper [1].

38.2   Outline of Method

The remainder of this paper outlines a fully classical approach to measurement theory. Both the system and the meter are classical, and the treatment is fully classical. We hope that this classical treatment will help the reader to better understand the methods used in Lamb's quantum theory papers [2–5]. At a suitable point, we will enrich the discussion by taking into account such effects as classical noise in the reading of the meter. The results of such a calculation bear a striking similarity to results found in the above-mentioned papers on the quantum theory of measurements of quantum systems. We find that this fully classical approach does indeed prove to be helpful in understanding a fully quantum theory of measurement.

We take the simplest possible configuration for our discussion. The system is a point particle of mass m and coordinate $x(t)$ which is constrained to move along the x–axis, without friction. The particle may be considered entirely free, save for two exceptions: The system particle itself is being used as a meter in some kind of seismographic manner. For this, it may be acted on by a spatially uniform, and time-dependent, force field $F(t)$, due perhaps to a passing seismic disturbance, or possibly a gravitational wave, whose force-time profile is to be determined. In addition, there may be other forces acting on the system when it is disturbed by the classical-mechanics measurement processes. These involve interaction between the system and the measuring instruments or "meters" similar to those of von Neumann's Chapter VI and of Arthurs and Kelly.

The von Neumann meter [28] suffers from several major defects. Von Neumann arranged his device so that it would introduce a meter-like characteristic into the calculation of the combined meter-system wave function. It was stated above that the wave function of the combined system–meter was initially $u(x)v(X)$, and at the end of the interaction time τ, was changed into a wave function $u(x)v(X - \theta x)$. The x dependence of the wave function of the combined system-meter is then changed by the system-meter interaction. Unfortunately, von Neumann did not spell out what to do next. Arthurs and Kelly came close to supplying an answer. However, they were primarily interested in the simultaneous measurement of x

and p, and corresponding uncertainty products, and hence did not extract some very important features of the theory of measurement of one variable, which might be either x or p.

Our object is different. We want to learn something from a single measurement on the x system by putting it into a brief encounter with an X meter. We only make an observation on the X meter at a time when there is no longer any coupling between the X meter and the x system. From the result of this observation, we hope to learn something about the x system. What can we learn? We can learn enough about the classical state of the system particle, i.e., its position and velocity, to be able to follow its motion after the measurement.

Beginning in 1985, Lamb [2–5], studied the quantum mechanical problem of successive measurement of position. At first he used the von Neumann, and the Arthurs-Kelly meter model. At a certain point in the discussion, he had a two-body wave function of the form $u(x)v(X - x)$. The conventional wisdom at that time was that once one had a two-body wave function for which entanglement had occurred, there could only be a statistical mixture of the corresponding one-body states. Density matrices were required to describe the separated systems.

Suppose, however, that the meter coordinate X were somehow measured, and one got the result $X = B$. In that case, it was tempting to think that the wave function of the system would be the pure-case one-body wave function $u(x)v(B - \theta x)$, where B has a well-defined value. B is now a *parameter* in the system wave function, but no longer a dynamical variable for the meter. Then we have a pure wave function of the system particle, and can expect it to evolve under the action of the force F until it is time for another measurement, etc., etc.

This looks like the "collapse of the wave function" of Dirac and von Neumann. However, it is not that sort of thing at all. Dirac and von Neumann would have been writing about the measurement of the position x of the system particle, and we are writing about measurement of the meter particle coordinate X, hoping to learn something about the system by looking at the meter. In the former case, the wave function was supposed to undergo (a very) substantial collapse to an eigenfunction of the x operator. In our case, the system wave function changes, but not catastrophically. Furthermore, the amount of change can be controlled at the will of the measurement manager by taking a different coupling constant θ.

Despite Lamb's dislike of the von Neumann interaction, it seemed to be a useful thing to explore. He used this model in his 1986 papers [2–4]. It was stated that a new postulate was needed for the kind of quantum mechanics mentioned in the previous paragraph. However, Lamb now realizes that the procedure does not require a new postulate, but follows rigorously from the fundamental principles of conventional quantum mechanics. This will be discussed in the future paper.

There are still other serious difficulties with the von Neumann meter. The assumed square wave temporal pulse interaction energy $(\theta/\tau)xP$ between the system (x, p) and meter (X, P) cannot be realized in any physically reasonable manner. This interaction purports to be some kind of velocity-dependent interaction involving a product of the system x and the meter particle's momentum P. Several view points are possible, but we feel that it is highly undesirable to base a measurement scheme on an interaction energy that cannot be realized in some imagined, but realistic, manner. (Note that there is a big difference between saying that one can have an arbitrary Hamiltonian $H(x, p)$, and saying that one can have an arbitrary potential $V(x, t)$ in a Hamiltonian $T + V(x, t)$.) Of course, it might be held that no measurement facility can exist. In that case, our question would be: "Why talk about measurement in quantum mechanics at all?"

Fortunately, instead of the von Neumann interaction, we can use a different kind of interaction potential, very familiar in both classical and quantum physics, which involves the square of the distance $x_1 - x_2$ between two particles. This Hooke's law potential can have a time dependence. For simplicity, we could take the time dependence to be given by a rectangular pulse. It will be still simpler for the calculations to use a Dirac δ–function pulse. Of course, von Neumann would never have used a delta function, but, as indicated above, he came pretty close to it. For a measurement made at time t_m, we have an interaction potential given by $(1/2)\kappa\tau(x_1 - x_2)^2\delta(t - t_m)$, where κ is a Hooke's law spring constant, and τ is now used as some convenient constant with dimensions of time. Note that τ is not the width of the pulse. That is zero. We have an *impulsive* force. The factors κ and τ are there for dimensional reasons, and only their product matters.

Another very serious difficulty with the von Neumann system-meter interaction based on the product $x P$ is that the location of the meter system makes no difference at all. Except for the difficulty of reading it, the meter could as well be on the planet Mars. When we use our more realizable interaction energy between meter and system, we must specify where the meter system is to be put along the x-axis when its appointed time of interaction arrives. We denote that x value by D, and later give a recipe for choice of a D value for each measurement.

Our object is to follow the motion of the system as a function of the time. Hence, we must have a sequence of measurements at fairly closely spaced time intervals. Several alternative procedures can be considered. We could try to use just one meter continuously connected to the system. However, for the extension to quantum mechanics, the use of a single meter would prove very inconvenient. The meter must be treated by quantum mechanics, before and during the interaction, and we must know its initial wave function at the beginning of each measurement time. Once a meter reading has been made, we want the quantum state of the meter to be known. The meter state is irreversibly "messed up" by the measurement. We have two choices: (1) we simply throw the meter away after its reading and use a new meter, carefully prepared in a known initial state, or, alternatively, (2) we could use just one meter for the whole series of measurements, and re-initialize it in some way after each measurement. This could prove to be a little difficult in practice. Therefore, in order to make a later connection with the quantum theory, we will consider, for our purely classical treatment, that we have prepared a large number of meters, and a new one is put into interaction with the system at the right time and place.

Thinking about, talking about [34], [36], looking at, listening to, touching, tasting, smelling the particle or talking to Wigner's friend [35] about the system will not be considered an accurate way of making measurements on a particle. As physicists, we must stick to a mechanical model, as nearly as possible. We will take a meter to be a particle of mass M. The notation suggests that $M \gg m$, but it need not be. Its initial x–axis coordinate is D and its coordinate after the interaction is $D + X$. We will treat the interaction between the system and meter particle using classical Newtonian mechanics.

It is a convenient fiction of classical mechanics that we can simultaneously make accurate measurements of position like x or X and momentum like p or P on a particle, whether it is a system or meter particle. In that case, we could measure the position of the system particle at any time, and would not need a meter. However, to prepare the ground for the subsequent quantum calculation, we will not try to make such measurements directly on the system, but only on a series of meters at equally spaced times. After each meter has finished its brief coupling to the system, and is no longer coupled to the system, we make a measurement on the meter.

In summary, we would like to determine $F(t)$ by making an observation on each meter which is connected for a brief time to the system at regular time intervals. For now, we will assume a *known* form for $F(t)$ and confirm that the motion of the system particle $x(t)$ determined by the meter measurements agrees with the result expected from Newton's laws of motion under the assumed force $F(t)$. There is a well-defined classical procedure for arranging that experimental errors in the determination of $x(t)$ will lead to errors in $F(t)$. This is called the classical theory of measurement errors, and is easily implemented in our program.

38.3   Discussion of the Measurement Process

The meter particle will be taken to be a "free" particle of mass M. The system is also acted on by a uniform force $F(t)$. At present, we will assume a plausible known form and magnitude for $F(t)$. The system particle is observed by using meters at regular time intervals. These meters have the effect of disturbing the system. This disturbance must be taken into account in the subsequent time development of the system particle.

The system will be a classical free particle of mass m and, for certain brief times, there will be a Hooke's law force between system particle at x and a meter particle at X. We choose the Hooke's law force for simplicity only; other interactions could also be treated using more complicated computational techniques. Unlike von Neumann [25], we assume that the position of our meter does have an effect on the measurement process, so we pay attention to where we place the meter with respect to the system particle. Each meter system is initially placed at rest at the appropriate position D along the x–axis. Hence, initially $X = 0$. After the interaction, the meter particle acquires a momentum P and may possibly be displaced from D to $D + X$ along the x–axis.

The initial position of the meter D has to be chosen for each measurement. At $t = 0$, we set the system particle to be initially at $x = 0$. For the first measurement, which is at some finite time later, we take the starting position of the meter to be $D = 0$, which was the starting position of the system. By this time the system particle may have moved away from the origin. For the subsequent measurements we will choose D to be the system coordinate $x(t)$ determined in the previous measurement. This is a very crude approximation, and it could be easily improved upon. The meter particle coordinate will be represented as $D + X$. Initially, the meter has $X = 0$. The offset X will not change much during the interaction pulse if, as assumed, the time duration of the meter-system interaction is very small.

With the δ-function interaction, the position of the meter particle is not changed at all during the pulse. However, the meter particle does acquire a momentum, and therefore a velocity, and therefore a displacement X at a later time. Then X could be measured, and used to calculate the system coordinate x. It is simpler to measure the momentum P just after the impulsive interaction. Remember that classical mechanics, as we define it, permits either position or momentum to be measured and it is not necessary to go into details about that choice. In later work, the interaction will be allowed to have a finite duration, and the displacement of the meter particle will be considered. Thus we measure P, and use it to calculate the system coordinate, making use of the Hooke's law of interaction between the system and meter particles. After the interaction has terminated, we read the meter momentum. The system particle gets a change in momentum equal and opposite to the one measured by the meter. The current meter is then discarded, and in due course, the next meter is coupled to the system, and so on.

38.4   Computational Details

Before its use, a meter will be prepared in a known classical state. In our case, that means the meter particle is placed at $x = D$ with zero velocity. After the interaction with the system, the meter is no longer coupled to the system. Since the meter is classical, its momentum can be read and this can be done without disturbing either particle. When a measurement is to be made, a meter is allowed to interact with the system particle for a very short time. It helps to have some specific numbers in mind. We take the classical system particle to have a mass of $m = 10^3$ Kg. Under present conditions, the mass M of the meter does not enter, since its momentum is measured. As an example, we can assume a rectangular square wave pulse for $F(t)$ of height 2.5×10^{-17} Newton, which has a duration of 4000 seconds, starting at time 3000 seconds and ending at time 7000 seconds. The motion of the system will be followed for a total of $10,000$ seconds. After every 100 seconds we will make a very short measurement as described below. All the numbers specified here are variables in the computer program. We took $\kappa = 1.0$ Newton/meter and $\tau = 1.0 \times 10^{-12}$ seconds. All the numbers specified here are variable in the computer program.

The potential energy of the interaction, $V(x, X, t)$, between the system particle at x and the meter particle at $D+X$ is dependent on the square of the distance between the particles. The time dependence of V is sharply peaked, and is represented by a Dirac δ-function. We write,

$$V(x, X, t) = \frac{\kappa \tau}{2} (x - D - X)^2 \delta(t - t_m).\qquad(38.1)$$

This interaction potential energy implies an impulsive force on the system particle at time t_m of

$$F(t) = -\frac{\partial V}{\partial x} = -\kappa \tau (x - D - X)\delta(t - t_m),\qquad(38.2)$$

which gives the system particle a momentum change of $-\kappa \tau (x - D - X)$ The change in velocity v of the system particle caused by the measurement is $-\kappa \tau (x - D - X)/m$. With the δ-function interaction, the changes in the instantaneous positions of the system and meter particle are zero. After each interaction, the momentum of the meter particle is measured, and this is equal and opposite to the change in the system particle's momentum. The momentum changes thus determined are proportional to the distance $x - D$. This immediately gives the value of the coordinate x of the system particle. In the quantum measurement paper [5] where the interaction duration τ was small but finite, Lamb "measured" the position of the meter after the interaction. This is one possibility, but we now believe that it is simpler to use the momentum transfer to the meter instead of the coordinate measurement of the meter during the interaction. This will be further investigated in a future paper on the quantum theory of measurement to be published elsewhere.

The meter particle interaction is "switched on" at 100 second intervals. At the end of the 10,000 seconds data run we will have 100 successive values of x which can approximate the desired function $x(t)$ of the system particle. Each measurement has made a disturbance on the system, but a known one. After allowance for the disturbance, application of Newton's second law of motion,

$$F = m\frac{d^2 x}{dt^2},\qquad(38.3)$$

will give a rather good calculation of the desired force $F(t)$, which is the ultimate goal of our enterprise. For higher precision we would need to make a larger number of measurements. As mentioned above, the measurements disturb the system. This effect can be made arbitrarily small by reducing the coupling constants κ and τ. The recoil momentum of the meter particle will become smaller. In an ideal classical mechanics, the smallness of a quantity would make no difficulty for its measurement.

Even in purely classical physics, there are some errors of measurement. A plausible way to take them into account is to bring in, from the classical theory of measurement errors, the idea that when a momentum P is being measured, the result is shifted by a noise momentum P_n selected at random from the abscissae of a gaussian probability distribution $W(P_n)$. This distribution has a mean value 0 and a standard deviation σ. The standard deviation σ is taken to be some small fraction $\beta \ll 1.0$ of a momentum P_0 somehow characteristic of the problem. We take $P_0 = F(\text{peak})\tau$.

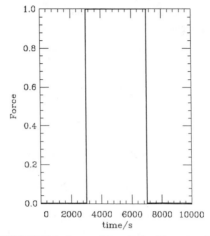

FIGURE 38.1. Square Wave Pulse Force Profile.

38.5   Results

As a simple example, we will take the applied force to be a square wave pulse of the type shown in Fig. 38.1. The actual strength of the pulse will be set by use of a suitable numerical factor. Our computerized measurement program is also applied to obtain the positions (Fig. 38.2a) and velocities (Fig. 38.2d) of the system particle as a function of time. Note that these results for $\beta = 0$ are just what we would expect. Classical measurements can be used to follow classical motion. The error from the simple D recipe is hardly noticeable. We also show a number of curves for the time dependent system particle velocities, (Figs. 38.2b, 38.2c), and positions, (Figs. 38.2e, 38.2f), when the momentum measurements are subject to Gaussian random errors. As described above, the errors are characterized by a fractional β. We take two β values equal to 1×10^{-3} and 1×10^{-2}. Note that the curves are similar to quantum mechanical curves shown in references [2]–[5].

38.6   Toward a Quantum Theory

Our meters must have quantum properties in order for them to adequately make measurements on a quantum mechanical system. (See Ref. [6] for a discussion of classical measurements on a quantum system.) Instead of using Newton's equation of motion we will be dealing with the Schrödinger equation for the specific problem. Fast Fourier transform techniques may be employed to integrate the Schrödinger equation using a split operator approach. Other methods are also available.

   Uncertainties in the measurement will now arise due to the quantum nature of the problem. There may be further uncertainties (or errors of measurement) due to meter reading, in the sense of the above β variables. In fact we would like to

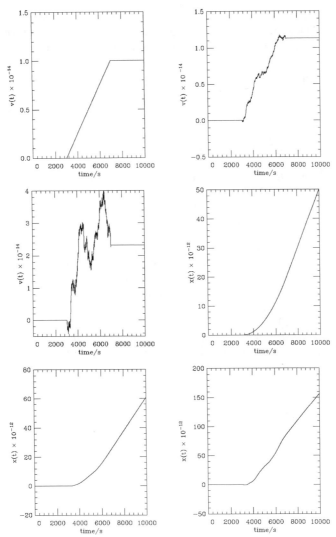

FIGURE 38.2. (a) Velocity of System Particle for Square Wave Force, no error $\beta = 0.000$. (b) Velocity of System Particle for Square Wave Force, error $\beta = 0.001$. (c) Velocity of System Particle for Square Wave Force, error $\beta = 0.010$. (d) Position of System Particle for Square Wave Force, no error $\beta = 0.000$. (e) Position of System Particle for Square Wave Force, error $\beta = 0.001$. (f) Position of System Particle for Square Wave Force, error $\beta = 0.010$.

suggest that the introduction of the β fractional error of momentum measurement, in the classical theory outlined above, has taken us one step in the direction of a quantum theory of measurement. We *are not* suggesting that classical errors in measurement are the same as quantum mechanical uncertainties due to the interaction between system and meter, but rather that a classical theory with errors (or imperfect measurements) takes us one step closer to a theory of quantum mechanical measurement than might have been thought possible. We hope that this work will allow for a better understanding of the quantum theory of measurement papers in references [1–8], and the forthcoming treatment of the fully quantum mechanical problem.

References

[1] W. E. Lamb Jr., "Interpretation of Nonrelativistic Quantum Mechanics", Physics Today **22**, 23 (1969).

[2] W. E. Lamb, Jr., "Theory of Quantum Mechanical Measurements", Ann. N. Y. Acad. Sci. **480**, 407 (1986).

[3] W. E. Lamb, Jr., "Theory of Quantum Mechanical Measurement", *Proceedings of the 2nd International Symposium on the Foundations of Quantum Mechanics in the Light of New Technology, August, 1986*, edited by M. Namiki, Y. Ohnuki, Y. Murayama and S. Normura (Physical Society of Japan, Tokyo, 1987), pp. 185–192.

[4] W. E. Lamb, Jr., "Sequential Measurements in Quantum Mechanics". Lecture at Como NATO Advanced Research Workshop in Quantum Optics, September, 1986. *Quantum Measurement and Chaos*, ed. E. R. Pike and S. Sarkar, pp. 183-193 (Plenum Press, New York and London, 1987).

[5] W. E. Lamb, Jr., "Quantum Theory of Measurement," *Proceedings of the Nato Advanced Research Workshop on Noise and Chaos in Nonlinear Dynamical Systems*, edited by F. Moss, L. Lugiato and W. Schleich (Cambridge University Press, 1990).

[6] W. E. Lamb, Jr., "Classical Measurements on a Quantum Mechanical System" in *Proceedings of the International Symposium on Spacetime Symmetries*, eds. Y. S. Kim and W. M. Zachary, pp. 197-201, reprinted from Nuclear Physics B (Proc. Suppl.) 6 (North-Holland, Amsterdam, 1989).

[7] W. E. Lamb, "Suppose Newton had Invented Wave Mechanics," Am. J. Phys. **62**, 201 (1994).

[8] W. E. Lamb Jr., "Schrödinger's Cat," in *Paul Adrien Maurice Dirac*, eds. B. N. Kursunoglu and E. P. Wigner, pp. 249–261 (Cambridge University Press, 1987).

[9] L. de Broglie, *Thesis* (Masson et Cie, Paris, 1924).

[10] D. Bohm, and B. Hiley, *The Undivided Universe*, (Routledge, London, 1993), and P. R. Holland, *The Quantum Theory of Motion; an Account of the DeBroglie-Bohm Causal Interpretation of Quantun Mechanics* (Cambridge University Press, 1993).

[11] W. Heisenberg, "Quantum Theoretic Re-interpretation of Kinematic and Mechanical Relations," (1925), translated in [13].

[12] B. L. van der Waerden, *Sources of Quantum Mechanics* (Dover, 1968).

[13] J. A. Wheeler and W. H. Zurek, editors, *Quantum Theory and Measurement* (Princeton University Press, 1983).

[14] W. Heisenberg, "The Physical Content of Quantum Kinematics and Mechanics" (1927), translated in [13].

[15] E. Schrödinger, *Collected Papers on Wave Mechanics* (Blackie, London, 1928).

[16] W. Heisenberg, *The Physical Principles of the Quantum Theory* (Chicago University Press, 1930).

[17] E. Schrödinger, "The Uncertainty Principle," Sitzungberichte, Preuss. Akad. Wiss., 296 (1930).

[18] H. P. Robertson, "The Uncertainty Principle," Phys. Rev. **34**, 163-164 (1929).

[19] E. Schrödinger, a translation of the "cat paradox" papers appears in Proc. Am. Phil. Soc. **124**, 323-338 and in [13].

[20] E. Schrödinger, *What is Life?*, (Cambridge University Press, 1944).

[21] K. B. Mullis, "The Polymerase Chain Reaction," pp. 102-117, *Les Prix Nobel, 1993* (Almqvist and Wiksell, Stockholm, 1994).

[22] See reference [12].

[23] A. Einstein, M. Born and H. Bohr, *The Einstein-Born Letters, 1916–1955* (Macmillan, London, 1971).

[24] P. A. M. Dirac, *The Principles of Quantum Mechanics*, 1st ed. (Oxford University Press, London 1930).

[25] J. von Neumann, *Mathematische Grundlagen der Quantenmechanik* (Springer, Berlin, 1932). English translation by R. T. Beyer (Princeton University, Princeton, N.J., 1955).

[26] E. Arthurs and J. Kelly, "On the Simultaneous Measurement of a Pair of Conjugate Variables," Bell Syst. Tech. J. **44**, 725–729, (1965).

[27] M. A. Bunge, *Foundations in Physics*, especially Chapter 5, Section 8 (Springer, New York, 1967).

[28] See reference [2] for a discussion of the classical version of von Neumann's X-meter.

[29] W. Pauli, "General Principles of Wave Mechanics," *Handbuch der Physik*, vol. 24/1 (Springer, Berlin, 1933).

[30] V. B. Braginsky, K. Khalili and K. Thorne, *Quantum Measurements*, (Cambridge University Press, 1992). Also, M. B. Menski, *Continuous Quantum Measurements and Path Integrals* (IOP Pub., London, 1992).

[31] C. M. Caves, "Quantum Mechanics of Measurements Distributed in Time," Phys. Rev. D **33**, 1643 (1986).

[32] S. Weinberg, "Testing Quantum Mechanics," Annals of Physics, **194**, 336–386 (1989).

[33] M. Gell-Mann and J. B. Hartle, "Classical Equations for Quantum Systems," Phys. Rev. D **47**, 3345–3382 (1993).

[34] R. Omnes, *The Interpretation of Quantum Mechanics* (Princeton University Press, 1994).

[35] E. P. Wigner, "Remarks on the Mind-Body Question," in *The Scientist Speculates* eds I. J. Good (Heineman, London, 1961).

[36] See, for example, R. Penrose, *The Emperor's New Mind* (Oxford University Press, 1989); R. Penrose, *Shadows of the Mind* (Oxford University Press, 1994); and H. P. Stapp, *Mind, Matter and Quantum Mechanics* (Springer, N.Y., 1993).

39

The Physics of Nerve Excitation

Harold Lecar

As a graduate student, when I worked on molecular-beam masers, I was struck by how the maser, a state-selection device, resembled the usual picture of the Maxwell Demon. An electrostatic focuser spatially separated molecules in different energy states to direct the higher energy members of a random mixture into a resonant cavity wherein they constituted a negative temperature medium. This Demon is, of course, not paradoxical. It does not violate the second law of thermodynamics because work is done by the focusing electrode's field at the expense of entropy increase of some power supply.

In biology, many wondrous molecular machines employ similar sorting demons to perform various energy transductions. Nerve cells, for example act as amplifiers, whose power supplies are ion concentration cell batteries. These batteries are charged by ion pumps—protein molecules in the cell membrane which are able to extrude Na^+ ions from a mainly K^+ ion mixture in the cell interior and take up K^+ ions from a mainly Na^+ ion mixture in the cell exterior in order to maintain an ion concentration gradient which in turn supports an electrical potential difference across the cell membrane. The pumps are also non-paradoxical demons because during their cycle of activity, they extract energy stored in high energy phosphate bonds to effect a structural change which alters the affinity of ion binding sites for Na^+ and K^+.

In the last fifteen years, research into the biophysics of nerve excitation and sensory transduction has made a transition from the phenomenological understanding of bioelectric phenomena in terms of ionic currents across cell membranes to the beginnings of fairly detailed pictures of how the underlying molecular gadgetry functions. Thus the story of nerve excitation is a bit like that of a branch of solid-state physics run backwards. In semiconductor physics, the quantum mechanical understanding of electronic conductance mechanisms led to the development of the technology of junctional electronic devices whose miniature size (now almost as small as neurons) was essential to the development of modern complex computers. In biology, the brain exists as a complex computer which evolved naturally, utilizing vast numbers of sophisticated ionic-conducting devices—neurons. The biophysics question was to find out the physical mechanisms which underlie the operation of these preexisting devices made of living matter.

The neuronal equivalent of the semiconductor junction is the excitable cell membrane, a bilayer of lipid molecules about 50 Å thick which forms an insulating barrier separating two conducting media—the intra- and extra-cellular electrolytes. Analogous to the electronic currents of semiconductor junctions, the ionic currents across the membrane are mediated by impurities. The membrane impurities are specialized protein molecules which are imbedded in the insulating matrix and act as channels for ion current. When activated, the ion channels are selectively permeable pathways which can distinguish between different ionic species. These ion-specific pathways allow ion concentration gradients between the inside and outside of the cell to set up transmembrane electromotive forces. The cell then acts as a concentration battery to supply the power for generating transmembrane potentials—the electrical signals of the nervous system.

As a topic in physics, the study of the electrical origin of nerve excitation is intertwined with the early history of electricity itself. Galvani's discovery of animal electricity and Volta's development of the battery represent the earliest scientific studies of electricity in the 18th century. In earlier times, speculation about the wave of information that intercedes between thought and action regarded the phenomenon as something occurring in the spiritual realm and hence infinitely rapid. As the anatomy of nerve tracts became known, nerve signals were envisioned by Descartes as winds of animal spirits blowing through the tube-like nerves. This pneumatic mechanism did not seem right even in the 17th century, because such a wind seemed not to be able to blow fast enough, and no anatomists had ever seen hollow channels in nerves. In his *Optiks*, Newton [1] suggested something akin to light wave propagation through the nerve fibers. One of the questions in that book asks, "Is not animal motion performed by the vibrations of this medium excited in the brain by the power of the will, and from thence through the solid pellucid and uniform capillamenta of the nerves into the muscles for contracting and dilating them?"

Helmholtz [2], in 1850, first actually measured the speed of nerve impulse transmission in a frog sciatic nerve. His ingenious apparatus used an inductive switch both to initiate a nerve impulse and to start a ballistic galvanometer running as a clock. When the nerve impulse reached its target muscle, the muscle twitch broke a mercury contact in the circuit which terminated the time measurement. The measured speed was 27 meters/second, certainly finite, an electrical wave moving at less than 60 miles per hour.

A nerve fiber, surrounded by its insulating membrane and immersed in a conducting electrolyte has the same electrical configuration as a submarine coaxial cable. By the 1850s, around the time that technologists were attempting to lay down the first Atlantic cable, the physiologists understood that the nerve impulse was essentially a pulse of electric current transmitted down a leaky coaxial cable. It was also understood that the cable could not act merely as a passive conduit for a signal generated at its input end. A comparison of the physical parameters of the submarine cable and the nerve cable illustrates the design problem encountered in transmitting signals via leaky electrical cables. The first man-made submarine cables, had they not broken while being laid down, would probably not have functioned anyway. The insulating walls used initially were too leaky and the core wires too resistive for transmission across the ocean. Kelvin's analysis of the submarine cable showed that a potential imposed at one end of the cable spreads like heat flow. The signal amplitude decays exponentially with a characteristic length

$$\lambda = \hat{E}\sqrt{\frac{r_m}{(r_i + r_o)}} \qquad (39.1)$$

where r_m is the resistance per unit length of the insulating wall, r_i is the resistance per unit length of the core conductor and r_o is the resistance per unit length of the external medium.

Using Eq. (39.1), we can estimate the cable length for a transatlantic cable made of the materials available in 1850 and compare its function to a nerve axon considered as a cable made of "flesh and blood." The original Atlantic cable had a core of 7 strands of 1.7 mm copper wire, an insulating "membrane" of gutta percha, 1 cm thick and seawater (essentially 0.5 M NaCl) as its external conducting medium. From Eq. (39.1), such a cable has a characteristic length of 1.7×10^4 km, which is long compared to the ocean. Indeed, it was Kelvin [3] who used this analysis to suggest making purer copper wire and more perfectly insulating gutta percha in order that the cable be capable of transmitting a signal across the ocean.

A long tubular cell, such as a nerve axon, forms a coaxial cable, whose operation is much more severely hindered by its materials than the Atlantic cable. The conducting core of an axon is a tiny elongated cylinder of electrolyte gel instead of a copper wire, and the insulating sheath is a cell membrane of molecular dimensions. The core of the biggest known nerve, the squid giant axon, is less than 1 mm in diameter, but the cytoplasmic electrolyte gel has a specific resistivity 10^7 greater than copper. The membrane has a very high specific resistance, but still such a thin insulating layer is many orders of magnitude leakier than the macroscopic cable sheath. With these values, transmission over many centimenters, from head to toe of an animal, is equivalent in terms of losses to having a transmission line across the entire solar system. The input dendrites of neurons often do act as passive cables for conduction over short distances of the order of hundreds of micrometers; however, transmission of information over centimeter- to meter-distances via nerve axons requires an active process amplifying the signal continuously to make up for cable losses.

How this amplification is provided by the excitable membrane, which acts as a negative conductance device, is the central question of classical biophysics. Nerve impulses propagate down the cable as pulses of transmembrane potential which create and are coupled to propagating changes in the membrane conductance itself. The crowning achievement of pre-World War II biophysics was the measurement by K. S. Cole and H. J. Curtis [4] of the impedance of a length of nerve axon membrane as a wave of action potential excitation passed through it. There was a 50-fold decrease in the resistance with only a less than 2% change in capacitance. Thus the membrane exhibits something like a controlled electrical breakdown. Furthermore, the two-order-of-magnitude change in electrical conductance is not accompanied by a substantial change in the membrane's dielectric properties.

The current-voltage characteristics of the cell membrane could only be determined with the development of the voltage clamp by Cole, a method of using a multielectrode feedback arrangement to control the voltage across the cell membrane in order to measure the dynamic currents in this highly nonlinear element. In the early '50s, Hodgkin and Huxley [5] used the voltage clamp to establish a complete theory of the ionic currents responsible for excitation. The excitable membrane has two separate ionic current pathways—one that was selectively permeable to sodium ions and is transiently activated by voltage, and the other a potassium pathway which is activated with a delay and can give a steady-state current.

In the Hodgkin-Huxley theory [6], the distribution of membrane current along the axon is described by a partial differential equation—a nonlinear form of the cable equation—in which the nonlinear characteristics of the membrane are described by subsidiary differential equations which govern the dynamic voltage-dependent conductance changes:

$$\left(\frac{a}{2R}\right)\frac{\partial^2 V}{\partial x^2} = C\frac{\partial V}{\partial t} + J_i(V, m, n, h) \tag{39.2}$$

$$J_i = G_{Na}m^3h(V - V_{Na}) + G_K n^4 (V - V_K)$$
$$+ G_L (V - V_L) \tag{39.3}$$

$$\frac{dm}{dt} = \frac{(m_\infty(V) - m)}{\tau_m(V)} \tag{39.4}$$

$$\frac{dh}{dt} = \frac{(m_\infty(V) - h)}{\tau_h(V)} \tag{39.5}$$

$$\frac{dn}{dt} = \frac{(m_\infty(V) - n)}{\tau_n(V)}. \tag{39.6}$$

$$\tag{39.7}$$

The ionic current J_i has three components: a sodium ion current J_{Na}, a potassium ion current J_K, and a leakage current J_L, which is essentially the resting membrane current. Each current component is given as the product of a driving force (potential minus the equilibrium potential for the ions permeating that pathway) and a conductance. The Na$^+$ and K$^+$ pathways have conductances which can be activated by a change in the transmembrane electric field. The auxiliary variables m, n, and h describe the instantaneous state of the voltage-dependent conductances; they are essentially the probabilities that certain gates are in the open position. The variable m describes the activation of Na$^+$ conductance with positive voltage change; n describes a similar but slower activation of the K$^+$ conductance; h describes the inactivation of the Na$^+$ conductance, which is only activated transiently. When voltage changes, these gate functions relax to new voltage-dependent steady-state values $m_\infty(V), n_\infty(V)$, and $h_\infty(V)$ following the relaxation equations, Eqs. (5-7). The relaxation times for these processes, $\tau_m(V)$, $\tau_n(V)$ and $\tau_h(V)$ are also voltage dependent. The empirical functions for the gate probabilities contain the physics of excitability. The steady-state probabilities are sigmoid functions of voltage and the corresponding relaxations are bell-shaped functions—all deriving as we shall see from the relaxation of charged groups in response to changes in the transmembrane electric field.

One of the great successes of the Hodgkin-Huxley theory was the prediction of the propagated action-potential speed by searching for a traveling-wave solution of Eqs. (2-7). Constant speed propagation requires that the nonlinear reaction-diffusion system have a traveling wave solution of the form

$$V(x, t) = V(x - st) .\tag{39.8}$$

Substituting Eq. (8) into Eq. (3) leads to the higher-order ordinary differential equation

$$\frac{a}{2R}s^2\frac{d^2V}{dt^2} + C\frac{dV}{dt} - J_i(V, m, n, h) = 0 .\tag{39.9}$$

Equation (9) is over-constrained and only has a stable solution for a particular value of s, which turns out to be the observed nerve conduction speed. Thus conduction speed is obtained as a nonlinear eigenvalue. Hodgkin and Huxley computed the value of s which gave a stable solution, and it agreed with the observed speed. Physically, these conditions describe how the circulating currents through an excited region of axon act as a source to stimulate adjacent unexcited regions to threshold in a process that is repeated continuously down the cable. The process is similar to how a burning candle shortens at a constant speed when it dissipates power at a rate equal to the chemical energy stored per unit length in the candle times the velocity of propagation. Mathematically, it is interesting to see how postulating a traveling wave solution leads to a system in an $n+1$ dimensional phase space in which stable singular points of the membrane equations become saddle-point singularities. The propagated action potential trajectory which then proceeds through the saddle point must be a homoclinic (or singular) solution which is stable for only one value of the velocity.

By the late 1960s it was clear that phenomenological entities of the Hodgkin-Huxley theory corresponded with the molecular conducting structures in the membrane. The problem then became that of identifying the molecular mechanisms underlying these voltage-dependent and ion-selective membrane conductances. Although numerous models had been proposed, some involving membrane phase transitions, others involving discrete ion carriers or channels, it was clear that the sigmoid voltage-dependent conductance functions of the Hodgkin-Huxley description resembled polarization functions. This suggested that gating probabilities were related to a Boltzmann distribution of charges or dipoles moving or reorienting within the membrane. There was also much evidence that transport occurred at a limited number of sites sparsely distributed in the membrane. For example, the Na^+ and K^+ ion currents could be blocked by very small quantities of different neurotoxins which were themselves small molecules. Thus one molecule of toxin could act like a cork stoppering a channel. Furthermore, if conductance occurred at such a small number of sites, then each site must be the locus of a rather large ionic flux—a condition most likely to be satisfied by pores through the membrane.

The Hodgkin-Huxley theory could thus be given a physical interpretation in terms of discrete conducting channels having voltage-operated gates. The variables of the theory represented the probabilities of gates being in the open position. One way to produce physical evidence for the existence of the putative channels was to study fluctuations in steady-state ionic currents. The thermal transitions of conductance must produce fluctuations about equilibrium which would give a novel form of electrical noise, channel-gating noise, whose intensity is related to the conductance magnitude of the channels (much like shot noise) and whose spectrum is related to the channel opening—and closing—kinetics. In the early '70s, channel noise was observed and the random transitions of discrete channels were verified [7]. The channel noise explained the characteristics of a classical neurophysiological phenomenon—threshold fluctuations—the probabilistic firing of nerves in response to a constant stimulus [8].

A more direct demonstration of the existence of gated channels came out of the study of artificial membranes [9,10]. Synthetic lipid bilayer membranes were introduced as a model system having many of the physical properties of natural cell membranes. The synthetic bilayers form insulating barriers between electrolyte solutions, as would be expected of a medium of low dielectric constant. However when the bilayers are doped with certain antibiotics and proteins they become conductive. In small amounts, the dopants increase the membrane conductance in discrete steps—each step resulting from the insertion of a single pore structure into the membrane.

In some instances the doped membranes exhibited the type of voltage-dependent relaxing conductance associated with excitability. This allowed the direct observation of the channel gating process at the single-channel level. The fluctuating current of a single channel held at a fixed voltage looks exactly like a random telegraph signal—a series of rectangular current pulses of more or less constant amplitude but with variable durations. The duration distributions of the pulses represent the dwell times of the channels in the open and closed states. The pulses are the stochastic events which underlie channel activation. Current-jump amplitude measured as a function of voltage gives the current-voltage relations of individual open molecular channels. These relations turn out to be fairly linear, indicating that the nonlinearities of gating reside in the statistics of channel opening and not in the transport process through the channel.

The open- and closed-state lifetime distributions change with voltage in such a way that the probability of a single channel being in the open state exactly parallels the macroscopic voltage-dependent conductance. This is a bit like experimentally verifying the ergodic hypothesis. You actually measure the time-average statistics of an individual channel and show that it coincides with the average conductance for an ensemble of channels.

The random-duration jumps seen in single-channel recordings represent the thermal transitions of channel structures between nonconducting and conducting conformational states. The intervals between transitions are generally distributed as sums of exponentials characteristic of an underlying Markov process. Thus the channel structure appears to make memory-free transitions among discrete states. Typical dwell times in the discrete states are of the order of milliseconds, a time scale for the stochastic transition many orders of magnitude slower than the molecular-dynamics time scale for small fluctuations of the channel protein about some equilibrium configuration. The kinetics of conductance change thus represents the biasing of an ongoing random process by the transmembrane electric field.

The important channels for action-potential generation are the voltage-gated Na^+, Ca^{++} and K^+ channels. Voltage-gated channels account for the sharp neural thresholds because of their steep voltage-dependent conductance (e-fold change per $4-12$ mV). The transition probability for gating is drastically altered by very small changes in membrane potential. The modest membrane potentials of the order of 100 mV represent transmembrane electric fields of the order of 100,000 V/cm. In such an intense transmembrane electric field, charged and polar groups experience enormous stresses, which bias the energy difference between different stable states of the channel. Individual channels literally switch between open and closed conformations, and an ensemble of channels relaxes to new conductance values with relaxation times that are just the reciprocal of the sum of the single-channel transition rates.

Thus if the energy difference between the open and closed states is linear in the transmembrane potential, the two conformations in equilibrium are populated according to a Boltzmann distribution in the electric field energy. The probability of a two-state channel being in the open state would then be given by

$$p(V) = \left[1 + e^{(Q(V-V_0)/kT)}\right]^{-1} , \tag{39.10}$$

which is the observed voltage-dependent conductance function of the channel ensemble.

The principle of detail balance tells us that the transition rates must also be exponentially voltage dependent. This leads to the characteristic bell-shaped relaxation time,

$$\tau(V) = \frac{1}{\alpha + \beta} = \operatorname{sech}\left(\frac{Q(V - V_0)}{2kT}\right). \tag{39.11}$$

Thus the voltage-dependent permeability and its concomitant voltage dependent relaxation time come about because the transmembrane electric field and the effective charge Q are sufficiently great that the energy $Q(V - V_0)$ needed to move a charge across the membrane can be much greater than kT.

The quantity Q represents the amount of charge that must be transferred across the membrane potential as the gating subunit undergoes its structural change. For reversible transitions it is easier to think of a group of charges each moving a small distance. The reality of the gating charge movement was demonstrated in experiments that measured the gating current, a displacement current across the membrane resulting from charge movement within the channel structure that could be unmasked when the ion-conduction currents were blocked. Recent measurements of gating current fluctuations [11] have established the gating charge to be 2.3 e per subunit, carried by 3 subunits. The gating currents provide experimenters with an alternative kinetic measurement of the underlying molecular process. The voltage-clamp and single-channel currents monitor the conductance changes that occur when concerted action of channel subunits leads to the opening of a channel. The gating displacement currents monitor the charge movements of the individual subunits.

By the mid-'70s, single-channel gating had been studied for a number of protein and antibiotic channels in synthetic membranes, and channel noise had been observed for many of the important physiological channels. These results provided the impetus for the measurement of single-channel currents in living cells. In natural membranes, the background level of electrical noise is much higher than for bilayers, and the channels are likely to gate faster, requiring experimenters to use broader bandwidth for their detection. To see ion channels in living cells, experimenters had to overcome a more difficult signal-to-noise problem of seeing subpicoampere current jumps of millisecond duration in a living membrane. This obstacle was surmounted by the development of the patch-electrode technique [12]. In the patch clamp, a polished microelectrode is used to seal off a patch of membrane of micrometer dimensions in which only a small number of channels are operative. The background current noise is minimized by making a high shunt resistance between the electrode walls and the cell membrane.

As patch-clamp methods were perfected [13], all the gated channels hypothesized for different excitable membranes were observable *in situ*. With the patch-clamp technique applied to cells grown in tissue culture, gated channels show up everywhere, in all kinds of cells. Not only are there voltage-gated channels of axonal excitation, but chemically gated channels responsible for synaptic excitation between nerves, stretch-activated channels probably responsible for the detection of mechanical stimuli, and a variety of more indirectly activated channels which underlie all the sensory transductions. As examples of a sensory mechanisms, we can now see that vision and smell have a very similar but more indirect transduction mechanism. Receptor molecules absorb light or bind with odorants. The result of this interaction is that the receptors activate a catalytic site; this in turn modulates the concentration of an intracellular messenger substance which regulates the opening of channels responsible for the sensory transduction current.

Molecular biology in the meantime has led to an identification of the channel proteins [14]. From genetic technology, the amino-acid sequence of channel proteins can be discerned. From these sequences we can begin to see a pretty suggestive molecular structure model of the channel. Regions of the protein having long runs of hydrophobic amino acid side chains form transmembrane helices. The basic folding motif of the protein is that of helical columns winding back and forth across the membrane. The Na^+ and K^+ channels are made of four identical domains, each formed of six helical columns. The domains thus correspond to the hypothetical subunits needed to give the Hodgkin-Huxley kinetics their cooperative character.

The challenge now for the molecular structural research is to identify the part of the structure corresponding to the actual voltage-sensitive gate. The amount of charge predicted from the Boltzmann factor requires the transfer of six electron charges across the entire cell membrane in order to get the e-fold per 4 mV change in conductance. Equivalently there might be 30 charges each moving 20% of the way. The latter is more plausible for such a rapidly reversible change as gating. This is a surprising amount of charge for the interior of a low-dielectric-constant membrane, and prompts the search for a rather special distribution of charged amino-acid groups in the makeup of the channel.

In the evolution of proteins, such as the voltage-gated channels, the most conserved sequences of amino acids are likely the ones most essential for function. Since the voltage-gated channels belong to a family of homologous proteins, and all have similarly steep voltage-dependent conductances, there might be a voltage-sensing region common to all. In fact, there is a particular helical column in each domain, called the S4 helix, which has charged argenine and lysine groups in every third position—giving 5 to 8 charges for the entire membrane-spanning helix. The S4 groups are thus compelling candidates to be the molecular concomitants of the Hodgkin-Huxley m and n gates. Experiments in which site-directed mutagenesis is used to modify these groups indeed show that gating is modified.

Furthermore, the channel has a molecular counterpart to the inactivation gate h which could be cleaved by internal enzymes. Each subunit has a small protein ball able to swing freely in solution, but tethered to the channel. These tethered balls act like corks to close channels after they have been activated. When the tethered ball is cleaved by an enzyme, the channel operates without inactivation; when a similar sized protein ball is added to solution, inactivation resumes.

At present, research is beginning to produce a detailed picture of the protein channels and of molecular motions involved in the gating process. In the current state of reductionism, the goal of many investigators is to use this new structural knowledge and a whole panoply of electrical techniques to measure different aspects of the gating process in order to develop an exact picture of channel gating as a molecular stochastic process. Such a description would replace the Hodgkin-Huxley kinetics by kinetics coming from molecular first-principles. When all of the fine details are worked out, the Hodgkin-Huxley equations may not embody the exact literal truth of how the channels operate, but they are amazingly close.

References

[1] I. Newton, *Optiks* (based on 4th ed., London, 1730) (Dover Publications, Inc., New York, (1952).

[2] A. C. Scott, Rev. Mod. Phys. **47**, 487 (1975).

[3] B. Dibner, *The Atlantic Cable* (2d ed.) (Blaisdell Pub. Co., New York, 1964).

[4] K. S. Cole and H. J. Curtis, J. Gen. Physiol. **22**, 649 (1939).

[5] A. L. Hodgkin, *The Conduction of the Nervous Impulse* (Charles C. Thomas, Springfield, Ill., 1964).

[6] A. L. Hodgkin and A. F. Huxley, J. Physiol. **117**, 500 (1952).

[7] H. Lecar and F. Sachs in Excitable cells in *Tissue Culture*, edited by M. Lieberman and P. G. Nelson (Plenum Publ. Corp., New York, 1981).

[8] H. Lecar and R. Nossal, Biophys. J. **11**, 1069 (1971).

[9] G. Ehrenstein, H. Lecar and R. Nossal, J. Gen. Physiol. **55**, 119 (1970).

[10] G. Ehrenstein and H. Lecar, Quart. Rev. Biophys. **10**, 1 (1975).

[11] F. Conti and W. Stuhmer, Eur. Biophys. J. **17**, (1989).

[12] E. Neher and B. Sakmann, Nature **260**, 799 (1976).

[13] B. Sakmann and E. Neher, *Single-Channel Recording* (Plenum Publ. Corp., New York, 1983).

[14] B. Hille, *Ionic Channels in Excitable Membranes* (2d ed.) (Sinauer Assoc., Sunderland Mass., 1992).

40

The Future of Science Education

Leon M. Lederman

40.1 Preamble: The Influence of Science and Technology

In the 17th and 18th centuries, technology, whose origins go back to pre-history, was largely invention-based. Inventors did not have a basic training in scientific fundamentals—they thrived by gifted intuition, by trial and error and by a heritage of handed-down experience. But even in this period, and much more so in the 19th century, the driving force for technology was scientific understanding. Thus, Faraday's (1820) invention of the electric motor and generator came directly out of the drive to understand the physics of electromagnetism. Faraday did not even take the time to patent his discoveries. In our own times the pace of technological discoveries continues to increase. These are based upon understanding of basic scientific principles, but additionally, the creation of new technologies provides a powerful tool for conducting basic research. Thus, we have an accelerating pace of change: science begets technology and technology enables new science which begets more technology. An example helps: In the 1920's, experimental data from the atom required an entirely new theory—which became known as the quantum theory. The quantum theory, applied to electrons in metals and semi-conductors, led to the invention of the transistor. The transistor revolutionized electronic engineering and gave rise to microelectronics and high-speed digital computers. Modern physical instruments and particle accelerators are based upon these inventions and provide the tools for further advances in all fields of basic research.

We live in a world which is characterized by change. The pace of change, in fact, is increasing so that what used to evolve over a period of fifty years now takes place in ten years. The driving engines for this spiralling change is science and science-based technology. The change has to do with increased longevity because of improved sanitation and health care, revolutionary improvements in communications, transportation and access to information and entertainment. Of course, not all of these changes are positive, but they are a facet of our times and they influence economics, politics, modes of living and thinking. The world has become so entwined, continent with continent, region with region. The fate of nations is welded together into what is aptly named "the global village." *As a world society, it seems clear that we have arrived at a point in our history when there must be a major increase in the capability of ordinary people to cope with the scientific and technological culture that is shaping their lives and the lives of their children.* In a world in which illiteracy is the shame of societies in which it is found, the *science illiteracy* is increasingly disastrous. And wherever it is measured, one finds illiteracy rates of 90-95%. The quality of primary school science education is similarly abysmal and these conditions are common in developed and developing nations.

Our world is full of brilliant potentialities and menacing threats. For the past hundred years our science and our technology have been driven largely by military and economic forces. We stand at a crucial juncture at the end of millennia: whether to apply our science with humanistic wisdom for the advance of humankind or to succumb to the base forces and epic tragedies that weave our history. The global village image raises the stakes enormously in this age-old conflict.

Education in the science tradition must take place on all fronts. We must begin in the earliest grades of schooling or even pre-schoolings, but we must not overlook the need to reach the parents and citizens who oversee the schools. It is this wasteland of education that we must address.

40.2   Science Education

Historians of education and especially science education have filled volumes with the waxing and waning of educational reform—in nations over the globe. In the U. S., reforms received great boosts after successive wars— most dramatically after World War II when the famous Vannevar Bush report, "Science, the Endless Frontier," gave a high priority to science education as an essential component of a new relationship between the State and the scientific and technological community. Some of this activity in reform of science education began to fade in the U. S. only to be shocked into greater frenzy by the 1957 success of the Soviet Sputnik. I recall many of my colleagues and teachers pausing their research activities to write splendid high school books in all fields of science. These books were unfortunately aimed for future scientists and little attention was paid to elementary school work. This was left to the educational psychologists, led by the Swiss psychologist Piaget and the succeeding generations that founded the subject of cognitive science in the mid-1950's. A blending of disciplines—psychology, philosophy, linguistics, anthropology, computer sciences and neuroscience—were marshaled to try to explain human behavior in terms of internal cognitive processes and structures of the human mind.

In engaging in science education—science for children— education for investment—I naturally relate it to science itself. There is a deep connection between the future of science education and the future of science, with the way children learn about the world and the way scientists learn about the world. There is more here than mere metaphor. Let me try to be more specific.

Science is a process of observation, measurement and synthesis. And this is the sequence that has been modeled in many of the *hands-on* science education programs that we know can be so successful. But what scientists observe and what they choose to measure are constrained by what we already know and think we understand. The *creative* insight comes from recognizing the limitations of intuition—very similar to intuitions which children acquire and accumulate in their explorations. Consider the far from trivial example of Galileo's great discovery (immortalized as Newton's First Law) "An isolated body will continue its state of motion forever." Boy, if that isn't counter-intuitive! The creative act was to realize that our experience is irrelevant because in our normal experience, objects are never *isolated*. Balls stop rolling, horses must pull carts to continue the motion.... However, Galileo's deeper intuition suspected simplicity in the law governing moving bodies and his insightful surmise was that if one *could* isolate the body, it would indeed continue moving forever. So he polished the block and he polished the table and the block moved much further....he knew he could not succeed in complete isolation but perhaps enough to expose the underlying simplicity. Galileo and his followers for the past 400 years have demonstrated how scientists must construct new intuitions in order to know how the world works. Now listen to educational psychologist Howard Gardner in 1994: "We argue that during the early years of life children form extremely powerful theories or sets of beliefs about how the world works—theories of mind, theories of matter, theories of life....these become so deeply entrenched in the human mind that they prove very difficult to eradicate in favor of more comprehensive and more valid views that have been painfully constructed in and across disciplines."

What Gardner says about children can be said about scientists, Congressmen, judges....and elementary school teachers! Replacing children's powerful misconceptions is the art of science education. Children need the same "intuitionÐ modifying" experiences that the scientists need—but where the scientist's device is a synchrotron light source, a dye laser, a mass spectrometer or a particle accelerator, the children need opportunities to use their own hands, to consort with their own small collective of other children in order to confront artfully contrived experiential processes.

As *they* polish the block and the table, the block will go further and further and it is in the accumulation of what must be a large number of such examples that *"science as a way of thinking"* will begin to crystallize. Let me simply assure you that the new way of teaching science to young children, ages 4-12, can be fantastically successful—can open the door to the joy of learning. But let us all realize that this is a difficult, time-consuming process, nothing less than a change in the culture of the classroom. To change ways of thinking of children, we must first change the ways of thinking of teachers. However, you think about education, bureaucracy, politics, technology it is, fundamentally, the teacher in the classroom with children that is the *key priority*. For teachers who love children and who love teaching, it is still a major problem-changing culture is extraordinarily difficult. And teachers who must teach science are poorly educated to do this—true in Rio, in Nairobi, in Milano and in Chicago. The role of scientists in this task should be obvious. It is critical that we get involved.

After spending the past five years immersed in the school reform business in the city of Chicago, I can certainly understand the pessimism that most people have about the future of our schools—especially the large public school systems that serve the large fraction of disadvantaged children. But we do have a model in which scientists and private industry can collaborate to intervene in the public schools and retrain teachers to teach science and math in the new way.

What makes this seeming near-impossibility of improving large numbers of schools so frustrating is that we actually know how to make the schools work better—we know what makes a good school and that they do exist and share common attributes. These are the necessities that emerge from the literature of school reform:

- A *belief* that all children can learn, although they may learn differently.

- An environment that is caring, personal, considerate and respectful of both children and adults.

- A shared vision–an education mission shared by the *entire* school community: parents, administrators, mayors, head-masters and teachers.

- A prioritization that places children's learning needs at the center of every activity.

- High expectations of everyone–children, teachers, parents and principals.

- A competent, well-trained teaching staff which is *given time* for *collegial interactions* and professional development. These teachers must be empowered to make decisions based on sound professional judgement.

- A basic collaboration between the school, as described, the parents, the principal, the local community; including those human resources that can be marshalled to the school's purposes: local industry, university, research laboratories.....

I am aware that I am not including many important ingredients such as the educational technologies, the capability of distance learning and the need for continuous education as the new ground rules of the work-place, which, from the factory floor to research lab, are changing almost hourly. I do want to focus on the most important issues.

If the primary mission of our schools is to prepare students to cope in the changing world of the Global Age, then educational systems designed 100 years ago cannot succeed. To succeed in implementing this radical reform of the style and culture of classroom teaching is not easy. It requires the concurrence of willing and enthusiastic parents and citizens and this leads me to the second crucial component– the science literacy of the general public.

40.3   Scientific Literacy

As we have emphasized, science and technology, which is increasingly based upon science, have transformed the way humans live and think. One has only to note the rate at which new words enter our vocabulary: consider the verb "to fax." I believe it is the same in Italian. We also "jet" to Rome or Tokyo. Then there is CAT SCAN and INFO SUPERHIGHWAY. These are examples of technologies that are ushering in the new century—the Global Age. The nature of these transformations and the interplay of science, technology and education are poorly understood by the great majority of people...there is, among the citizens of the world, the expectation of change, of improvement, of continued relief from disease, protection from natural disasters, continued extension of longevity but there is also a new awareness of the dark sides of technology, a mistrust which eventually turns to fear if it is ignored. Fear is born of ignorance, and this leads us to education and the need for science literacy for all peoples.

Why should we worry about this? Of course, the future scientists and engineers who will practice science and work in technology must have a solid science education but why is it necessary for the rest?

Some reasons for the need for science literacy are less than convincing. Your home is full of objects of modern technology and we need to know how they work! Why? Most of my physics colleagues cannot program their own VCR's and God forbid that you should try to repair the TV set or the vacuum cleaner.

It is argued that our politicians and CEO's need to understand science in order to make informed decisions on scientific matters or at least to know enough to support research. In a recent conversation, President Clinton convinced me that he is probably the only U.S. President who really understands what a neutrino is—but research funding in the U.S. is in deep trouble anyway. The science base for public policy decisions are usually provided by others—by science advisors with the help of commissions and panels. If you have the power, experts are always available.

Sir George Porter, the British Nobel Prize winning chemist, identified three reasons why the public understanding of science is important:

First, science is part of our culture, like the arts, like literature and music, it can enrich the life of the individual who engages in it, however vicariously. Children usually love science and if we define a scientist as someone who asks the right questions, they are all scientists, before schools destroy that natural curiosity. There is a tremendous response when efforts are made to make science and technology palatable. The long TV series entitled "The Ascent of Man" narrated by Joseph Bronofski, was attended by over 10 million viewers, as was Carl Sagan's "Cosmos." Stephen Hawking's book, *A Brief History of Time* has sold over five million copies and has spawned sequels, book-length keys and cult groups wearing Hawking t-shirts. More people visit science museums in the U.S. than attend baseball, football and basketball combined. Polls continually expose a hunger for science news among the general public. In any bookstore, there are, in recent times, always a few science books among the "best sellers." Perhaps science is replacing the frontiers between civilization and the vanishing wildernesses to satisfy the basic human yearning for romance.

The second reason is economics—increasingly, occupations of all kinds need some mastery of mathematics and understanding of science and technology. Industry spends billions of dollars training their work force and the winners in the quest for location of job-producing industry are regions where the educational level is high.

But the most urgent reason has to do with the need for citizens to be involved in decision making–in engaging in the discussions of public policy where science and technology are important factors. We have already noted the need for citizens to insist on a reform of the early education of children. Without the active concern and political insistence of the public, we will not have the kind of new education the 21st century demands.

A fourth reason or better, an extension of Porter's argument, is gleaned from the newspapers and magazines; from the front page headlines and from the intellectual back pages. We learn that we are stuck with a world which is addicted to technology and the science which drives it—not only to keep us healthy, to make us rich, to provide for our culture—but also to solve some of the many problems which were generated by technology.

We need a huge advance of science and technology to provide a reasonable prospect for a world with a population heading for 8-10 billion people. To provide the energy and sustenance in an environmentally safe way is beyond our present capabilities. To provide the resources, human and material, to address these problems requires a popular consensus.

But we have the dilemma that ignorance of science all too often goes over to fear—and fear is the end of rational discourse. We see today a growth of anti-science—to the point of hatred.

The worldwide emergence of radical fundamentalism, from Teheran to Hebron, from Algeria to the skinheads of Germany and California....all constitute the extreme counterpoint to the enlightenment, a 500-year old commitment to rationality. Another menace to the orderly and compassionate advance of humanity is a rising mood of immediacy. In business and in government, the short-term outcome seems to weigh increasingly over investment for the long-term. The business chief too often wants an increase of dividends to appear during his short tenure. The politician faces re-election *next year*—so cut taxes now! As a consequence, research and education suffer. These are long-term investments. In the industrialized world, we see a trend in which children are no longer better off than their parents. Nation after nation is now re-examining faltering educational systems.

Finally, there is a growing anti-science movement that originates in the Academy itself. Featuring such contentious phrases as "end of progress" or "post-modernism," it condemns science and its vaunted objectivity as outmoded. There are, in our universities, well-placed and articulate scholars who promote the view that western science is biased to its deepest roots and that nothing but its whole-sale destruction and subsequent rebuilding will rid western science of its parasitic chauvinism, and this includes Newton's principles, as well as the quantum theory and relativity.

In 1995, we have parliamentary critics (and their staffs) who sit on powerful committees and who don't understand how science works.

40.4 Action Plan

The commitment to rationality and the advance of mankind is under attack by extreme dogmatists, by excessive greed, fear, superstition, hatred and ignorance. Using the existing international organization and, in particular, UNESCO and ICSU (International Council of Scientific Unions), a unified working group should be assembled to address the linked issues of science education in primary and secondary schools and the scientific level of understanding of the general public. Structures exist to do this, experts abound. A groundswell of demand for action is needed.

There has never been a coordinated, international effort to increase the public understanding of science. This is needed in the underdeveloped nations as well as in the developed nations. This conforms to the new maxim: "think globally, act locally." The equally abysmal ignorance of science creates a symmetry between the North and the South in the U.S. It is a problem that must be addressed at all levels of education. In the primary schools the issue is sharp because poorly educated teachers there begin the process of turning young people away from science and mathematics. However, a strong improvement in primary schools requires the support and insistence of parents so we must wage the campaign at all levels: the children, the teachers, the parents, the ministers. Pedagogy and entertainment aspects of this new educational program must be understood. The assistance of behavioral psychologists, popularizers, advertising professionals must be enlisted if we are to have a successful educational program. A world-wide collaboration is an exciting prospect. The tools exist, TV is a superb medium for raising the level but, in fact, it is too often a negative influence. The Internet and the rapid, world-wide increase of connectivity is the bright and affordable hope of addressing all these issues. Use should also be made of cinema, radio, magazines, etc.

In recent years the world has witnessed two spectacular events: the Rio de Janeiro Conference of 1992 on the environment and the Cairo Conference of 1994 on the population. World leaders assembled to discuss these seminal issues for the future of the planet. But the future success or failure of these bold ventures in global collaboration will hinge on the public acceptance, the public endorsement, the public understanding of the issues involved. A similar conference on public science literacy would seem to be a necessity.

41

Noncoherent Feedback in Space Masers and Stellar Lasers

Vladilen S. Letokhov

The discovery of quantum electronics was an inevitable and logical stage in the practical application of the laws of quantum physics that is rightfully considered one of the most important scientific technological achievements of the 20th century.

41.1   From Quantum Physics to Quantum Electronics

In 1900, Max Planck introduced into physics the concept of a quantum—an indivisible portion of energy that can be absorbed or given up in the process of emission. In 1905, Albert Einstein introduced the notion of the quantum of light, or photon, as a real particle of the electromagnetic field. Then Niels Bohr analyzed the photon emission process and arrived at the following fundamental conclusion: An atom is characterized by a set of certain energy levels, i.e., certain values of its total energy, E_1, E_2, E_3, and so on. Each of these levels corresponds to a stationary atomic state in which the atom does not emit; emission only takes place when the atom makes a transition from one state and into another; the emission frequency is related to the difference in energy between the initial and final energy levels: $\omega = (E_2 - E_1)/\hbar$, where $\hbar$ is Planck's constant.

The next and most important step forward was made by Einstein in 1916 in his famous work [1] entitled "The Emission and Absorption of Radiation by the Quantum Theory," wherein he introduced two types of quantum transitions of an atom or molecule between discrete quantum states:

- the *spontaneous transition* of an excited atom to a state with a lower energy, accompanied by the emission of a quantum of light. As Einstein puts it [1]: "This transition takes place without any external influence. One can hardly imagine it to be like anything else but radioactive decay."

- the *induced (stimulated) transition* of an atom between energy states. This transition is due to radiation acting on the atom, and its probability is proportional to the radiation intensity. If the atom that is induced to make a transition is excited, it then gives an additional light quantum to the radiation, and vice versa, if it is not excited, the atom then absorbs a similar quantum from the radiation.

Einstein introduced the concepts of the spontaneous and stimulated transitions, based on his astonishing intuition: The only proof was the fact that these notions helped him easily derive Planck's formula for the thermal radiation spectrum.

Einstein proved absolutely right. In 1925-1926, *quantum mechanics* was developed by the efforts of a brilliant international community of physicists, who described the necessary interaction between light and atoms and molecules. And soon (in 1927) the outstanding English physicist Paul Dirac constructed the *quantum radiation theory* [2], leaning upon the principal equation of quantum mechanics, namely, the Schrödinger equation, and the Einsteinian idea of the light quanta. The Dirac radiation theory gave a rigorous proof of Einstein's hypothesis for spontaneous and stimulated emission of radiation, and predicted the main properties of these phenomena. A photon whose emission was induced by another photon was demonstrated to have absolutely the same characteristics as the latter, i.e., the same propagation direction, energy or radiation frequency, and polarization. It is precisely on this *identity of photons* of stimulated radiation that the whole of quantum electronics is based, specifically including lasers—coherent light sources. And the term "coherence" itself means that the elementary quantum emitters—atoms and molecules—emit quanta of identical characteristics, i.e., they emit in synchrony.

But the phenomenon of the stimulated emission of radiation cannot by itself help one either control the emission process or, the more so, obtain any powerful light emission. Indeed, as postulated by Einstein, there exist two types of stimulated transitions: one accompanied by the emission of photons and the other involving the absorption of photons. Under ordinary, equilibrium conditions, the number of atoms residing at any excited energy level is always smaller than that at any lower energy level. For that reason, the number of stimulated transitions attended by the absorption of photons always prevails in matter over that of such transitions involving the emission of photons. This manifests itself in an obvious fashion: all bodies around us absorb radiation to some or another extent.

By 1930 ideas were conceived that this situation might have changed cardinally if a *nonequilibrium distribution* of atoms among excited states had been produced in some other way, so that the number of atoms in one of the excited states had proved greater than in a lower quantum state (see chronology in [3]). Such a nonequilibrium distribution of atoms throughout the available energy levels is now referred to as the *population inversion* of the levels.

The idea of the possibility of *light amplification* by stimulated emissions in a medium with inverted population had opened the way a little to the production of flows of identical photons, but subject to one condition: a flow of equally identical photons had to be fed to the input of the amplifying medium. Otherwise, all the imperfection of the light beam at the input of the amplifying medium would be repeated in an amplified form at its output. All common light sources emit chaotic radiation from which it is impossible to isolate a flow of identical photons of any intensity for their subsequent amplification. It took one more decisive idea, namely, to place the amplifying medium in a *resonant cavity* [4–6] wherein the photon loss factor was lower than the gain factor, i.e., to establish a *positive resonance feedback*. One then could start the amplification process with a single or a few spontaneous photons that would give birth, while repeatedly passing through the amplifying medium, to a great many photons just like themselves. This chain process of undamped multiplication of identical photons is usually called generation, and such a device itself is known as the *maser* in the case of microwaves, and the *optical quantum generator* or *laser* in the case of light. However, the idea of the coherent light quantum generator came to optics from radio. In substance, the discovery of the laser occurred through the synthesis of the key ideas of optics and radio. For this reason, it is interesting to discuss why it was the maser and not the laser that was the first to be discovered.

The experimental verification by Heinrich Hertz of the existence of electromagnetic waves predicted by James Maxwell to have the same nature as light waves marked the beginning of the intense mastering of the radio-frequency band. The development of radio engineering followed, raising the power output of radio generators, improving the sensitivity of radio receivers, and gradually shortening the wavelength of radio waves being mastered. The development of radar led to the mastering of the centimetric wave band. Microwave generators allowed the start of systematic research on the absorption spectra of molecules in the microwave band, and *microwave spectroscopy* [7] made its appearance. It was perhaps here that the first synthesis took place of the concepts of quantum mechanics that clearly explained microwave molecular spectra, and the concepts of experimental radio engineering.

Microwave spectroscopic studies were similar to investigations of the optical spectra of atoms and molecules, but the experimenters now had at their disposal radiation sources of a much higher spectral brightness than in optics. Since spontaneous transitions in the microwave band have a very low probability, the main part in the interaction between radio waves and molecules was played by stimulated transitions. Along with the direct (absorbing) stimulated transitions, reverse (radiative) stimulated transitions in excited molecules predicted were also observed by Einstein. This was the second fruitful synthesis of the predictions of quantum mechanics and the experimental capabilities of radio engineering. Using a nonuniform electric field in experiments with molecular beams, one could sort out molecules in different quantum states. Based on such experiments, C.H. Townes and co-workers in the U.S.A., and N.G. Basov and A.M. Prokhorov in the U.S.S.R. suggested and implemented the idea of the *molecular generator* or *maser* [4–6]. This in-principle new type of electromagnetic generator relies for its operation on the preparation of "active" (excited) molecules, and the stimulated emission by them, of energy in a resonant cavity sustaining undamped electromagnetic oscillations.

The development of microwave quantum electronics led to the creation of low-noise quantum amplifiers built around paramagnetic crystals [8, 9] and a quantum generator using hydrogen atoms [10] that featured a very high frequency stabil-

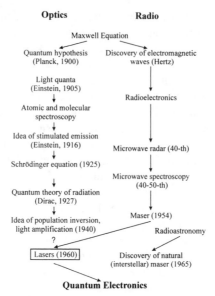

FIGURE 41.1. Sequence of discoveries in optics and radio which led to the advent of quantum electronics.

ity. But the most impressive achievements in quantum electronics still lay ahead, on the way to extending the maser principle to the optical band [11]. In 1958, A.L. Schawlow and C.H. Townes [11] and A.M. Prokhorov [12] suggested using in the short-wavelength region an open Fabry-Perot cavity instead of a closed cavity. And soon, in 1960-61, the first *optical masers* or *lasers* were developed [13, 14].

Figure 41.1 illustrates the sequence of events in the development of concepts, ideas, and experiments that led to the advent of quantum electronics. One can clearly see the potential possibility, not implemented in reality, of discovering the laser on the way to the development of the ideas of light amplification in a medium with an inverted population.

The design of the laser is much simpler than that of the maser, and it now seems surprising that science had to zigzag so much on the road to the coherent light source. This can be explained. The first lasers used perfect crystals or highly reflecting mirrors that were unavailable in the 1920s-30s. But this is hardly of principal importance, for now it is possible to build a laser, for example, a pulsed CO_2 laser in the IR region or an N_2 laser in the UV region, from the handy materials that were available at the physical laboratory even in the '20s. The fact cannot be ruled out that some experimenters would accidentally come across phenomena close to the laser effect, but would pay no attention to them. If one had managed to observe stimulated emission of radiation in an experiment of this sort, it would have probably been explained because the necessary theoretical basis was already there. Such a discovery would be a matter of chance, but it would not have any part in the advent of quantum electronics.

To this must be added the following historical factor. In the '30s-40s, the majority of the active physicists were engaged in the field of nuclear physics and technology. This led to a "brain drain" from the other fields of physics, atomic-molecular and optical physics in particular. A certain role in this process was played by the accelerated development of nuclear physics and radar technique during World War II. This made for the predominant development of the radio trend compared to the optical one. This may perhaps also explain the "zigzag" in the discovery of quantum electronics distinctly seen in Fig. 41.1.

Irrespective of the above explanations, the advent of quantum electronics is a good example of the development of a new domain of science and technology along a logical, consistent path preordained by the course of evolution of science and technology as a whole. The interconnection and intertwining of the different branches of the "tree of science" make such a discovery inevitable, even though chance had not played a part. So, with quantum electronics in mind, it was the theoretical basis, namely, the stimulated radiation theory [1, 2], that was developed first. Then experimental techniques and instruments were worked out to observe the phenomenon predicted theoretically. In our case these were microwave generators that made possible the microwave spectroscopy of molecules and the observation of stimulated transitions. Thereafter, the concept of positive feedback universally adopted in radio engineering was naturally combined with the phenomenon of stimulated emission of radiation by excited molecules, and the result was the development of the maser [4–6]. The creation of the laser that soon followed was a logical, natural stage in the evolution of quantum electronics whose discovery and further development was commended by the 1964 Nobel Prize in physics awarded to Townes, Prokhorov, and Basov.

To conclude this brief discussion of the "genesis" of quantum electronics, let us present one more proof of the inevitability of its discovery on the verge of the 50s-60s as a result of the interplay between various domains of science and technology. Ten years after the development of the laboratory maser, natural space masers were discovered in certain regions of interstellar ionized hydrogen (HII regions) emitting on quantum transitions in the OH radical ($\lambda = 18.5$ cm) [15] and the H_2O molecule ($\lambda = 1.35$ cm) [16, 17]. The exceptionally high temperature of the radio emission of the hydroxyl molecules ($T_{bright} \gg 10^{10}$ K) was at once explained on the basis of the maser principle—amplification in a medium of molecules with an inverted population of the energy levels involved in the transition corresponding to the frequency of the spectral line being amplified. Imagine if no laboratory maser had been developed by that time. In that case, it would be inevitably discovered thanks to the radio-astronomical observations.

And finally, consider the remarkable fact that Professor C. H. Townes was a pioneer in the discovery of the *maser* [6], the *laser* [11], and the *space maser* [16]. In the text below, we will discuss some specific features of interstellar masers and stellar lasers with a scattering feedback which convert them from amplifiers to generators of an unusual type.

41.2   Noncoherent and Nonresonant Scattering Feedback

Masers are based upon placing the amplifying medium in a resonance cavity [4, 5], and lasers, upon placing this medium in an open Fabry-Perot cavity [11, 12]. In both these cases, generation takes place on a limited number of resonant modes, and so the generated electromagnetic field is spatially coherent. Such a feedback may be called *coherent* and *resonant* (Fig. 41.2a). This type of feedback fulfills two functions: (1) it returns some of the electromagnetic energy emitted by the active medium and (2) forms a stable electromagnetic field configuration in the cavity because of the constructive interference between the incident and reflected waves.

However, other types of positive feedback are also possible, where there occurs the return of electromagnetic energy into the amplifying medium without any stable phase retention, i.e., without formation of a small number of resonant modes. This was demonstrated in [18] using as an example ruby crystals with a high amplification and scattering surfaces or cells containing a scattering medium instead of the mirrors (Fig. 41.2b).

Escape of radiation from the system by scattering is then the predominant loss mechanism for all modes. As a result, instead of individual high-Q resonances there appear a large number of low-Q resonances which overlap and form a continuous flat spectrum. The absence of resonant feedback means that the spectrum of the radiation being generated tends to be continuous, i.e., it does not contain discrete components at selected resonant frequencies [18] (see review [19]). The only resonant element left in the laser is the amplification line of the active medium. Therefore, after reaching a threshold the oscillation spectrum narrows continuously towards the center of the amplification line. It was found [20] that the process of spectral narrowing is much slower than in an ordinary laser with resonant feedback, and that this limits the width of the spectrum in the pulsed mode operation. In the steady-state or continuous-wave (cw) mode the spectrum width is determined by fluctuations. The statistical properties of the radiation in the case of nonresonant feedback are somewhat unusual for a laser. It has been shown [21] that it possesses the statistical properties of the radiation of an extremely bright "black body" in a narrow range of the spectrum. The radiation of such a laser has no spatial coherence and is not stable in phase. Since the only resonant element in a laser with nonresonant feedback in the amplification line of the active medium, the mean frequency of the radiation generated does not depend on the dimensions of the laser but is determined only by the frequency of the center of the resonant amplification line. If this frequency is sufficiently stable the radiation of this kind of laser has a stable frequency.

The above-described type of laser with a nonresonant scattering feedback may be more properly referred to as a laser with a *nonresonant feedback* (Fig. 41.2b). Such a laser can be built around an amplifying medium with *nonresonant scattering particles* inside it, i.e., on the basis of a scattering medium with amplification [22].

A. Resonant Coherent Feedback

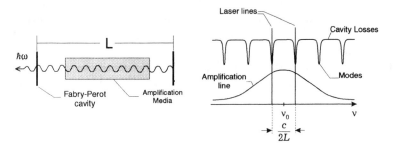

B. Nonresonant Noncoherent Feedback

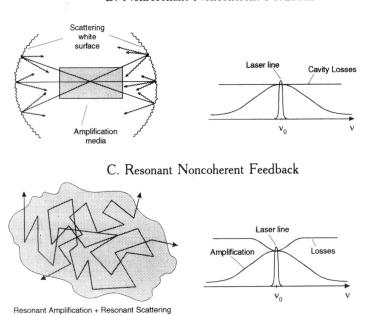

C. Resonant Noncoherent Feedback

Resonant Amplification + Resonant Scattering

FIGURE 41.2. Types of feedback in lasers: (a) resonant coherent feedback effected by means of the Fabry-Perot resonator; (b) nonresonant noncoherent feedback effected by means of scattering surfaces or by particles in the bulk of the active medium that scatter radiation in a nonresonant fashion; (c) resonant noncoherent feedback effected by means of atoms or molecules that scatter radiation in a resonant fashion.

There is a wide diversity of nonresonant scattering mechanisms—microparticles, Rayleigh scattering, etc. Let us consider briefly the specific features of such a laser with nonresonant noncoherent feedback.

Take for the sake of simplicity an ensemble of identical dielectric particles with an average number per unit volume N_0 and a complex permittivity $\epsilon = \epsilon_0 + i\epsilon''_\omega$, where $\epsilon''_\omega > 0$ in the vicinity of the frequency ω_0. Let Q_s be the scattering cross section, and Q_ω the cross section for amplification of light of a frequency ω on the particle. Here we take the case when each particle scatters and amplifies. Naturally the results obtained can easily be extended to the case of different amplifying and scattering particles, and also to the case of scattering particles immersed in a uniform amplifying medium.

Consider the case when the average dimension of the region occupied by the cloud R, the mean-free path of a photon due to scattering $\Lambda_s = 1/Q_s N_0$, and the wavelength of the emission λ satisfy the inequalities

$$R \gg \Lambda_s \gg \lambda, \tag{41.1}$$

while the mean distance between the scattering particles $N_0^{-1/3}$ is much greater than λ:

$$N_0^{-1/3} \gg \lambda. \tag{41.2}$$

Then the change in flux density $\Phi_\omega(\mathbf{r}, t)$ of photons of frequency ω at the point $\mathbf{r}$ can be described as follows in the diffusion approximation:

$$\frac{1}{c}\frac{\partial \Phi_\omega(\mathbf{r}, t)}{\partial t} = D\Delta\Phi_\omega(\mathbf{r}, t) + Q_\omega(\mathbf{r}, t)N_0\Phi_\omega(\mathbf{r}, t), \tag{41.3}$$

where D is the diffusion coefficient, Δ is the Laplace operator, and c is the mean velocity of light in the region occupied by the particles. The frequency dependence of the photon density is connected with the resonant nature of the amplification. The cross section $Q_\omega(\mathbf{r}, t)$ does not remain constant, since the imaginary part of the permittivity $\epsilon_\omega''(\mathbf{r}, t)$ depends on the photon flux density due to the saturation effect.

The process of diffusion multiplication of photons described by Eq. (41.3) is in many ways reminiscent of the diffusion of neutrons in a homogeneous nuclear reactor. In the solution we can therefore make use of certain results known for the diffusion of neutrons [3]. At the same time in our case there are a number of essential differences connected with the resonant and nonstationary nature of the amplification cross section $Q_\omega(\mathbf{r}, t)$.

The diffusion coefficient in an absorbing medium with anisotropic scattering takes the form [23]:

$$D = \frac{\chi_s}{3\Sigma(\Sigma - \bar{\mu}\chi_s)}, \tag{41.4}$$

where $\chi_s = Q_s N_0$; $\alpha_a = Q_\omega N_0$ are the macroscopic coefficients per unit length for scattering and amplification; $\Sigma = \chi_s + \alpha_a$; $\bar{\mu}$ is the mean cosine of the scattering angle. In the case we are discussing $\chi_s \gg \alpha_a$; and therefore

$$D \approx \frac{1}{3\chi_s(1 - \bar{\mu})} = \frac{\Lambda_s}{3(1 - \bar{\mu})}. \tag{41.5}$$

For axisymmetrical scattering (Rayleigh type) $\bar{\mu} > 0$. In the case of preferential forward scattering ($\bar{\mu} = 0$), the diffusion coefficient increases.

The imaginary part of the particles' permittivity $\epsilon''_\omega(\mathbf{r}, t)$ in the case of homogeneous line broadening can be represented in the form

$$\epsilon''_\omega(\mathbf{r}, t) = a(\omega)\epsilon''(\mathbf{r}, t), \qquad (41.6)$$

where $a(\omega)$ is the form of the normalized absorption lineshape. The quantity $\epsilon''(\mathbf{r}, t)$ is determined by the pumping power and depends on the photon flux intensity. The equation for $\epsilon''(\mathbf{r}, t)$ can be written in the form:

$$\frac{\partial \epsilon''(\mathbf{r}, t)}{\partial t} + \frac{1}{T_1}\epsilon''(\mathbf{r}, t) = -2\sigma_a\epsilon''(\mathbf{r}, t)\int a(\omega)\Phi_\omega(\mathbf{r}, t)\,d\omega + \frac{1}{T_1}\tilde{\epsilon}''(\mathbf{r}),$$
$$(41.7)$$

where $\sigma_0 = \sigma(\omega_0)$ is the cross section for radiative transitions in the ions in the dielectric particles responsible for resonance amplification; T_1 is the longitudinal relaxation time describing the spontaneous decay of amplification; $\tilde{\epsilon}''(\mathbf{r})$ is a term proportional to the pumping power. In the steady-state case in the absence of generation, $\tilde{\epsilon}''_\omega(\mathbf{r}, t) = \tilde{\epsilon}''(\mathbf{r})$.

The system of Eqs. (41.6) and (41.7) completely describes the stimulated emission of a cloud of scattering and amplifying particles. Boundary conditions must be set in order to solve it. The photon density is very low on the outer boundary of the medium. However, it cannot be zero, since photons diffuse from the cloud and can pass through the boundary surface. We can introduce the distance d at which the photon flux extrapolated linearly outwards from the boundary of the medium would be zero. This distance, which is called the linear extrapolation distance, is defined by the expression [23]:

$$d = \frac{2}{3}\Sigma^{-1} = \frac{2}{3}\Lambda_s. \qquad (41.8)$$

The right-hand side of Eq. (41.3) describes the attenuation of the emission due to diffusion "spreading" and amplification. There is obviously a threshold at which the emission losses are compensated for by the amplification. At the threshold the amplification saturation can be neglected, and it can be taken that $\tilde{\epsilon}''_\omega(\mathbf{r}, t) = \tilde{\epsilon}''(\mathbf{r})$. We shall limit ourselves to the case of uniform pumping $\epsilon''(r) = \epsilon''_0$, when the complex permittivity is the same for all scattering particles. Then the solution of the problem is reduced to the solution of Eq. (41.3) with a cross section $Q_\omega(\mathbf{r}, t) \equiv Q_\omega$ that is constant in ensemble and time. The general solution of Eq. (41.3) takes the form:

$$\Phi_\omega(\mathbf{r}, t) = \Sigma a_n \Psi_n(r)\exp\left[-\left(DB_n^2 - Q_\omega N_0\right)ct\right], \qquad (41.9)$$

where $\Psi_n(\mathbf{r})$ and B_n are the eigenfunctions and eigenvalues of the equation

$$\Delta\Psi_n(\mathbf{r}) + B_n^2\Psi_n(\mathbf{r}) = 0 \qquad (41.10)$$

with the boundary condition $\Psi_n = 0$ at a distance d from the boundary. a_n are arbitrary constants determined by the initial distribution $\Phi(\mathbf{r}, t)$ when $t = 0$.

The threshold condition follows at once from (9):

$$DB^2 - N_0Q_0 = 0, \qquad (41.11)$$

where B is the lowest eigenvalue B_n (usually $B = B_1$), $Q_0 = Q_{\omega_0}$, ω_0 is the frequency of maximum amplification.

If the region occupied by the ensemble of scattering particles takes the form of a sphere of radius R, then [23]

$$\Psi_n(r) = \frac{1}{r} \sin \frac{n\pi r}{R}, \quad B_n = \frac{\pi n}{R}, \quad B = \frac{\pi}{R}. \tag{41.12}$$

The cross sections Q_s and Q_0 are determined by the geometry and value of the complex permittivity $\epsilon = \epsilon_0 + i\epsilon''(\omega_0)$ of the scattering particles. Their calculation for an arbitrary relation between the particle size and the wavelength is a very complex problem even for a definite form of particle [24]. Therefore we shall limit ourselves to examples of spherical particles with a radius a much less or much more than the wavelength λ.

The case $ka \ll 1$ can be calculated exactly [24]:

$$Q_s = \frac{8}{3}(ka)^4 \left| \frac{\epsilon - 1}{\epsilon + 1} \right| S; \quad Q_0 = 4ka \operatorname{Im} \left(\frac{\epsilon - 1}{\epsilon + 2} \right) S; \quad \tilde{\mu} = 0, \tag{41.13}$$

where $S = \pi a^2$ is the geometrical cross section. The imaginary part of the permittivity $\epsilon_0'' = \alpha_o/k \ll 1$, where α_0 is the amplification factor per unit length in a substance of scattering particles, and k is a wave number. Then we have:

$$Q_s = \frac{8}{3}(ka)^4 \left| \frac{\epsilon - 1}{\epsilon + 1} \right| S; \quad Q_0 = \frac{4}{\epsilon_0 + 2} a\alpha_0 S. \tag{41.14}$$

In the case $ka \gg 1$ the calculation of the precise values of the cross sections and diagrams is cumbersome. However, for approximate estimates in the region $(\epsilon_0 - 1) \approx 1$ it can be taken that:

$$Q_s \approx S; \quad Q_0 \approx 2\eta a\alpha_0 S; \quad \tilde{\mu} = 0, \tag{41.15}$$

where η is the mean transmission factor of the boundaries of a dielectric sphere. Then the expression for the critical size becomes:

$$\frac{\pi}{B_{thr}} \approx g^3 \sqrt{\frac{32a}{3\eta\alpha_0}} \tag{41.16}$$

where $g = N_0^{-1/3}/2a$ is the ratio of the average distances between particles to their diameter. For example, for a spherical distribution of ruby particles ($\lambda = 7 \times 10^{-5}$ cm) with a radius $a = 2 \times 10^{-4}$ cm, $\eta \approx 1$, $g = 2$ and amplification $\alpha_0 \approx 1$ cm^{-1} (at 77 K the critical radius of the region is $R_{thr} \approx 4$ mm).

The emission spectrum of the laser with a nonresonant scattering feedback also has interesting features due to scattering. Consider them as previously in the diffusion approximation. Note that with the steady-state value for the integral intensity $\Phi(r, t)$ the spectral density $\Phi_\omega(r, t)$ may be non-stationary. The non-stationary nature of the spectrum consists of continuous narrowing of the emission spectrum after the start of generation. This is easy to check by examining the equation for the spectral density (41.3) for a stationary value of the amplification cross section Q_0 and stationary spatial distribution of the photon density. For the sake of simplicity we shall neglect the distortion of Q_0 due to saturation, so $\Delta\Phi_\omega = -B^2\Phi_\omega$. Equation (41.3) is then reduced to the following:

$$\frac{1}{c}\frac{\partial\Phi_\omega}{\partial t} = -DB^2\Phi_\omega + a(\omega)Q_0 N_0\Phi_\omega. \tag{41.17}$$

Since $a(\omega_0)Q_0 N_0 = DB^2$,

$$\Phi_\omega(t) \sim \exp\{-[a(\omega) - a(\omega_0)]Q_0 N_0\, ct\}. \tag{41.18}$$

By expanding the function of the line form $a(\omega)$ near the center of the line ω_0 we find the law for spectrum narrowing in the stationary generation state:

$$\Delta\omega(t) = \frac{\Delta\omega_0}{\sqrt{(Q_0 N_0 ct/\ln 2)}}, \tag{41.19}$$

where $\Delta\omega_0$ is the width of the amplification line at half-height.

Spectral narrowing proceeds, as usual, up to a certain fluctuation limit. In practice the most important part is played by the unavoidable fluctuation (Brownian) motion of the scattering particles, which leads to a random change (wandering) in the photon frequency by the Doppler effect from the scattering particles. If at each scatter the frequency of a photon changes by a mean amount $\delta\omega \ll \omega_0$, then the change in the spectrum can be found by examining only the frequency diffusion of the photons, i.e., by examining the spatial and frequency diffusion separately. In this case we can use the frequency diffusion equation,

$$\frac{1}{c}\frac{\partial\Phi_\omega}{\partial t} = D_\omega\frac{\partial^2\Phi_\omega}{\partial\omega^2} + Q_\omega N_o\Phi_\omega, \tag{41.20}$$

where D_ω is the frequency diffusion coefficient defined in the model used by the expression

$$D_\omega = \frac{(\delta\omega)^2}{2\Lambda_s}. \tag{41.21}$$

The stationary solution of Eq. (41.20) takes the form

$$\Phi_\omega = A\exp\left[-\frac{(\omega-\omega_0)^2}{\Delta\omega_0}\sqrt{\frac{Q_0 N_0}{D_\omega}}\right] \tag{41.22}$$

or, by substituting the value of D_ω and Λ_s, we obtain

$$\Phi_\omega = A\exp\left[-\frac{(\omega-\omega_0)^2}{\Delta\omega_0\delta\omega}\sqrt{2\frac{Q_0}{Q_s}}\right]. \tag{41.23}$$

The expression for the stationary width of the spectrum at half-maximum is of the form:

$$\Delta\omega_{st} = \left[4\ln 2\Delta\omega_0\delta\omega\sqrt{\frac{Q_s}{2Q_0}} \right]^{1/2}. \tag{41.24}$$

The first experiments on the observation of the laser effect in an amplifying medium with scattering were conducted in [25]. In these experiments, amplification was high (a dye solution) and scattering (Rayleigh) was weak.

The laser action has recently been observed to occur in strongly scattering media [26]. An active dye laser medium was charged with TiO_2 microparticles (250 nm in average diameter). It has been found that emissions from such a system can exhibit spectral and temporal properties characteristic of a multimode laser oscillator, even though the system contains no external cavity. Thanks to the short photon mean-free path in comparison with the size of the system, the diffusion losses are small, and accordingly the threshold excitation energy for the laser action is surprisingly low. This is a direct implementation of the laser with a noncoherent nonresonant scattering feedback suggested in [22]. The authors, however, note that it is not in all cases that such experiments can be explained within the framework of the simple diffusion model. Such is the case, for example, with a thin cell where condition (1) is not satisfied. It may well be that account should be taken of the effect of random interference of photons. This was already emphasized by C. H. Townes (private communications). And this has recently become real in connection with the development of the theory of propagation of random fields in heterogeneous media. Specifically, the role has been noted of weak localization in backscattering from the amplifying medium [27]. The author of this work predicts a very narrow peak in reflection when the system is close to the lasing threshold. Thus, the theory of the laser with a noncoherent nonresonant scattering feedback is still to be constructed with due regard for the effects of random interference of light waves.

A scattering feedback may have a *resonant* character as a result of the resonant scattering by the amplifying particles themselves (Fig. 41.2c). The maximum resonant scattering cross section, hence the minimum photon mean-free path, and accordingly the minimum scatter loss, coincides with the maximum amplification. For that reason, the lasing frequency coincides with the amplification line center.

The cross section for resonant scattering $\sigma_{scat}(\omega)$ is related to the resonant radiative transition cross section $\sigma_{abs.em}(\omega)$. The cross section for a resonance radiative transition between two levels at the frequency ω is given by the expression [28]

$$\sigma_{abs.em}(\omega) = \sigma_0\frac{\Gamma^2}{\Gamma^2 + (\omega - \omega_0)^2}; \tag{41.25}$$

$$\sigma_0 = \frac{\lambda^2}{2\pi}\frac{A_{21}}{\Gamma}, \tag{41.26}$$

where A_{21} denotes the probability of a radiative transition between the two levels considered, and Γ is the homogeneous half linewidth, determined by the full

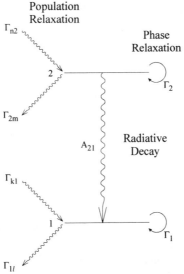

FIGURE 41.3. Various mechanisms for relaxation of population (longitudinal relaxation) and relaxation of phase (transversal relaxation) of two levels.

probability of relaxing the phase of the two levels, i.e., the combined operation of longitudinal relaxation (relaxation of the populations) and transverse relaxation (relaxation of the phase of the wave function without relaxation of the populations), as illustrated schematically in Fig. 41.3

$$2\Gamma = 2\Gamma_{\text{hom}} = \frac{2}{T_2} + \frac{1}{T_1}, \tag{41.27}$$

where rate of phase relaxation is defined by the expression

$$\frac{2}{T_2} = \Gamma_1 + \Gamma_2, \tag{41.28}$$

and where Γ_i is the phase relaxation rate of i-th level, a rate of population relaxation is given by

$$\frac{1}{T_1} = (\Gamma_{n2} + \Gamma_{2m}) + (\Gamma_{k1} + \Gamma_{1l}) + A_{21}. \tag{41.29}$$

The coefficient of the resonance amplification per unit length is given by

$$\alpha(\omega) = \sigma_{em}(\omega)N_2 - \sigma_{abs}(\omega)N_1, \tag{41.30}$$

or

$$\alpha(\omega) = \sigma(\omega)(N_2 - N_1), \tag{41.31}$$

if for the sake of simplicity no account is taken of the difference in degeneracy of the two levels. Where N_1 and N_2 represent the number of particles per unit volume in the lower and upper levels, with the transition frequency between the two levels falling inside a spectral interval of width 2Γ, corresponding to homogeneous broadening. If a line has an inhomogeneous broadening $\Delta\omega_D$, say, because of the Doppler effect, then

$$\alpha_{nonhom} = \alpha_{hom} \frac{2\Gamma}{\Delta\omega_D} . \qquad (41.32)$$

To determine the resonance scattering coefficient per unit length one must know the cross section for resonance scattering of radiation by a single particle. The cross section for resonance scattering of radiation of frequency ω into 4π steradians is related to the cross section of the radiative transition by

$$\sigma_{sc}(\omega) = \sigma_{abs.em}(\omega) \left(\frac{A_{21}}{2\Gamma} \right). \qquad (41.33)$$

The resonance scattering cross section is always smaller than the resonance emission or resonance absorption cross section.

The coefficient for resonance scattering per unit length into a unit solid angle by particles in either the lower or the upper level is equal to

$$\chi(\omega) = \sigma_{sc}(\omega)(N_1 + N_2). \qquad (41.34)$$

Equations (41.31), (41.32), and (41.33) imply that the coefficient for resonance scattering into the solid angle $\Delta\Omega$ is related to the resonance amplification coefficient by the simple expression

$$\chi(\omega) = q \left(\frac{\Delta\Omega}{4\pi} \right) \alpha(\omega), \qquad (41.35)$$

where the factor q is defined as

$$q = \frac{A_{21}(N_1 + N_2)}{\Gamma(N_2 - N_1)} = \frac{A_{21}}{2\Gamma} \frac{N_0}{\Delta N}. \qquad (41.36)$$

The coefficient χ_0 for resonance scattering per unit length into all directions will be

$$\chi_0 = q\alpha(\omega). \qquad (41.37)$$

The relationship between the resonance amplification and resonance scattering coefficients is defined by the product of the factors $(A_{21}/\Gamma) \ll 1$ and $q \gg 1$. At $\alpha \gg \chi$, the lasing threshold in a laser with a noncoherent resonant scattering feedback is only reached with a high amplification per the scattering length $A_{sc} = 1/\chi_{sc}$. It is precisely such a case that can be realized in space masers (Sec. 3). Resonant scattering is very weak in the long-wavelength region, for $\Gamma \gg A_{21}$, but becomes quite substantial in the visible and UV regions, where A_{21} and Γ may be commensurable of the levels.

41.3   Space Masers with Scattering Feedback

In 1965 the scientists of the Radio Astronomy Laboratory of the University of California at Berkeley, reported that they had discovered radio frequency emission lines near the wavelength $\lambda = 18$ cm in the direction of certain regions of interstellar ionized hydrogen ("HII regions") [29]. The frequencies of the observed lines corresponded to transitions between sublevels of the Λ-doubled ground state of the OH radical. At that time, the observations revealed the striking properties of this radiation: unusually high intensity, strong polarization, and variability in time.

Four years later, a group of radio astronomers at the same university headed by Prof. Townes discovered in a number of HII regions emission sources in the 1.35 cm line, which correspond to a rotational transition of H_2O molecules [30]. It was soon found that these sources exhibit the same remarkable features as the OH sources.

The brightness temperature at certain OH and H_2O lines is as high as $10^{13} - 10^{17}$ K, whereas the Doppler width of these lines corresponds to kinetic temperatures of 10-100 K that are usual for neutral clouds of interstellar gas. Researchers found a solution to this problem by means of hypothesis of amplification of the OH and H_2O lines by a maser mechanism in clouds having an inverted population of the corresponding molecular levels (see reviews and books [31–37]).

The universally accepted theoretical approach to the construction of space maser models is the traveling-wave amplifier model [38, 39] (see [31–37]). This approach has until recently allowed in a sufficiently satisfactory manner the main emission characteristics of many observed maser sources. But thanks to the development of radio interferometry techniques and the creation of very-long-base radio interferometer systems (VLBI), powerful H_2O maser outbursts have lately been observed experimentally (see [37]). For example, interstellar masers have long been known to be variable on time scales from several years to a few days, but recent observations [40] have suggested that OH maser sources may show variability on time scales of a few *minutes*. The microwave emission of these powerful space masers exhibit new properties that make it greatly different from the emission of the earlier H_2O masers [41]. Attempts made to explain them within the framework of the standard space maser model in the form of a traveling wave amplifier have run into serious difficulties. Thus, the problem of interpreting the compact OH and H_2O radio emission sources is still far from a final solution.

However, a generator mechanism has long been suggested for the production of intense narrow-band radiation in populated-inverted space maser media [42, 43]. Feedback in this case can be formed on account of nonresonant scattering of radio emission by electrons or microscopic dust particles [42], or as a result of resonance scattering by the active molecules themselves [43]. And scattering in the maser model with a noncoherent feedback can naturally explain many unusual properties of compact space masers.

Resonance scattering over the characteristic amplification distance $1/\alpha(\omega)$ will return into the solid angle $\Delta\Omega$ according to Eq. (41.36) approximately the propor-

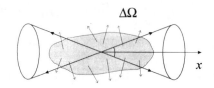

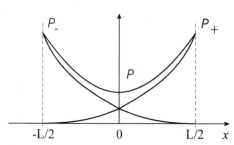

FIGURE 41.4. Amplifying cloud with weak backscattering: (a) angular distribution of radiation; (b) intensity distribution along cloud P_+ and P_- being propagated in two opposite directions and total intensity $P = P_+ + P_-$.

self-excited for an amplification $G_0 = \exp[\alpha(\omega)L]$ on passage through the cloud, equal to

$$G_0 \cong \frac{\alpha(\omega)}{\chi(\omega)} = \frac{4\pi}{q\,\Delta\Omega}, \tag{41.38}$$

where the factor q is defined by Eq. 41.36.

This relation can readily be derived for a simple configuration of an elongated amplifying cloud, in which one may neglect the angular dependence and consider the propagation of radiation beams of flux P_+ and P_- in two opposite directions along the x axis (Fig. 41.4), with allowance for backscattering:

$$\frac{1}{c}\frac{\partial P_+}{\partial t} + \frac{\partial P_+}{\partial x} = (\alpha_0 - \chi_0)P_+ + \chi_{bk}P_- \tag{41.39}$$

$$\frac{1}{c}\frac{\partial P_-}{\partial t} - \frac{\partial P_-}{\partial x} = (\alpha_0 - \chi_0)P_- + \chi_{bk}P_+.$$

Here, χ_{bk} is the coefficient for resonance back-scattering per unit length into the solid angle $\Delta\Omega$ of the cloud, and χ_0 is the coefficient for resonance scattering per unit length in all directions. By solving the above systems of equations, we can find a condition for exponential growth of power with time (a threshold condition):

$$G_0 = e^{(\alpha_0 - \chi_0)L} > \frac{\alpha_0 - \chi_0}{\chi_{bk}} \tag{41.40}$$

or

$$G_0 > \frac{1-q}{q}\frac{4\pi}{\Delta\Omega}, \tag{41.41}$$

where L is the length of the cloud and $\alpha_0 - \chi_0$ is the effective amplification per unit length corrected for losses by resonance scattering. If the medium contains additional nonresonant losses γ per unit length, the amplification α_0 in the condition should be replaced by an effective amplification $\alpha_0 - \gamma$.

One can obtain the value of the threshold amplification for other geometries of the amplifying cloud by considering the scattering of radiation in all directions in the presence of amplification, as has been shown in [22] (see Sec. 2). But it is not possible to apply these previous results [22] directly, because in the space maser case $\alpha_0 \gg \chi_0$; i.e., we have weakly scattering active media. The generation threshold for the case of a weakly scattering medium may be written in the more simple form

$$q\left(\frac{\Delta\Omega}{4\pi}\right) \cdot \exp(\alpha L) > 1 \qquad (41.42)$$

$$\qquad\qquad \uparrow \qquad\qquad \uparrow$$
probability of resonant   amplification (gain)
scattering   per pass

Table 1 lists the main tentative parameters of the OH, H_2O, and SiO space masers, borrowed from [31–37], which are necessary for the estimation of the masing threshold.

TABLE 41.1. Some approximate parameters of space masers

	OH	H_2O	SiO
λ [cm]	18	1.35	$0.07 - 0.23$
$A_{21}[s^{-1}]$	$0.8 \cdot 10^{-10}$	$2 \cdot 10^{-9}$	$3 \cdot 10^{-6}$
$\Gamma[s^{-1}]$	10^{-2}	1	5
$\Delta N/N$	$\sim 10^{-2}$	$\sim 10^{-1}$	$\sim 10^{-1}$
q	10^{-10}	$2 \cdot 10^{-10} - 10^{-9}$	10^{-5}
$G_0 = \exp(\alpha L)$	$10^8 - 10^{12}$	$10^{11} - 10^{14}$	$> 10^{10}$

We shall now obtain estimates for amplifying OH and H_2O clouds. We first consider the case of OH molecules. The transition probability for the strongest Λ-doublet lines of OH is $\Gamma_{12} = 8 \cdot 10^{-11}$ sec^{-1}. The homogeneous width 2Γ of the transition depends, according to Eq. (41.42), on the combined contribution of the pumping rate (longitudinal relaxation) and the broadening effects of collision with atoms (transverse relaxation). The contribution of the pumping rate is quite different for different pumping mechanisms. In the model of pumping by ultraviolet radiation the contribution to the quantity Γ is $10^{-6} - 10^{-8}$ sec^{-1}, and the relative inversion reaches $\Delta N/N \approx 0.1 - 0.01$ [44]. For pumping by infrared radiation this contribution to the homogeneous width, according to Litvak's estimates [45], would be 10^{-4} and 10^{-2} sec^{-1} for far- and near-infrared radiation, respectively. In the case of collisions with H atoms the contribution to the homogeneous width depends on the density of atoms. If we adopt the quite high value 10^5 sec^{-3} for the density of hydrogen atoms in OH clouds, the contribution from collisions would yield a broadening of about 10^{-4} sec^{-1}. This contribution to Γ is decisive for the mechanism of pumping by ultraviolet radiation, and is also important for pumping by far-infrared radiation. In these cases the quantity $\Gamma \approx 10^{-4}$ sec^{-1}, and with $\Delta N/N_0 \simeq 0.1$ a cloud having a geometrical solid angle of, say, $\Delta\Omega = 0.2$ sr, would require an amplification upon passage of $G_0 \lesssim 10^8$ in order to be self-excited. Although many of the values adopted for this estimate are admittedly rather arbitrary ($\Delta N/N_0$, the density of H atoms, the solid angle), it seems entirely possible that the threshold amplification upon passage through the maser-generator model with feedback as a result of resonance scattering could explain the observations with less amplification than is required in the maser-amplifier model ($G_{amp} \approx 10^{11} - 10^{12}$ [46]). If the homogeneous width of the transition has a higher value (for example, in the case of pumping by near-infrared radiation $\Gamma \approx 10^{-2}$ sec^{-1} [45]), the necessary amplification would increase ($G_0 > 10^{10}$ if we adopt the same estimates as before for $\Delta\Omega$ and $\Delta N/N_0$), and the proposed model will no longer offer an appreciable advantage over the model of amplification of a traveling wave.

In the case of galactic H_2O clouds [47] the situation is more definite. Amplification will take place for the transition $6_{16} \rightarrow 5_{23}$ between high rotational sublevels of the ground vibrational state (the energy of excitation is $\Delta E = 447$ cm^{-1}) [30]. The homogeneous width of the transition will be relatively large here, and will be determined by spontaneous decay of the 6_{16} and 5_{23} states to the lower rotational states $5_{05}, 5_{14}, 4_{14}, 4_{32}$:

$$\Gamma_{10} + \Gamma_{20} \simeq \Gamma_{6_{16}-5_{05}} + \Gamma_{5_{73}-4_{14}} = 0.77 + 0.42 \simeq 1.2 \text{ sec}^{-1}.$$

Since the radiative transition $6_{16} \rightarrow 5_{23}$ has a probability $\Gamma_{12} = 2 \cdot 10^{-9}$ sec^{-1}, the factor q is given by

$$q = \frac{\Gamma_{12}}{\Gamma_{10} + \Gamma_{20}} \frac{N_0}{\Delta N} \simeq 10^{-8},$$

provided that $N_0/\Delta N$ is of the order of several units, and accordingly the threshold amplification will be $G_0 > 10^{10}$. Such an amplification is of the order of, or higher than, the amplification required for passage through the maser-amplifier model [47].

A choice between the maser-amplifier model and the model of a maser with feedback can be made for galactic maser regions on the basis of special supplementary observations. For example, it would be important to measure the characteristic periods T_{var} of intensity variation when measuring the size L of compact emission regions by means of radio interferometry with a very long baseline [46]. In the

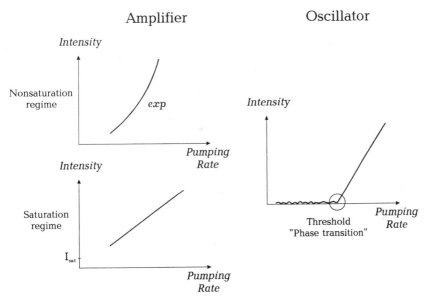

FIGURE 41.5. Radiation intensity of a maser amplifier (left) and a maser generator (right) as a function of the pumping rate.

maser-amplifier model the condition

$$T_{var} \leq \frac{L}{c}, \tag{41.43}$$

should always be satisfied, since saturation of the amplification in a traveling-wave amplifier cannot take place in more than one pass of radiation through the amplifier. Furthermore, in an amplifier with a large gain G_0 upon passage ($\ln G_0 = \alpha_{eff} L \gg 1$, where $\alpha_{eff} = \alpha_{(0)} - \gamma$ is the effective amplification factor per unit length), at the moment the amplification becomes saturated, an intensity variation would, strictly speaking, be possible within the time required for the radiation to traverse the distance L_e over which the intensity rises by a small factor, such as a factor e: $L_e \simeq \frac{1}{\alpha_{eff}}$, so that the variation would exhibit a characteristic time scale

$$T_{var} = \frac{L_e}{c} < \frac{L}{c}. \tag{41.44}$$

Of course, to observe short intensity variation times $\tau_{var} \approx L_e c$, it is necessary to have a high angular resolution capable of resolving as small a size as $L_e \ll L$.

Figure 41.5 (left) presents a qualitative relationship between the emission intensity of a maser amplifier and the pumping rate under saturated and unsaturated conditions. Figure 41.6 (left) illustrates qualitatively the dynamic response of the maser amplifier to a fast increase of the pumping rate.

But in the maser-generator model with feedback, growth of intensity and thereby saturation of the amplification can take place over many passes (in principle, tens

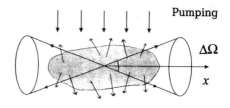

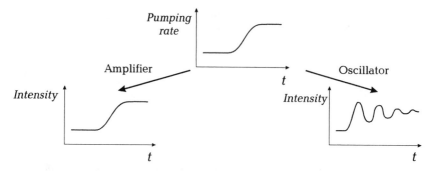

FIGURE 41.6. Dynamical response of maser amplifier (left) and maser-generator (right) to a fast increase of the pumping rate.

or hundreds), and it would be possible for variations to develop with

$$T_{var} \gg \frac{L_{th}}{c},$$
(41.45)

where L_{th} is the threshold length of the masing region that can, in principle, be much shorter than the length L of the amplifying region.

Figure 41.5 (right) shows qualitatively the emission intensity of a maser generator as a function of the pumping rate, which in contrast to the maser amplifier has a distinct phase-transition-type threshold. Figure 41.6 (right) presents a qualitative response of the maser generator to a fast change of the pumping rate. An oscillating response is possible in that case. The pertinent numerical calculations were presented in [22].

But the most essential distinction between the maser amplifier and maser generator models is the difference between their emission spectrum widths. For the maser amplifier, the narrowing of the emission spectrum is due to the predominant amplification at the center of the amplification line, $\alpha(\omega)$. As a result, the emission linewidth $\Delta\omega$ is smaller than the Doppler width $\Delta\omega_D$ by a factor of $1/\sqrt{\alpha_0 L}$.

$$\frac{\Delta\omega}{\Delta\omega_D} = (\alpha_0 L)^{-1/2} = (\ln G_0)^{-1/2}.$$
(41.46)

The observed 4- to 5-fold narrowing of the emission spectrum agrees well with the estimate of Eq. (41.46). Under amplification saturation conditions, the radiation intensity $I \propto \alpha_0 L$ and so $\Delta\omega/\Delta\omega_D \propto I^{-1/2}$ which also agrees with the experimental data of [41].

The calculation made in [3] for a maser generator model with a strong resonance scattering yielded the following expression for the emission linewidth under steady-state conditions:

$$\Delta\omega/\Delta\omega_D = \left(\frac{N_2}{N_2 - N_1} \frac{1}{\langle n_0 \rangle} \right)^{1/2}, \tag{41.47}$$

where $\langle n_0 \rangle$ is the emission power in terms of the number of photons per degree of freedom of a multiple-mode maser generator field. Equation (41.47) may be represented in the form

$$\Delta\omega/\Delta\omega_D = \left[\frac{c\sigma_0 \Omega_{osc}}{L_{th}\lambda^2(\eta - 1)A_{21}} \right]^{1/2}, \tag{41.48}$$

where Ω_{osc} is the solid angle of the radiation being emitted, $\eta = P/P_{th}$ is the excess of the pumping rate over the threshold value. Equation (41.48) has the same intensity dependence as Eq. (41.46). But as distinct from Eq. (41.46), it allows for a much greater narrowing of the emission spectrum relative to the amplification linewidth $\Delta\omega_D$ [3]. This can explain a narrow feature in the spectrum of the strong OH source W3(OH) with a linewidth of only 0.1 km/sec or an equivalent temperature of 5 K [49].

Finally, let us estimate the threshold gain G_0^{th} for the case of nonresonant scattering, for example, by free electrons [42]. The sources of anomalous microwave emission of OH lie in the regions where the hydrogen is highly ionized [29]. These regions have high densities of electrons and interstellar dust, which are capable of scattering the microwave emission.

The coefficient of Thomson back-scattering by electrons is given by

$$\chi = N_e \Omega_{osc}(e^2/m_e c^2)^2 \left(\frac{\Delta\omega_D}{\Delta\omega_{sc} + \Delta\omega_D} \right) \tag{41.49}$$

where N_e is the electron density, Ω_{osc} is the generation solid angle, determined by the shape of the generation region, and the last factor takes into account the broadening of the emission spectrum due to back-scattering by the electrons. For the Doppler broadening $\Delta\omega_D$ of the amplification line and $\Delta\omega_{sc}$ of the scattered-radiation spectrum we can assume that $\Delta\omega_{sc} \gg \Delta\omega_D$ (the mass of the amplifying molecule is $M \gg m_e$) and

$$\frac{\Delta\omega_D}{(\Delta\omega_{sc} + \Delta\omega_D)} = \frac{1}{2} \left(\frac{m_e T_M}{M T_e} \right)^{1/2}, \tag{41.50}$$

where T_M and T_e are the kinetic temperatures of the amplifying molecules and the electrons, respectively. For example, for OH molecules and for $T_e = 10 T_M$ we obtain

$$\chi \simeq 0.5 \cdot 10^{-9} N_e \Omega_{osc}(\text{parsec}^{-1}) = 1.7 \cdot 10^{-28} N_e \Omega_{osc}(\text{cm}^{-1}). \tag{41.51}$$

For compact OH emission regions with a size of $L \approx 10^{12} - 10^{13}$ cm [36], the proportion of radiation scattered by the electrons χ is very small. At $N_e \approx 10^6$ cm^{-3} and $\Omega_{osc} \approx 1$ sr, $\chi L \approx 1.7 \cdot (10^{-9}$ to $10^{-10})$. Thus the threshold gain is $G_0 = \exp(\alpha L_{th}) = \alpha_0/\chi \approx 10^{10} - 10^{11}$ with $\alpha_0 L_{th} \approx 20 - 25$. This very approximate estimate agrees with observations.

The emission statistics of a generator with a noncoherent feedback was theoretically studied in [50]. The distribution function of the number of quanta in the kth emission mode, n_k was found to be given by the expression

$$P(n_k) = \frac{\langle n_k \rangle^{n_k}}{(1 + \langle n_k \rangle)^{1 + n_k}}, \tag{41.52}$$

i.e., the distribution was found to be of the Bose-Einstein type. At $\langle n_k \rangle \gg 1$, Eq. (41.52) yields a Gaussian distribution. It was precisely this type of distribution that was experimentally observed in the case of the OH maser [51].

And finally, let us note the recent work [52] which has also analyzed a space maser generator model. It has been noted in this work that the model suggested in [42, 43] for the establishment of masing conditions in space media and the powerful space maser outbursts observed in [40, 41] may be one of the first examples of the manifestation of cooperative nonsteady-state processes occurring in astrophysical conditions.

41.4   Stellar Lasers with a Resonance Scattering Feedback

The concept of a maser with a noncoherent resonance scattering feedback can be extended to optical quantum transitions in atoms, for example, in stellar atmospheres [53]. There exist stellar atmospheres conditions reminiscent of those that exist in a low-pressure gas laser. There occur in stellar spectra, besides absorption lines, bright emission lines whose appearance is explained by the excitation of atoms during collisions [54], by the recombination of ions, as well as by fluorescent excitation based on the accidental coincidence of a bright emission line of one element with the absorption line of another and the subsequent successive transition of the excited atom.

However, in a number of cases these mechanisms do not give a satisfactory account of the anomalies observed in stellar spectra. Furthermore, it can be seen from the analysis of the anomalous emission lines presently known in the astrophysics literature [55] that certain schemes used by astrophysicists in attempts to explain these anomalies are essentially astrophysical analogs of the three- and four-level optical-pumping schemes used at present in lasers. Such an analogy suggested the possible existence in certain stellar atmospheres of conditions that facilitate the creation of population inversion at the corresponding transitions.

As an example of one of the still unexplained anomalies, we can point out the behavior of the 7774 Å and 8446 Å OI-oxygen lines in the spectra of Be

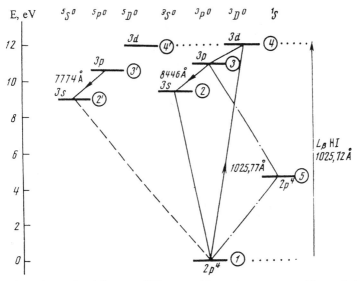

FIGURE 41.7. Energy-level diagram of OI oxygen. The spontaneous-transition probabilities are equal (in $\sec^{-1}$) to: $A_{21} = 5 \times 10^8$, $A_{32} \approx A_{43} \approx A_{41} \approx A_{4'1'} \approx A_{3'2'} = 3 \times 10^7$; $A_{2'1'} = 10^5$; $A_{4'3'} = 2 \times 10^6$. The multiplicities of the degeneracies of the levels are: $g_1 = 9$, $g_2 = 3$, $g_3 = 9$ and $g_4 = 15$.

stars [55]. The ratio of the intensities of the 8446 Å $(3p^3 P - 3s^3 S)$ and 7774 Å $(3p^5 P - 3s^5 S)$ lines differ from the laboratory value. Bowen [56] has suggested that this is connected with the phenomenon of fluorescence caused by the coincidence of the bright HI-hydrogen line L_β 1025.72 Å with the 1025.77 Å $(2p^3 P - 3d^3 D)$ OI absorption line. On the other hand, it can be seen from the transition scheme (Fig. 41.7) under consideration that the fluorescence process coincides with a four-level laser scheme and that, owing to a serendipitous ratio between the probablities A_{ij} of spontaneous transitions between levels (3)-(2) and (2)-(1), (as designated in Fig. 41.7), a population inversion can arise at the (3)-(2) transition. The considered example is of even greater interest in that generation in the continuous regime has been realized under laboratory conditions at the 8446 Å line [57].

Let us consider the specific case of the appearance of inversion under stellar atmosphere conditions at the (3)-(2) OI transition owing to optical pumping, according to the Bowen scheme, of level (4) with a subsequent cascade transition to level (3) (Fig. 41.7).

Let us assume that a particle gas of density N_0 and temperature T_0 is located in the field of radiation of spectral density U and frequency $\omega \approx \omega_{41}$. One can write balance (rate) equations for level populations in the steady-state case and obtain the following expression for the difference in population between levels (3) and (2):

$$\Delta N = \frac{N_3}{g_3} - \frac{N_2}{g_2} = \frac{N_0 \, \mu (p_{14} - p_{14}^{thr})}{\rho} \left[1 - \frac{1}{p_{21}} \left(\frac{g_3}{g_2} p_{32} - p_{23} \right) \right], \quad (41.53)$$

where

$$p_{14}^{thr} = \frac{p_{12}}{p_{21}} \frac{(p_{32}g_3/g_2 - p_{23})(p_{41} + p_{43}) + p_{41}p_{43}g_3/g_2}{p_{43}\left[1 - p_{21}^{-1}(p_{32}g_3/g_2 - p_{23})\right]},$$

$$\rho = \mu p_{14}\left(1 + \frac{p_{32} + p_{34}}{p_{43}}\right) + \left(p_{32} + \mu p_{41}\frac{p_{34}}{p_{43}}\right)\left(1 + \frac{p_{12}}{p_{21}}\right)$$

$$+ \frac{p_{12}}{p_{21}}p_{23}\left(1 + \mu\frac{p_{34}}{p_{43}}\right)$$

$$\mu = p_{43}/(p_{41} + p_{43}). \tag{41.54}$$

The probabilities of the transitions between levels i and j can be represented with the aid of the Einstein coefficients A_{ij} and B_{ij} in the form

$$p_{ij} = A_{ij} + B_{ij}U_{ij}^0 + d_{ij} + B_{ij}U_{ij} = p_{ij}^0 + B_{ij}U_{ij} \tag{41.55}$$

$$p_{ji} = B_{ji}U_{ij}^0 + d_{ji} + B_{ji}U_{ij} = p_{ji}^0 + B_{ji}U_{ij}.$$

Here the d_{ij} are the nonradiative-transition probabilities,

$$p_{ji}^0/p_{ij}^0 = (g_i/g_j)\exp(-\hbar\omega_{ij}/kT_0), \tag{41.56}$$

g_i is the multiplicity of the degeneracy of the level i, and U_{ij}^0 is the equilibrium radiation density corresponding to the temperature T_0.

Besides the quantity ΔN, we shall need below the quantity

$$N_2 + N_3 = \frac{N_0}{\rho}\left[\mu p_{14}\left(1 + \frac{p_{32} + p_{23}}{p_{21}}\right)\right.$$

$$\left. + \frac{p_{12}}{p_{21}}\left(p_{23} + p_{32} + \mu p_{41}\frac{p_{34}}{p_{43}}\right)\right]. \tag{41.57}$$

It should be noted that in four-level generators a significant influence on the threshold pump level is exerted by the population of level (2), and that this population depends on the temperature of the radiation actuating the (1)-(2) transition (the nonradiative transitions contribute little to the case under consideration). If the gas were in equilibrium with the radiation, then the radiation temperature would coincide with the temperature T_0 of the gas. However, there exists in stellar atmospheres hotter radiation emanating from the photosphere and characterized by a temperature T considerably exceeding T_0. Therefore, below we shall assume that the (1)-(2) transition is actuated by radiation of temperature T.

Let us now estimate the possible value of the amplification factor $\alpha(\omega)$ for the 8446 Å OI line in the atmospheres of the Be stars. The amplification factor per unit length in a medium with a given inverse-population value ΔN is equal to

$$\alpha(\omega) = \frac{\lambda^2}{2\pi}\frac{(\pi\ln 2)^{1/2}}{2}\frac{\gamma}{\Delta\omega_D}\exp\left[-\frac{(\omega - \omega_0)^2}{\Delta\omega_D^2}\ln 2\right]g_3\Delta N; \tag{41.58}$$

$$\Delta\omega_d = \left[\frac{2kT_0}{M_{OI}}\ln 2\right]^{1/2}\frac{\omega_0}{c}$$

is the Doppler halfwidth of the line and $\gamma = A_{32}$.

The density of the OI atoms in the region of the atmosphere close to the outer boundary of the photosphere (i.e., in the inverting layer) is $N_0 \sim 10^5 - 10^6$ cm^{-3}, the gas temperature here is $T_0 \sim 1.5 \times 10^4$ K, and the effective temperature of the photosphere is $T \sim 3 \times 10^4$ K [58, 59]. If we assume that the temperature of the radiation emitted in the 1-2 transition is of the order of T, then, as follows from Eqs. (41.53) and (41.54), the requisite brightness temperature of the fluorescent pump at the threshold should be equal to 4×10^4 K, while in order to obtain an inverse population of, say, $\Delta N \sim 10^{-6} N_0$, it must exceed this value.

Besides optical pumping, the contribution made by collisions with electrons to the inversion of levels (3) and (2) (the excitation of level (3) via the metastable level (5)) was estimated. Inversion is not attainable through this mechanism when the (1)-(2) transition is stimulated by radiation of temperature T in the absence of stimulation by radiation of the other transitions.

The line widths necessary for the estimate are: $\gamma \approx 3 \times 10^7$ sec^{-1}, $\Delta \omega_D \approx 5 \times 10^{10}$ sec^{-1}. Substituting these values and $\Delta N = 10^{-6} N_0$ into Eq. (41.58), we obtain the following value for the amplification factor at the line center: $\alpha_0 \equiv \alpha(\omega_0) \sim 10^{-12}$ cm^{-1}.

It was shown in Sect. 3 that in space sources of microwave radiation of anomalously high brightness temperature, a generation regime maintained by part of the radiation returning to the amplifying volume as a result of resonance scattering at the amplifying transition is, in principle, possible. In the optical region resonance scattering can play a more important role because of the larger value of the factor γ / Γ where 2Γ is the homogeneous line width Eq. (41.27).

The coefficient per unit length of resonance scattering into a solid angle of 4π in the medium given by Eq. (41.33) has (for a Doppler line shape) the form

$$\chi(\omega) = \frac{\lambda^2}{2\pi} \frac{(\pi \ln 2)^{1/2}}{2} \frac{\gamma^2}{\Delta \omega_D \Gamma} \exp\left[-\frac{(\omega - \omega_0)^2}{\Delta \omega_D^2} \ln 2\right] (N_2 + N_3). \quad (41.59)$$

For the considered conditions $N_2 + N_3 \approx 10^{-2} N_0$ and $\Gamma \approx 5 \times 10^8$ sec^{-1} and we obtain the estimate: $\chi_0 \equiv \chi(\omega_0) \approx 10^{-10}$ cm^{-1}.

The attractiveness of the generator model lies in the fact that a high gain per pass is not required when the scattering is sufficiently effective, since above the threshold, when the gain per pass exceeds the loss per pass, the intensity grows in the course of many passes.

The problem consists in the determination of the so-called threshold or (in the spherically symmetric model) of the critical radius r_0 at which the system is on the threshold of self-excitation. For the case when $\alpha(\omega) \ll \chi(\omega)$ the solution of the problem is known [22]; however, as can be seen from Eqs. (41.58) and (41.59), depending on the values of the quantities determining $\alpha(\omega)$ and $\chi(\omega)$, the relation between the latter quantities can be arbitrary, and it is necessary to consider the problem in its general form. We can, in determining the threshold, restrict ourselves to the linear approximation, neglecting saturation effects.

The equation satisfied by the radiation intensity J_ω at the frequency ω in the case of low particle densities N_0 and coherent light scattering (scattering without a change in frequency) with a spherical indicatrix has the form [60]

$$\cos\Theta \frac{\partial J_\omega}{\partial r} - \frac{\sin\Theta}{r}\frac{\partial J_\omega}{\partial\Theta} + \frac{1}{c}\frac{\partial J_\omega}{\partial t} = [\alpha(\omega) - \chi(\omega)]J_\omega\frac{\chi(\omega)}{4\pi}\int J_\omega d\Omega \quad (41.60)$$

where $d\Omega = \sin\Theta d\Theta d\varphi$ and c is the velocity of light. The solution of Eq. (41.60) yields the threshold condition [42]:

$$\tan(qr_0) = 2\alpha_0 s r_0/(2\alpha_0 - s^2 r_0), \quad (41.61)$$

where $s^2 = 3\alpha_0(\chi_0 - \alpha_0)$ and $\chi_0 \gg \alpha_0$. In principle, all the solutions q_n for which $\alpha > \alpha_{thr}(r_0)$ (where $\alpha_{thr}(r_0)$ is the threshold value) are possible. Physically, however, it is clear that the solution possessing the least threshold value $\alpha_{thr}(r_0)$ does not allow the appearance of the other possible solutions (at least before the onset of saturation).

Let us now estimate the critical dimension r_0 for OI laser, using the estimates for α_0 and χ_0:

$$r_0 \approx \pi(3\alpha_0\chi_0)^{-1/2} = \frac{2\pi^2}{\lambda^2}\frac{\Delta\omega_D}{\gamma} \times \frac{2}{(\pi\ln 2)^{1/2}}$$
$$\times\left[3\frac{\Gamma}{\gamma}g_3\Delta N(N_2 + N_3)\right]^{-1/2} \approx 10^{11}\text{ cm} \quad (41.62)$$

Taking into account the fact that the dimension of the atmosphere of a Be star is $10^{13} - 10^{14}$ cm, we can conclude that if there exists a population inversion in the stellar atmosphere, then the appearance of the generation regime is more practicable than that of the amplification regime.

The considered generator could have been called a generator with a nonresonant feedback, but such a term would have been a poor one in the present case, since the scattering cross section has a resonance character. Therefore, we employ the term "incoherent resonance feedback," which reflects the fact that the radiation emitted by such a generator does not possess spatial coherence.

Let us now consider the difference between the stimulated radiation of a generator with an incoherent feedback and the spontaneous radiation emitted by the same medium as a result of the cyclic process connected with the fluorescence mechanism. In the case when generation occurs at the 8446 Å line the relation between the integral intensities in the 7774 Å and 8446 Å lines does not change in comparison with the fluorescence mechanism. This happens because the ratio of the number of transitions giving rise to the 7774 Å line depends only on the rates of occupation of the $3p^3P$ and $3p^5P$ levels—rates which are determined only by the external conditions and the values of the probabilities of transition from the $3d^3D$ and $3d^5D$ levels respectively to the $3p^3P$ and $3p^5P$ levels. However, the emission lines due to the stimulated-emission mechanism and those due to spontaneous emission should differ in their spectral line widths. A narrowing of the spectrum occurs during generation because the gain at the line center is different from the gain at the edges. Such a narrowing can serve as a criterion for the unambiguous selection of the mechanism responsible for the appearance of the emission lines. Notice that the contribution of spontaneous emission in generators with incoherent feedbacks operating in the large solid-angle regime (in the present case $\Omega_{gen} = 4\pi$) is greater by many orders of magnitude than in ordinary generators, in which $\Omega_{gen} \lesssim 10^{-6}$ sr.

Let us consider generation by a system of two-level atoms of density N_0 contained in a spherical volume of diameter d. The feedback is effected by resonance scattering at the amplifying transition. Owing to the scattering, the conventional "mode" concept is not applicable to such a generator. Therefore, in the example below, the word "mode" means the normal mode of the closed volume V corresponding to the volume occupied by the generator. The number of "modes" coupled by scattering is determined by the expression

$$M = \Omega_{gen}(d/\lambda)^2. \tag{41.63}$$

Solving the set of equations describing such a generator yields expressions (Eqs. (41.47) and (41.48)) for the spectral width of radiation emitted under continuous generation conditions. In our case, where the generator emits within a solid angle of 4π sr, these expressions may be reduced to the form [48]:

$$\frac{\Delta\omega}{\Delta\omega_D} = \left(\frac{3c}{2\pi^2(\eta - 1)\Delta\omega_D r_0}\right)^{1/2}. \tag{41.64}$$

The line width is—up to a constant factor—the geometric mean of the amplification line width $\Delta\omega_0$ and the reciprocal of the time it takes light to pass through the medium (without allowance for scattering):

$$\Delta\omega = f(\Delta\omega_D c/2r_0)^{1/2}, \quad f = \left[3/\pi^2(\eta - 1)\right]^{1/2}. \tag{41.65}$$

For the estimate of $\Delta\omega$, let us take the values $\eta - 1 \approx 3.5 \times 10^{-4}$ (a 0.035% excess pump power over the threshold value); $\Delta\omega_0 \approx 10^{10}$ sec^{-1}; $r_0 \approx 10^{11}$ cm. We then find that $\Delta\omega = 10^{-4}\Delta\omega_0$, i.e., that the generation line should be 10^4 times narrower than the spontaneous-radiation line.

Consideration should be given to the fact that the inhomogeneously Doppler-broadened amplification line $\Delta\omega$ interacts with isotropic monochromatic radiation as a homogeneously broadened one. This is explained by the fact that for any section of the Doppler contour, there always exists a light wave with an appropriated propagation direction that is in resonance with it (Fig. 41.8(b)) [61]. Therefore, as distinct from the case of directional laser radiation (Fig. 41.8(a)), no "hole-burning"-type effects occur here.

Because of the narrowing of the spectrum, the brightness temperature T_b of the line can significantly differ from the temperature of the spontaneous radiation. Thus, if for the spontaneous line the brightness temperature $T \approx 3 \times 10^4$ K, then

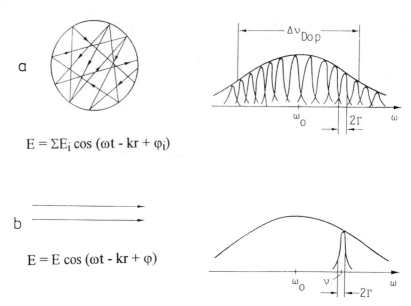

FIGURE 41.8. Interaction of (a) a monochromatic isotropic light field and (b) a directional light field with a Doppler-broadened amplification line.

for the line of width Δv it will be equal to

$$T_b \approx T \frac{\Delta\omega_0}{\Delta\omega_D} \frac{\hbar\omega_{32}/kT}{\exp[\hbar\omega_{32}/kT] - 1} \approx T \frac{\Delta\omega_0}{\Delta\omega} \text{ for } (\hbar\omega_{32} \ll kT_b). \quad (41.66)$$

Observation of the narrowing of some spectral lines emitted from a stimulated emission zone would be the best proof of the occurrence of the laser effect. However, the critical size of a stimulated emission zone r_0 may be much smaller ($r_0 > 10^6$ km) than a typical diameter L of the atmosphere of a hot star ($L \sim 10^8 - 10^9$ km) [48]. (Moreover, a large number of separate stimulated emission zones may appear in a stellar atmosphere.) The rotation of a star causes variation of the projection of the velocity of a stimulated emission zone v_{eff} along the direction of observation, i.e., an alterating Doppler shift of the center of the stimulated emission line v_0 should be observed. The magnitude of this shift $\Delta\omega_{\text{eff}} \approx \omega_0 v_{\text{eff}}/c$ (c is the velocity of light) may not only be much greater than the width of the stimulated emission line but also greater than the Doppler width of the gain profile; thus, this shift may mask the narrowing of the spectrum emitted from a given stimulated emission zone.

Thus, the problem of reconstruction of the true local width of a line emitted by a distant source is encountered in the case of strong spatially inhomogeneous broadening of the spectrum (for example, due to the rotation of an extended source). Naturally, the stimulated emission zone and the source have such very small angular dimensions that it is impossible to resolve the separate zones by ordinary optical methods and a radiation detector receives the signal from the whole source with the motion-broadened spectrum. In this formulation, the problem is more general

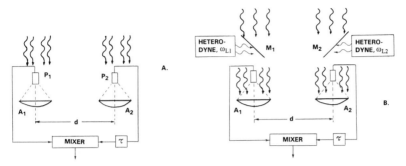

FIGURE 41.9. Schematic diagrams showing intensity (a) and independent-heterodyne (b) interferometers. Here, P_1 and P_2 are photodetectors; A_1 and A_2 are receiving mirrors (antennas); M_1 and M_2 are semitransparent mirrors.

and it has applications other than the laser effect in stellar atmospheres. Possible methods of detection of narrow "laser" lines masked by spatially inhomogeneous broadening in radiation emitted from stellar atmospheres have been considered in [62].

Narrow lines in the radiation emitted from local regions of angular size (at the point of observation) much smaller than the angular size of a stellar atmosphere can be detected using the Brown-Twiss intensity interferometer [63] (Fig. 41.9(a)), in which radiation is received by two spatially separated detectors and the correlation function of the intensity fluctuations of the received radiation is measured. We may expect a narrow component in a stellar spectrum to give rise to a low-frequency maximum in the correlation function if the receivers are separated by a distance corresponding to the resolution of a local "laser" zone in the stellar atmosphere. An exact answer can be obtained by calculating the mutual correlation function of the radiation $\gamma_{12}(P_1, P_2, \tau)$ at the points of location of the two receivers P_1 and P_2. One can also use a heterodyne interferometer (Fig. 41.9(b)), in which fluctuations of the radiation at points P_1 and P_2 are detected by heterodyning the received fields with radiation from two independent lasers of frequencies ω_{L1} and ω_{L2}. In this case measurements are made of the correlation between fluctuations of the signal amplitude at difference frequencies at the inputs of mixers, and one may expect an increase in the signal/noise ratio. In estimating the capabilities of this method once again, one has to calculate the mutual correlation function $\gamma_{12}(P_1, P_2, \tau)$.

Radiation emitted by a noncoherent feedback laser is characterized, like the radiation emitted from thermal sources in a narrow spectral range, by Gaussian statistics [21, 50]. Therefore, in a full analysis of the correlation method of the detection of a narrow emission line it is sufficient to find the normalized mutual

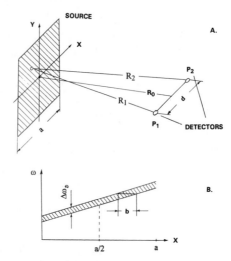

FIGURE 41.10. Representation of the relative positions of a source and detectors adopted in the calculation of the correlation function $\gamma_{12}(P_1, P_2, \tau)$ (a) and a diagram showing the spectrum of a source with a narrow local line and strong broadening of the spectrum (b).

correlation function of the fields at the points of reception (Fig. 41.10(a)):

$$\gamma_{12}(P_1, P_2, \tau) = \frac{\langle V(P_1, t + \tau)V(P_2, t)\rangle}{[\langle V^2(P_1)\rangle\langle V^2(P_2)\rangle]^{1/2}} \tag{41.67}$$

where $V(P_1, t)$ is the optical field at a point P_1 measured at a moment t.

For simplicity, we shall consider a source in the form of a square of side a which is located very far from the observer ($R_0 \gg a$). (Each point of this source emits light with a narrow Lorentzian spectral line

$$I(\omega, x, y) = I_0 \left\{ 1 + \left[\frac{\omega - \omega_0(1 - \beta x)}{\Delta\omega} \right]^2 \right\}^{-1}, \tag{41.68}$$

where ω_0 is the frequency of the line center at the origin of the coordinates located at the center of the square; β is the spatial broadening parameter; the axes are x and y (Fig. 41.10(a)); $\Delta\omega$ is the local width of the emission line.)

In the case of a Gaussian statistics source the expression, Eq. (41.67), for γ_{12} was derived in [62]. Since the inequalities $a/R_0 \ll 1$, $\Delta\omega/\omega_0 \ll 1$, and $\Delta\omega_{\mathrm{eff}}/\omega_0 \ll 1$ are valid in the case under consideration, we may assume that $R_1 R_2 \simeq R_0^2$ and $R_1 - R_2 = -xd/R_0$ where d is the interferometer base. The expression for γ_{12} may then be represented in the following simple form [62]:

$$\gamma_{12}(P_1, P_2, \tau) = e^{-i\omega_0\tau} e^{-\Delta\omega\tau} \frac{\sin\left\{\pi \frac{a}{\ell}\left(1 - \frac{\ell}{b}\Delta\omega\tau\right)\right\}}{\pi \frac{a}{\ell}\left(1 - \frac{\ell}{b}\Delta\omega\tau\right)}. \tag{41.69}$$

Where $\ell = \lambda R_0 d$ is the linear size of the region of the source resolved by the interferometer, $b = \Delta\omega/\beta\omega_0$ (Fig. 41.10(b)), λ is the wavelength of light.

Let us consider how the function $|\gamma_{12}(P_1, P_2, \tau)|$ behaves as a function of τ in the range $0 \ll \tau \ll \Delta\omega^{-1}$ for various relationships between a, ℓ, and b.

1. The source is not resolved by the interferometer ($\ell \gg a$):

   a) the spectrum exhibits weak spatial broadening ($b \gg a$),

   $$|\gamma_{12}(P_1, P_2, \tau)| \approx \exp(-\Delta\omega\tau); \qquad (41.70)$$

   b) the spectrum shows strong spatial broadening ($a \gg b$),

   $$|\gamma_{12}(P_1, P_2, \tau)| \simeq \exp(-\Delta\omega\tau)\frac{|\sin \pi a \Delta\omega\tau/b|}{(\pi a \Delta\omega\tau/b)} \qquad (41.71)$$

   $$\text{for } \tau \gg \frac{b}{\ell}\frac{1}{\Delta\omega}.$$

   We can see that in the latter case the function $|\gamma_{12}|$ oscillates rapidly and the oscillation period is $\Delta\tau \simeq b/a\Delta\omega \ll 1/\Delta\omega$.

2. The function $|\gamma_{12}|$ behaves similarly if $a \gg \ell \gg b$ (provided $\tau \gg b/\ell\Delta\omega$).

3. Finally, if $a \gg \ell \gg b$, i.e., when the interferometer resolves a region which is small compared with the characteristic spectral broadening scale b, we have

   $$|\gamma_{12}(P_1, P_2, \tau)| \quad \sim \quad \exp(-\Delta\omega\tau)\frac{\ell}{a}$$

   $$\times \left|\sin\left\{\pi\frac{a}{\ell}\left(1 - \frac{\ell}{b}\Delta\omega\tau\right)\right\}\right|. (41.72)$$

This behavior of the function $|\gamma_{12}|$ can be understood on the basis of the following qualitative description of fluctuations of radiation in the plane of the two detectors. The fluctuations of the radiation emitted by a source can be divided into two types:

   a) fluctuations in space and time due to each part of a star of size smaller than or of the order of b and with a spectral width $\Delta\omega$; these fluctuations give rise, in the plane of the detectors, to large-scale correlations of size $\tilde{d} \geq R_0\lambda/b$ and are characterized by correlation times $\tau_{corr} \sim \Delta\omega^{-1}$;

   b) fluctuations in space and time due to various parts of a star of size ℓ lying within the range $a \geq \ell > b$; they correspond to spatial correlations in the detector plane whose size is $\tilde{d}$ ($R_0\lambda/a < \tilde{d} < R_0\lambda/b$) and temporal correlations with a time τ_{corr} ($\Delta\omega_{eff}^{-1} \lesssim \tau_{corr} < \Delta\omega^{-1}$).

If the detectors are separated by $d > R_0\lambda/a$ all the small-scale fluctuations in the reception plane are detected in the frequency interval $\Delta\omega_{\text{eff}}$. In this case the signal due to fluctuations is strongest, but it is then distributed over a wide spectral range (case 1b). However, if the detectors are separated by a distance $d > R_0\lambda/b$, then only the large-scale fluctuations of type 1a) in a spectrum of width $\Delta\omega$ are detected. The small-scale fluctuations are uncorrelated and they produce a general background. This corresponds to case 3.

If $R_0\lambda/a < d < R_0\lambda/b$, we have the intermediate variant (case 2). It should also be noted that in all three cases we have oscillations of period $\tau \sim \Delta\omega_{\text{eff}}^{-1}$ (in case 1a) $\Delta\omega_{\text{eff}} \gg \Delta\omega$ and, therefore, this term is omitted, which corresponds to beats of signals from different receivers, each of which "sees" the spectral interval $\Delta\omega_{\text{eff}}$.

It follows from the estimates [62] that the condition for the observation of narrow spectral lines by means of an interferometer with independent heterodynes, particularly in the case of strong spatial broadening of the spectrum, are much more favorable than in the case of an intensity interferometer.

Note that in the optical region it is rather difficult to have independent heterodynes with a stable optical oscillation phase (Fig. 41.9(b)). It is much simpler to have a single stable local laser heterodyne, the emission of which can be transported any distance to the receiving mirrors by means of a low-loss optical fiber. The linear resolution attainable by this method at a distance of R_0 will be

$$\ell = \frac{\lambda R_0}{d}. \tag{41.73}$$

For $R_0 = 10^{20}$ cm (10^2 light years), a wavelength of $\lambda = 500$ nm, and a distance between the receiving mirrors of $d = 1$ km, the resolution ℓ will be equal to 5×10^{10} cm $= 5 \times 10^4$ km. This corresponds to an angular resolution of $\varphi = \ell/R_0 = \lambda/d = 5 \times 10^{-10} = 0.1$ milliarcsecond. Such a resolution is quite sufficient to resolve individual "laser regions" in stellar atmospheres without any fear that narrow spectral lines will be broadened as a result of stellar rotation. To make observations at great distances R_0, one may increase the distance d between the receiving mirrors.

41.5   Conclusions and Outlook

Noncoherent feedback (resonant or nonresonant) in lasers and masers is of interest in studying the behavior of high-gain active media under laboratory and natural conditions, when no use is made of mirrors, providing for a coherent resonant feedback.

Under laboratory conditions, this effect enables one to study the properties of random light fields, specifically the interference properties of backscattering. An interesting possibility of investigating noncoherent nonresonant feedback in a single transverse radiation mode was opened up after the observation of the laser effect in a semiconductor amplifying medium coupled with a long (10 km) low-loss optical fiber without any optical isolation [66]. The Rayleigh backscattering in the fiber provides for the return of some radiation into the amplifying medium and the establishment of the generation threshold with the correspondent dramatic narrowing of the radiation linewidth. The nature of the origination of very narrow lines in such a laser is still to be theoretically analyzed. This is apparently associated with random interference effects and a weak Anderson localization in the backscattered radiation.

Under natural space conditions, a distinct dichotomy is observed in space masers. Large and relatively low-density ($\leq 10^4$ cm^{-3}) clouds are observed, which lead to an unsaturated and very weak maser action in media with a gain much smaller than unity. Such intergalactic masers with a huge luminosity but without any narrowing of the radiation spectral width are referred to as mega-masers [67, 68]. These space megamasers apparently operate under unsaturated amplification conditions. Space masers of the second type (interstellar masers) are active in compact and dense regions with densities usually in excess of at least 10^5 cm^{-3}. They operate in conditions of high and saturated amplification per pass. It is exactly in this type of compact space maser that one can expect *maser generation* with a scattering noncoherent feedback as shown in Sec. 3.

Finally, it would be extremely interesting to detect the laser effect in stellar atmospheres. However, the compactness of such regions compared to the size of stars and the rotation of the latter require that experiments should be conducted with a very high spatial resolution (better than 10^{-7} arcsecond) and a very high spectral resolution (of the order of $10^6 - 10^5$ Hz). The progress in tunable lasers, fiber optics, and optoelectronics, makes such experiments quite real.

References

[1] A. Einstein, Verh. Dt. Phys. Ges., Bd **18**, 5 318 (1916).

[2] P. A. M. Dirac, *The Principles of Quantum Mechanics*, 3rd ed, Oxford (1947).

[3] W. E. Lamb, Jr., *Impact of Basic Research and Technology*, ed. B. Kursunogly, A. Perlmutter, Plenum Press (1975).

[4] J. P. Gordon, H. J. Zeiger, and C. H. Townes, Phys. Rev. **95**, 282 (1954).

[5] N. G. Basov and A.M. Prokhorov, Zh. Eksp. Teor. Fiz. (in Russian) **27**, 431 (1954).

[6] N. G. Basov and A. M. Prokhorov, Zh. Eksp. Teor. Fiz. (in Russian) **28**, 249 (1955).

[7] C. H. Townes and A. L. Schawlow, *Microwave Spectroscopy*, McGraw-Hill (1955).

[8] N. Bloembergen, Phys. Rev. **104**, 324 (1956).

[9] H. E. Scovil, G. Feher, and H. Seidel, Phys. Rev. **105**, 762 (1957).

[10] H. M. Goldenberg, D. Kleppner, and N. F. Ramsey, Phys. Rev. Lett. **5**, 361 (1960).

[11] A. L. Schawlow and C. H. Townes, Phys. Rev. **112**, 1940 (1958).

[12] A. M. Prokhorov, Zh. Eksp. Teor. Fiz. (in Russian) **34**, 1658 (1958).

[13] T. H. Maiman, Nature **187**, 493 (1960).

[14] A. Javan, W. R. Bennett, Jr., and D. R. Herriott, Phys. Rev. Lett. **6**, 106 (1961).

[15] H. F. Weaver, D. R. W. Williams, N. H. Dieter, and W. T. Lum, Nature **208**, 29 (1965).

[16] A. C. Cheung, D. M. Rank, C. H. Townes, D. D. Thornton and W. J. Welch, Nature **221**, 626 (1971).

[17] D. M. Rank, C. H. Townes, and W. J. Welch, Science **174**, 1083 (1971).

[18] R. V. Ambartzumian, N. G. Basov, P. G. Kryukov, and V. S. Letokhov, Pis'ma ZhETF **3**, 261 (1966) [JETP Letters, **3**, 167 (1966)]; Zh. Eksp. Teor. Fiz. **51**, 274 (1966)] [Sov. Phys. JETP **24**, 481 (1966)]; JEEE J Quantum Electronics, **QE-2**, 442 (1966).

[19] R. V. Ambartzumian, N. G. Basov, P. G. Kryukov, and V. S. Letokhov, Progress in Quantum Electronics, ed. J. H. Sanders and K. W. H. Stevens, (Oxford, Pergamon Press) Vol. **1**, 107 (1971).

[20] R. V. Ambartzumian, P. G. Kryukov, and V. S. Letokhov, Zh. Eksp. Teor. Fiz. **51**, 1669 (in Russian) (1966).

[21] R. V. Ambartzumian, P. G. Kryukov, V. S. Letokhov, and Yu. A. Matveets, Pis'ma ZhETF **5**, 378 (1967) [JETP Letters **5**, 312 (1967)]; Zh. Eksp. Teor. Fiz. **53**, 1955 (1967) [Sov. Phys. JETP **26**, 1109 (1967)].

[22] V. S. Letokhov, Pis'ma ZhETF **5**, 262 (1967); Zh. Eksp. Teor. Fiz. **53**, 1442 (1967) [Sov. Phys. JETF **26**, 835 (1968)].

[23] A. M. Weinberg and E. P. Wigner, *Physical Theory of Neutron Chain Reactors*, Univ. of Chicago (1958).

[24] H. C. van de Hulst, *Light Scattering by Small Particles*, Wiley (1975).

[25] G. Marovsky, Private communication.

[26] N. M. Lawandy, R. M. Balachandran, A. S. L. Gomes, and E. Sauvain, Nature **368**, 436 (1994).

[27] A. Yu. Zyuzin, Europhys. Lett. **26**, 517 (1994).

[28] W. Heitler, *The Quantum Theory of Radiation*, Oxford Univ. Press (1954).

[29] H. Weaver, D. R. W. Williams, N. H. Dieter, and W. T. Lum, Nature **208**, 29 (1965).

[30] A. C. Cheung, D. M. Rank, C. H. Townes, D. D. Thornton, and W. J. Welch, Nature **221**, 626 (1969).

[31] D. ter Haar and M. A. Pelling, Rep. Prog. Phys. **37**, 481 (1974).

[32] W. H. Kegel, Problems in Stellar Atmospheres and Envelopes, ed. B. Baschek, W. H. Kegel and G. Traving. Springer-Verlag (1975).

[33] V. S. Strelnitskii, Usp. Fiz. Nauk (in Russian) 113, 463 (1974) [Sov. Phys.-Usp. **17**, 507 (1975)].

[34] A. H. Cook, *Celestial Masers*, Cambridge University Press (1977).

[35] M. J. Reid and J. M. Moran, Masers. Ann. Rev. Astron. Astrophys. **19**, 231 (1981).

[36] M. Elitzur, Rev. of Modern Phys. **54**, 1225 (1982).

[37] A. W. Clegg and G. E. Nedoluha, eds., *Astrophysical Masers*, Springer-Verlag (1993).

[38] M. M. Litvak, Phys. Rev. A **2**, 2107 (1970).

[39] L. Allen and G. I. Peters, Nature, Physical Sciences **235**, 143 (1972).

[40] A. W. Clegg and J. M. Cordes, Ap J. **374**, 150 (1991).

[41] P. R. Rowland and R. J. Cohen, Cephens A. Mon. Not. R. Astr. Soc. **220**, 233 (1986).

[42] V. S. Letokhov, Pis'ma ZhETF (in Russian) 4, 477 (1966) [JETP Letters. **4**, 321 (1966)].

[43] V. S. Letokhov, Astronomicheskii Zhurnal (in Russian) **49**, 737 (1972) [Soviet Astronomy-A.J. **16**, 604 (1973)].

[44] M. M. Litvak, A. L. McWhorter, M. L. Meeks, and H. J. Zeiger, Phys. Rev. Lett. **17**, 821 (1966).

[45] M. M. Litvak, Ap J. **156**, 471 (1969).

[46] M. H. Cohen, D. L. Jancey, K. I. Kellerman, and B. G. Clark, Science **162**, 88 (1968).

[47] S. H. Knowles, C. H. Mayer, W. T. Sullivan III, and A. C. Cheung, Science **166**, 221 (1969).

[48] N. N. Lavrinovich and V. S. Letokhov, Zh. Eksp. Teor. Fiz. (in Russian), **67**, 1609 (1974) [Sov. Phys.-JETP, **40**, 800 (1975)].

[49] A. H. Barrett and A. E. E. Rogers, Nature **210**, 188 (1966).

[50] V. S. Letokhov, Zh. Eksp. Teor. Fiz. (in Russian) **53**, 2210 (1967) [Sov. Phys.-JETP **26**, 1246 (1968)].

[51] N. J. Evans II, R. E. Hills, O. E. H. Rydbeck, and E. Kollberg, Phys. Rev. **76**, 1643 (1972).

[52] D. Ishankuliev, Zh. Eksp. Teor. Fiz. (in Russian) **102**, 731 (1992).

[53] V. S. Letokhov, Paper presented at the 7th Intern. Conf. on Quantum Electronics, Montreal, May 1972; Preprint, Institute of Spectroscopy, USSR Academy of Sciences, N9 (1972).

[54] V. G. Gorbatskii, *The Theory of Stellar Spectra* (in Russian), Publ. House "Nauka" (1966).

[55] P. Merrill, Lines of the Chemical Elements in Astronomical Spectra. Washington (1956).

[56] J. S. Bowen, Publ. Astron. Soc. Pac. **59**, 196 (1947).

[57] W. R. Bennett, Jr., W. L. Faust, R. A. McFarlane, and C. K. N. Patel, Phys. Rev. Lett. **8**, 470 (1962).

[58] C. W. Allen, *Astrpophysical Quantities*, University of London, Athlone Press (1963).

[59] A. A. Boyarchuk, Astron. Zh. (in Russian) **34**, 193 (1957) [Sov. Astron.-AJ. **1**, 192 (1957)].

[60] V. V. Sobolev, Radiative Energy Transfer in Stellar and Planetary Atmospheres. Van Nostrand, Princeton, N.J. (1963).

[61] V. S. Letokhov and V. P. Chebotayev, Nonlinear Laser Spectroscopy, Ch.1, Berlin, Springer-Verlag (1977).

[62] N. N. Lavrinovich and V. S. Letokhov, Kvantovaya Elektron. (in Russian) **3**, 1948 (1976) [Sov. J. Quantum Electron. **6**, 1061 (1976)].

[63] R. Hanbury Brown and R. Q. Twiss, Proc. Roy. Soc. **A242**, 300 (1957); **A243**, 291 (1958); **A248**, 199 (1958); **A248**, 222 (1958).

[64] M. A. Johnson, A. L. Betz, and C. H. Townes, Phys. Rev. Lett. **33**, 1617 (1974).

[65] H. Gamo, J. Opt. Soc. Am. **56**, 441 (1966).

[66] L. Goldberg, H. F. Taylor, and J. F. Weller, Electron. Lett. **18**, 353 (1982).

[67] J. B. Whiteoak and F. F. Gardner, Astrophys. Lett. **15**, 211 (1973).

[68] V. V. Burdjuza, Uspekhi Fiz. Nauk (in Russian) **155**, 703 (1988).

42

Application of Millisecond Pulsar Timing to the Long-term Stability of Clock Ensembles

**Demetrios N. Matsakis
and Roger S. Foster**

42.1   Introduction

The 1982 discovery of the 1.6 millisecond pulsar B1937+21 [1] has provided an object with rotational stability comparable to the best atomic time standards over periods exceeding a few years [2]. The microwave beams of millisecond pulsars, which are rigidly anchored to rotating neutron stars, can act as precision celestial clocks owing to the combined effects of their large rotational energies, 10^{50-52} erg, and low energy loss rates. Millisecond pulsars provide an opportunity for astronomical observations to define International Atomic Time (TAI) over long periods.

Because each pulsar has a different spin and spin-down rate, timing measurements of millisecond pulsars cannot provide a more *accurate* timescale than current atomic timescales. Since the best foreseeable daily-averaged pulsar timing measurements will be accurate to 0.1 microsecond, two observations separated by 5×10^7 seconds would be needed to reach the 2.5×10^{-15} s/s TAI monthly *precision* [3] (defined as the internal rms of the data set, ignoring systematic errors), although more frequent pulsar observations would reduce this considerably.

The existence of a number of millisecond pulsars distributed across the sky leads to the possibility of timing several pulsars against each other, with terrestrial clocks providing merely the endpoint times and a means of interpolating observations [4-6]. Such a program can smooth out variations in TAI, better determine the masses of the outer planets [7], place limits on perturbations produced by a primordial spectrum of gravitational radiation [8, 9], constrain the nature of interstellar turbulence [10], and provide an astrometric tie for the planetary and radio reference frames [11].

The search for new millisecond pulsars is a matter of intense activity and steady progress. More than 30 millisecond pulsars have been found in globular clusters, and over 25 have been found in the galactic disk, over a wide range of galactic latitudes and longitudes. Unlike slow pulsars, more than 75% of the millisecond pulsars have stellar companions; millisecond pulsars are often termed recycled pulsars because it is believed they were spun-up by their companions [12]. It is estimated that over 30,000 millisecond pulsars in the galaxy are beaming in our direction [13]. Non-cluster millisecond pulsars are discovered at a rate of about 1 per 200-300 square degrees [14]. Assuming a factor of two improvement in search efficiency, and ignoring serendipity, perhaps up to 400 field millisecond pulsars could be discovered within the next few decades.

In this work we summarize the problems and status of terrestrial timescales and pulsar observations, and explore the contributions pulsars can make to terrestrial timescales.

42.2   Terrestrial Clocks

According to the International System (SI), the second is defined so that the frequency of the $(F, m_f) = (3, 0)$ to $(4,0)$ hyperfine transition of cesium 133 is exactly 9.192631770 GHz, in the absence of a magnetic field and located on the geoid [15]. As recognized in the definition, the great majority of the timing data used to compute TAI are from cesium-based frequency standards [16]. Cesium atoms in the $(3,0)$ state are created in an oven and isolated from cesium atoms in other states by their deflection in an inhomogeneous magnetic field. They are then allowed to travel through a magnetically shielded cavity, in which they are exposed to 9 GHz microwaves at each end. The probability that the 9 GHz microwaves will induce a transition to the $(4,0)$ state has a Ramsey-pattern dependence upon frequency of the microwaves [17]; the number of atoms which have made the transition to the $(4,0)$ state is measured by the ionic current created after deflection by an inhomogeneous field at the exit point of the cavity. The principal error sources are the magnetic field correction and the phase difference between the microwave fields at the two interrogation regions [18]. Recently, Hewlett-Packard introduced their model 5071 cesium standards, with a significantly improved electronics package [19]. This cesium clock has demonstrated superior performance both in the laboratory [20] and in the field [21]. The world's most accurate laboratory standards are maintained by the Physikalisch-Technische Bundesanstalt (PTB) in Germany and the National Institute of Standards and Technology (NIST), which achieve long-term stability through use of special beam-path and field switching techniques and the use of a longer cesium travel path [22].

A second frequency standard used for high precision time keeping is the hydrogen maser. Hydrogen masers are based upon the 21-cm transition of atomic hydrogen and can achieve much greater precision in the short-term [23]. However, their long-term accuracy is limited by variations in the dimensions of the confining cavity and interactions with the cavity walls [24]. Special Teflon coatings have been developed which minimize the wall-shift. Recently the Sigma Tau Standards Corporation has begun marketing masers whose cavity capacitance is stabilized and optimized through an attached varactor diode and a tuning circuit

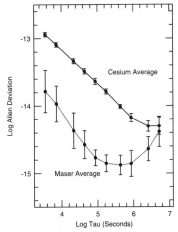

FIGURE 42.1. Observed frequency stability of USNO standards as function of averaging time. Plotted are the average Allan deviation (rms of differences between adjacent averages, divided by $2^{1/2}$) of the frequencies of 6 cavity-tuned masers and 12 model 5071 cesiums as a function of averaging time τ. The error bars indicate the ensemble rms.

which measures the maser amplitude at 7.5 kHz on each side of the microwave resonance [25].

The precision of the best types of frequency standards at the U.S. Naval Observatory (USNO) as a function of the averaging time are shown in Fig. 42.1, which was adapted from Breakiron [20]. Note that masers have significantly better performance on short timescales, but on the longest scale, 60 days, their performance is similar to the cesiums. Following the methodology of Barnes et al. [26], we summarize in Table 42.1 the error performance of the clocks as a sum of white and random-walk frequency components. Since these clocks have not been in existence long enough to gather data on still longer timescales, it is likely that future analyses will identify other components to the variation of the cesium standards. A long-term analysis of maser data will be more difficult to perform since they undergo discontinuous changes in frequency and frequency drift (the time derivative of the frequency) on timescales of a few months, which are removed by comparison with other clocks.

Within the next decade, we can expect further improvements in terrestrial standards. The closest to implementation are optically-pumped mercury stored-ion devices. Their advantage is based upon the high frequency of the observed transition (40.5 GHz) and coherence times which allow line-widths as low as 17 mHz. Although the accuracy of prototypes was limited by vacuum contamination [27], recent designs based upon a linear ion trap have achieved frequency stabilities better than 10^{-15} s/s on timescales of a few hours [28] (frequency is defined hereafter as the time derivative of the timing error in units of time, not phase). Somewhat further away are clocks based upon trapped yttrium [29], single trapped mercury ions [30], and cesium atomic fountains, which achieve long coherence times by

TABLE 42.1. Simple Noise Model

	White Phase Noise nanosecond	White Frequency Noise 10^{-15}s/s (t/days)$^{-0.5}$	Random Walk Frequency Noise 10^{-15}s/s (t/days)$^{+0.5}$
cesium *	0	25	0.5
hydrogen maser *	0	2	1.0
PSR B1937+21 **	100	0	3

* It is assumed that the times of the phase, rate, and frequency drift discontinuities have been identified, and their effects removed by least-squares methods. Such methods are required on monthly scales for masers, and much less frequently for cesiums. It has the effect of masking higher order variations.

** Although today's best Arecibo data for PSR B1937+21 have formal TOA errors of 200 nanoseconds, higher precision will be achievable with the GBT and up-graded Arecibo. The random walk noise, computed from Monte Carlo simulations, is numerically larger than modeled by Petit et al. [65]

first cooling atoms in "optical molasses" to 700 nanokelvins, and then allowing them to rise out of the measurement region until gravity brings them back [31].

42.3 Global Time Transfer

The Global Positioning System (GPS) has within the last few years become the chief instrument for comparing clocks of different institutions, as well as for determining position on the Earth. It consists of a constellation of 24 satellites, run by the U.S. Air Force Second Satellite Operations Squadron, each containing at least one cesium frequency standard and circling the Earth twice per sidereal day [32]. These satellites constantly broadcast their coded position and timing information, in both classified and unclassified channels, enabling anyone with the proper equipment to determine, by triangulation, the position and time at the receiver. Depending upon the kind of GPS receiver, the time can be communicated through an output signal and/or compared digitally to the time input from a standard at the site. The individual timing characteristics of the satellites are monitored by the USNO, which forwards the information to the Air Force for use in determining the frequency and time offset corrections, and in measuring orbital variations. Although the specified GPS time precision is 1 microsecond, since January 1993 GPS time has kept within 260 nanoseconds of the USNO Master Clock #2 (UTC(USNO)), and the rms difference over that period was 60 nanoseconds. Satellites also broadcast on-line predictions of GPS−UTC(USNO), which are specified to be less than

90 nanoseconds, but are usually below 20 nanoseconds (which is close to the measurement error). Evaluations of GPS−UTC(USNO) are published as USNO Series 4, and are also available via modem, Internet, and the World Wide Web home page http://tycho.usno.navy.mil.

While the received GPS signal is distorted and delayed by unmodeled tropospheric, ionospheric, relativistic, and multi-path effects, the largest error contribution to unclassified use is the intentional distortion of up to 340 nanoseconds imposed by the military for security reasons and termed "selective availability" (SA). SA consists of both the insertion of a variable time delay and a misrepresentation of satellite orbital parameters, which are used to determine the propagation time for the signal to reach the observer.

If one is merely interested in time difference between remote clocks, it is possible to eliminate the error due to the variable time delay through the "Common View" method. In Common View, observations are scheduled so that both sites observe the same satellite at the same time. While the difference neatly cancels satellite timing errors, it cannot eliminate the errors due to the incorrectly broadcast orbital parameters. It is possible to correct for this after the fact, using satellite orbital elements routinely disseminated by the International GPS Service for Geodynamics, which relies upon multiple observations for its analysis. Common View techniques routinely provide time-transfer precision of 5 to 15 nanoseconds [33, 34].

While it is possible, in principle, to correct any GPS measurement in real time for SA using Common View, such scheduling tends to result in fewer observations which are also conducted at lower elevation angles, so that tropospheric and ionospheric modeling errors are more significant. At the USNO, unclassified GPS transfers are achieved using the "melting-pot" method, which forsakes Common View in order to obtain a straight-forward average of as many satellites as possible. Even in the presence of SA, a precision of 20 nanoseconds is attainable if one is able to average at least seventy 13-minute satellite passes over a 48-hour period [35]. The advantage of this method is even greater for receivers which can observe more than one satellite at the same time. Another form of time-transfer consists of using available non-GPS satellites to send signals to, and from, the two remote sites. Aside from achieving sub-nanosecond time transfer precision, such studies often reveal systematic and perhaps seasonal differences with conventional GPS measurements, of the order of 10 nanoseconds [36].

42.4   Terrestrial Timescales and the BIPM

The goal of timescale formulation is to approximate "true" coordinate time, from an average of differential clock timing measurements. In essence, the clocks serve as frequency standards to create a heterodyned tone of known or defined value. Individual timing events are realized either through summation of cycles or use of "zero-crossings" of the sinusoidal output voltage. In practice, timescales are

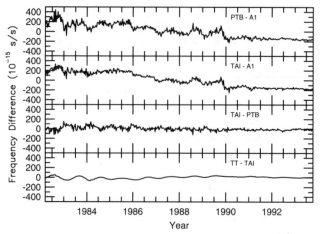

FIGURE 42.2. Frequency difference between free-running timescales of TAI, PTB, and USNO (A.1) in fs/s. The USNO and PTB timescales are independent, but TAI is gently steered towards the PTB, whereas the USNO clock's contribution to TAI has increased to 40%. Nevertheless, improvement in the last few years is obvious. The lowest plot shows the difference between TT and TAI, where TT represents TAI recomputed using hindsight corrections to clock weights, offsets, and drifts. A constant frequency offset has been removed from all plots. Data were obtain from the BIPM by anonymous ftp to address 145.238.2.2 The time range shown is from January 1982 to December 1993.

usually generated in the frequency domain, in which a mean frequency is computed and used to define the time from an arbitrary offset.

Internationally, the official responsibility for timescale generation lies with the International Bureau of Weights and Measures (BIPM) in Paris, France, as first established by the Convention du Mètre in 1875. The BIPM produces TAI by independently evaluating the individual characteristics of approximately 200 clocks maintained by almost 50 institutions [37]. Evaluations are made at 10-day intervals, and reported monthly in their Circular T. The BIPM also evaluates timescales generated by many national institutions, including GPS time. Some of these timescales are free-running, such as the USNO "A.1" series, which is currently comparable to TAI in accuracy. Others, such as UTC(NIST) and UTC(USNO), are steered so as to track Coordinated Universal Time (UTC) closely. The UTC(USNO), which is the national standard for legal and military matters, is available on-line and is in practice based upon the output of a single maser that is steered daily through the addition of small frequency offsets so as to approach the computed mean of all contributing USNO clocks, which itself is adjusted towards extrapolations of the BIPM-computed TAI−UTC(USNO). As a result of this steering TAI and UTC(USNO) have since 1991 kept within 200 nanoseconds (see Figs. 42.2 and 42.3).

Once all the time comparisons have been assembled, the problem of timescale generation reduces to calculating weights and offsets for each standard. The BIPM first generates a free-running timescale, the Echelle Atomique Libre (EAL). The

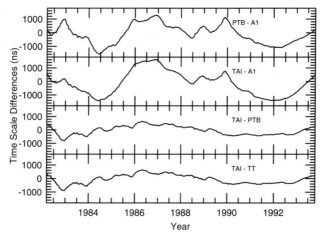

FIGURE 42.3. As in previous figure, except that the vertical axis shows the timing difference, with a quadratic term removed. Pulsars cannot contribute to the quadratic terms as their true periods and spin-down rates cannot be determined.

EAL is computed by evaluating clock weights each 60 days. The weights are determined from the variance of the frequency offsets over the past year, after allowing for any intentional rate adjustments. A maximum weight is chosen so that the timescale will not depend upon just a few of the clocks. While both the general and specific procedures are periodically under review, in 1994 about half of all clocks used had this maximum weight, while most of the rest were weighted less than 20% of the maximum. The BIPM does not subtract offsets from the clock frequencies, and this can lead to minor discontinuities as clocks are added, subtracted, or reweighted. The TAI timescale is created by gently steering the EAL so that its long-term behavior conforms to the "primary" frequency standards, which are currently the two "long-tube" primary standards of the PTB [22] and NIST-7 [37]. This is a practical way for a real time system to take into account the fact that the primary standards have greater long-term accuracy than accounted for by the precision-based weighting system. Again, there is a minor problem in that no allowance is made for fact that the NIST-7 frequency is corrected by about 1.9×10^{-14} s/s for the AC Stark effect due to the microwave background experienced by the cesium atoms [38], whereas the PTB frequencies are not.

At the USNO, the A.1 timescale is currently determined using a weighting function which weights clocks by their type. Recent maser data are given a much higher weight than recent cesium data, but older maser data are progressively down-weighted, with zero weight assigned to maser data more than 60 days old [39]. Also, data from the older model cesium standards are included at 0.65 the weight of model 5071 cesium clocks. Unlike the BIPM, each clock is introduced to the mean after subtracting an initial frequency offset. This has the effect of reducing discontinuities, but the drawback of subtly over-weighting initial clocks is soon lost in the random-walk noise, as long as no systematic errors are present. For

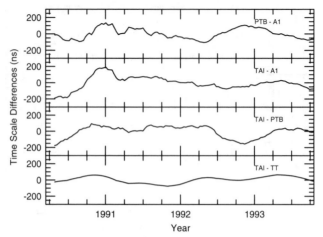

FIGURE 42.4. As in previous figure, except the time range of data is shortened to include only the latest and best data, from 1990 to 1994.

historical reasons, the A.1 is intentionally steered a constant 19 ns/day [40]. Using data from 1993 and 1994, the BIPM estimates the 100-day error of the EAL as slightly less than the TAI, and both to be about 2.5×10^{-15} s/s [3]; the errors in the 1980s were above 10^{-14} s/s. The recent improvement in both TAI and the USNO free-running timescale A.1 is evident in Fig. 42.4, which gives their difference as a function of time, as compared with Figs. 42.2 and 42.3.

Part of this apparent improvement is due to the fact that the contribution of the USNO clocks to TAI has varied (from 30% in the 1980s to a low near 15% in 1990 to almost 40% in 1994), but most is due to the world-wide increase in number, quality, and environmental isolation of the individual frequency standards. Similar increases in precision are shown in the comparisons with the independent PTB timescale.

For practical reasons, other timescales are often used which differ from TAI by specific and calculable amounts. Coordinated universal time (UTC) differs from TAI only by an integral number of leap seconds, which crudely take into account variations in the rate of Earth rotation [41]. These leap seconds are extra seconds included in UTC at UT midnight on either December 31 or June 30 so as to minimize UT1−UTC, where UT1 is time related to the rotational angle of the Earth after allowing for polar motion [41-43]. They have lately been inserted at a rate approaching once a year, and on July 1, 1994, TAI−UTC= 29 seconds. Measurements and predictions of UT1, polar motion, and nutation parameters [44] are available on-line from the home page http://maia.usno.navy.mil. In order to allow for continuity with ephemeris time (ET), Terrestrial Time (TT, also called Terrestrial Dynamical Time, TDT) is defined to be offset by exactly 32.184 seconds from the "ideal" form of TAI, which is what TAI would be if no instrumental noise were present. Although in the future TT may be realized from an average of many sources, including pulsars, currently TT is realized only from TAI, by post-processing the raw clock data and improving upon the initial weighting and steering corrections [45]. Since 1982, the BIPM-computed numerical difference TT−TAI has varied by over 3 microseconds absolutely, and over 2 microseconds after removing the quadratic component of the variation. In order to provide continuity in the interpretation of old data and allow for the effects of general relativity in a consistent manner, the IAU in 1991 passed resolutions recognizing timescales which distinguish between centers of reference at the solar system barycenter (Barycentric Coordinate Time, TCB), the center of the Earth (Geocentric Coordinate Time, TCG), and the geoid (TT, TAI, UTC, ET) [46-48].

42.5   The Techniques of Pulsar Timing

The precision of an individual pulsar time-of-arrival (TOA) measurement depends upon its pulse period and the signal-to-noise ratio as determined from the particular instrumental setups [49, 50]. The fastest pulsar known today is still PSR B1937+21, whose 1.6 ms period is close to the theoretical limit determined from balancing the centripetal and gravitational forces, assuming a typical neutron star equation of state at densities of greater than 10^{14} g cm^{-3} [51].

The simplest rotation model for a solitary neutron star can be given in terms of phase residuals as a power series in time:

$$\phi(t) = \phi_0 + \Omega(t - t_0) + \dot{\Omega}(t - t_0)^2/2 + \ddot{\Omega}(t - t_0)^3/6 + \ldots \qquad (42.1)$$

where ϕ is the pulse phase, Ω is the rotation frequency, $\dot{\Omega}$ is the spin-down rate. The spin frequency is related to the pulse period by $\Omega = 2\pi/P$ ($\Omega = 2\pi\nu$), while $\dot{\Omega} = -2\pi\dot{P}/P^2$. If the spin-down of the pulsar is driven by magnetic dipole radiation due to misalignment of the magnetic and rotational axes, then the braking index n, defined by $\Omega\ddot{\Omega}/\dot{\Omega}^2 \equiv 2 - P\ddot{P}/\dot{P}^2$, is 3. If the initial pulsar period was P_0, the pulsar age can be computed from

$$\tau = \frac{P}{(n-1)\dot{P}} \left[1 - \left(\frac{P_0}{P} \right)^{n-1} \right]. \qquad (42.2)$$

For most pulsars $\ddot{P}$ measurements yield only upper limits, but in those pulsars for which $\ddot{P}$ can be measured above the noise, n is found to be between 2 and 3. Inverting the equations above, we derive a theoretical value

$$\ddot{\Omega} = \frac{n\dot{\Omega}^2}{\Omega} = \frac{2n\pi\dot{P}^2}{P^3} . \qquad (42.3)$$

In the case of PSR B1937+21 the theoretical slowdown would result in a 0.1 microsecond residual from a constant $\dot{P}$ model after 11 years. The relativistic time-of-flight effects and acceleration due to the galactic gravitational field can also make small, but non-negligible contributions to $\dot{\Omega}$ and the theoretical $\ddot{\Omega}$ [52].

Also, for pulsars in globular clusters, contributions to $\ddot{\Omega}$ from Doppler acceleration can be important where the gravitational force due to the cluster accelerates the pulsar, influencing the rotation period over several-year timescales [53, 54].

Pulsar timing requires an explicit definition of both a steady atomic time standard and accurate monitoring of the motion of the Earth with respect to the background stars. This task is accomplished by using a clock whose offset is frequently calibrated against a real time-steered UTC service, such as UTC(NIST) or UTC(USNO). This real time UTC realization is later corrected to a more accurate timescale using after-the-fact determinations published by the BIPM. Although TAI is often used, TT is recommended as the most accurate terrestrial scale available today [5]. Once a terrestrial timescale has been chosen, it must be converted to the barycentric frame (TCB). The omission of this time conversion would, among other things, result in a sinusoidal monthly error of about a microsecond as the Moon pulls the the Earth back and forth through the solar gravitational potential well. In most software, the time conversion happens automatically as part of the four-dimensional general-relativistic transformation which accounts for the Earth's orbit. The IAU has recommended that all users avoid transformations to Barycentric Dynamical Time (TDB) [46], which is scaled to remove secular differences with TT; as a result it diverges from TCB at a constant rate of 1.55×10^{-8}, which would be absorbed into the fitted parameters.

The geometric, general relativistic, and astrometric effects due to the position and velocity of the antenna with respect to the solar system barycenter must also be removed. In most cases this is done using the International Earth Rotation Service (IERS) values for UT1−UTC and the DE200 ephemeris [55]. Since most observations used for computing the Earth's orbit are Earth-based they are relatively insensitive to the uncertainties in the masses of the outer planets. The most uncertain long-term component of the ephemeris for pulsar work is the motion of the Sun and known planets about the solar system barycenter. The largest error contribution is due to the uncertainty in the mass of Saturn, which could result in a TOA error of order 7 microseconds with a 30-year period [49]. This and other such errors would be identifiable through their different and systematic projections upon the pulsars observed [4]. Fukushima has examined the differences between the DE200, DE102, and DE245 ephemerides and found the differences between them (including their effects on the relativistic time correction TCB−TT) to be negligible for pulsar work [48].

Pulsar signals must be corrected for dispersion caused by the intervening plasma using the frequency dependence of the delay. For a homogeneous and isotropic medium the group velocity is given by $v_g = c(1 - \omega_p^2/\omega^2)^{1/2}$, where ω_p is the plasma frequency and ω is the wave frequency. The plasma frequency in Gaussian units is $\omega_p^2 = 4\pi n_e e^2/m$, where n_e is the mean electron density, and e and m are the electron charge and mass. The delay is proportional to the column density of electrons in the line of sight towards the pulsar. This column density is called the dispersion measure and is defined as $DM = \int_0^z n_e dl$, where z is the pulsar distance. The dispersion measure is a directly measurable quantity determined from the differential pulse arrival time between two frequencies. To first order, observing v_g gives a time delay between two frequencies as

$$t_2 - t_1 = \frac{2\pi e^2}{mc}(\omega_2^{-2} - \omega_1^{-2})DM .$$ (42.4)

Observations of dispersion measures toward 706 radio pulsars give values that range from 1.8 to ~ 1000 cm^{-3} pc [56]. For some lines of sight the turbulent electron population may cause different wavefronts to travel along slightly different paths, which can lead to delays that are not removable by the two-frequency formula, although some improvement can occur with multi-frequency observations. The unmodeled delay, which depends upon the total electron column density and the turbulent properties on the interstellar electron population, has been estimated in the worst cases to be at the sub-microsecond level for radio frequency observations at 1.4 GHz and above [10].

Another important consideration is that millisecond pulsars have steep radio spectra and are generally weak enough that observations at the largest telescopes are required in order to achieve an adequate signal-to-noise (SNR) ratio. An estimate of the SNR achievable at current and future sites is provided in Table 42.2.

TABLE 42.2. Antenna Parameters at 1.4 GHz

	T_{sys} K	Gain K/Jy	Bandwidth MHz	SNR 2 Polarizations	T_{int}* Hours
Arecibo, upgraded	20	11	200	150	0.3
Arecibo, current	40	8	40	24	10
GBT	10	1.9	200	50	2
Effelsberg	20	1.5	200	20	14
Nancay	40	1.5	200	10	55
Parkes	20	.64	320	11	50
Jodrell Bank	20	.64	320	11	50
NRAO 140′	20	.26	120	3	800

* Time required to observe PSR J1713+0747 with a TOA error of 100 nanoseconds assuming a flux of 3 mJy and a pulse width of 18 degrees.

Once the TOA's from a pulsar are assembled and iteratively corrected for known effects, several pulsar parameters are typically solved. Uncertainty in the apriori pulsar period and spin-down rate are identified through their quadratic dependence. The pulsar parallax is identified from the TOA by its semiannual periodicity, position errors through their annual periodicity, and proper motion by a growing annual periodicity. Pulsar parameters associated with orbits about companions are

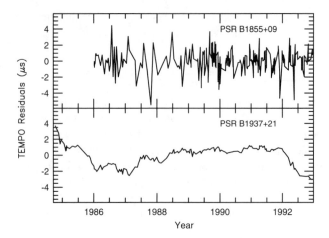

FIGURE 42.5. Timing residuals to pulsar parameter solutions for PSR B1937+21 and PSR B1855+09, from October 1984 to December 1992.

more complex [57, 58]. More exotic error sources, such as gravitational radiation and unknown planets, can be identified through comparison of delays with other members of the pulsar ensemble. If the TOA errors are modeled as a multi-pole expansion over the sky, timing errors are related to the monopole term, ephemeris errors result in a dipole dependence, and gravitational radiation would cause a quadrupole and higher order effect [4].

42.6   Stability of Pulsar Time Standards

Using data on the millisecond pulsars B1937+21 and B1855+09 made available by anonymous FTP from Princeton University [2], the program TEMPO [59] was used to solve for the pulsar periods, spin-down rates, position, and proper motion. The residuals to the solutions are shown in Fig. 42.5. As Kaspi et al. [2] found with the same data, the residuals of PSR B1937+21 appear to display red noise (long-period correlations, in this case well-modeled as a non-zero $\ddot{P}$), whereas the residuals of PSR B1855+09 do not. Since these two pulsars are close together in the sky, the red noise is considered to be intrinsic to the pulsar B1937+21, and not due to the timescale or ephemeris errors. The timing noise for millisecond pulsars seems to follow the correlation with $\dot{P}$ found for slower pulsars by Arzoumanian et al. [2, 60, 61]:

$$\Delta_8 = 6.6 + 0.6 \log \dot{P} , \qquad (42.5)$$

where the "stability parameter" $\Delta_8 = \log\left(\frac{1}{6\nu}|\ddot{\nu}|(10^8 \text{ s})^3\right)$ with a scatter of $\pm$ one decade. Pulsar B1937+21 has a $\Delta_8 = -5.5$. If this inverse correlation is valid for millisecond pulsars, then the PSR J1713+0747 would be a particularly promising

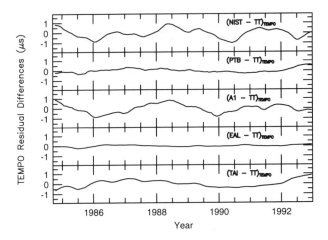

FIGURE 42.6. Difference between timing residuals of PSR B1937+21. Solutions differ only in their reference timescale, and each plot is the difference between the solutions with the indicated timescale and with TT. The data are from October 1984 to December 1992.

candidate to have a very low Δ_8 stability parameter [62], with a red noise less than that of PSR B1937+21 by a factor of about 5.

The effects of different input timescales are shown in Fig. 42.6, which plots the difference between the residuals of parameter fits to PSR B1937+21, which differ only in the input timescale assumed. For these and subsequent solutions, the parameters are the pulsar period, period derivative, position, parallax, and proper motion. In Fig. 42.7 these difference solutions are double-differenced with the differences between the timescales, as made available by the BIPM, with a second-order term removed. The double-differences indicate how much timescale variation is soaked up into the fitted parameters, and hence is indeterminable from pulsar TOA data. The annual signature evident in the plots, for example, shows how pulsar timing data alone will never be able to identify strictly periodic annual variations in terrestrial standards. In the presence of an independent means of determining pulsar positions, proper motion, or parallax (such as astrometric Very Long Baseline Interferometry, or VLBI), such variations would be recognizable as displacements in the fitted parameters. Table 42.3 shows the current positional precision of a number of millisecond pulsars, derived from timing data. The coupling of the dynamic and radio reference frames will, along with VLBI observations of millisecond pulsars, aid in removing two degrees of freedom from the pulsar fitting models and improve our ability to constrain other sources of errors.

Slightly extending the analysis of Blandford et al. [63], it is possible to apply a Weiner filter to the residuals by Fourier-transforming the residuals, and then transforming back to the time domain after multiplication by a "transfer function," which corrects for the distortion caused by fitting to pulsar parameters. The exact correction factor would also include a factor incorporating the clock and pulsar stabilities as a function of frequency. Since pulsar data are much less accurate on

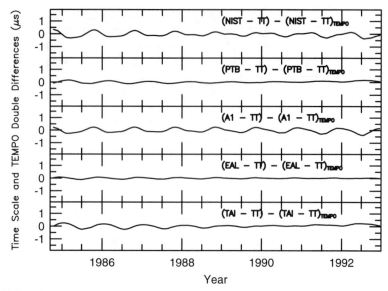

FIGURE 42.7. Double-difference plots subtracting the TEMPO residual differences of previous figure from the terrestrial timescale differences. Since pulsars cannot determine variations below second order, a quadratic term was removed from the terrestrial differences. The data are from October 1984 to December 1992.

TABLE 42.3. Current Pulsar Astrometric and Timing Precision

Pulsar Name	Observation Duration (years)	Timing Precision (μsec)	RA Precision (mas)	DEC Precision (mas)
PSR B1257+12 [58]	2.6	2.3	0.4	1.0
PSR J1713+0747 [66]	1.8	0.4	0.2	0.3
PSR B1855+09 [2]	6.9	1.0	0.07	0.12
PSR B1937+21 [2]	8.2	0.2	0.03	0.06
PSR J2019+2425 [67]	2.7	3.0	0.6	0.9
PSR J2322+2057 [67]	2.3	2.9	1.0	2.0

short time-scales, compared to terrestrial standards, pulsar data would be given insignificant weights at short periods. For the generation of the plots in Fig. 42.8, a simpler analysis was used in which all periods less than 400 days were discarded. Figure 42.9 shows the "transfer function" for two different time ranges, which indicate what frequencies in the raw data are absorbed by the fitted parameters [63]. The low values at periods of one year and six months indicate that any noise in

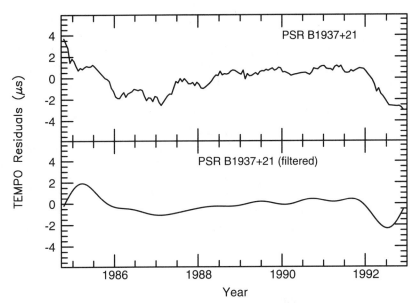

FIGURE 42.8. Upper part shows the residuals for a fit to pulsar parameters for PSR B1937+21. Lower part shows the same residuals after using a Fourier analysis to extract all periodicities shorter than 400 days. No corrections have been made for spectral leakage or aliasing. The data are from October 1984 to December 1992.

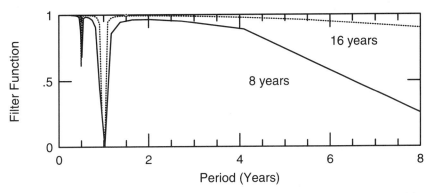

FIGURE 42.9. Transfer functions for 165 observations equally spaced over 8 years, and for 330 observations equally spaced over 16 years as a function of period (1/frequency). In this plot, a value of 1 indicates no noise is absorbed in the fit to pulsar parameters, and a value of 0 indicates that all noise is absorbed.

the data with those frequencies will be absorbed into the fitted pulsar parameters. It is evident that the accumulation of another decade of data will bring about a considerable improvement in the ability of pulsar timing data to discern long-term variations.

The contribution millisecond pulsar data can make to a timescale is entirely dependent upon the quantity and quality of the pulsar TOA's compared to the terrestrial standards. Included along with the simple clock model in Table 42.1 are parameters for the timing stability of PSR B1937+21, which is characterizable as a random walk in frequency, or perhaps an integrated random walk, which is even redder [64]. These numbers were chosen to be consistent with the residuals plotted in Fig. 42.5, and they show that PSR B1937+21 is less accurate than a single model 5071 cesium standard. If the relation between timing stability and $\dot{P}$ is valid, then the pulsar J1713+0747 would be five times more accurate, and an ensemble consisting of several such pulsars could approach the accuracy of today's ensemble of terrestrial frequency standards. More important than the numerical evaluations is the fact that pulsars can provide an independent means of evaluating clock performance, and can serve to constrain the amount of extremely red noise in terrestrial clocks, which would be undetectable on the short timescales available so far.

42.7   Acknowledgments

We would like to thank D. C. Backer, L. A. Breakiron, R. T. Clarke, H. Chadsey, J. DeYoung, T. M. Eubanks, F. J. Josties, V. M. Kaspi, W. J. Klepczynski, S. Lundgren, D. J. Nice, P. K. Seidelmann, F. Vannicola, and G. M. R. Winkler for many helpful discussions.

References

[1] D. C. Backer, S. R. Kulkarni, C. E. Heiles, M. M. Davis, and W. M. Goss, Nature, **300**, 615 (1982).

[2] V. M. Kaspi, J. H. Taylor, and M. F. Ryba, ApJ, **428**, 713 (1994).

[3] C. Thomas, Proc. 26th Annual PTTI, in press (1995).

[4] R. S. Foster and D. C. Backer, ApJ **361**, 300 (1990).

[5] B. Guinot, G. Petit, A&A **248**, 292 (1991).

[6] G. Petit, C. Thomas, and P. Tavella, Proc. 24th Annual PTTI, 73 (1993).

[7] J. D. Mulholland, ApJ **165**, 105 (1971).

[8] D. C. Backer, Proc. Marcel Grossmann Meeting, in press (1995).

[9] D. R. Stinebring, M. F. Ryba, J. H. Taylor, and R. W. Romani, Phys. Rev. Lett., **65**, 285 (1990).

[10] R. S. Foster and J. M. Cordes, ApJ **364**, 123 (1990).

[11] J. F. Lestrade, A. E. E. Rogers, A. R. Whitney, A. E. Niell, R. B. Phillips, and R. A. Preston, AJ **99**, 1663 (1990).

[12] P. Ghosh and F. K. Lamb, In: *X-ray Binaries and Formation of Binary and Millisecond Pulsars*, eds. E. P. J. van den Heuvel & S. A. Rappaport, (Dordrecht: Kluwer, p.487, 1992).

[13] M. Bailes and D. R. Lorimer, in: *Millisecond Pulsars: A Decade of Surprise*, ed. Fruchter, Tavani and Backer, 17 (1995).

[14] R. S. Foster, B. J. Cadwell, A. Wolszczan, and S. B. Anderson, ApJ **454**, Dec. 1, in press (1995).

[15] Thirteenth General Conference of Weights and Measures (1967).

[16] N. F. Ramsey, J.Res., N.B.S. **88**, 301 (1983).

[17] N. F. Ramsey, *Molecular Beams* (Oxford, Clarendon Press, 1963).

[18] A. Bauch, K. Dorenwendt, B. Fischer, T. Heindorff, E. K. Muller, and R. Schroder, IEEE Trans. Instrum. and Meas. IM-36 **No. 2**, 613 (1987).

[19] L. S. Cutler and R. P. Giffard, Proc. 1992 IEEE Frequency Control Symposium, 127 (1992).

[20] L. Breakiron, Proc. 26th Annual PTTI, in press (1995).

[21] W. J. Wheeler, D. N. Chaulmers, A. D. McKinley, A. J. Kubik, and W. Powell, (1995) Proc. 26th Annual PTTI, in press.

[22] A. Bauch, K. Dorenwendt, and T. Heindorff, Metrologia **24**, 199 (1987).

[23] C. H. Townes, J. Appl. Phys., **22** , 1365 (1951).

[24] H. E. Peters, Proc. Fifth Annual PTTI, 283 (1974).

[25] H. B. Owings, P. A. Koppang, C. C. MacMillan, and H. E. Peters, Proc. 1992 IEEE Frequency Control Symposium, 92 (1992).

[26] J. A. Barnes, R. C. Andrew, L. S. Cutler, D. J. Healey, D. B. Leeson, T. E. McGunigal, J. A. Mullen, W. L. Smith, R. L. Sydnor, R. F. C. Vessot, and G. M. R. Winkler, IEEE Trans. Instr. and Meas. IM-20 **2**, 105 (1971).

[27] D. Matsakis, A. T. Kubik, J. DeYoung, R. Giffard, and L. Cutler, Proc. 1995 IEEE Frequency Control Symposium, in press (1995).

[28] R. L. Tjoelker, J. D. Prestage, G. J. Dick, and L. Maleki, Proc. 1994 IEEE Frequency Control Symposium, 739 (1994).

[29] P. T. H. Fisk, M. J. Sellars, M. A. Lawn, C. Coles, A. G. Mann, and D. G. Blair, Proc. 1994 IEE Frequency Symposium, 731 (1994).

[30] D. J. Wineland, D. J. Heinzen, and C. S. Weimer, Proc. 22nd Annual PTTI, 53 (1991).

[31] A. Kastberg, W. D. Phillips, S. L. Rolston, R. J. C. Spreeuw, and P. S. Jessen, Phys. Rev. Letters **74**, 1542 (1995).

[32] T. H. Dixon, Reviews of Geophysics, **29**, 249 (1991).

[33] W. J. Klepczynski, Proc. IEEE **71**, No. 10, 1193 (1983).

[34] W. Lewandowski, C. Thomas, Proc. IEEE **79**, No. 7, 991 (1991).

[35] H. Chadsey, Proc. 25th Annual PTTI, 317 (1994).

[36] J. A. DeYoung, W. J. Klepczynski, A. D. McKinley, W. Powell, P. Hetzel, A. Bauch, J. A. Davis, P. R. Pearce, F. Baumont, P. Claudon, P. Grudler, G. de Yong, D. Kirchner, H. Ressler, A. Soring, C. Hackman, and L. Veenstra, Proc. 26th Annual PTTI, in press (1995) .

[37] BIPM Annual Report (1993).

[38] R. E. Drullinger, J. P. Lowe, D. J. Glaze, and J. Shirley, Proc. IEEE Frequency Symposium, 71 (1993).

[39] L. Breakiron, Proc. 23rd Annual PTTI, 297 (1992).

[40] G. M. R. Winkler, R. G. Hall, and D. B. Percival, Metrologia **6**, 126 (1970).

[41] T. M. Eubanks, *Contributions of Space Geodesy to Geodynamics: Earth Dynamics*. Geodynamics **24**, 1 (1993).

[42] J. D. Mulholland, PASP **84**, 357 (1972).

[43] S. Aoki, B. Guinot, G. H. Kaplan, H. Kinoshita, D. D. McCarthy, and P. K. Seidelmann, A&A **105**, 359 (1982).

[44] U.S. National Earth Orientation Service Annual Report (1995).

[45] B. Guinot, Metrologia **24**, 195 (1987).

[46] IAU Resolution A4 Proc. 21st General Assembly Transactions of the IAU, XXI. Kluwer, Dordrecht (1992).

[47] P. K. Seidelmann and T. Fukushima, A&A **265**, 833 (1992).

[48] T. Fukushima, A&A **294**, 895 (1995).

[49] D. C. Backer, IAU Symposium **165** (1994).

[50] D. C. Backer, Springer-Verlag Lecture Notes in Physics **418**, 193 (1991).

[51] J. L. Friedman, J. R. Ipser, and L. Parker, ApJ, **304**, 115 (1986).

[52] F. Camilo, S. E. Thorsett, and S. R. Kulkarni, ApJ, **421**, L15 (1994).

[53] R. D. Blandford, R. W. Romani, and J. H. Applegate, Mon. Not. R. Astron. Soc. **225**, 51P (1987).

[54] A. Wolszczan, S. R. Kulkarni, J. Middleditch, D. C. Backer, A. S. Fruchter, and R. J. Dewey Nature, **337**, 531 (1989).

[55] E. M. Standish, A&A **233**, 252 (1990).

[56] J. H. Taylor, D. N. Manchester, A. G. Lyne, and F. Camilo, in preparation (1995).

[57] T. Damour and N. Deruelle, Ann. Inst. H. Poincare (Physique Theorique) **44**, 263 (1986).

[58] A. Wolszczan, Science, 264, 538 (1994).

[59] J. H. Taylor and J. M. Weisberg, ApJ, **345**, 434 (1989).

[60] I. Cognard, G. Bourgois, J. F. Lestrade, and F. Biraud, in: *Millisecond Pulsars: A Decade of Surprise*, eds. Fruchter, Tavani and Backer, Astron. Soc. of the Pac., p.372 (1995).

[61] Z. Arzoumanian, D. J. Nice, J. H. Taylor, and S. E. Thorsett, ApJ, **422**, 671 (1994).

[62] R. S. Foster, F. Camilo, and A. Wolszczan, Proc. Marcell Grossmann Meeting, in press (1995).

[63] R. Blandford, R. Narayan, and R. W. Romani, J. Astrophys. Astr. **5**, 369 (1984).

[64] V. M. Kaspi, *Ph.D. Thesis*, Princeton University (1994).

[65] G. Petit, P. Tavella, and C. Thomas, Proc. 6th European Frequency and Time Forum, 57 (1992).

[66] F. Camilo, R. S. Foster, and A. Wolszczan, ApJ **437**, L39 (1994).

[67] D. J. Nice and J. H. Taylor, ApJ **441**, 429 (1995).

43

Some Security Implications of Growing Electricity Demand for the Use of Nuclear Power in East Asia

Michael May

It gives me particular pleasure to contribute this modest essay to Charles Townes' 80th birthday Festschrift. I am not the one best qualified to expound on the excellence of his scientific work over a career spanning at least six decades. Over the three and more decades that I have known Charlie, he has been a model to me of what serious, mature, effective science statesmanship should be. From chairing seminars at which all sides of the then-contentious Cold War issues could be presented, to chairing Department of Defense studies on equally contentious issues in a very different context, his leadership has introduced insight, balance and rationality into situations where those qualities were direly needed and not often present in sufficient supply. It has always been a privilege to work with him. I wish him well in his 80th year and, I hope, many more.

43.1   The Global Context

Over the past many decades, the use of electricity world-wide has been very tightly coupled with economic growth (see Fig. 43.1)

Even assuming a high level of conservation in the future, together with some increase in energy fuel costs, it is hard to see how the coupling can be broken. For one thing, in several important cases, conservation itself leads to increased use of electricity at the expense of other forms of energy. For another, rising fuel costs matter less to electricity costs than to transportation or heating costs, because a larger fraction of electricity costs reflects the cost of fixed plant.

As a result, assuming that world population grows to about nine billion people, a rate of population growth on the low side of the expected mean, that conservation is used to the maximum degree which present technology makes economical, a level of conservation on the high side of what can be expected given past experience, and a moderate world economic growth of about 2.3%, energy consumption and with it installed electric capacity is nevertheless estimated to increase two-to-four fold in the next half-century.

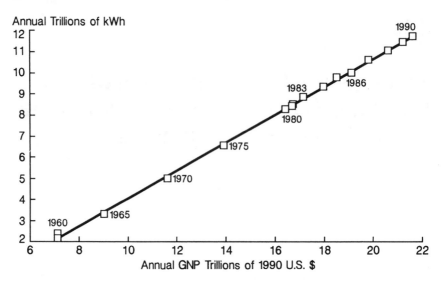

FIGURE 43.1. World electricity use vs. GNP 1960-1990.(Source: Starr, *Energy* **18**, 227, (1993).)

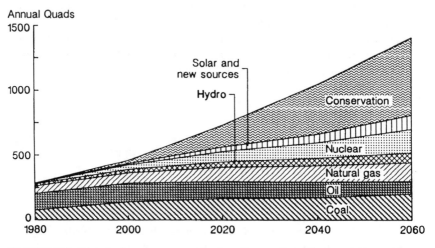

FIGURE 43.2. Global primary energy production. (Source: Starr, Searl, and Alpert, *Science* **256**,1992.)

Anything significantly less than what Fig. 43.2 shows would mean a serious setback to growth in the developing countries, and/or a depression lasting decades in the developed countries. Such setbacks would probably be accompanied by considerable political unrest and attendant security risks, and should be avoided if possible on both humanitarian and prudential grounds.

Thousands of Quads (1 quad=1.055 EJ)

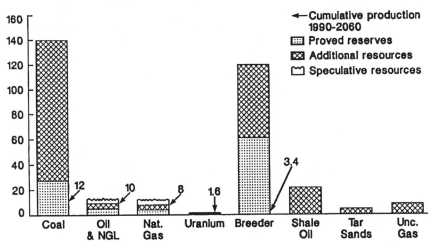

FIGURE 43.3. Mineral energy resources and cumulative production. (Source: Starr, Searl, and Alpert, *Science* **256**,1992.)

If the growth is to occur, however, a number of concerns also arise. The first, and the most easily disposed of, is whether enough fuels of all kinds will be available for such growth, as well as for the growth in other forms of energy consumption besides electricity, at prices that will not raise the cost of electricity significantly. The answer is now generally believed to be yes, although some of the more important resources—coal, oil, gas—are not evenly distributed and may not be near the points of consumption or within the major consuming countries. Since most of the next half century's economic growth is slated to occur in the developing countries, patterns of international distribution are likely to change. Thus, the growth of China's oil imports starting this decade is likely to affect market conditions throughout the world.

A second question, to which much attention is being devoted today, concerns the environmental consequences of the anticipated growth in energy consumption, of which electricity generation will probably provide an increasing fraction. These questions are principally associated with combustion of carbon-containing fuels such as oil, coal and gas, although the storage and disposal of spent nuclear fuel, and the land use associated with hydroelectric, biomass, and other diffuse forms of energy also raise concerns.

A third question concerns the availability of capital to build the electrical generating plants. This question will most likely continue to receive different answers in different locales. Thus, East Asian countries are likely to be able to finance approximately the desired rate of build-up, if present savings and investment patterns continue. Other areas may experience more difficulty. The differential availability of capital in different parts of the developing world, and the extent to which it can (or should) be remedied, and by whom, is a question deserving of further attention from policy makers.

A fourth question has to do with the security consequences attendant upon this growth. Again these consequences will depend on the area under consideration. The OECD countries, in addition to sizable if insufficient reserves of their own, have well-established import agreements with suppliers all over the world. The more rapidly growing developing countries, on the other hand, as noted, will need to establish new supply arrangements with the regions on earth which will have an exportable surplus of energy fuels, such as the Middle East and possibly Central Asia. While there is probably no economic reason why such new supply arrangements should not be accommodated by the markets in an efficient way, the anticipation of new large calls on limited resources can have a negative impact on perceived security if the growth in demand is not planned for in a joint or at least transparent manner. The establishment of new supply arrangements will therefore require careful political and economic coordination.

The different energy fuels also pose different kinds of security problems. We have noted the likely increased competition for inexpensive oil. The use of coal poses international pollution problems. All hydrocarbon fuels increase the emission of carbon dioxide into the atmosphere, and the presence of different, even contradictory views among the world's countries, evident at Rio and now at Berlin, is well-known. Nuclear power avoids these problems, but makes for some others. They arise mainly because of the possible connection of nuclear power with nuclear weapons. In this piece, these problems will be discussed briefly, particularly in the context of East Asia, where a great deal of the world's economic growth is anticipated to take place. First, what is the likely growth in nuclear power? Second, what problems with international implications may be associated with this growth? Third, what kind of arrangements will be needed to deal with these problems? We turn to these questions next.

43.2   Growth in Nuclear Power

The energy consumption growth shown in Fig. 43.2 would lead to a growth in nuclear power from the present 340 GWe (17% of total world electric consumption) to a range of perhaps 600-3,000 GWe, depending on whether the nuclear part of electric power generation remains the same, falls or increases. Clearly, this range of numbers cannot be forecast accurately, since it depends in the first place on the difference between two larger and poorly known quantities, to wit, the difference between the total electrical power used and that part provided by sources other than nuclear, and, in the second place, on economic and political developments which are also poorly predictable.

Nevertheless, major departures from the technical, economic and political assumptions usually made would be required for the numbers to fall outside the range shown.

Uranium Reserves

Total Estimated Supply: 6 - 7 Mtonnes

•Recoverable at less than $80 / kg (1990) : 1.4 Mtonnes

•Seawater Reserves: Estimated at 4000 Mtonnes at a concentration of approximately 3 parts per billion.

Material Cost

•Uranium ore purchase is now 1-2 % of electric busbar cost

•A 50% increase in ore prices would increase busbar cost by a few percent.

FIGURE 43.4. Options for nuclear fuel supply: more cheap uranium ore.

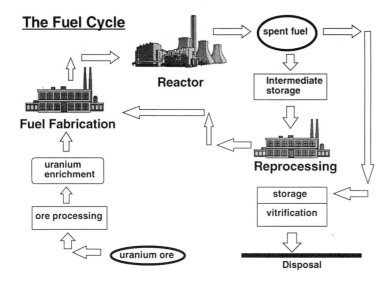

FIGURE 43.5. Options for nuclear fuel supply: reprocessing.

Of this total, if we take it that, in forty to fifty years, the proportion of nuclear power in East Asia will be about the same as it is world-wide, and assuming further that East Asia produces and consumes between 25 and 35% of the world's electric power, versus 20% now, the nuclear electric installed capacity there will be of the order of 150-1,000 GWe, or approximately that number of standard-size 1 GWe nuclear reactors. The amount of uranium ore required will be 36,000-240,000 tonnes if the reactors are all fueled solely with uranium fuel, and not much

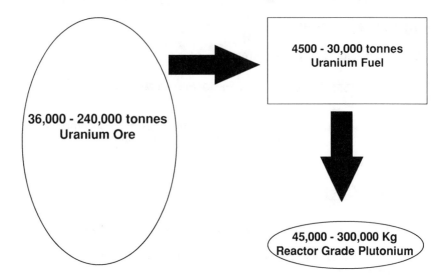

FIGURE 43.6. Typical material flows for a gigawatt of electricity.

less at first even if plutonium recycle is used. A total of 45,000-300,000 kilograms of reactor-grade plutonium will be generated per year, using figures for present-day reactor designs.

43.3   Some Problems With International Implications

Material which is usable, at least in theory, for nuclear weapons is created and handled at several stages of the nuclear fuel cycle, depicted in Fig. 43.6 above. The usability of the material varies quite widely over the cycle, however, from a physical state where the material could, if not adequately guarded, be diverted or stolen by a comparatively few people with relatively simple equipment, to physical states where major and quite noticeable operations would be needed to get at the material.

Nuclear reactors, even though they may be originally fueled with uranium that is not sufficiently enriched in the prompt fissionable isotope U-235 to make a weapon, and therefore poses little diversion or theft problem, nevertheless generate (from neutron capture in the more abundant isotope U-238) and partially burn plutonium in their core as they are operated. In fact, a significant fraction of the power generated by the reactor comes from burning this plutonium. The plutonium left over—usually a few percent of the heavy metal in the core—is "reactor-grade plutonium," a category which includes a variety of isotopic mixtures, none of them ideal for nuclear weapons, but all of them usable for that purpose, and all of them therefore posing various levels of risk of diversion or theft.

Weapons-grade plutonium contains over 90% Pu-239. It is obtained by with-drawing the fuel from a reactor much earlier than would be economical for

•Fuel Fabrication and Shipment

•Reactor Operations

•Spent Fuel Shipment and Storage

•Reprocessing

FIGURE 43.7. Some problems with international implications.

generating electricity. Reactor-grade plutonium contains from 40 to 60% Pu-239, depending on the reactor type and operating history. The balance in both cases consists mainly of the higher isotopes of plutonium, together with some Pu-238. These isotopes are much more radioactive than Pu-239, and one, Pu-241, has a gamma-radioactive daughter, Am-241. Because of this radioactivity, of the consequent exposure to personnel and heat generated, and because of the higher spontaneous fission rate of these isotopes, reactor-grade plutonium is more difficult to handle in a military or production setting than is weapon-grade plutonium, and requires a more sophisticated design to give a well-determined yield. No nuclear state has used it. Nevertheless, explosives made out of reactor-grade plutonium can give a nuclear yield and, if available to a terrorist group for instance, could constitute a serious danger.

This plutonium is relatively difficult to get at as long as it remains embedded in the highly radioactive spent fuel in which it originated. If it is separated, however, as it must be if it is to be recycled into new fuel, it becomes easier to handle, especially by a group that does not dispose of the facilities, time and protection available to a government-sponsored plant or laboratory.

The major problems with international implications stemming from this situation and others associated with the use of nuclear power are listed in Fig. 43.7.

The question of fuel fabrication and shipment arises because fewer fuel fabrication plants are needed than nuclear reactors. As a result, when a number of geographically proximate countries use nuclear power, as in Europe for instance, it has been economical for some but not all of these countries to invest in fabrication of nuclear fuel plants, resulting in the need for international shipment of fuel, both new and spent. Spent fuel is highly radioactive initially, so much so that it must remain for some years at the site of the utility which generated it where it is generally protected by shielding in water pools. By the time it is shipped for reprocessing or disposal, it is generally much less radioactive, and can be handled with ordinary heavy equipment when encased within large, multi-ton casks. The fuel remains quite dangerous for decades when outside the casks. New fuel, while it still must be handled with care, can be handled safely without radiation shielding.

The negative publicity associated with the recent shipment of reprocessed plutonium from France to Japan, though it was long-planned, approved by all concerned, and legal, has called attention to what is normally a routine event within Europe. This publicity, while it may delay Japanese plans to use reprocessed plutonium in the short run, will almost surely reinforce Japanese desires to reprocess their own plutonium in the longer run. Nuclear operations in Europe are carried out within the framework of a European pact backed by an organization, Euratom. One question deals with what arrangements will be made for such large-scale shipments in the now-developing areas, such as much of East Asia.

Reactor operations themselves can pose problems with international implications, as we have seen some years ago with the Chernobyl accident, and as we are continuing to see as Soviet-made reactors with inadequate provision for safety by modern standards continue to operate both in the former Soviet Union and in Eastern Europe in order to provide needed electricity. This problem has so far been avoided in East Asia. Provision for international cooperation may be needed to continue to avoid it.

Spent fuel storage and shipment is not handled in a fully satisfactory way now. Most of the spent fuel is stored at utility sites, and provision for longer term disposal is tied up in political difficulties in many countries. The problem is not so much a technically insoluble one as it is a politically difficult one. The question has arisen whether an Internationally Monitored Retrievable Surface (or near surface) Storage (IMRSS) regime is needed, both in order to validate individual states' plans for dealing with the situation, and in order to assist states which might not have a well-developed system for handling spent fuel. Such a regime is now under very preliminary and informal investigation by a group of academics and utility representatives in Europe and the U.S. As now envisaged, it would comprise, not only a set of standards, but also an agreed international management system for shipping and storing spent fuel and other material of potential concern. One suggestion is that it might function like an international bank for nuclear material. In any case, as and if nuclear power use grows, the need for such a system implementation will also grow.

Finally, as discussed above, reprocessing poses special problems associated with the manufacture and shipment of separated reactor-grade plutonium. For that reason, the US government has opposed reprocessing spent fuel in order to recycle the plutonium, at least while low-cost uranium ore is abundant. Most US analyses show that using fresh uranium fuel is cheaper than using fuel that includes recycled plutonium. On the other hand, nations such as France, Britain, Germany, Japan, which have little in the way of fuel resources and wish to develop a technology that could fully use the uranium resource (theoretically, continued recycling, coupled with the use of breeder reactors, would multiply the amount of energy available from a given amount of uranium by a factor of 100 or so) are using or plan to use recycled plutonium in their civilian fuel cycle. So does Russia.

Even in the countries that have adopted it, some of the impetus for recycling has been lost since the plans for it were originally laid in the 1970s because of the slower-than-expected growth of nuclear power and the added cost of using recycled plutonium in the reactor fuel. If, however, the world in general and East Asia in particular are facing a period of extensive growth in nuclear electricity generation, either much larger reserves of low-cost uranium (that is, sufficiently low-cost not to affect the cost of electricity seriously) than have been identified so far must be developed and made available, or plutonium recycling will be practiced widely within the area, or both.

The problems of safeguarding separated reactor plutonium have been dealt with successfully in Europe, where such manufacture and shipments have mainly taken place. It should be noted that all nuclear material in the non-nuclear weapons states party to the Nuclear Non-Proliferation Treaty, and in some nuclear weapons state parties as well, must be safeguarded according to international standards promulgated by the International Atomic Energy Agency (IAEA) or, as in the case of Euratom, standards compatible with IAEA standards. Where safeguards meeting these standards have been applied to the nuclear power reactor fuel cycle, they have been effective. The question is more to ensure that safeguards will be adhered to world-wide than to improve the standards.

43.4   What Arrangements Will Be Needed?

What arrangements will be needed to ensure that adequate standards of transparency, accountability and security are met so that nuclear power can do its part in meeting the coming vastly increased world demand for electricity at affordable cost and with a minimum of environmental damage? While a detailed answer must await further study, three kinds of questions for further analysis may be distinguished at this time.

1) What kind of regional arrangements will be made for meeting the problems outlined above? For reasons of economics and safety, much of the routine handling and shipment of nuclear materials will be carried out on a regional basis, where a region will be large enough to afford all needed facilities, but compact enough to minimize costs and risks associated with shipments. This is what has happened in the past. Thus, we may expect to see regional nuclear arrangements develop in East Asia as we have seen them develop in Europe. While this is not a foregone conclusion, and political relations may militate for other arrangements, nevertheless, the nature of possible regional compacts, their financing, the standards and systems they will adopt, their enforcement mechanisms, all these things will help define the energy world we and our descendants will live in, and deserve further study

2) What global arrangements will be needed, if any, beyond what the IAEA provides today? While regional arrangements will be driven by considerations of economics, safety and political relationship, global arrangements will also be continue to be needed to ensure that diversion and theft, which may be sponsored from outside the region, are avoided, and the flow of components, technology, and services is appropriately regulated. Something more complete than the present Nuclear Suppliers Guidelines will be needed if the number of states using nuclear power and the number of nuclear facilities increase significantly. It may also be the case that some facilities, perhaps longer-term storage for spent fuel, will require trans-regional arrangements. In that case, a management system which will ensure that agreed standards are observed will also be needed. The present IMRSS studies may provide a start on answering these questions.

3) What new technologies will be needed or helpful for a wider exploitation of nuclear power? At present, most of the world's commercial nuclear power experience has been with light-water reactors. These reactors have many virtues, but their adoption stems from a combination of budgetary and political exigencies in the US in the fifties. They do not lend themselves to a broader exploitation of nuclear power if widespread reprocessing of spent fuel to make full use of the uranium resource is to take place. At what time, and on what scale should research and development on different fuel cycles and reactors be carried out? And under whose sponsorship? These questions can be usefully studied at the present time, before there is a need for immediate action.

This short essay is a preliminary attempt to understand what security-related questions may be associated with the future utilization of nuclear power in a world where demand for electricity is all but sure to grow, and environmental restrictions on carbon-based fuels are likely to become more stringent. Several questions regarding both economic-political arrangements and the need for new technology arise in this connection. While fifty years is a very long time for planners and an eternity for political leaders, it is approximately equal to the operating lifetime of an electrical generating plant, and not much longer than the length of time which it takes for new technology to penetrate the power grid. Given that shifts in the locales of rapid economic growth are usually accompanied by shifts in relative political power, and can easily be traumatic if not foreseen and handled with care, it may well be time to give the shape of this particular aspect of our future some thought.

43.5   Acknowledgments

I thank Celeste Johnson for research assistance and the Carnegie Corporation of New York for support of the program which made this work possible.

44

Dynamic Control of the Photon Statistics in the Micromaser and Laser

Georg M. Meyer
Tserensodnom Gantsog
Marlan O. Scully
and Herbert Walther

It is a pleasure to participate in a Festschrift for Professor Charles Townes, but for whom physics would be much poorer.

44.1   Introduction

Quantum coherence and interference can completely change the optical properties of atomic systems. They have led to many surprising innovations in quantum optics, including quantum noise quenching [1, 2], lasing without inversion [3, 4], and new optical materials with an enhanced index of refraction [5]. On the other hand, a new kind of maser, the micromaser [6–11], has been used to investigate nonclassical aspects of the interaction between radiation and matter such as sub-Poissonian photon statistics [6, 7], trapping states [7, 9], and the production of number states [7, 8]. Recently the problem has been extended to the study of one- and two-mode micromasers operating on three-level atoms [12].

In this paper we are interested in the question of how the micromaser and laser photon statistics are affected when we drive three-level atoms resonantly by an external field so that the Autler-Townes effect [13] comes into play. Emission properties of a laser operating on driven Λ-type atoms have previously been studied by Manka *et al.* [14]; trapping states in the corresponding case of the micromaser have been investigated by Orszag and Ramírez [15]. Here we focus on V-type three-level atoms. As we will show in the present paper, the addition of a coherent field to the traditional setting of the micromaser and laser enables us to control the photon statistics of the cavity mode.

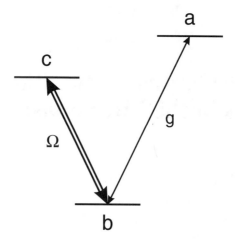

FIGURE 44.1. Schematic representation of a driven three-level atom in V configuration.

44.2   Master Equation for the Photon Distribution

We consider a micromaser or laser cavity into which three-level atoms are injected according to a Poissonian process. The atoms have the V-type configuration of transitions as shown in Fig. 44.1. After entering the cavity, the atoms start to interact with both the cavity mode and the external coherent field, which are assumed to be resonant with the atomic $a \leftrightarrow b$ and $b \leftrightarrow c$ transitions, respectively.

The interaction of the atom with the quantized cavity mode and the external classical field is described in the interaction picture by the Hamiltonian

$$H = \hbar\Omega \left(|b\rangle\langle c| + |c\rangle\langle b| \right) + \hbar g \left(a\, |a\rangle\langle b| + a^{\dagger}\, |b\rangle\langle a| \right) . \qquad (44.1)$$

Here a and $a^{\dagger}$ are the annihilation and creation operator of the cavity mode, respectively; Ω is the Rabi frequency of the driving field; g is the atom-cavity mode coupling constant. The corresponding time-evolution operator can be expressed in the form

$$U(t) \equiv \exp\left(-iHt/\hbar\right) = \left(1 + \frac{K}{\Lambda^2}\right)\cos(\Lambda t) - i\frac{H}{\hbar\Lambda}\sin(\Lambda t) - \frac{K}{\Lambda^2}, \qquad (44.2)$$

where $\Lambda^2 = \lambda^2 + \Omega^2$. Here the operators

$$\lambda^2 \; = \; g^2(a^{\dagger}a + |a\rangle\langle a|) , \qquad (44.3)$$

$$K \; = \; g\Omega \left(a\, |a\rangle\langle c| + a^{\dagger}\, |c\rangle\langle a| \right) - \lambda^2\, |c\rangle\langle c| - \Omega^2\, |a\rangle\langle a| \qquad (44.4)$$

are constants of motion. The atoms are injected into the cavity in an incoherent superposition $\varrho_A = \varrho_{aa}\, |a\rangle\langle a| + \varrho_{bb}\, |b\rangle\langle b| + \varrho_{cc}\, |c\rangle\langle c|$ with an average rate r.

44.2.1   Micromaser

In the micromaser, atoms are injected into a high-Q cavity at such a low rate that at most one atom is present inside the resonator at a time. We assume that the relaxation of the cavity field can be ignored while an atom is inside the cavity. For simplicity, we suppose that all atoms have the same velocity so that they interact with the cavity mode for the same time.

As discussed in [7], the time evolution of the density matrix ϱ of the cavity field is governed by the equation

$$\dot{\varrho} = r\,\delta_\tau\varrho + L\varrho , \qquad (44.5)$$

where the change $\delta_\tau\varrho$ due to an atom interacting with the field for the time τ, and the Liouvillian operator L, which describes the coupling of the cavity mode to a thermal bath, respectively, are given by

$$\delta_\tau\varrho = \mathrm{Tr}_{(A)}\left\{U(\tau)\,\varrho_A \otimes \varrho\, U^\dagger(\tau)\right\} - \varrho ,$$

$$L\varrho = \frac{1}{2}C(n_b + 1)(2a\varrho a^\dagger - a^\dagger a\varrho - \varrho a^\dagger a)$$
$$+ \frac{1}{2}Cn_b(2a^\dagger\varrho a - aa^\dagger\varrho - \varrho aa^\dagger) . \qquad (44.6)$$

Here the subscript A denotes the atomic degrees of freedom, n_b is the average number of thermal photons, and C is the cavity decay rate.

The master equation for the photon distribution $P(n) = \varrho(n, n)$ is obtained from Eqs. (44.5) and (44.6)

$$\dot{P}(n) = a_{n-1}P(n-1) - (a_n + b_{n-1})P(n) + b_n P(n+1) , \qquad (44.7)$$

where the coefficients a_n and b_n are defined by

$$a_n = r\varrho_{aa}\left[\sin^2(\Lambda_n\tau) + \frac{4\Omega^2}{\Lambda_n^2}\sin^4\left(\frac{\Lambda_n\tau}{2}\right)\right]\frac{\lambda_n^2}{\Lambda_n^2} + Cn_b(n+1) ,$$

$$b_n = \left[r\varrho_{bb}\sin^2(\Lambda_n\tau) + r\varrho_{cc}\frac{4\Omega^2}{\Lambda_n^2}\sin^4\left(\frac{\Lambda_n\tau}{2}\right)\right]\frac{\lambda_n^2}{\Lambda_n^2}$$
$$+ C(n_b + 1)(n+1) \qquad (44.8)$$

with $\lambda_n^2 = g^2(n+1)$ and $\Lambda_n^2 = \lambda_n^2 + \Omega^2$. The various terms on the right-hand side of Eq. (44.7) can be interpreted as inflow and outflow of probabilities.

44.2.2 Laser

In the laser, the levels a, b, and c can decay to other levels. For simplicity, we assume that all three levels decay with the same rate γ. Following the standard quantum theory of the laser [16], we can write the coarse-grained time derivative of the field density matrix in the form

$$\dot{\varrho} = r\int_0^\infty \delta_\tau\varrho\,\mathcal{P}(\tau)d\tau + L\varrho , \qquad (44.9)$$

where $\mathcal{P}(\tau) = \gamma\exp(-\gamma\tau)$ is the distribution of the atomic lifetime τ.

By averaging Eq. (44.7) over τ, we obtain the master equation for the laser

$$\dot{P}(n) = a_{n-1} P(n-1) - (a_n + b_{n-1}) P(n) + b_n P(n+1) , \qquad (44.10)$$

where

$$a_n = r\varrho_{aa} \left(1 + \frac{3\Omega^2}{\gamma^2 + \Lambda_n^2}\right) \frac{2\lambda_n^2}{\gamma^2 + 4\Lambda_n^2} + Cn_b(n+1) ,$$

$$b_n = \left(r\varrho_{bb} + r\varrho_{cc} \frac{3\Omega^2}{\gamma^2 + \Lambda_n^2}\right) \frac{2\lambda_n^2}{\gamma^2 + 4\Lambda_n^2} + C(n_b + 1)(n+1) . \qquad (44.11)$$

44.3   Transient and steady-state photon statistics

The steady-state solutions of Eqs. (44.7) and (44.10) can be calculated from a detailed-balance argument to be

$$P(n) = P(0) \prod_{k=0}^{n-1} \frac{a_k}{b_k} , \qquad (44.12)$$

where $P(0)$ is determined by normalization. From Eqs. (44.8) and (44.11), it is obvious that the photon statistics strongly depend on the value of Ω.

The transient evolution to steady state cannot be solved analytically. However, the master equations (44.7) and (44.10) can be solved if the Rabi frequency of the driving field is so large that for all relevant values of the photon number n

$$\Omega^2 \gg g^2(n+1) \qquad \text{for the micromaser,}$$
$$\gamma^2 + \Omega^2 \gg g^2(n+1) \qquad \text{for the laser.} \qquad (44.13)$$

Under these conditions, Eqs. (44.7) and (44.10) reduce to

$$\dot{P}(n) = A [n P(n-1) - (n+1) P(n)] - B [n P(n) - (n+1) P(n+1)] \, (44.14)$$

with the coefficients

$$A = r\varrho_{aa} \frac{4g^2}{\Omega^2} \sin^2\left(\frac{\Omega\tau}{2}\right) + Cn_b ,$$

$$B = \left[r\varrho_{bb} \cos^2\left(\frac{\Omega\tau}{2}\right) + r\varrho_{cc} \sin^2\left(\frac{\Omega\tau}{2}\right)\right] \frac{4g^2}{\Omega^2} \sin^2\left(\frac{\Omega\tau}{2}\right)$$
$$+ C(n_b + 1) \qquad (44.15)$$

for the micromaser and

$$A = r\varrho_{aa} \frac{2g^2}{\gamma^2 + \Omega^2} + Cn_b ,$$

$$B = \left(r\varrho_{bb} + r\varrho_{cc} \frac{3\Omega^2}{\gamma^2 + \Omega^2}\right) \frac{2g^2}{\gamma^2 + 4\Omega^2} + C(n_b + 1) \qquad (44.16)$$

for the laser. The mean number of photons satisfies the simple equation

$$\langle \dot{n}(t) \rangle = (A - B)\langle n(t) \rangle + A , \tag{44.17}$$

whose solution is

$$\langle n(t) \rangle = \left(\langle n(0) \rangle + \frac{A}{G} \right) \exp\left(Gt \right) - \frac{A}{G} . \tag{44.18}$$

Here the parameter $G \equiv A - B$ plays the role of the small signal gain. Note that a positive value of G leads to the build-up of maser and laser radiation under the condition (44.13).

44.3.1 Micromaser

The steady-state mean photon number $\langle n \rangle$ is shown in Fig. 44.2 as a function of the scaled interaction time $g\tau$ for various values of the Rabi frequency Ω. The atoms are injected into the cavity in the excited state a with the rate $r/C = 50$. Figure 44.2(a) for $\Omega = 0$ corresponds to the ordinary micromaser operating on two-level atoms. From Fig. 44.2, where Ω ranges from zero to $5g$, we observe that at the first threshold the mean photon number does not depend on Ω. By contrast, for increasing Ω the subsequent peaks become higher and sharper, and their positions move closer to the origin $g\tau = 0$. At a certain value of Ω, the first and second peak combine to form a single peak.

For larger Ω, the heights of the peaks start to decrease until they practically disappear. Using a gain/loss argument, we can estimate the value of Ω for which the maser action stops. In the regime $\Omega^2 \gg g^2(n+1)$, the net signal gain is given by

$$G = r \frac{4g^2}{\Omega^2} \sin^2\left(\frac{\Omega\tau}{2} \right) - C . \tag{44.19}$$

The first term is the gain due to the atomic $a \to b$ transition. Masing occurs when this term is greater than the cavity loss. For $\Omega/g \geq 2(r/C)^{1/2}$, the first term in Eq. (44.19) cannot compensate the loss term, and as a result we have an example of inversion without masing. This can be explained by the dynamic Stark shift of the driven levels which gives rise to the Autler-Townes doublet [13]. In effect, for large Ω, the lower masing level falls outside the response spectrum of the cavity mode, and hence emission is suppressed.

One of the most intriguing features of the conventional micromaser is the existence of trapping states at zero temperature [7, 9]. In our case, the trapping condition must be slightly modified. For atoms injected in the excited state a, the trapping states are given for $\Omega \neq 0$ by

$$g\tau\sqrt{n_q + 1 + \Omega^2/g^2} = 2q\pi , \tag{44.20}$$

where n_q and q are integers. If Eq. (44.20) is fulfilled for some integers $n_q \geq 0$ and $q \geq 1$, then $P(n) = 0$ for $n > n_q$, i.e., the photon distribution is truncated

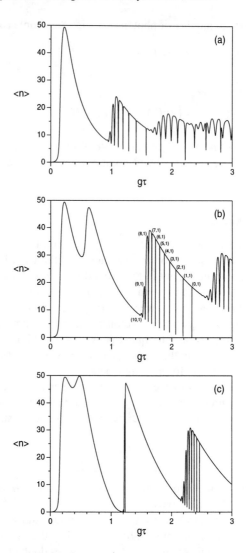

FIGURE 44.2. The mean photon number versus the reduced atom-field interaction time $g\tau$ for the micromaser with the parameters $\varrho_{aa} = 1$, $r/C = 50$, $n_b = 0$, and (a) $\Omega/g = 0$, (b) $\Omega/g = 2.5$, and (c) $\Omega/g = 5$.

at the number state $|n\rangle$ with $n = n_q$. The existence of trapping states leads to the resonances in the curves of Fig. 44.2. For example, in Fig. 44.2(b) for $\Omega/g = 2.5$ the successive resonances corresponding to $q = 1$ and to n_q going down from 10 to 0 are clearly visible and, in contrast to Fig. 44.2(a), separated from the trapping series for $q > 1$. Because of the cavity loss the mean photon number at each trapping resonance is smaller than the corresponding value of n_q. With increasing Ω the resonances move closer to the position of the phase transition. As

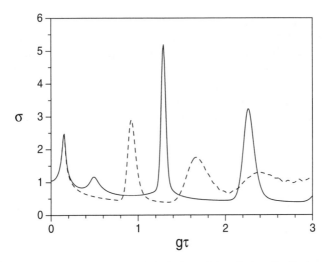

FIGURE 44.3. The normalized standard deviation of the photon distribution versus the reduced atom-field interaction time $g\tau$ for the parameters $\varrho_{aa} = 1, r/C = 50, n_b = 0.1$, $\Omega/g = 0$ (dashed curve) and $\Omega/g = 4$ (solid curve).

a consequence, the number of trapping resonances is reduced because in the region where these resonances should have appeared, the photon number of the cavity field is smaller than the corresponding value of n_q. This behavior is illustrated in Fig. 44.2(c).

In Fig. 44.3, the normalized standard deviation $\sigma = (\langle n^2 \rangle / \langle n \rangle - \langle n \rangle)^{1/2}$ is plotted as a function of $g\tau$ for $n_b = 0.1$ and two values of Ω. The initial atomic variables and the other parameters are the same as in Fig. 44.2. We see that the presence of the driving field has a considerable effect on the photon number fluctuations, e.g., for $g\tau = 0.9$ the change from $\Omega = 0$ to $\Omega = 4g$ leads to sub-Poissonian instead of strong super-Poissonian statistics.

It was previously shown that a conventional micromaser, operating on two-level atoms, can exhibit masing without inversion due to the fact that micromasers operate in a regime where the interaction between atoms and field is coherent [10]. Atoms that are not strictly in their ground state do carry some excitation, and for a suitable atom-field interaction time, this energy can be deposited into the cavity mode. However, because the upward atomic transitions nearly balance the downward transitions, the net gain is almost negligible. If we manage to exclude the unwanted absorptive transitions, any amount of population in the upper masing level is sufficient to amplify the cavity field.

In our arrangement, we achieve this goal by adding the coherent driving field, which induces the transition $b \to c$. From Eq. (44.15), we obtain for $\varrho_{cc} = 0$

$$G = \left[r\varrho_{aa} - r\varrho_{bb} \cos^2 \left(\frac{\Omega\tau}{2} \right) \right] \frac{4g^2}{\Omega^2} \sin^2 \left(\frac{\Omega\tau}{2} \right) - C . \qquad (44.21)$$

It is clear from Eq. (44.21) that we can have positive net gain even if $\varrho_{aa} < \varrho_{bb}$ but $\varrho_{aa} > \varrho_{bb} \cos^2(\Omega\tau/2)$ and if the cavity loss is small enough. As Ω increases, the transition probability from b to c becomes larger, and the population in the lower level starts to oscillate between the levels b and c. Consequently, the transition probability from $b \to a$ is reduced and the net gain on the operating transition is enhanced. At this point one should note that, as a consequence of the conservation of the total excitation $a^\dagger a + |a\rangle\langle a|$ on the $a \leftrightarrow b$ transition during the coherent atom-field interaction (see Eq. (44.3)), the energy of the maser field comes from the atoms as in a normal maser, and not via parametric conversion from the external field.

44.3.2 Laser

Now we proceed to investigate the photon statistics of the laser operating on driven V-type atoms. In Fig. 44.4, we plot the steady-state mean photon number $\langle n \rangle$ and the normalized standard deviation σ versus Ω/g for $\varrho_{aa} = 1, r/C = 200, \gamma/g = 1$, and $n_b = 0$. As seen from Fig. 44.4, the photon statistics changes drastically with increasing Rabi frequency Ω. Initially the mean photon number increases, while the normalized standard deviation falls below unity indicating sub-Poissonian photon statistics. After reaching its maximum value, the mean photon number decreases until almost complete suppression of laser emission occurs.

Population inversion between the lasing levels is necessary for the operation of a conventional laser. In the present model, however, lasing is possible even without inversion. To see this most clearly, we write from Eq. (44.16) the net signal gain for $\varrho_{cc} = 0$ as

$$G = \left(r\varrho_{aa} - r\varrho_{bb} \frac{\gamma^2 + \Omega^2}{\gamma^2 + 4\Omega^2} \right) \frac{2g^2}{\gamma^2 + \Omega^2} - C . \tag{44.22}$$

The first term is the gain due to the atomic $a \leftrightarrow b$ transition. Lasing can be achieved even when $\varrho_{aa} < \varrho_{bb}$ but $\varrho_{aa} > \varrho_{bb}(\gamma^2 + \Omega^2)/(\gamma^2 + 4\Omega^2)$ and the cavity loss is small enough. In Fig. 44.5, we plot the distribution $P(n)$ of photons in the cavity for three values of the Rabi frequency Ω. The other parameters are $\varrho_{aa} = 0.3, \varrho_{bb} = 0.7, r/C = 2000, \gamma/g = 1$, and $n_b = 0$. Note that in this case $\varrho_{aa} < \varrho_{bb}/2$. The curve for $\Omega/g = 10$ has a single peak and corresponds to the case of lasing without inversion. The physical mechanism is the same as in the case of the micromaser discussed above. The inset of Fig. 44.5 shows two limits. The behavior of $P(n)$ for $\Omega = 0$ is similar to a thermal distribution, corresponding to the fact that the ordinary laser does not operate without population inversion. On the other hand, when Ω is very large (e.g., $\Omega/g = 10^3$), complete suppression of emission occurs, and the cavity field remains in the vacuum state.

Finally, we return to the transient evolution. In Fig. 44.6, we illustrate the time evolution of the mean photon number (lower plot) of the laser when the atoms are driven by an external periodic field, the Rabi frequency which varies as shown in the upper plot. The two values of the Rabi frequency have been chosen before

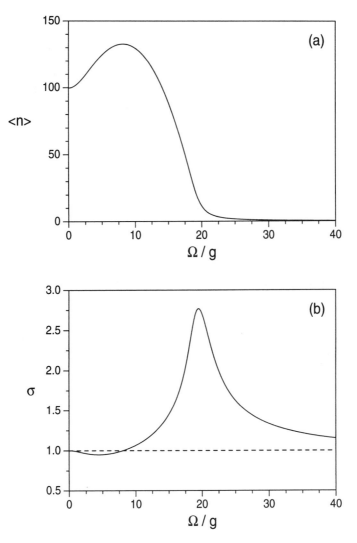

FIGURE 44.4. The mean photon number (a) and the normalized standard deviation (b) for the laser versus the Rabi frequency Ω for the parameters $\varrho_{aa} = 1, r/C = 200, \gamma/g = 1,$ and $n_b = 0$.

and after a steep decrease of the steady-state mean photon number. As seen from Fig. 44.6, our model operates in this case as an optical inverter, i.e., an optical realization of a gate which produces the logical complement of its input [17]. This is an example of an all-optical switch: light controls light with the help of a nonlinear optical medium, whose photon emission rate is altered directly by the presence of the driving field. We have thus demonstrated that using an external field we can either enhance or suppress the radiation, i.e., control the laser emission.

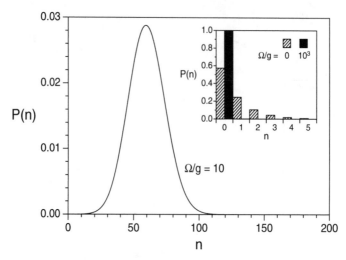

FIGURE 44.5. The photon distribution for the Rabi frequencies $\Omega/g = 0$, $\Omega/g = 10$, and $\Omega/g = 10^3$. The other parameters are $\varrho_{aa} = 0.3$, $\varrho_{bb} = 0.7$, $r/C = 2000$, $\gamma/g = 1$, and $n_b = 0$.

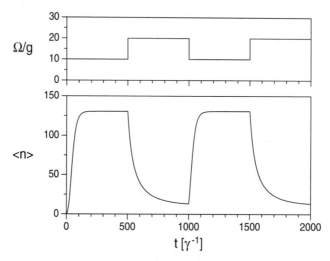

FIGURE 44.6. The time evolution of the mean photon number (lower plot) from initial vacuum when the atoms are driven by an external periodic field (upper plot) for the parameters $\varrho_{aa} = 1$, $g/\gamma = 1$, $r/\gamma = 10$, $C/\gamma = 0.05$, and $n_b = 0$.

44.4   Acknowledgments

One of us (Ts. G.) would like to acknowledge the Alexander von Humboldt Foundation for the award of a Humboldt Research Fellowship. This work has been supported partly by the Office of Naval Research, the Welch Foundation, and the Texas Advanced Research Program.

References

[1] M. O. Scully, Phys. Rev. Lett. **55**, 2802 (1985); M. O. Scully and M. S. Zubairy, Phys. Rev. A **35**, 752 (1987).

[2] M. P. Winters, J. I. Hall, and P. E. Toschek, Phys. Rev. Lett. **65**, 3116 (1990).

[3] O. A. Kocharovskaya and Ya. I. Khanin, Pis'ma Zh. Eksp. Teor. Fiz. **48**, 581 (1988); S. E. Harris, Phys. Rev. Lett. **62**, 1033 (1989); M. O. Scully, S.-Y. Zhu, and A. Gavrielides, *ibid.* **62**, 2813 (1989).

[4] A. Nottelmann, C. Peters, and W. Lange, Phys. Rev. Lett. **70**, 1783 (1993); E. S. Fry, X. Li, D. E. Nikonov, G. G. Padmabandu, M. O. Scully, A. V. Smith, F. K. Tittel, C. Wang, S. R. Wilkinson, and S.-Y. Zhu, *ibid.* **70**, 3235 (1993).

[5] M. O. Scully, Phys. Rev. Lett. **67**, 1855 (1991); M. Fleischhauer, C. H. Keitel, M. O. Scully, C. Su, B. T. Ulrich, and S.-Y. Zhu, Phys. Rev. A **46**, 1468 (1992).

[6] The first masing using a single-atom "micromaser" was reported in D. Meschede, H. Walther, and G. Müller, Phys. Rev. Lett. **54**, 551 (1985). For a review of the work of H. Walther and co-workers, see G. Raithel, C. Wagner, H. Walther, L. M. Narducci, and M. O. Scully, in *Cavity Quantum Electrodynamics*, edited by P. R. Berman (Academic, Boston, 1994), p. 57.

[7] P. Filipowicz, J. Javanainen, and P. Meystre, Phys. Rev. A **34**, 3077 (1986); L. A. Lugiato, M. O. Scully, and H. Walther, *ibid.* **36**, 740 (1987).

[8] J. Krause, M. O. Scully, and H. Walther, Phys. Rev. A **36**, 4547 (1987); J. Krause, M. O. Scully, T. Walther, and H. Walther, *ibid.* **39**, 1915 (1989).

[9] P. Meystre, G. Rempe, and H. Walther, Opt. Lett. **13**, 1078 (1988).

[10] Fam Le Kien, Opt. Commun. **82**, 380 (1991).

[11] For a review of the work of S. Haroche and co-workers, see S. Haroche and J. M. Raimond, Adv. At. Mol. Phys. **20**, 350 (1985), and for the first realization of a two-photon micromaser, see M. Brune, J. M. Raimond, P. Goy, L. Davidovich, and S. Haroche, Phys. Rev. Lett. **59**, 1899 (1987).

[12] Fam Le Kien, G. M. Meyer, M. O. Scully, H. Walther, and S.-Y. Zhu, Phys. Rev. A **49**, 1367 (1994); Fam Le Kien, G. M. Meyer, M. O. Scully, and H. Walther, *ibid.* **51**, 1644 (1995). A two-channel Raman micromaser has been studied in R. R. Puri, C. K. Law, and J. H. Eberly, *ibid.* **50**, 4212 (1994).

[13] S. H. Autler and C. H. Townes, Phys. Rev. **100**, 703 (1955).

[14] A. S. Manka, C. H. Keitel, S.-Y. Zhu, M. Fleischhauer, L. M. Narducci, and M. O. Scully, Opt. Commun. **94**, 174 (1992).

[15] M. Orszag and R. Ramírez, Opt. Commun. **101**, 377 (1993).

[16] M. O. Scully and W. E. Lamb, Jr., Phys. Rev. **159**, 208 (1967).

[17] For an introduction to photonic switching and computing, see B. E. A. Saleh and M. C. Teich, *Fundamentals of Photonics* (Wiley, New York, 1991), Chap. 21.

45

Sgr A* - A Starving Black Hole?

Peter G. Mezger
Wolfgang J. Duschl
Robert Zylka
and Thomas Beckert

45.1   The Early Days

The Galactic Center was actually the first radio source identified after Carl Guthe Jansky's discovery of cosmic radio radiation in 1933. At the long wavelength of λ 14.5 m Jansky nearly exclusively observed only synchrotron emission. Although the contribution of free-free emission becomes stronger at decimeter wavelengths, in later surveys with 25 m telescopes at λ 20 cm [1] and λ 11 cm (Altenhoff *et al.* [2]), the Sagittarius A (Sgr A) radio complex was still observed as a single compact but predominantly nonthermal radio source, well above the level of the diffuse emission from the Galactic Disk.

Only in 1971 Downes and Martin resolved with the Cambridge 1-mile telescope the Sgr A complex into the compact Hii region Sgr A West and the more extended synchrotron source Sgr A East [3]. Their search had actually been aimed for a much more compact and weaker source (0.5 Jy, 1 mas), whose possible presence at the Center of the Galaxy had been predicted by Lynden-Bell and Rees [4]. These latter authors had investigated a black hole/accretion disk model as "the engine" which may power quasars. Knowing that our Galaxy does not harbor an active galactic nucleus (AGN) they suggested the presence of a massive ($\sim 10^7 - 10^8$ M$_\odot$) but underfed black hole with a maximum luminosity of $\sim 10^8$ L$_\odot$. They suggested searching for a compact synchrotron source, which Balick and Brown [5] discovered three years later at λ 11 cm, using the National Radio Astronomy Observatory (NRAO) Green Bank interferometer combined with a 14 m telescope located about 35 km southwest of the other telescopes. Moreover, Lynden-Bell and Rees had also suggested searching for radio recombination lines in the thermal plasma which they presumed to surround the starving black hole, arguing that the width of these lines could be used to estimate the mass of the black hole.

Observations of radio recombination lines [6, 7] and the [Neii] 12.8 μm forbidden line emission [8, 9] of Sgr A West got the Max Planck Institut für Radio Astronomie astronomers of Peter Mezger's group into a friendly competition with Charles Townes' group at Berkeley. Thanks to the much higher angular resolution of the Neii spectroscopy, the Townes group arrived about a decade earlier than we did with the 100-m telescope in Effelsberg [10] at the conclusion that there is a mass of $\sim 4 \cdot 10^6 \, M_\odot$ enclosed within the central parsec. It is remarkable that—with the exception of Downes and Martin—none of the early papers on Sgr A* and the dynamics of the ionized gas in Sgr A West refers to Lynden-Bell and Rees. Apparently, most observers outside the United Kingdom had not read this paper.

Since then, discussions and exchanges of preprints between Peter Mezger's group at Bonn and Charles Townes' group at Berkeley have been concerned with the enigmatic nature of Sgr A* and its environment. It is therefore more than appropriate that we devote our latest paper on the radio/IR spectrum of Sgr A* (Beckert *et al.* [11]) to Charles Townes. The results of this paper will be summarized here as part of a more general review of the central $\sim 4 \, \mathrm{kpc}$ of the Galaxy.

45.2   From the Nuclear Bulge to the Circumnuclear Disk

We begin with a general overview of the Galactic Center but refrain from overloading this paper with detailed references which may be found in the recent reviews by Genzel, Hollenbach and Townes [12], in the NATO ASI Conference Proceedings edited by Genzel and Harris [13] and in a forthcoming review by Mezger, Duschl and Zylka [14].

Modeling of the Hi, CO and CS emission in the inner $10°$ and of the near-infrared (NIR) $\lambda 2.4 \, \mu m$ emission in the inner $12°$ indicates the presence of a bar structure which is tilted relative to the line of sight by $\sim 16°$ with the near side of the bar lying at positive longitudes. While the stellar bar may extend close to the inner Lindblad resonance (ILR) at $R \sim 4 \, \mathrm{kpc}$, there exists a zone of avoidance for interstellar matter (ISM) between the galactocentric radii $R \sim 1.5 - 3.5 \, \mathrm{kpc}$ since there exist no stable orbits there. Within $R \sim 0.3 - 1.5 \, \mathrm{kpc}$ atomic hydrogen moves on very eccentric orbits which follow approximately the stellar bar structure. For $R < 0.3 \, \mathrm{kpc}$ the now predominantly molecular hydrogen moves on only slightly elliptical orbits (for a recent review, see Gerhard [16]). This is the region we refer to as the Nuclear Bulge. Figure 45.1 shows the morphology of our Galaxy. For $R \geq 3 \, \mathrm{kpc}$ it is dominated by the spiral structure outlined here by the distribution of Hii regions [15]. Inside the ILR both near-infrared (NIR) and radiospectroscopic observations suggest a bar-structure as outlined here by the distribution of the surface density of gas (according to model calculations; see, e.g., [16]).

The stellar population of the Nuclear Bulge consists of old population ii stars which probably form the inner part of the Galactic Halo, and somewhat older but very metal-rich stars, which could be the result of a giant star burst during the early evolution of the Galaxy. Much younger massive stars exist in the thin ($h \sim 30 - 50 \, \mathrm{pc}$) layer of molecular ISM which extends to radii $R \sim 250 - 300 \, \mathrm{pc}$. Zylka *et al.* [17] find from free-free and far-infrared (FIR) emission (see Fig. 45.2) that $\sim 10\%$ of all high and medium mass stars of the Galaxy are formed within a

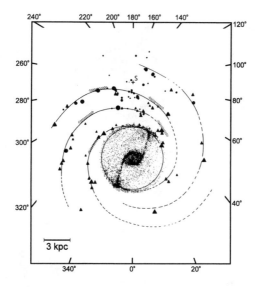

FIGURE 45.1. The morphology of our Galaxy [15,16].

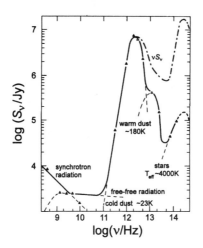

FIGURE 45.2. Integrated radio/far-infrared (FIR) spectrum of the central 500 pc [18].

volume, which is only $\sim 1/1000$-th of that occupied by ISM in the Galactic Disk, but which contains $\sim 10\%$ of all molecular gas in the Galaxy. This total mass of $\sim 10^8 \, M_\odot$ of molecular hydrogen accounts for $\sim 2 - 3\%$ of the estimated total stellar mass of the Nuclear Bulge of $\sim 3.5 - 5 \cdot 10^9 \, M_\odot$. The stellar and radiation density increases dramatically towards the Galactic Center thus creating rather extreme physical conditions in the central parsec.

The Sgr A giant molecular cloud (GMC) complex extends over the central 50 pc. This region is filled with very compact, massive ($\sim 10^5 \, M_\odot$) clouds which

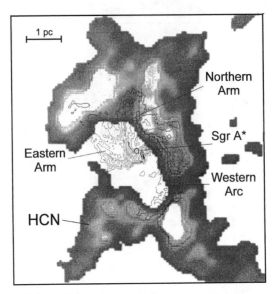

FIGURE 45.3. The inner edge of the CND overlayed on a Very Large Array (VLA) radio image of the minispiral. The area inside the CND is referred to as Central Cavity (adapted from Güsten *et al.* [20]).

are in part located in front and in part behind the central star cluster. Fortunately, no GMC covers the central 4 pc surrounding Sgr A*, otherwise any observations short of 100 μm would be blocked. Even without a GMC within the line-of-sight the interstellar dust between Sun and Galactic Center accounts for a visual extinction of $A_V \sim 31^m$, thus rendering any observations between the NIR at $\lambda \sim 1\,\mu$m and the soft X-ray regime $E \leq 1$ keV impossible.

The Sgr A radio complex contains the extended (~ 10 pc) synchrotron source Sgr A East, which is the remnant of an explosion that occurred $5 \cdot 10^4$ yr ago in the massive Sgr A East cloud core. The energy input corresponds to that of ~ 40 supernova explosions. Sgr A West is an Hii region of size ~ 3 pc which is located in front of Sgr A East [19].

An irregular cloud of $\sim 10^4\,M_\odot$, referred to as Circumnuclear Disk (CND), rotates around Sgr A*, the compact synchrotron source mentioned in Sect. 45.1. The CND has a rather sharp inner edge which limits the Hii region Sgr A West. Duschl [21] and Yi, Field, and Blackman [22] have shown that this sharp inner edge can be explained as the consequence of a kink in the rotation curve, which must occur at that galactocentric distance where the gravitational potential of a compact central source begins to dominate over that of the central star cluster. Hence, the sharp inner edge of the CND is another indicator of the presence of a compact central object of a few million solar masses. Since gas column densities drop by factors of greater than ten at the transition from the inner edge of the CND to the Hii region one refers to the central ~ 3.4 pc also as the Central Cavity (CC).

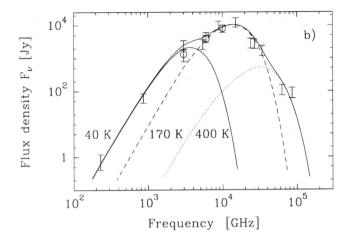

FIGURE 45.4. The infrared (IR)-spectrum of the central 30″ ($\sim$ 1.2 pc) corrected for a visual extinction of $A_V \sim 31^m$ [17].

The geometry and dynamics of the CND have been thoroughly investigated by Güsten *et al.* [20] (see also Fig. 45.3). The disk is inclined towards the line of sight by $\sim$ 70°. NIR observations by Gatley *et al.* [23] indicate that the southwestern side of the CND lies in front of (and the northeastern behind) the central star cluster. The position (tilt) angle of the CND decreases from the inner edge outward while its full-width-half-maximum (FWHM) increases from $\sim$ 0.5 to 1.5 pc. The gas is very clumpy (filling factor $\sim$ 5 − 10%; typical clump mass $\sim$ 30 M$_\odot$). For a review of the physical state of the CND see Genzel [24], who also discusses arguments that the CND/Cavity configuration cannot be much older than about 10^5 yr. The rather similar ages of CND and extended synchrotron source Sgr A East (see above) could link the formation of the CND to the explosion that created Sgr A East.

45.3   The Central Cavity, its Morphology and Luminosity

Most of the volume of the CC of radius $R \sim 1.7$ pc and depth $l \sim 0.5 - 1.5$ pc is filled with $\sim$ 250 M$_\odot$ of ionized gas, a small mass fraction ($\sim$ 10%) of which forms a pattern of arms and arcs, referred to as "minispiral" (see Fig. 45.3; for the morphology of Sgr A West see [11] and references therein). One also recognizes Sgr A* as the most intense radio source in the CC, located at the northern edge of the "bar" which connects northern and eastern arms with the western arc.

Figure 45.4 from [17] shows the IR-spectrum of the central 30″ ($\sim$ 1.2 pc), corrected for an extinction of $\sim 31^m$. It has been decomposed into three components of temperatures 40 K, 170 K and 400 K. Other fit parameters are given in [17].

The 40 K dust component appears to be a blend of dust associated with the extended ionized gas within the CC and (probably somewhat cooler) dust contained in the Sgr A East core GMC against which Sgr A West is seen [11].

Hot dust which dominates the emission at $\lambda \leq 60\,\mu$m (see Fig. 45.4) is mainly associated with ionized gas contained in the minispiral, as the beautiful high-resolution mid-infrared (MIR) images by Gezari [25] and Telesco, Davidson, and Werner [26] show. From a correlation between $12.4/7.8\,\mu$m dust color temperatures and the position of the more prominent Hei emission line stars detected by Krabbe et al. [27], Gezari shows that these stars—and not a central source associated, e.g., with Sgr A*—are the prime heating sources of the hot dust. Actually, if these stars have typical luminosities of $\sim 10^6\,L_\odot$ they can heat dust to temperatures ≥ 170 K only within distances of $\leq 10^{17}$ cm, corresponding to $\leq 1''$ at the distance of the Galactic Center.

The systematic decrease of the $30/19.2\,\mu$m dust color temperature with galactocentric radius found by Telesco, Davidson and Werner [26] suggests that most of the stars responsible for the heating of the hot dust are located within the central $\sim 16''$. The existence of a cluster of luminous stars within the central $10''$ surrounding Sgr A* was recently demonstrated by high-resolution NIR imaging (Eckart et al., [28]). The IR-luminosity integrated over the central 1.2 pc is $L_{IR} \sim 5\cdot10^6\,L_\odot$. On the assumption that the stellar heating sources are uniformly distributed within the CC, Zylka et al. [17] arrive at a total luminosity of $2\cdot10^7\,L_\odot$.

On the other hand, a comparison of the IR spectra integrated over the central 500 pc (Fig. 45.2) and the central 1.2 pc (Fig. 45.4), respectively, shows that much cooler dust with temperatures ~ 23 K is the main contributor to the IR-luminosity of $\sim 1.3\cdot10^9\,L_\odot$ of the Nuclear Bulge [17]. Such temperatures are typical for dust heated by the interstellar radiation field which is dominated by medium mass A stars [18]. The luminosity provided by hot (~ 180 K) dust in the Nuclear Bulge is about an order of magnitude lower.

About 10^6 yr is an upper-age limit for the massive stars in the central star cluster. Hence, most of these stars must have formed within the CC. The detection of a "Tongue" of $\sim 200\,M_\odot$ of neutral gas sandwiched between the ionized Northern and Eastern Arms [29] and its recent confirmation by Telesco, Davidson, and Werner [26] gives a clue as to how neutral matter is transported from the CND to the central cluster of massive stars and thus can sustain their formation.

45.4   A Summary of Recent Observations of Sgr A*

In the Introduction we have summarized how the compact synchrotron source Sgr A* has been detected and identified with an object of a few $10^6\,M_\odot$. We have also mentioned that a layer of gas and dust between the solar system and the Galactic Center prevents any observations in the NIR/optical wavelength range $\lambda \leq 1\,\mu$m and that this layer becomes transparent again only in the X-ray regime $E \geq 1$ keV.

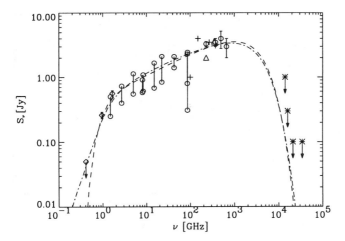

FIGURE 45.5. The observed spectrum of Sgr A* as given in [17], but supplemented by low-frequency observations (shown as rhombs) obtained by Davies, Walsh and Booth [30]. Symbols "o——o" refer to upper and lower flux densities observed for the variable radio source; "*——>" are upper limits obtained in the mid-infrared (MIR). Curves relate to model computations (see text).

45.4.1   The Radio through MIR Source Characteristics: Spectrum and Morphology

Figure 45.5 shows the observed spectrum of Sgr A* as given in [11]. It is characterized by a low turnover frequency $\nu_t \sim 0.8\,\text{GHz}$, a high cut-off frequency $\nu_c \sim 2 - 4 \cdot 10^3\,\text{GHz}$, and an inverted spectrum $S_\nu \propto \nu^{1/3}$ in between. Note, however, that for reasons discussed in [17], $\lambda \sim 450\,\mu\text{m}$ is still the shortest wavelength at which Sgr A* has actually been detected. At shorter wavelengths, especially in the MIR, there exist to date only upper—although significant—limits. The variability of the radio spectrum is well-established. Zhao *et al.* [31] suggest that for $\nu \geq 8.5\,\text{GHz}$ intrinsic variations on time scales of a few months determine the variability. The investigation of variability at frequencies $\nu \geq 100\,\text{GHz}$ has not yet led to unambiguous results [17].

As first noted by Davies, Walsh and Booth [30] the apparent size of Sgr A* depends on the wavelength $\propto \lambda^2$ (see Fig.45.6). This is the sign of source broadening due to electron scattering. Very Long-Base Interferometer (VLBI) observations have now been successfully extended to $\lambda \sim 1\,\text{mm}$ [32, 33]. The shortest wavelength at which Sgr A* has been detected to date with VLBI is $\lambda \sim 3\,\text{mm}$ [34] and $\sim 7\,\text{mm}$ [35, 36] yielding source sizes $\sim 0.33 - 0.13\,\text{mas}$ ($\hat{=} 4 - 1.6 \cdot 10^{13}\,\text{cm}$) as compared to a size $2R_S \sim 1.2 \cdot 10^{12}\,\text{cm}$ of a $2 \cdot 10^6\,\text{M}_\odot$ black hole. The corresponding brightness temperatures are $T_b \sim 2 \cdot 10^9 - 1.4 \cdot 10^{10}\,\text{K}$. The elongated jet-like structure detected by Krichbaum *et al.* [35] has yet to be confirmed.

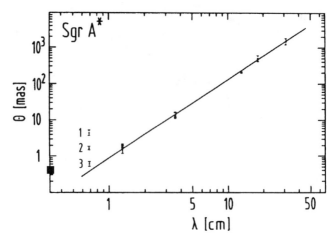

FIGURE 45.6. The apparent angular diameter θ of Sgr A* as a function of the wavelength λ.

45.4.2   The NIR through X-ray Source Characteristics

Eckart *et al.* [37] using high-resolution imaging, detected an apparently extended NIR source in the direction of Sgr A*. Assuming optically thick free-free emission and adopting electron temperatures of $2 - 4 \cdot 10^4$ K one obtains optical/UV luminosities of $0.34 - 2.8 \cdot 10^5$ $L_\odot$ [38]. More recently, however, with the remarkable K-band resolution of 0.15″, Eckart *et al.* [28] resolved the extended source into a cluster of six (probably stellar) sources. The relative accuracy of radio and NIR frames does not yet allow us, however, to decide if any of these sources actually coincides with the position of the radio source Sgr A*.

Predehl and Trümper [39] detected with ROSAT (Röntgen satellite) a soft X-ray source within 10″ of the radio position of Sgr A*. However, comparison with X-ray images obtained with ASCA (Advanced Satellite for Cosmology and Astrophysics) clearly indicate that this source is not identical with the much stronger source detected at higher energies with the ART-P telescope [40]. As in the case of the NIR sources, the weak X-ray emission could be well explained by the presence of a Hei star in the ROSAT beam.

45.5   The Nature of Sgr A*

At present Sgr A* has been unambiguously detected only in the radio range $1 \leq \nu/\text{GHz} \leq 670$. The integrated radio/IR spectrum yields luminosities of $\sim 250 - 600 L_\odot$. Sgr A* could be a strong ($\leq 2 \cdot 10^7 L_\odot$) but undetected optical/UV emitter with all its radiation absorbed by surrounding dust. However, heating of the dust by extended (stellar) sources concentrated within the central parsec and an upper limit of the Sgr A* luminosity of $\leq 10^5 L_\odot$ appears to be much more likely (see Sect. 45.3). Hence, the radio observations summarized in Subsect. 45.4.1 may be the only clue for an understanding of this enigmatic source.

45.5.1   The Radio Spectrum above 1 GHz

The Sgr A* radio/IR spectrum is shown in Fig. 45.5 and has been qualitatively discussed in Sect. 45.4.1. Beckert *et al.* [11] offer the following interpretation of this spectral behaviour. The $S_\nu \propto \nu^{1/3}$ for $\nu > 1$ GHz as well as the high-frequency cut-off at a few 10^3 GHz is best explained by a quasi-monoenergetic distribution of relativistic electrons which emit optically thin synchrotron radiation [17, 41, 42]. A best fit of this model to the observations shown in Fig. 45.5 as dashed curve is obtained for the following source parameters: the electron energies are in the range between 50 and 150 MeV with a steep increase of electron densities towards the higher energies. Other fit parameters are the number density of relativistic electrons of $\sim 4 \cdot 10^4$ cm^{-3}, the source size of a few 10^{13} cm, and a magnetic field strength of ~ 13 G. This magnetic field and the electron number densities are well within the range of what one expects as being potentially due to an accretion disk around a black hole of a few 10^6 M$_\odot$ with an accretion rate of $10^{-7\ldots-6}$ M$_\odot$/yr [41]. It is especially encouraging that the independently deduced source size of Sgr A* agrees very well with recent VLBI determinations (see Subsect. 45.4.1). The acceleration process for the electrons is not yet fully clear.

45.5.2   The Low-frequency Cut-off at $\nu < 1$ GHz

Beckert *et al.* [11] elaborate on two explanations for the low-frequency cut-off: Free-free absorption due to the ionized gas of Sgr A West, and synchrotron self-absorption. Model fits for the two alternatives are shown as dashed (synchrotron emission + free-free absorption) and dash-dotted (synchrotron emission + self-absorption) curves in Fig. 45.5. The model parameters for free-free absorption place Sgr A* in the center of the extended component Sgr A West. The parameters for the compact synchrotron source are given above. It is clear that both solutions (or even a combination of the two) are fully compatible with the presently available data. However, future observations below 1 GHz could resolve this uncertainty as the free-free cut-off is steeper than the one due to self-absorption.

45.5.3   The Accretion Disk around Sgr A*

If Sgr A* is a black hole of a few 10^6 M$_\odot$ with an accretion disk around it through which matter is accreted at a rate of a few 10^{-7} M$_\odot$/yr, as suggested above, one expects a luminosity of around 10^5 L$_\odot$, well below its Eddington limit of a few 10^{10} L$_\odot$. The exact value of the luminosity depends on both the accretion rate and on the efficiency of transforming particle rest energy into accretion luminosity (typically $\sim 10\%$). A luminosity of $\sim 10^5$ L$_\odot$ would not exclude the possibility that one of the now resolved NIR objects seen in the direction of Sgr A* (Subsect. 45.4.2) is identical with Sgr A* and a black hole/accretion disk combination.

As discussed in [17], the standard accretion disk model can also explain the observed high formation rate of massive stars close to Sgr A*. It suggests a disk mass of approximately 1 (10) M$_\odot$ within $10^3 (10^4) R_S$ depending on the value of the viscosity [17]. Then for radii of a few $10^4 R_S$—depending again on viscosity—the disk becomes gravitationally unstable. At larger radii the gravitational instability and hence fragmentation of the disk followed by star formation will continue. The observations reviewed in this paper suggest that we are currently in a phase where most of the accretable disk material is not fed into the black hole but rather consumed by star formation, thus leaving the black hole starving. In such a situation, it would be no surprise that a sizeable fraction, or even most, of the luminosity of the CC region comes from young massive stars rather than from the black hole at the center. Of course, if one wants to explain the formation of the Hi/Hei star cluster $\sim 10^6$ yrs ago by this process, one has to argue that the disk then was more massive with accordingly higher mass flow and energy release rates. Details of this have to be clarified by disk model calculations. The same is true for another instability of such a disk which our model calculations predict for radii $\leq 250 R_S$, which might be responsible for the observed variability in the Sgr A* spectrum.

45.6   Summary and Conclusions

There is converging observational evidence—based on the dynamics of gas and stars in the central ~ 3 pc—of a massive $(2 - 3 \cdot 10^6 \, M_\odot)$ compact object close to or at the dynamical center of our Galaxy. Radio observations suggest that this object coincides with the compact synchrotron source Sgr A* and represents a starving black hole. The detection of Sgr A* as a NIR or a X-ray source is yet highly uncertain.

Millimeter VLBI observations yield an upper limit to the size of Sgr A* of $\sim 0.33 - 0.13$ mas corresponding to a linear size of $4 - 1.6 \cdot 10^{13}$ cm, which should be compared with the Schwarzschild radius of a $2 \cdot 10^6 \, M_\odot$ black hole of $2R_S \sim 1.2 \cdot 10^{12}$ cm.

The compact source is surrounded by a rotating disk which is ionized at galactocentric distances ≤ 1.7 pc. While differential rotation and viscosity can account for mass flow within $R \leq 200$ pc, a bar structure which extends out to $R \sim 4$ kpc appears to maintain mass inflow from regions farther out. The molecular disk itself extends out to $R \sim 250 - 300$ pc and contains $\sim 10^8 \, M_\odot$ of gas and dust. Model computations indicate the onset of gravitational instability of the disk which—depending on the accretion rate and the viscosity—begins at radii $\sim 10^4 R_S$ and may continue further out. This instability could eventually lead to the formation of massive stars close to Sgr A*.

While all of the elements of an accretion disk/black hole configuration are present at the Center of the Galaxy, it appears that this configuration during the past $10^{6...7}$ yrs has been and still is in a phase, where nuclear fusion in massive stars dominates the energy production in the central 1.2 pc.

The contributions by Charles Townes and his associates were essential to arrive at the picture of the Galactic Center given here. Discussions with him have always been a source of stimulation and encouragement to us.

References

[1] G. Westerhout G., Bull. Astron. Inst. Neth., Vol. **XIV** Nr. 488 (1958) p. 215.

[2] W. J. Altenhoff, P. G. Mezger, H. Strassl, H. Wender, and W. Westerhout, Veröff. d. Universitäts-Sternwarte zu Bonn Nr. **59** (1960).

[3] D. Downes D. and A. H. M. Martin, Nature **233,** 112 (1971).

[4] D. Lynden-Bell and M. J. Rees, MNRAS **152,** 461 (1971).

[5] B. Balick and R. L. Brown, ApJ **194,** 265 (1974).

[6] T. Pauls, P. G. Mezger, and E. Churchwell, A&A **34,** 327 (1974).

[7] P. G. Mezger and T. Pauls, IAU Symp. No. 84 (W.B. Burton, ed.), (1979) p. 357.

[8] E. R. Wollman, T. R. Geballe, J. H. Lacy, C. H. Townes, and D. M. Rank, ApJ **218,** L103 (1977).

[9] E. R. Wollman, IAU Symp. No. 84, W. B. Burton, ed., (1979) p. 367.

[10] P. G. Mezger and J. E. Wink, A&A **157,** 252 (1986).

[11] T. Beckert, W. J. Duschl, P. G. Mezger, and R. Zylka, A&A (1995) *submitted.*

[12] R. Genzel, D. Hollenbach, C. H. Townes, 1994, The Nucleus of our Galaxy, Rep. Prog. Phys. **57,** 417 (1994).

[13] R. Genzel, A. I. Harris (eds.), The Nuclei of Normal Galaxies – Lessons from the Galactic Center, NATO-ASI Series Vol. **445** (1994).

[14] P. G. Mezger, W. J. Duschl, and R. Zylka, *in preparation.*

[15] Y. M. Georgelin and Y. P. Georgelin, A&A **49,** 57 (1976).

[16] O. E. Gerhard, IAU-Symp. No. 169, L. Blitz, ed. (1995) *in press.*

[17] R. Zylka, P. G. Mezger, D. Ward-Thompson, W. J. Duschl and H. Lesch, A&A (1995) *in press.*

[18] P. Cox and P. G. Mezger, A&A Rev. **1,** 49 (1989).

[19] P. G. Mezger, R. Zylka, C. J. Salter, *et al.*, A&A **209,** 337 (1989).

[20] R. Güsten, R. Genzel, and M. C. H. Wright, *et al.*, ApJ **318** 124 (1987).

[21] W. J. Duschl, MNRAS **240,** 219 (1989).

[22] I. Yi, G. B. Field, and E. G. Blackman, ApJ **432,** L31 (1994).

[23] I. Gatley, R. Joyce, A. Fowler, D. DePoy, and R. Probst, IAU Symposium No. 136, M. Morris, ed. (1989) p. 361.

[24] R. Genzel, IAU Symp. No. 136, M. Morris, ed. (1989) p. 393.

[25] D. Y. Gezari, in: *The Center, Bulge and Disk of the Milky Way* L. Blitz, ed., (Kluwer, Dordrecht, 1992). p. 23.

[26] C. A. Telesco, J. A. Davidson, and M. W. Werner, ApJ (1995) *in press.*

[27] A. Krabbe, R. Genzel, S. Drapatz, and V. Rotaciuc, ApJ **382,** L19 (1991).

[28] A. Eckart, R. Genzel, R. Hofmann, B. J. Sams, and L. E. Taccconi-Garman, ApJ, (1995) *in press.*

[29] J. M. Jackson, N. Geis, R. Genzel, *et al.*, ApJ **402,** 173 (1993).

[30] R. D. Davies, D. Walsh D., and R. S. Booth, MNRAS **177,** 319 (1976).

[31] J.-H. Zhao, W. M. Goss, K. -Y. Lo, and R. D. Ekers, in: *Relationship between Active Nuclei and Starburst Galaxies* A. Filippenko, ed., (1991) p. 295.

[32] A. Greve, *et al.*, A&A (1995) *in press.*

[33] M. Grewing, IRAM Newsletter **19,** 1 (1995).

[34] T. P. Krichbaum, C. J. Schalinski, A. Witzel, *et al.*, in: *The Nuclei of Normal Galaxies – Lessons from the Galactic Center* NATO-ASI Series Vol. **445,** (1994) p. 411.

[35] T. P. Krichbaum, J. A. Zensus, A. Witzel, *et al.*, A&A **274,** L37 (1993).

[36] A. E. E. Rogers, S. Doeleman, M. C. H. Wright, *et al.*, ApJ **434,** L59 (1994).

[37] A. Eckart, R. Genzel, A. Krabbe, *et al.*, Nature **355,** 526 (1992).

[38] R. Zylka, P. G. Mezger, and H. Lesch, A&A **261,** 119 (1992).

[39] P. Predehl and J. Trümper, A&A **290,** L29 (1994).

[40] P. Predehl and Y. Tanaka, *private communication.*

[41] W. J. Duschl and H. Lesch, A&A **286,** 431 (1994).

[42] T. Beckert, Diploma thesis, University of Heidelberg, Germany (1995).

46

Infrared Semiconductor Laser by Means of J × H Force Excitation of Electrons and Holes

Takeshi Morimoto
Meiro Chiba
and Giyuu Kido

46.1 Introduction

In the earlier stage of the investigation of lasing action in solids, especially in semiconductors, the possibility of utilizing Landau levels was proposed by several investigators [1,2]. While these attempts were not successful, the first achievements of a ruby laser was reported by Maiman in 1960 [3]. In semiconductors, after the subsequent developments of injection-type semiconductor lasers [4], the research shifted to the lasing action by use of the quantum size effect in low-dimensional materials, such as quantum wires [5,6] and dots [5]: The principle is based on utilizing the singularity in the density of states of electrons associated with quantization, which produces discrete energy states due to the finite-size effect.

In this paper, we shall describe a new type of quantum effect in bulk semiconductor lasers brought about by applying a high magnetic field $\mathbf{H}$ perpendicular to the current flow $\mathbf{J}$ passed through the bulk sample. The inverted population, being a self-organized state of the system, is caused by the transverse excitation of electrons from valence to conduction bands by the action of $\mathbf{J} \times \mathbf{H}$ force in narrow-gap semiconductors, such as n-InSb and $Hg_{1-x}Cd_xTe$ (see Fig. 46.1). The condition for inverted population is easily satisfied at low temperatures. The resulting stimulated infrared emissions are tunable by varying the applied magnetic field. The application of high magnetic fields brings about a one-dimension-like density of states having a divergently high value of the gain for stimulated emissions at the band edge.

46.2 Principle of Lasing Action

In narrow-gap semiconductors having almost equal numbers of electrons and holes in the configuration of $\mathbf{J} \perp \mathbf{H}$ as shown in Fig. 46.1, electrons and holes are driven in the −y-direction with almost equal drift velocity at high magnetic field limit by

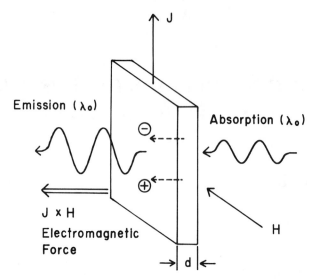

FIGURE 46.1. Classical concept of the magnetoelectric photoeffect in intrinsic narrow-gap semiconductors, indicating a pumping action of the $\mathbf{J} \times \mathbf{H}$ force. ($\mathbf{H}$ is applied in the z-direction; $\mathbf{J}$ is in the x-direction; $\mathbf{J} \times \mathbf{H}$ is in the $-\mathbf{y}$ -direction.) The dotted arrows indicated the drift motion of electrons $\ominus$ and holes $\oplus$ at high magnetic fields. Note that in compensated electron-hole plasmas the Hall electric field, $\mathbf{E}_{y,H}$ vanishes as $\mathbf{H} \to \infty$. It causes the transverse drift motion of electrons and holes. The total system constitutes a driven system which produces stimulated emission, and the amplification of background radiation.

the action of $\mathbf{J} \times \mathbf{H}$ force, or the equivalent transverse electric field E_y^*, as given by $E_y^* \sim JH/n^* |e| c$. Here n^* is the effective concentration of electrons. Note that in high magnetic fields the value of E_y^* becomes much higher than the longitudinal electric field E_x originally applied [7]. Then a part of the excess electrons will be excited along the y-direction by the action of E_y^*, which lasts during the travel time through the mean free path l^* until annihilation by recombination with excess holes, without suffering severe damping by collisions [7-9]. We call such an electron a "lucky electron" [10]. The concentration of electrons and holes n and p then deviate from the thermal equilibrium values n_0 and p_0. The excess electrons and holes annihilate by radiative and nonradiative recombinations. The emitted power by the band-to-band recombination per unit volume of the sample is then written as

$$P(\omega)d\omega = \hbar\omega \cdot B(\omega) \cdot (n \cdot p - n_i^2)d\omega , \qquad (46.1)$$

for the frequency interval $d\omega$, with $n_0 \cdot p_0 = n_i^2$. Here $B(\omega)$ is the probability of radiative recombination at frequency ω. The above is the classical concept of the phenomenon at the intrinsic region [7,11,12].

   In the extrinsic region at low temperatures, there would be no holes in the originally n-type materials at thermal equilibrium. However, at low temperatures highly nonequilibrium-like generation of holes was observed under high magnetic fields due to the transverse excitation by hot electrons [8,9,13]. At the quantum

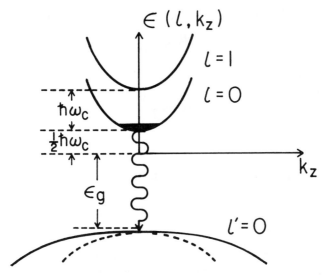

FIGURE 46.2. Population of electrons at the quantum limit and Landau emission, neglecting the spin splitting of electrons. In the valence bands heavy holes play a much more dominant role for emissions than light holes due to their higher density of states [7-9].

limit, these excited electrons and holes will populate mostly the lowest Landau levels, $\ell = \ell' = 0$, ℓ and ℓ' being Landau's orbital quantum numbers of electrons and (heavy) holes, respectively, as schematically illustrated in Fig. 46.2. At high excitation it causes an inverted population of electrons and holes at the band edge $k_z = 0$, k_z being the wave vector of electrons along the magnetic field direction.

It has been found that the peak frequencies $\omega_\pm$ of the strongest emissions α_1 $(0_+ \rightarrow 0')$ and α_2 $(0 \rightarrow 0')$, which correspond to the transitions between the lowest Landau levels of electrons and heavy holes, obey the relation [8,9,13]

$$\hbar\omega_\pm = \varepsilon_g + \frac{1}{2}\hbar\omega_{c1}\left[1 \pm \frac{m_{*1}\,g_1}{2m_0}\right] + \frac{1}{2}\hbar\omega_{c2,h} - |e|\,E_y^*l^*, \qquad (46.2)$$

with $E_y^* \sim JH/n^*\,|e|\,c$. Here the $\pm$ signs correspond to the up- and down-spin states of conduction electrons; ω_{c1} and $\omega_{c2,h}$ are the cyclotron frequencies of electrons and heavy holes, respectively, ε_g is the energy gap, m_1^* is the effective mass of electrons, g_1 is the effective g-factor, m_0 is the rest mass of electrons, and l^* is the mean free path of the lucky electrons mentioned earlier. This phenomenon would be the inverse of the interband magneto-optical absorption (IMO) [1, 14], if the last term of Eq. (46.2) were not involved: This term is added in order to explain the reduction of the virtual energy gap in crossed electric and magnetic fields [15,16], as will be seen below.

It should be noted that at the quantum limit, the gain $g(\omega, H)$ of the stimulated emission becomes divergently high at the band edge $k_z = 0$ for α_1 $(0_+ \rightarrow 0')$ and

$(0_- \rightarrow 0')$ emissions, since it has singularities at $\omega = \omega_\pm$ of the form [17]:

$$g(\omega, H) = A\frac{eH}{c\omega}\sum_\pm(\hbar\omega - \hbar\omega_\pm)^{-1/2}. \qquad (46.3)$$

Here $\omega_\pm$ are the peak frequencies of the two strongest emissions α_1 and α_2 given by Eq. (46.2). Therefore, stimulated emission can take place at an extremely low value of the critical current density such as $J_c \sim 17\,A/cm^2$ for $H = 10\,T$ at 43 K, for example.

46.3   Characteristics of the Emission

The first discovery of the quantum effect was done at 80 K in a magnetic field up to 7 T [8]. The present data were taken in magnetic fields extended up to 22 T at a temperature around 20 K using a hybrid-type magnet at the high field laboratory of Tohoku University. As the infrared emitter, a chemically etched thin plate of n-InSb sample was used. The dimensions were 4.06 mm long, 31.18 mm wide, and 0.22 mm thick. The electron concentration n_0 and the mobility μ at 80 K were 5×10^{14} cm^{-3} and 3.7×10^5 cm^2/Vs, respectively. The magnetic field $\mathbf{H}(\|\mathbf{z})$, was applied along the [1 1 0] direction parallel to the sample surface. A constant current $\mathbf{J}(\|\mathbf{x})$, chopped with a 50% duty factor, was passed through the sample using a constant current supply. The photo-signal was guided into a monochromator through a light pipe of length ~ 5 m and synchronously detected by a HgCdTe infrared detector.

The typical examples of the emission spectrum and the magnetic field variation of the emission peaks are shown in Figs. 46.3 and 46.4. In Fig. 46.3, it will be seen that the α_1 and α_2 spectral components in emission show a remarkable line narrowing, within the spectral resolution $\Delta\lambda = 0.08\mu$m. This emission spectrum is much sharper than the spectra [8] observed first at $H \le 7\,T$. The line narrowing is caused by repeated stimulated emissions with a divergently high value of the gain at $\omega = \omega_\pm$, despite the fact that no Fabry-Perot cavity is formed by the etched surfaces of the sample.

As seen from Fig. 46.4, the peak frequencies of the α_1 and α_2 emissions fit well with Eq. (46.2), except for slight deviation from the linearity with respect to H at higher magnetic fields due to the non-parabolicity [1] in the conduction band, in spite of the complicated structures of the magnetic energy levels in the valence band [16,18]. At higher magnetic fields $H > 11\,T$, as shown in Fig. 46.3, the β_1 $(1_+ \rightarrow 1')$ component of the emission disappears, and the α_1 $(0_+ \rightarrow 0')$ emission becomes dominant. No fine structures were observed in the emission spectra in high magnetic fields, whereas the continuous wave (cw) emission spectra at room temperature involve many fine structures at high magnetic fields [16]. This suggests that almost all electrons and holes are degenerate in the cw excitation at low temperatures.

In Fig. 46.4 the extrapolation of the $\hbar\omega$ vs. H curves to zero magnetic field gives the values of the virtual energy gap $\varepsilon_g(0)$ to be 0.200 eV at 20 K in the

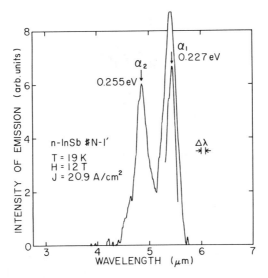

FIGURE 46.3. Emission spectrum of n-InSb for $H = 12\,\mathrm{T}$ and $J = 20.9\,\mathrm{A/cm^2}$ at 19 K.

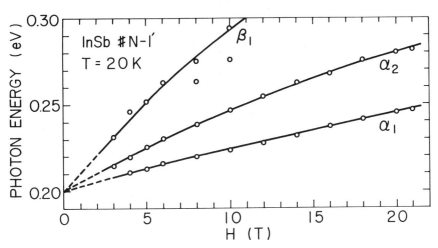

FIGURE 46.4. Magnetic field variation of photon energy $\hbar\omega$ of emission peaks in n-InSb at 20 K.

present cw excitation under the crossed fields $\mathbf{E}^*$ and $\mathbf{H}$, whereas the value of ε_g in the absence of external fields has been evaluated as 0.225 eV [8]. The last term in Eq. (46.2) gives such a reduction in the energy gap of about 25 meV, if we take the reasonable values of $\mathbf{E}^*_y \sim 250$ V/cm and $l^* \sim 10^{-4}$ cm, respectively. The details are to be presented elsewhere [16]. The reduction in the energy gap in the crossed fields is closely related to the onset of stimulated emission and the excitation of electron-hole pairs at low temperatures, as will be discussed below.

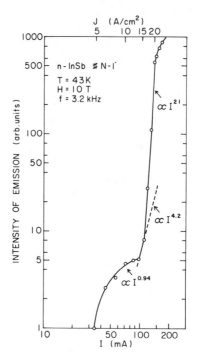

FIGURE 46.5. Current variation of the total output of the emission from n-InSb for $H = 10\,\text{T}$ at 43 K.

Figure 46.5 shows the current dependence of the total output of the emission of n-InSb at 43 K in a magnetic field of 10 T. The $V - I$ curves, measured for the magnetic fields $H = 8, 10$, and 14 T, are shown in Fig. 46.6. In Fig. 46.5, we can see that the generation of electron-hole pairs by the action of $\mathbf{J} \times \mathbf{H}$ force, or the equivalent $\mathbf{E}_y^*$, is taking place at a current density as low as $J \sim 5\,\text{A/cm}^2$ for $H = 10$ T at 43 K, as can be understood from the increasing emitted intensity with the current. At a critical current density, $J_c \approx 17.1\,\text{A/cm}^2$, the emitted intensity steeply increases, with a current dependence proportional to J^{21}, suggesting the occurrence of stimulated emission. The output is estimated to be more than 20 mW at the source point.

In the $V - I$ curve, on the other hand, the increase in the number of electron-hole pairs is reflected to the super-linear increase in the differential conductivity $\partial I / \partial V$ at the first stage. However, at a critical current density J_c, the differential conductivity $\partial I / \partial V$ suddenly diminishes to a value approaching to zero, associated with the steep increase in the emitted intensity proportional to J^{21}. A distinct increase in the voltage drop across the sample at $J = J_c$ would be needed for keeping the value of the constant current flow against the rapid decrease in the number of excess electron-hole pairs caused by stimulated emission.

The subsequent decrease in the voltage drop, as seen at higher currents $J > 20\,\text{A/cm}^2$ for $H = 10$ T, for example, is due to the increase in the number of

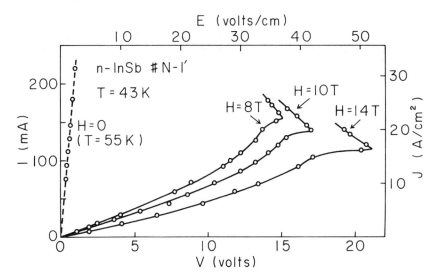

FIGURE 46.6. $V - I$ characteristics of n-InSb under the constant current operation ($J \perp H$) at 43 K.

thermal carriers generated by Joule-heating of the sample at higher currents. The increase in the numbers of thermal electrons n_0 and of thermal holes p_0 will lead to saturation of the emitted power, as can be understood from Eq. (46.1). Indeed, such a saturation of the output is seen at higher currents $J > 20$ A/cm^2 for $H = 10$ T, in Fig. 46.6. Such a characteristic behavior is enhanced by increasing the magnetic field, as shown here, or by forming a Fabry-Perot cavity between parallel surfaces, presumably due to the increase in the rate of stimulated emission [8].

It is worthwhile noting the condition for the inverted population at the quantum limit at the band edge (refer to Fig. 46.1). Taking into account the degeneracy of Landau levels proportional to H and the energy fluctuation kT of the width of the lowest Landau level, we have the critical concentration of electrons n_T for realizing the inverted population:

$$n_T = eH(2m_1^* kT)^{1/2}/2ch^2 \tag{46.4}$$

where k is the Boltzmann constant and T the temperature. The value of n_T is estimated to be $n_T = 5.24 \times 10^{15}$ cm^{-3} for $H = 10$ T at 20 K with $m_1^* = 0.0164 m_0$, which is a realizable concentration even for the cw excitation at low temperatures.

The application of high magnetic fields in semiconductors brings about a one-dimension-like singularity in the density of states of electrons, being similar with that of quantum wires. This singularity drastically lowers the value of the critical current density J_c for lasing, as shown here. However, the value of the critical concentration of electrons n_T goes up in proportion to H due to the Landau degeneracy, so that the operation at low temperatures will be needed. From this point of view, the application of too high magnetic fields will not be so effective for producing population inversion, as long as the field strength H sufficiently satisfies the quantum limit condition $\hbar\omega_{c1} >> kT$. For p-type InSb [19] the generation of excess electron-hole pairs at low temperatures by the $\mathbf{J} \times \mathbf{H}$ force excitation will be difficult because of the twice-higher value of the threshold energy for hole-initiated impact ionization [10].

46.4   Conclusions

We have demonstrated that tunable stimulated emission, controlled by magnetic field strength, is realized in n-InSb bulk samples being excited by $\mathbf{J} \times \mathbf{H}$ force at low temperatures. The critical current density for lasing becomes as low as 17 A/cm^2 due to the extremely high value of the gain $g(\omega, H)$ in the quantum limit. Further investigations on this highly nonequilibrium system involving the details of the excitation mechanism are left as a fruitful task in the future.

46.5   Acknowledgments

We are indebted to S. Ueda, M. Inoue and T. Hamamoto for their technical assistance. Thanks are also due to Dr. S. Akai for offering high quality InSb samples.

References

[1] B. Lax and K. J. Button, *Physics of Solids in Intense Magnetic Fields*, E.D. Haidemenakis (ed.), p. 145 (Plenum, New York, 1969).

[2] B. Lax, J. Mag. & Mag. Matter **11**, 1 (1979); P. Phelan, A. Calawa, R. H. Rediker, R. J. Keyes, and B. Lax, Appl. Phys. Lett. **3**, 143 (1963).

[3] T.H. Maiman, Nature **187**, 493 (1960).

[4] A. Yariv, *Introduction to Optical Electronics*, Chap. 15 (Holt, Rinehart and Winston, New York, 1985).

[5] C. Weisbuck, Proc. of 22nd International Conference on Physics of Semiconductors, E. J. Lockwood (ed.), (World Scientific, Singapore, 1955); W. Wegscheider, L. N. Pfeiffer, M. M. Dignam, A. Pinczuk, K. W. West, S. L. McCall, and R. Hull, Phys. Rev. Lett. **71**, 4071 (1994).

[6] R. Cingolani, Proc. of 22nd International Conference on Physics of Semiconductors, E. J. Lockwood (ed.), (World Scientific, Singapore, 1955).

[7] T. Morimoto and M. Chiba, Infrared Phys. **29**, 371 (1989).

[8] T. Morimoto and M. Chiba, J. Phys. Soc. Jpn. **60**, 2446 (1991).

[9] T. Morimoto, M. Chiba, G. Kido, and A Tanaka, Semicond. Sci. Technol. **8**, S417 (1993).

[10] F. Capasso, Semiconductors and Semimetals **22**, Part D, 1, R. K. Willardson and A. C. Beer (eds.) (Academic Press, New York, 1985).

[11] T. Morimoto and M. Chiba, Phys. Lett. A **85**, 395 (1981); A **95**, 343 (1983).

[12] T. Morimoto and M. Chiba, Jpn. J. Appl. Phys. **23**, L821 (1984); V. K. Malyutenko, S. S. Bolgov, and E. I. Yablonovsky, Infrared Phys. **25**, 115 (1985).

[13] T. Morimoto, M. Chiba, S. Ueda, G. Kido, and M. Inoue, Physica B **184**, 123 (1993).

[14] B. Burstein, G. C. Picus, R. F. Wallis, and F. Blatt, Phys. Rev. **113**, 15 (1959).

[15] W. Zawadzki, Surface Sci. **37**, 218 (1973).

[16] T. Morimoto, M. Chiba, S. Ueda, G. Kido, and T. Hamamoto, Proc. of 11th International Conference on High Magnetic Fields in Semiconductor Physics, D. Heiman (ed.), (World Scientific, Singapore, 1995).

[17] T. Morimoto and M. Chiba, Semicond. Sci. Technol. **7**, B652 (1992).

[18] C. R. Pigeon and R. N. Brown, Phys. Rev. **146**, 575 (1966).

[19] F. R. Kessler, P. Paul, and R. Nies, Phys. Stat. Sol. (b) **167**, 349 (1991).

47

From Laser Beam Filamentation to Optical Solitons: The Influence of C. H. Townes on the Development of Modern Nonlinear Optics

Elna M. Nagasako
and Robert W. Boyd

Self-action effects began to be studied soon after the advent of the laser. Suddenly sources were available that could create beams of light which not only could influence the materials in which they were propagating, but could also, through this material interaction, act back on themselves. These self-action effects could alter the size and shape of the laser beam and introduce new spectral components. In this article, we single out self-focusing, and the associated phenomena of filamentation and the generation of spatial solitons, and describe how recent research has evolved from the pioneering contribution of C. H. Townes.

Self-focusing occurs in materials with an intensity-dependent index of refraction $n = n_0 + \bar{n}_2 \langle E^2 \rangle = n_0 + n_2 I$, where E is the electric field strength and I is the field intensity [1]. In a material with $n_2 > 0$, regions of higher intensity will see a larger index of refraction than regions of lower intensity, causing the material to behave in many ways as a positive lens (Fig. 47.1). A defocusing effect is also possible and

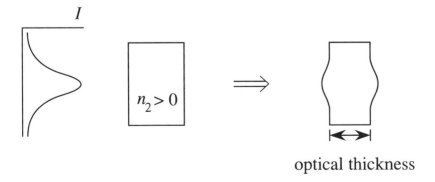

optical thickness

FIGURE 47.1. The intensity-dependent refractive index.

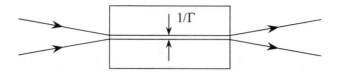

FIGURE 47.2. The self-trapping of light.

will occur in a material with $n_2 < 0$. Investigations of this nonlinear optical effect began in the 1960's as an attempt to explain the newly discovered laser-induced damage tracks in solids [2], but has since expanded to encompass many topics of research, including the field of spatial solitons, which show promise in the areas of optical switching and optical couplers, and fundamental studies of the interaction of light with matter.

In 1964, it was discovered by Hercher that a Q-switched laser beam focused into a transparent solid would cause damage in the form of long tracks, possibly over a few centimeters in length and of a few microns in diameter [2]. Without the consideration of self-action effects, this phenomenon was inexplicable. In an attempt to understand this effect, Chiao, Garmire, and Townes [3] proposed that a beam could become self-trapped [4] in a material with a positive nonlinear index of refraction, and that this confined, intense beam could cause damage of the form previously observed. They suggested that this self-trapping could arise because the beam's transverse intensity profile would induce a waveguide-like index profile in the material (Fig. 47.2). Their steady-state solution for the field envelope for the case in which the field amplitude is allowed to vary in one transverse dimension has the form

$$A(x, y) = A_0 \operatorname{sech}(\Gamma x) \tag{47.1}$$

where the peak amplitude A_0 and beam width $1/\Gamma$ are related by

$$1/\Gamma = (kA_0)^{-1}\sqrt{2n_0/n_2} . \tag{47.2}$$

where $k = n_0 k_0 = n_0 \omega_0 / c$. The cylindrically symmetric solution (variation in two transverse dimensions) was found numerically and the power contained in this solution was determined to be [5]

$$P_{cr} = \frac{5.763 \lambda^2 c n_{eff}}{16 \pi^3 \bar{n}_2 n_0} \tag{47.3}$$

where $n_{eff} = n_0$ for beams with diameters larger than a few wavelengths.

Kelley [6] further investigated the behavior of a cylindrically symmetric beam in a self-focusing material. He found numerically that, for a Gaussian beam with sufficiently high power, nonlinear effects would overcome diffraction and estimated that the beam would come to a focus in the material after a distance

$$z_f = \frac{n_0 a^2}{4} \sqrt{\frac{c}{n_2}} \frac{1}{\sqrt{P} - \sqrt{P_{cr}}} \tag{47.4}$$

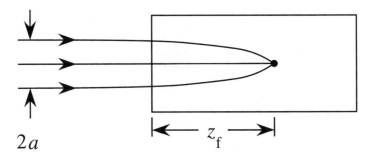

FIGURE 47.3. The self-focusing of light.

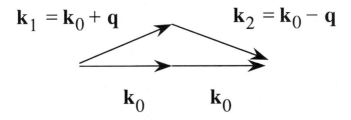

FIGURE 47.4. Self-focusing as a form of forward four-wave mixing.

where a is the beam radius (Fig. 47.3). He speculated that this process could also be responsible for the anomalously high Raman gain that had been observed in certain geometries [7]. Subsequent numerical work [8] on beam propagation in self-focusing media showed that a Gaussian input beam with a power near P_{cr} would evolve into a shape similar to the cylindrically symmetric self-trapped solution and behave temporarily as a trapped beam. However, these beams did not remain trapped, but eventually diffracted ($P < P_{cr}$) or came to a focus ($P > P_{cr}$). It was later found that a full explanation of Hercher's damage tracks would involve a time-dependent version of this self-focusing effect [9].

Townes and co-workers also performed experimental investigations of the self-focusing process. In 1966 they published experimental observations of a Q-switched beam collapsing into a diameter as small as 50 microns [10] and demonstrated the collapse of a single beam into many small (few micron) filaments [11]. The connection between the appearance of multiple filaments and the transverse mode structure of the beam was made by Loy and Shen in 1969 [12]. They showed that a single-mode laser beam would tend to focus into a single filament, indicating that deviations from this single mode structure form the origin of the multiple filaments. Chiao, Kelley and Garmire [13] gave a physical explanation of the growth of the off-axis modes in terms of stimulated light scattering. They showed that the nonlinear index of refraction allows the coupling of a strong incident wave with weak waves having different transverse wavevectors. The four-photon interaction of a photon with wavevector $\mathbf{k}_1 = \mathbf{k}_0 + \mathbf{q}$, a photon at $\mathbf{k}_2 = \mathbf{k}_0 - \mathbf{q}$ and two pump photons with $\mathbf{k} = \mathbf{k}_0$ can allow the two weak waves to see gain (Fig. 47.4). This gain is dependent on the fulfillment of a phase-matching relation involving the participating waves. Because of this condition, certain transverse wavevectors will see more gain than others and will come to dominate the spatial evolution of the total field. This picture is consistent with the model of Bespalov and Talanov [14], who predicted the existence of a dominant instability size $\Lambda = (\pi/k)\sqrt{n_0/2n_2 I}$. It also emphasizes the ability of the self-focusing process to allow small perturbations to have a large effect on the overall beam evolution. Experimental confirmation of this effect, by Carman *et al.* [15] and others followed.

These issues—the interplay of diffraction and self-focusing and the initiation and stability of filaments—are all still of interest twenty years later. In the area of spatial solitons, the balance of the nonlinear index of refraction with diffraction gives new tools to those interested in all-optical switching. The extension of these theories into the quantum domain allows the investigation of the properties of the quantum vacuum. What began as a hindrance, causing damage to optical materials, has come to provide unforeseen opportunities.

The authors of the present article have recently become interested in the initiation of the filamentation process by quantum mechanical fluctuations in the electro-magnetic vacuum field. According to traditional theories, the filamentation [14] of a laser beam as it passes through a nonlinear material is a consequence of the spatial growth [13,16] of weak perturbations initially present on the beam's wave-front. Consequently, one would expect that if these perturbations were removed (e.g. by passing the beam through a spatial filter) filamentation would not occur. However, quantum fluctuations [17] in the field amplitudes of the beam's trans-verse side modes impose perturbations that cannot be removed by spatial filtering. These quantum fluctuations can lead to the filamentation of a beam with an other-wise perfect wavefront. For any given nonlinear optical medium, this phenomenon imposes a fundamental limit on the maximum intensity of a laser beam that can propagate without the occurrence of filamentation [18].

This effect can be modeled by using a paraxial quantum statistical theory to investigate the interaction of a plane wave with the other modes that are in the vacuum state. These vacuum side modes can couple with the populated mode in a four-wave mixing interaction. It can be shown that side-modes with a transverse wavevector magnitude above $q_{max} = k\sqrt{4n_2 I/n_0}$ simply display oscillatory be-havior. Those with transverse wavevector magnitudes below that value experience exponential growth. These are the side modes which will lead to filamentation. The value of the nonlinear phase shift B_{thr} at which this quantum filamentation process produces significant changes in the wavefront is determined by the size of the non-linear index of refraction and the intensity and wavelength of the initial light. The shorter the wavelength and the larger the nonlinearity, the smaller the phase shift at which quantum-initiated filamentation develops (Fig. 5). This is simply a return to the stimulated scattering picture of Chiao, Kelley, and Garmire [13], but with perturbations caused by the vacuum field rather than by wavefront imperfections.

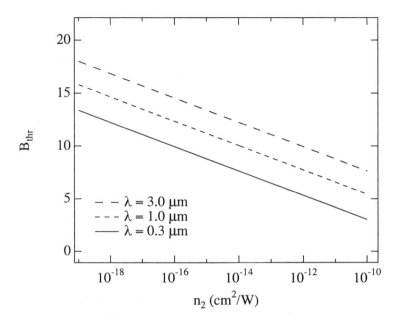

FIGURE 47.5. The value B_{thr} of the nonlinear phase shift at which filamentation occurs is shown as a function of n_2 for various wavelengths.

The area of self-trapped filaments [3] was to reappear in a different guise, that of spatial solitons, when in 1973 Zakharov and Shabat [19] pointed out the connection between the propagation of a pulse in a medium with both group-velocity dispersion and a Kerr nonlinearity, and the variation in the transverse profile of a beam undergoing both diffraction and self-focusing. For the case in which only one transverse dimension is allowed to evolve, the mathematics of the two problems are identical. Diffraction is analogous to anomalous group-velocity dispersion; self-focusing is analogous to self-phase modulation. Hasegawa and Tappert [20] demonstrated that group-velocity dispersion and self-phase modulation could be balanced to produce pulses whose shape did not vary with propagation. Consequently, one would expect to find that diffraction and self-focusing could be balanced to produce beams whose profiles did not vary. These beams were called spatial solitons, making their connection to temporal solitons explicit.

Solitons may exist in any of three dimensions and may be either bright or dark. Bright spatial solitons [19,28-30] are the analogue of bright temporal solitons [20]. They have a hyperbolic-secant shaped transverse profile, which is invariant under propagation. In addition to the $(1+1)D$ case (i.e. the (1 transverse + 1 longitudinal) dimensions case, where there is one dimension in which the soliton evolves and one dimension in which the soliton propagates) there also exist $(2+1)D$ [21] and $(3+1)D$ [21] generalizations. These solutions are confined in two and three dimensions, respectively. In the bright $(1+1)D$ and $(2+1)D$ cases, these spatial solitons are none other than the solutions proposed by Chiao, Garmire and Townes [3,22]; see Eqs. (47.1-47.3) above. Dark spatial solitons [23,25-27] are the expected analogue of their temporal counterparts. In the temporal case [20], they are possible in regimes of positive n_2 and normal GVD. In the spatial soliton case, it is necessary for the material to have a negative (self-defocusing) nonlinear index of refraction. Dark solitons in one transverse dimension are characterized by a dip in the field amplitude centered on a bright background and a phase jump of π across their center. Solitons that are circularly symmetric ($(2 + 1)D$) [23] may be characterized as optical vortices [24] and consist of a dark "hole" in a bright background with a helical phase variation across their azimuthal transverse profile.

Much of the experimental work to date has focused on dark solitons since they are stable when confined in one ($(1 + 1)D$), two ($(2 + 1)D$), or three ($(3 + 1)D$) dimensions. Both $(1 + 1)D$ [25,26] and $(2 + 1)D$ [23] dark spatial solitons have been demonstrated in a variety of materials. The methods of creating the necessary spatial profile vary, from using phase masks [23] to wire meshes [27]. Because of the instability of the $(2 + 1)D$ and $(3 + 1)D$ bright solitons in Kerr media, only the $(1 + 1)D$ bright solution has been demonstrated [28]. This has been done in a range of materials including nonlinear glass waveguides [28], liquid carbon disulfide [29] and semiconductor materials [30]. In some experiments, one-dimensional behavior was induced by using a waveguide geometry to confine one transverse dimension [28,30]. In other experiments with bulk materials, one-dimensional behavior was obtained by using a cylindrical lens to create a highly elliptical spot [29]. The smaller dimension evolves over a much smaller distance than the larger dimension, allowing effectively one-dimensional propagation.

Researchers have also broadened their scope to include non-Kerr law materials. An area of recent activity has been the study of photoreactive materials. The propagation of both bright [31] and dark [32] spatial solitons has been studied. Because the mechanism involved is photorefractive two-wave mixing rather than the Kerr effect, there are some significant differences in the properties of these solitons as compared to conventional solitons. The photorefractive solitons are independent of absolute light intensity and may be observed at powers as low as 10μW. Additionally, in both the bright and dark cases the beam can be trapped in both transverse dimensions. This contrasts with Kerr $(2 + 1)D$ bright solitons, which are unstable. Another alternative material class is that of cascaded $\chi^{(2)}$ materials [33,34]. In these materials the interaction of the fundamental with the generated second-harmonic is used to confine the waves. The solitary wave solutions found for these materials are also stable in the higher dimensions.

The demonstration of "light bullets" or $(3 + 1)D$ bright solitons has been a goal of many researchers. In a Kerr-law material, these solitons are unstable and undergo a collapse. Silberberg [21] proposed in 1990 that if the material were saturable, a symmetric collapse could occur, leading to the formation of stable light bullets. Other researchers have gone on to consider the formation of bistable light bullets [35] and have proposed types of nonlinear materials which would support this class of solitary waves [36].

In addition to simply demonstrating the existence of various types of solitons, researchers have begun to show the utility of spatial solitons in the area of optical couplers and all-optical switches. The waveguide-like profile induced in a nonlinear material by bright [37] or dark [38] $(1 + 1)D$ spatial solitons can be exploited as a waveguide for another pulse. The steerable nature of these waveguides has been demonstrated [29,38]. Additionally, soliton interactions may provide a mechanism for optical switching [39]. Consider an aperture arranged so that it passes a signal soliton propagating alone. An asymmetric collision with a properly directed control soliton will deflect the signal so that it is blocked by the aperture. In this way, the soliton dragging effect can be used to create an optical switch.

In conclusion, self-action effects are a mixed blessing, proving to be a hindrance in some settings while providing opportunities for new science in others. Even air has a nonlinear refractive index sufficiently large [40] for self-trapping effects to be important for the propagation of femtosecond pulses [41]. Self-action effects can provide a test bed for fundamental studies of the interaction of light with matter and also provide hope for uses such as optical switching. We are grateful to Professor Townes for his work in establishing this area of research. This work was supported by the U. S. Army Research Office and by the National Science Foundation.

References

[1] A pedagogical discussion of self-focusing and self-trapping is presented in Chapter 6 of R. W. Boyd, Nonlinear Optics (Academic, San Diego, 1992).

[2] M. Hercher, J. Opt. Soc. Am. **54**, 563 (1964).

[3] R. Y. Chiao, E. Garmire, and C. H. Townes, Phys. Rev. Lett. **13**, 479 (1964); Erratum, *ibid.* 14, 1056 (1965).

[4] G. A. Askar'yan, Zh. Eksp. Teor. Fiz. **42**, 1567 (1962); English translation: Sov. Phys. JETP **15**, 1088 (1962).

[5] The correction factor of 1/2 introduced by H. A. Haus, Appl. Phys. Lett. **8**, 128 (1966) has been included.

[6] P. L. Kelley, Phys. Rev. Lett. **15**, 1005 (1965); Erratum, *ibid.* **16**, 384 (1966).

[7] N. Bloembergen and Y. R. Shen, Phys. Rev. Lett. **13**, 720 (1964); P. Lallemand and N. Bloembergen, Appl. Phys. Lett. **6**, 210 (1965); F. J. McClung, W. G. Wagner, and D. Weiner, Phys. Rev. Lett. **15**, 96 (1965).

[8] E. L. Dawes and J. H. Marburger, Phys. Rev. **179**, 862 (1969); See also J. H. Marburger, Prog. Quantum Electron. **4**, 35 (1975).

[9] For a discussion of the moving focus model, see J. H. Marburger, Prog. Quantum Electron. **4**, 35 (1975) and Y. R. Shen, Prog. Quantum Electron. **4**, 1 (1975).

[10] E. Garmire, R. Y. Chiao, and C. H. Townes, Phys. Rev. Lett. **16**, 347 (1966).

[11] R. Y. Chiao, M. A. Johnson, S. Krinsky, H. A. Smith, C. H. Townes, and E. Garmire, IEEE J. Quantum Electron. **QE-2**, 467 (1966).

[12] M. M. T. Loy and Y. R. Shen, Phys. Rev. Lett. **22**, 994 (1969).

[13] R. Y. Chiao, P. L. Kelley, and E. Garmire, Phys. Rev. Lett. **17**, 1158 (1966).

[14] V. I. Bespalov and V. I. Talanov, Zh. Eksp. Teor. Fiz. Pis'ma **3**, 471 (1966); English translation: JETP Lett. **3**, 307 (1966).

[15] R. L. Carman, R. Y. Chiao, and P. L. Kelley, Phys. Rev. Lett. **17**, 1281 (1966).

[16] E. S. Bliss, D. R. Speck, J. F. Holzrichter, J. H. Erkkila, and A. J. Glass, Appl. Phys. Lett. **25**, 448 (1974).

[17] For extensive work on other systems in nonlinear and quantum optics where quantum noise initiates the process, see I. A. Walmsley and M. G. Raymer, Phys. Rev. Lett. **50**, 962 (1983); R. Glauber and F. Haake, Phys. Lett. **68A**, 29 (1978); G. S. Agarwal and R. W. Boyd, Phys. Rev. A **38**, 4019 (1988); M. D. Reid and D. F. Walls, Phys. Rev. A **34**, 4929 (1986); and M. Sargent III, D. A. Holm, and M. S. Zubairy, Phys. Rev. A **31**, 3112 (1985).

[18] E. M. Nagasako, R. W. Boyd, and G. S. Agarwal, submitted for publication. A related paper "Quantum fluctuations in nonlinear optical self-focusing," by E. M. Wright has appeared in *Chaos, Solitons, & Fractals* **4**, 1805 (1994).

[19] V. E. Zakharov and A. B. Shabat, Zh. Eksp. Teor. Fiz. **61**, 118 (1971); English translation: Sov. Phys. JETP **34**, 62 (1972).

[20] A. Hasegawa and F. Tappert, Appl. Phys. Lett. 23, 142 (1973); *ibid.* **23**, 171 (1973).

[21] Y. Silberberg, Opt. Lett. **15**, 1282 (1990).

[22] Silberberg (Ref. 21) notes that the self-trapped beam power he calculates differs by 1.5% from the value given in Ref. 3.

[23] G. A. Swartzlander, Jr. and C. T. Law, Phys. Rev. Lett. **69**, 2503 (1992).

[24] P. Coullet, L. Gil, and F. Rocca, Opt. Comm. **73**, 403 (1989).

[25] D. R. Andersen, D. E. Hooten, G. A. Swartzlander, Jr., and A. E. Kaplan, Opt. Lett. **15**, 783 (1990).

[26] G. R. Allan, S. R. Skinner, D. R. Andersen, and A. L. Smirl, Opt. Lett. **16**, 156 (1991).

[27] G. A. Swartzlander, Jr., D. R. Andersen, J. J. Regan, H. Yin, and A. E. Kaplan, Phys. Rev. Lett. **66**, 1583 (1991).

[28] J. S. Aitchison, Y. Silberberg, A. M. Weiner, D. E. Leaird, M. K. Oliver, J. L. Jackel, E. M. Vogel, and P. W. E. Smith, J. Opt. Soc. Am. B **8**, 1290 (1991).

[29] M. Shalaby and A. Barthelemy, Opt. Comm. **94**, 341 (1992).

[30] A. Villeneuve, J. S. Aitchison, J. U. Kang, P. G. Wigley, and G. I. Stegeman, Opt. Lett. **19**, 761 (1994).

[31] G. C. Duree, Jr., J. L. Shultz, G. J. Salamo, M. Segev, A. Yariv, B. Crosignani, P. Di Porto, E. J. Sharp, and R. R. Neurgaonkar, Phys. Rev. Lett. **71**, 533 (1993).

[32] G. Duree, M. Morin, G. Salamo, M. Segev, B. Crosignani, P. Di Porto, E. Sharp, and A. Yariv, Phys. Rev. Lett. **74**, 1978 (1995).

[33] K. Hayata and M. Koshiba, Phys. Rev. Lett. **71**, 3275 (1993).

[34] L. Torner, C. R. Menyuk, W. E. Torruellas, and G. I. Stegeman, Opt. Lett. **20**, 13 (1995).

[35] D. E. Edmundson and R. H. Enns, Opt. Lett. **17**, 586 (1992).

[36] R. H. Enns and S. S. Rangnekar, Phys. Rev. A **45**, 3354 (1992).

[37] R. de la Fuente, A. Barthelemy, and C. Froehly, Opt. Lett. **16**, 793 (1991).

[38] B. Luther-Davies and X. Yang, Opt. Lett. **17**, 1755 (1992).

[39] S. Blair, K. Wagner, and R. McLeod, Opt. Lett. **19**, 1943 (1994).

[40] D. M. Pennington, M. A. Henesian, and R. W. Hellwarth, Phys. Rev. A. **39**, 3003 (1989).

[41] A. Braun, G. Korn, X. Liu, D. Du, J. Squier, and G. Mourou, Opt. Lett. **20**, 73 (1995).

48

Industrial Research in Today's World

A. Penzias

As luck would have it, I decided to learn optics at just the same time that Charlie Townes did—he by teaching the course at Columbia, me by taking it. Almost ready to quit physics in a moment of panic during that course's uniquely-imaginative open-book final, I managed to garner a respectable grade together with a lifelong admiration for the scientist whom I am privileged to call my teacher.

While engaged in research which would revolutionize optics—and several other fields besides—Charlie also pursued his long-standing interest in radio astronomy. Assigned by him to adapt his maser technology to 21-cm hydrogen line research, I thought of astronomy as a momentary diversion from my intended career path in applied science. Charlie advanced my career along both these dimensions by introducing me to astronomy as my thesis advisor, as well as by introducing me to Bell Labs—his former employer.

Like Charlie, I got my first job at Bell Labs, but unlike him, I have yet to take the next step. My "momentary diversion" in astronomy lasted for over twenty years— highlighted by the use of a low-noise amplifier based upon Charlie's invention to pinpoint a cosmic source of background radiation, as well as the detection of several species of interstellar molecules predicted by Charlie in a prescient paper written during a sabbatical in Paris.

With almost thirty-five years of Bell Labs service to my credit, the respective lengths of time spent on astrophysics and applied sciences have now come into rough parity. In that spirit, I offer the following item, from the latter phase in my research career, to Charlie's Festschrift.

Call it poetic justice. Year after year, we scientists and engineers have transformed life and work around the globe through a dazzling flow of technological advances. But now, like most of our fellow citizens, we find a rapidly-changing environment pulling us along—rather than the other way around.

We live in interesting times, to say the least. Small wonder that many look back upon the past with nostalgia. In retrospect, the '50s, '60s and '70s look like a golden era. Having emerged from the Second World War with a booming economy when everyone else's lay in ruins, the U.S. commanded fully half of the gross product of our planet, and the seemingly-assured prospect of further growth.

Full speed ahead seemed the obvious course in those days. More of the same, in other words, only better because of past experience. Even with a half-century's hind-sight, we can hardly call that optimism unfounded. If a properly organized and financed team could build an A-Bomb, what might post-war America accomplish with a whole slew of smaller-scale Manhattan Projects?

And early experience proved the optimists right. What a list: the transistor, computers, television, jet engines, antibiotics, and a host of lesser-known successes. Furthermore, this chain of successes persisted for more than a quarter of a century, culminating in the success of Project Apollo. Well into the 1970's many an American's world introduced a rehearsal of their favorite gripe with the words: "If we can put a man on the moon, what can't we....?"

What did it take to produce a practical transistor, or Neil Armstrong's footprint on a dusty lunar landscape? A well-led and properly supported team of specialists focused on a challenging goal. Time after time this model proved a sure-fire recipe for success.

But nothing lasts forever—in part because most organizations last too long. The massive internally-focused corporate hierarchies which blossomed for sound economic reasons in the fifties and sixties persisted intact into the nineteen eighties, long after the rules of the game had changed.

Since it's easier for humans to see the shortcomings of internal focus in others, few scientists questioned the value of research for its own sake, and neither did the folks who funded them. With another transistor just around the corner, leaving researchers alone to do their thing seemed a profitable course of action in many cases—especially since few outsiders could really understand the specialized concepts and language involved.

Looking ahead however, organizing industrial research around disciplines as was done in the 1960's—the physicists in one group and the electrical engineers in another, say—cannot support the external focus which future challenges demand. Today, we see much of industrial research organized by technology, such as electronics or software engineering, for instance. Far better than discipline-based compartmentalization, but still not good enough. These days, our Business Unit colleagues and their customers think and speak in terms of applications—devices, products and systems—rather than the technologies which underlie them.

In that spirit, managing industrial research on the basis of applications, rather than by technology or discipline, can help to strengthen the customer-focused partnership so vital to success in a competitive world. But how does this affect the people involved? Originally trained in some discipline, and experienced in one or more technologies, how can individual researchers find meaningful careers in today's fast-changing environment?

Herein lies a great challenge to research management. In the past, an industrial researcher could aspire to promotion to management just by being a top notch scientist—leading by example and rating the organization's performance after the fact. But no longer. Tomorrow's research managers and the people they lead, cannot wait to find how some project turns out before assessing its worth. Instead, research leaders must become skilled in evaluating the potential value of research projects, and engaging both customers and associates in that evaluation process. That way, each member of the group not only works on the "right stuff," but also understands the risks and rewards associated with that work.

In this more-open work environment, individual researchers should be able to "vote with their feet" by transferring to whatever task best matches their own needs and skills. Contrast this with the "freedom" offered by old-style discipline-based organizations. In the old days, newly-hired researchers were left free to find their own projects, and could look forward to more of the same as long as they continued to meet an implicit standard of excellence. From the point of view of the survivors, this hidden-rules game worked pretty well. Of course, managers insured success by recruiting people like themselves—frequently on the recommendation of their own thesis advisors. A practice more suited to publishable results than entry into risky new fields.

By contrast, the more-open work environment I have in mind depends upon three complementary elements:

a. Management prioritization: In concert with colleagues from corporate Business Units, as well as input from individual researchers, research managers conduct an ongoing evaluation of all the work in their organization, ensuring an appropriate balance between the needs of radically new ventures and the healthy growth of existing applications.

b. Open dialogue: Researchers and their managers build a shared understanding of the value of each undertaking, its likely chances of success, and how the risks involved will be apportioned. For example, management might ask one individual to undertake a risky project for the good of the company, while another might insist on pursuing a long-shot with only reluctant support from above.

c. Job posting: All openings get advertised within the research organization. So that individuals can make appropriate career moves as smoothly as possible, management helps by providing training and other support. In this environment, a management-sponsored program of regular career-planning dialogues helps individuals to aim in desired directions and gain whatever management aid might be available.

As a result, the organization, and the individuals within it can adapt to a changing environment.

In the past, research managers might concern themselves with technical issues associated with a particular project and leave business issues to others. In the new environment however, managers must ask themselves questions such as:

Who are the customers for this work? Do they exist within AT&T? If so: What is the degree of their buy-in? What role did they play in project selection? If not: How did you determine the need for this work? How do you expect to bring it to market?

When do you expect results? When will these results be needed and/or most profitably applied? Who will be your likely competitors? What will they offer the market in the same time frame?

What will define success? What is the probability of that success? What benefits will it bring? How much will it cost to produce? What else might one have done with these resources? What would it have cost to obtain the results by other means?

While some may see such constraints as an impediment to the highest levels of creativity, I claim that quite the contrary holds true. Anyone can solve problems in an environment in which investigators can set boundary considerations to please themselves, after all. Personally, I prefer the challenges which real-world problems offer, but I'll have to admit that my enthusiasm does get me into trouble some times.

A couple of years ago, I got a lot of flack for a "formula" I created in response to the following question in a session on personal advancement: "How can I tell if the project I'm working on is likely to advance my career?" Exercising my lifelong habit of estimating answers to life's puzzles, I produced the following formula:

$$(\text{Figure of merit}) = (\text{Reward}) \times (\text{Probability of Success})/(\text{Cost})$$

and advised my listeners to seek activities which appear to reward their investments of time and resources to the greatest degree.

I went on to explain "Reward" as the value that our corporation would gain from products or services based upon a project's output, and "Cost" as number of person-years and other resources needed—all denominated in dollars.

"Probability of success" is the product of the likelihood: a) that the research yields the desired outcome; b) that a Business Unit adopts the resulting technology; and c) That the Business Unit succeeds in bringing the technology to market.

Even as a thought experiment, few in my audience seemed willing to entertain this idea. The task seemed too daunting. Even the ones intrigued enough to discuss the idea with their Department Heads generally came away discouraged. Find customers in advance, get them to opine on their future technology needs, and evaluate the success of offerings they had not even begun to think about yet? No way!

No one else seemed to like it, either. While admitting that individual researchers needed to be aware of such matters, most of my managerial colleagues feel that asking everyone to pursue business issues to this level of detail would exact too great an overhead upon individual time and energy. One hires researchers to conduct research, after all.

As of this writing, however, I note a grass roots revival of such notions. Dialogues with customers concerning needs, long-range plans and the risks involved no longer seem quite as far-fetched as they did only a couple of years ago. While most such efforts center upon management-driven prioritization of work portfolios, a small but growing number of individual scientists and engineers are using the back of an odd envelope or two as a means of assessing their career paths. Quite a contrast to the days when an "Organizations Man" could merely cede such concerns to the corporation.

In the final analysis, individuals own ultimate responsibility for their careers, and should assure themselves of an optimal mix of their own aspirations and larger business needs of the corporation. Having made the world a more "interesting" place with our technology, we researchers can't help living in interesting times ourselves, after all.

49

Far-infrared Imaging of the HII Region-Molecular Cloud Complex W51A with a Balloon-borne Telescope

Francesco A. Pepe
Roland Brodbeck
Daniel Huguenin
and Fritz K. Kneubühl

49.1   Introduction

Star formation is a field of high interest in astrophysics. In recent years, considerable effort has been made in order to understand the processes of star formation. Molecular clouds, in which star formation takes place, and dust associated with these clouds extinguish all short-wavelength radiation up to the near-infrared. Consequently, protostars and young stars cannot be observed directly and, therefore, one is restricted to study the dust and gas surrounding them. By means of continuum radio astronomy free-free emission from compact or more extended sources can be mapped. These maps trace regions in which gas is highly ionized by ultra-violet (UV) radiation emitted by nearby or embedded stars. Quantitative measurements imply restrictions to the number and the spectral type of the stars. Studies on molecular line emission provide additional information concerning physical conditions, composition and dynamics of the clouds. These lines are most frequent in the submillimeter- and millimeter-wavelength region. By contrast, dust emission is continuous and is observed in the far-infrared (FIR). Its broad-band spectrum can be approximated by a gray-body function with an emissivity which varies slightly with wavelength. The bulk of this radiation falls into the wavelength range between 30 μm and 300 μm.

Unfortunately, far-infrared measurements suffer serious limitations due to the Earth's atmosphere. Water vapour absorbs the far-infrared radiation, and emits strongly at these wavelengths. Balloon-borne and air-borne telescopes on the one hand, and satellites on the other hand, solve this problem. As a consequence, telescopes employed for this purpose are restricted in diameter. Typical spatial resolutions achieved are > 1 arcmin. This is in contrast to ground-based radio observations. Large telescopes and interferometers map HII regions with arcsecond or even subarcsecond resolution. Far-infrared measurements do not provide such detailed analyses of small-scale structures. Nevertheless, one can determine the mass and total luminosity of the cloud with high accuracy. These are calculated on the basis of the measured temperature and optical depth. In addition, comparison with measurements of the extended free-free and molecular line emission reveals whether there is any spatial association among dust, ionized gas, and molecular gas at the medium scale.

In this paper we report on the results of the balloon flight MIL 1 of our far-infrared 60 cmØ telescope equipped with liquid-helium cooled spectral filters and a composite silicon bolometer [1–3]. This flight took place in Aire-sur-l'Adour, France, on September 20-21, 1993. Intensity maps of the star-forming region W51A measured at 130 μm and at 260 μm wavelength as well as several source parameters of W51-I are presented. None of these wavelengths has been covered by the US satellite IRAS in 1983 [4]. Therefore, there is still a lack of data between 100 μm, i.e., the longest IRAS wavelength taken into account by the spectral filters of IRAS, and the atmospheric window at 350 μm wavelength. Our measurements fill this gap in the specific case of W51A.

49.2   The HII Region—Molecular Cloud Complex W51

W51 is one of the most luminous giant HII region—molecular cloud complexes in the Milky Way. It is situated in the Sagittarius spiral arm at a galactic longitude of 49.5°. Its mean distance to the sun is 7.5 kpc. It is located at a galactocentric distance of 8 kpc and 53 pc above the galactic plane. The radio-continuum observations by Goss and Shaver [5] shown in Fig. 49.1 demonstrate that the HII-region complex consists of the two major groups, W51A and W51B, which are embedded in an extended low-density ionized gas region. Their 5 GHz map reveals in W51A two strong sources labeled G49.5-0.4 and G49.4-0.3 and two others in W51B, G49.2-0.3 and G48.9-0.3. In contrast to W51C, which is a non-thermal source, all of these HII regions find their counterpart at far-infrared wavelengths. Rengarajan et al. [6], who mapped the entire region in the band ranging from 40 μm to 120 μm wavelength, denoted them W51-I, W51-II, W51-III and W51-IV, respectively. Since there is good spatial agreement between the measured far-infrared emission and radio-continuum emission, Rengarajan et al. [6] conclude that ionized gas and FIR-emitting dust are closely associated at large scale and powered by the same stellar association. In addition, the similarities of velocities in recombination lines and molecular emission lines indicate that the various HII regions and the molecular clouds observed are not a mere superposition, but are spatially associated. Because at this galactic longitude the line of sight is nearly parallel to the Sagittarius arm, superposition does not necessarily mean that these clouds are spatially associated with each other. In order to improve the analysis of the structure of the molecular clouds, Pankonin et al. [7] have investigated the radial

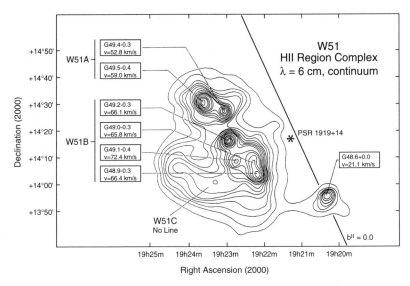

FIGURE 49.1. Overview of the W51 HII-region complex. The 6 cm-wavelength continuum contours are from Goss and Shaver [5]. The H109α recombination line velocity shown at each continuum peak has been taken from Wilson *et al.* [1970]. The position of PSR 1919+14 is from Hulse and Taylor [1974]. The map is adapted from Mufson and Liszt [33].

velocity of radio recombination lines at several positions in the W51 complex. They came to the conclusion that W51 can be separated in three velocity groups: W51B (66 kms^{-1}), G49.5-0.4 (58 kms^{-1}) and G49.4-0.3 (53 kms^{-1}). Although these groups do not appear to be directly related spatially, it has been suggested [8] that there is a high-velocity molecular gas stream, of the same velocity as W51B and probably associated with it, passing between G49.5-0.4 and G49.4-0.3.

With respect to the two major components of W51, the complex W51A has been studied more extensively. By centimeter-continuum observations, Martin [9] has found various compact radio sources in G49.5-0.4, which he labeled W51a-h. The diameters of these compact HII regions range from 10$''$ to 50$''$. Lightfoot *et al.* [10] have discovered far-infrared counterparts to most of these HII regions. By taking into account diameter, radio, and far-infrared emission they assigned these sources to different stages of star formation, which seem to evolve with increasing distance to the east of W51d. Observations made by Gaume *et al.* [11] with the Very Large Array (VLA) telescope of the National Radio Astronomy Observatory (NRAO) at its highest resolution reveal that each of the compact HII regions mentioned above is in turn composed of several ultracompact HII regions of subarcsecond size.

W51A is an active region of massive star-formation. OH and H$_2$O masers, which trace active regions within star-forming complexes have been found near W51d and e [12, 13]. W51 d and e are also associated with the extremely bright near-infrared sources IRS2 and IRS1, respectively. These are probably regions of star formation at an early stage. Moving to the east, star formation seems to evolve progressively to later stages. Lightfoot *et al.* [10] interpreted this situation as an evidence for sequential star formation. There are various mechanisms known which may cause massive star formation to follow a sequential path. On the one hand, star formation could be activated by shocks produced by multiple radiation-driven implosions arising from massive O stars [14]. On the other hand, Pankonin *et al.* [7] proposed that star formation in W51A may have been triggered by the collision of the molecular clouds with the cold gas stream blowing through the complex. Star formation would then take place at the interface of the high-velocity gas stream and the molecular clouds. Both scenarios would lead to the formation of massive stars while inhibiting the formation of low-mass stars.

There are structural differences between W51A and W51B, yet they may be due to different evolutionary states of the groups. W51B appears to be more evolved than the other two groups (G49.5-0.4 and G49.4-0.3) of W51A. It does not show the usual indicators of current star formation which occur in the other two groups, i.e., OH and H_2O maser sources, compact HII regions, and near-infrared sources.

49.3   Instrumentation

The experimental equipment employed in our experiments is that used by Holenstein *et al.* [1] and Schenker *et al.* [2] in previous flights. It consists of a 60 cmØ Cassegrain telescope integrated in a balloon-borne gondola, which has been designed and built in collaboration with the Geneva Observatory [3]. The telescope control system provides a pointing accuracy $\Delta\psi$ of 17 arcsec and a pointing stability $\delta\psi$ of about 3 arcsec rms, both considerably smaller than the telescope's field of view. The theoretical field of view FOV_{th} has been selected to match the diffraction limit of our telescope at 260 μm wavelength. It amounts to 118 arcsec, if an effective primary diameter $D_E = 550$ mm is assumed when taking into account the undersized secondary mirror.

A liquid-helium cooled photometer attached to this telescope serves as detector of the incoming FIR radiation. This radiation is filtered by one of three selectable metal-mesh band-pass filters with central wavelengths λ_{center} at about 80 μm, 130 μm and 260 μm. The precise wavelengths are incorporated in Table 49.2. These filters have relative bandwidths $\Delta\lambda/\lambda$ of 11%, 13% and 25%, respectively. These are considerably narrower than those of the filters used in other experiments, e.g., with the satellite IRAS [4] or balloons [6, 10]. The radiation is detected by a composite silicon bolometer working at a temperature of 1.7 K during measurements. It exhibits a noise equivalent power (NEP) of about $1.4 \cdot 10^{-14}$ WHz$^{-1/2}$. This value is significantly higher than the background NEP from the warm instrument and the atmosphere, which was estimated by Schenker [15] as $NEP_{bg} \approx 8 \cdot 10^{-16}$ WHz$^{-1/2}$. Thus, the detectivity is limited definitively by the detector.

In order to remove the high offset of the detector signal produced by background radiation from sky, atmosphere, and telescope it is necessary to chop the signal by beam switching. This switching was achieved by wobbling the secondary mirror

TABLE 49.1. Telescope Parameters

Effective Primary Mirror Diameter	D_E	550 mm
F-Number of the Telescope	F	15
Theoretical Field of View	FOV_{th}	118 arcsec
Measured Field of View	FOV_m	117 arcsec
Diffraction Limit at 250 μm	Θ_D	118 arcsec
Pointing Accuracy	$\Delta\psi$	17 arcsec
Pointing Stability	$\delta\psi$	4 arcsec
Chopping Frequency	ν_{ch}	8 Hz
Chopper Throw	α_{ch}	$\pm$ 5 arcmin

TABLE 49.2. Photometer Parameters

System	NEP_{SYS}	$1.4 \cdot 10^{-14}$ WHz$^{-1/2}$
Central Wavelengths:		
Filter No. 1	λ_{center}	141.0 μm
Filter No. 2	λ_{center}	250.8 μm
Band Widths:		
Filter No. 1	$\Delta\lambda$	18 μm
Filter No. 2	$\Delta\lambda$	61 μm

in the scan direction at a frequency of $\nu = \hat{E}8$ Hz. The chopping angle between the two beam directions and the optical axis of the telescope was $\alpha_{ch} = \hat{E}\pm5$ arcmin. A detailed description of the complete instrumentation is given by Holenstein *et al.* [1] and by Schenker *et al.* [2]. All relevant telescope and photometer parameters are summarized in the Tables 49.1 and 49.2.

49.4   Balloon Flight MIL 1

The balloon flight MIL 1 took place in the night of September 20/21, 1993 at the balloon facility of the "Centre Nationale d'Etudes Spatiales" (CNES) in Aire-sur-l'Adour, France. The gondola was launched on September 20 at 17.45 UT. A 380,000 m^3 stratospheric hydrogen balloon carried our telescope during an ascension time of 3 hours to a height of 3.5 HPa, which corresponds to about 39 km. Since the stratospheric winds were strong, the balloon drifted rapidly eastward. We could not expect an observation time as long as during the flight Cygnus III [2], where the wind conditions were exceptionally good. Our measurements started at 21:00 UT. After some initial positioning problems caused by a slight misalignment of the telescope and the star tracker, we were able to point the telescope on Saturn. We mapped Saturn at both wavelengths, 260 μm and 130 μm. These observations allowed us to determine the precise direction of the optical axis of the telescope, as well as to verify the chopping angle α_{ch} and whether the measured field of view corresponded to the theoretical field of view FOV_{th}. In addition, this operation served to perform the photometric calibration. The Saturn measurements took about 50 minutes. After calibration, at about 22:15 UT, we mapped W51A at 260 μm wavelength during 45 minutes. This measurement was followed by that on W51A at 130 μm wavelength. At 1:00 UT the telescope was turned in its vertical position and the gondola was separated from the balloon, which was drifting towards the Mediterranean. The gondola landed smoothly half-an-hour later near Limoux, about 20 km from Carcassone. It was recovered and returned to CNES on the same day. No visible damage resulted from the landing, except for the primary mirror, which was wet because of rain.

49.5   Calibration on Saturn

Since we cannot perform absolute intensity measurements with our instrument we have to calibrate it on a well-known source. Planets are the best candidates for this purpose. The only visible planet during the flight MIL 1 was Saturn. The diameter of Saturn is much smaller than our telescope's field of view. Thus we can treat Saturn as a point source.

In order to perform the calibration we have observed the continuum emission of Saturn at both wavelengths, 130 μm and 260 μm. Fig. 49.2 shows a single line scan over Saturn at 260 μm wavelength. The detector signal is plotted as a function of the horizontal beam position in the startracker coordinate system. Since we chop the signal by means of the beam-switching technique described in Section 49.3, the response is proportional to the difference of the flux at the left and the right beam position. When the right beam position is passing over the source we have maximum negative signal and vice versa. Therefore, the zero position of the telescope axis is given by the center between the negative and the positive peak. The distance between the two peaks determines the total chopping angle. By taking into account that one startracker step is equivalent to 17 arcsec we can calculate that the in-flight total chopping angle was 9′38″, slightly less than the 10 arcmin expected. By means of a Gaussian fit to this observed profile we are also able to evaluate the FWHM of the beam. The measured 117 arcsec matches well the theoretical value of 118 arcsec. Figure 49.3a represents a surface plot of the measured intensity of Saturn at 260 μm wavelength. We have fitted this map by assuming a Gaussian beam profile described by the function

$$p(x, y) = A \cdot e^{-k[(x-x_0)^2+(y-y_0)^2]} , \tag{49.1}$$

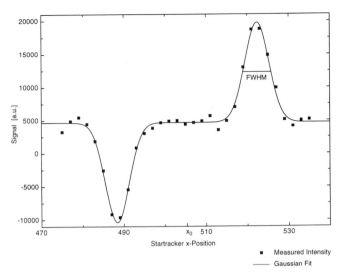

FIGURE 49.2. Single scan of the telescope over Saturn at 260 μm wavelength. The star-tracker x-direction of the scan is parallel to the horizon. A negative amplifier signal is measured when the right beam is passing over the source and vice versa. In order to evaluate the "real" signal a spatial integration has to be performed.

where x_0 and y_0 are the startracker coordinates of the telescope zero position and k determines the width of the Gaussian profile. The point spread function (PSF) results by setting A equal unity. This PSF (Fig. 49.3b) is essential when deconvolving the measured W51A maps by means of the deconvolution techniques of Lucy [16] which have been described by Schenker [15].

The intensity calibration relates the known Saturn flux density to the measured signal when pointing on Saturn. For the latter we have used the far-infrared data reported by Hildebrand *et al.* [17], which include the flux density of Saturn when the ring inclination to the earth was $< 1.7°$. During our measurements the ring inclination was only about 12°, a low value which does not affect the calibration. For the intensity calibration we define the converting factor CF as the quotient of the intensity to the deconvolved amplifier signal V_A of the measured object at a given telescope position:

$$I_A = CF \cdot V_A. \qquad (49.2)$$

For a known Saturn intensity $I_{\lambda_{Sat}}$ and the filter transmission function $t(\lambda)$ of the employed spectral filters [2] the converting factor CF results in

$$CF = \frac{G_{Sat}}{G_{Source}} \cdot \frac{\Omega_{Sat} \cdot \int I_{\lambda_{Sat}} t(\lambda) d\lambda}{\Omega_{FOV} \cdot \int t(\lambda) d\lambda} \cdot \frac{DF}{V_{Sat}}, \qquad (49.3)$$

where G_{Sat} and G_{Source} indicate the amplifier gains during the Saturn and the W51A measurements, respectively. Ω_{Sat} represents the solid angle subtended by Saturn, while Ω_{FOV} is the telescope's field of view. The signal of the on-source

a)

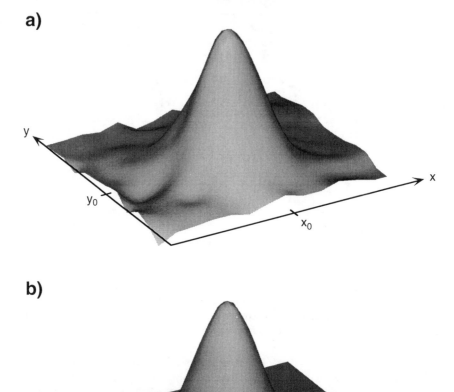

b)

FIGURE 49.3. Surface plot (a) of the measured 260 μm Saturn flux and (b) of the resulting point spread function (PSF) in startracker x- and y-coordinates. x_0 and y_0 denote the zero position of the telescope axis.

position when the telescope is centered on Saturn is given by V_{Sat}. This expression for CF needs to be corrected by the diffraction factor DF, which corresponds to the portion of the diffraction disk of a point source falling inside the field stop. The DF is about 0.47 at 260 μm and 0.94 at 130 μm wavelength.

The calculated converting factor is $CF_{260} = 0.63 \cdot 10^7$ Jy sterad^{-1} signal^{-1} for the 260 μm map and $CF_{130} = 2.4 \cdot 10^7$ Jy sterad^{-1} signal^{-1} for the 130 μm map. The intensity at each telescope position can be calculated by simply multiplying

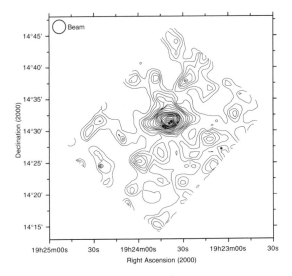

    • 80 μm peaks [Rengarajan et al. 1984]
    • 800 μm peak FIR1 [Sievers et al. 1991]
    • 400 μm peak [Jaffe et al. 1984]
    • 158 μm peak [Nikola 1993]

FIGURE 49.4. Intensity contour plot of the continuum emission from the W51A region at 130 μm wavelength. The contour levels are at 2.5, 5, 7.5, 10, 12.5, 15, 17.5, 22.5, 27.5, ..., 87.5 × 1000 MJy sterad^{-1}.

the deconvolved measured amplifier signal by the converting factor at the observed wavelength.

49.6   Imaging and Radiometry of W51A

49.6.1   Intensity Maps

The intensity maps of W51A have been made by observing in the raster scanning mode. The imaged field extends over a rectangle of 19×28 arcmin^2. Neighbouring telescope positions are separated by about 1 arcmin. The integration time at each position was about 4 sec. The differential maps have been spatially integrated along the wobbling direction of the secondary mirror which corresponds to the star tracker x-direction and subsequently deconvolved with the data reduction techniques described by Schenker [15]. Contour plots of the resulting intensity maps are shown in the Figs. 49.4 and 49.5. The circles in the upper left edge are 117 arcsec in diameter, marking the beam width of the telescope.

As expected, the strongest source in the W51A complex is the central peak denoted W51-I, at the position $\alpha(2000) = 19$h 23m 39s, $\delta(2000) = 14°\ 31'\ 32''$. This position corresponds well to that of the peak flux density measured by Rengarajan et al. [6], who mapped the continuum emission from the entire W51 cloud complex

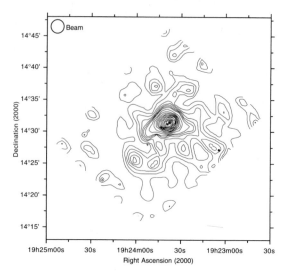

FIGURE 49.5. Intensity contour plot of the continuum emission from the W51A region at 260 μm wavelength. The contour levels are at 0.6, 1.2, 1.8, 2.4, 3, 3.6, 4.2, 5.4, 6.6, ..., 21 × 1000 MJy sterad^{-1}.

in a relatively broad spectral band ranging from 40 μm to 120 μm, with a central wavelength of about 80 μm. Good agreement with the position of peak emission is also found in the 400 μm map of Jaffe *et al.* [18], the 158 μm continuum map of Nikola [19] and with the millimeter-emission center FIR1 which has been mapped by Sievers *et al.* [20].

The other FIR source W51-II at $\alpha(2000)$ = 19h 23m 08s, $\delta(2000)$ = 14° 27′ 30″ is at both observed wavelengths much weaker than W51-I. While at 260 μm wavelength the peak position of W51-II observed by Rengarajan *et al.* [6] fits well in our map, there is a discrepancy at 130 μm between the two peak positions. Probably this is due to the lower spectral resolution of Rengarajan's maps, which results in an integration over different dust-temperature components of this source.

Two additional strong components are the sources lying southeast to W51-I. They do not occur in the IRAS map because of its lower spatial resolution. In this map they appear as an extended bump (Fig. 49.6). Their dust emission is as strong as in W51-II at 130 μm and even stronger at 260 μm.

It is evident that the 260 μm emission traces primarily colder regions and the 130 μm emission warmer regions in the complex. By comparing some features in the 130 μm map with those in the 260 μm map (Fig. 49.7) we can conclude that most of the contours do not represent noise but real sources. This is an additional indication for the morphological complexity of this star-forming region. Nevertheless, some sources appear to be at slightly different positions in the 260 μm and

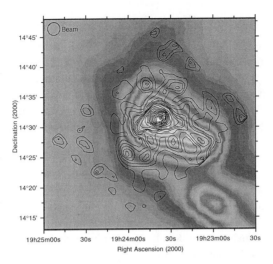

FIGURE 49.6. W51A. The intensity contour plot of the 260 μm-continuum emission is superimposed on a gray-scale image of the 100 μm-continuum emission derived from IRAS data.

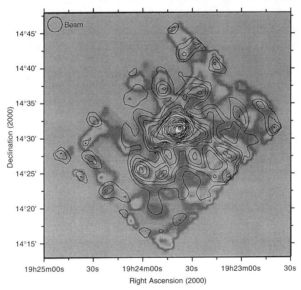

FIGURE 49.7. W51A. The intensity contour plot of the 260 μm-continuum emission is superimposed on a gray-scale image of the 130 μm-continuum emission.

the 130 μm map. This displacement may partially have its origin in the pointing error of the telescope, but as mentioned in a previous chapter the pointing error is less than 17 arcsec. More probably, the different temperature components within the complex cause a shift of the emission peaks at the analyzed wavelengths.

49.6.2   Dust Emission Spectrum

The W51-I dust-emission spectrum can be described by a three-component model, in which each component is associated with dust of a different temperature. The bulk of the dust emission has its origin in warm and compact sources. Typically, the emission maximum occurs near 100 μm. These compact sources are supposed to be embedded in a cold and more extended envelope, which is predominantly heated by the interstellar radiation field. Since both components are optically thin at the wavelengths where they emit the main portion of their radiation, the resulting emission corresponds to the sum of the emission from both components. This argument does not hold for the hot component, because for radiation with wavelengths shorter than 100 μm the dust becomes optically thick. Nevertheless, the assumption that the emission from the hot component is not influenced by the cold and warm dust is valid for one of the two following scenarios in W51-I: a) The emission originates in hot dust in the vicinity of the heating star. In this case the cooler dust envelope has to be highly fragmented, for the photons to escape. b) As suggested by Sievers *et al.* [20] the emission may not arise from "normal" dust but from very small grains located in the outer, tenuous layers of the molecular clouds. Thus, the grains are temporarily heated by energetic photons of the ISRF to temperatures of several hundred degrees.

In order to describe the total flux density of W51-I, we approximate the spectral emission from each component by a Planck function $B_\lambda(T)$, modified by a wavelength-dependent emissivity $\varepsilon(\lambda)$. Assuming that all components emit independently from each other, the total flux density S_λ at the wavelength λ from the whole cloud can be written as linear combination

$$S_\lambda = \sum_{i=1.3} \Omega_i \cdot \varepsilon(\lambda) \cdot B_\lambda(T_i) , \tag{49.4}$$

where $\varepsilon(\lambda) = 1 - e^{-\tau_\lambda}$. Ω_i denotes the solid angles of the dust components, T_i their temperatures and $\varepsilon(\lambda)$ the wavelength-dependent grain emissivity. $\varepsilon(\lambda)$ is related to the optical depth τ_λ. The wavelength dependence of the optical depth of most star-forming regions is described [21, 22] by

$$\tau_\lambda = \left(\frac{\lambda_c}{\lambda} \right)^\beta , \text{ with } \beta \approx 2.0 , \tag{49.5}$$

where λ_c represents the critical wavelength where the optical depth equals unity.

We have fitted Eq. (4) to our data and to those of other authors [20]. The measured flux densities of W51-I as well as the calculated spectrum are shown in Fig. 49.8. Since there are several fit parameters involved in this calculation which are not independent of each other, there may be large uncertainties in the results. Therefore, we first fitted the data in the wavelength region ranging from 50 μm to 160 μm. In this range the spectrum is expected to be dominated by the emission of the warm dust component, so that the other components can be neglected in this fit. Subsequently, we fitted the short-wavelength and the long-wavelength regions by including in the fit function the hot and the cold component, respectively, while

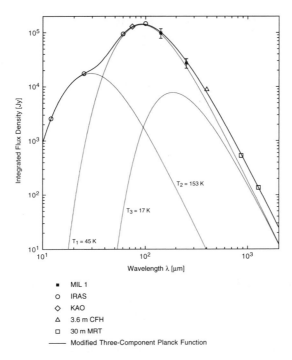

FIGURE 49.8. Total flux density of W51-I. Our measurements labeled MIL 1 and those listed by Sievers et al. [20] are plotted. The uncertainty of our measurements is about 20%. The three-component fit to the spectrum is indicated by the solid line. The dotted lines represent the spectral emission of cold, warm and hot dust.

now keeping the parameters of the warm component fixed. The results confirm that the contribution of hot and cold dust is indeed extremely low in the wavelength range from 50 μm to 160 μm.

The warm dust component has a temperature of $T_1 = 45$ K. It is emitted from a region with an effective solid angle of about $\Omega_1 \approx 9.8 \cdot 10^{-8}$ sterad. The hot component with $T_2 = 153$ K fills a considerably smaller volume. The corresponding solid angle amounts to $\Omega_1 = 2.8 \cdot 10^{-10}$ sterad. Finally, the cold envelope exhibits a temperature of about $T_3 = 17$ K and has a solid angle $\Omega_1 = 4.9 \cdot 10^{-6}$ sterad. The filling factor is defined as the ratio of the fitted solid angle to the apparent source solid angle. It represents a measure for the degree of fragmentation of the cloud if the considered source is spatially resolved in the sense that the source diameter is larger than the field of view of the telescope. This is evidently not the case for the hot dust component. Table 49.3 summarizes these results. The critical wavelength λ_c of each component is also shown in this table.

TABLE 49.3. Fitted Source Parameters of W51-I

Dust Component	Temperature T [K]	Solid Angle Ω [arcsec^2]	Critical Wavelength λ_c [μm]	Filling Factor
Warm	45	4200	133	0.012
Hot	153	12	50	$3.3 \cdot 10^{-5}$
Cold	17	$2.1 \cdot 10^5$	104	0.58

49.6.3   Cloud Mass and Luminosity

By means of the previously calculated total flux density we can estimate the total infrared luminosity of W51-I. For the mean source distance of 7.5 kpc the result is $L_{IR} \approx 1.4 \cdot 10^7 L_{\odot}$. This is in good agreement with the data of Rengarajan *et al.* [6] and Sievers *et al.* [20], who made the estimates $L_{IR} \approx 1.5 \cdot 10^7 L_{\odot}$ and $L_{IR} \approx 1.8 \cdot 10^7 L_{\odot}$, respectively. The resulting infrared excess $IRE = L_{IR}/Ly_{\alpha}$ as well as the luminosities of the single dust components are listed in Table 49.4.

In order to calculate the total cloud mass we had first to determine the hydrogen column density N_H. Mezger *et al.* [23] relate N_H directly to the optical depth τ_{λ} by the equation

$$\tau_{\lambda} = N_H \cdot \sigma = N_H \cdot (Z/Z_0) \cdot b \cdot 7 \cdot 10^{-21} \lambda_{\mu m}^{-2} \qquad (49.6)$$

for $\lambda_{\mu m} \geq 100$, where σ represents the optical cross section of the dust grains. The relative metallicity is assumed unity, since W51 has nearly the same galactocentric distance as the sun. The parameter b depends on the type of dust grains which are responsible for far-infrared extinction in the observed cloud. In particular, b is larger in cold and dense clouds, because the dust grains are covered by ice mantles, which increase the extinction. Mezger [24] suggests that in molecular clouds of medium hydrogen density, i.e. $n_H < 10^6$ cm^{-3}, the parameter $b = 1.9$ yields a good estimate of the column density. For denser and colder clouds $b = 3.4$ should be adopted.

The hydrogen mass of a single dust component is evaluated from its column density and the area subtended by its effective solid angle. The total hydrogen mass is then just the sum of the masses of all components. For W51-I this mass is $M_H \approx 2.4 \cdot 10^5 M_{\odot}$. The total cloud mass M_c including the dust is 1.36 times larger [25]. Column density and mass of each component of W51-I are summarized in Table 49.4.

TABLE 49.4. General Source Parameters of W51-I

Dust Component	Warm	Hot	Cold	Total
H$_2$ Column Density $N_H[\text{cm}^{-2}]$	$1.3 \cdot 10^{24}$	$1.9 \cdot 10^{23}$	$8.1 \cdot 10^{23}$	$8.5 \cdot 10^{23}$
H$_2$ Mass $M_H[M_\odot]$	$5.9 \cdot 10^4$	23	$1.8 \cdot 10^5$	$2.4 \cdot 10^5$
Mean H$_2$ Density $n_H[\text{cm}^{-3}]$	$2.4 \cdot 10^5$	$6.4 \cdot 10^5$	$6.6 \cdot 10^4$	$8.1 \cdot 10^4$
Cloud Mass $M_c[M_\odot]$	$8.0 \cdot 10^{24}$	32	$2.4 \cdot 10^5$	$3.2 \cdot 10^5$
Infrared Luminosity $L_{IR}[L_\odot]$	$9.8 \cdot 10^6$	$4.1 \cdot 10^6$	$2.5 \cdot 10^5$	$1.4 \cdot 10^7$
Lyman-α Luminosity $Ly_\alpha[L_\odot]$				$2.5 \cdot 10^6$
Infrared Excess IRE				6
Lumin.-to-Mass $L_{IR}/M_c[L_\odot/M_\odot]$	123	3430	1.2	44

49.6.4   Determination of Temperature and Optical Depth

Warm dust provides the strongest contribution to the total infrared luminosity of W51-I. Its emission is most significant in the wavelength region from 50 μm to 300 μm. At these wavelengths, the contribution from cold and hot dust are at least one order of magnitude lower. Consequently, our maps at 130 μm and 260 μm trace mainly the warm dust component. In addition, the emission in this wavelength range can be approximated by a single Planck function. Since we know the intensity at two different wavelengths, we can determine the temperature T as well as the critical wavelength λ_c, and consequently the optical depth τ_λ, by solving the following system of equations:

$$I_{260} = \left(1 - e^{(\lambda_c/260\mu m)^2}\right) \cdot B_{260}(T), \qquad (49.7)$$

$$I_{130} = \left(1 - e^{(\lambda_c/130\mu m)^2}\right) \cdot B_{130}(T) .$$

The results for W51-I and its surroundings are shown in Figures 49.9a-b and 49.10a-b, where we have superimposed the continuum emission and the optical depth, as well as the continuum emission and the temperature at the respective wavelengths. The calculated temperature is generally lower than that expected from the fitted emission spectrum, because the cold dust component, which is not considered here, contributes to the 260 μm intensity more than it does to the

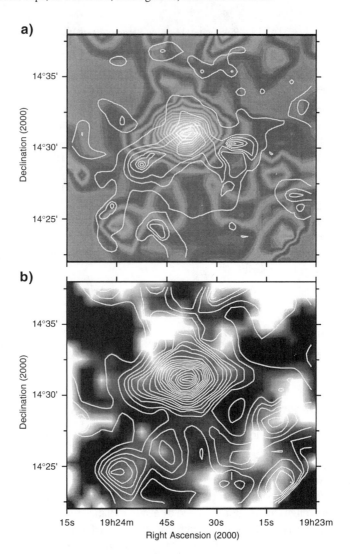

FIGURE 49.9. W51A. (a) The contours of the optical depth at 130 μm wavelength are superimposed on a gray-scale image of the continuum emission at 130 μm wavelength. The contour levels are at 0.015, 0.03, 0.06, 0.09, 0.12, 0.15, 0.18, 0.21.   (b) The contours of the continuum emission at 130 μm wavelength are superimposed on a gray-scale image of the temperature varying between black, corresponding to $T \leq 16$ K, and white with $T \geq 70$ K.

130 μm intensity. Consequently, in order to produce the same intensity the optical depth has to be overestimated. However, this calculation of the optical depth does not take into account that dust may be clumped together in regions which are not resolved by our telescope. This systematically results in beam-averaged values.

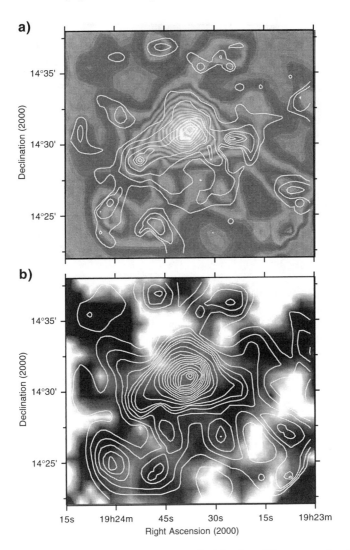

FIGURE 49.10. W51A. (a) The contours of the optical depth at 260 μm wavelength are superimposed on a gray-scale image of the continuum emission at 260 μm wavelength. The contour levels are at 0.005, 0.01, 0.02, 0.03, 0.04, 0.05, 0.06.   (b) The contours of the continuum emission at 260 μm wavelength are superimposed on a gray-scale image of the temperature varying between black, corresponding to $T \leq 16$ K, and white with $T \geq 70$ K.

In Figs. 49.9a and 49.10a white regions represent a temperature of about 70 K or more while the lowest temperatures of about 16 K correspond to the black regions. At the position of the peak intensity the temperature is about 30 K. The hydrogen column density at the same position is $N_H = 6.7 \cdot 10^{23}$ cm^{-2}. The maximum hydrogen column densities are reached at about 3 arcmin to the west

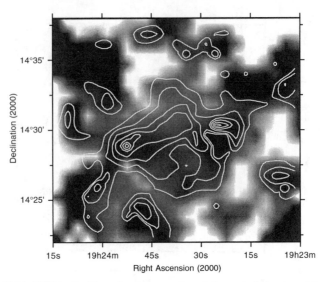

FIGURE 49.11. W51A. Overlay of a contour plot of the optical depth at 260 μm and a gray-scale image of the temperature varying between black, corresponding to $T \leq 16\,\mathrm{K}$, and white with $T \geq 70\,\mathrm{K}$. The contour levels of the optical depth are at 0.005, 0.01, 0.02, 0.03, 0.04, 0.05, 0.06.

and to the east of this position. There the temperature appears to be about 17 K and the hydrogen column density amounts to $N_H = 1.2 \cdot 10^{24}\ \mathrm{cm}^{-2}$. This value is in excellent agreement with that measured by Rudolph *et al.* [26], who found $N_H \approx 1 \cdot 10^{24}\ \mathrm{cm}^{-2}$.

In general, regions of high column density coincide well with those of low temperature. This fact is illustrated by Fig. 49.11 which shows an overlay of the optical depth and the temperature. It seems that in dense regions the emission is dominated by cold and warm dust. This is not the case in the low-density regions where small grains are probably temporarily heated by energetic photons of the interstellar radiation field to higher temperatures than one would expect if the dust grains were in thermodynamical equilibrium with the environment.

Furthermore, we are able to calculate the total hydrogen mass by integrating the column density over the area of interest. For W51-I the result is $M_H = 5.1 \cdot 10^5 M_\odot$, i.e., about a factor of 2 higher than that given in Section 49.6.3. There is a simple explanation for this discrepancy, namely, the method we employed in this chapter to calculate the temperature and the optical depth. As mentioned above, the temperature is underestimated by this method. Consequently, in order to produce the same intensity, the beam-averaged optical depth and, therefore, the column density is increased. This implies a higher cloud mass.

49.7 Conclusion

We have mapped the continuum emission from W51A at 130 μm and at 260 μm wavelength. The observations have been performed by means of a balloon-borne telescope of 60 cmØ equipped with a chopping secondary mirror and a liquid helium-cooled radiometer at CNES in Aire-sur-l'Adour, France, on September 20/21, 1993. We succeeded in mapping the dust emission from the HII region-molecular cloud complex W51A at 130 μm and 260 μm. In order to perform instrumental and radiometric calibration we have also observed Saturn at this wavelength. On the basis of these measurements, we have determined several source parameters of W51-I, such as the color temperature, the hydrogen mass and the total infrared luminosity. In addition, we have studied the emission spectrum, the surface brightness, the column density and the temperature distribution. In general, our results agree well with those of other authors, who have studied W51-I at other wavelengths [6, 10, 18, 20].

The wavelength region from 20 μm to 350 μm is poorly studied because strong atmospheric absorption and emission does not allow ground-based observations. In 1983 the Infrared Astronomical Satellite IRAS made for the first time a complete sky survey at far-infrared wavelengths up to 100 μm. It did not cover the wavelength region from 100 μm to 350 μm where the continuum emission from cool dust is strongest. As a consequence, it is difficult to calculate cloud mass and infrared luminosity of molecular clouds exclusively on the basis of IRAS data. In general, they would be strongly underestimated. Other previous measurements of the far-infrared continuum emission are restricted to individual objects or are of poor spatial and spectral resolution, e.g., by Lightfoot *et al.*, Rengarajan *et al.*, Nikola [6, 10, 19]. Especially in the 170 μm to 350 μm range there is a serious lack of measurements with medium or high spectral resolution. For a better understanding of the processes involved in star formation a higher resolution is required. This would permit study of compact objects of cloud complexes individually. Astrophysicists are also interested in the observation of line emission which can provide additional information on chemistry and dynamics and the cooling mechanism in molecular clouds. There are several line emissions in the far-infrared which are sufficiently strong to be observed. Examples are the [OI] lines at 63 μm and at 145 μm, the [CII] line at 158 μm and the [NII] lines at 122 μm and 205 μm wavelength. While the [CII] line was detected for the first time in 1980 [27] the [NII] line at 205 μm was discovered only recently by the Cosmic Background Explorer (COBE) [28]. Since the COBE measurements are restricted to spatially extended emission (7° beam) efforts are now being made to study the [NII]-line emission from molecular clouds with higher spatial resolution.

In this context we have recently developed a new liquid-helium cooled spectrometer, which operates in the 145 μm to 220 μm wavelength region with a spectral resolution of $R \approx 2000$. This resolution is achieved by two liquid-helium cooled scanning Fabry-Perot interferometers, which have been developed and tested in our laboratory [29]. A stressed Ge:Ga photoconductor detector with a low NEP of about 10^{-17} WHz$^{-\frac{1}{2}}$ guarantees that the detectivity is limited by the background radiation of the telescope. This spectrometer will be incorporated into our balloon-borne 60 cmØ telescope on the occasion of our next flight MIL 2 which will take place in autumn, 1995.

Another instrument, the Infrared Space Observatory (ISO) of the ESA, will also be launched in autumn, 1995. It is equipped with spectrometers and an infrared camera and will also possess polarimetric capabilities. However, ISO will not cover the wavelength region from 195 μm to 350 μm. Also the spatial resolution is lower than 1 arcmin for wavelengths longer than 100 μm [30]. Two other projects, which will extend the measurements to the submillimeter wavelength region, are the Far-Infrared and Submillimeter Space Telescope (FIRST) and the Stratospheric Observatory for Infrared Astronomy (SOFIA). FIRST will be equipped with a 3 mØ Cassegrain telescope and possess spectroscopic capabilities in the wavelength region from 100 μm to 1 mm [31]. SOFIA is planned as the successor of the Kuiper Airborne Observatory (KAO). It is planned as a 2.5 mØ telescope housed in a Boeing 747 aircraft to cover the 0.3 μm to 1600 μm wavelength range [32].

49.8   Acknowledgments

We are very grateful to A. Benz, ETH, Zürich, G. G. Fazio, Harvard Smithsonian Center for Astrophysics, Cambridge, Mass., U.S.A., M. C. E. Huber, ESTEC/ESA, Noordwijk, Netherlands, P. G. Mezger, MPI für Radioastronomie, Bonn, Germany, G. N. Schenker, ETH Zürich, G. Winnewisser and J. Stutzki, Universität Köln, Germany, and H. Zinnecker, Universität Würzburg, Germany, for valuable advice, suggestions and discussions. We also wish to express our gratitude to D. Bhend, W. Herrmann, H. Scherrer and E. Zimmermann, Physics Departement, ETH, Zürich, as well as to M. Georges, M. Fleury and Ch. Maire, Geneva Observatory, for technical assistance. The balloon launch and recovery operations were performed by the staff of the Balloon Facility of the CNES, Aire-sur-l'Adour, France, whose assistance and hospitality are gratefully acknowledged.

This study was supported by the Swiss National Science Foundation, by the ETH, Zürich and the Geneva Observatory.

References

[1] A. P. Holenstein, G. N. Schenker, D. Huguenin and F. K. Kneubühl, Astron. Astrophys. Suppl. Ser. **96**, 115 (1992).

[2] G. N. Schenker, A. P. Holenstein, F. A. Pepe, D. Huguenin, and F. K. Kneubühl, Infrared Phys. Technol. **35**, No. 2/3, 221 (1994).

[3] D. Huguenin, Infrared Phys. Technol. **35**, No. 2/3, 195 (1994).

[4] G. Neugebauer, H. J. Habing, R. van Duinen, H. H. Aumann, B. Baud, C. A. Beichmann, T. A. Beintema, N. Bogess, P. E. Clegg, T. de Jong, J. P. Emerson, T. N. Gautier, F. C. Gillett, S. Harris, M. G. Hauser, J. R. Houck, R. E. Jennings, F. J. Low, P. L.

Marsden, G. Miley, F. M. Olnon, S. R. Pottasch, E. Raimond, M. Rowan-Robinson, B. T. Soifer, R. G. Walker, P. R. Wesselius, and E. Young, Astrophysical J. Lett. **278**, L1 (1984a).

[5] W. M. Goss and P. A. Shaver, Australian J. Phys. Ap. Suppl. **14**, 1 (1970).

[6] T. N. Rengarajan, L. H. Cheung, G. G. Fazio, K. Shivanandan, and B. McBreen, Astrophysical J. **286**, 573 (1984).

[7] V. Pankonin, H. E. Payne, and Y. Terzian, Astron. Astrophys. **75**, 365 (1979).

[8] J. Bieging, *HII Regions and Related Topics* **42**, 443, Eds. T. L. Wilson and D. Downes, Springer Verlag, Berlin (1979).

[9] A. H. M. Martin, Mon. Not. R. astr. Soc. **157**, 31 (1972).

[10] J. F. Lightfoot, W. Cudlip, I. Furniss, W. M. Glencross, R. E. Jennings, K. J. King, and G. Poulter, Mon. Not. R. astr. Soc. **205**, 653 (1983).

[11] R. A. Gaume, K. J. Johnston, and T. L. Wilson, Astrophysical J. **417**, 645 (1993).

[12] R. A. Gaume and R. Mutel, Astrophysical J. Suppl. **65**, 193 (1987).

[13] R. Genzel *et al.*, Astrophysical J. **247**, 1039 (1981)

[14] R. I. Klein, M. T. Sandford, and R. W. Whitaker, Astrophys. J. Lett. **271**, L73 (1983).

[15] G. N. Schenker , PhD. Thesis No. 10060, Federal Institute of Technology, Zürich (1993).

[16] L. B. Lucy, Astron. J., **79**, 745 (1974).

[17] R. H. Hildebrand, R. F. Loewenstein, D. A. Harper, G. S. Orton, J. Keene, and S. E. Whitecomb, Icarus **64**, 64 (1985).

[18] J. T. Jaffe, E. E. Becklin, and R. H. Hildebrand, Astrophysical J. **279**, L51 (1984).

[19] T. Nikola, Diploma Thesis, Max-Planck Institut für extraterrestrische Physik, Ludwig-Maximilians-Universität München (1993).

[20] A. W. Sievers, P. G. Mezger, M. A. Gordon, E. Kreysa, C. G. T. Haslam, and R. Lemke, Astron. Astrophys. **251**, 231 (1991).

[21] Gordon M. A., Astrophysical J. **316**, 258 (1987).

[22] Gordon M. A., Astrophysical J. **331**, 509 (1988).

[23] P. G. Mezger, J. E. Wink, and R. Zykla, Astron. Astrophys. **228**, 95 (1990).

[24] P. G. Mezger, Infrared Phys. Technol. **35**, No. 2/3, 337 (1994).

[25] R. H. Hildebrand, Q. Jl R. Astr. Soc. **24**, 267 (1983).

[26] A. Rudolph, W. J. Welch, P. Palmer, and B. Dubrulle, Astrophysical J. **363** 528 (1990).

[27] R. W. Russell, G. Melnick, G. E. Gull, and M. Harwit, Astrophysical J. **238**, L99 (1980).

[28] E. L. Wright, J. C. Mather, C. L. Bennett, E. S. Cheng, R. A. Shafer, D. J. Fixsens, R. E. Eplee Jr., R. B. Isaacman, S. M. Read, N. W. Boggess, S. Gulkis, M. G. Hauser, M. Janssen, T. Kelsall, P. M. Lubin, S. S. Meyer, S. H. Moseley Jr., T. L. Murdock, R. F. Silverberg, G. F. Smoot, R. Weiss, and D. T. Wilkinson, Astrophysical J. **381**, 200 (1991).

[29] F. A. Pepe, R. Brodbeck, D. P. Scherrer, G. N. Schenker, D. Bhend, E. Zimmermann, and F. K. Kneubühl, Infrared Phys. Technol. **35**, No. 7, 863 (1994).

[30] ISO INFO, Newsletter on the Infrared Space Observatory, No. 4, March 1994, ESA (1994).

[31] G. Pilbratt, Infrared Phys. Technol. **35**, No. 2/3, 407 (1994).

[32] P. M. Harvey and E. F. Erickson, Infrared Phys. Technol. **35**, No. 2/3, 153 (1994).

[33] S. L. Mufson and H. S. Liszt, Astrophysical J. **232**, 451 (1979).

50

Charles Townes: The Scientist and the Person

Alexander M. Prokhorov

In connection with the 80th birthday of Professor Charles Townes I would like to express my regards to him as a scientist and as a person. I take upon myself this responsibility because our common scientific interests and friendly relations span many decades.

Charles Townes and I first became personally acquainted exactly 40 years ago in 1955, during a conference in Cambridge, England. Although we had not met before, we had come to know about each other somewhat earlier (from the beginning of the fifties) through our publications in the field of radio spectroscopy. At that time it was a young, fast-developing area of spectroscopy. Our mutual interests were greatly enhanced by our independent publications [1-4], in which the new principle of amplification and generation of electromagnetic radiation by stimulated emission of molecules had been proposed. This new principle was first demonstrated in the microwave region, when the first quantum oscillators (1954) and amplifiers (1957-1958) [4-6], called "masers," were created. Six years later this principle was demonstrated in the optical range of the spectrum, when the first optical quantum oscillator, the "laser," was created (1960) [7]. Thus was born a new area of science and engineering, i.e., quantum electronics. This new area developed rapidly, and it continues to develop more than 40 years after the creation of the first maser. It shows how fruitful these ideas of quantum electronics were,[1] and how universal the new principle of amplification and generation was over essentially the entire electromagnetic spectrum: at present quantum amplifiers and oscillators have been demonstrated in a wide spectral range from microwaves up to X-rays. Quantum electronics has now become one of the most dynamic and broad areas of scientific activity, and has had a revolutionary effect on many fields of science and engineering.

Modern optics, in essence, is not conceivable without lasers. Many areas in physics, chemistry, biology, and other sciences have been advanced in important

---

[1]I do not discuss here a history of the occurrence of these ideas, in particular, the history of the transition from masers to lasers. These were discussed rather in detail in the Nobel lectures [8,9].

new directions by the application of laser methods. One new direction has given rise to the field of "laser physics"—or more broadly—"laser science." The range of applications of devices and methods of quantum electronics is huge: extremely long-distance space radio communication and radio astronomy based on the usage of quantum amplifiers with an extremely low intrinsic noise, fundamental metrology based on the usage of extremely precise maser and laser standards of frequency and time, the technology of material processing based on the usage of high power lasers, fiber-optic communications, ultra-short light pulses, surgery and medicine, and many other applications.

It should be noted that the great potential for application of masers and lasers, in general, became clear rather soon after their discovery. However, results of the subsequent rapid and wide development of quantum electronics exceeded the most optimistic expectations. In this respect the evaluation of achievements and forecasts of future developments made by Charles Townes on the 10th and 20th anniversaries of quantum electronics [8,10], proved, perhaps, to be most accurate.

Certainly, during the progress of quantum electronics over the past four decades, there have been contributions made by hundreds and even thousands of scientists from many countries. But doubtless, the contribution of Prof. Townes in this progress is huge, not only because of his first basic research work mentioned above, but also because of his subsequent work, and his stimulating reports at many conferences.

In this connection, spectroscopy-based investigations of interstellar space are especially noteworthy. This work, carried out by Townes, his coworkers, and colleagues from various astrophysical laboratories from 1960 to the present, require a special analysis, but I shall comment on only some of them here, i.e., those which are closely connected to quantum electronics. The work done by Townes and coworkers in radio- and infrared-spectroscopy of interstellar molecules is very important for the understanding of the structure and dynamics of the interstellar medium. Using advanced experimental methods in radio astronomy and infrared spectroscopy of space, including those based on the achievements of quantum electronics (for example, the application of masers as high sensitivity amplifiers and high-stability frequency standards), they obtained some very important data for astrophysics on the absorption and emission spectra of interstellar atomic and molecular gases. Many molecules, including complex organic ones, were discovered; their spatial localization and density, as well as the sizes and the motions of molecular formations (clouds) in interstellar media have been determined (the details of these investigations and their significance for astrophysics can be found in the excellent review [11]).

An essentially new result was the detection of maser action in space: microwave radiation amplified by stimulated emission of H_2O molecules at a wavelength of the 1.35 cm was observed [12]. This discovery stimulated a search for other maser sources in space at shorter wavelengths. The search has been successful: maser radiation at a millimeter wavelength has been discovered in the spectra of the star MWC349, and quite recently [13], there has been the discovery of laser amplification in the far-infrared range (at 169.4 μ and 52.53 μ), corresponding to the hydrogen recombination lines.[2]

The work by Townes and coworkers on the detection of maser and laser sources in the interstellar medium, on the one hand, confirms the universality of the quantum amplification principle and, on the other hand, has a significant impact especially on astrophysics (permitting, in particular, the determination of an effective temperature of radiation sources and excitation mechanisms of molecules, which are important for the understanding of processes in the interstellar medium).

Analyzing Townes' research in interstellar molecular spectroscopy, its genesis can be delineated: it is a logical continuation of his work in laboratory molecular radio spectroscopy and quantum electronics. As pointed out above, the work in radio spectroscopy resulted in the birth of quantum electronics ideas, and they in turn stimulated a new direction in space research, namely, spectroscopic radio astronomy, which has been developing in many countries over the past four decades. This direction, enriched with ideas of quantum electronics, resulted in very important new data on the contents and properties of the interstellar medium. Speaking of the connection of quantum electronics with space research, it is worth noting one more important direction—the application of masers as ultimately low intrinsic noise microwave receivers in radio reception systems of radio astronomy and long-distance space communication stations.

The most successful development and application of effective masers in the decimeter, centimeter and millimeter range of the spectrum was achieved during the mid-'70s in the U.S.S.R. and the U.S.A. [15]. Due to the applications of masers, important results have been obtained in studies of galactic neutral hydrogen atoms emitting radiation at the 21 cm line, and of water vapor at the 1.35 cm line; there were numerous emission lines discovered in the centimeter and millimeter wavelength ranges originating from highly excited hydrogen atoms as well. Other very interesting results were obtained from the joint Soviet-American program on compact astronomical sources (quasi-stellar objects, and the nuclei of galaxies) with the very long base interferometer (U.S.S.R.-U.S.A.) at 3 cm and 1.35 cm wavelengths. Also, masers have been successfully used in radar studies of planets Mars, Venus, Jupiter, and Mercury.

Besides the outstanding, concrete contributions made by Townes to radio astronomy and interstellar spectroscopy, the experts in this field place enormous importance on his monograph written jointly with Prof. Schawlow [16], which has become a handbook for all radio astronomy spectroscopists.

In giving a brief analysis of the scientific activity of Charles Townes, it is impossible not to mention his contribution to nonlinear optics. In particular, his work [17] on the investigation of self-focusing of laser beams in nonlinear media stimulated

---

[2]Hydrogen recombination radio lines in space were predicted, discovered and investigated for the first time by the Soviet scientists [14]. At the present time this direction of radio astronomy is undergoing significant development. It is interesting to note that in the recombination emission spectra highly excited states of hydrogen atoms (up to $n = 800$, where n is the principal quantum number) have been observed in which the atomic size reaches the macroscopic value 0.1 mm.

wide experimental and theoretical studies of this important phenomenon in the propagation of high-power laser radiation in optical media.

Though the personal scientific interests of Charles Townes in recent years have been connected mainly with astrophysical problems, i.e., with the studies of interstellar media by radio astronomy methods and astronomical infrared (IR) spectroscopy, he has maintained a continued interest in developments in quantum electronics. He participates regularly at the International Quantum Electronics Conference, which was founded by him in 1958, and at other conferences on quantum electronics. His talks at these conferences, with his analyses of the achievements and trends of quantum electronics, have aroused enormous interest in the participants, and stimulated the development of many important directions, since the authority of his analyses and forecasts is very great.

Charles Townes is well-known to the scientific community not only as an outstanding scientist, but also as a person deeply interested in the global problems of mankind, the social processes of his country, and the world. He supports the wide cooperation of scientists of different countries, understanding well the universal and international character of science. Connected with this, I would like to note his interest in my country and its difficult problems at the present time. He appreciates the true value of the contribution of Soviet scientists to the world's scientific progress, and especially emphasizes the fundamental character of their research. He is sincerely concerned with those difficulties which our scientists have experienced in recent years. He has visited the Soviet Union (Russia) many times as the guest of the Academy of Sciences under my invitation and the invitations of other scientists. During his trips, he participated in conferences and visited many laboratories. He was my personal guest, and a guest at the homes of many colleagues. Many of his colleagues and I look forward to his forthcoming participation as the chairman of the Laser Optics '95 Conference, which will be held in June-July 1995, at St. Petersburg. His participation at conferences gives them high authority. Moreover, the opportunity for personal dialogue and discussions is always very pleasent and useful. There is no doubt that the American and the international scientific communities will commemorate the birthday of Professor Townes, eminent scientist and person.

In conclusion, I would like to note the significant role played by Charles Townes's wife Frances. My friend Charles found a good match when he married such a charming, intelligent and well-educated woman, who renders, I believe, a very positive influence on all his many-sided activities. I send to both of them my congratulations and best wishes.

References

[1] N. G. Basov and A. M. Prokhorov, JETP **27**, 431 (1954).
[2] J. P. Gordon, H. J. Zeiger, and C. H. Townes, Phys.Rev. **95**, 282 (1954).
[3] C. H. Townes, J. Inst. Elec. Commun. Eng. (Japan) **36**, 650 (1953).
[4] N. G. Basov and A. M. Prokhorov, JETP **28**, 249 (1955).
[5] H. E. Scovil, G. Feher, and H. Seidel, Phys. Rev. **105**, 762 (1957).

[6] G. M. Zverev, L. S. Kornienko, A. A. Manenkov, and A. M. Prokhorov, JETP **34,** 1660 (1958).

[7] T. H. Maiman, Nature **187,** 493 (1960).

[8] C. H. Townes, IEEE Spectrum **2,** 30 (1965).

[9] A. M. Prokhorov, "Quantum Electronics," Preprint from Les Prix Nobelen (1964) Stockholm 1965, Kungl. Boktryckeriet P. A. Norstedt Soner.

[10] C. H. Townes, "Quantum Electronics - Where We Have Been and Where We May Go," Plenary Paper at CLEO/IQEC Joint Plenary Session, June 18-21, 1984, Anaheim, California.

[11] D. M. Rank, C. H. Townes, and W. J. Welch, Science, **174,** 1083 (1971).

[12] S. H. Knowles, C. H. Mayer, A. C. Cheung, D. M. Rank, and C. H. Townes, Science **163,** 1055 (1969).

[13] V. S. Strelnitski, H. A. Smith, M. R. Haas, S. W. J. Colgan, E. F. Erickson, N. Geis, D. J. Hollenbach, and C. H. Townes, "A Search for Hydrogen lasers in MW349 from KAO," Proc. of the Airborne Astronomy Symp. on the Galactic Ecosystem: From Gas to Stars to Dust, edited by M. R. Haas, J. A. Davidson, and E. F. Erickson, p. 305 (1994).

[14] R. L. Sorochenko, "Postulation, Detection and Observations of Radio Recombination Lines," Radiorecombination Lines: 25-Years of investigation, edited by M. A. Gordon and R. L. Sorochenko, (Kluwer Academic Publishers, 1990).

[15] A. A. Manenkov and V. Shteinshleiger, "Quantum Amplifiers and Their Usage in Radio Reception Systems of Far-space Communication and Radioastronomy," 1977 Annual Comprehensive Soviet Encyclopedia, **21,** 566, (Moscow, 1977).

[16] C. H. Townes and A. L. Shawlow, *Microwave Spectroscopy* (McGraw-Hill, New-York, 1955).

[17] R. Y. Chiao, E. Garmire, and C. H. Townes, Phys. Rev. Lett. **13,** 479 (1964).

51

Neutron Spin Reorientation Experiments

Norman F. Ramsey

It is an honor to write an article in celebration of Charles Townes' 80th birthday. His tremendous contributions to physics and astronomy include masers and lasers, as well as many ingenious and fundamental experiments on microwave spectroscopy and molecules in outer space. For my contribution to the celebration, I am reviewing a related, but different field of research that spans most of his highly productive career.

51.1   Introduction

Experiments which reorient neutron spins have been fruitful sources for both fundamental and applied physics, including the measurement of the neutron magnetic moment [1–3] and the setting of a limit to the neutron electric dipole moment as a test of time reversal symmetry [4, 5]. "Dressed" neutrons [6] and Berry phases [7, 8] have been studied with reoriented neutron spins. In neutron interaction condensed matter experiments, small changes in neutron velocity have been sensitively measured by the changes in the precession angle in magnetic fields before and after the interaction [9], or by comparing the direction of the neutron spin with a rotating magnetic field whose phase serves as a clock [10].

Heckel, Ramsey, Forte, and their associates [11] found that the spins of polarized neutrons, moving at approximately 200 m/s through several centimeters of various solid materials, are rotated by parity nonconserving interactions to the right or to the left by a few microradians.

The success of these widely different neutron spin reorientation experiments is dependent upon the availability of intense neutron beams and an effective method for polarizing them. Therefore, before reviewing the experiments, the means for producing, transporting, polarizing and confining neutrons will be described [12].

51.2   Neutron Sources

Slow neutrons by convention are usually classified as follows according to their approximate temperature T, kinetic energy E, velocity v and wave length λ:

Hot neutrons: $T \approx 2100\,\text{K}$, $E \approx 200\,\text{meV}$, $v \approx 5700\,\text{ms}^{-1}$ and $\lambda \approx 0.7\,\text{Å}$

Thermal neutrons: $T \approx 300\,\text{K}$, $E \approx 25\,\text{meV}$, $v \approx 2200\,\text{ms}^{-1}$ and $\lambda \approx 1.8\,\text{Å}$

Cold neutrons: $T \approx 40\,\text{K}$, $E \approx 7\,\text{meV}$, $v \approx 200\,\text{ms}^{-1}$ and $\lambda \approx 20\,\text{Å}$

Ultracold neutrons (UCN): $T \approx 1\,\text{mK}$, $E \approx 10^{-7}\,\text{eV}$, $v \approx 5\,\text{ms}^{-1}$ and $\lambda \approx 800\,\text{Å}$.

Most of the spin reorientation experiments have been done with thermal, cold and ultracold neutron beams from reactors at Grenoble, Brookhaven, Oak Ridge, Argonne, Munich, and Gatchina (St. Petersburg, Russia) with the most powerful and most frequently used neutron source for research being the one at the Institut Laue-Langevin (ILL) at Grenoble. The cold and ultracold neutrons at Grenoble are cooled in a liquid D_2 moderator near the core of the reactor. Many of the cold neutrons enter a horizontal pipe in which the neutrons are guided by glancing angle reflections to the experiments. The UCN on the other hand, leave the moderator in an upward directed pipe with a gradual bend so only ultracold neutrons are transmitted. The neutrons then strike and are reflected by a rotating turbine which further slows the longitudinal velocity components before the neutrons enter the guide pipe to the UCN experiment.

Neutron fluxes Φ near the ILL reactor core are about $1.5 \times 10^{15}\,\text{cm}^{-2}\text{s}^{-1}$ and at the end of a guide $3 \times 10^9\,\text{cm}^{-2}\text{s}^{-1}$. The UCN fluxes, on the other hand, are approximately $3 \times 10^4\,\text{cm}^{-2}\text{s}^{-1}$ and the density of UCN is about $60\,\text{cm}^{-3}$.

Ultracold neutrons can also be obtained from pulsed spallation sources, such as ISIS in England and those at Argonne, Los Alamos, and the European Spallation Source. The spallation sources usually provide time of flight velocity selection.

51.3   Reflection, Transport Polarization and Trapping of Neutrons

Due to the coherent forward scattering of neutrons by nuclei, neutrons upon entering a medium are reflected and refracted in accordance with the following index of refraction:

$$n = \left(1 - \frac{\lambda^2 N a_{coh}}{\pi} \pm \frac{\mu_n B}{\frac{1}{2}Mv^2}\right)^{\frac{1}{2}}, \tag{51.1}$$

where N = number of nuclei for one volume, a_{coh} = coherent forward scattering amplitude and μ_n = the magnetic moment of the neutron and B the magnetic induction with the $\pm$ sign depending on whether the neutron spin is parallel or antiparallel to B. For neutrons, unlike light, the sign of a_{coh} is positive with some materials and negative with others so the index of refraction may be either greater than one or less than one. If the neutrons in a vacuum strike a surface such as Cu, quartz, or BeO with index of refraction less than one, the neutrons will be totally reflected, as in optical fiber optics, if the glancing angle θ in striking the surface is less than the critical angle θ_c, which is given by $\cos\theta_c = n$.

For thermal neutrons at 2200 m/s incident on a surface of BeO, $\theta_c = 0.15°$, but this increases for slower neutrons and for UCN at 5 m/sec there is total reflection at normal incidence. Consequently, hollow pipes of such materials can be used to transport such neutrons from the reactor moderator to the experiment.

Since the index of refraction in Eq. (51.1) and consequently the critical angle θ_c depend on the orientation of the neutron spins, total reflection can be used to polarize neutrons. Although polarization of thermal neutrons by a single mirror reflection provides only a narrow beam, Mezei and Dagleisch [9] have developed multiple "supermirrors" which combine total and Bragg reflection techniques to provide polarizers of $4 \times 6 \, \text{cm}^2$ cross-section and 98% polarization. For ultracold neutrons the polarization is even easier since a magnetized iron-cobalt foil 0.5 μm thick will transmit neutrons of one spin orientation and totally reflect the opposite polarization. Polarization has also been produced by transmission through polarized magnetic materials, magnetic Bragg scattering from single crystals, and by spin selective absorption in polarized proton on ^{3}He targets, but the various total reflection techniques are so effective that they are now used in almost all cold and UCN experiments.

When UCN slower than 7 m/s impact surfaces of Be, BeO, or quartz, $\theta_c = 90°$ so there is total reflection even at normal incidence. Such neutrons can therefore be stored or trapped in a box with such walls for 100 s or more. A neutron moving at 7 m/s velocity has several unusual properties due to its kinetic energy being only 2×10^{-7}eV, its effective temperature being 0.002 K, and its wave length being 670 Å. Since nuclei in the reflector are about 1 Å apart, this means that the neutrons interact with more than 670^2 nuclei and the effective reflecting mass is so large compared to the neutron mass that no energy is exchanged and the neutrons in such a storage bottle remain effectively at 0.002 K while the storage bottle is at room temperature.

51.4   Neutron Magnetic Moment

Bloch and Alvarez [1] made the first measurement of the neutron magnetic moment with a beam of neutrons from Ra-Be source in a paraffin moderator. The neutrons were polarized by transmission through magnetically saturated iron, were subjected to an oscillating magnetic field perpendicular to a constant field, and then analyzed by transmission through saturated iron. Later, Corngold and Ramsey [2] improved the measurement by using a more intense reactor neutron beam, and by using Ramsey's separated oscillatory field method [16]. The latest and most accurate measurement of the neutron magnetic moment combines the previous features with the use of slower cold neutrons and with a calibration of the magnetic field by the proton magnetic resonance in a stream of water that traverses the same pipe and is subject to oscillating fields from the same coils used for the neutrons. In this experiment the magnetic moment was measured to be 1.91304275(45) nuclear magnetons [3].

51.5   Neutron Electric Dipole Moment

In 1950 Purcell and Ramsey [14] pointed out that there was no experimental evidence for assumed parity conservation in nuclear forces and that the assumption should be tested. They proposed to test the assumption by searching for a neutron electric dipole moment which should not occur if parity is conserved. With Smith [15] they searched for such an electric dipole but found none. Later maximal parity non-conservation was discovered in the weak interaction, but not in the strong forces which were the ones tested by the neutron electric dipole moment [16].

Many theorists then argued [16] that, although parity no longer prevented a neutron electric dipole moment, time reversal symmetry would do so, but Ramsey [16, 17] and Jackson [16] argued that time reversal symmetry was also an untested assumption, and that a continued search for an electric dipole moment would test the assumption of time reversal symmetry [16]. No electric dipole moment was found in the next search, but a few years later, Fitch, Cronin, Christenson and Turlay [16, 17] discovered a failure of charge conjugation and parity (CP) conservation in the decay of the long lived neutral kaon, which implied a failure of T symmetry if time reversal, charge conjugation, and parity (TCP) were conserved. From this time onward, the most favored theoretical predictions were usually just below the experimental limit, and many of these theories were later eliminated by further lowerings of the experimental limits on the neutron electric dipole moment [16, 17].

The present lowest experimental limit to the neutron electric dipole moment comes from the combination of the latest experiments from an international group working at the Institut Laue-Langevin [4] and a Russian group at St. Petersburg Nuclear Physics Laboratory [5].

In the experiment of the Grenoble group, ultracold neutrons from the liquid deuterium moderator at the reactor core are further slowed by reflection from the blades of a rotating turbine and are transmitted to the resonance apparatus shown in Fig. 51.1. The magnetized Fe-Co foil polarizes the incident beam and serves as an analyzer for the returning beam. The neutrons are stored by total reflection in a cylinder 25 cm in diameter and 10 cm long with the sides made of BeO and the end

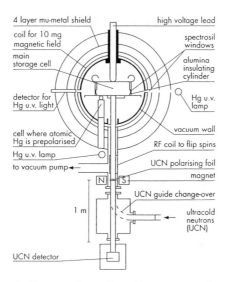

FIGURE 51.1. Schematic diagram of new Grenoble apparatus. The apparatus used in the most recently completed experiment was similar to that in the diagram except the Hg magnetometer was not included.

plates of Be. After filling for about 10 s, a shutter is closed and the neutrons are stored for about 80 s. The neutron density is about 10 cm^{-3}. The oscillating field is supplied in short initial and final pulses [13, 17]. After the shutter is opened, the neutrons again pass through the magnetized foil which serves as an analyzer, and transmits only neutrons whose orientations have not been changed by the coherent separated oscillatory fields. Consequently, the numbers of neutrons reaching the UCN detector provide a measure of the neutron magnetic resonance. The neutrons are counted at the steepest part of the resonance curve as the directions of the static electric and magnetic fields are successively reversed. The strengths of the static magnetic and electric fields were 1 μT and 1.6 MV m^{-1}, respectively. A neutron electric dipole moment would produce a change in the neutron counts when the relative orientations of the electric and magnetic fields were reversed.

The St. Petersburg neutron experiment is similar in principle except for the use of a double storage bottle with electric fields in opposite directions in the two halves. The combination of the Grenoble [15] and St. Petersburg experiments [5] places a 95% confidence upper limit on the magnitude of the neutron electric dipole moment of

$$|d_n| < 9 \times 10^{-26} e \, \text{cm} . \tag{51.2}$$

Since an electric dipole moment requires a failure of both parity and time reversal symmetry, the above result directly limits forces that are T-odd and P-odd. However, Khriplovich and Conti [18] have recently pointed out that through radiative corrections from the T-even, P-odd weak interaction, the neutron electric dipole moment experiments also set the most sensitive limit to the T-odd, P-even interaction parameters $\beta_{qe} < 10^{-6}$ and $\beta_{qq} < 0.3 \times 10^{-6}$.

Both the St. Petersburg and the Grenoble groups are beginning a new series of experiments to increase the sensitivity of their experiments for which the Grenoble group has made major changes in its apparatus. The inner magnetic shield has been removed to increase the size of the neutron storage bottle, and ^{199}Hg will be used as a magnetometer in the same storage vessel along with the neutrons, to be sure that there is no systematic error from a possible magnetic field associated with the direction of the electrostatic field. A later increase of the neutron sensitivity is planned by Golub and Lamoreaux [19] using neutrons stored in superfluid liquid ^{4}He with the neutrons being polarized and analyzed by small amounts of dissolved polarized ^{3}He.

In recent years, there have been a number of atomic and molecular searches for an atomic electron dipole moment, as recently reviewed by the author [17]. None of these experiments has yet found an electric dipole moment, but the different atomic and molecular experiments are complementary to each other and to the neutron experiments since different experiments provide the most sensitive tests [17] for different CP or time-reversal (T) violating theories. The neutron electric dipole moment, for example, provides the most sensitive limit, $\theta < 10^{-19}$, to the quantum chromodynamic parameter θ and to some of the possibilities for T non-conservation in supersymetric (SUSY) theories and in some of the T-odd, P-even theories. On the other hand, the atomic and molecular experiments are more sensitive to T-odd effects in the leptonic and semi-leptonic parameters [17]. The very small value of θ from the neutron electric dipole moment gives rise to what is often called the "Strong CP Problem." Quantum Chromodynamics (QCD) naturally contains a CP odd term and an extreme fine-tuning is required to make its contribution so small. Peccei and Quinn have introduced a new symmetry to solve this problem, but it implies the existence of axions that have not yet been observed. In the future, the atomic and molecular electric dipole moment limits, as well as the neutron limits, should be markedly lowered.

51.6   Dressed Neutrons

In analogy with the studies of C. Cohen-Tannoudji, and S. Haroche [20], and others on "dressed atoms," Muskat, Dubbers, and Schärpf [22] have studied neutrons "dressed" with a large number of radiofrequency quanta. The energy levels of a neutron in relatively strong oscillatory fields vary with magnetic field B_0 applied parallel to a z-axis as shown by the dashed lines in Fig. 51.2(a) provided the interaction between the neutron and the oscillatory field can be neglected. But with the inclusion of the neutron-photon magnetic interaction the energy levels of the "dressed neutron" are distorted to the solid curves in the figure. The energy levels were measured by passing velocity-selected neutrons through a polarizer perpendicular to the z-axis, after which they go through a region with a uniform magnetic field of strength B_0 followed by a spin analyzer oriented opposite to the first one. Then for $B_0 = 0$ no neutrons in principle get through the filter, but when

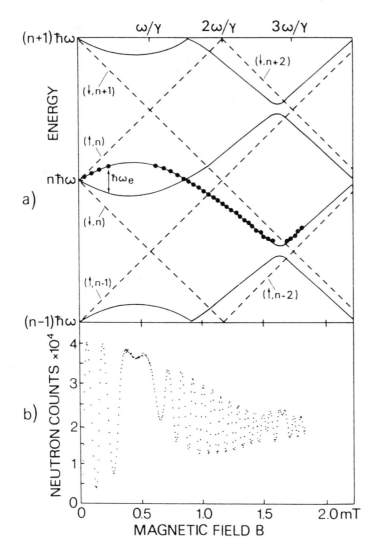

FIGURE 51.2. (a) Energy level diagram for the dressed-neutron system for $B_1 = 0.1$ mT. Dashed lines: Zeeman splitting (from H_M) omitting the interaction with photons. Solid lines: The energy levels repel each other because of coupling (from H_{int}) between the neutron and the photon field with energy $n\hbar\omega$. (b) Dressed-neutron spin-rotation pattern; from the measured eigenfrequencies the experimental energy levels in (a) are reconstructed. The error bars are of the same order as the size of the points.

B_0 is increased so that the neutron spin precesses $180°$ between the polarizer and analyzer, the number transmitted increases to a maximum and then diminishes as shown in Fig. 51.2(b).

From the number of oscillations as a function of B_0 and from the time B_0 acts on the neutron, the oscillation frequency, the effective magnetic moment, and the energies of the energy levels as a function of B_0, can all be inferred as shown by interrelation between parts (a) and (b) of Fig. 51.2. The dots on part (a) show the experimental energies inferred in this way and it can be seen that they are in agreement with the "dressed neutron" calculations.

51.7   Berry Phase

Bitter and Dubbers [7] have observed the "Berry" or "geometric" phase by observing the neutron polarization after passage through a helical magnetic field. Richardson and his associates [8] have also observed the Berry phase in a rotating frame with a rotating magnetic field. They note that in this case the results can also be explained by standard magnetic resonance theory [8].

51.8   Detection of Small Velocity Changes

A powerful technique for studying condensed matter is to measure small losses of neutron energy when neutrons interact with matter. Originally such experiments either had poor resolution or required the selection of a very narrow initial velocity range so that a small change in velocity could be detected. Mezei [21] pointed out that a more intense broad velocity band of neutrons could be used with the slight changes in velocity being detected by changes in the times required for the neutrons to pass through two magnetic field regions in which the times are measured by the angles precessed by the neutron spins. In particular, the polarized neutrons first pass through a region with a magnetic field in one direction, followed by the interaction region, which in turn is followed by a magnetic field in the opposite direction, and then an analyzer and detector. If the neutrons lose no kinetic energy in the interaction region the second magnetic field will exactly compensate the first, and a maximum number of neutrons will be detected. If, on the other hand, the neutrons lose some velocity in the interaction region, the detected neutron intensity will be reduced. From the change in one magnetic field required to restore the maximum, the loss of neutron velocity can be determined. A number of successful variants of this method have been devised, including one which measures the velocity change by a resonance experiment comparing the direction of the neutron spin with a rotating field whose phase serves as a clock [22].

51.9   Parity Non-conserving Spin Rotations

Heckel, Ramsey, Forte and their associates [11] observed for the first time, parity nonconserving rotations of spin-polarized neutrons passing through various materials. Since the parity-nonconserving interaction $\sigma \cdot \mathbf{v}$ is proportional to the velocity v while the durations of the collisions are inversely proportional to v, the angles of rotation are velocity independent. In subsequent experiments they showed that the rotation angles with various isotopes of tin, lead, and lanthanum ranged from -217 to $+2.25$ micro-radians per cm of material traversed. In some cases this was much larger than originally expected, due, presumably, to near degeneracy with a P state of the complex nucleus. Although this enhancement was helpful in providing a more easily measurable rotation angles, it also meant that the results depended on both particle and nuclear physics phenomena. Presently, Heckel and his associates are preparing an experiment to observe the parity-nonconserving spin rotation of neutrons passing through ^4He and $^1H_2$. In these light nuclei the parity-nonconserving particle interactions should be more easily distinguished from nuclear effects.

References

[1] F. Bloch and L. Alvarez, Phys. Rev. **57**, 111 (1940).

[2] N. Corngold and N. F. Ramsey, Phys. Rev. **104**, 283 (1956).

[3] G. L. Greene, N. F. Ramsey, W. Mampe, J. M. Pendlebury, K. Smith, W. B. Dress, P. D. Miller and P. Perrin, Phys. Rev. D **20**, 2139 (1979).

[4] K. F. Smith, N. Cramptin, J. M. Pendlebury, D. J. Richardson, D. Shiers, K. Green, A. I. Kilvington, J. Moir, H. B. Prosper, D. Thompson, N. F. Ramsey, B. R. Heckel, S. K. Lamoreaux, P. Ageron, W. Mampe, and A. Steyrl, Phys. Lett. B **234**, 191 (1990).

[5] I. S. Altarev, Uy. V. Barisov, N. V. Borovikova, E. A. Kolomensky, M. Lasakov, A. P. Serebov, Yu. V. Sobolev, E. V. Shulgina, and A. I. Yegerov, Phys. Lett. B **276**, 242 (1992).

[6] G. L. Greene, N. F. Ramsey, W. Mampe, J. M. Pendlebury, K. Smith, W. B. Dress, P. D. Miller and P. Perrin, Phys. Rev. D **20**, 2139 (1979).

[7] T. Bitter and D. Dubbers, Phys. Rev. Lett **59**, 251 (1987).

[8] D. J. Richardson, A. I. Kilvington, K. Green and S. K. Lamoreaux, Phys. Rev. Lett. **61**, 2030 (1988).

[9] F. Mezei and P. A. Dagleisch, J. Phys. France **45**, C3 (1984) and Comm. Phys. **2**, 41 (1977).

[10] R. Golub and R. Gahler, Z. Phys. **B65**, 269 (1987) and Phys. Lett. A **123**, 43 (1987).

[11] B. Heckel, N. Ramsey, M. Forte, K. Greene and G. L. Green, Phys. Rev. Lett **45**, 2088 (1980); Phys. Lett. B **119**, 298 (1982); Phys. Rev. C **29**, 2489 (1984); J. de Physique **45**, C3, 89 (1984).

[12] D. Dubbers, Prog. in Part. and Nucl. Phys. **26**, 173 (1990).

[13] N. F. Ramsey, Phys. Rev. **78**, 695 (1950).

[14] E. M. Purcell and N. F. Ramsey, Phys. Rev. **78**, 807 (1950).

[15] J. H. Smith, E. M. Purcell and N. F. Ramsey, Phys. Rev. **108**, 120 (1951).

[16] N. F. Ramsey, Ann. Rev.Nucl. Part. Science **32**, 211 (1982) and **40**, 1 (1990). These articles review the field and gives references to the many experimental and theoretical papers.

[17] N. F. Ramsey, Proceedings of the 14th Intl. Conf. on Atomic Physics **14**, 3 (1995).

[18] I. B. Khriplovich, Nucl. Phys. **B352**, 385 (1991) and I. B. Khriplovich and R. S. Conti, Phys. Rev. Lett. **68**, 3262 (1992).

[19] R. Golub and S. K. Lamoreaux, Physics Reports **237**, 1 (1994).

[20] C. Cohen-Tannoudji and S. Haroche, J. Phys. France **30**, 125 and 153 (1969).

[21] F. Mezei, Z. Phys. **225**, 146 (1972).
[22] E. Muskat, D. Dubbers and O. Scharpf, Phys. Rev. Lett. **58**, 2047 (1987).

52

An Appreciative Response to Townes on Science and Religion

Robert J. Russell

52.1 Introduction

With the elegance and seasoned wisdom of one supremely distinguished in his field, Charles Townes in the chapter "Spiritual Views from a Scientific Base" of his book *Making Waves* [1] recounts the astonishing story of one of humanity's most important technologies—the laser—and his pivotal role in its discovery and development. In the process he suggests how technology serves a source of basic science, how difficult it is to foresee and evaluate new technologies on the horizon, and how crucial it is that the public become educated about, and involved in decisions concerning, science and technology.

Still, in what may be this book's most important contribution, Townes argues strongly for the similarities between science and religion and their eventual convergence, urging us to see beyond the apparent conflicts between them or the attempt to compartmentalize them. In this short essay, I want to summarize the reasons Townes gives for the similarity between science and religion and then place these arguments within the context of the growing interdisciplinary research. Finally I want to suggest that, while we do not yet know for sure what lies on the distant horizon, an important step forward is now being taken through the ongoing creative mutual interaction of science and religion.

52.2 Similarities between Science and Religion

Townes describes four broad categories in which science and religion are much more similar than they are different. We can summarize them as follows:

a. The role of faith in science. Townes claims that faith plays a role, not only in religion, but also in science. "(F)aith is essential to science...Faith is necessary for the scientist to even get started, and deep faith necessary for him to carry out his tougher tasks." There are at least two central beliefs which scientists hold by faith: belief in "an objective and unique reality which is shared by everyone" and belief that "there is order in the universe and that the human mind...has a good chance of understanding this order." Without such faith, one would never undertake the grueling efforts of scientific research. With these beliefs, science has been able to make the kind of progress that distinguishes our age from "an age of superstition." Townes compares the role of faith in science to its role in religion, citing Constantine's famous claim: "I believe so that I may know."

b. The Role of Revelation. Granting that the discoveries of religion come primarily from revelation, what about science? Although the popular view stresses deduction from raw data, Townes sees this as a "travesty on the real thing." The discovery of the structure of benzene by Kekule is more akin to the religious revelation of a Moses confronted by the burning bush or Gautama the Buddha sitting under a Bo tree. Scientific insight, though something which we cannot yet describe 'scientifically,' does not come according to the "scientific method" of data analysis. Instead scientific insight, like religious inspiration, is laced with intuition, accident, and the joy of a "wonderful idea."

c. The problem of proof. Are our scientific ideas absolute and universal, since they have been proven, while our religious claims are limited, being based in faith and revelation? As Townes points out, science is subject to Gödel's theorems which prove that a set of postulates can never be both self-consistent and complete. Secondly, Townes reminds us that science can disprove but never prove its theories. Religion deals with much more complex problems than science. Perhaps we should regard it, too, as a set of "working hypotheses, tested and validated by experience."

d. Limits of paradox and uncertainty. Townes lists the following: the tentative nature of knowledge in both fields; the occurrence of paradoxes and limited knowledge (the nature of light in physics? the presence of suffering given a loving God in religion?); the uncertainty principle and with it complementarity (position and momentum in physics; the human person involving justice and love).

In the end, Townes believes that science and religion should "at some time clearly converge. I believe this confluence is inevitable." [1] In the end, the basic question comes down to this: "How well we can commit our lives, effort, and devotion to ideas which we recognize in principle as only tentative represents a real test of mind and emotions...(W)e must be willing to live and act on our conclusions." [1]

52.3   The Broader Context of Townes' Claims

In this chapter, Townes is tapping into a series of key arguments under development now for several decades by scientists, philosophers and theologians who broadly support the claim that science and religion are remarkably similar. One of the clearest analyses of these developments has been given by Ian Barbour [1], and I shall be following it here.

First we focus on the role of faith in scientific epistemology. One of the earliest figures to point this out was Oxford physicist C. A. Coulson. In 1955, Coulson argued that scientists hold precisely the faith presuppositions that Townes cites, namely that the world is orderly, lawful and intelligible, and Coulson claims that they are essential to science [2]. This view has been stressed recently by British physicist and Anglican priest, John Polkinghorne, who writes that "Without (belief in the intelligibility of nature) science would be impossible." [3] According to physical chemist and Anglican priest Arthur Peacocke, the search for intelligibility is part of the drive through which both science and religion makes sense out of the experimental data they share [4]. Twentieth century science, according to Peacocke, brings together three perspectives: the search for intelligibility, the questions concerning our existence as such, and our search for personal meaning [5].

Structuring much of this conversation was the watershed text, *Personal Knowledge* [6], by Michael Polanyi. Published in 1958, Polanyi explored in detail what he called the "fiduciary component" of science. "We must recognize belief once more as the source of all knowledge. Tacit assent and intellectual passions, the sharing of an idiom and of a cultural heritage, affiliation to a like-minded community: such are the impulses which shape our vision of the nature of things on which we rely for our mastery of things. No intelligence, however critical or original, can operate outside such a fiduciary framework." [7] Polanyi cites St. Augustine as giving us the maxim which encapsulates this view and by which the West made its break with Greek philosophy: *nisi credideritis, non intelligitis* ("unless you believe, you shall not understand") [8]. It is through the participation in a set of assumptions and key questions that we are equipped to explore the evidence and relate it to these assumptions. Through faith we gain an understanding of self and world.

What about the construction and proof of scientific theories? It was in the explosive period of 1958-1962 that a variety of philosophers and scientists launched the strongest attacks yet on the empiricism of the first half of this century and its strictly rationalist account of science. During these pivotal years, Thomas Kuhn [9], Norwood Hanson [10], Paul Feyerabend, Stephen Toulmin [11] and Imre Lakatos [12], and others developed several further arguments against the earlier neo-positivism: theories color the data by telling us what categories of data are relevant by how best to interpret them [13]; theories resist direct (Popperian) falsification; when theories are replaced their terms are radically redefined [14]; and science must be seen in a sociological perspective to be best understood. Kuhn's development of the concept of paradigms in science is particularly important. A paradigm involves a broad set of conceptual and methodological presuppositions held by a community of scientists and represented by a standard example. It focuses research on certain types of questions, provides acceptable modes of explanation, and determines the appropriate form of solutions [15]. Paradigms condition the way scientists see the world and determine what can count as data from the infinite plethora of phenomena in nature. Moreover no single criterion is used for choosing between competing paradigms. Scientists may appeal to simplicity, elegance, or mathematization, instead of the empiricist's preference for predictive power, as a reason for staying within a paradigm. For example, P. A. M. Dirac preferred aesthetics over prediction [16] as did Einstein [17]. The problem is that, since none of these criteria dominate the rest, none can serve to induce a unanimous choice between competing paradigms.

How does this relate to religion? Ian Barbour has brought many of these ideas together in his detailed analysis of similarities and dissimilarities between science and religion [18]. Barbour first identifies the subjective elements and the objective elements at work in both science and religion. The subjective elements are (quoting Barbour): (1) all data are theory-laden; (2) comprehensive theories are highly resistant to falsification; (3) there are no rules for choice between research programmes. The objective elements are (again quoting Barbour): (1) rival theories are not incommensurable; (2) observation exerts some control over theories; (3) there are criteria of assessment independent of particular research programmes. Barbour then claims that while the objective elements are predominant in science and the subjective ones in religion, there is a degree of subjectivity in science and of objectivity in religion. Thus science and religion are similar, but certainly not identical. The lack of lower-level regularities and the difficulty of reaching consensus in religion are significant features which differentiate it strongly from science.

Arguments for both the similarities and differences between science and religion are being developed further by theologians, philosophers and scientists such as Philip Clayton, Philip Hefner, Mary Hesse, Chris Kaiser, Hans Kung, Sallie McFague, Nancey Murphy, Wolfhart Pannenberg, Ted Peters, Janet Soskice and Bill Stoeger. With Townes, they underscore the limitations and hypothetical nature of all knowledge, the presence of paradox, complementarity and uncertainty in theology, the pivotal role of insight, and the complex ways in which data from the wealth of human experience and nature interact in the testing of theological proposals. With Townes, they point beyond the present dialogue to a richer understanding of the enormously complicated but crucially important relations between these diverse fields.

52.4   Looking Ahead

Unlike so many of his peers who dismiss religion, Townes has thought deeply about his faith, and out of this he points towards a convergence of science and religion as the goal looming on the distant horizon. I applaud him for the courage it takes to hold this view when science is so often elevated to ultimacy and religion so often relegated to the ghetto of subjective distortion. Indeed, something is 'definitely afoot,' something which will take us far beyond the naive and often dogmatic confines of those holding that science and religion are utterly independent or in inevitable and total conflict. At present we do not really know for sure what lies in the future, though I believe we are going to find out. For now, its shape shimmers tantalizingly on the horizon. It may indeed involve the convergence Townes envisions, converging to a common discipline of shared presuppositions, insights, methods of inquiry, and a wider calling to human service.

What I find particularly exciting about the time we live in, however, is that, in order to move towards the horizon and discover what lies there, scholars from around the world are now increasingly engaged in research aimed at the mutual creative interaction of the natural sciences and religion. A variety of centers and institutions located throughout the world sponsor conferences, education, and public service on such scientific issues as cosmology, physics, evolution, genetics, technology and the environment. The programs include the publication of books and journals, development of courses in university, college and seminary contexts, sponsoring of colloquia within existing professional societies in each field, and the creation of new international interdisciplinary societies. Religious leaders and organizations as well as scientists and technologists are active in each of these programs. Moreover, these programs include forums for the wider public discussion of the ethical, legal, social and spiritual implications of science and technology. Perhaps most importantly, this interaction seeks to honor the diversities within religion, within science, and between religion and sciences. The interaction requires of all participants a radically self-critical stance in with other dialog partners are heard in their own authenticity, integrity, and perspective on truth [19].

And so Townes is not only a pioneer in physics but in helping to create this interdisciplinary interaction. In *Making Waves*, Townes acknowledges—with characteristic humility—being fortunate for the colleagues he has worked with and the opportunities he has received. Here I must beg to differ. It is we who are indeed fortunate to have someone whose pioneering genius in physics has reached beyond the spheres of the scientific community and government service to shed light on the wider cultural significance of science and its growing unity with religion, and who is taking an active role personally in supporting and shaping the mutual creative interaction between these fields.

References

[1] C.H. Townes, *Making Waves*, AIP Press (1995).

[2] Ian G. Barbour, *Issues in Science and Religion*, Harper Torchbook (1966), and *Myths, Models and Paradigms: A Comparative Study in Science & Religion*, Harper & Row (1974).

[3] C. A. Coulson, *Natural Religion and Christian Theology*, Vo. 2.10, Cambridge University Press (1953).

[4] John Polkinghorne, *Science and Creation: The Search for Understanding*, p. 20, New Science Library (1989). See also John Polkinghorne, *One World: The Interaction of Science and Theology*, p. 45, SPCK (1986).

[5] A. R. Peacocke, *Creation and the World of Science*, p. 33, Claredon Press (1979).

[6] *Ibid.*, p. 74.

[7] Michael Polanyi, *Personal Knowledge,* University of Chicago Press (1958).

[8] *Ibid.*, p. 266.

[9] Perhaps an even clearer statement of the central role of faith in the epistemic process is Augustine's pivotal claim: *fides quaerens intellectum* ("faith seeking understanding").

[10] Thomas S. Kuhn, *The Structure of Scientific Revolutions*, University of Chicago Press (1962).

[11] See in particular Norwood R. Hanson, *Patterns of Discovery*, Cambridge University Press (1958).

[12] Stephen Toulmin, *Foresight and Understanding,* Hutchinson's University Library and Indiana University Press (1961).

[13] Imre Lakatos, "Falsification and the Methodology of Scientific Research Programmes," in I. Lakatos and A. Musgrave, *Criticism and the Growth of Knowledge*, Cambridge University Press (1970).

[14] As Hanson put it, all data are "theory-laden."

[15] As Feyerabend put it, they are "incommensurable."

[16] Classical physics was dominated by Newtonian mechanics, epitomized by the falling apple story, in which questions about nature were restricted to the categories of matter in motion, explanations were given in terms of efficient causes represented as forces producing accelerations, and solutions were stated as predictions found through mathematical solutions to the appropriate differential equations.

[17] P.A.M. Dirac, "The evolution of the physicist's picture of nature," Scientific American (May, 1963). I am grateful to Paul Davies for this reference, cf. *God and the New Physics*, p. 221, Simon & Schuster (1983).

[18] See Einstein's response to reporters about the eclipse data of Sir Arthur Eddington.

[19] See in particular Ian G. Barbour, *Myths, Models and Paradigms*, especially p. 118, 144-145.

[20] From where I stand in the theological world, I work diligently to convince one side of the interaction—the religious community—that it cannot do its job without a serious and self-critical interaction with science—cosmology, quantum physics, molecular genetics, population evolution, and so on—and that it can do its job better if it takes science on board. I also write that scientists can benefit in their research by engaging in long-term conversations with philosphers and theologians who can articulate the wider social and historical signifance of their work and its theological root in the history of the West.

53

The Academic Ivory Tower Under Siege

Bernard Sadoulet

53.1 The Crisis of Science and Society

Science, and physics in particular, is clearly in crisis. There are many symptoms: the cancellation of the Superconducting Super Collider (SSC), the technological vulnerability of NASA's large projects (e.g., the Challenger accident, the flawed Hubble Space Telescope before its repair), the proposed abolition of the Department of Energy, the questioning of the role of the national laboratories, and so forth. At the same time, we are facing a painful unemployment crisis in physics, with many of our postdoctoral fellows unable to find tenure-track academic jobs [1]. The budgetary outlook is grim, with civilian basic research slated by the Republican House for a 35 percent decrease in purchasing power between 1996 and 2000 [2], and we are witnessing a deep unhappiness among scientists, see, e.g., [3]. The burden of proposal writing, the numerous evaluation and bureaucratic requirements, the intense competition for inadequate resources, the increasing time pressure which jeopardizes deep thinking and originality, and the impression of being misunderstood by society all contribute toward dampening our enthusiasm.

To a large extent, the stresses of our profession mirror a more profound crisis in our society at large. This crisis is first one of goals. With the end of the Cold War, we seem to be missing a sense of mission. We are increasingly troubled by the difficulties of living in the global world in which we find ourselves: an embattled American hegemony, a serious loss of economic competitiveness in a few key sectors, and the rise of other economic giants (Japan, Europe, China). This is also a crisis of resources, both for the federal government and industry. The national debt keeps rising, the budget share of discretionary spending is under severe pressure, and we have less and less the means of carrying out our policies. Simultaneously, confronted by harsh international competition, companies have chosen to scale back any long term investments (including basic research). There is finally a crisis of government: the debate between socialized and entrepreneurial solutions, the tensions between pluralism and centralization, the need to streamline bureaucracy and "reinvent government," the failure of our educational system, and so forth. This societal crisis deeply affects the conditions under which science is done. Not only have we lost the prevailing justification to fund basic research (just in case the Soviets might have discovered something militarily crucial before we did), but the impact of scientific research on the long term welfare of the nation is being questioned. The changing majorities in Congress and the successive administrations seem trapped in large oscillations in their attempts to redefine the role of science in this new environment. And out of fear of losing support, our funding agencies may not have helped in their haste to appease Congress with such clumsy responses as labeling more than half of the National Science Foundation's activities "strategic research" to be sure to appear to distance them from "curiosity-driven research."

What a perfect target this latter characterization of basic research offers, as the public at large seems to be increasingly skeptical about science [4]! True, at a deeper level, some enthusiasm remains: awe in the face of revealed mysteries of the universe, appreciation for spectacular technological achievements, a sense of prestige. However, a number of clouds hover over the scene. Clearly the general public has no idea what science is about, how it works, and how it is related to everyday life. By and large, our schools fail to provide a sufficient scientific background at least basically to ground our citizens in the many relationships between science, technology and society. The media usually portrays science as some strange activity performed by "nerds" of uncertain genius, isolated in their separate laboratories. The person in the street (or, in most cases, his or her Congressperson) has no idea about the way "scientific truth" is arrived at, *a fortiori* about the peer review system or what we mean by serendipity in basic research. Costs seem enormous, even if we claim that the returns will be proportionally much greater than the investments. The timescale for payoffs is totally underestimated. Worse, there is clearly a negative perception of the scientific community. We appear to be increasingly irrelevant, trapped in old and abstruse problems, an arrogant cast of priests living in elite isolation, only able to clone themselves. We are also seen as a community given to bickering, unable to agree about the priority of projects or the mode of funding. And one of the sad legacies of the SSC story has been that physicists are now perceived as another special interest group, undermining any attempts to explain our impact upon society. To make things even worse, ethical problems have come to the surface: the more prevalent being our perceived reluctance to teach undergraduates in favor of doing research, the more spectacular being well-advertised affairs of scientific misconduct or fraud with federal funds. Perhaps more fundamentally, these doubts come from deep worries about a world which becomes more complex every day, and about vaguely threatening technologies which are increasingly difficult to evaluate. These anti-science reactions are not only those of a New Age fringe, nor of marginal reactionary movements. They testify to a deep rejection of the cold "objectivity" of the technical world and of decision mechanisms dominated by "experts" which cannot be democratically controlled, see, e.g., [5]. I am convinced that this anti-science movement is first an anti-technocratic movement.

53.2   The Responsibility of Scientists

Unfortunately, as scientists we are seen as part of this technocracy. I believe that we bear part of the responsibility for such a state of affairs. I would submit that a primary factor is that we have let our scientific activity increasingly drift away from society, by unconsciously isolating ourselves from the world. Of course, a certain amount of isolation is part of the intellectual process, which requires stepping back from the distractions of everyday life; scientists and philosophers have always been somewhat remote. What is new since World War II [6], however, is that science has become a true profession, with a large number of practitioners and an increasingly self-defined agenda. The necessary intellectual abstraction is becoming esotericism and hyperspecialization. In the large expansion of the number of scientists, we have tended largely to replicate the nineteenth century scientific community, with a very small number of women and even fewer minority members in our ranks. Moreover, the evolution of a scientific field takes on an existence of its own when the survival of large corps of scientists and engineers becomes a primary consideration in the decisions of new directions for the field. The debate in the scientific community about the construction of the SSC is a case in point. It may also be argued that the federal funding of science established after the war has contributed to separating the scientific enterprise from its base [7], as it was no longer necessary for the scientific community and the university to maintain close links with local industry, which used to be their major benefactor. Finally, the primacy of research—partially because it is a source of income and prestige for the university—has led to a clear de-emphasis on undergraduate education [8]. Combined with the difficulty of bridging the "two cultures" of C.P. Snow [9], our schools tend to produce a scientifically illiterate citizenry (and we have ourselves become humanistically illiterate), not to mention the implicit message that we give to our students that any career outside academia is a failure.

These factors have led to the general perception of science as an ivory tower. The abstraction of our fields makes it very difficult to explain the frontier scientific questions to the public at large. We have broken most of our links to the humanities. Our agenda seems increasingly foreign to the problems of the nation, and in our gender and race composition we are deemed not representative of the society at large. The impact that this can have on the scientific process is frightfully exemplified by the lack of studies on women's illnesses by a predominantly male medical profession. Moreover, it certainly contributes to the perception of science as a vaguely hostile microcosm in which differences are not welcomed (and therefore do not correct themselves). Academic scientific research is seen by many industries as irrelevant and the Ph.D. graduates that we train difficult to assimilate in the industrial sector. Finally, some members of Congress would like us to quit "splitting hairs" to finally become "useful"!

I would add that because of our isolation, not only are we not understood, but also we do not understand society and have developed what I am obliged to call a fair amount of arrogance. Intellectual arrogance, first: I am speaking here not only of the vigorous debates between ourselves, with the all too frequent violent clashes of strong egos, but of our general pretensions based on our absolute faith in the scientific method. Rationalism soon becomes reductionism and we tend to legislate in the name of science, forgetting the multiplicities of human and cultural factors. In that sense, we are responsible for the technocratic ideology that technological fixes are always the solution. As a group, we also poorly understand political realities and are often also politically arrogant. Unfortunately we gave too many examples of this tendency during the campaign defending the SSC! We consider federal funding as an entitlement and tend to blast any opponent as reactionary and illiterate. Any arguments are deemed good if they justify the support we receive, even if sometimes they are at the limits of honesty and overstate the impact of our activity. How many of us have not argued about the impact on the medical sector of our physics research? Is this really our primary motivation? Even our attempts to defend the long term nature of basic research take on the tone of a demand to be protected from the corrupt nature of the world. Our call for the opportunity to set the direction of our research ourselves and our priorities is interpreted as a childish demand for the "freedom to play."

53.3   Opportunities for Change

I realize that so far I have been very critical of our community, and may have upset a large number of my colleagues reading these lines, by appearing to side with the adversaries of science at a time where our funding is at risk. Yet I passionately believe in science, I hold that our research institutions still have the noblest of missions, and . . . I certainly support a strong federal funding of science. But in my mind there is no point in presenting a hagiography of science. It does not help our cause to attempt to appear superhuman and beyond criticism. We should acknowledge that we have not yet fully assimilated the professionalization of science and that major changes are on the way. And the ivory tower may indeed have to be leveled. The main question for me is the following: In this time of crisis, do we, as scientists, want to cling to old descriptions, steeling ourselves in a futile attempt to let the storm pass over us, or do we want to look squarely at the reality and these changes instead of resisting them?

Let us return for a moment to what I believe to be the fundamental mission of basic research. In my mind, this is first to revolutionize our vision of the world and of our place in it. Such cultural shifts [10] are rare—the Copernican revolution, the evolution of Darwin, the genetic code—but they are the ultimate goals of our activities. This is what scientific research is tinkering about. Our quest for intellectual rigor and excellence, our insistence on the scientific method, and our debates are tools for us to extricate ourselves from presuppositions or ideology. Our search for a more accurate description of our environment is directed at finding the detail that does not fit in the standard model which could trigger illumination and paradigm shift. Our insistence upon allowing scientists to freely pursue their intuition comes from the fact that this process is highly uncertain and individual, and it is impossible to guess with certainty what will lead to breakthroughs. Fundamental research is part of this exploration of the possible and of the future, which is so basic to human nature, as is art or literature. A civilization which gives up on this exploration is in the long run doomed. Of course the question is whether in the short run we can afford these activities and what share of resources to devote to them. The marvelous mystery of science is that over time, this noble cultural quest for another vision of the world profoundly influences our lives and increases our economic wealth. I would submit that this transformation is the second fundamental goal of our activity. Not only does our attitude with respect to nature and the environment change, but in the process we learn a number of skills which totally alter our economy and our culture. Take the example of the transistor and the integrated circuit, which emerged from apparently esoteric research on the quantum nature of matter and is now revolutionizing our civilization and represents a large fraction of our gross national product. This elaboration of another world model cannot, however, take place in isolation. I would suggest that we tend to entertain an inadequate description of the scientific process, one that is too pyramidal and linear—with scientific research at the top and applications at the bottom—that ignores the fact that creativity emerges first from interactions. And this may be a profound cause for the ivory tower we build around ourselves. We still have a myopic vision of the many connections between basic research and society, which are complex, multidimensional, and formed by history. Our questions emerge from our experience [11], while the answers we obtain modify this experience. We scientists, while we can profoundly influence this culture, at the same time are also the product of the surrounding culture. In other words, the standard description of the scholar isolated from the world is inaccurate; worse, it is both an impediment to the desired grounding of our pursuits in reality, and an obstacle to society's being able to benefit from its investment in research.

This is why I hold it important that we vigorously engage ourselves in society. This is not an external requirement, foreign to our mission; it is part of it, a necessity for our fundamental goal to transform society [12]. In the rest of this essay, I would like to identify two directions in which we should get more involved. The first one is industry. Although the present majority in Congress de-emphasizes contact between academia and industry, we would be foolish not to actively pursue such contacts. It is a matter of moral obligation with respect to our students who will increasingly end up in industry instead of academia. How can we train them effectively for a world we do not know? It is a matter of ethics to pay back our debt to the taxpayer who foots the bill. It is also a matter of survival, as irrelevant activities will not be supported very long. What could we contribute to the general welfare in the medium term? First, of course, our students. They are the best "technology transfer" from universities to industry. The problem which develops is that, because of the lack of contacts, our curriculum and our degrees may not be best adapted to the needs of industry. In particular, the Ph.D. appears increasingly obsolete in the economic sector: too time-consuming and too specialized. We should have the courage to invent in collaboration with industry other degrees if they are needed. At a second level, I believe we could also contribute more efficiently to the progress of technology. The traditional distinction between fundamental and applied research is artificial and misleading. In my mind, it reflects more our demand for autonomy than the reality of our experimental laboratories. As a cosmologist, for instance, I know only too well that our science critically depends on advanced sensors, and although it is difficult to imagine a science more fundamental than cosmology, most of what I (or more precisely my students) do every day in the laboratory is indeed very applied in our attempts to invent or improve our instruments. Although this sensor development involves industry in some cases, for instance in a mutual effort to reach the demanding specifications needed for our application, there may not be enough contacts. This leads to inefficiency in our own research. As physicists during and immediately after World War II, we had to invent many tools ourselves, for instance our own computer programs, because computers were too new yet for industry to be involved. In too many cases, we tend to still behave in the same way and "reinvent the wheel," expending enormous effort in the development of products which are already commercially available but of which we simply do not know. Conversely, some of the clever designs or innovative sensors we come up with that could be interesting in the economic sector never make their way to industry, simply because we do not have the appropriate contacts. At a third level, our fundamental studies, particularly in condensed matter physics and materials science, could readily lead to interesting practical applications. In the occasion of this Festschrift, it is interesting to meditate on the example of the work of Townes in molecular spectroscopy and his discovery of the maser, which led to the laser and, among other applications, fiber optics telecommunications (in spite of the questioning of his superiors at Bell Laboratories wondering, before the discovery, what molecular spectroscopy could possibly have to do with telecommunications). The obstacle encountered here again in too many cases is the lack of an efficient flow of information, of organizations like Bell Labs or a start-up firm where connections with practical applications can be established. It is not the calling of most of us in academia to spot new entrepreneurial opportunities and to develop new products, but we should think seriously how to improve linkages to dynamic firms which could run with some of our ideas. Finally, as academics, we could also help industry to master the overwhelming flow of information by appropriate reviews, for instance, at the occasion of the meetings of professional societies: Where are the various technologies going, what promise do they hold, what are the fundamental limits (e.g., quantum limits to sensitivity)? But for these reviews to be useful, we are also required to know what the trends are in industry. In summary, there are a number of ways we could be useful to industry on a much shorter timescale than that of the radical concept which will revolutionize our economic production and our lives, but this impact is directly proportional to the efficiency of our knowledge exchange with industry. I prefer the phrase "knowledge exchange" to "technology transfer" which does not carry the bi-directional character of the needed flow of information. It relies on relationships between individual scientists and engineers in academia and industry and is truly a "contact sport"! We need to be more creative on how to improve these linkages: for example, informal contacts at "industrial liaison programs" inviting members of industry to visit our campuses, courses in our universities by industrial scientists, graduate student internships, or professor sabbatical leaves in industry, consulting, and so forth.

The second natural direction in which we can step out of our ivory tower is education, conceived in a more general sense than just teaching our own students at the undergraduate, graduate and postgraduate levels. As would many of my colleagues, I would claim that it is our duty to explain to the public at large the science that it has paid for. Whether this is done through the media or through public lectures, this is always an enthusing but delicate exercise. Not only is it not easy to describe in simple terms the basic concepts and the purported breakthroughs, but it is also difficult to present them in a way which does not perpetuate the prevalent outdated models. How do we at the same time share the excitement about a result and the deep uncertainties we may still have about the whole picture? How do we describe the teamwork at the basis of most scientific discoveries while the news media want to focus on a single hero? Another difficulty arises when the general media are used to publicize a result before it is peer reviewed, in order to improve our chances of funding or to claim priority. Although general standards of behavior are well defined in the scientific environment, we often witness serious lapses of those standards, in particular when funding is at stake. Another role that we can assume in the scientific education of the surrounding society is to participate in the political debates of issues involving a large technological component: for instance, nuclear energy, information networking, or genetic engineering. If we succeed in explaining the technical issues and possible scenarios while not presenting ourselves as infallible experts, we could contribute to an evolution beyond a technocratic mode of government. In yet another direction, many of our younger colleagues are enthusiastically involving themselves in K–12 education, helping locally to overcome the difficulties of a bankrupt education system and improve the scientific literacy of our children. Moreover, contact at an early age with a scientist is the best vehicle to demystify science, to improve understanding of its process, and to provide role models for a potential career in science or technology. Many programs are also allowing primary and secondary school teachers to spend time in research laboratories, to rekindle their enthusiasm and develop a new understanding of current science. I would even add that we may want to involve undergraduate and graduate students—whom our primary duty is to educate—in these activities, as they not only represent a large pool of enthusiastic manpower but will also tremendously benefit from this community service. There is no better way to assimilate material than to explain it to others, no better way to train an interactive scientist or engineer than to immerse him or her early on in diverse environments.

In summary, the relation between science and society is much more complex than our usual descriptions imply. Compounded with a number of historic circumstances—such as the inheritance of the Manhattan Project and the ensuing confidence in science, the Cold War, and a strong growth in federal funding—the limits of our paradigm may have isolated us in a comfortable ivory tower. For better or worse, those times are gone; the atmosphere is now one of crisis and our survival is at stake. The question for us is whether we cling to the past, defending our privileges under siege, or take the initiative and creatively adapt to the new environment. I submit that as a community, we should take the second line of action and depart willingly from the ivory tower. There is a lot to be done, both in improving linkages with industry and in being more effective educators in the society at large. And many of our colleagues are spearheading innovative efforts in those directions. Yes, there will always be tension between our academic work and our engagement in society; this will require profound changes in our organization and our tenure and reward system. But this is the price we must pay if we are determined not only to survive but to reconnect to the deeper mission of our endeavor: to be a humble but effective ferment in society, to help it to adapt to the present and to explore the future, and to constantly rebuild its vision of the world.

References

[1] K. Kirby and R. Czujko, "The Physics Job Market: Bleak for Young Physicists," *Physics Today* **22,** (Dec. 1993).
[2] *New York Times* (May 22, 1995).
[3] L. Lederman, "Science: The End of the Frontier," *Science* **256** (1992).
[4] See, e.g., G. Holton, "The Value of Science at the 'End of the Modern Era,'" presented at the 1993 Sigma Xi Forum on Ethics, Values, and the Promise of Science, Harvard preprint (1993) and Science and Anti-Science" (Cambridge, Harvard University Press, 1993).
[5] H. Vaclav, "The End of the Modern Era," *New York Times*, March 1, 1992 and the reflections it inspired in George E. Brown, Jr. "The Objectivity Crisis," *Am. J. Phys.* **60**, 779 (1992).
[6] This was largely due to the influential report of Vanevar Bush: "Science, The Endless Frontier" which defines the basic concepts of the current system.
[7] N. Rosenberg and R. R. Nelson, "American Universities and Technical Advance in Industry," *CEPR Pub. #342,* Center for Economic Policy Research, Stanford (1993).
[8] This is changing rapidly; see, e.g., the UCLA conference on "Reinventing the Research University," 1994, reported on, e.g., by J. Maddox, *Nature* **369**, 703 (1994).
[9] C. P. Snow, *The Two Cultures and the Scientific Revolution* (Cambridge, 1959).
[10] T. S. Kuhn, *The Structure of Scientific Revolutions* (Univ. of Chicago Press, 1962).
[11] John Von Neuman wrote that "As a mathematical discipline travels far from its empirical source, or still more, if it is second or third generation only indirectly inspired by ideas coming from reality, it is beset with very grave dangers. It becomes more and more aestheticising, more and more purely *l'art pour l'art*. . . . At a great distance from its source, or after much 'abstract' inbreeding, a mathematical subject is in danger of degeneration." Quoted in P. A. Griffiths, "Science and the Public Interest," *The Bridge*, National Academy of Engineering, **23**, 3 (1993).
[12] Numerous calls to action have been made recently. Of particular interest are the texts of P. A. Griffiths, "Science and the Public Interest," *The Bridge*, National Academy of Engineering **23**, 3 (1993) and L. S. Wilson, "The Scientific Community at a Crossroads; Discovery in a Cultural and Political Context," Keynote Address, AAAS Meeting (1993).

54

The Correlated Spontaneous Emission Maser Gyroscope

Marlan O. Scully
Wolfgang Schleich
and Herbert Walther

We dedicate this paper to our hero Charles Townes in recognition of his pioneering work in maser and laser physics as one of the many spin-offs of his great inventions.

In the present paper we propose and analyze a new type of Sagnac gyroscope [1], the correlated spontaneous emission [2] maser (CEM) gyro, which has a potential sensitivity superior to the best state-of-the-art optical laser gyro. The proposed CEM device utilizes a holographic [3] (striated gain medium) micromaser [4] configuration, a high-Q cavity such as that used in current micromaser experiments, and a novel intracavity detection scheme based on earlier work [5]. We present the CEM gyro concept, and compare it with a conventional laser gyro in the next few paragraphs. We then outline a quantum Langevin sensitivity analysis for the proposed device and conclude with a numerical example.

The envisioned CEM gyro involves a single-standing wave field, whose nodal and antinodal positions vary with rotation rate. These positions are located via well-defined atomic beams passing through the cavity, or striated thin photo-detectors as in Fig. 54.1. By concentrating on locating the position of minimum photo excitation (the nodes) we determine the relative phase between two traveling wave modes of the gyro with a minimum disturbance of the maser fields. Thus the present detection scheme would constitute a "quantum non-back-action" [6] or at least a "minimally perturbative" measurement scheme [7].

Such a device would be particularly well-suited for general relativistic frame-dragging and cosmological preferred-frame studies [8]. In these experiments the effective rotation rate (due to general relativistic effects) is constant over a long period of time, and a non-back-action measurement would be of special interest.

Conventional laser gyro operation is governed by the (time dependent) phase difference between the counterpropagating waves $\Delta\phi_s(t) = \Delta\Omega t_m$, where t_m is the measurement time, and the Sagnac frequency difference $\Delta\Omega = S\Omega_{\rm rot}$ where

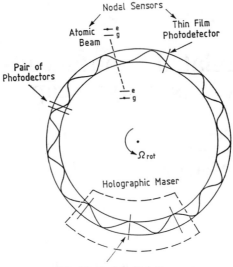

Thin "Film" of Excited Atoms $\rightleftharpoons$

FIGURE 54.1. CEM gyro in which maser field is generated by holographic (striated gain) maser. Rotation rate Ω_{rot} determines position of nodes which are probed by atomic beam or thin film photodetectors. On left side of gyro a pair of detectors is placed on both sides of a nodal point. As the device rotates, one detector will be more frequently excited than the other, thus indicating the rotation rate.

Ω_{rot} is the rotation rate and S is the gyro scale factor which may be written in terms of the gyro area A, perimeter p and the reduced wavelength $\lambda_l = c/\nu_l$ as $S = 4A/(\lambda_l p)$. The limiting noise source in such a gyro is spontaneous-emission phase diffusion [9] which yields a phase noise $\Delta\phi_n = \sqrt{\gamma_l \cdot t_m / \bar{n}}$, where the cavity decay rate γ_l is given by the sum of the absorption and transmission losses as $\gamma_l = \gamma_{\text{abs}} + \gamma_{\text{tr}}$ and $\bar{n}$ is the average photon number in the cavity. Equating the spontaneous emission phase noise to the signal phase, and solving for the resulting minimum detectable rotation rate, we obtain the standard quantum limit for the laser gyro

$$\Omega_{\text{rot}}^{\text{min}} \cong S^{-1}(\gamma_{\text{abs}} + \gamma_{\text{tr}})^{1/2}/\sqrt{\bar{n}t_m}. \tag{54.1}$$

When we express $\bar{n}$ in terms of the photon energy $\hbar\nu_l$ and laser power P as $\bar{n} = P/[\hbar\nu_l(\gamma_{\text{abs}} + \gamma_{\text{tr}})]$, this standard quantum limit reads

$$\Omega_{\text{rot}}^{\text{min}} = S^{-1}(\gamma_{\text{abs}} + \gamma_{\text{tr}})\sqrt{\frac{\hbar\nu_l}{Pt_m}}. \tag{54.2}$$

Next we consider the CEM gyro. In this case the radiation modes are locked due to the CEM action, and the average relative phase angle ϕ obeys the equation of motion

$$\dot{\phi} = \Delta\Omega - \gamma\sin\phi. \tag{54.3}$$

From Eq. (54.3) we see that in the case of strong locking, where $\gamma \gg \Delta\Omega$, the phase angle will be small so that $\sin\phi \cong \phi$ and in steady state $\phi = \Delta\phi_s = \Delta\Omega/\gamma_{abs}$, since γ_{tr} is "negligible" in the present measurement scheme.

In order to determine its sensitivity, we must calculate the noise signal $\Delta\phi_n$ in our CEM gyro. We model the photodetection process, which may in fact involve solid state devices, by considering a beam of atoms which can be photoexcited. A thin beam of such atoms is injected into the maser perpendicular to the propagation direction of the radiation in order to locate the maser nodes. The probability, $P(T)$, of exciting a detector atom [10] is given by

$$P(T, s) = \kappa \int_0^T dt' \langle \hat{E}^\dagger(t', s) \hat{E}(t', s) \rangle \tag{54.4}$$

where κ is a constant depending on matrix elements and density of state factors, T is the interaction time, i.e., the time a detector atom needs to pass through the cavity, $\hat{E}^\dagger(t', s)[\hat{E}(t', s)]$ is the negative [positive] frequency creation [annihilation] operator part of the maser field at position s in the cavity. We write the field operator $\hat{E}$ in terms of the clockwise $\hat{E}_c$ and counterclockwise $\hat{E}_{cc}$ operators as

$$\hat{E}(t, s) = \hat{E}_c(t, s) + \hat{E}_{cc}(t, s) = E_0[\hat{a}_c(t)e^{i(ks-vt)} + \hat{a}_{cc}(t)e^{-i(ks+vt)}] \tag{54.5}$$

where E_0 is the electric field per photon, $\hat{a}_c(t)[\hat{a}_{cc}(t)]$ is the annihilation operator for the clockwise [counterclockwise] mode, and k, v is the wave vector and frequency for these modes. When we substitute Eq. (54.5) into Eq. (54.4) we find

$$P(T, s) = \eta \int_0^T dt' \langle \hat{a}_c^\dagger(t')\hat{a}_c(t') + \hat{a}_c^\dagger(t')\hat{a}_{cc}(t')e^{-i2ks} + c \leftrightarrow cc \rangle \tag{54.6}$$

where $\eta = E_0^2 \kappa$.

At this point we note that above threshold, amplitude fluctuations are not important for the present problem, i.e., only fluctuations in the phase are of interest here. Therefore, we may to a good approximation [11] write our field operators as

$$\hat{a}_c = \sqrt{\bar{n}}e^{i\widehat{\phi_c}} \tag{54.7}$$

and

$$\hat{a}_{cc} = \sqrt{\bar{n}}e^{i\widehat{\phi_{cc}}} \tag{54.8}$$

where $\bar{n}$, $e^{i\widehat{\phi_c}}$ and $e^{i\widehat{\phi_{cc}}}$ denote the average number of photons, and the London phase operators for the c and ec modes, respectively, and we have assumed $\bar{n}_c = \bar{n}_{cc} \equiv \bar{n}$. Inserting Eqs. (54.7) and (54.8) into Eq. (54.6) we have

$$P(T, s) = \eta\bar{n} \int_0^T dt' [1 + e^{-2iks} \langle e^{-i\widehat{\phi_c}} e^{i\widehat{\phi_{cc}}} \rangle] + c.c.$$

Now we adjust our detectors so as to locate the nodes, i.e., inject the detector atoms or move the thin film photodetectors such that $2ks = (2n + 1)\pi - \phi$ where

ϕ is the average relative phase. In this case the probability of photo excitation at a node

$$P(T, s = \text{node}) = \eta\bar{n} \int_0^T dt' \delta\varphi^2(t') , \tag{54.9}$$

governed by the deviation from the null signal, i.e., the phase noise

$$\delta\varphi^2(t') \equiv 1 - e^{i\phi}\langle \widehat{e^{-i\phi_c}} \widehat{e^{i\phi_{cc}}}\rangle + c.c.,$$

will be minimized. Indeed, when the two modes are in phase eigenstates $|e^{i\phi_c}\rangle$ of the London phase operator with

$$\widehat{e^{i\phi_c}}|e^{i\phi_c}\rangle = e^{i\phi_c}|e^{i\phi_c}\rangle$$

and analogously for the counterclockwise mode, we find [12] as a result of $\phi = \phi_c - \phi_{cc}$, a vanishing phase noise $\delta\varphi^2$. Hence in the absence of phase fluctuations such a choice of s will result in a null reading. However, the two modes are not in phase states. Consequently the noise $\delta\varphi^2$ is non-zero and we find a non-vanishing probability. In the appendix we show that the phase noise $\delta\varphi^2$ can be expressed by

$$\delta\varphi^2 = 1 - \left(1 - \frac{1}{4\bar{n}}\right)e^{i\phi}\int_{-\pi}^{\pi} d\varphi e^{-i\varphi} P(\varphi, t) + c.c. \tag{54.10}$$

where $P = P(\varphi, t)$ is the distribution for the phase difference φ. In the case of a CEM this distribution in steady state, i.e., for $T \geq 1/\gamma$ reads [13]

$$P_{ss}(\varphi) \equiv P(\varphi, t \to \infty) = (2\pi\mathcal{D}/\gamma)^{-1/2}\exp[-(\varphi - \phi)^2/(2\mathcal{D}/\gamma)] \tag{54.11}$$

where [2]

$$\mathcal{D} \equiv D(1 - \cos\phi)$$

denotes the diffusion constant of the CEM and D is the Schawlow-Townes phase diffusion rate. When we substitute Eq. (54.11) into Eq. (54.10) and perform the integration we arrive at

$$\delta\varphi^2 = 2\left\{1 - \left(1 - \frac{1}{4\bar{n}}\right)\exp[-\mathcal{D}/(2\gamma)]\right\}$$

which for $\mathcal{D}/(2\gamma) \ll 1$ reduces to

$$\delta\varphi^2 \cong 2\left(\frac{1}{4\bar{n}} + \frac{\mathcal{D}}{2\gamma}\right) = 2\left(\frac{1}{4\bar{n}} + \frac{D}{2\gamma}(1 - \cos\phi)\right) . \tag{54.12}$$

The first term is the shot noise contribution to the phase noise, whereas the second term contains the phase diffusion due to spontaneous emission. When we substitute the expression for the phase noise $\delta\varphi^2$ into Eq. (54.9) the probability of photo excitation at the node is given by

$$P(T, s = \text{node}) = \frac{1}{2}\eta\bar{n}T\left[\frac{1}{\bar{n}} + 2\frac{D}{\gamma}(1 - \cos\phi)\right] . \tag{54.13}$$

We note for detector atoms passing at an antinode of the field we would like to have $P(T, s = \text{antinode}) \cong 1$ so that $\frac{1}{2}\eta\bar{n}T \cong 1$. Moreover, for the ultra-small rotation rates envisioned, that is for $\phi \cong \Delta\Omega/\gamma \ll 1$, we find

$$2\frac{D}{\gamma}(1 - \cos\phi) \approx D\Delta\Omega^2/\gamma^3 \approx \frac{1}{\bar{n}}\Delta\Omega^2/\gamma^2 ,$$

i.e., the phase noise due to spontaneous emission is very small. Hence we may write

$$P(T, s = \text{node}) \cong \frac{1}{\bar{n}} . \tag{54.14}$$

Next we note that our detector arrangement should be such that a rotation rate is readily observed. To this end we could envision symmetrically placed "detector pairs" as in Fig. 54.1. In order that the detector not "load" the cavity we require that $\gamma_{\text{det}} < \gamma_{\text{maser}}$. Now the probability of exciting a detector atom at δs is

$$P(T, \delta s) = \frac{1}{2}\eta\bar{n}T\mathcal{N}\sin(k\delta s) \simeq \frac{1}{2}\eta\bar{n}T\mathcal{N}k\delta s \tag{54.15}$$

where $\mathcal{N}$ is the number of atoms in the beam during a time T. Since $\frac{1}{2}\eta\bar{n}T \cong 1$ we have the probability of atomic excitation

$$P(T, \delta s) \cong \mathcal{N}k\delta s . \tag{54.16}$$

Hence the rate of "photon loss" due to detector atoms is

$$\gamma_{\text{det}} = r_{\text{det}}P(T, \delta s) = r_{\text{det}}\mathcal{N}k\delta s , \tag{54.17}$$

where r_{det} is the rate of injection of detector atoms. Now for optimum operation we should take $\gamma_{\text{det}} \cong \gamma_m$, where γ_m is the maser absorption loss rate, then from Eq. (54.17) the error $k\delta s$ is of order

$$\delta\phi_{\text{det}} = k\delta s = \frac{\gamma_m}{r_{\text{det}}}\frac{1}{\mathcal{N}} = \frac{\gamma_m T}{\mathcal{N}^2} , \tag{54.18}$$

where we have taken $r_{\text{det}} = \mathcal{N}/T$. We can clearly arrange Eq. (54.18) to be small compared to the vacuum fluctuation error $\delta\phi \sim \sqrt{1/\bar{n}}$. Further, as noted earlier, we may set the transmission losses to zero, that is $\gamma_{\text{tr}} = 0$ for the maser, since no light is required for "external" photodetectors.

Returning to the signal, we recall that the CEM gyro yields the phase shift

$$\delta\phi_{\text{sig}} = \frac{S\Omega_{\text{rot}}}{\gamma_m} . \tag{54.19}$$

However, we are allowed a time t_m in which to accumulate signal so that we could accumulate a signal phase for a time $\tau = \gamma_m^{-1}$, next "spin-up" the gyro so as to "null out" the signal Eq. (54.19); then shut off the spin-up bias and allow the system to again come to its steady state with an average phase shift given by

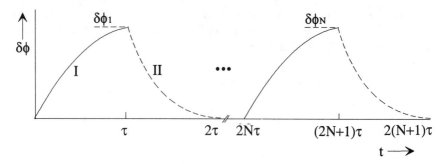

FIGURE 54.2. Figure summarizing measurement strategy of CEM gyro. At $t = 0$ Sagnac phase shift begins to develop towards a steady state (locked) value at $\tau = \gamma_m^{-1}$, region I. At this time, the phase shift $\delta\phi_1$ is recorded and nonreciprocal element (e.g., rotational "spin-up" bias or Faraday effect type bias, etc.) is activated nulling out Sagnac phase shift by time $t \geq 2\tau$, region II. This process is repeated many times yielding a phase shift signal $\delta\phi_N$ in the Nth such sequence, N measurements made in time interval t_m.

Eq. (54.19). This we could do for N times during the measurement interval t_m where $N = t_m/(\gamma_m)^{-1}$. Upon adding the results of these N measurements we have $\Delta\phi_{\text{sig}} = N\delta\phi_{\text{sig}}$ and the measurement-averaged phase shift

$$\langle\delta\phi_{sig}\rangle_{N\text{ meas}} = \frac{S\Omega_{\text{rot}}}{\gamma_m}N = S\Omega_{\text{rot}}t_m . \tag{54.20}$$

The preceding measurement sequence is summarized in Fig. 54.2. However, the same multiple measurement sequence will only increase the noise by $\sqrt{N}$, i.e.,

$$\langle\delta\phi_{\text{err}}\rangle_{N\text{ meas}} = \sqrt{\frac{N}{\bar{n}}} = \sqrt{\frac{\gamma_m t_m}{\bar{n}}} . \tag{54.21}$$

The minimum observable rotation rate as found by equating the signal Eq. (54.20) to the noise Eq. (54.21), is thus

$$\Omega_{\text{rot}}^{\min} \cong S^{-1}\gamma_m^{1/2}/\sqrt{\bar{n}t_m} = S^{-1}\gamma_m\sqrt{\frac{\hbar\nu}{Pt_m}} , \tag{54.22}$$

where in the last step we have used the relation $\bar{n} = P/(\hbar\nu\gamma_m)$.

Comparing the sensitivity of the CEM gyro as given by Eq. (54.22) to the usual laser gyro, given by Eq. (54.2) we see they are superficially similar. But in the present case $\gamma_{\text{tr}} = \gamma_{\text{det}} = 0$, and γ_{abs} (maser) $\ll \gamma_l$ (laser). Calling γ_m and γ_l the total decay rate of the present CEM and the conventional laser gyro, respectively, and denoting the maser and laser wavelengths by $\lambda_m = c/\nu$ and λ_l, respectively, the minimum detectable rotation rates are related by the expression

$$\Omega_{\text{rot}}^{\min}(\text{CEM}) = \frac{\gamma_m}{\gamma_l}\sqrt{\frac{\lambda_l}{\lambda_m}}\Omega_{\text{rot}}^{\min}(\text{laser}) . \tag{54.23}$$

Today's best laser gyros have $\gamma_l \cong 10^4$ Hz and the best maser cavities have lifetimes of several seconds so that $\gamma_m \cong 10^{-1}$ Hz. Taking $\lambda_m \cong 1$ cm and $\lambda_l \cong 10^{-4}$ cm we see that Eq. (54.23) holds promise for substantial enhancement of sensitivity.

We conclude by summarizing our main results: We have investigated the sensitivity of a correlated spontaneous emission maser gyroscope. In contrast to a conventional laser gyro this device is free of spontaneous emission phase noise. Moreover we have proposed a minimally perturbative measurement scheme based on an atomic beam or thin film detector probing the nodes of the standing light wave. This strategy eliminates transmission losses. Due to the high Q-values of microwave cavities this CEM gyro holds promise for substantial enhancement of sensitivity.

54.1   Acknowledgments

It is a pleasure to thank the Office of Naval Research for its support of this work, and M. Freyberger, J. Gea-Banacloche, J. Hall, and W.E. Lamb for useful and enjoyable discussions.

Appendix: Phase Diffusion and Shot Noise from a Phase Operator Approach

In this appendix we calculate the phase noise

$$\delta\varphi^2 = 1 - e^{i\phi}\langle \widehat{e^{-i\phi_c}}\,\widehat{e^{i\phi_{cc}}}\rangle + c.c. \tag{54.1}$$

where

$$\langle \widehat{e^{-i\phi_c}}\,\widehat{e^{i\phi_{cc}}}\rangle = \int d^2\alpha_c d^2\alpha_{cc} P(\alpha_c, \alpha_{cc}, t)\langle\alpha_c|\widehat{e^{-i\phi_c}}|\alpha_c\rangle\langle\alpha_{cc}|\widehat{e^{i\phi_{cc}}}|\alpha_{cc}\rangle \tag{54.2}$$

and $P = P(\alpha_c, \alpha_{cc}, t)$ denotes the Glauber-Sudarshan P-distribution of the two modes. The coherent states $|\alpha_c\rangle$ and $|\alpha_{cc}\rangle$ correspond to the clockwise and counterclockwise modes, respectively. We show that this expression does not only contain the phase diffusion (via the P-distribution) but also the shot noise. The latter results from the operator nature of the phase.

We start our calculation of $\delta\varphi^2$ by first evaluating the expectation value $\langle\alpha|\widehat{e^{i\varphi}}|\alpha\rangle$ of the London phase operator

$$\widehat{e^{i\varphi}} \equiv \sum_{n=0}^{\infty} |n\rangle\langle n+1| $$

in a coherent state $||\alpha|e^{i\varphi_\alpha}\rangle$ of large amplitude $|\alpha| \gg 1$. With this definition the expectation value reads

$$\langle\alpha|\widehat{e^{i\varphi}}|\alpha\rangle = \sum_{n=0}^{\infty}\langle\alpha|n\rangle\langle n+1|\alpha\rangle = e^{i\varphi_\alpha}\sum_{n=0}^{\infty}\frac{|\alpha|^{2n}}{n!}e^{-|\alpha|^2}\frac{|\alpha|}{\sqrt{n+1}}.$$

When we recall the asymptotic expansion

$$\frac{|\alpha|^{2n}}{n!}e^{-|\alpha|^2} \cong \frac{1}{\sqrt{\pi}}\frac{1}{\sqrt{2}|\alpha|}\exp\left[-\left(\frac{n-|\alpha|^2}{\sqrt{2}|\alpha|}\right)^2\right]$$

of the Poisson distribution, valid for $|\alpha|^2 \gg 1$, and replace summation by integration we arrive with the help of the relation

$$\frac{|\alpha|}{\sqrt{n+1}} = \left(1 + \frac{\sqrt{2}}{|\alpha|}\frac{n-|\alpha|^2}{\sqrt{2}|\alpha|} + \frac{1}{|\alpha|^2}\right)^{-1/2}$$

at

$$\langle\alpha|\widehat{e^{i\varphi}}|\alpha\rangle \cong e^{i\varphi_\alpha}\frac{1}{\sqrt{\pi}}\int_{-\frac{|\alpha|}{\sqrt{2}}}^{\infty}dy\left(1 + \frac{\sqrt{2}}{|\alpha|}y + \frac{1}{|\alpha|^2}\right)^{-1/2}\exp(-y^2).$$

Here we have also introduced the new integration variable $y \equiv (n-|\alpha|^2)/\sqrt{2}|\alpha|$. Due to the Gaussian the dominant contribution to the integral arises for $|y| < 1$ and since $1 \ll |\alpha|$ we can expand the square root in powers of $y/|\alpha|$, i.e.,

$$\left(1 + \frac{\sqrt{2}}{|\alpha|}y + \frac{1}{|\alpha|^2}\right)^{-1/2} \cong 1 - \frac{1}{\sqrt{2}|\alpha|}y - \frac{1}{2|\alpha|^2} + \frac{3}{4|\alpha|^2}y^2.$$

When we substitute this expression into the above integral, extend the lower limit to $-\infty$, and perform the remaining Gaussian integral, we find for the expectation value of the London phase operator in a coherent state

$$\langle\alpha|\widehat{e^{i\varphi}}|\alpha\rangle \cong e^{i\varphi_\alpha}\left(1 - \frac{1}{8}\frac{1}{|\alpha|^2}\right). \tag{54.3}$$

This result has also been obtained in the Appendix A of Ref. 14.

We apply Eq. (54.3) to the c- and cc-mode and substitute these expressions into Eq. (54.2) which yields

$$\langle\widehat{e^{-i\phi_c}}\,\widehat{e^{i\phi_{cc}}}\rangle \cong \int_{-\pi}^{\pi}d\phi_c\int_{-\pi}^{\pi}d\phi_{cc}e^{-i(\phi_c-\phi_{cc})}\bar{P}(\phi_c,\phi_{cc},t) \tag{54.4}$$

where

$$\begin{aligned}
\bar{P}(\phi_c,\phi_{cc},t) &\cong \int_0^{\infty}d|\alpha_c|\int_0^{\infty}d|\alpha_{cc}||\alpha_c||\alpha_{cc}|\left[1 - \frac{1}{8}\left(\frac{1}{|\alpha_c|^2} + \frac{1}{|\alpha_{cc}|^2}\right)\right]\\
&\quad\times P(|\alpha_c|e^{i\phi_c},|\alpha_{cc}|e^{i\phi_{cc}},t)\\
&\cong \left(1 - \frac{1}{4\bar{n}}\right)\int_0^{\infty}d|\alpha_c|\int_0^{\infty}d|\alpha_{cc}||\alpha_c||\alpha_{cc}|\\
&\quad\times P(|\alpha_c|e^{i\phi_c},|\alpha_{cc}|e^{i\phi_{cc}},t).
\end{aligned}$$

Here we have used the fact that the Glauber-Sudarshan distribution for a maser or laser above threshold is sharply peaked around $|\alpha_c|^2 = |\alpha_{cc}|^2 = \bar{n}$, which allows us to evaluate the slowly varying contribution $1 - \frac{1}{8}\left(1/|\alpha_c|^2 + 1/|\alpha_{cc}|^2\right)$ at these values and factor it out of the integral. Hence Eq. (54.4) reads

$$\langle e^{-i\widehat{\phi_c}} e^{i\widehat{\phi_{cc}}}\rangle = \left(1 - \frac{1}{4\bar{n}}\right)\langle : e^{-i\widehat{\phi_c}} e^{i\widehat{\phi_{cc}}} :\rangle \qquad (54.5)$$

where

$$\langle : e^{-i\widehat{\phi_c}} e^{i\widehat{\phi_{cc}}} : \rangle \equiv \int_{-\pi}^{\pi} d\phi_c \int_{-\pi}^{\pi} d\phi_{cc} e^{-i(\phi_c - \phi_{cc})} \int_0^{\infty} d|\alpha_c|$$
$$\times \int_0^{\infty} d|\alpha_{cc}||\alpha_c||\alpha_{cc}| P(|\alpha_c|e^{i\phi_c}, |\alpha_{cc}|e^{i\phi_{cc}}, t) \quad (54.6)$$

i.e., the right-hand side defines the average of the normally ordered operator : $e^{-i\widehat{\phi_c}} e^{i\widehat{\phi_{cc}}}$: via the Glauber-Sudarshan distribution. We note that the integration over the phases ϕ_c and ϕ_{cc} only involves the phase difference $\varphi \equiv \phi_c - \phi_{cc}$. Hence we can express this average Eq. (54.6) by

$$\langle : e^{-i\widehat{\phi_c}} e^{i\widehat{\phi_{cc}}} : \rangle = \int_{-\pi}^{\pi} d\varphi e^{-i\varphi} P(\varphi, t) \qquad (54.7)$$

where the distribution $P(\varphi, t)$ for the phase difference φ reads

$$P(\varphi, t) \equiv \int_{-\pi}^{\pi} d\Phi \int_0^{\infty} d|\alpha_c| \int_0^{\infty} d|\alpha_{cc}||\alpha_c||\alpha_{cc}|$$
$$\times P\left(|\alpha_c|e^{i(\Phi + \varphi/2)}, |\alpha_{cc}|e^{i(\Phi - \varphi/2)}, t\right)$$

which is the Glauber-Sudarshan distribution of the two-mode field integrated over the two amplitudes and the average phase $\Phi \equiv \frac{1}{2}(\phi_c + \phi_{cc})$ of the two modes. When we substitute Eqs. (54.5) and (54.7) into Eq. (54.1) we arrive at

$$\delta\varphi^2 \cong 1 - \left(1 - \frac{1}{4\bar{n}}\right) e^{i\phi} \int_{-\pi}^{\pi} d\varphi e^{-i\varphi} P(\varphi, t) + c.c.$$

We conclude by noting that for an ordinary two-mode laser with Schawlow-Townes diffusion rate D we have [9]

$$P(\varphi, t) = (2\pi)^{-1} \sum_{n=-\infty}^{\infty} e^{in(\varphi - \phi)} e^{-n^2 Dt}$$

and hence

$$\delta\varphi^2 = 1 - \left(1 - \frac{1}{4\bar{n}}\right) e^{-Dt} + c.c.$$

In the limit $D \cdot t \ll 1$ we arrive at

$$\delta\varphi^2 = 2\left(\frac{1}{4\bar{n}} + Dt\right).$$

The first term—the shot noise contribution—results from the operator of the phase, or more precisely, from the fact that the expectation values $\langle \alpha_c | e^{-i\widehat{\phi_c}} | \alpha_c \rangle$ and $\langle \alpha_{cc} | e^{i\widehat{\phi_{cc}}} | \alpha_{cc} \rangle$ are not just $e^{-i\phi_c}$ or $e^{i\phi_{cc}}$ but are corrected by the term $\frac{1}{8} |\alpha|^{-2}$. This term results from the fact that the coherent states $|\alpha_c\rangle$ and $|\alpha_{cc}\rangle$ have a phase distribution of finite width $\sim \bar{n}^{-1/2}$. Indeed, when we insert a complete set of London phase states into the expectation value

$$\langle \alpha | \widehat{e^{i\varphi}} | \alpha \rangle = \int_{-\pi}^{\pi} d\varphi \langle \alpha | \widehat{e^{i\varphi}} | \varphi \rangle \langle \varphi | \alpha \rangle = \int_{-\pi}^{\pi} d\varphi e^{i\varphi} |\langle \varphi | \alpha \rangle|^2$$

and recall that the phase distribution of a coherent state of large amplitude reads [15]

$$W(\varphi) = (2|\alpha|^2/\pi)^{1/2} \exp\left[-2|\alpha|^2(\varphi - \varphi_\alpha)^2\right]$$

we arrive at

$$\langle \alpha | \widehat{e^{i\varphi}} | \alpha \rangle \cong e^{i\varphi_\alpha} \exp\left(-\frac{1}{8|\alpha|^2}\right) \cong e^{i\varphi_\alpha} \left(1 - \frac{1}{8|\alpha|^2}\right)$$

in agreement with Eq. (54.3).

References

[1] For a review of laser gyro physics, see W. Chow, J. Gea-Banacloche, L. Pedrotti, V. Sanders, W. Schleich, and M. Scully, Rev. Mod. Phys. **57**, 61 (1985).

[2] The CEL concept is presented in: M. Scully, Phys. Rev. Lett. **55**, 2802 (1985) and M. O. Scully and M. S. Zubairy, Phys. Rev. A **35**, 752 (1987). Experimental observation of the effect has been reported by M. P. Winters, J. L. Hall, and P. E. Toschek, Phys. Rev. Lett. **65**, 3116 (1990) and most recently by I. Steiner and P. E. Toschek, Phys. Rev. Lett. **74**, 4639 (1995).

[3] CEL action in a ring cavity utilizing a striated gain medium was discussed in M. Scully, Phys. Rev. A **35**, 452 (1987). Experimental success in achieving a holographic, striated gain medium semiconductor laser is reported in M. Raja, S. Brueck, M. Osinski, C. Schaus, J. McInerney, T. Brennan, and B. Hammons, Appl. Phys. Lett. **53**, 1678 (1988).

[4] See for example, D. Meschede, H. Walther, and G. Müller, Phys. Rev. Lett. **54**, 551 (1985); for a recent review see G. Raithel, C. Wagner, H. Walther, L. Narducci, and M. O. Scully, Adv. At. Mol. and Opt. Phys., supplement 2, edited by P. R. Berman (Academic Press, New York, 1994) p.57.

[5] M. Scully and W. Lamb, Phys. Rev. **166**, 246 (1968).

[6] C. Caves, K. Thorne, R. Drever, V. Sandberg, and M. Zimmerman, Rev. Mod. Phys. **52**, 341 (1980).

[7] We emphasize that the scheme proposed in the present paper uses the atoms to locate the nodes of the radiation field. This is different from the idea of using the light field to channel the atoms at the nodes, as discussed in C. Salomon, J. Dalibard, A. Aspect, H. Metcalf, and C. Cohen-Tannoudji, Phys. Rev. Lett. **59**, 1659 (1987); for a review of atomic motion in laser light, see C. Cohen-Tannoudji in *Fundamental Systems in Quantum Optics*, edited by J. Dalibard, J. -M. Raimond, and J. Zinn-Justin (North-Holland, Amsterdam, 1992). The concept of a minimally perturbative measurement

as introduced in the present paper has been applied to determine the photon statistics of a quantized field by scattering atoms off the nodes of the standing light wave, see A. M. Herkommer, V. M. Akulin, and W. P. Schleich, Phys. Rev. Lett. **69**, 3298 (1992).

[8] M. Scully, M. Zubairy and M. Haugan, Phys. Rev. A **24**, 2009 (1981); see also W. Schleich and M. O. Scully in *Modern Trends in Atomic Physics*, edited by G. Grynberg and R. Stora (North-Holland, Amsterdam, 1984).

[9] See for example, M. Sargent, M. Scully, and W. Lamb, *Laser Physics*, (Addison Wesley, Reading, 1974) and M. Lax in *Statistical Physics, Phase Transitions and Superfluidity*, eds. M. Chretien, E. P. Gross, and S. Deser (Gordon and Breach, New York, 1966).

[10] For a good treatment of the problem, see R. Glauber in *Quantum Optics and Electronics*, eds. C. DeWitt, A. Blandin, and C. Cohen-Tannoudji (Gordon and Breach, New York, 1964).

[11] For an early discussion of the phase operator concept, see F. London, Z. Phys. **37**, 915 (1926) and **40**, 193 (1927) and P. A. M. Dirac, Proc. R. Soc. London **A114**, 243 (1927). See also the review articles by P. A. Carruthers and M. M. Nieto, Rev. Mod. Phys. **40**, 411 (1968); D. Pegg and S. Barnett, Phys. Rev. A **39**, 1665 (1989) and J. H. Shapiro and S. R. Shepard, Phys. Rev. A **43**, 3795 (1991). For a recent review see the Topical Issue T48 of Physica Scripta *Quantum Phase and Phase Dependent Measurements* edited by W. Schleich and S. M. Barnett (1993).

[12] Strictly speaking Eqs. (7) and (8) are not valid for a phase state since in this case the amplitude fluctuations are infinite and Eq. (6) cannot be simplified by the phase operator ansatz, Eqs. (7) and (8), to Eq. (9).

[13] W. Schleich and M. O. Scully, Phys. Rev. A **37**, 1261 (1988) and W. Schleich, M. O. Scully, and H. -G. von Garssen, Phys. Rev. A **37**, 3010 (1988). See also J. Bergou, M. Orzag, M. Scully, and K. Wódkiewicz, Phys. Rev. A **39**, 5136 (1989); for a calculation of the CEL noise containing phase diffusion and shot noise using the Pegg-Barnett phase operator, see for example M. Orszag and C. Saavedra, Phys. Rev. A **43**, 554 (1991) and *ibid*. A **43**, 2557 (1991).

[14] A. Bandilla, H. Paul, and H. -H. Ritze, Quantum Optics **3**, 267 (1991).

[15] See for example, W. P. Schleich, J. P. Dowling, R. J. Horowicz, and S. Varro in *New Frontiers in Quantum Electrodynamics and Quantum Optics*, edited by A. Barut (Plenum Press, New York, 1990).

55

Astronomical, Atmospheric, and Wavefront Studies with a Submillimeter-wavelength Interferometer

Eugene Serabyn

55.1 Introduction

Over the course of the last half century, astrophysical investigations have expanded into ever wider sections of the electromagnetic spectrum. One of the most difficult wavebands to access, both for atmospheric and technological reasons, is the submillimeter band, between wavelengths of roughly 1 mm and 100 μm. Home to a large set of molecular rotational transitions [1, 2], these wavelengths mark the boundary between radio and infrared waves, and in terms of the atmosphere, coincide with the transition between the high-transparency radio 'windows' and the opaque far-infrared sky (Fig. 55.1).

To either side of the submillimeter band, detection techniques differ in character. At longer wavelengths, i.e., for microwave and millimeter-wave frequencies, the standard detection system is the heterodyne receiver, in which the incoming signal is first coherently down-converted to lower frequencies, and then processed electronically. By contrast, at infrared wavelengths the incoming signal is usually first processed optically, after which direct detection of the signal photons is employed. Both of these approaches are now being actively extended into the 'crossover' submillimeter region, with spectroscopic observations currently pursued primarily with heterodyne systems [3, 4], and broadband continuum observations predominantly with direct bolometric detectors [5–7].

The reasons for this duality are not fundamental, as insofar as submillimeter wavelengths can be considered to be in the Rayleigh-Jeans limit, ideal coherent and incoherent detectors should share the same ultimate sensitivity [8]. Rather, the split between spectroscopic and continuum observing techniques are due to the more practical issues of non-ideal detector performance, and spectral resolution capabilities. Heterodyne mixers are currently much closer to ideal sensitivity limits than are continuum bolometers, but it is in their resolution capabilities that the inherent differences between these systems are most apparent. Heterodyne receivers have arbitrarily high spectral resolution, with typical values (in frequency units) ranging from $\Delta \nu \approx 0.1$ MHz to 1 MHz, and corresponding resolving

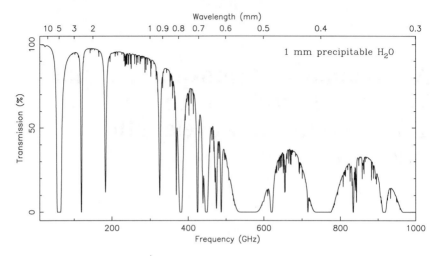

FIGURE 55.1. Model of the zenith millimeter and submillimeter wavelength atmospheric transmission on Mauna Kea. The plotted model is for the case of 1 mm of precipitable H_2O.

powers, $R = \nu/\Delta\nu \approx 10^5 - 10^6$. However, these systems in general have only limited bandwidths, with typical instantaneous frequency coverages of approximately 1 GHz. On the other hand, continuum systems typically have only a small number of very wide spectral channels, of typical width ≈ 100 GHz, and $R \leq 10$.

Although much effort has gone into improving the sensitivities of these two classes of systems, at submillimeter and longer wavelengths astronomical observations with a resolving power intermediate to these two cases are rare. Likewise, wideband spectroscopy is also rarely attempted (the COBE FIRAS spectrometer [9] being a notable exception). However, it is possible to access this large region of parameter space with Fourier transform spectroscopy, a common laboratory technique at submillimeter wavelengths. Indeed, since submillimeter wave telescopes at high, dry sites can access close to half a terahertz of bandwidth through the available submillimeter atmospheric windows (Fig. 55.1), broadband spectra would clearly be a valuable probe of these vast spectral regions. As is discussed in the following, such observations can uniquely address several astronomical and atmospheric issues, and, with appropriate instrumental modifications, can also be used to measure the shape of the telescope surface itself.

55.2   The Interferometer

At the Caltech Submillimeter Observatory (CSO), the Fourier transform technique [10] was first made use of in studies of the telescope surface, for which a novel interferometer was constructed [11]. This interferometer has the general layout of a reflective Twyman-Green interferometer (Fig. 55.2), which is used to

TO TELESCOPE

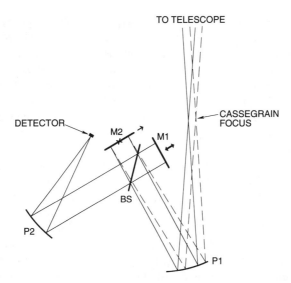

FIGURE 55.2. Schematic diagram of the interferometer. P1 and P2 are off-axis paraboloids, BS is a dielectric beamsplitter, and M1 and M2 are flat mirrors. M1 is translatable along its normal as in an FTS, while M2 is gimbal-mounted to enable it to rotate about two perpendicular axes in its plane. The solid rays show the on-axis beam, while the dashed rays show an off-axis beam.

measure wavefront interference patterns at optical wavelengths, but the lack of appropriate detector arrays at submillimeter wavelengths makes this straightforward approach impractical. Instead, a single detecting element is used to measure the electic field distribution in the focal plane pattern produced by a point source in the far-field of the telescope. The measured diffraction pattern is related to the desired aperture plane field by a spatial Fourier transform.

The measurement of the field in the Cassegrain focal plane is accomplished [11] by using plane mirror M2 to steer the beam in its arm of the interferometer off axis. If we imagine a diffraction pattern situated in the Cassegrain focal plane of Fig. 55.2, centered on the convergence of the on-axis (solid) rays, it becomes clear that the beamsplitter (BS) combines the radiation from the center of the pattern with that from an off-axis point (dashed rays). By steering M2 over a regular grid pattern, but leaving the orientation of M1 fixed (to provide a common reference), the focal plane electric field can be mapped out. The detector could in principle be of either the heterodyne or direct types, but a substantial sensitivity advantage emerges from the broad bandwidth available with direct detection. However, to preserve phase information across a wide band, the frequencies across the band must be processed separately, which can be accomplished by scanning mirror M1 as in a Fourier transform spectrometer (FTS), for each focal plane point under inspection. The resultant modulation of the different frequencies allows for their post-detection separation.

The interferometer described is now used quite regularly to measure the wavefront quality of the CSO, and provides the required data both for the proper alignment of the 84 panels which make up the telescope primary surface, and also for the correct positioning of the secondary mirror. The most recent series of such 'holographic' measurements (to use the radio parlance) were made in September of 1994, just after the installation of a chopping secondary mirror at the CSO. The top part of Fig. 55.3 shows the final beam map acquired on Jupiter after several measurement/adjustment cycles. Evidence for at least two, and possibly three Airy rings beyond the central lobe is present, implying a high surface quality. By spatially Fourier transforming this distribution, the surface map shown in the bottom half of Fig. 55.3 results. Each contour represents a deviation of 10 μm above (solid lines) or below (dashed lines) the ideal parabolic surface. The region near the right-hand edge of the dish is excessively noisy, due to low illumination levels there, and so should be ignored. The illumination-weighted root-mean-square (rms) surface error of this map is 15 μm, or roughly $\lambda/20$ at the shortest wavelength the atmosphere transmits. The true rms surface error is likely slightly higher due to the limited spatial resolution of the data, but not much so, as high beam efficiencies measured independently constrain the surface rms from above.

55.3 Spectroscopy

Since dish measurements with this instrument require only a low resolving power [11], and the spectral resolution of an FTS is inversely proportional to the maximum optical path difference provided by the translating mirror, the initial interferometer employed only a short translation stage. However, as the only instrumental modifications needed to provide a serviceable Fourier transform spectrometer were a longer translation stage and position sensor, these were realized at a later date. Now, with a one-sided maximum travel of 0.45 m, broadband spectroscopy with a resolution of up to 200 MHz is possible. This provides for a maximum resolving power which rises linearly with frequency from 1000 at 200 GHz, to 5000 at 1 THz. Five bandpass filters, with individual bandwidths of 100 to 150 GHz, are used to match to the available atmospheric transmission windows in this range [12].

Spectroscopic measurements obtained thus far with the FTS include observations of the interstellar medium, the planets, and our own atmosphere, as described in the following sections. Astronomical spectra are put onto the standard T_A^* scale (antenna temperature corrected for atmospheric absorption and hot spillover) by taking scans on- and off-source, as well as observing an ambient temperature 'hot' load, differencing and dividing the resultant spectra to form the (On-Off)/(Hot-Off) spectrum, and then scaling by the ambient temperature [13]. The sky emission spectra are converted to an absolute transmission scale by observing both hot and cold (liquid N_2 temperature) loads in addition to a blank sky spectrum, forming

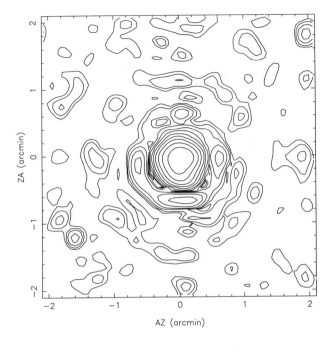

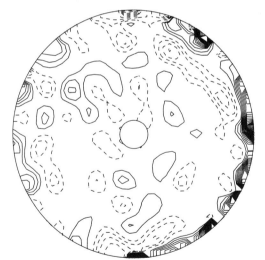

FIGURE 55.3. CSO beam map at 300 GHz (top) and dish map (bottom). The contour intervals are 3 dB (top), and ± 10 μm (bottom).

the (Hot-Sky)/(Hot-Cold) spectrum, and then scaling by $(T_{hot} - T_{cold})/T_{sky}$. Here T represents the physical temperatures of the bodies denoted by the subscripts.

55.3.1   The Orion Molecular Cloud Core

The richest spectrum acquired to date is that of the Orion Molecular Cloud core [12], the nearest region of massive star formation. This spectrum, measured toward the position of the embedded source IRc 2, with a beam of half power width $\approx 30''$, is presented in Fig. 55.4. The top panel shows the data in its entirety, with the detected CO transitions labelled, and the bottom panel shows a blowup of the lowest frequency portion, with SO triplets labelled. A high density of spectral lines is observed, with rotational transitions of 17 molecular species (including isotopomers) detected. In total, over 180 spectral features have been detected between 190 and 900 GHz, a large number of which are first detections in the interstellar medium. Although all of the molecular species have been detected previously in the Orion Molecular Cloud, the broadband coverage afforded by the FTS observations provides for an immediate overview of the dominant radiators, and of the cloud core energetics. Another important aspect of this data set is its uniform calibration over a wide frequency range, which allows for well determined line ratios. Several results emerge from a detailed examination of the data.

By far the brightest features in the spectrum are the five CO transitions which fall within the observing passbands. Interestingly, the observed (unresolved) CO antenna temperatures rise approximately linearly with frequency. Because these lines are unresolved, and lines of fixed velocity width and brightness temperature increase in area in proportion to the frequency, this behavior is consistent with optically thick emission in lines of constant brightness temperature. The implied brightness temperatures exceed 100 K, suggesting that the bulk of the emission in *all* of the CO rotational lines detected arises in rather warm layers, such as UV-excited gas [14, 15], and not in ambient cloud material.

Also seen throughout the spectrum are transitions of the linear rotors SO, CS and HCN. Emission is seen from these molecules from levels as high as 447 K, 402 K, and 234 K above the ground state, respectively, implying substantial excitation temperatures. Rotation diagrams for ^{32}SO and ^{34}SO give a temperature of roughly 90 K. Furthermore, the collisional excitation requirements are quite high. Because of the cubic frequency dependence of the Einstein A-coefficients, the highest frequency transitions seen have the highest critical densities for excitation, roughly 10^9 cm^{-3} [16]. Radiative trapping and vibrational pumping may lower these requirements somewhat, but clearly, hot and dense gas is called for, such as is produced in shocks. Indeed, due to the FTS's moderate resolution, narrow lines from the ambient cloud are not detectable, but broad linewidth components are, suggesting, along with comparisons to heterodyne observations [17–19], that the observed emission arises in the outflowing 'plateau' cloud component [14, 20].

The most unique aspect of these observations is their ability to address the overall energetics. As Fig. 55.4 shows, although the CO transitions are the strongest present, the SO molecule has many more transitions, both because of its $^3\Sigma$ ground state, and because of its smaller rotational constant. A detailed energy flux summation, including interpolations for lines falling between the observed passbands, indicates that below 1 THz, the SO molecule emits more energy into our beam

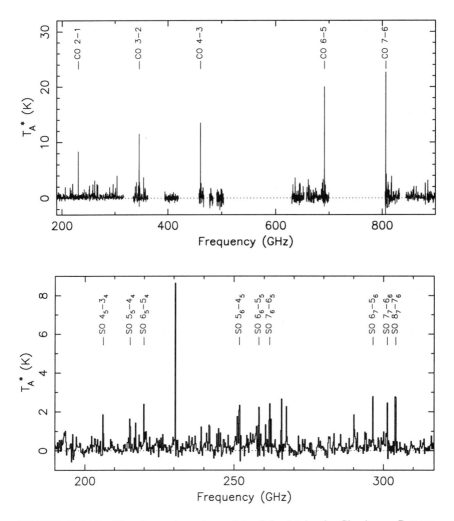

FIGURE 55.4. Top: The observed spectrum of the Orion Molecular Cloud core. Bottom: The lowest frequency portion of the spectrum.

than does CO. The asymmetric rotor SO_2 is an even more extreme case, showing the largest number of lines (just over 90), and the highest energy flux. In total, SO_2 emits slightly over twice the energy of CO below a THz. Indeed, the combined radiation from SO and SO_2 transitions accounts for roughly half of the detected line flux from the cloud core at sub-THz frequencies. These S-bearing species likely owe their existence to the passage of shock waves, and their presence thus substantially alters the radiation budget.

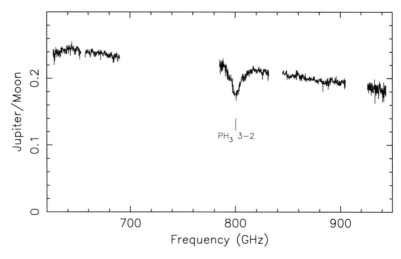

FIGURE 55.5. The submillimeter spectrum of Jupiter normalized to that of the Moon. The PH$_3$ J=3-2 triplet at 800.5 GHz is identified.

55.3.2   Planetary Spectroscopy

At submillimeter wavelengths, spectroscopic and continuum observations of the planets are still quite few, e.g. [21–23], even though several molecules are predicted to have observable transitions in this frequency band [24, 25]. In particular, in the giant planets, saturated and hydride molecules are expected, as both ammonia, NH$_3$, and phosphine, PH$_3$, have their lowest rotational transitions at submillimeter or near-millimeter wavelengths, as do HCl and HBr. In addition, the H$_2$S, HCN, and HCP molecules have submillimeter rotational transitions. In the terrestrial planets, CO is expected to be detectable. The main reason for the paucity of observational data is that collisionally broadened tropospheric lines tend to be broader than the capabilities of heterodyne receivers. Thus, planetary spectroscopy is one area where broadband spectral coverage is essential.

A program to measure the broadband submillimeter spectra of all of the planets with the FTS is underway, but to date the observations have concentrated on the gas giants Jupiter and Saturn. In both planets, the FTS has been used to detect very wide ($\approx$ 10 GHz FWHM) absorption features due to low-J rotational transitions of phosphine. An example, the 3-2 rotational transition of PH$_3$ in Jupiter is shown in Fig. 55.5 [26]. Unfortunately, the lowest rotational transition of NH$_3$, at 572.5 GHz, is obscured by our atmosphere.

The large linewidths of the detected PH$_3$ transitions indicate that these lines arise in the tropospheres of the planets in question. Both the depths of the lines, and their lack of stratospheric emission cores further suggest that PH$_3$ does not extend into the stratosphere with constant abundance, but instead cuts off at some level. This is consistent with photodestruction of PH$_3$ in the stratosphere by the incident solar UV flux [27]. As PH$_3$ is not an equilibrium species at low temperatures, its presence in the upper troposphere must arise by convective transport from lower, hotter regions where it is stable [28]. The equilibrium distribution of PH_3 thus probes both the vertical mixing rate, and the UV photolysis rate.

Detailed analysis of the $J = 1 - 0\,PH_3$ line profile in Saturn yields a tropospheric phosphine abundance, relative to H_2, of 3 ppm [13], implying an atomic P/H abundance ratio roughly five times solar. In Jupiter on the other hand [26], the 3-2 absorption profile (Fig. 55.5) is approximately consistent with a solar P abundance deep in the troposphere. This difference is likely due to a combination of a greater gas retention by Jupiter in its formation stage, and a greater convective transport rate from the deep atmosphere for Saturn. Modelling of the phosphine falloff at high atmospheric levels is the next step, for which use of $J = 1 - 0$ and 3-2 data in concert will ultimately be necessary.

Surprisingly, no other lines have yet been detected with the FTS in the jovian planets, yielding upper limits on the abundances of several minor atmospheric constituents, such as HCN, HCP, HBr, HCl, and H_2S. Absolute calibration of the planetary submillimeter continuum emission is also planned, as this emission level relates to the far-wing lineshape of the NH_3 molecule. The latter project is awaiting additional observations of Mars, which will provide beam coupling information.

55.3.3 Atmospheric Spectroscopy

One of the most severe limitations to submillimeter wavelength observations of astronomical sources is the transmission of the terrestrial atmosphere. At submillimeter wavelengths, the transmission of our atmosphere is highly variable, both as a function of frequency and time. Both of these factors are determined primarily by the amount of water vapor present along the line of sight to the source, because of numerous H_2O transitions at these (and higher) frequencies. Even between the atmospheric H_2O (and O_2) lines, there is considerable absorption, caused by the low frequency wings of the ensemble of far-infrared H_2O lines [29]. These line wings add together to produce a 'quasi-continuum' absorption, which theoretically rises with frequency as ν^2. Although attempts have been made to model the atmospheric transmission at submillimeter wavelengths empirically, very few quantitative measurements exist which can constrain the models. Of particular interest is the actual lineshape which the water vapor absorption assumes.

A program to quantify the atmospheric transmission as a function of water vapor column was therefore undertaken at the CSO. Initial results of FTS observations in the lower half of the frequency range accessible from the CSO are now in hand, some of which are shown in Fig. 55.6. Simultaneous opacity readings, τ_{225}, from a fixed frequency (225 GHz) NRAO water vapor radiometer are also recorded. The data are then modelled with AT, a single-temperature atmospheric transmission model [30], modified by the inclusion of an empirical power law continuum term. So far, the best fit to the continuum requires a power law exponent of 1.9, consistent with the theoretical value. The optimal lineshape is not yet well constrained, but extrapolations suggest that it may be possible to discriminate between the van Vleck-Weisskopf and Zhevakin-Naumov lineshapes [29] with the extension of

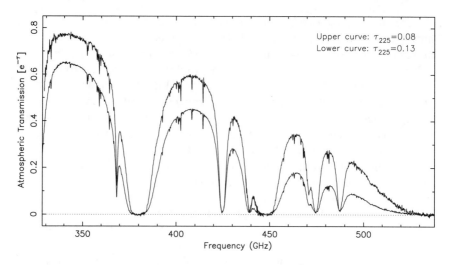

FIGURE 55.6. Measured atmospheric transmission spectra for two water vapor columns. The corresponding opacities at 225 GHz, as measured by the NRAO radiometer, are provided in the top right corner.

these measurements to 1 THz. The ultimate goal is a model which can accurately predict the atmospheric transmission at any frequency, from the current 225 GHz water vapor radiometer reading.

Acknowledgments

This work has progressed with the assistance of many parties. E. Weisstein is involved in all aspects of this project, but most specifically in the planetary modelling area. W. Schaal, B. Force, and K. Young have assisted with the mechanical, electronic, and communications aspects, respectively. A. Schinckel and D. Woody are integral to the holography effort, and D. C. Lis is involved with the atmospheric modelling effort. I would also like to express my deep appreciation to Charles H. Townes, both for providing the initial impetus for my work in the areas of radiation detection and astrophysics, and for the near-infinite patience he showered upon this erstwhile student. This work is supported by NSF grant AST-9313929 and NASA grant NAGW-3303.

References

[1] G. J. Melnick, Int. J. IR and mm Waves **9**, 781 (1988).
[2] T. G. Phillips and J. Keene, Proc. IEEE **80**, 1662 (1992).
[3] T. G. Phillips, in *Astronomy with Millimeter and Submillimeter Wave Interferometry*, eds. M. Ishiguro and W. J. Welch (ASP, San Francisco, 1994), p. 68.

[4] A. I. Harris, K. F. Schuster, R. Genzel, B. Plathner, and K. H. Gundlach, Int. J. IR and mm Waves **15**, 1465 (1994).

[5] W. D. Duncan, E. I. Robson, P. A. R. Ade, M. J. Griffin, and G. Sandell, MNRAS **243**, 126 (1990).

[6] W. K. Gear and C. R. Cunningham, in *Multi-feed Systems for Radio Telescopes*, edited by D.T. Emerson and J.M. Payne (ASP, San Francisco, 1995), p. 215.

[7] N. Wang, T. R. Hunter, D. J. Benford, E. Serabyn, T. G. Phillips, and S. H. Moseley, Proc. SPIE **2198**, 749 (1994).

[8] J. W. V. Storey, Infrared Phys. **25**, 583 (1985).

[9] E. L. Wright *et al.*, ApJ **381**, 200 (1991).

[10] H. W. Schnopper and R. I. Thompson, in *Meth. Exp. Phys. 12A: Astrophysics*, edited by N. Carleton (Academic, New York, 1974), p. 491.

[11] E. Serabyn, C. R. Masson, and T. G. Phillips, Appl. Opt. **30**, 1227 (1991).

[12] E. Serabyn and E. Weisstein, ApJ **451**, 238 (1995).

[13] E. Weisstein and E. Serabyn, Icarus **109**, 367 (1994).

[14] R. Genzel and J. Stutzki, Ann. Rev. A&A **27**, 41 (1989).

[15] A. Castets, G. Duvert, A. Dutrey, J. Bally, W. D. Langer, and R. W. Wilson, A&A **234**, 469 (1990).

[16] C. M. Walmsley and R. Güsten, in *The Structure and Content of Molecular Clouds*, editor T.L. Wilson (Springer, Berlin, 1994).

[17] E. C. Sutton, G. A. Blake, C. R. Masson, and T. G. Phillips, ApJ Supp. **58**, 341 (1985).

[18] G. A. Blake, E. C. Sutton, C. R. Masson, and T. G. Phillips, ApJ Supp. **60**, 357 (1986).

[19] T. D. Groesbeck, Ph. D. Dissertation (California Institute of Technology, 1994).

[20] G. A. Blake, E. C. Sutton, C. R. Masson, and T. G. Phillips, ApJ **315**, 621 (1987).

[21] R. H. Hildebrand, R. F. Loewenstein, D. A. Harper, G. S. Orton, J. Keene, and S. E. Whitcomb, Icarus, **64**, 64 (1985).

[22] G. S. Orton, M. J. Griffin, P. A. R. Ade, I. G. Nolt, J. V. Radostitz, E. I. Robson, and W. K. Gear, Icarus **67**, 289 (1986).

[23] A. Marten, D. Gautier, T. Owen, D. B. Sanders, H. E. Matthews, S. K. Atreya, R. P. J. Tilanus, and J. R. Deane, ApJ **406**, 285.

[24] E. Lellouch, T. Encrenaz, and M. Combes, A&A **140**, 405 (1984).

[25] B. Bezard, D. Gautier, and A. Marten, A&A **161**, 387 (1986).

[26] E. Weisstein and E. Serabyn, in preparation.

[27] D. F. Strobel, ApJ Lett. **214**, L97 (1977).

[28] R. G. Prinn and J. S. Lewis, Science **190**, 274 (1975).

[29] J. W. Waters, in *Meth. Exp. Phys. 12B: Astrophysics*, edited by M. L. Meeks (Academic, New York, 1976), p. 142.

[30] E. Grossman, AT user's manual (Airhead Software, Boulder, 1989).

56

Theory of an Optical Subharmonic Generator

Koichi Shimoda

56.1 Introduction

Techniques for optical harmonic generation, sum and difference frequency generation have been developed since the beginning of nonlinear optics. Optical frequency division or subharmonic generation, however, has not yet been realized. A device for generating optical subharmonics has recently been proposed in order to produce a train of subfemtosecond pulses [1]. A semiclassical theory of the steady-state and build-up of optical subharmonic oscillations in the device is presented here.

56.2 Steady-State Operation

The proposed optical subharmonic generator is composed of two nonlinear elements and two amplifiers as shown in Fig. 56.1. It is driven by a CW laser at an input frequency ω_0. The difference frequency output of the mixer is amplified and fed to a harmonic generator, giving the nth harmonic. This harmonic is amplified and fed back to the mixer, which gives a beat with the input frequency. Then the beat frequency is

$$\omega_1 = \omega_0 - \omega_n$$

and the nth harmonic frequency becomes

$$\omega_n = n\omega_1.$$

Therefore we obtain subharmonic frequencies as given by

$$\omega_n = \frac{n}{n+1}\omega_0, \quad \omega_1 = \frac{1}{n+1}\omega_0. \tag{56.1}$$

A practical scheme with $n = 2$ generates one third of the input frequency has been proposed [1] as shown in Fig. 56.2. The input radiation from a single-mode 514.5 nm Ar laser is represented by

$$E_0(t) = \tilde{E}_0 e^{i\omega_0 t} + c.c.,$$

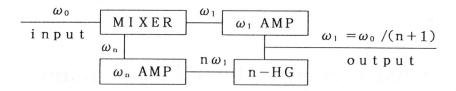

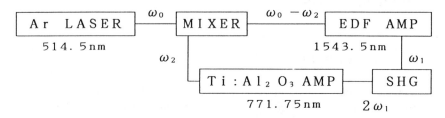

FIGURE 56.1. Schematic diagram of proposed subharmonic generator. AMP denotes an amplifier; n-HG, n^{th} harmonic generator.

FIGURE 56.2. An example for generating one-third and two-thirds of the input frequency of the 514.5 nm Ar laser. EDF denotes an erbium doped fiber; AMP, amplifier; SHG, second harmonic generator.

where $\widetilde{E}_0$ is the amplitude. Another input to the mixer is similarly expressed as

$$E_2(t) = \widetilde{E}_2 e^{i\omega_2 t} + c.c. \tag{56.2}$$

and $\omega_2 = (2/3)\omega_0$ from Eq. (56.1).

Then we obtain the complex amplitude of the beat to be

$$\widetilde{E}_1 = \chi_m \widetilde{E}_0 \widetilde{E}_2^* \tag{56.3}$$

where χ_m is the effective nonlinear coefficient of the mixer. This beat signal is amplified and drives the second harmonic generator. Then the second harmonic is amplified to become

$$\widetilde{E}_2 = A_2 \chi_2 (A_1 \widetilde{E}_1)^2, \tag{56.4}$$

where A_1, A_2 are complex amplification factors of the amplifiers and χ_2 is the nonlinear coefficient of second harmonic generation.

Self-consistency in the steady-state of waves from Eqs. (56.3) and (56.4) before and after the round trip gives

$$\widetilde{E}_2^3 = A_1^2 A_2 \chi_m^2 \chi_2 \widetilde{E}_0^2 |\widetilde{E}_2|^4. \tag{56.5}$$

Therefore the subharmonics are automatically phase-locked to the input wave, if the phase shift in the optical loop is stable.

56.3   Build-up of Subharmonic Oscillation

Subharmonic oscillation in the device does not start from small noise, because the conversion efficiency for harmonic generation is too low at low levels. In order

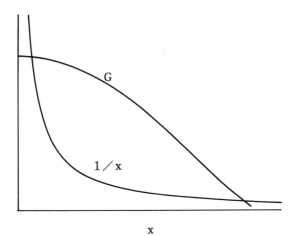

FIGURE 56.3. Schematic diagram showing gain coefficient as a function of the amplitude x and the inverse amplitude $1/x$.

to investigate the build-up and frequency locking of subharmonics we consider plane waves traveling along the optical loop. We shall assume that a trigger wave as given by

$$E_2(t) = \tilde{E}_2(t)e^{i\omega_2 t} + c.c. \tag{56.6}$$

is injected initially and its complex amplitude varies slowly with time and space, while the input radiation is steady.

Similar calculations to those in Sec. 2 give

$$\tilde{E}_2(t + \tau) = G\tilde{E}_2^*(t)^2 \exp[i(2\omega_0 - 3\omega_2)t] \tag{56.7}$$

where τ is the round-trip time around the loop and

$$G = A_1^2 A_2 \chi_m^2 \chi_2 \tilde{E}_0^2 \tag{56.8}$$

is the complex gain coefficient which saturates with the light intensity.

To find the development of the build-up, however, we first assume that the amplitudes and gain coefficient are real quantities and frequencies are locked in the initial state as will be shown later. Then the condition for build-up can be expressed as

$$\tilde{E}_2(t + \tau) > \tilde{E}_2(t)$$

and it is rewritten by using Eqs. (56.7) and (56.8) as

$$G > 1/\tilde{E}_2.$$

The gain saturation as a function of the wave amplitude is schematically shown in Fig. 56.3. If the gain is high enough, curves for G and $1/\tilde{E}_2$ have two crossing points: the lower one corresponds to the threshold and the upper one to the steady state. Above the threshold, waves grow as they circulate along the loop until they reach the steady state.

56.4   Locking Condition

Finally we consider the temporal evolution of the complex amplitude and discuss the condition for phase locking. This problem is similar to a classical van der Pol oscillator with an injected signal. Let us rewrite the complex amplitude as

$$\tilde{E}_2(t) = x(t)e^{i\theta(t)}, \quad G = ge^{i\phi}.$$

Then Eq. (56.7) can be approximated in the form

$$x + \tau \frac{dx}{dt} + ix\tau \frac{d\theta}{dt} = gx^2 \exp[i(2\omega_0 - 3\omega_2)t - 3i\theta + i\phi].$$

By using

$$\psi = (2\omega_0 - 3\omega_2)t - 3\theta + \phi, \tag{56.9}$$

we find the real and imaginary parts to be

$$x + \tau \frac{dx}{dt} = gx^2 \cos \psi \tag{56.10}$$

$$\tau \frac{d\theta}{dt} = gx \sin \psi. \tag{56.11}$$

Differentiation of Eq. (56.9) with time yields

$$\frac{d\psi}{dt} = 2\omega_0 - 3\omega_2 - 3\frac{d\theta}{dt}.$$

Substituting Eq. (56.11), we obtain

$$\frac{d\psi}{dt} = 2\omega_0 - 3\omega_2 - \frac{3gx}{\tau} \sin \psi.$$

This may be integrated using the slowly varying amplitude approximation in the form

$$t - t_0 = \int_{\psi_0}^{\psi} \frac{d\psi}{2\omega_0 - 3\omega_2 - 3gx\tau^{-1} \sin \psi}. \tag{56.12}$$

Investigation of Eq. (56.12) shows that t becomes infinite with a finite value of the upper limit of integration, if

$$|3\omega_2 - 2\omega_0| < 3gx\tau^{-1}. \tag{56.13}$$

In order that Eq. (56.9) takes a finite value with increasing t, we must have

$$3\omega_2 - 2\omega_0 = 0. \tag{56.14}$$

Thus the frequency is locked to exactly 2/3 of the input frequency. Then Eq. (56.12) diverges at $\psi = 0$ for an infinite value of t, so that the phase can be locked to

$$\theta = \phi/3,$$

where ϕ is the phase of the gain coefficient of the loop.

At threshold and at steady-state, we have $gx = 1$. Thus the frequency-locking condition is expressed as

$$\left| \omega_2 - \frac{2}{3}\omega_0 \right| < \frac{1}{\tau} \tag{56.15}$$

from Eq. (56.13). The growth rate of the amplitude is small near the threshold, and the phase is initially locked. While the amplitude grows, the phase will be hard to unlock. If the trigger is a kind of amplified broadband noise, its spectral components outside the locking range of (56.15) are rejected, and those within the locking range are concentrated (due to growth) onto the locked component.

56.5 Conclusion

Optical subharmonic oscillations in the proposed system have been analyzed and their reliable performance has been found. The proposed subharmonic generator is capable of delivering substantial powers at integral multiples of frequencies. It may be used to produce a train of subfemtosecond pulses by Fourier synthesis. Utilization of subharmonics is more practical than the method of locking phases of separate lasers as proposed by Hänsch [2].

Optical pulse shaping will become easier and more convenient with this method. It will also be useful for optical frequency measurements and frequency comparisons. Progress in semiconductor lasers and nonlinear optics in the infrared will allow versatile applications of this method. Thus we should be able to find a coherent connection between masers and lasers by making use of such subharmonic generators.

This work is dedicated to C. H. Townes who has inspired me ever since his first visit to Japan in 1953.

References

[1] K. Shimoda, presented at the International Workshop on Femtosecond Technology, Tsukuba, Japan, February 1995.
[2] T. W. Hänsch, Optics Commun. **80**, 71 (1990).

57

Hydrogen Masers and Lasers in Space

Howard A. Smith
Vladimir S. Strelnitski
and Victor O. Ponomarev

57.1   Introduction

Atomic hydrogen appears to be a natural candidate for population inversions. The increasing rate of spontaneous decay with decreasing energy level suggests the common astrophysical pumping conditions in HII regions—ionization followed by recombination and radiative cascading—efficiently produce population inversions. Indeed, this was found in the early calculations of the hydrogen level populations, which only used these processes and assumed low density regions (e.g., J. C. Baker *et al.* [1]). Seaton [2] and, in a more general way, Brocklehurst [3] then included into the calculations collisional redistribution of the level populations (especially important for the high n transitions) for densities up to about 10^4 cm^{-3}, and they found inversions still occur because the cascading processes are so fast. Since then many approaches have been tried to make the calculations even more complete, and extend them to a wider range of different environments of density, temperature, radiation fields, etc. (e.g., Krolik and McKee [4]).

Goldberg [5] first suggested interpretating radio recombination lines (RRLs) from HII regions by taking into account deviations of the hydrogen populations from local thermodynamic equilibrium (LTE). Since these RRL observations only involved optically thin HII regions, there was never a question of real amplification in the inverted transitions, only the need to introduce some intensity corrections. Smith, Larson, and Fink [6] observed near-IR lines from much denser, potentially optically thick regions around new stars. They observed an abnormally bright infrared line in the recombination line spectrum of the BN object in Orion, and interpreted it using level calculations by W. Clarke [7] which considered densities $\sim 10^6$ cm^{-3} and which implied inversions persisting up to levels $n \leq 15$. They noted that should the inversions exist at these higher densities they could have gains ~ 1 and produce lasing in the infrared recombination lines, but cautioned that at the conditions needed to achieve optically thick lines, Clarke's assumptions break down.

It was therefore somewhat of a surprise when millimeter observations of RRLs discovered real, high-gain masers in several RRLs at 1-3 mm in the emission line star MWC349 [8]. The lines are very bright and time varying, double-peaked (as characteristic of emission from a Keplerian disk seen edgewise) with some weaker broad pedestal of thermal-like emission present between the peaks. Interferometric observations of the dimensions of the region ($\simeq$ 0.06 arcsec) make it fairly certain that it is indeed a maser. The densities in the disk are thought to reach high values $\geq 10^9$ cm^{-3} based on the near-IR lines like CaII [9], stimulating interest in new population calculations for the regime where collisions might drastically change the population distributions. One of the most accurate and comprehensive calculations is due to Storey and Hummer [10, 11]. Following the approach of Brocklehurst, these authors extended the calculations of the hydrogen level populations to levels up to $n = 50$, and for conditions with $2 \leq \log N_e \leq 14$ and $2.7 \leq \log T_e \leq 5$. They assume Menzel's Case B applies, for which the Lyman lines are infinitely optically thick, but all other lines are thin. They also derive the conditions under which their models break down because of the contribution of transitions from $n = 1$ and 2 to other levels; their conclusions are that for the regions under discussion here these effects are small. Their results strikingly confirm quantitatively (to densities even higher than $N_e = 10^{10}$ cm^{-3}) the general tendencies of hydrogen noted earlier: the populations are inverted over a wide range of levels and densities.

57.2   Population Inversions and Masing in Atomic Hydrogen

We plot the values of the "departure coefficients" b_n versus n at five densities and $T_e = 10^4$ K in Fig. 57.1, using data from the machine readable files of Storey and Hummer [11]. This parameter measures the deviation of the population for the nth-level from its LTE value which occurs when collisions dominate (and where $b_n = 1$); the dotted curve labeled "Rad" shows the purely radiative limit when the electron density N_e is zero. It is important to note in the figure both the regions where the slopes of these curves are positive—the regions where masing is possible—but also the non-LTE regions where the slope is negative, where the phenomenon of "overcooling" (see below) can occur.

The critical parameter in determining an inversion and the possibility of masing is the *relative* populations of neighboring levels, given conveniently by the quantity $\beta_{nn'}$,

$$\beta_{nn'} = \frac{1 - (b_{n'}/b_n)e^{(-h\nu_0/kT_e)}}{1 - e^{(-h\nu_0/kT_e)}} \tag{57.1}$$

first introduced by Brocklehurst and Seaton [12]. $\beta_{nn'} < 0$ indicates a population inversion. $\beta_{nn'}$ measures the deviation of the absorption coefficient κ from its LTE value κ_{LTE}.

$$\kappa = b_n \beta_{nn'} \kappa_{\text{LTE}}. \tag{57.2}$$

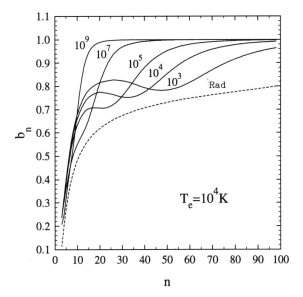

FIGURE 57.1. Plot of the deviation b_n from LTE *vs.* the principal quantum number n, for various values of electron density N_e.

The optical depth (or gain when $\kappa < 0$) is:

$$\tau_{nn'} = \kappa L. \tag{57.3}$$

Figure 57.2 plots the behavior of $\beta_{nn'}$ for several alpha transitions (labeled by their principal quantum number n) as a function of electron density, at an electron temperature $T_e = 10^4$ K. It is seen that the populations are heavily inverted over a range of N_e and n, that the peak of the inversion drops both in magnitude and in n location as the density increases, but that some inversion is present even at the highest densities.

It is important to keep in mind two additional points when examining Fig. 57.2. The first is that although the size of the inversion in $\beta_{nn'}$ decreases at the high densities, the gain (or absorption) may actually increase substantially because the LTE absorption coefficient κ_{LTE}, and also κ, is proportional to the density squared. For comparison, in Fig. 57.2 the bar labeled $\kappa_l(36\alpha)$ shows the location and width of the peak of the inverted H36α line absorption coefficient. The second point is that the effect of free-free absorption must also be included, so that the *net* absorption κ_{net} is what finally matters. The bar labeled $\kappa_n(36\alpha)$ shows the net absorption, when free-free absorption is included. Figure 57.3(a) plots κ_l and κ_{net} versus density for three α-lines which have been observed in maser emission in MWC349; Fig. 57.3(b) plots κ_{net} versus N_e for some far-infrared α-lines, including the H21α already detected at 452 μm.

The observed maser emission in a Keplerian disk in MWC349 provides a clear example of just such a high-density environment, and confirms that the high-density

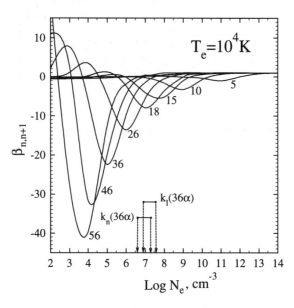

FIGURE 57.2. Plot of the relative population difference of neighboring levels $\beta_{nn'}$ vs. the logarithm of electron density N_e.

hydrogen level population calculations of Storey and Hummer are realistic. In fact, our analysis of the kinematic behavior of such a disk provides the fine details of the line emitting regions. Comparing the observed double-peaked profile with a model profile calculated for an amplifying Keplerian edge-on disk, we concluded that the line is at the threshold of saturation, and that its maser gain is ~ 6 [13]. We were also able to set limits to the temperature, turbulent activity, radial density distribution, and free-free absorption in the disk. Figure 57.3(a) shows that to obtain $|\tau| \approx 6$ in the H30α transition, with the optimum density being 3×10^7 cm^{-3}, an amplification length of about 2×10^{15} cm is needed at $T_e = 10^4$ K. At $T_e = 5000$ K this requirement drops to about 4×10^{14} cm. These size estimates are in close agreement with the interferometric scale of the masing region [14].

57.3   Hydrogen Lasers

Figure 57.2 shows that the inversions persist to even higher density regimes and lower n transitions, even down to the H5α line at 4 μm where the process could more appropriately be called "lasing." The weighting effect of $N_e N_+$ on κ shifts the regions of optimum gain to these higher density/lower $n\alpha$ transition cases, as seen in Figs. 57.3(a) and (b), in spite of the smaller magnitude of the population inversion. For example, at 10^4 K the 30α transition at 1.29 mm peaks around $N_e = 3 \times 10^7$ cm^{-3} with $\tau = (3-10) \times 10^{-15}$ (L/cm), depending on T_e, while the infrared 8α transition at 27.8 μm, with a much smaller maximum inversion, nevertheless

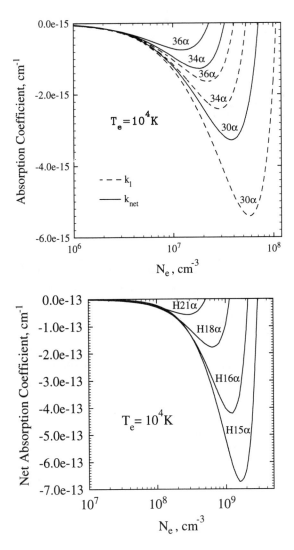

FIGURE 57.3. (a) Absorption coefficients κ_l and κ_{net} vs. the logarithm of the electron density N_e. (b) Net absorption coefficient κ_{net} vs. the logarithm of the electron density N_e.

has a *much greater* $\tau = 2 \times 10^{-11}$ (L/cm) because its κ_{net} peaks at a much higher density, $N_e \approx 10^{10}$ cm^{-3}. In fact, the unsaturated maser gain is a sharp function of N_e, much sharper than that for the population inversion. Furthermore, the rapidly decreasing importance of free-free absorption with decreasing wavelength enhances this trend toward short wavelengths. We also emphasize that Hnβ, Hnγ, and other transitions are inverted.

We have searched for evidence of such laser action, with mixed results to date. Using both the present generation of the Townes' group Fabry-Perot interferometer

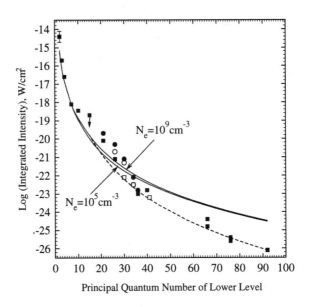

FIGURE 57.4. The logarithm of the integrated intensity *vs*. the principal quantum number *n* of the lower level.

on the Kuiper Airborne Observatory ("FIFI") to observe the H15α line at 169.4 μm, and the KAO grating spectrometer ("CGS") to observe the H10α line at 52.5 μm, we obtained a marginal detection of the latter, and an upper limit to the former [15]. Using the Irshell echelle spectrometer of Lacy on the IRTF, we also observed the H7α line at 19.06 μm and the H6α line at 12.4 μm [16]. Our results are included in Fig. 57.4 below, which summarizes most of the α-transitions measured from MWC349 to date; dots give the fluxes in narrow components, and squares in broad/non-masing components (filled/open symbols represent single/averaged data, cf. [17]).

The figure also shows as solid lines the "Case B" spontaneous emissivity ratios for two extreme densities, and those same ratios corrected for free-free absorption (important for α-lines with $n > 30$) as a dotted line (the lines' ordinates are arbitrarily placed). The non-masing radio and near-IR transitions are distinguished by their close correspondence to this line. More sensitive and complete infrared and far-infrared observations are clearly warranted, because it seems extremely unlikley that the phenomenon should suddenly terminate at the strongly masing H21α (452.6 μm) line.

57.4 "Dasars"

A relatively unnoticed phenomenon—population *overcooling*, which may result in stimulated *absorption* of radiation—occurs in the region where the slope of

b_n versus n is negative and $\beta_{nn'}$ is greater than unity. This region is seen clearly in Fig. 57.1 at the low densities, and in Fig. 57.2. In fact this phenomenon has been known for some time in interstellar molecules. The observational effect it produces, an abnormally strong absorption in the line, was named *dasar* by Townes and Schawlow, for "darkness amplification by the stimulated absorption of radiation." In hydrogen the negative slope of b_n versus n occurs over a range of n where the angular momentum 1 structure is only partly affected by collisions, before proton collisions at higher densities completely blur the l-levels. When the optical depth of such a transition is comparable to unity, it produces a recombination line which is seen *in absorption* on the free-free continuum of the cloud itself. Because this occurs when the density is relatively modest, however, significant optical depth can only occur with a very large path length, and so the phenomenon is most likely to be observed directly only in peculiar extragalactic sources. A more complete discussion of the phenomenon is presented in Strelnitski, Ponomarev, and Smith [18].

57.5   Is MWC349 Unique?

The requirement of high (negative) optical depth in the masing lines might normally imply high positive optical depths in many other transitions, besides the Lyman lines, invalidating the optically thin "Case B" assumptions of Storey and Hummer, and requiring the line radiation itself be included into the population calculations. Furthermore saturation effects must also be included. Despite these theoretical obstacles, hydrogen masers are seen, and we conclude that any adjustments to the population calculations are of second-order, and that collisional and stimulated radiative non-Lyman effects are small in comparison with the main Lyman continuum pumping and natural cascading processes, at least for MWC349. But then the inversion is predicted over a wide-range of conditions: Why does MWC349 remain a unique source?

The first reason may be that finding an object with the right values of ionization, density, and path length is actually not so easy. MWC349, which M. Cohen *et al.* [19] place at a distance of 1.2 kpc, is a very bright 26 $M_\odot$ star ($L \approx 3 \times 10^4 L_\odot$) with high ionization flux ($\Phi \approx 1 \times 10^{49}$ photons/sec) and a mass loss rate of 10^{-5} $M_\odot$ yr^{-1}. The observed masers arise from an ionized disk with a zone of densities $\approx 10^{7-8}$ cm^{-3} over rather significant distances of perhaps 5×10^{14} cm = 30 AU [20]. Even if the other conditions are favorable, the gain width for masing lines of a factor of two around in density (see Figs. 57.3(a) and (b)) may mean a simple dense cloud ($N_e \approx r^{-2}$) does not have a very large zone of density constant to within the required factors. A complementary feature is suggested by the particular maser geometry in MWC 349: a flattened circumstellar disk seen edge-on. We show [20] that all of the observed masing lines originate from this disk, and not (as has been suggested) in the strong associated bipolar wind from the star.

A second critical parameter is the *column density per thermal linewidth* [20]. A search for hydrogen masers in Wolf-Rayet stars was unsuccessful [21]. Perhaps this was because the high-velocity dispersions there, $\Delta V \approx 10^3$ km s^{-1}, as compared with the thermal linewidth of only about 20 km s^{-1}, greatly reduce the effective optical depth. But the highest-frequency-lasing regime samples not the wind but the stellar atmosphere whose densities are in the range of $\geq 10^{11}$ cm^{-3}, and there the situation may be more favorable. Similar comments apply to the atmosphere of Be stars. These are, however, more complex to analyze because of the strong continuum emission present. In addition, we show [20] that, because of saturation effects and the strong background, small solid angles of lasing might be needed for high-frequency hydrogen lasers to be observable.

Another prospective category for searches is that of active galactic nuclei (AGN), especially those with a "broad line region" (BLR) where high densities and very large dimensions are promising. If they should contain high-density ionized cloudlets of dimensions around 1 AU, maser and laser emission is possible. At the same time AGNs seem to be the most likely place to find dasars, as they may also have regions of $N_e \approx 10^4$ cm^{-3} over large dimensions as required.

57.6 *Notes added in proof*

In August 1995, the KAO CGS program team reobserved MWC349 in the H1α (169 μm) line, and positively detected the line at an intensity six times higher than expected from non-amplified, spontaneous emission—consistent with it lasing. Also detected (but with lower confidence levels) were the H12α (89 μm) and H10α (52μm) lines, consistent with them lasing as well.

We further note that Cox et al. [22] recently report the detection of masing in a set of millimeter hydrogen recombination lines in η Carinae.

57.7 Acknowledgments

We thank Prof. Townes, not only for his many physical insights, creativity, and guidance, but also for his optimism. Nature is rich in variety and imagination, and we are optimistic that new and exciting cases of masers, lasers and even dasars will be discovered in the future. We would also like to thank the Smithsonian Institution's Scholarly Studies Program, which supported much of this research, and which enabled the Laboratory for Astrophysics to host V.S.S. for a Research Fellowship and V.O.P. for a Short-Term Visitorship.

References

[1] J. C. Baker and D. H. Menzel, ApJ Lett. **88**, 52 (1939).

[2] M. J. Seaton, MNRAS **127**, 117 (1964).

[3] M. Brocklehurst, MNRAS **148**, 417 (1970).

[4] J. H. Krolik and C. F. McKee, ApJ Supp. **37**, 459 (1978).

[5] L. Goldberg, ApJ **144**, 1225 (1966).

[6] H. A. Smith, H. P. Larson, and U. Fink, ApJ **233**, 132 (1979).

[7] W. Clarke, Ph.D. thesis, UCLA (1965), unpublished.

[8] J. Martin-Pintado, R. Bachiller, C. Thum, and M. Walmsley, A&A **215**, L13 (1989).

[9] F. Hamann and M. Simon, ApJ **327**, 876 (1988).

[10] P. J. Storey and D. G. Hummer, MNRAS **254**, 277 (1992).

[11] P. J. Storey and D. G. Hummer, MNRAS **272**, 41 (1995).

[12] M. Brocklehurst and M. J. Seaton, MNRAS **157**, 179 (1972).

[13] V. O. Ponomarev, H. A. Smith, and V. S. Strelnitski, ApJ **424**, 976 (1994).

[14] P. Planesas, J. Martin-Pintado, and E. Serabyn, ApJ Lett. **386**, L23 (1992).

[15] V. S. Strelnitski, H. A. Smith, M. R. Haas, S. W. J. Colgan, E. F. Erickson, N. Geis, D. J. Hollenbach, and C. H. Townes, in *Proceedings of the Airborne Astronomy Symposium on the Galactic Ecosystem: From Gas to Stars to Dust*, M. R. Haas, J. A. Davidson, and E. F. Erickson, eds. (ASP, San Francisco, 1995) p.271.

[16] H. A. Smith, V. S. Strelnitski, J. Lacy, D. Kelly, and J. Miles (in preparation).

[17] C. Thum, J. Martin-Pintado, and R. Bachiller, A&A **256**, 507 (1992).

[18] V. S. Strelnitski, V. O. Ponomarev, and H. A. Smith, ApJ (1995) submitted.

[19] M. Cohen, J. H. Bieging, J. W. Dreher, and W. J. Welch, ApJ **292**, 249 (1985).

[20] V. S. Strelnitski, H. A. Smith, V. O. Ponomarev, ApJ (1995) (submitted).

[21] H. E. Matthews, (private communication).

[22] P. Cox *et al.*, A&A, **295**, L39, (1995).

58

Beyond the South Pole

John W. V. Storey

58.1 Why Antarctica?

The popular image of Antarctica is of a harsh, wind-swept continent, totally un-suited to any form of astronomy. However, this image is appropriate only to the coast, where the fierce katabatic winds can create devastating blizzards. On the plateau, conditions are relatively benign. Cloud cover and precipitation are mini-mal, with wind speeds averaging a mere 3 m/s. It is, nevertheless, extremely cold: typically $-60°$C at the South Pole in winter, and falling to $-90°$C at Dome A.

There are many reasons why the Antarctic Plateau is an exceptional site for astronomy, but three reasons stand out in particular. First, the atmospheric wa-ter vapour content is extremely low. At the Pole in summer the precipitable water vapour averages 0.2 to 0.5 mm, and falls below 0.1 mm in winter [1]. This dramati-cally improves the atmospheric transmission in the far-infrared and sub-millimetre windows over any other observing site on the surface of the Earth. For example, from the summit of Mauna Kea in Hawaii (the premier observing site currently in use) the precipitable water vapour is only rarely as low as 1 mm. From the Antarctic Plateau several windows throughout the far-infrared are opened up to observation from the ground, allowing detailed study of many spectral lines from fine-structure transitions of common interstellar atoms and ions [2].

Second, the Antarctic Plateau may also provide substantially better "seeing" at optical wavelengths than even the best mountain-top sites currently operat-ing [3]. Seeing is the degradation of the spatial resolution of a telescope caused by inhomogeneities in the refractive index of the atmosphere through which the light is travelling. These inhomogeneities result principally from tiny tempera-ture fluctuations in the air temperature, or "microthermal turbulence." On the Plateau, microthermal turbulence is minimised by the combination of the high altitude ($\sim$ 4000 m), extreme cold (creating an equivalent pressure altitude of over 5000 m), small diurnal temperature changes and complete uniformity in the emis-sivity of the terrain. In addition to this is the possibility of significantly enhanced UV transmission because of the high altitude and reduced aerosol concentration in the atmosphere.

Third, the thermal background radiation from the atmosphere and telescope is considerably reduced, leading to increased sensitivities. This is most dramatic in the near-infrared "K" band at 2.2 μm, where the thermal background is predicted to be reduced by a factor of 220 over that at Mauna Kea [4, 5]. In the spectral range from 2.27 to 2.45 μm no airglow emission has been measured from OH radicals or aurorae. Hence, this drop in thermal background translates directly into a signal-to-noise gain of about 15 if detector performance is background-limited, rising to 220 if sensitivity is limited by the ability to flat-field an array detector. Thus for near-IR observations, a telescope on the Antarctic Plateau of 2 to 3 meter aperture could out-perform the largest telescopes being built (with diameters of 10 to 16 meters) anywhere else on the Earth.

At near-infrared wavelengths the zodiacal background radiation from Solar System dust reaches a minimum, with scattered sunlight dominating at shorter wavelengths and thermal radiation from the dust dominating at longer wavelengths. The Antarctic Plateau may therefore provide a site with as dark a background as can be found anywhere within the inner Solar System, offering us a "cosmological window" through which we may peer back with the greatest clarity into the early universe.

Additional factors which make Antarctica attractive for astronomy include:

- It is well-positioned to complete VLBI north-south baselines to countries such as Australia.

- It offers the possibility of continuous monitoring (24 hours/day) of astronomical sources.

- There is a very low level of radio interference.

These issues have recently been addressed by the Australian Working Group on Antarctic Astronomy. Their report [6] was very positive, and concluded that *"...the case for development of Antarctic astronomy is overwhelming."*

The potential of the Antarctic Plateau as a site for astronomy has of course been recognised for a number of years. Early work by Martin Pomerantz at the South Pole has led to the establishment there of a thriving group of experiments under the umbrella of CARA (see next section). On the international scene, interest in Antarctica is rapidly gaining momentum, as the following extracts show:

- 1991: The International Astronomical Union (IAU) General Assembly passed a resolution recognising the importance of Antarctica for astronomy, and set up a working group.

- 1994: A special session of the XXIII STAR/SCAR meeting was devoted to Antarctic astronomy. A resolution was subsequently passed by STAR (the Solar-Terrestrial and Astrophysics Research working group) in support of further astronomical research in Antarctica, and urged: *"The acquisition and analysis of directly comparable site testing data (preferably with identical instrumentation) at high altitude sites including the South Pole, Vostok, Dome C and Dome A."*

- 1995: The National Committee for Astronomy of the Australian Academy of Science, in its report "Australian Astronomy beyond 2000," ranked an International Observatory on the Antarctic Plateau amongst the highest priorities for Australia's next generation facilities, and specifically recommended that *"Australia should participate in site testing for an international observatory on the Antarctic Plateau, with a view to participation in the development of such an observatory should it prove feasible."*

58.2   South Pole Station

Pre-eminent in Antarctic astronomy is the Chicago-based Center for Astrophysical Research in Antarctica (CARA). CARA currently operates three telescopes at the South Pole. These are a cosmic background experiment (COBRA), a submillimeter telescope (ASTRO) and a near-infrared telescope (SPIREX). CARA's Advanced Telescope Project (ATP) is also carrying out site testing at the Pole, measuring (among other things) the sky brightness and the optical seeing. There are another three experiments under development at the Pole to measure cosmic rays, gamma rays and neutrinos, funded by individual grants from the National Science Foundation, and involving collaborators from Sweden and the United Kingdom.

A team from the University of Nice, CARA and the University of New South Wales (UNSW) is carrying out measurements of microthermal turbulence as a function of height in the atmosphere. These data can be used to derive the astronomical seeing. Measurements carried out from a fixed mast over the 1994 winter demonstrated an extremely low level of near-ground turbulence. In the second stage of this experiment, a series of balloon flights will take place in 1995 to carry the experiment to higher altitudes. The first three of a planned 28 such balloon flights were made from the South Pole in February 1995.

Another experiment being conducted by UNSW and CARA at the South Pole makes use of an infrared photometer recycled from the Anglo Australian Observatory: the Infrared Photometer Spectrometer (IRPS). This instrument uses a liquid-nitrogen cooled InSb detector to measure the infrared sky brightness in the 1 to 5 μm region. Of particular importance is establishing whether there are weak airglow lines in the 2.27–2.45 μm "cosmological window," and what effect aurorae may have on observations.

During 1994 IRPS obtained the first confirmation of the existence of the predicted cosmological window, though the level of residual sky brightness was somewhat higher than expected. In January, 1995 IRPS was moved to the roof of the new M. A. Pomerantz Building. The instrument was overhauled, recalibrated, and will winter-over again in 1995 to gather a second, more detailed data set.

58.3   Other High Plateau Sites

At an altitude of 2835 m, the South Pole is by no means the highest nor the driest location on the Plateau. While it currently is the only location with logistical support sufficient for a major observatory, astronomers always seek the ultimate observing site. Possibilities include Vostok (3488 m), Dome C (3425 m), and the highest point on the Plateau, Dome A ($\sim$ 4200 m).

Data on the suitability of these sites are scarce. The best studied is Vostok [7], and data from there have already been analysed by Townes and Melnick to show its potential for far-infrared observations [2]. Other information comes from automatic weather stations, and from the U.S. Plateau Station which operated from 1965 to 1969.

Construction of a major observatory at any of these sites represents a daunting logistical challenge. Before any decision can be made, comprehensive site testing data are required. In order to do this we are taking advantage of the pioneering efforts of our colleagues in the geophysical community. Several years ago they were faced with a similar challenge—how to gather data from remote, uninhabited sites throughout the Antarctic winter.

58.4   The Automated Geophysical Observatory

Beginning in 1981, the National Space Foundation (NSF) funded an eight-year program with Lockheed to develop a self-powered, self-heated field station to enable geophysical data to be obtained from the remote Antarctic interior. This station, of which six were built, is called "the Automated Geophysical Observatory (AGO)" [8].

The AGO is a very well-insulated, small portable laboratory. It provides accommodation for up to four people, who accompany the AGO to its destination, set up the experiments over a period of about one week, then leave the AGO to operate automatically for the next 12 months. A major challenge that Lockheed had to face was how to provide heat and electrical power for the full twelve months that the AGO would sit on the ice in sub-zero temperatures. There is little wind on the plateau, it is dark for six months of the year, and environmental and cost considerations rule out radioisotope generators. The solution chosen is a propane-fueled catalytic oxidiser, which produces 2.5 kW of heat, plus 50 watts of electrical power via a thermoelectric generator.

Data taken by the AGO are recorded on optical disk, and retrieved at the end of the 12-month deployment. Only the health and status of the experiment can be monitored in close to real time—via transmissions to polar-orbiting Argos satellites.

The AGO is designed to fit exactly into the cargo hold of a ski-equipped LC-130 "Hercules" transport plane. In the initial "put-in" flight, the AGO is placed on the snow, together with the science team of three or four persons and enough fuel for three weeks. Approximately one week later the Hercules returns with a year's supply of fuel and retrieves the science team. The AGO remains on the ice for the next twelve months, gathering data automatically until the next service visit.

58.5   The AASTO

Although developed for the geophysical community, an AGO is equally suited to powering astronomical instruments. We are therefore in the process of purchasing a seventh "AGO," built to our specifications, and which we call the Automated Astronomical Site-Testing Observatory, or AASTO.

In May 1994 the Automated Astronomical Site Testing Observatory Working Group (AASTOWG) was established to coordinate Australian and U.S. plans for site testing from the AASTO. Subsequently, the University of New South Wales and the Australian National University signed a Memorandum of Understanding establishing JACARA (the Joint Australian Centre for Astrophysical Research in Antarctica) and agreed to purchase the AASTO from internal funds. Instruments will be developed by interested parties in the U.S. and Australia. At the present time (April 1995) several groups have expressed interest in supplying submillimeter and millimeter radiometers, and the ANU is to develop a low-power DIMM (Differential Image-Motion Monitor) for direct measurements of the seeing.

Many of the site-testing instruments proposed are considerably more complex than the "all-electronic" instruments currently installed in AGOs. Small telescopes, scanning mirrors, and detectors operating at cryogenic temperatures will all be required. Designing these instruments to operate unattended for a full 12 months, and within the very low power budget available (a total of 50 W), is a major challenge. Three of the instruments are under development at UNSW, and are briefly described below.

58.5.1   Near-infrared Sky Monitor

Precise measurement of the background flux in the 2.35 μm "cosmological window" is a primary goal of the AASTO program. To achieve this, a low power, Stirling-cycle cooled instrument is under development for the near-infrared. Containing only a single, cold, 2.35 μm filter, this instrument will be optimised for accurate measurement of the very low background levels expected. A tilt mirror will enable observations to be made as a function of zenith distance.

58.5.2   Mid-infrared Transmission and Sky Brightness Monitor

Mid-infrared (7–17 μm) astronomy also stands to gain from an Antarctic Plateau observatory [9]. Although the drop in thermal background due to the lower temperature is smaller than in the near-infrared, it is still significant. The most important gains, however, will be from the improved atmospheric transmission, especially at the edges of the commonly used windows. Two very important spectral lines which may become accessible are those of [NeVI] at 7.75 μm and molecular hydrogen at 17.0 μm.

One concern, however, is the possible background contribution of thermal noise from ice crystals, whose abundance and significance is not well determined. A simple sky monitor is therefore proposed, with a mercury cadmium telluride detector, cooled via a second Stirling engine to 77 K. The detector will look out at the sky in a 1 arcminute field of view through a selection of ambient-temperature filters and a tilt mirror.

58.5.3   UV/visible Sky Monitor

This experiment will monitor the atmospheric transmission from 300 to 1100 nm, and will also allow an assessment of the possible impact of the auroral emission lines throughout this range. Because there is negligible atmospheric thermal emission at these wavelengths, the atmospheric transmission must be measured directly by taking spectra of bright stars with a small telescope. This will be a simple prime-focus instrument with a 20 cm diameter primary mirror. At the focus, a small bundle of optical fibres will direct the light back to the AASTO where a spectrometer with a passively cooled CCD array will cover the UV/visible region at a resolution of approximately 1 nm.

58.5.4   Future Plans

The AASTO will be built at Lockheed during 1995, and shipped to McMurdo at the end of the year. In the meantime, a test facility is being developed at UNSW to allow the testing of full-scale instruments at $-85°C$, while operating them from an AASTO power supply and data acquisition system.

In late 1996 it is planned to set the AASTO up at the Amundsen-Scott South Pole station, where it will operate unattended for the next 12 months. There it will not only acquire the necessary "baseline" data on conditions at the Pole, but will also allow the instruments to be fully field tested at a serviced location.

Each following austral summer, the AASTO can be redeployed to a new location. Dome C is the obvious next choice. Not only is an AGO already operating there, but France and Italy have ambitious plans to build a permanent scientific facility on the Dome. Eventually it is hoped to take the AASTO to the summit of Dome A. This is likely to be the very best submillimeter and far-infrared site on Earth, but also the least accessible.

According to the current timetable, this site-testing program should be nearing completion at the end of 1999. It seems appropriate that, as we enter a new millenium, humankind will be in the process of choosing the ultimate site on Earth on which to build the next generation of instruments to unravel the mysteries of the universe.

58.6   Acknowledgments

The Antarctic astronomy program at the University of New South Wales would not be possible without the major contributions of Michael Burton and Michael

Ashley. Nor would it be possible without the inspiration of earlier workers in the field such as Charlie Townes, whose 1990 paper and enthusiastic tales of his visit to the frozen continent were the spark that ignited my interest in Antarctic astronomy.

References

[1] J. Bally, in *Astrophysics in Antarctica,* edited by D. J. Mullan, M. A. Pomerantz and T. Stanev (American Institute of Physics Conference Proceedings, 1989), p. 100.

[2] C. H. Townes and G. Melnick, Pub. Astron. Soc. Pac. **102,** 357 (1990).

[3] P. R. Gillingham, *ANARE Research Notes # 88,* edited by M. Duldig and G. Burns (ANARE, 1992), p. 285.

[4] D. Lubin, Masters Thesis, University of Chicago (1988).

[5] D.A. Harper, in *Astrophysics in Antarctica,* edited by D. J. Mullan, M. A. Pomerantz and T. Stanev (American Institute of Physics Conference Proceedings, 1989), p. 123.

[6] M. G. Burton *et al.*, Proc. Astron. Soc. Aust. **11,** 127 (1994).

[7] L. P. Burova, V. D. Gromov, N. I. Luk'yanchikova, and G. B. Sholomitskii, Sov. Astron. Lett. **12,** 811 (1986).

[8] J. H. Doolittle and S. Mende, Antarctic Journal of the United States (preprint).

[9] C. H. Smith, *ANARE Research Notes # 88,* edited by M. Duldig and G. Burns (ANARE, 1992), p. 301.

59

Townes and Nonlinear Optics

Boris P. Stoicheff

It was in the early 1960s, immediately after the development of the ruby and the helium-neon lasers that the first observations were made of optical harmonics, two-photon absorption, stimulated Raman and Brillouin scattering, and self-trapping and focusing of laser beams. This was an exciting period for nonlinear optics, and Charles Townes and his students and collaborators were major players in the exploration of this field of research.

In reviewing briefly some aspects of this group's contributions, I should like to begin by recounting the theme of the introductory paper given by Townes at the first Quantum Electronics Conference [1] held in September of 1959 at Shawanga Lodge in High View, New York. He began with: "The subject matter of this Symposium lies near the interface between two highly developed disciplines—the field of electronics and that of spectroscopy." He went on to say: "Here we are concerned with resonance spectroscopy and resonant interactions, not the resonances dependent on macroscopic properties such as mass or size of a piece of material ... but the resonances which are primarily determined by characteristics of individual atoms or molecules. ... In fact, the spectroscopist can probably produce from atomic and molecular resonances almost any type of circuit element or phenomenon to which the electronics engineer is normally accustomed."

Clearly, this was an early prognosis of the effects of mixing of spectroscopy and electronics—one might say, the anticipation of nonlinear phenomena. A pertinent early example came with the realization that the interferometer of the spectroscopist was the cavity or resonator of the electronics engineer—and this juxtaposition of meanings was instrumental in the development of the laser. Later, at the first conference, Townes suggested the possibility of phonon maser action using paramagnetic crystals [2].

I first encountered Charles Townes when as a graduate student I attended my first Molecular Spectroscopy Conference at Ohio State University in 1949, where Townes, Van Vleck, Gordy, and Dennison, among others, were thrilling us with the wonders of the new field of microwave spectroscopy. Next, we met briefly when he visited Gerhard Herzberg's Laboratory at the National Research Council in Ottawa, where I was involved in high-resolution Raman spectroscopy. So you can imagine my surprise when attending the Second Quantum Electronics Conference at Berkeley in March of 1961, my telephone rang at 6:00 a.m. on Monday morning before the opening of the conference: "Hello, this is Charlie Townes; I hope I didn't waken you. I would like to talk to you about some ideas on Brillouin scattering. How about breakfast in 15 minutes? I'd like to bring along Joe Giordmaine a graduate student of mine." What piqued my interest was that he had mentioned Brillouin scattering, a topic he was to discuss later in his paper "Some Applications of Optical and Infrared Masers," [3] which opened the conference. Today, Brillouin scattering is widely known and well advanced! But at that time, while numerous studies [4] had been carried out in the USSR and India, only two papers had been published in North America, one by David Rank in 1948 [5], and the other, a decade later by colleagues and myself [6] which apparently Charlie had read.

At breakfast, we discussed light scattering in general, and he raised the point that Doppler broadening would be minimized for scattering in the forward direction, since the Doppler width would then be dependent on frequency shift rather than on the much higher exciting frequency. Thus it should be possible to use heterodyne spectroscopy, that is, the mixing of scattered light with the exciting radiation on a fast detector to produce the beat or difference frequency. This would allow very high resolution and high precision for measurement of rotational Raman frequencies and Brillouin doublets. It is well-known that this method is now one of the standard techniques for Brillouin spectroscopy.

In his paper, Charlie discussed the above more fully, along with four other experiments, representing exciting areas of fundamental interest to physical scientists. He considered the problem of communication between our civilization and one possibly located on some planet associated with a star many light-years away; he emphasized the possibility of frequency multiplication from the radio frequency and microwave range into the visible and ultraviolet regions, and the measurement of frequency to high precision; and he noted the paucity of high resolution spectroscopy in the far infrared, suggesting that tunable oscillators throughout the relatively unworked region of 10 to 500 μm would be highly desirable. Finally, he reviewed the importance of the monochromaticity of infrared and optical masers to fundamental experiments, and discussed as an example a Michelson-Morley experiment using the frequency comparison of two optical masers. Most of these experiments have been carried out successfully, well beyond the early expectations, and thus have provided advanced experimental techniques and a large body of new knowledge. During this early period, Townes was in Washington much of the time, serving as Vice-President and Director of Research at the Institute of Defense Analyses, while guiding the research of graduate students at Columbia University. At the same time, he was preparing to take up new responsibilities as Provost and Professor of Physics at the Massachusetts Institute of Technology, a post he held from 1961 to 1966. In spite of these onerous commitments, his insight and dedication inspired the research of a group of young scientists working in nonlinear optics and laser spectroscopy. It is worth noting some of their achievements during the early years of quantum electronics.

The stage was set for nonlinear optics with the announcement of the observation of optical second-harmonic generation (SHG) by Franken, Hill, Peters, and Weinreich in 1961 [7], followed by the classic theoretical paper of Armstrong,

Bloembergen, Ducuing, and Pershan in 1962 [8]. Soon after this first observation of SHG in crystalline quartz, and of two-photon excitation [9] of impurity levels in a crystal of $CaF_2:Eu^{++}$, Isaac Abella at Columbia University compared the wavelengths of second harmonic and laser radiation, and found that the ratio was the expected 1:2 to 1.5 ppm [10]. He also observed two-photon absorption in Cs vapor [11]. He used the transition $^9D_3/2 \leftarrow {}^6S_1/2$ at 28,828 cm^{-1} pumped by a ruby laser, thermally tunable in the region of $\sim 14,410$ cm^{-1}, and monitored the process through the fluorescence decay of $^9D_3/2 \rightarrow {}^6P_3/2$ at 584.7 nm. Heterodyne detection of light by driven ultrasonic waves in water was accomplished by Cummins, Knable, Gampel, and Yeh [12]. They generated travelling waves of 5, 10, and 15 MHz, and measured the resulting frequency shifts of a cw He-Ne laser beam. Cummins and Knable [12] also observed Bragg reflection of light by travelling ultrasonic waves, showing this to be a simple method for shifting light frequencies and spatially separating the shifted components from the incident forward beam. With Yeh [12] they examined the broadening of the Rayleigh line from scattering by dilute solutions of polystyrene molecules using the heterodyne detection technique to achieve an instrumental resolution of 6 Hz. Earlier, Cummins and Abella were involved in an unsuccessful attempt to achieve laser action with alkali vapors, following the suggestion described in the original paper of Schawlow and Townes [13].

The year 1962 saw the discovery of stimulated Raman emission by Woodbury and Ng [14] in the nitrobenzene-filled Q-switch cell of the ruby laser, and then with colleagues, in a variety of organic liquids placed in the laser cavity. Investigations with liquids external to the cavity, revealed emission of several orders of anti-Stokes radiation, in concentric cones around the incident beam [15], as well as the unusual characteristics of stimulated Raman radiation in comparison with the corresponding normal Raman scattering [16]. The first contribution to this effect by the Townes group was the development of a classical theory of coherently driven molecular vibrations [17] which explained many of the observations, including threshold requirements, and in particular, the directional properties of the Stokes and anti-Stokes emission. The mechanism suggested by Garmire, Pandarese, and Townes was the excitation of intense, coherent oscillations at infrared frequencies, which then modulate the incident and Raman scattered light to produce Stokes and anti-Stokes radiation of many orders. A more detailed presentation of the theory and its consequences, along with the first prediction of stimulated Brillouin scattering, was given by Chiao, Garmire, and Townes [18] at the Enrico Fermi summer school "Quantum Electronics and Coherent Light", organized by Townes in 1963. Several other papers at the school dealt with theories and experimental results in stimulated Raman scattering [18]. Shortly afterwards, I was pleased to spend a year's leave with Townes and his students at M.I.T. We began a detailed study of the angular dependence of the stimulated Raman process, and confirmed theoretical predictions with the observation [19] of sharply-defined anti-Stokes emission cones and Stokes absorption cones from calcite crystals.

Next came the observation of stimulated Brillouin scattering [20] of an intense laser beam involving coherent amplification of an acoustic wave (a hypersonic lattice vibration) in the forward direction, and a light wave scattered in the back-direction. The first experiments were carried out with quartz and sapphire crystals, with the spectra resolved by Fabry-Perot interferometers, and detected on photographic plates. Each pulse of the focused 694 nm incident radiation caused extensive internal fractures in the crystals, resulting from high local stresses produced by the intense acoustic waves or from local heating produced by damping of these acoustic waves. The relevant theory had been discussed earlier [18, 21]. A short time later, stimulated Brillouin scattering was reported in a variety of liquids [22].

During these early investigations of stimulated Raman and Brillouin scattering, a recurring problem was the lack of reproducibility of observations from lab to lab, and even in the same laboratory. These anomalies seemed to stem from the intensity and quality of the incident laser beam, and from the media being studied. Clarification of these problems came rapidly once it was realized that an intense laser beam can produce its own waveguide and propagate without spreading, as discussed by Chiao, Garmire, and Townes in 1964 [23]. They showed that such self-focusing and self-trapping can occur in materials whose refractive index (dielectric constant) increases with electric field intensity. With the refractive index $n = n_0 + n_2 E^2$, they found that the coefficient n_2 for many liquid or solid dielectric media is such that the threshold power of $\sim 10^6$ W for trapping was easily reached with the ruby lasers used in such experiments. In a later work [24], they demonstrated the evolution of an optical waveguide in liquid carbon disulfide from a 1 mm diameter incident beam to a bright filament of ~ 50 μm, and to propagation for 10-50 cm without diffraction.

Along with these contributions to nonlinear optics, Charlie was involved in examining, along with Ali Javan, the frequency stability of He-Ne lasers [25], and testing the special theory of relativity [26], using heterodyne techniques to detect any variation in the frequencies of two lasers. He then developed these methods further, for use in astronomy, another challenge for Charlie, and one where his ideas and innovative techniques have opened vast new horizons for exploration.

59.1   Note

This essay is based on remarks at the Birthday Symposium in honor of Charles H. Townes on May 22, 1995 at CLEO/QELS 95, in Baltimore, Maryland.

References

[1] C.H. Townes, in Quantum Electronics, edited by C.H. Townes (Columbia University Press, New York, 1960) pp. vi-viii.

[2] C.H. Townes, ibid., pp. 405-409.

[3] C.H. Townes, in Advances in Quantum Electronics, edited by J.R. Singer (Columbia University Press, New York, 1961) pp. 3-11.

[4] I. L. Fabelinskii, *Molecular Scattering of Light* (Plenum Press, New York, 1968) and references therein.

[5] D. H. Rank, J. M. McCortney, and J. Szaz, J. Opt. Soc. Amer. **38**, 287 (1948).

[6] P. Flubacher, A. J. Leadbetter, J. A. Morrison, and B. P. Stoicheff, J. Phys. Chem. Solids **12**, 53 (1959).

[7] P. A. Franken, A. E. Hill, C. W. Peters, and G. Weinreich, Phys. Rev. Lett. **7**, 118 (1961).

[8] J. A. Armstrong, N. Bloembergen, J. Ducuing, and P. S. Pershan, Phys. Rev. **127**, 1918 (1962).

[9] W. Kaiser and C. G. B. Garrett, Phys. Rev. Lett. **7**, 229 (1961).

[10] I. D. Abella, Proc. IRE **50**, 1824 (1962).

[11] I. D. Abella, Phys. Rev. Lett. **9**, 453 (1962).

[12] H. Cummins, N. Knable, L. Gampel, and Y. Yeh, Appl. Phys. Lett. **2**, 62 (1963); H.Z. Cummins and N. Knable, Proc. IEEE **51**, 1246 (1963); H.Z. Cummins, N. Knable, and Y. Yeh, Phys. Rev. Lett. **12**, 150 (1964).

[13] A. L. Schawlow and C.H. Townes, Phys. Rev. **112**, 1940 (1958).

[14] E. J. Woodbury and W.K. Ng, Proc. IRE 50, 2367 (1962); G. Eckhart, R. W. Hellwarth, F. J. McClung, S. E. Schwarz, D. Weiner, and E. J. Woodbury, Phys. Rev. Lett. **9**, 455 (1962).

[15] R. W. Terhune, Bull. Am. Phys. Soc. **8**, 359 (1963).

[16] B. P. Stoicheff, Phys. Lett. **7**, 186 (1963).

[17] E. Garmire, F. Pandarese and C. H. Townes, Phys. Rev. Lett. **11**, 160 (1963).

[18] R. Y. Chiao, E. Garmire, and C. H. Townes in Proc. Int. School of Phys. "Enrico Fermi," Course XXXI, edited by P.A. Miles (Academic Press, New York, 1964) pp. 326-338. See also: N. Bloembergen, *ibid.*, pp.247-272; A. Javan, *ibid.*, pp.284-305; B.P. Stoicheff, *ibid.*, pp 306-325.

[19] R. Y. Chiao and B. P. Stoicheff, Phys. Rev. Lett. **12**, 290 (1964).

[20] R. Y. Chiao, C. H. Townes, and B. P. Stoicheff, Phys. Rev. Lett. **12**, 592 (1964).

[21] N. M. Kroll, Proc. IEEE, **51**, 110 (1963).

[22] R. G. Brewer and K. E. Rieckhoff, Phys. Rev. Lett. **13**, 334 (1964); E. Garmire and C. H. Townes, Appl. Phys. Lett. **5**, 84 (1964).

[23] R. Y. Chiao, E. Garmire, and C. H. Townes, Phys. Rev. Lett. **13**, 479 (1964).

[24] E. Garmire, R. Y. Chiao, and C. H. Townes, Phys. Rev. Lett. **16**, 347 (1966).

[25] T. S. Jaseja, A. Javan, and C. H. Townes, Phys. Rev. Lett. **10**, 165 (1963).

[26] T. S. Jaseja, A. Javan, J. Murray, and C.H. Townes, Phys. Rev. 133, A1221 (1964).

60

Spectral Observations of the Molecular Cloud Orion S

Edmund C. Sutton
and Goran Sandell

60.1   Introduction

Approximately 1.5′ south of IRc2 in the molecular cloud OMC-1 there is a pro-tostellar condensation known as Orion S. This source exhibits strong continuum emission at 400 μm, indicative of a large column density of warm dust. In the map of Keene et al. [1] this "southern source" had a flux density approximately 2/3 of that from the Orion-KL region. Higher resolution continuum maps at 3 mm and 1.3 mm showed the source structure in additional detail [2, 3]. Observations of metastable NH_3 out to (J,K) = (7,7) revealed strong emission, which is interpreted as arising from a high temperature core which has been designated "S6" [4, 5]. This region also exhibits strong CS $J = 2 \rightarrow 1$ emission, which is generally an indicator of high density. The CS emission was mapped by Mundy et al. [2, 6], who designated this source as both "CS3" and "condensation 4". A variety of estimates give the density of this region to be of order 10^6 cm^{-3} or greater and the mass to be of order 50 $M_\odot$. Evidence for a bipolar molecular outflow for Orion S has been seen in SiO and CO emission [7-9]. Recently McMullin et al. [10] made a number of interferometric maps of this source with the Berkeley-Illinois-Maryland Association (BIMA) array to develop a picture of its chemical composition. They interpret Orion S as an object in a relatively early stage of chemical evolution. Overall the chemistry is rather similar to that of quiescent regions in which much molecular material remains deposited on grain surfaces and has not yet been released back into the gas phase.

In this paper we report a spectral survey of molecular emission from Orion S between 334 and 342 GHz. This survey was undertaken as a companion project to the five position survey of the Orion-KL region by Sutton et al. [11]. Although the spectral coverage reported here is somewhat limited, the data are nicely complementary to the survey of Orion S by Groesbeck [12], who has observed a slightly higher frequency band with coverage which partially overlaps our own. Between these two surveys sufficient data exist to develop a rather good picture of the chemical composition of Orion S.

60.2   Observations

The observations reported here were obtained at the James Clerk Maxwell Telescope (JCMT) during 1990 March using the equipment and techniques described by Sutton *et al.* [11]. The receiver system was a 345 GHz superconducting tunnel junction (SIS) mixer which operated in a double-sideband mode. Atmospheric conditions were generally good during these observations, with overall system temperatures (including atmospheric losses) of order 1100 K single sideband. The position of Orion S was taken to be $\alpha(1950) = 5^h 32^m 46^s.1$, $\delta(1950) = -5°26'02''$. At these frequencies the beam width of the JCMT was 13''.7 FWHM, larger than the Orion S continuum source size of approximately 6 arcsec but smaller than the extent seen in several molecular distributions (approximately 20 arcsec). The data were obtained using a position-switched observing mode and calibrated using standard "chopper-wheel" techniques. All data reported here have been converted to the T_R^* antenna temperature scale, in which corrections have been made for atmospheric losses and the antenna forward spillover and scattering.

A total of 13 double-sideband spectra were obtained covering bandwidths of 500 MHz in each sideband. These spectra were partially overlapping, allowing us to reconstruct the underlying single-sideband spectrum through an application of maximum-entropy analysis. The degree of redundancy (spectral overlap) in these data on average is less than that which was present in the data reported by Sutton *et al.* [11]. This tends to make the maximum entropy analysis somewhat less reliable, although this is partially compensated by the fact that the spectrum of Orion S is rather sparse and to a large extent unaffected by blending of overlapping lines. There are a few gaps in spectral coverage between 334 and 342 GHz. Our derived single-sideband spectrum is presented in Fig. 60.1. Most of the frequency range displayed was covered by between two and six independent observations (that is, between 2 and 6 double-sideband observations, each with a different local oscillator setting). In these regions the spectrum appears to be generally reliable. Near the ends of the spectrum and near gaps, the spectrum is somewhat less reliable. The only extensive region in Fig. 60.1 thus affected is the band between 335.35 and 335.85 GHz.

60.3   Analysis

A total of 23 lines were detected in Orion S, as shown in Fig. 60.1 and listed in Table 60.1. All of these lines also were detected in one or more of the Orion-KL positions studied by Sutton *et al.* [11]. Six molecules are seen in just single transitions: CO, CS, OCS, HCS$^+$, HDCO, and H$_2$CS. In the cases of CO and CS these are detections of rare isotopic species (C^{17}O and C^{34}S). The detection of OCS is somewhat questionable, since the line is weak and the energy of the $J = 28$ level is rather high ($E = 237$ K). Also the observed frequency for OCS implies a v_{lsr} of 5.3 km s^{-1}, somewhat blueshifted with respect to the other species seen in Orion S.

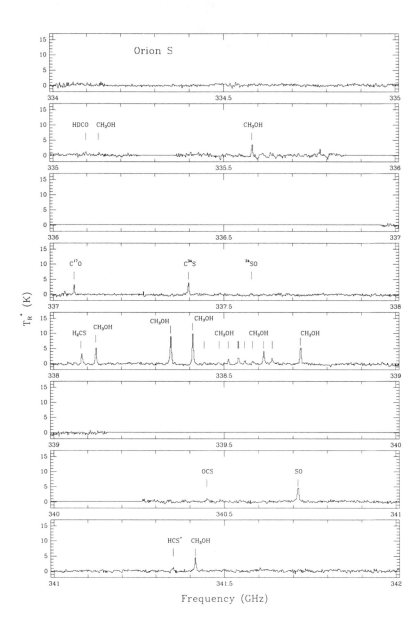

FIGURE 60.1. Single sideband spectrum of Orion S at 2 MHz resolution, with the continuum subtracted. The rest frequency scale is based on an assumed v_{lsr} of 6 km s^{-1}.

TABLE 60.1. Observed features in Orion S

Species	ν (MHz)	E_u (K)	J, N_J, J_K or J_{K_p,K_o}	T_R^*(pk) (K)	$\int T_R^*\, dv$ (K km s^{-1})
$C^{17}O$	337061.1	32.4	$3-2$	3.3	11.4
$C^{34}S$	337396.6	64.8	$7-6$	4.2	15.0
^{32}SO	340714.2	81.2	8_7-7_6	5.4	30.3
^{34}SO	337580.2[a,b]	86.1	8_8-7_7	0.7	2.2
OCS	340449.3	237.0	$28-27$	1.0	7.5
HCS$^+$	341350.8[c]	73.7	$8-7$	1.6	4.8
HDCO	335096.5	56.2	$5_{1,4}-4_{1,3}$	1.0	3.2
H_2CS	338080.8[b]	102.4	$10_{1,10}-9_{1,9}$	3.5	13.2
CH_3OH	335133.5	44.7	$2_2-3_1\ A^-$	1.5	3.6
	335582.0	79.0	$7_1-6_1\ A^+$	4.6	10.6
	338124.5	78.1	$7_0-6_0\ E$	5.4	19.1
	338344.6	70.5	$7_{-1}-6_{-1}\ E$	9.8	41.4
	338408.7	65.0	$7_0-6_0\ A^+$	10.5	41.3
	338442.3[d]	258.7	$7_6-6_6\ A^\pm$	0.6	0.9
	338486.3	202.9	$7_5-6_5\ A^\pm$	0.4	0.6
	338512.7	145.3	$7_4-6_4\ A^\pm$	1.9	5.0
	338512.9[e]	102.7	$7_2-6_2\ A^-$	—	—
	338540.8	114.8	$7_3-6_3\ A^+$	2.1	8.7
	338543.2[e]	114.8	$7_3-6_3\ A^-$	—	—
	338559.9	127.7	$7_{-3}-6_{-3}\ E$	1.0	3.6
	338583.2	112.7	$7_3-6_3\ E$	0.9	3.2
	338615.0	86.0	$7_1-6_1\ E$	4.3	18.1
	338639.9	102.7	$7_2-6_2\ A^+$	2.0	5.7
	338721.6	87.2	$7_2-6_2\ E$	6.0	23.8
	338722.9[e]	90.9	$7_{-2}-6_{-2}\ E$	—	—
	341415.6	80.1	$7_1-6_1\ A^-$	5.1	23.4

Notes: [a]Frequency from A. Salek [13].
[b]Frequency uncertainty $\pm$ 0.7 MHz.
[c]Frequency uncertainty $\pm$ 1.8 MHz.
[d]Emission in this line appears to be anomalously strong.
[e]Blended with preceeding line; treated as a composite.

This line of OCS was not seen in the "extended ridge" spectrum of Sutton *et al.* [11], although it was seen towards the Orion-KL core regions. Sulfur monoxide is seen in the form of both ^{32}SO and ^{34}SO. Methanol is seen in 15 resolvable features, all of them transitions of the main isotopic form ($^{12}CH_3OH$) in its ground torsional state. Overall the Orion S spectrum is most similar to the previously studied Orion "extended ridge" source, although some chemical differences are discussed below.

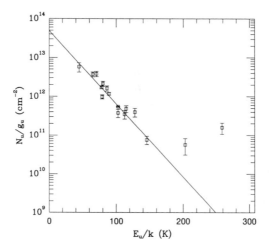

FIGURE 60.2. Rotation diagram for methanol in Orion S. The data are fit with a rotational excitation temperature of 26.5 K.

We have used these data to calculate molecular column densities and fractional abundances, based on estimates of the molecular excitation and line opacities. Our methods are discussed in Sutton *et al.* [11]. For CO we have assumed thermal excitation at a kinetic temperature of 80 K and an isotopic ratio of $^{16}O/^{17}O = 2400$. The remaining species have significantly larger dipole moments than CO and are expected to have sub-thermal excitation. We have assumed excitation temperatures for these species in the range from 25 K to 60 K, as listed in Table 60.2. For methanol we have constructed a rotation diagram, which is shown in Fig. 60.2. This shows that the bulk of the methanol emission from Orion S is well described by a rotational temperature of 23 K and a column density of 7.0×10^{15} cm^{-2}. Our derived beam averaged molecular column densities are summarized in Table 60.2.

To determine fractional abundances it is necessary to have a value for the molecular hydrogen column density. The correct value for this is somewhat uncertain, but we will adopt $N_c(H_2) = 10^{24}$ cm^{-2} for the average over our 13″.7 beam. This is generally consistent with the 1.3 mm fluxes of Mezger *et al.* [3], assuming a dust opacity law, of $\lambda^{-1.5}$ and a standard relationship between the dust opacity and the gas [14]. Larger column densities may be derived if one assumes a steeper opacity law or if one averages only over the apparent size of the source. The H_2 column density is also somewhat dependent on the assumed temperature. Continuum maps at 450 μm, 800 μm, and 850 μm obtained at the JCMT suggest an opacity law of approximately $\lambda^{-1.7}$ and give estimates for the H_2 column density as large as 5 $\times 10^{24}$ cm^{-2}. Our adopted value of 10^{24} cm^{-2} gives a CO fractional abundance of 1.7×10^{-5}, which is rather low compared with typical values for molecular clouds (although a similar CO depletion has been seen in NGC 1333/IRAS4 [15]). The CO fractional abundance will be even lower if the H_2 column density is higher than 10^{24} cm^{-2}.

TABLE 60.2. Beam Averaged Molecular Column Densities and Fractional Abundances

Species	T_{rot}	N_c	f
	(K)	(cm^{-2})	
H_2	—	1.0×10^{24}	—
CO	80[a]	1.7×10^{19}	1.7×10^{-5}
CS	25[a]	1.4×10^{15}	1.4×10^{-9}
SO	60[a]	5.9×10^{14}	5.9×10^{-10}
OCS	60[a]	2.7×10^{15}	2.7×10^{-9}
HCS^+	25[a]	4.0×10^{13}	4.0×10^{-11}
HDCO	30[a]	2.4×10^{13}	2.4×10^{-11}
H_2CS	40[a]	3.1×10^{14}	3.1×10^{-10}
CH_3OH	23	7.0×10^{15}	7.0×10^{-9}

Notes:[a] Assumed value for the rotational temperature.

One interesting feature of these data is the observation that Orion S has an active sulfur chemistry. We seem to have detected five sulfur-bearing species (CS, SO, OCS, HCS^+, and H_2CS), most of which have column densities comparable to those observed for the Orion "extended ridge". Groesbeck [12] has also observed weak lines of SO_2. Sulfur chemistry is often considered an indicator of warm gas, confirming the picture of Orion S as a warm, dense, and somewhat active region.

The inferred methanol column density is also large, in fact somewhat larger than the "extended ridge" methanol column density. In cold dark clouds, methanol is expected to have rather high abundances on grain surfaces but rather low gas phase abundances. A large gas phase methanol abundance can then be produced when the grains are heated and material is returned into the gas phase [15, 16]. The release of methanol and other species into the gas phase can then drive other molecular reactions, some of them leading to the formation of more complex molecular species. Methanol is the only molecule we have observed with more than four atoms. Groesbeck [12] has observed in addition several weak lines of CH_3OCH_3 in this source. These data may be interpreted as suggesting that grain mantle evaporation has begun rather recently, and that the time elapsed since that onset has been insufficient for the synthesis of the more complex molecular species, a process requiring on the order of 10^5 years. Thus Orion S may be viewed as a very young stellar or protostellar object.

A comparison of our data with the survey of Orion S at similar frequencies undertaken at the Caltech Submillimeter Observatory (CSO) [12] shows similar intensities for most lines observed in common. Where different lines were observed (e.g., for CS, SO, and H_2CS), the two surveys imply similar beam-averaged column densities, assuming reasonable values for the isotopic ratios and excitation conditions (and after taking proper account of spin degeneracies). These results suggest that the molecular material in this region is spatially extended. If the emission were from a compact region it would undergo different amounts of beam dilution in the CSO and JCMT beams, contrary to what is observed. This is consistent with source sizes of order 20″, as seen in interferometer maps of most molecular species. We agree with the conclusion of McMullin *et al.* [10] that the molecular abundances do not appear to be peaked on the small scale of the millimeter continuum source.

The kinematics of Orion S as seen in these lines shows a rather quiescent cloud. Line widths are typically $\Delta v \approx 3.8 \, \mathrm{km \, s^{-1}}$ (FWHM), similar to but slightly broader than the Orion extended ridge. Some of the strong, low energy methanol lines have broad wings extending $\pm 10 \, \mathrm{km \, s^{-1}}$ from line center, suggesting there may be some involvement of methanol with the molecular outflow. The average radial velocity of all lines seen in Orion S is $v_{\mathrm{lsr}} = 6.6 \, \mathrm{km \, s^{-1}}$, following the trend towards progressively more blueshifted material to the south of IRc2. Little variation is seen in the radial velocities of the different molecules, except for species with significant uncertainties in line frequencies (such as H_2CS).

60.4   Summary

We have surveyed molecular line emission from Orion S between 334 and 342 GHz. The rather sparse spectrum (compared with Orion-KL) appears to reflect rather low intrinsic abundances for many molecular species. However methanol is reasonably abundant, suggesting that some evaporation of grain mantles has begun to occur. Combined with other information about the density, temperature, and kinematics of this region, this indicates that Orion S is a very young stellar object.

60.5   Acknowledgments

We would like to thank R. Peng for his assistance in preparing this manuscript. This work has been supported in part by the National Science Foundation under grant AST-9117740. The James Clerk Maxwell Telescope is operated by the Royal Observatories on behalf of the Particle Physics and Astronomy Research Council of the United Kingdom, the Netherlands Organization for Scientific Research, and the National Research Council of Canada.

60.6   Concluding Personal Remarks (by Ed Sutton)

When I first joined the Townes group in 1974, the field of interstellar molecules was still relatively new. The discovery of ammonia emission had occurred and astronomers and physicists were well on their way to discovering the multitude of polyatomic molecules which few people other than Professor Townes had ever imagined would be present in interstellar space. As a member of that group at Berkeley I had a front row seat for many of those events, although at that time

I was not a radio astronomer myself. I vividly remember the discussions of such topics as maser mechanisms, isotopic ratios, and formaldehyde excitation at the noon-hour group lunches and the 290I seminars. Some of that must have sunk in, since I have gone on to spend much of my subsequent career in millimeter and sub-millimeter radio astronomy.

Yet my debt to Professor Townes goes far beyond this mere "background". Of all the influence that he has had on my career and personal development, I can think of four areas which seem most important. In each of these Professor Townes taught first and foremost by the way in which he exemplified these ideals in his own life. The first of these is breadth of scholarship. He liked to start to work in a new branch of physics each decade. He was uniquely equipped to do this with his breadth of knowledge of physics (and beyond), but it is an example for all of us and a warning against intellectual stagnation and complacency. I also remember that he liked to teach a given class at most three times, for similar reasons. The second principle is hard work. While hard work by itself is no guarantee for success, it certainly is necessary. And I cannot think of any individual I know who has worked harder (and with fewer hours of sleep per night) than Charles Townes. Thirdly, he has an outstanding intellectual flexibility. He always seems prepared to shift gears and discuss virtually *any* topic which comes up in conversation, and is usually able to draw upon an impressive storehouse of relevant knowledge in doing so. And finally and most importantly I have learned to respect his values and personal integrity. Those of us who have been privileged to work and study with him will always be grateful for what he has given us.

References

[1] J. Keene, R. H. Hildebrand, and S. E. Whitcomb, ApJ **252**, L11 (1982).
[2] L. G. Mundy, N. Z. Scoville, L. B. Bååth, C. R. Masson, and D. P. Woody, ApJ **304**, L51 (1986).
[3] P. G. Mezger, J. E. Wink, and R. Zylka, A&A **228**, 95 (1990).
[4] W. Batrla, T. L. Wilson, P. Bastien, and K. Ruf, A&A **128**, 279 (1983).
[5] R. Mauersberger, C. Henkel, T. L. Wilson, and C. M. Walmsley, A&A **162**, 199 (1986).
[6] L. G. Mundy, T. J. Cornwell, C. R. Masson, N. Z. Scoville, L. B. Bååth, and L. E. B. Johansson, ApJ **325**, 382 (1988).
[7] L. M. Ziurys and P. Friberg, ApJ **314**, L49 (1987).
[8] L. M. Ziurys, T. L. Wilson, and R. Mauersberger, ApJ **356**, L25 (1990).
[9] J. Schmid-Burgk, R. Gusten, R. Mauersberger, A. Schulz, and T. L. Wilson, ApJ **362**, L25 (1990).
[10] J. P. McMullin, L. G. Mundy, and G. A. Blake, ApJ **405**, 599 (1993).
[11] E. C. Sutton, R. Peng, W. C. Danchi, P. A. Jaminet, G. Sandell, and A. P. G. Russell, ApJ Suppl. **97**, 455 (1995).
[12] T. D. Groesbeck, Ph.D. thesis, California Institute of Technology (1995).
[13] A. Salek, private communication (1994).
[14] R. H. Hildebrand, Quarterly J. Roy. Astron. Soc. **24**, 267 (1983).

[15] G. A. Blake, G. Sandell, E. F. van Dishoeck, T. D. Groesbeck, L. G. Mundy, and C. Aspin, ApJ **441**, 689 (1995).
[16] P. Caselli, T. I. Hasegawa, and E. Herbst, ApJ **408**, 548 (1993).

61

A Visit to America

David H. Whiffen

Visits from Britain to America these days are two a penny and certainly nothing to write about for a largely American readership. Nevertheless, I hope that some reminiscences and some consequences of my visit, 1946-47, may be of interest.

The visit was the result of being awarded the Harkness Fellowship which included full funding and much detailed help. Normally the Fellowships were held at a university. In my own case matters took a variant turn. My doctorate work [1] at Oxford included the measurement of rotational relaxation of polar molecules in liquids, and I expressed a wish to work at Cornell under Peter Debye. But by 1946 he had changed his research to polymers and had no microwave equipment; he suggested and arranged that I work at the Bell Telephone Laboratories at Murray Hill, New Jersey where he was a consultant and could see me monthly.

Not joining a university, I required a work visa which I acquired just in time to join a ship with five other Harkness Fellows. One night in late September we drove over 200 miles from London to Fowey, a china clay port, to board the *S.S. Thomas Pinkney*, a war-time freighter called a liberty boat, as its first and probably last commercial passengers. The conditions were rough, and it was typical that we went the longer way via the Azores as this area had not yet been declared free of mines, so that the crew could claim danger money, even though danger was extremely improbable. Next there was a drunken brawl within the crew and one member was in real need of help from one of the Fellows, a qualified doctor. Then we moved into the edge of a hurricane which we failed to avoid as the telegraphist had dropped the warning message on the floor and it was not seen by the captain until a day or so later. After a week or more at sea we hove to on the mid-coast of New Jersey and were met by one of the Harkness staff who had ordered a cutter to take us ashore. We were then driven to New York. As a contrast I returned in September 1947 on the *Queen Elizabeth*.

Although parts of New Jersey did not have a good reputation, I found Summit, where I had lodgings, to be a pleasant town. It was near enough to New York for day visits which I often made by myself or with friends. The one trouble was that the train in the subway had a tendency to halt under the Hudson River with an ear-piercing screech leaving us wondering if we would catch the last train to Summit; the alternative, which did not arise, would have been to spend an unarranged night in Hoboken.

FIGURE 61.1. From left to right Ralph Merritt, Ben Wright—a vacation visitor—Bill Hewitt, and Charles Townes, Summer 1947.

The arrangement at Murray Hill was that I should continue work on dielectric loss. I was to share a room for both working laboratory and desk space with Charles H. Townes, his technician Ralph Merritt, and another staff member Bill Hewitt (see Fig. 61.1). At the time Charlie was working on gas microwave spectroscopy. With hindsight, I wish I had joined his team. However, as I was not a paid member of the Bell staff this would have been awkward in many ways, and would have meant jettisoning the research I had intended. As an uninvolved individual I used to countersign Townes' notebook confirming the date in case of an argument over a patent. Bell Telephone was very generous in permitting me to work at their expense and in allowing help from their staff, especially Townes, in assembling the microwave and other equipment I needed.

My work resulted in a publication [2] concerning the rotational relaxation of camphor in solutions. But it showed that the difficulties of a microwave cavity full of liquid exceeded its advantages. The project's real intention was to confirm the loss arising from collisions in non-polar liquids; this I was able to do on my return [3] to Oxford and show that the relaxation was more rapid than for rotations. Many privately thought I was using impure non-polar liquids, but my work was fully legitimated [4, 5] when spectroscopy was available in the range 20 to 1 cm^{-1}.

The staff at all levels was very friendly; the more senior such as J. Bardeen, W. H. Brattain, C. Davisson, L. H. Germer, C. Kittel, W. Shockley, would join in at lunch or in the class for learning Russian. Although far from fluent, I learned enough Russian to understand papers in my field and write short accounts for

FIGURE 61.2. From left to right: Mrs. Ellen Townes (mother of Charles H. Townes), Mary Townes (sister), and Henry Townes (father).

Physics Abstracts. Others who became friends included Morris Fine, Harry Ham, Cyril Harris, Conyers Herring, Alan Holden, Bob Hull, Warren Matthews, Jim Menard, Stan Morgan, Ed Murphy, Bob Newton, John Richardson, Morgan Sparks; many invited me home, and the younger bachelors invited me to join in visits to New York, the beaches, etc., often with female company. These lists are notable for the very distinguished scientific careers of their members.

As required by the terms of the Harkness Fellowship, I toured the U. S. A., choosing the early summer before university vacations. Traveling alone, using the Greyhound coaches for most of the trip, I visited most states. Both the Harkness staff and those at Murray Hill, especially Townes, provided introductions to scientists, the best-known of whom included F. Bloch, R. R. Brattain, B. L. Crawford, G. Glockler, W. Gordy, G. Herzberg (then at the Yerkes Observatory), R. C. Lord, L. Pauling, C. P. Smyth, E. B. Wilson, and R. W. Wood. There were also visits to relatives and friends of those I knew at Murray Hill. The hospitality I especially remember was from Charlie's father Henry, mother Ellen, and sister Mary in Greenville, South Carolina (see Fig. 61.2). Mrs. Townes arranged for me to visit a local junior school and talk to the pupils, all blacks, and answer their questions about England. I also spent a week at Cornell, arranged by Debye, to sample an American university.

An occasion which greatly impressed me occurred one day late in 1946 when Townes said I should come with him to a lecture by E. M. Purcell on nuclear magnetic resonance (NMR), which was my introduction to the subject. As well as the phenomenon itself and the measurement of the magnetic moment of the hydrogen nucleus, he had slides of "the-eating-of-the-hole-in-the-curve effect." The viscous paraffin sample and inhomogeneity of the magnet meant that the peak width was broad as demonstrated by a magnetic field sweep. However, if the magnet was fixed and a strong oscillating field power was used, saturation removed the signal. If a normal sweep was immediately applied, the narrow missing signal produced a hole in the middle of the unsaturated signal. The saturation disappeared exponentially with a relaxation time of seconds which could be measured. When I returned to Oxford in September 1947, my great friend, Rex E. Richards, had just acquired a secure position and was looking for a new research topic and was considering NMR. I strongly encouraged this, hoping that his nuclear relaxation times might have related features. However, other features were being discovered including crystal structures, chemical shifts, and spin-spin coupling, so that relaxation was less interesting: Richards' scientific career was devoted to NMR throughout. For myself I was keen to follow the subject and teach it, but I have only a few publications which relate to the determination of the relative signs of spin-spin coupling in molecules [6]. I preferred electron spin resonance where my knowledge of microwave circuitry was useful.

Of wider interest are the differences I perceived between American industrial scientific research closely linked with technology and British university scientific research linked with teaching. The Bell Laboratories at Murray Hill had a very large staff, from Nobel prize winners to bottle-washers. But everyone seemed to recognise that they were part of the same organisation. And each programme accumulated the appropriate staff in respect to their backgrounds and abilities. The contrast to universities where the staff is chosen for the subject to be taught and where research students are commonly restricted to their first academic discipline was striking. And Bell Labs seemed to be able to identify scientists who became good administrators; many scientists, often older and better, were content to let the administrators look after their needs. And I strongly suspected that the scientists often had higher salaries than their overlords. Another feature of the Bell enterprise was its attitude to staff who felt in need of a change. Such individuals, however eminent, were happily released, sometimes with a consultant position to retain contact; thus able younger staff could be appointed by Bell Telephone, and stultification avoided.

The company ran a canteen which many of the staff patronised. I often joined Townes and colleagues for lunch, but on many occasions he apologised for absence because somebody in the organisation, sometimes previously unknown to him, wanted to pick his brains. And similarly he had telephone requests. On one occasion he was nearly caught out, when he had a query regarding the protection of wooden telephone poles from fungal decay; he at least knew the man who did know. I tell this story to illustrate the atmosphere that employees understood they were all members of one large enterprise. And I remember an inverse case when Townes had difficulty with the microwave spectra of molecules which had more than one quadrupolar nucleus: Bardeen became a frequent visitor to our room for several weeks and the problem was resolved [7].

How was my own career influenced by this visit? Firstly, a small item: it was Townes who explained to me the nature of stimulated emission in spectroscopy; consequently I easily recognised a mistake in the literature which needed to be adjusted [8]. He and colleagues increased my grasp of microwave components and their use which was valuable when I was building spectrometers. Unfortunately, I could not work on gas microwaves when I first had a tenured post in Birmingham University in 1949. Unknown to me another staff member, J. Sheridan, had already adopted this subject and was already learning the trade with W. Gordy at Duke University. My colleagues and I did construct a very effective gas microwave spectrometer [9] at the University of Newcastle-upon-Tyne in 1969.

Between these two university posts I spent 1959-68 at the National Physical Laboratories at Teddington in its Basic Physics Divison. I would not have been interested in such a post if I had not sensed that it had some of the advantages for research that I knew in Bell Laboratories. In particular, the division staff included mathematicians, computer specialists, physicists, electrical engineers, chemists, and first-class technicians. My specific interest at this time was Electron Spin Resonance and included two spectrometers, the second of which [10] at 35 GHz was especially valuable for ENDOR measurements of N in diamond [11] and also for the relative and absolute sign determinations [12] of the electron-nuclear couplings; this needed double ENDOR.

As one who has worked in three universities and two establishments committed to scientific research and its applications, I have a view of their respective roles. Firstly, it is that research and undergraduate teaching are different matters, not least in financial requirements, and that, at least in Britain, this has not been fully realised. Teaching should be funded from public funds plus contributions from philanthropic sources and, in some cases such as students from other countries, from finance for individual students. The research institutions might be funded by international cooperation, national support, philanthropic sources, individual firms who pay for advice, patents, etc., sponsor-specific projects, or, like Bell, may even be sole owner and provider of the laboratory. With essentially separated financial matters, it is easier to arrange for the research staff to give some university course lectures and for the university staff to spend some vacations in a research environment. Research training should also be arranged in the research laboratories, doctorate degrees should be monitored by an independent body. Geography would play some part in relationships that develop between specific teaching and research bodies, but the nature and location of such bonds should be as varied as is realistic. Such a fairly sharp division should benefit the good teachers who no longer need to publish to retain their position or promotion. Conversely, researchers who are incompetent teachers would not be expected to teach.

To close this account of my visit to America and its consequences, I must thank the Harkness Foundation—now the Commonwealth Fund—for the opportunity it provided, and all the friends who made the visit so successful, and especially Charlie and Frances Townes who supported my scientific and social well-being, respectively.

References

[1] D. H. Whiffen and H. W. Thompson, Trans. Far. Soc. **42A**, 114 and 122 (1946).

[2] D. H. Whiffen, J. Am. Chem. Soc. **70**, 2452 (1948).

[3] D. H. Whiffen, Trans. Far. Soc. **46**, 124 (1950).

[4] J. E. Chamberlain, E. B. C. Warner, H. A. Gebbie, and W. Slough, Trans. Far. Soc. **63**, 2605 (1967).

[5] S. K. Garg, J. E. Bertie, H. Kilp, and C. P. Smyth, J. Chem. Phys. **49**, 2551 (1968).

[6] R. Freeman and D. H. Whiffen, Mol. Phys. **4**, 321 (1961); K. A. McLauchlan and D. H. Whiffen, Proc. Chem. Soc., p. 144 (1962); A. D. Cohen, R. Freeman, K. A. McLauchlan and D. H. Whiffen, Mol. Phys. **7**, 45 (1963).

[7] J. Bardeen and C. H. Townes, Phys. Rev. **73**, 647 and 1204 (1948).

[8] I. M. Mills, J. Chem. Phys. **30**, 1619 (1959).

[9] J. H. Carpenter, J. D. Cooper, J. B. Simpson, J. G. Smith, and D. H. Whiffen, J. Phys. E. **7**, 678 (1974).

[10] R. J. Cook, J. R. Rowlands, and D. H. Whiffen, Mol. Phys. 7, 32 (1963); R. J. Cook, J. Sci. Inst. **43**, 548 (1966).

[11] R. J. Cook and D. H. Whiffen, Proc. Roy. Soc. **295**, 99 (1966).

[12] R. J. Cook and D. H. Whiffen, Proc. Phys. Soc. 84, 845 (1964); R. J. Cook and D. H. Whiffen, J. Chem. Phys. **43**, 2908 (1965).

62

Review of Some Photothermal Effects

John R. Whinnery

62.1 Introduction

When a laser beam passes through a lossy material, the absorbed energy produces thermal gradients, which in turn produce gradients of refractive index. There may then occur a variety of effects including focusing or defocusing (thermal lens effects), beam distortion, self-phase modulation, or generation of acoustic waves. These are simple effects, easy to understand in principle, but the surprising aspect is their sensitivity to loss. Important effects, some useful but most undesirable, may occur with low-power beams passing through materials of a high degree of transparency. Because of the variety and importance of these effects, and because Charles Townes made a key suggestion in the early stages of their observations, it seems appropriate for a brief review in this Festschrift. An earlier review was given [1], but there have been later developments, especially in spectroscopic applications. The emphasis will be on the effect in liquids, but most of the points apply to gases and solids also.

S.P.S. Porto in 1963 placed cells of organic liquids inside the resonator of a helium-neon laser to increase the power in his study of Raman spectra of these materials. A series of transients was observed when the cell was inserted, moved, or the laser turned on, with time constants of a few seconds, reaching steady state with an expanded beam [2]. Because of the time constants, it seemed likely that these were thermal effects, but the first rough calculations appeared to rule these out because of the low absorptivities of the liquids (order of 10^{-4} cm^{-1}). Other explanations were sought until a more careful analysis was made [3], showing that thermal effects could indeed explain the phenomena. At this point, Townes, who was told about the observations, suggested measuring an absorbing solution by standard photometric means, and then diluting it and checking the absorption by the thermal lens method. Solutions of copper sulphate in water of tenth and hundredth molar were measured by photometric means. A diluted solution to thousandth molar could not be measured accurately by that method, but measurements by the thermal lens method yielded a result consistent within 10% of the photometric measurements. That was about the limit of accuracy in measuring the thermally expanded beam, so this was considered a key experiment in verifying the thermal hypothesis.

62.2   The Thin Thermal Lens

The analysis of [3] assumed a Gaussian laser beam, substantially constant in power and beam diameter over the length of the absorbing cell. This required that the cell length be small compared with confocal distance of the beam $\pi w_0^2/\lambda$ (where w_0 is minimum beam radius and λ is wavelength), and also small compared with focal length of the resulting lens. Only radial thermal conduction was considered and the cell radius was assumed large in comparison with beam radius so that the thermal boundary conditions of the cell played no role during the transient phase. Solution of the heat conduction equation then gave a temperature increase as a function of radius and time as

$$\Delta T(r, t) = \frac{\alpha P}{4 J \pi k} \left[\text{Ei}\left(-\frac{2r^2}{w^2}\right) - \text{Ei}\left(-\frac{2r^2}{8Dt + w^2}\right) \right] \qquad (62.1)$$

with α the absorption coefficient (cm^{-1}), P the power of the laser beam (watts), k the thermal conductivity (cal/cm/sec °C), w the "beam radius" (cm), defined as the radius for which field amplitude has fallen to e^{-1} of its axial value, $J = 4.184$ Joule/cal, and thermal diffusivity $D = k/\rho c_p$ where ρ is density (gm/cm^3) and c_p is specific heat (cal/g °C).

For small temperature changes, the refractive index change is proportional to ΔT through the rate of change of index with temperature, dn/dT. The exponential integrals of Eq. (62.1) may be expanded in a power series with terms up to second order retained in the paraxial approximation. The refractive index as a function of radius and time is then

$$n(r, t) \approx n_0 + \frac{dn}{dT} \frac{\alpha P}{4 J \pi k} \left[\ln\left(1 + \frac{2t}{t_c}\right) - \frac{2(r^2/w^2)}{1 + (t_c/2t)} \right]. \qquad (62.2)$$

where t_c is a characteristic thermal time constant,

$$t_c = \frac{w^2}{4D} \text{ sec }. \qquad (62.3)$$

For a cell of length l, the quadratic terms produce a simple lens with focal length

$$f = \frac{\pi J k w^2}{P \alpha l (dn/dT)} \left[1 + \frac{t_c}{2t} \right]. \qquad (62.4)$$

For most materials, dn/dT is negative, producing a diverging lens leading to a beam expansion (thermal blooming). This is usually undesirable but has been used for absorption measurement.

Figure 62.1 shows curves of temperature increase as a function of normalized radius for different values of t/t_c. Included is a steady state curve calculated with a fixed temperature boundary at $r/w = 10$, and the quadratic approximation near the axis.

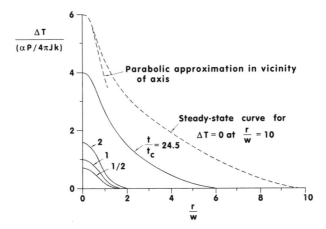

FIGURE 62.1. Curves of normalized temperature increase as a function of normalized radius for various values of normalized time. Also included is a steady-state curve with fixed temperature boundary at $r/w = 10$ and the parabolic approximation near the axis.

62.3   Uses for Absorption Measurement

The usefulness in measurement of materials with low absorptivities comes from the remarkable sensitivity of the effect. Immediately after its identification, Leite *et al.* [4] used the effect to measure absorptivities of five organic liquids with values in the 10^{-4} range. As with the first observations, the cell was placed within the resonator of the helium-neon laser, the beam diameters measured at the mirrors by scanning, and the focal lengths found from Gaussian mode theory.

Although placement of the cell within the resonator increases power in the beam, it is inconvenient and may be impossible in many commercial lasers. Although other measurements were made that way [5], it was found that comparable sensitivities could be obtained with the test cell outside the resonator, provided it is placed at the position of maximum curvature [6]. This is a confocal distance beyond the waist (i.e., the position of minimum beam radius). It was also found more convenient to find the expanded beam radius by monitoring the intensity on the axis of the beam rather than by scanning. The intensity at beam center as a function of time is [6]

$$\frac{I_{bc}(t)}{I_{bc}(0)} = \left[1 - \frac{\Theta}{(1 + t_c/2t)} + \frac{\Theta^2}{2(1 + t_c/2t)^2} \right]^{-1} \tag{62.5}$$

where $\Theta = (\alpha Pl/J\lambda k)(dn/dT)$. A curve of the normalized intensity change as a function of time,

$$[I_{bc}(t) - I_{bc}(\infty)] [I_{bc}(0) - I_{bc}(\infty)]^{-1} \tag{62.6}$$

is shown in Fig. 62.2. Although calculated for $\Theta = -0.3$, it is nearly the same over wide ranges of Θ. Dovichi and Harris made many refinements to this method,

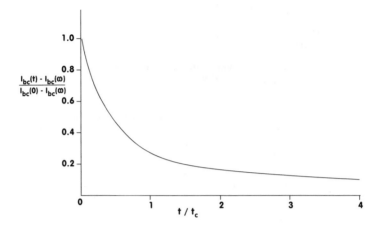

FIGURE 62.2. Normalized intensity differential at beam center as a function of normalized time. Curve is calculated for $\Theta = -0.3$, but is substantially the same over a wide range Θ.

including a two-cell arrangement in which comparison was made between the test cell and a standard cell [7]. The standard, with carefully chosen absorber and solvent, had been measured by photometric means and then diluted, as in the suggestion by Townes mentioned earlier. This work extended sensitivities into the 10^{-6} cm^{-1} range.

The variation of refractive index with radius produces a phase variation with radius of the beam passing through it, which of course is the wave point-of-view for explaining beam divergence or convergence. But it also makes possible interference effects, some of which may be useful for absorption measurement. One of the simplest is an interference between the ray propagating along the axis and rays from the wings of the Gaussian beam, all of which propagate parallel to the axis. Observation of the transient interference signal can give a measure of absorption [8]. Another interesting interference was that from two transverse modes of the laser beam, one forming the thermal lens and the other too weak to disturb it, but strong enough to give a beat frequency with the other [9]. Stone used two beams from the same laser, laterally displaced in passing through the cell, one strong enough to form a thermal lens with its induced phase shift, and the other weak with negligible thermal effect. Comparison was then made in an interferometer [10].

All of the above were limited to specific frequencies for which lasers were available, and for many spectroscopic purposes one would wish measurement over a range of frequencies. Long, Swofford and Albrecht [11] used a dual-beam arrangement in which a tunable dye laser formed the thermal lens and a low power continuous-wave (CW) laser served as probe. The signal was monitored on the axis and the dye laser beam chopped so that synchronous detection could be used. Stone [12] used an incoherent light from a high pressure xenon arc lamp, with filters to obtain a variety of frequencies for the lens-forming beam, and a He-Ne laser in his interferometric arrangement for induced phase measurement. Fang and Swofford describe still other arrangements and their limits of accuracy in a comprehensive review article [13]. They note that an important source of uncertainty comes from the imprecise knowledge of thermal conductivity and dn/dT for many materials.

62.4   Longer and Shorter Cells: Thermal Self-focusing

For cells which are not short compared with confocal distance of the Gaussian beam, or with the focal length of the induced lens, two factors enter. Beam power decreases with distance because of absorption and scattering, and beam radius changes because of the thermal lens effect and diffraction. There is thus a distributed lens system with each longitudinal element producing an amount of convergence (positive or negative) depending upon local power and beam radius. Analytic formulations have been made [14-16], but numerical integration of the distributed effect may be necessary. Carman and Kelley made interesting observations of the time development of thermal blooming in a long cell [17].

A particularly interesting example of long cells is one in which thermal self-focusing has been demonstrated. Lead glasses and some crystals have positive values of dn/dT, so that the thermal effect leads to converging lenses in such materials. In [18], long cylinders of crown glass with $dn/dT \approx 10^{-5}/\,^\circ\text{C}$ demonstrated very clear self-focusing effects for the beam from an argon ion laser. At the appropriate power (about 3 watts) beam trapping was observed, with the diameter maintaining a nearly constant 50 μm over the length of the cylinder. This is the condition in which focusing effects and diffraction effects essentially balance. Akhmanov et al. report similar self-focusing effects in lithium niobate crystals doped with chromium [15].

Short cells may be needed if it is desired to maintain very small diameter beams over the interaction region in order to obtain short thermal time constants. In that case longitudinal heat conduction may be important. In [19] cell lengths of about 20 μm and beam diameters of the same order produced thermal time constants less than a millisecond.

62.5   Higher Power Effects

As power is increased for a given absorptivity, the beam spreads out within the cell so that the paraxial idealization is no longer valid. Additional terms beyond the quadratic in the expansion of the Ei functions in Eq. (62.1) are then necessary and lead to interference rings resulting from the spherical aberration which these represent [15, 20, 21]. At still higher powers (or higher intensities resulting from focusing to very small diameter beams), the heating may cause a free-convection effect in fluids, causing an upward motion of the fluid and a general blurring at the top [15, 20]. The effect of forced convection on thermal blooming in gases and liquids has been extensively studied by Gebhart and Smith [22]. As expected, the flow of fresh material into the path of the optical beam decreases the overall blooming, but causes a shift into the "wind."

Thermal absorption of a laser beam also causes expansion, which converts some of the power into acoustic waves [23–25, 26]. Although a somewhat inefficient process, it has been shown to provide very sensitive absorption spectroscopy, with absorptivities as low as 10^{-9} measured. Excellent reviews of this art, called photoacoustical (PA) or optoacoustical (OA) spectroscopy, are available [27, 28].

62.6   Conclusion

The effects of this relatively simple phenomenon, easy to understand but somewhat surprising in its sensitivities, are mostly undesirable and can seriously disturb a variety of experiments if not accounted for. It has proved a useful effect in absorption spectroscopy and has been proposed for use in power limiters [29], as a transducer from modulated optical waves to undersea acoustic waves [30], and as a technique for determining fluorescence quantum yield of dyes [6]. So far its greatest use has been that of absorption measurement.

References

[1] J. R. Whinnery, Acc. Chem. Res. **7**, 225 (1974).
[2] J. P. Gordon, R. C. Leite, R. S. Moore, S. P. S. Porto, and J. R. Whinnery, Bull. Am. Phys. Soc. **9**, 501 (1964).
[3] J. P. Gordon, R.C. Leite, R. S. Moore, S. P. S. Porto, and J. R. Whinnery, J. Appl. Phys. **36**, 3 (1965).
[4] R. C. Leite, R. S. Moore, and J. R. Whinnery, Appl. Phys. Lett. **5**, 141 (1964).
[5] D. Solomini, J. Appl. Phys. **37**, 3314 (1966).
[6] C. Hu and J. R. Whinnery, Appl. Opt. 12, **72** (1973).
[7] N. J. Dovichi and J. M. Harris, Anal. Chem. **52**, 2338 (1980).
[8] Y. Kohanzadeh, K. W. Ma, and J. R. Whinnery, Appl. Opt. **12**, 1584 (1973).
[9] Y. Kohanzadeh and D. H. Auston, IEEE J. Quant. Elect. **6**, 475 (1970).
[10] J. Stone, J. Opt. Soc. Am. **62**, 327 (1972).
[11] M. E. Long, R. L. Swofford, and A. C. Albrecht, Science **191**, 183 (1976).
[12] J. Stone, Appl. Opt. **12**, 1828 (1973).
[13] H. L. Fong and R. L. Swofford, in *Ultrasensitive Laser Spectroscopy*, David S. Kliger, ed., (Academic Press, New York, 1983).
[14] P. K. Tien, J. P. Gordon, and J. R. Whinnery, Proc. IEEE **53**, 129 (1965).
[15] S. A. Akhmanov, D. P. Krindach, A. V. Migulin, A. P. Suchorukov, and R. V. Khoklov, IEEE J. Quant. Elect. **4**, 568 (1968).
[16] F. W. Dabby and H. A. Haus, J. Appl. Phys. **40**, 439 (1969).
[17] R. L. Carman and P. L. Kelley, Appl. Phys. Lett. **12**, 241 (1968).
[18] F. W. Dabby and J. R. Whinnery, Appl. Phys Lett. **13**, 284 (1968).
[19] F. W. Dabby, R. W. Boyko, C. V. Shank, and J. R. Whinnery, IEEE J. Quant. Elect. **5**, 516 (1969).
[20] J. R. Whinnery, D. T. Miller, and F. Dabby, IEEE J. Quant. Elect. **3**, 382 (1967).

[21] M. Born and E. Wolf, *Principles of Optics*, 6th ed. (Pergamon Press, Oxford, 1980), Sec. 9.4.1.

[22] F. G. Gebhart and D. C. Smith, IEEE J. Quant. Elect. **7**, 63 (1981).

[23] A. G. Bell, Am. J. Sci. **20**, 305 (1880).

[24] R. M. White, J. Appl. Phys. **34**, 3559 (1963).

[25] I. B. Kreuzer, J. Appl. Phys. **42**, 2934 (1971).

[26] Y. Kohanzadeh, J. R. Whinnery, and M. M. Carroll, J. Acoust. Soc. Am. **57**, 67 (1975).

[27] C. K. N. Patel and A. C. Tam, Rev. Mod. Phys. **53**, 517 (1981).

[28] A. C. Tam, in *Ultrasensitive Laser Spectroscopy*, David S. Kliger, ed. (Academic Press, New York, 1983).

[29] R. C. C. Leite, S. P. S. Porto, and T. C. Damen, Appl. Phys. Lett. **10**, 108 (1967).

[30] P. J. Westerveldt and R. S. Larson, J. Acoust. Soc. Am. **54**, 121 (1973).

63

Dark Matter and Faint Galactic Halo Light

Eric R. Wollman

63.1 Introduction

Sackett *et al.* [1] have reported the detection of faint R-band light from the halo of the edge-on spiral galaxy NGC5907. On a line of sight passing 6.5 kpc above the plane of the galaxy, the R-band intensity is roughly 27 mag arcsec^{-2}. The intensity of this light decreases with distance above the plane less steeply than any known luminous component of spiral galaxies.

It is natural to suspect that the faint light comes from halo stars. Based on its intensity and spatial distribution, Sackett *et al.* suggest that the light is emission from a halo population of late M subdwarfs and that these stars may constitute the halo dark matter. On the other hand, recent measurements of Bahcall *et al.* [2] strongly constrain the number of faint red stars in the halo of the Milky Way. If the halo stellar population of NGC5907 is similar to that of the Milky Way, it is unlikely that the faint light from the halo of NGC5907 is due primarily to faint red stars. Counts of halo stars in the solar neighborhood rule out hotter stars as the source of the NGC5907 halo light (again assuming similarity to the Milky Way) [3]. Red giants cannot be ruled out [4], but the fact that the intensity profile of the faint light differs from that of known stellar components suggests the possibility that the faint light comes from something other than halo stars.

What else might explain the faint light from the halo of NGC5907? The possibility considered here is that the faint light is disk starlight scattered by halo dark matter. Commonly considered forms of dark matter do not scatter light significantly. However, it has been proposed [5] that the galactic dark halo is composed of solid pieces of normal elements and compounds several to tens of centimeters in size. An estimation of the intensity of disk starlight scattered by a dark halo of this form is straightforward (Sec. 2). The significant result is that in the case of NGC5907, the expected intensity of scattered light is similar to the observed intensity.

63.2   The Intensity of Scattered Light

Consider a simplified model in which the galactic luminosity is a point source and the dark halo consists of a spherically symmetric distribution of identical scatterers. The scatterers are assumed to be solid pieces of normal elements and compounds. As will be seen, the required size of a scatterer is much greater than an optical wavelength. We further simplify the problem by assuming that the scattering is isotropic. Then as seen from outside the galaxy, the intensity of scattered R-band light is [6]

$$I_R = \frac{L_R \sigma \overline{Q}_s}{16\pi^2} \int \frac{n(r)}{r^2} dl \tag{63.1}$$

where L_R is the galactic R-band luminosity, σ is the geometrical cross section of a scatterer, $n(r)$ is the number density of scatterers at distance r from the center of the galaxy, and the integration is along the line of sight. The quantity $\overline{Q}_s$ is the spectrally averaged and weighted reflective scattering efficiency factor, given by

$$\overline{Q}_s \equiv \frac{1}{L_R} \int_R L_\lambda Q_s(\lambda) d\lambda .$$

where $Q_s(\lambda)$ is the scattering efficiency factor (in this case the albedo) at wavelength λ, L_λ is the galactic spectral luminosity density, and the integration is over the R-band.

For a dark halo composed of identical scatterers of mass m_s, a standard representation of $n(r)$ is

$$n(r) = \frac{v^2}{4\pi G m_s(r^2 + r_c^2)} \tag{63.2}$$

where v is the large-r limit of the galactic orbit speed, G is the gravitational constant, and r_c is the dark halo core radius. Now consider a line of sight that passes at distance b from the center of the galaxy. If the scatterers are spheres of radius r_s composed of matter of density ρ_s, the R-band intensity along this line of sight is

$$I_R = \left(\frac{3}{256\pi^2 G}\right) \frac{L_R v^2 \kappa \overline{Q}_s f(r_c, b)}{r_c^2 b} \tag{63.3}$$

where $\kappa \equiv (r_s \rho_s)^{-1}$ and

$$f(r_c, b) \equiv \left[1 - \frac{1}{\sqrt{1 + (\frac{r_c}{b})^2}} \right].$$

In the above expression (3) for the scattered light intensity, the only unconstrained parameter is κ. For NGC5907, $v \approx 220$ km/sec [7]. If NGC5907 is more-or-less normal, L_R is approximately $6 \times 10^9 L_\odot$ [1]. The core radius r_c

is not well-known; but analyses of spiral galaxies, including NGC5907, invoke values between about one kiloparsec and a few tens of kiloparsecs [1, 8-16]. We refer r_c to 10 kpc. In anticipation of the final result, we refer κ to $1\,\mathrm{cm^2/g}$. Then the R-band intensity expected from the halo of NGC5907 is

$$I_R \sim 7 \times 10^{-5}\,\overline{Q}_s\,f(r_c, b)\left(\frac{\kappa}{1\,\mathrm{cm^2/g}}\right)\left(\frac{10\,\mathrm{kpc}}{r_c}\right)^2$$
$$\times \left(\frac{1\,\mathrm{kpc}}{b}\right)\frac{\mathrm{ergs}}{\mathrm{cm^2\,sec\,ster}}. \tag{63.4}$$

As noted earlier, Sackett *et al.* detect faint emission out to about 6.5 kpc above the plane, where the R-band intensity is roughly 27 mag arcsec^{-2} or

$$\text{observed } I_R \approx 2.6 \times 10^{-6}\,\frac{\mathrm{ergs}}{\mathrm{cm^2 sec\,ster}}. \tag{63.5}$$

By comparison, for $r_c = 10\,\mathrm{kpc}$, the scattered light intensity expected at $b = 6.5\,\mathrm{kpc}$ is, according to equation (63.4),

$$\text{expected } I_R \sim 5 \times 10^{-6}\,\overline{Q}_s\left(\frac{\kappa}{1\,\mathrm{cm^2/g}}\right)\frac{\mathrm{ergs}}{\mathrm{cm^2 sec\,ster}}. \tag{63.6}$$

If r_c does not differ very greatly from 10 kpc and $\overline{Q}_s$ is not very much less than unity, the expected and observed R-band intensities are similar if $\kappa \sim 1\,\mathrm{cm^2/g}$. For low-density solids, the corresponding diameter of a scatterer is in the range of a few to a few tens of centimeters. Observations at other wavelengths and of other galaxies are consistent with this picture [12-14].

The significance of the requirement $\kappa \sim 1\,\mathrm{cm^2/g}$ is that the same requirement emerged from considerations of other seemingly unrelated galactic and cosmological phenomena. These other considerations are reviewed below.

63.3 Discussion

An earlier paper [5] explored various consequences of the hypothesis that the galactic halo dark matter is solid pieces of normal elements and compounds. According to this model, a galaxy evolves from a relaxed self-gravitating condensation of these solid pieces of matter. Within the galaxy, direct collisions vaporize the solids. The hot vapor radiatively cools and settles toward the center of the system to form the luminous core of the galaxy. Regardless of the size of the solid pieces, the galactic luminosity L and orbit speed v are related according to $L \propto v^{3.5}$, in agreement with the observed Tully-Fisher relation. It was found that for $\kappa \approx 0.5\,\mathrm{cm^2/g}$, the extent of hot halo gas, the rate of star formation, and the total mass within the luminous core are roughly consistent with what is observed in galaxies.

The condition $\kappa \sim 1\,\mathrm{cm^2/g}$ also has interesting cosmological consequences. In a critical-density universe dominated by solid matter for which $\kappa \sim 1\,\mathrm{cm^2/g}$, the extinction optical depth is unity at redshift $z \sim 3$, thus accounting for the absence of detectable high-z quasars. Moreover, large solids are normally good microwave emitters. Consequently, radiant energy introduced significantly prior the epoch corresponding to redshift $z \sim 3$ will have been thermalized more or less thoroughly, depending on the wavelength, epoch of introduction, and optical properties of the solids. In this picture, the microwave background need not be entirely primordial. (An interesting early discussion of the possibility that solid matter both thermalizes cosmic radiation and obscures the distant universe is given by Layzer and Hively [15].)

Because of warming by galactic radiation, a dark halo of solid matter is a source of thermal microwave radiation. This microwave emission may be detectable if the solids are reasonably good optical absorbers [5, 16]. The present proposal (that scattered light may be detected in the case of NGC5907) suggest high scattering efficiency and low absorption efficiency, in which case the microwave emission may be difficult to detect. On the other hand, the possible detection of scattered disk starlight offers an alternative test. If a significant fraction of the faint light from the halo of NGC5907 is disk starlight scattered by halo dark matter, then the faint light should be partially polarized.

Sackett *et al.* show that the intensity profile of the faint halo light is roughly consistent with emission from a flattened r^{-2} source distribution with a small core radius. Scattering of disk starlight by halo dark matter gives qualitatively similar results if the halo core radius is relatively large. In this case, the density of scatterers in the region of interest is roughly constant but the flux of disk radiation approaches r^{-2} well above the plane. Flattening of the halo intensity contours is expected, since the halo is being illuminated by a flattened source (i.e. the disk). Preliminary numerical calculations confirm this qualitative picture.

63.4   Summary

Sackett *et al.* [1] have detected faint visible light from the halo of the edge-on spiral galaxy NGC5907. Since the distribution of the faint light is unlike that of any known stellar component, it is possible that the light comes from something other than halo stars. It was shown previously [5] that a model in which the galactic dark halo is composed of solid pieces of normal elements and compounds may explain a number of galactic and cosmological phenomena. For low-density solids, the model requires that the pieces be several to tens of centimeters in size. The foregoing analysis shows that scattering of disk starlight by halo dark matter in this form may also account for a significant fraction of the faint light coming from the halo of NGC5907. If scattering is significant, the faint halo light should be partially polarized.

63.5   Acknowledgments

It is a great pleasure to contribute this paper in honor of Charles H. Townes, who has been a kind, generous, and inspiring teacher. Thank you, Professor Townes. I also wish to thank Penny Sackett, Harley Thronson, and William Hubbard for helpful discussions.

References

[1]   P. D. Sackett, H. L. Morrison, R. Harding, and T. A. Boroson, Nature **370**, 441 (1994).

[2]   J. N. Bahcall, C. Flynn, A. Gould, and S. Kirhakos, ApJ **435**, L51 (1994).

[3]   J. N. Bahcall and S. Casertano, ApJ **308**, 347 (1986).

[4]   H. L. Morrison, AJ **106**, 578 (1993).

[5]   E. R. Wollman, ApJ **392**, 80 (1992).

[6]   D. E. Osterbrock, *Astrophysics of Gaseous Nebulae*, W. H. Freeman and Company (1974).

[7]   R. Sancisi and T. S. van Albada, in *Int. Astr. Un. Symp. No. 117* (eds. J. Kormendy and G. R. Knapp), p. 67, Reidel (1985).

[8]   J. N. Bahcall and S. Casertano, ApJ **293**, L7 (1985).

[9]   T. S. van Albada, J. N. Bahcall, K. Begeman, and R. Sancisi, ApJ **295**, 305 (1985).

[10] S. M. Kent, AJ **93**, 816 (1987).

[11] R. Flores, J. R. Primack, G. R. Blumenthal, and S. M. Faber, ApJ **412**, 443 (1993).

[12] D. Barnaby and H. A. Thronson, Jr., AJ **107**, 1717 (1994).

[13] P. C. van der Kruit, in *Int. Astr. Un. Symp. No. 117* (eds. J. Kormendy and G.R. Knapp), p. 415, Reidel (1985).

[14] M. F. Skrutskie, M. A. Shure, and S. Beckwith, ApJ **299**, 303 (1985).

[15] D. Layzer and R. Hively, ApJ **179**, 361 (1973).

[16] E. R. Wollman, B. R. Tuttle, and C. E. Parrish, Bull. Amer. Astron. Soc. **24**, 1248 (1992).

64

Optical Pump-Probe Experiments and the Higgs Field

Herbert J. Zeiger

I feel honored to participate in this Festschrift in celebration of Charles Townes's 80th birthday. Ever since my days as a graduate student at Columbia University, Charlie has been my most admired model of a physicist. But, like the Platonic Ideal, he is a model to aspire to, although impossible to match. Even his wild speculations turn out to be right on target—take the maser, for example. The rest of us have to accept our park-bench musings as nothing more than rambling daydreams. As the reader will see later in this paper, I cannot resist some wild speculation of my own on this happy occasion.

Since the development of colliding pulse mode-locked (CPM) laser technology in the early 1980s [1], a host of experiments using sub-picosecond, intense optical pulses have been performed. In one class of such experiments, a strong pump pulse of light is incident on a sample, and a delayed, weaker pulse probes the effect produced by the pump. Because of the shortness of the pulses, phenomena produced by the pump can be studied with subpicosecond resolution. Optical pump-probe studies of metal surfaces have shown a fractional disturbance of reflectivity of $\sim 10^{-4}$ which decays with a roughly exponential behavior on a time scale of a few picoseconds. Since the decay of electronic excitation is mainly due to phonon emission, such measurements allow estimates of elecron-phonon coupling strengths which can be compared with both transport determinations [2], and with determinations from superconducting transition temperatures [3].

I report here on more recent experiments using optical pump-probe techniques in which I have collaborated with colleagues at MIT. In these studies of semimetals and semiconductors using 60 fs pulses, this group has observed strong oscillatory signals superposed on the exponentially decaying background, with fractional changes in reflectivity as large as $\sim 10^{-2}$ [4]. An example of such a spectrum is shown in Fig. 64.1 for the case of Sb. The oscillation periods were found to agree with the totally symmetric A_1 or A_{1g} optical mode frequencies of the material, and thus far only crystal structures having these modes have given such strong signals. For reasons which will be discussed below, we have called the excitation mechanism for these oscillations "displacive excitation of coherent phonons" (DECP).

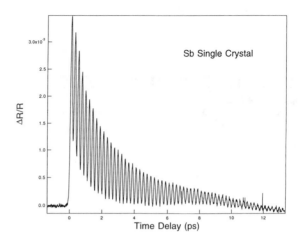

FIGURE 64.1. DECP spectrum of single crystal Sb at room temperature.

The excitation mechanism for DECP is based on the unique properties of A_1 vibrational modes. Crystal structures with A_1 modes can be pictured as derived from "virtual" crystal structures of higher symmetry by a displacement of some of the atomic positions. For example, the Sb structure can be thought of as essentially arising from a cubic structure by small displacements q of (111) planes along a cubic body diagonal to produce pairing of adjacent planes. The equilibrium pairing displacement q_{eq} is determined by the minimum free energy of the system. The A_1 mode of the system is due to the vibration of q about q_{eq}. (In the case of Sb, which has inversion symmetry, this is an A_{1g} mode.) As the temperature of Sb is raised, one might expect q_{eq} to move toward its "virtual" crystal value of zero. However, long before q_{eq} can change appreciably, Sb melts. There are materials such as GeTe which at room temperature have a trigonal structure related to the Sb structure, and which go over to a cubic NaCl structure at higher temperatures (670 K for GeTe).

Since the value of the equilibrium A_1 coordinate q_{eq} is determined by the minimum of free energy, one might expect that electronic excitation of a crystal by a sub-picosecond optical pump pulse would displace the equilibrium value of q. If the excited electron system comes to quasi-equilibrium in a time short compared to an A_1 vibrational period, the ion system should be set into coherent A_1 vibration about the new displaced equilibrium. This mechanism, of course, requires a pump pulse of duration less than an A_1 vibrational period. The coherent A_1 vibration modulates the dielectric properties of the crystal, producing a DECP modulation of the optical reflectivity. Since other modes, such as the E_g mode in Sb, cannot change their equilibrium value without changing the equilibrium crystal symmetry, they cannot be excited by a DECP process except in most unusual circumstances.

These ideas on the origin of the DECP process for A_1 modes have been used to produce a phenomenological model which gives an excellent fit to experiment. The model assumes that the electronic excitation due to the pump pulse decays exponentially. This excitation modulates the reflectivity directly, but also produces a displacement of the equilibrium A_1 coordinate, which then produces the DECP modulation of reflectivity. The theoretical expression obtained for the reflectivity can be written [4]

$$
\begin{aligned}
\frac{\overline{\Delta R(\tau)}}{R} &= A \int_0^\infty G(t - \tau) e^{-\beta \tau} d\tau \\
&\quad + B \int_0^\infty G(t - \tau) \left[e^{-\beta \tau} - e^{-\gamma \tau} \cos(\omega_{A_1} \tau) \right] d\tau \\
&\quad + C \int_0^\infty G(t - \tau) d\tau.
\end{aligned}
\tag{64.1}
$$

Here, $\beta = (1/\tau_{el})$ is the effective electronic relaxation rate, $\gamma = (1/\tau_{ph})$ is the effective phonon relaxation rate, $G(t)$ is the optical pulse auto-correlation function, and $\omega_{\hat{A}_1}$ is the A_1 phonon angular frequency. Equation (1) assumes that $\omega_{\hat{A}_1} \gg \beta, \gamma$. The A term in Eq. (1) represents the direct effect of electronic excitation, while the vibrational B term is the indirect effect due to the change in equilibrium A_1 coordinate produced by the electronic excitation. The C term represents a background shift due to heating or a long-time electronic excitation effect. Equation (64.1) has given a good description of DECP experiments in Sb, Bi, Bi-Sb alloys, Ti_2O_3, and GeTe.

In order to understand the underlying origin of Eq. (64.1), we have examined a thermodynamic free energy model of the equilibrium condition for the A_1 coordinate q. Since, in a DECP experiment, degrees of freedom of the crystal other than the A_1 are presumably constrained, our free energy will be a function only of q. In the materials we have studied, the Fermi level lies in the vicinity of a conduction band (1) and a valence band (2). We assume that at a given temperature the free energy has already been minimized with respect to all lattice coordinates other than the A_1 coordinate q. For small changes in q about the equilibrium value, the free energy per unit volume can then be written,

$$
F = -n_0 \frac{a}{2} q^2 + n_0 \frac{b}{4} q^4 + F_1 + F_2 + F_{ph}
\tag{64.2}
$$

where

$$
\begin{aligned}
F_1 &= (-kT) \int_0^\infty \rho_1(\epsilon) \ln \left[1 + e^{-(\epsilon + \epsilon_1 - \mu)/kT} \right] d\epsilon \\
F_2 &= (-kT) \int_0^\infty \rho_2(\epsilon) \ln \left[1 + e^{-(\epsilon - \epsilon_2 + \mu)/kT} \right] d\epsilon \\
F_{ph} &= \sum_n (kT) \int_0^\infty \rho_n(\hbar\omega) \ln \left[1 - e^{-\hbar(\omega_{0n} + \omega)/kT} \right] d(\hbar\omega)
\end{aligned}
\tag{64.3}
$$

In Eq. (64.2), the a and b terms on the right-hand side represent q-dependent, symmetry-breaking elastic terms, and n_0 is the inverse of the volume of a crystal unit cell. For simplicity, we assume that in the experiments where DECP has thus far been observed, a and b are only weakly dependent on temperature. F_{ph} is the free energy due to the lattice phonons, F_1 comes from the electronic free energy of electrons in the conduction band (1), and F_2 comes from holes in the valence band (2). The sum over n in F_{ph} represents contributions from different branches of the phonon spectrum, and ω_{0n} is the lowest angular frequency of that branch. The phonon free energy may be q-dependent if the atomic interactions depend on the q coordinate. The phonon density of states $\rho_n(\hbar\omega)$ as well as the phonon angular frequencies ω_{0n} may depend on q. The electronic free-energy terms are q-dependent by virtue of the q-dependence of the band edge energies and carrier densities of state $\rho_1(\epsilon)$ and $\rho_2(\epsilon)$. The electronic densities of state may also include factors for the multiplicity of band edges. The energy ϵ is measured up from the bottom of the conduction band in F_1, and down from the top of the valence band in F_2. The energy ϵ_1 is that of the conduction band edge, ϵ_2 is that of the valence band edge energy, and μ is the Fermi energy. In writing the electronic free energy in terms of electrons and holes, we have left out a filled valence band contribution, which is independent of electron temperature, and can thus be included in the elastic free energy. The condition that the free energy is a minimum with respect to A_1 coordinate q is

$$\frac{\partial F}{\partial q} = -n_0 a q + n_0 b q^3 + \left(\frac{\partial F_1}{\partial q}\right) + \left(\frac{\partial F_2}{\partial q}\right) + \left(\frac{\partial F_{ph}}{\partial q}\right) = 0 \qquad (64.4)$$

where

$$\left(\frac{\partial F_1}{\partial q}\right) = \int_0^\infty f_1(\epsilon)\left[\rho_1(\epsilon)\frac{\partial \epsilon_1}{\partial q} - \frac{\partial N_1(\epsilon)}{\partial q}\right] d\epsilon$$

$$\left(\frac{\partial F_2}{\partial q}\right) = \int_0^\infty f_2(\epsilon)\left[\rho_2(\epsilon)\frac{\partial \epsilon_2}{\partial q} - \frac{\partial N_2(\epsilon)}{\partial q}\right] d\epsilon \qquad (64.5)$$

$$\left(\frac{\partial F_{ph}}{\partial q}\right) = \sum_n \int_0^\infty f_{ph}(\hbar\omega)\left[\rho_n(\hbar\omega)\frac{\partial \hbar\omega_{0n}}{\partial q} - \frac{\partial N_n(\hbar\omega)}{\partial q}\right] d(\hbar\omega),$$

and

$$f_1(\epsilon) = \left[e^{(\epsilon+\epsilon_1-\mu)/kT} + 1\right]^{-1}$$

$$f_2(\epsilon) = \left[e^{(\epsilon-\epsilon_2+\mu)/kT} + 1\right]^{-1} \qquad (64.6)$$

$$f_{ph}(\hbar\omega) = \left[e^{\hbar(\omega_{0n}+\omega)/kT} - 1\right]^{-1}.$$

In Eq. (64.6) we have used integration by parts to obtain the electronic and phonon terms in the form shown. $N(\epsilon)$ is the integrated density of states up to energy ϵ, which may be a significant function of the A_1 coordinate q far into a band, but is likely to be nearly independent of q near a band edge in an indirect band gap material. The dependence of the Fermi energy on q has canceled from Eq. (64.4) because of the equality of numbers of holes and electrons. The term $(\partial F_{ph}/\partial q)$ has also been obtained using integration by parts. The contribution of different phonon branches depends on the properties of the particular system under consideration. The solution of Eq. (64.4) leads to the equilibrium value of q, q_{eq}. Neglecting the electronic and phonon terms in Eq. (64.4), the solution is $q = (a/b)^{\frac{1}{2}} = q_0$. Thus, from Eq. (64.4), the elastic free energy lowers the symmetry of the "virtual" crystal, and the electronic and phonon terms produce additional corrections to q_0.

Eq. (64.2) can be rewritten as an expansion about q_{eq}. To second order in $(q - q_{eq})$ the result is

$$
\begin{aligned}
F &= \frac{-n_0 a}{2} q_{eq}^2 \left[1 - \frac{q_{eq}^2}{2q_0^2} \right] + F_1(q_{eq}) + F_2(q_{eq}) + F_{ph}(q_{eq}) \\
&+ n_0 a \left[1 - \frac{3}{2q_{eq}n_0 a} \left\{ \left(\frac{\partial F_1}{\partial q} \right)_{eq} + \left(\frac{\partial F_2}{\partial q} \right)_{eq} + \left(\frac{\partial F_{ph}}{\partial q} \right)_{eq} \right.\right. \\
&+ \left.\left. \frac{1}{2} \left(\frac{\partial^2 F_1}{\partial q^2} \right)_{eq} + \frac{1}{2} \left(\frac{\partial^2 F_2}{\partial q^2} \right)_{eq} + \frac{1}{2} \left(\frac{\partial^2 F_{ph}}{\partial q^2} \right)_{eq} \right\} \right] (q - q_{eq})^2
\end{aligned}
$$
(64.7)

The second derivative terms in Eq. (64.7) involving F_1, F_2, and F_{ph} will contribute only if the electronic and phonon thermal relaxation times are short compared to the period of the A_1 vibration, so that these systems can adjust to the instantaneous equilibrium condition during the vibration. In our DECP experiments, there is some evidence that this is not the case. For simplicity we drop these terms.

The equation of motion of q in terms of the free energy is

$$
-\frac{\partial F}{\partial q} = n_0 \mu_0 \ddot{q}
$$
(64.8)

where μ_0 is an effective mass for the atomic motion associated with the A_1 vibration. Combining Eqs. (64.4), (64.7), and (64.8) we find the normal mode result,

$$
\omega_{A_1}^2 \mu_0 = 2a \left[\frac{3}{2} \left(\frac{q_{eq}^2}{q_0^2} \right) - \frac{1}{2} \right].
$$
(64.9)

Equations (64.2) through (64.9) are general within the context of the model. It should be kept in mind that Eq. (64.9) describes the $k = 0$ uniform mode of vibration of one branch of the optical mode spectrum, and that ω_{A_1} is the angular frequency of the $k = 0$ mode of $\omega(k)$.

The materials so far studied by our group in DECP experiments all have indirect band gaps with band edges at room temperature that are either nonoverlapping, as in Ti_2O_3 at room temperature, or slightly overlapping, as in Bi or Sb. For such materials, the behavior of the band edges with q can be described in terms of deformation potentials associated with the A_1 mode by the definition $d\epsilon/dq = D/q_0$, where q_0 is the equilibrium value of q due to the strain terms only in Eq. (64.4). Making the approximation that q can be replaced to lowest order in Eq. (64.9) by q_0, we obtain to first order in electronic and phonon terms,

$$
q \cong q_0 - \frac{1}{\omega_0^2 \mu_0} \left[\frac{n}{n_0} (1 + \eta_1) \frac{D_1}{q_0} - \frac{n}{n_0} (1 + \eta_2) \frac{D_2}{q_0} \right.
$$

$$+ \frac{1}{n_0} \left(\frac{\partial F_{ph}}{\partial q} \right)_{eq} \Bigg]$$

(64.10)

where

$$n = \int_0^\infty \rho_1(\epsilon) f_1(\epsilon) d\epsilon = \int_0^\infty \rho_2(\epsilon) f_1(\epsilon) d\epsilon \, ,$$

$$\eta_1 = \frac{-\int_0^\infty \frac{\partial N_1(\epsilon)}{\partial q} f_1(\epsilon) d\epsilon}{D_1/q_0} \, ,$$

(64.11)

$$\eta_2 = \frac{\int_0^\infty \frac{\partial N_2(\epsilon)}{\partial q} f_2(\epsilon) d\epsilon}{D_2/q_0} \, ,$$

and $\omega_0 = (2a/\mu_0)^{\frac{1}{2}}$.

Even though we do not calculate the electronic and phonon corrections to q, Eq. (64.10) predicts a relation between quantities, some of which can be determined empirically. We have applied Eq. (64.10) to the case of Ti_2O_3, where DECP spectra have been followed up from room temperature through a "soft" phase transition at $\sim T = 450$ K occurring over a temperature interval of ~ 50 K [5]. This phase transition is characterized by an increase in conductivity of an order of magnitude, fairly abrupt changes in lattice parameter, about a 10 percent decrease in Raman frequencies, and changes in many other physical parameters, but with no change in crystal structure [6]. It has been fairly well-established that the transition is due to an overlapping of indirect valence and conduction bands as the crystal temperature is raised [7]. Careful x-ray measurements of ion positions through the "soft" transition have been made [8], allowing an estimate of the changes in q with temperature. The assumption is made that η_1 and η_2 are small for small band overlap, and that D_1 and D_2 are nearly constant through the transition. Clearly, the phonon spectrum changes through the transition, but it is not clear how much of that change is due to a change in q. The decreases in Raman frequencies are consistent with a change proportional to n, so we assume that (dF_{ph}/dq), if not negligible, is proportional to n.

Then the x-ray measurements, combined with the knowledge that the gap energy is ~ 0.15 eV at room temperature and is 0 eV at about 450 K, allow a determination of energy gaps and n through the transition using Eq. (64.10). The band overlap after band crossing is found to be about 0.06 eV [9], with a relatively weak dependence of this result on the values of deformation potentials and band effective masses assumed. The effective electron relaxation rate is found experimentally to increase by about a factor of 3 as band crossing occurs. This can be understood as due to the opening of an additional interband channel for electron-phonon energy relaxation. A good fit of the electron relaxation rate as a function of temperature during band crossing is found [9] using the energy gap values obtained from Eq. (64.10), lending support to the argument that the phonon contribution to Eq. (64.10), if at all significant, is roughly proportional to n.

We have also used Eq. (64.10) for a description of the DECP phenomenon, assuming this equation can be applied if excited carrier distributions produced by the pump pulse can be characterized approximately by electron temperatures T_{el}. In the case of DECP in crystals with A_1 modes, the process is so rapid that the lattice phonon contribution to the change in q is negligible. The pump pulse changes the number of carriers in conduction and valence bands, as well as the energy distribution within the bands. In the presence of an energy gap, as in Ti_2O_3 at room temperature, the interband electron-phonon recombination process is slow, so that n changes slowly. Fast changes in q must be due to changes in η_1 or η_2. Thus, after the pump pulse, the magnitude of the η's can be large, but they will decay at essentially the same rate as the intraband relaxation rate of the electronic excitation. Hence, the first term in parentheses in the vibrational B term of Eq. (64.1), the change due to the shift in equilibrium q coordinate, has the same decay constant as the direct electronic excitation A term. If the bands overlap, as they do in Sb or in Ti_2O_3 above 450 K, then both n and the η's will change rapidly, and will also lead to shifts in equilibrium q-coordinate which decay at the same rate as the A term.

The description of the effect of carrier excitation on the equilibrium value of q given in terms of deformation potentials in Eq. (64.10) is appropriate for the samples in which we have observed DECP, since they are all indirect gap materials in the vicinity of the Fermi level. However, the energy gaps in these materials, which were produced by the symmetry lowering of the initial "virtual" crystal, do not lie near the Fermi level, and therefore do not play a role in changing q with carrier number or temperature. It is of interest to examine what would happen if the Fermi level *did* lie in the region of the symmetry-breaking energy gap. Carriers thermally excited across this gap would cause a decrease in q, given by Eq. (64.4), which would tend to drive the crystal toward its "virtual" state of higher symmetry. Of course, in real crystals, when a phase transition back to a structure of higher symmetry occurs with increasing temperature, the transition is dominated by other effects including the entire lattice vibrational spectrum.

Nevertheless, it is important for the discussion to follow, to consider the case of a model system described by Eq. (64.4) when the gap due to symmetry-breaking *does* lie near the Fermi level. For reasons which will become clear shortly, we consider a single isotropic pair of mirror conduction and valence bands at the Fermi level produced by symmetry breaking. For such a case, the energy gap Δ is even in q and therefore in lowest order is given by

$$\Delta = \kappa q^2 \tag{64.12}$$

where κ is a constant. The Fermi level lies in the middle of the gap, and the well-known result for the band energy as a function of lattice momentum $\mathbf{p}$ is

$$\epsilon = \left[\left(\frac{\Delta}{2}\right)^2 + \alpha^2 p^2\right]^{\frac{1}{2}} - \left(\frac{\Delta}{2}\right) \tag{64.13}$$

where α is a constant. For a fermion of spin $\frac{1}{2}$ the number of states per unit volume up to crystal momentum $\mathbf{p}$ is

$$N = \frac{8}{3}\frac{\pi p^3}{h^3} . \tag{64.14}$$

Making use of Eqs. (64.12–64.14), $(\partial F/\partial q)_{eq}$ in Eq. (64.7) for holes and electrons can be written,

$$\left(\frac{\partial F_1}{\partial q}\right) = \left(\frac{\partial F_2}{\partial q}\right) = \frac{2\pi\Delta}{h^3\alpha^3}\int_0^\infty \frac{\epsilon^{\frac{1}{2}}(\epsilon + \Delta)^{\frac{1}{2}}}{e^{(\epsilon + \frac{\Delta}{2})/kT} + 1}d\epsilon\left(\frac{\partial\Delta}{\partial q}\right). \qquad (64.15)$$

To model the phonon contribution to the free energy, we consider the simplified phonon spectrum consisting of only one optical phonon branch. The $\mathbf{k} = 0$ phonon of this spectrum is assumed to be the A_1 mode associated with the displacement q responsible for the lowering of the crystal symmetry, and the phonon frequency is taken to increase as k^2 for small $\mathbf{k}$. For reasons which will become clear shortly, we then take this simplified phonon spectrum to be of the form

$$\epsilon = \left[\epsilon_{A1}^2 + \alpha'^2 p^2\right]^{\frac{1}{2}} - \epsilon_{A1} \qquad (64.16)$$

where $\epsilon = \hbar\omega_{ph}$, $\epsilon_{\widehat{A_1}} = \hbar\omega_{\widehat{A_1}}$, $\mathbf{p} = \hbar\mathbf{k}$, and α' is a constant. The number of available states for phonons up to crystal momentum p is half the number for fermions of spin 1/2 given by Eq. (64.14). Since ω_{A_1} in Eq. (64.9) is not an explicit function of q, the phonons described by Eq. (64.16) do not contribute to Eq. (64.4). Making use of Eq. (64.12), Eq. (64.4) then becomes

$$\Delta = \Delta_0 - \frac{8\pi\kappa^2\Delta}{h^3\alpha^3 n_0 a}q_0^2\int_0^\infty \frac{\epsilon^{\frac{1}{2}}(\epsilon + \Delta)^{\frac{1}{2}}}{e^{(\epsilon + \frac{\Delta}{2})/kT} + 1}d\epsilon \qquad (64.17)$$

where $\Delta_0 = \kappa q_0^2$.

Now comes a great leap into the void: There is a suggestive analogy between the "displacement field" q which lowers the symmetry of a crystal, and the Higgs field which breaks the symmetry of the vacuum in the standard model [10]. It is intriguing to see how far this analogy can be pursued. Looked at in this light, if α and α' are taken to be the velocity of light c, then Eq. (64.13) describes a spin $\frac{1}{2}$ Klein-Gordon fermion of mass given by $mc^2 = \Delta/2$, and Eq. (64.16), describes a Klein-Gordon boson with mass given by $mc^2 = \epsilon_{A1}$. The fermion mass comes from its coupling to the symmetry-breaking Higgs field, while the boson owes its existence to that same field. At a temperature T, the value of Δ is determined by Eq. (64.17) which we rewrite in the form

$$\Delta = \Delta_0 - \left(\frac{T}{T'}\right)^2\Delta\int_0^\infty \frac{x^{\frac{1}{2}}\left(x + \frac{\Delta}{kT}\right)^{\frac{1}{2}}}{e^{(x + \frac{\Delta}{2kT})} + 1}dx \qquad (64.18)$$

where

$$\frac{1}{(kT')^2} = \frac{8\pi\kappa^2}{h^3 c^3 n_0 a}q_0^2. \qquad (64.19)$$

The boson mass, which changes with temperature through its coupling to the fermion field, is given by Eq. (64.9), now written in the form

$$\omega = \omega_0\left[\frac{3}{2}\left(\frac{\Delta}{\Delta_0}\right) - \frac{1}{2}\right]^{\frac{1}{2}} \qquad (64.20)$$

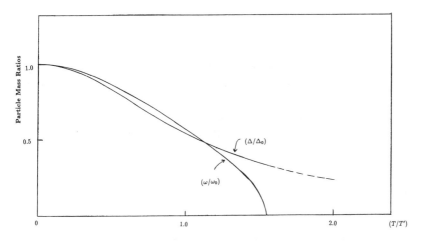

FIGURE 64.2. Plot of normalized boson mass (ω/ω_0) and normalized fermion mass (Δ/Δ_0) as a function of normalized temperature (T/T').

where $\omega_0 = (2a/\mu_0)^{1/2}$, and $M = \hbar\omega/c^2$. From Eq. (64.20) it can be seen that the system becomes unstable when $(\Delta/\Delta_0) = \frac{1}{3}$, signaled by $\omega \to 0$.

We have numerically solved Eq. (64.18) to obtain the ratio (Δ/Δ_0) and (ω/ω_0) as a function of normalized temperature (T/T') for values of the parameter (Δ_0/kT') ranging from 0.01 to 0.5. The results, which are almost independent of this parameter, are presented in Fig. 64.2. We find a gradual transition back toward the vacuum of unbroken symmetry with increasing temperature, until the instability is reached. Based on the present simplistic model, we cannot say anything about the nature of the system above that temperature.

What are we to make of all this? It could very well be that the analogy between the crystal displacement field and the Higgs field, although intriguing, is totally meaningless. After all, we are extrapolating the results obtained for a crystal with a periodic structure to the vacuum; buried in our definition of ω_0 is the mass μ_0 associated with the A_1 vibration. Furthermore, our results were obtained using an equation of motion, Eq. (64.8), which is both classical and non-relativistic. Yet it is tempting to pursue the analogy a bit further. Putting aside the enormous simplification of our model which ignores quarks, gluons, photons, and all their interactions, there is one implication which almost certainly remains: a heating of the vacuum will change the equilibrium displacement of the Higgs field. It has been argued that heavy ion-ion collisions could be energetic enough to produce quark-gluon plasmas [11]. This suggests that, in principle, a DECP process could excite bosons associated with the Higgs field. If the bosons discussed above are to be identified with the Higgs bosons of the standard model, then one theoretical estimate has given their mass as greater than 41 Gev [12]. This makes a DECP process highly unlikely, since the necessary plasma formation time for this would be less than $t = 10^{-25}$ sec. For comparison, the time for a relativistic particle to travel 10^{-13} cm is $\sim t = 3 \times 10^{-24}$ sec.

64.1   Acknowledgments

I wish to thank my colleagues at MIT for the pleasure of working with them on the DECP problem: E. P. Ippen, T. K. Cheng, J. Vidal, G. Dresselhaus, and M. S. Dresselhaus. I particularly thank G. Dresselhaus for a stimulating discussion of this paper.

Again, I congratulate Charlie Townes on his birthday. As always, he remains a model for us, both as an inspired scientist and as a human being.

References

[1] R. L. Fork, B. I. Greene, and C. V. Shank, Appl. Phys. Lett. **38,** 671 (1981).

[2] P. B. Allen, Phys. Rev. Lett. **59,** 1460 (1987).

[3] S. D. Brorson, A. Kazeroonian, J. S. Moodera, D. W. Face, T. K. Cheng, E. P. Ippen, M. S. Dresselhaus, and G. Dresselhaus, Phys. Rev. Lett. **64,** 2172 (1990)

[4] H. J. Zeiger, J. Vidal, T. K. Cheng, E. P. Ippen, G. Dresselhaus, and M. S. Dresselhaus, Phys. Rev. B **45,** 768 (1991).

[5] J. M. Honig and T. B. Reed, Phys. Rev. **174,** 174 (1968).

[6] S. H. Shin, G. V. Chandrashekhar, R. E. Loehman, and J. M. Honig, Phys. Rev. B **8,** 1364 (1973).

[7] G. Luckovsky, J. W. Allen, and P. Allen, Inst. Phys. Conf. Ser. No. **43,** 465 (1979).

[8] C. E. Rice and W. R. Robinson, Acta Cryst. **B33,** 1342 (1977).

[9] H. J. Zeiger, T. K. Cheng, E. P. Ippen, J. Vidal, G. Dresselhaus, and M. S. Dresselhaus, to be published.

[10] S. Weinberg, Rev Mod. Phys. **46,** 255 (1974).

[11] S. Das Gupta and G. D. Westfall, Physics Today **46,** 34 (1993).

[12] R. Hempfling, Phys. Lett. B **296,** 121 (1992).

Index